Lecture Notes in Computer Science 11414

More information about this series at http://www.springer.com/series/7412

Michel Couprie · Jean Cousty
Yukiko Kenmochi · Nabil Mustafa (Eds.)

Discrete Geometry for Computer Imagery

21st IAPR International Conference, DGCI 2019
Marne-la-Vallée, France, March 26–28, 2019
Proceedings

Springer

Editors
Michel Couprie
Université Paris Est, LIGM,
CNRS - ENPC - ESIEE Paris - UPEM
Noisy-le-Grand, France

Jean Cousty
Université Paris Est, LIGM,
CNRS - ENPC - ESIEE Paris - UPEM
Noisy-le-Grand, France

Yukiko Kenmochi
Université Paris Est, LIGM,
CNRS - ENPC - ESIEE Paris - UPEM
Noisy-le-Grand, France

Nabil Mustafa
Université Paris Est, LIGM,
CNRS - ENPC - ESIEE Paris - UPEM
Noisy-le-Grand, France

ISSN 0302-9743 ISSN 1611-3349 (electronic)
Lecture Notes in Computer Science
ISBN 978-3-030-14084-7 ISBN 978-3-030-14085-4 (eBook)
https://doi.org/10.1007/978-3-030-14085-4

Library of Congress Control Number: 2019932169

LNCS Sublibrary: SL6 – Image Processing, Computer Vision, Pattern Recognition, and Graphics

This Springer imprint is published by the registered company Springer Nature Switzerland AG
The registered company address is: Gewerbestrasse 11, 6330 Cham, Switzerland

Preface

This volume collects the articles presented at the 21st IAPR International Conference on Discrete Geometry for Computer Imagery (DGCI 2019), held at ESIEE Paris, France, during March 26–28, 2019, following, the 20th edition held in Vienna in 2017. We are very happy to highlight that this was the second time that a DGCI conference was held at ESIEE Paris: the 8th edition was already organized at the same place precisely 20 years earlier, in 1999.

This DGCI edition attracted 50 submissions by authors from 20 countries: Australia, Austria, Belgium, Brazil, China, France, Hungary, India, Iran, Italy, Japan, The Netherlands, Peru, Poland, Russia, Serbia, Spain, Sweden, UK, and USA. Out of these 50 submissions, 38 were selected for presentation at the conference after a review-and-rebuttal process where each submission received, on average, 4.1 reports, including a meta-review by a Program Committee member. These contributions focus on discrete geometric models and transforms, discrete topology, graph-based models, analysis and segmentation, mathematical morphology, shape recognition and analysis, and geometric computation.

In addition, three internationally well-known researchers were invited for keynote lectures:

- Professor Alexandre Xavier Falcão, from University of Campinas, Brazil, on "The Role of Optimum Connectivity in Image Segmentation: Can the Algorithm Learn Object Information During the Process?"
- Professor Longin Jan Latecki, from Temple University, USA, on "Computing Maximum Weight Cliques with Local Search on Large Graphs with Application to Common Object Discovery"
- Professor János Pach, from École polytechnique fédérale de Lausanne, Switzerland, on "The Blessing of Low Dimensionality"

Moreover, Professor Falcão and Professor Latecki contributed a written article that can be found in this volume.

Together with the main DGCI 2019 event, a one-day pre-conference workshop on Discrete Topology and Mathematical Morphology in honor of the retirement of Gilles Bertrand and a half-day post-conference workshop on Approximation Algorithms: Combinatorics, Geometry and Statistics were also organized, leading to a one-week program.

Following the tradition of DGCI, the DGCI 2019 proceedings appear in the LNCS series of Springer, and it is planned to produce a special issue of the *Journal of Mathematical Imaging and Vision* from extended versions of excellent contributions.

We are thankful to Springer for sponsoring a best student article award and to the International Association of Pattern Recognition (IAPR) for its sponsorship, DGCI being one of the main events associated with the IAPR Technical Committee on Discrete Geometry and Mathematical Morphology (TC18). We are also grateful to our institutions, namely, ESIEE Paris, Université Paris-Est, Université Paris-Est Marne-la-Vallée, Laboratoire d'Informatique Gaspard-Monge, Bézout Labex (grant ANR-10-LABX-58), and Centre National de la Recherche Scientifique (CNRS), and to Agence National de la Recherche (SAGA grant ANR-14-CE25-0016) that provided financial support.

We would like to thank all contributors, the keynote speakers, the Program and Steering Committees of DGCI, the Organizing Committee of DGCI 2019, and all those who made this conference happen. Last, but not least, we thank all participants and we hope that everyone had great interest in DGCI 2019.

January 2019

Michel Couprie
Jean Cousty
Yukiko Kenmochi
Nabil Mustafa

Organization

General Chairs

Michel Couprie	ESIEE Paris, Université Paris-Est, France
Jean Cousty	ESIEE Paris, Université Paris-Est, France
Yukiko Kenmochi	CNRS, Université Paris-Est, France
Nabil Mustafa	ESIEE Paris, Université Paris-Est, France

Local Organizing Committee

Michel Couprie	ESIEE Paris, Université Paris-Est, France
Jean Cousty	ESIEE Paris, Université Paris-Est, France
Yukiko Kenmochi	CNRS, Université Paris-Est, France
Camille Kurtz	Université Paris-Descartes, France
Nabil Mustafa	ESIEE Paris, Université Paris-Est, France
Vincent Nozick	Université Paris-Est Marne-la-Vallée, France
Pascal Romon	Université Paris-Est Marne-la-Vallée, France
Deise Santana Maia	Université Paris-Est, France
Lama Tarsissi	Université Paris-Est, France

Steering Committee

David Coeurjolly	CNRS, Université de Lyon, France
Éric Andrès	Université de Poitiers, France
Gunilla Borgefors	Uppsala University, Sweden
Srečko Brlek	Université du Québec à Montréal, Canada
Isabelle Debled-Rennesson	Université de Lorraine, France
Andrea Frosini	Università di Firenze, Italy
María José Jiménez	Universidad de Sevilla, Spain
Bertrand Kerautret	Université de Lyon, France
Walter Kropatsch	Technische Universität Wien, Austria
Jacques-Olivier Lachaud	Université de Savoie, France
Nicolas Normand	Université de Nantes, France

Program Committee

Éric Andrès	Université de Poitiers, France
Partha Bhowmick	Indian Institute of Technology Kharagpur, India
Isabelle Bloch	Telecom ParisTech, France
Srečko Brlek	Université du Québec à Montréal, Canada
Sara Brunetti	Università degli Studi di Siena, Italy
Guillaume Damiand	CNRS, Université de Lyon, France

Rocio Gonzalez-Diaz	Universidad de Sevilla, Spain
Yan Gérard	Université Clermont Auvergne, France
Thierry Géraud	EPITA, France
Atsushi Imiya	IMIT, Chiba University, Japan
Bertrand Kerautret	Université de Lyon, France
Reinhard Klette	Auckland University of Technology, New Zeland
Walter G. Kropatsch	Technische Universität Wien, Austria
Jacques-Olivier Lachaud	Université de Savoie Mont-Blanc, France
Loïc Mazo	Université de Strasbourg, France
Laurent Najman	ESIEE Paris, Université Paris-Est, France
Nicolas Passat	Université de Reims Champagne-Ardenne, France
Pawel Pilarczyk	Gdansk University of Technology, Poland
Gabriella Sanniti di Baja	ICAR, CNR, Italy
Isabelle Sivignon	GIPSA-lab, CNRS, France
Robin Strand	Uppsala University, Sweden
Akihiro Sugimoto	National Institute of Informatics, Japan
Imants Svalbe	Monash University, Australia
Michael H. F. Wilkinson	University of Groningen, The Netherlands

Additional Reviewers

Alpers, Andreas
Bai, Xiang
Balazs, Peter
Baudrier, Etienne
Berthé, Valerie
Biswas, Ranita
Borgefors, Gunilla
Boutry, Nicolas
Brimkov, Valentin
Brlek, Srečko
Bruckstein, Alfred M.
Brun, Luc
Buzer, Lilian
Carlinet, Edwin
Chazalon, Joseph
Coeurjolly, David
Comic, Lidija
Crombez, Loïc
Da Fonseca, Guilherme D.
Debled-Rennesson, Isabelle
Dokladal, Petr
Domenjoud, Eric
Dulio, Paolo
Escolano, Francisco
Even, Philippe
Falcão, Alexandre
Fernique, Thomas
Feschet, Fabien
Foare, Marion
Gérard, Yan
Glisse, Marc
Gonzalez-Lorenzo, Aldo
Grompone Von Gio, Rafael
Grélard, Florent
Géraud, Thierry
Hartke, Stephen
Hosseinian, Mohammad
Hubert, Pascal
Jacob-Da Col, Marie-Andrée
Jimenez, María Jose
Jonathan, Weber
Kerautret, Bertrand
Keszegh, Balázs
Kiselman, Christer
Kropatsch, Walter
Krähenbühl, Adrien
Kupavskii, Andrey
Kurtz, Camille
Labbé, Sébastien
Lachaud, Jacques-Olivier
Largeteau-Skapin, Geëlle
Latecki, Longin Jan
Lidayova, Kristina
Lindeberg, Tony
Loménie, Nicolas
Mari, Jean-Luc
Mazo, Loïc
Millman, David L.
Miranda, Paulo
Naegel, Benoît
Nagy, Benedek
Najman, Laurent
Ngo, Phuc
Oda, Masahiro
On Vu Ngoc, Minh
Orts Escolano, Sergio
Palagyi, Kalman
Passat, Nicolas
Pelillo, Marcello

Peri, Carla
Perret, Benjamin
Pilarczyk, Pawel
Pluta, Kacper
Prasad, Dilip K.
Pritam, Siddharth
Provençal, Xavier
Puybareau, Élodie
Reutenauer, Christophe
Romon, Pascal
Ronse, Christian
Rossi, Luca
Roussillon, Tristan
Salembier, Philippe
Sati, Mukul
Schnider, Patrick
Sivignon, Isabelle
Sladoje, Nataša
Stanko, Tibor
Strand, Robin
Svensson, Stina
Talbot, Hugues
Tamaki, Toru
Telea, Alexandru
Thiel, Edouard
Tijdemans, Robert
Tochon, Guillaume
Tougne, Laure
Vacavant, Antoine
Valette, Sébastien
Van Droogenbroeck, Marc
Veelaert, Peter
Velasco-Forero, Santiago
Vialard, Anne
Wagner, Hubert
Welk, Martin
Wendling, Laurent
Xu, Yongchao
Zrour, Rita

Contents

Graph-Based Models, Analysis and Segmentation

Mathematical Morphology

Shape Representation, Recognition and Analysis

Geometric Computation

Discrete Geometric Models and Transforms

Digital Two-Dimensional Bijective Reflection and Associated Rotation

Eric Andres[1(✉)], Mousumi Dutt[2], Arindam Biswas[3], Gaelle Largeteau-Skapin[1], and Rita Zrour[1]

[1] University of Poitiers, Laboratory XLIM, ASALI, UMR CNRS 7252, BP 30179, 86962 Futuroscope Chasseneuil, France
eric.andres@univ-poitiers.fr

[2] Department of Computer Science and Engineering, St. Thomas' College of Engineering and Technology, Kolkata, India
duttmousumi@gmail.com

[3] Department of Information Technology, Indian of Engineering Science and Technology, Shibpur, Howrah, India
barindam@gmail.com

Abstract. In this paper, a new bijective reflection algorithm in two dimensions is proposed along with an associated rotation. The reflection line is defined by an arbitrary Euclidean point and a straight line passing through this point. The reflection line is digitized and the 2D space is paved by digital perpendicular (to the reflection line) straight lines. For each perpendicular line, integer points are reflected by central symmetry with respect to the reflection line. Two consecutive digital reflections are combined to define a digital bijective rotation about arbitrary center, i.e. bijective digital rigid motion.

Keywords: Digital reflection · Bijective digital rotation · Digital rotation · Bijective digital rigid motion

1 Introduction

Reflection transformation is one of the most basic linear transforms [1]. There are however surprisingly few works that deal with such transforms in the digital world, although they are the key to defining n-dimensional rotations [2]. Digital rotations have many applications such as template matching [3], object tracking [4], etc.

The problem is that digital transforms on square grid are usually not bijective. The goal of this paper is to propose a digital bijective reflection on a two-dimensional image and to use this reflection to propose a digital bijective rotation algorithm that works for an arbitrary rotation center (i.e. bijective digital rigid motion). The main motivation behind this work in n-dimensional bijective digital rotation [2]. Classical methods based on shear matrices proved somewhat difficult to extend to higher dimensions so the idea was to explore digital reflections.

M. Couprie et al. (Eds.): DGCI 2019, LNCS 11414, pp. 3–14, 2019.
https://doi.org/10.1007/978-3-030-14085-4_1

The theoretical discussion on the subject of digital rigid motions is presented in [5]. While the reflection transform in the digital world has not been studied much, digital rotations and more precisely bijective digital rotations has been a point of interest of the digital community for some years now [6,7]. In [8] a discretized rotation is defined as the composition of a Euclidean rotation with rounding operations. An incremental approach to discretized rotations is presented in [9]. A link between Gaussian integers and the bijective digital rotation is proposed in [10]. In [11], bijective rigid motion in 2D Cartesian grid are discussed. Other mentionable work is the characterization of some bijective rotations in 3D space [12]. The bijective rotation has also been studied on the hexagonal grid [13,14] and triangular grid [15].

A continuous tranform applied to a digital image is not, in general, bijective, neither injective nor surjective for that matter, even for an apparently very simple transform such as the reflection transform (see Fig. 1). Our idea is to digitize the Euclidean reflection straight line as naive digital lines (i.e. 8-connected digital lines) and to partition the two-dimensional space into naive digital lines that are perpendicular to the reflection line. A central symmetry is performed on the points of each *Perpendicular Digital Straight Line* (PDSL) according to the *Digital Reflection Straight Line* (DRSL). This is always possible because we can compute the exact pixel to pixel correspondance on each side of the DRSL for such naive digital lines. The main problem comes from the fact that the digital reflection line and a given digital perpendicular line may or may not intersect and so two different cases have to be considered.

The organization of the paper is as follows: Preliminaries in Sect. 2. Section 3 presents the mathematical foundation of bijective digital reflection. The bijective digital rotation about an arbitrary center can be determined by applying digital reflection twice (see Sect. 4). An error criteria based on the distance between the continuous and the digital rotated points are discussed in Sect. 4.2. The concluding remarks are stated in Sect. 5.

2 Preliminaries

Let $\{\boldsymbol{i}, \boldsymbol{j}\}$ denote the canonical basis of the 2-dimensional Euclidean vector space. Let $\mathbb{Z}^2$ be the subset of $\mathbb{R}^2$ that consists of all the integer points. A *digital (resp. Euclidean) point* is an element of $\mathbb{Z}^2$ (resp. $\mathbb{R}^2$). Two integer points $p = (p_x, p_y)$ and $q = (q_x, q_y)$ are said to be 8-connected if $|p_x - q_x| \leqslant 1$ and $|p_y - q_y| \leqslant 1$.

For $x \in \mathbb{R}$, $\lfloor x \rfloor$ is the biggest integer smaller or equal to x and $\lceil x \rceil$ is the smallest integer greater or equal to x.

A *naive digital straight line* is defined as all the integer points verifying $-\frac{max(|a|,|b|)}{2} \leqslant ax - by < \frac{max(|a|,|b|)}{2}$, where $\frac{a}{b}$ represents the slope of the *Digital Straight Line* (DSL) and $\omega = max(|a|, |b|)$ the arithmetical thickness [16]. A naive DSL is 8-connected such that if you remove any point of the line then it is not 8-connected anymore [16]. There are no simple points.

A *reflection transformation* $\mathcal{R}_{\theta,(x_o,y_o)} : \mathbb{R}^2 \mapsto \mathbb{R}^2$ reflects (or flips) a continuous point, like in a mirror, on a continuous straight line called the reflection line.

The reflection line is defined as the line with the vector director $v = (\sin\theta, \cos\theta)$ passing through a point of coordinates $(x_o, y_o) \in \mathbb{R}^2$. The corresponding digital reflection is denoted $R_{\theta,(x_o,y_o)} : \mathbb{Z}^2 \mapsto \mathbb{Z}^2$. A continuous rotation $\mathcal{R}ot_{\theta,(x_o,y_o)}$ of center (x_o, y_o) and angle θ can be defined as the composition of two continuous reflections $\mathcal{R}_{\alpha,(x_o,y_o)}$ and $\mathcal{R}_{\alpha+\frac{\theta}{2},(x_o,y_o)}$. The angle α is arbitrary.

3 Digital Reflection

3.1 Principle

Let us consider the continuous reflection transform, $\mathcal{R}_{\theta,(x_o,y_o)}$. It is not difficult to note that if we simply compose the continuous reflection with a digitization transform (such as for instance $\mathcal{D} : (x, y) \mapsto (\lfloor x + 1/2 \rfloor, \lfloor y + 1/2 \rfloor)$) then we have a transform that is, in general, neither injective nor surjective (see the middle image in Fig. 1).

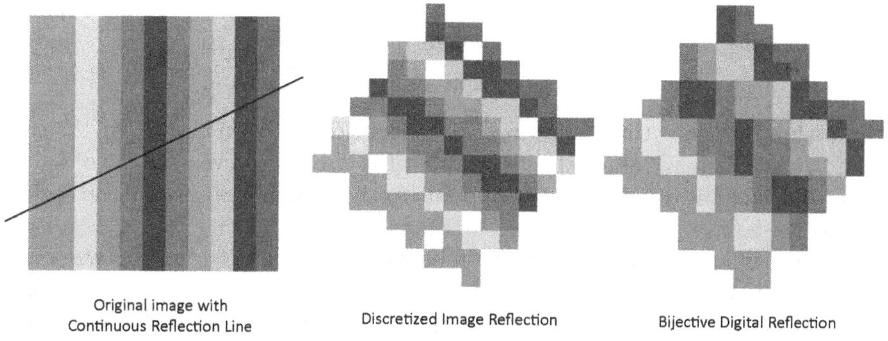

Fig. 1. Reflection transform applied to an image. On the left, the original image. In the center, the continuous reflection transform composed with a rounding function. On the right, our proposed bijective digital reflection

To avoid this problem and create a digital bijective reflection transform, a completely digital framework is proposed here based on the following digital primitives:

- A naive DSL called *digital reflection straight line* (DRSL);
- and a partition of the digital space with naive DSLs that are perpendicular to the digital reflection straight line. These digital lines are called *Perpendicular Digital Straight Lines* (PDSL).

The idea of the reflection method is the following. Let us consider a point p that belongs to a given digital perpendicular straight line P_k (k is a parameter identifying the PDSL in the partition). The point p may belong to the digital reflection line (in case there is an intersection) or lie on either side of that line. Let the digital reflection of p be determined in the following way (see Fig. 2):

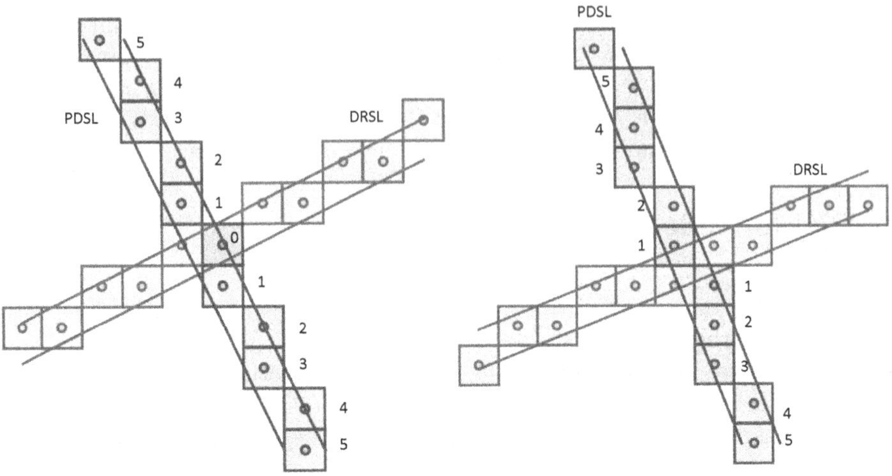

Fig. 2. The image of each point is the mirror image in the perpendicular line according to the digital reflection straight line. The image of a point numbered i is the opposite point numbered i. The point numbered 0 is its own image.

- let us first suppose that p lies in the intersection between the PDSL P_k and the DRSL. Such an intersection does not always exist between the two digital lines but when it does (in this case point p) then the image of p by the digital reflection is p;
- Let us now suppose that p does not belong to the intersection between the PDSL and the DRSL (whether such an intersection exists or not). The point p is then located on one side of the DRSL: the digital space is split by the DRSL into three regions, the DRSL and two regions on each side of it. All the integer points of P_k on each side of the DRSL can formally be ordered according to the distance to the DRSL with as first point the closest one to the DRSL. As we will see, there actually is no real ordering needed. Let us suppose that p is the nth point of the ordered list on its side and an integer point q is the nth point of the list on the other side of the DRSL. The digital reflection of p will then be the point q and vice-versa.

Figure 2 shows the two possible cases: on the left, the perpendicular digital line has an intersection point (the point marked as point 0) with the digital reflection line, while on the right there is no intersection point. The points marked by the number i (from 1 to 5) are images of each other. The image of the point marked 0 is itself. The idea is to create a digital transformation that is bijective and easily reversible. Let us now, in the next section, present the various definitions and mathematical details to make this work out.

3.2 Mathematical Details

Let us consider the continuous reflection $\mathcal{L}$ line defined by the point $(x_o, y_o) \in \mathbb{R}^2$ and the direction vector $v = (a, b) = (\sin\theta, \cos\theta)$. The analytical equation of the reflection line is given by: $\mathcal{L} = \{(x, y) \in \mathbb{R}^2 : a(x - x_o) - b(y - y_o) = 0\}$.

The *digital reflection straight line (DRSL)* the digitization (as naive DSL) of $\mathcal{L}$ defined as integer points verifying:

$$DRSL : -\frac{max(|a|,|b|)}{2} \leqslant a(x - x_o) - b(y - y_o) < \frac{max(|a|,|b|)}{2}$$

The *perpendicular digital straight lines (PDSL)* P_k are naive DSLs defined as all the integer points (x, y) verifying:

$$P_k : \frac{(2k-1)\max(|a|,|b|)}{2} \leqslant b(x - x_o) + a(y - y_o) < \frac{(2k+1)\max(|a|,|b|)}{2}$$

It is easy to see that all the PDSLs are perpendicular to the DRSL and that the set of PDSLs, as a set of naive digital lines, partitions the two dimensional digital space.

Let us now suppose for what follows, w.l.o.g, that $-\pi/4 \leqslant \theta \leqslant \pi/4$, then $0 \leqslant |a| \leqslant b$ and thus $\max(|a|,|b|) = b$. The DRSL is then defined by $-b/2 \leqslant a(x-x_o)-b(y-y_o) < b/2$ and the PDSLs are then defined by $P_k : (2k-1)b/2 \leqslant b(x - x_o) + a(y - y_o) < (2k + 1)b/2$.

The method supposes that we are able to formally order the points of a given perpendicular line that are located on either side of the reflection line. This is always possible because the PDSLs are naive lines and for an angle verifying $-\pi/4 \leqslant \theta \leqslant \pi/4$, there is one and only one point per integer ordinate y (For angles $\pi/4 \leqslant \theta \leqslant 3\pi/4$, we will have, symmetrically, one and only one point per integer abscissa x in a given PDSL).

More precisely, $(2k - 1)b/2 \leqslant b(x - x_o) + a(y - y_o) < (2k + 1)b/2$ means that $(2k - 1)/2 + x_o - \frac{a}{b}(y - y_o) \leqslant x < (2k + 1)/2 + x_o - \frac{a}{b}(y - y_o)$. Since $\frac{2k+1}{2} - \frac{2k-1}{2} = 1$, there is one and only one value x. For a given ordinate y, the abscissa x in the PDSL P_k is given by the following function:

$$\mathcal{X}(y) = \left\lceil \frac{(2k-1)}{2} + x_o - (a/b)(y - y_o) \right\rceil$$

Let us note that for a given integer point $p(x, y)$, it is easy to determine the PDSL P_k it belongs to: $(2k - 1)b/2 \leqslant b(x - x_o) + a(y - y_o) < (2k + 1)b/2$ leads to $2k \leqslant 2\frac{b(x-x_o)+a(y-y_o)}{b} + 1 < 2k + 2$ and thus:

$$k = \left\lfloor (x - x_o) + \frac{a}{b}(y - y_o) + \frac{1}{2} \right\rfloor$$

The next question that arises is the determination and the localization of the potential intersection point between the digital reflection line and a given

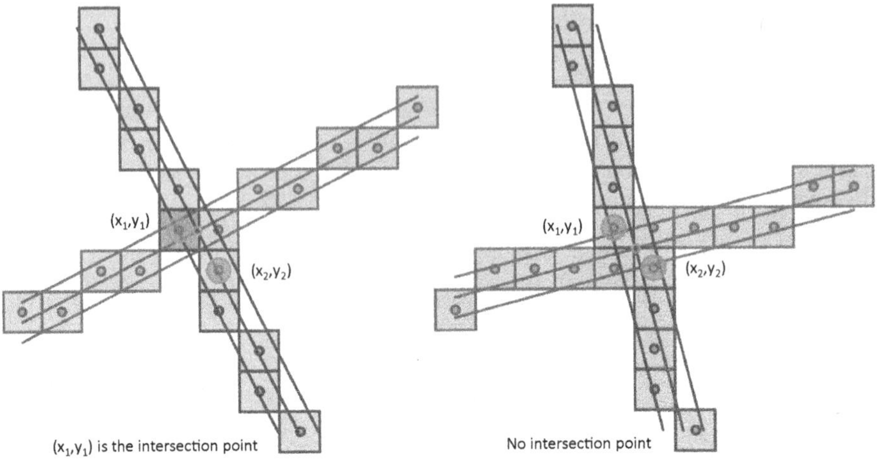

Fig. 3. The green point represents the intersection point between the continuous reflection line and the continuous perpendicular line. The orange disks mark the two candidate points for the digital intersection. On the left (x_1, y_1) is the digital intersection point between the DRSL and the PDSL. On the right there is no digital intersection point. (Color figure online)

digital perpendicular line. As already mentioned, there may be 0 or 1 digital intersection points between the DRSL and a given PDSL since both are naive digital lines. We propose here a simple criterion to determine the existence of such an intersection point and its localization, if it exists. If the point does not exist, it will yield the closest point of the PDSL to the DRSL on either side of the digital reflection line.

Firstly, let us note that the DRSL is the digitization of the continuous reflection line of equation: $a(x - x_o) - b(y - y_o) = 0$ which lies in the middle of the strip defining the DRSL. In the same way, the continuous straight line of equation $b(x - x_o) + a(y - y_o) = kb$ lies in the middle of the strip defining the perpendicular digital straight line P_k. The intersection point of those two continuous lines is given by $\left(kb^2 + x_o, abk + y_o\right)$ (since $a^2 + b^2 = 1$). It is easy to understand that the only digital intersection point, if it exists, has an ordinate value given by either $\lceil abk + y_o \rceil$ or $\lfloor abk + y_o \rfloor$. This is a direct consequence of the fact that the DRSL is a naive digital line of slope between -1 and 1 and thus with only one integer point per integer ordinate value. We have therefore a very simple test to determine the existence and coordinates of the intersection point (see Fig. 3):

- let us define $(x_1, y_1) = (\mathcal{X}(\lceil abk + y_o \rceil), \lceil abk + y_o \rceil)$ and $(x_2, y_2) = (\mathcal{X}(\lfloor abk + y_o \rfloor), \lfloor abk + y_o \rfloor)$
- If (x_1, y_1) belongs to the DRSL then there is an intersection point, (x_1, y_1) in this case. A central symmetry of the PDSL points can be performed around the ordinate value y_1. The digital reflection of a point $p(x_p, y_p) \in P_k$ is

Algorithm 1. REFLECTION TRANSFORM $R_{\theta,(x_o,y_o)}$

Input : $(x, y) \in \mathbb{Z}^2, (x_o, y_o) \in \mathbb{R}^2, -\pi/4 \leqslant \theta < \pi/4$
Output: $(x', y') \in \mathbb{Z}^2$
1 $k = \lfloor (x - x_o) + \frac{a}{b}(y - y_o) + \frac{1}{2} \rfloor$
2 Function $\mathcal{X}(y) : \mathbb{Z} \mapsto \mathbb{Z} : \mathcal{X}(y) = \left\lceil \frac{(2k-1)}{2} + x_o - (a/b)(y - y_o) \right\rceil$
3 $(x_1, y_1) = (\mathcal{X}(\lceil abk + y_o \rceil), \lceil abk + y_o \rceil)$
4 $(x_2, y_2) = (\mathcal{X}(\lfloor abk + y_o \rfloor), \lfloor abk + y_o \rfloor)$
5 If $-b/2 \leqslant a(x_1 - x_o) - b(y_1 - y_o) < b/2$ Then
6 $\quad (x', y') = (\mathcal{X}(2y_1 - y), 2y_1 - y)$
7 Elseif $-b/2 \leqslant a(x_2 - x_o) - b(y_2 - y_o) < b/2$ Then
8 $\quad (x', y') = (\mathcal{X}(2y_2 - y), 2y_2 - y)$
9 Else $(x', y') = (\mathcal{X}(y_1 + y_2 - y), y_1 + y_2 - y)$
10 **return** (x', y')

given by $(\mathcal{X}(2y_1 - y_p), 2y_1 - y_p)$. Let us note here that the central symmetry around the ordinate y_1 does not mean here that there is a central symmetry around the point (x_1, y_1) because space partition would not be guaranteed and thus bijectivity would be lost. The abcissa $\mathcal{X}(2y_1 - y_p)$ is computed so that the reflected point still belongs to P_k.
- Else, if (x_2, y_2) belongs to the the DRSL then there is an intersection point (x_2, y_2). A central symmetry of the PDSL points can be performed around the ordinate value y_2. The digital reflection of a point $p\,(x_p, y_p) \in P_k$ is given by $(\mathcal{X}(2y_2 - y_p), 2y_2 - y_p)$.
- Otherwise there is no intersection point and (x_1, y_1) and (x_2, y_2) are the first points of the PDSL on each side of the Reflection line. The central symmetry can be performed around the ordinate value $\frac{y_1+y_2}{2}$. The digital reflection of a point $p\,(x_p, y_p) \in P_k$ is given by $(\mathcal{X}(y_1 + y_2 - y_p), y_1 + y_2 - y_p)$.

The important point here is that, since there is only one point per ordinate y, a central symmetry on the ordinate leads directly to the reflection point. Algorithm 1 presents the digital reflection transform method.

3.3 Bijectivity of the Digital Reflection Transform

Algorithm 1 provides a digital reflection method for an integer point. It is not very difficult to see that this defines a bijective digital reflection transform. Let us briefly summarizes the arguments for this:

- The computation of the value k in line 1 of Algorithm 1 yields the same value for any point of a given PDSL (and only for those).
- Line 6, 8 and 9 ensure that the image of an integer point of a given PDSL P_k is an integer point belonging to P_k. Indeed the ordinate values $2y_1 - y, 2y_2 - y$ and $y_1 + y_2 - y$ are integers if y_1, y_2 and y are integers and $(\mathcal{X}(y), y)$ is, by construction, a point of P_k if y is an integer.

- The reflection of the reflection of an integer point $(x, y) \in P_k$ is the integer point (x, y). Let us consider $(x, y) \mapsto (\mathcal{X}(y'), y') \mapsto (\mathcal{X}(y''), y'')$ where $y' = 2y_1 - y$, $y' = 2y_2 - y$ or $y' = y_1 + y_2 - y$. Let us note first that $(x, y) = (\mathcal{X}(y), y)$. For all three cases, we have respectively $y'' = 2y_1 - y' = 2y - 1 - (2y_1 - y) = y$, $y'' = 2y_2 - y' = 2y_2 - (2y_2 - y) = y$ and $y'' = y_1 + y_2 - y' = y_1 + y_2 - (y_1 + y_2 - y) = y$ which proves the point.

4 Reflection Based Rotation

As mentioned in the preliminaries, a continuous rotation transform $\mathcal{R}ot_{\theta,(x_o,y_o)}$ of center (x_o, y_o) and angle θ can be defined as the composition of two reflections based on two reflection lines passing through (x_o, y_o) with an angle $\theta/2$ between the two lines. In the same way, we define the digital rotation by:

$$Rot_{\theta,(x_o,y_o)}(x, y) = \left(R_{\alpha+\frac{\theta}{2},(x_o,y_o)} \circ R_{\alpha,(x_o,y_o)}\right)(x, y)$$

Compared to previous digital rotations methods [6,9,13,17], there is an extra parameter that comes into play: the angle α. Each value of α defines the exact same continuous rotation although not the same digital rotation. Let us note as well that, since the digital reflection transform is bijective, the digital rotation based on the reflection transforms will be bijective as well. Furthermore, the inverse transform is easily defined. This last point may seem obvious but it is not because a digital transform is bijective that the inverse transform is easily computed.

4.1 Rotation Evaluation Criteria

In order to evaluate the "quality" of such a digital rotation, let us present some simple error measures [6]. Each grid point has one and only one image through a bijective digital rotation but that does not mean that the digital rotation is a good approximation of the continuous one. To measure how "wrong" we are by choosing the digital rotation over the continuous one, we are considering two distance criteria that were proposed in [6]. The considered distance is the Euclidean distance. Let us denote $\mathcal{R}ot_{\theta(x_o,y_o)}(p)$ the continuous rotation of center (x_o, y_o) and angle θ of a grid point $p \in \mathbb{Z}^2$ and $Rot_{\theta,(x_o,y_o)}(p)$ the digital rotation of center (x_o, y_o) of a grid point p. The *Maximum Distance quality criteria (MD)* consists in computing $\max_{p\in\mathbb{Z}^2}\left(d(\mathcal{R}ot_{\theta,(x_o,y_o)}(p), Rot_{\theta,(x_o,y_o)}(p))\right)$. The *average distance quality criteria (AD)* consists in computing $avg_{p\in\mathbb{Z}^2}\left(d(\mathcal{R}ot_{\theta(x_o,y_o)}(p), Rot_{\theta(x_o,y_o)}(p))\right)$, where $avg_{p\in\mathbb{Z}^2}$ is the average distance over the grid.

The idea is to measure what error is made by using the digital rotation instead of the continuous one in terms of (Euclidean) distance to the optimal position (to the continuously rotated point). Let us finish by noting that for the best known digital bijective rotations [10,11,17] and for the angles where

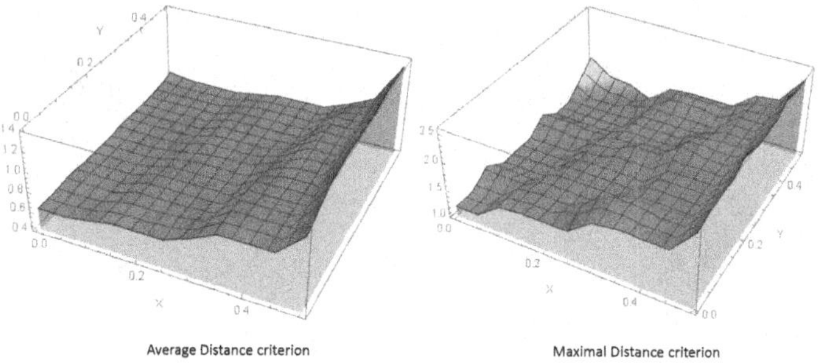

Fig. 4. Rotations of center (x, y), angle $\pi/8$ and $\alpha = 0$

the digitized rotations are bijective (except for the trivial $k\pi/2$ angles), the maximum MD values are $\frac{\sqrt{2}}{2} \approx 0.7$ and the AD values are ≈ 0.3. Note also that this method is not bijective for all angles. Other notable methods, that work for all angles, based on shear matrices [6,7], have MD values of ≈ 1.1 and AD values of ≈ 0.6.

4.2 Evaluation Analysis

It is difficult to give a completely detailed rotation evaluation here due to lack of space as there are four different parameters that can influence the outcome of the evaluation criteria: the center coordinates (x_o, y_o) and the angles θ and α.

At first, we wanted to have an idea on the effect of the rotation center on the rotation error measure with the angle $\alpha = 0$ and $\theta = \pi/8$. This means that the first reflection corresponds to a reflection with the continuous reflection line $y = y_o$, for $-1/2 \leqslant y_o < 1/2$ and the digital reflection $y = 0$. In Fig. 4 we represented the influence of moving the center on the interval $[0, 1/2]^2$. As can be seen, the error measures are almost not affected by the shift on the ordinate but greatly by a shift on the abscissa of the rotation center. This can be explained

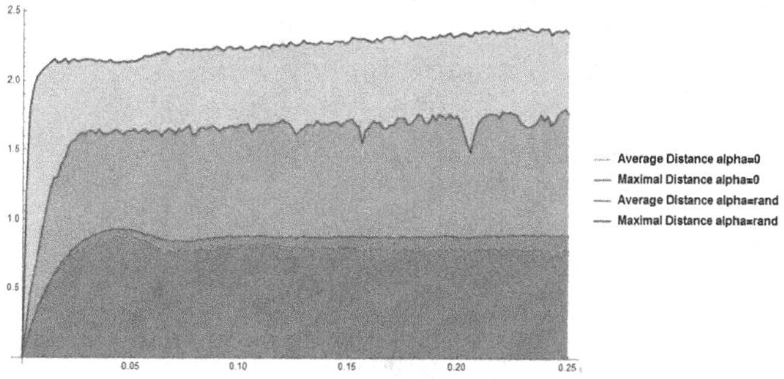

Fig. 5. For each rotation angle $0 \leqslant \theta \leqslant \pi/4$, the average error measures of 500 rotations with random center and respectively $\alpha = 0$ and random α (with $0 \leqslant \alpha \leqslant \pi/4$)

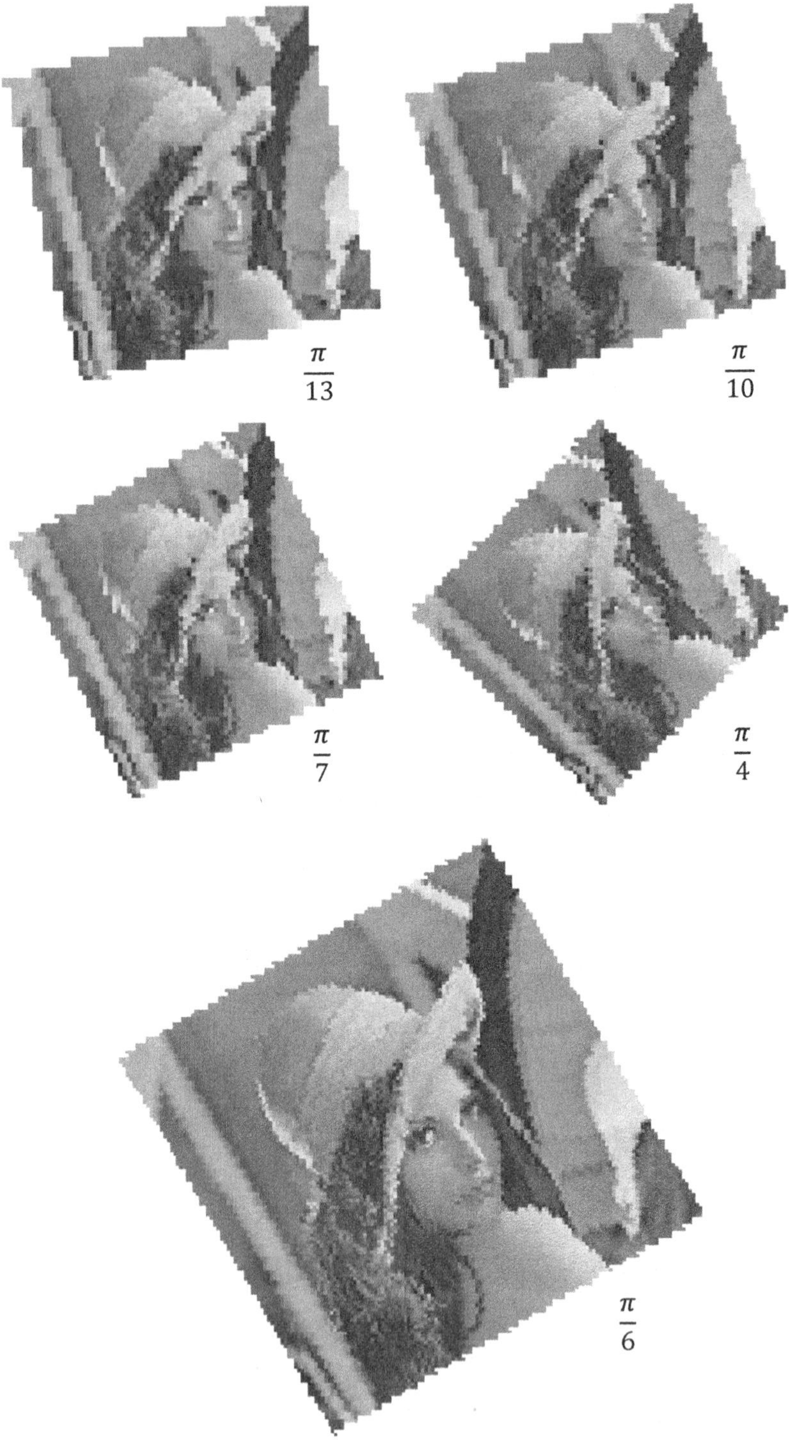

Fig. 6. Rotation of the Lena image by various angles and randomly chosen center.

by the fact that the shift on the perpendicular lines has a direct influence on the lateral error that adds up to the global error. Let us note that the surfaces are similar looking for other couple of angles (α, θ) even though the amplitude of the error measure may be different.

Next, we wanted to have an idea on the effect of the angle α on the rotation error measures. Although the angle α always defines the same continuous rotation, it changes the digital rotation. So, for each rotation angle $0 \leqslant \theta \leqslant \pi/4$, the average error measures for 500 rotations with randomly chosen center and respectively angle $\alpha = 0$ and a randomly chosen angle $0 \leqslant \alpha \leqslant \pi/4$. The result can be seen in Fig. 5. It can be seen that the angle α has only a minor influence on the average error but the maximal error is typically significantly increased for an angle $\alpha > 0$. It is interesting to note that this is not always so clear: the digital rotation of angle $\pi/6$ and center $(0,0)$ has an average / maximum error of $(0.6367, 1.3972)$ for $\alpha = 0$ and $(0.5726, 1.4186)$ for $\alpha = \pi/6$. What can be noticed as well is that the error ratio between $\alpha = 0$ and random chosen α is relatively stable over all the rotation angles. The error measures in general are significantly higher than those obtained for the shear based method [6] that is also defined for all angles and centers. The reflection based rotation is however easier to implement for arbitrary centers: see [6].

5 Conclusion

A novel bijective digital reflection transform in two dimensions has been proposed. The reflection line is defined by an arbitrary Euclidean point and a straight line passing through this point. A rotation can be defined by the composition of two reflections, and so an associated new bijective digital rotation transform has been proposed. This new bijective digital rotation is defined for all angles and for all rotation centers, defining a rigid motion transform (see Fig. 6 that illustrates the rotation of an image at various angles and centers). In average the distance between a continuously and a digitally rotated point is about 0.8 (for a pixel of side 1) which is more than for other known methods. However, this method is much simpler to implement for arbitrary centers and this reflection based rotation seems better suited for extensions to higher dimensions which is not easily done with previous methods. This is the main perspective for the future: exploring digital reflection based bijective rotations in three and higher dimensions.

References

1. Goodman, R.: Alice through looking glass after looking glass: the mathematics of mirrors and kaleidoscopes. Am. Math. Mon. **111**(4), 281–298 (2004). http://www.jstor.org/stable/4145238
2. Richard, A., Fuchs, L., Largeteau-Skapin, G., Andres, E.: Decomposition of nD-rotations: classification, properties and algorithm. Graph. Model. **73**(6), 346–353 (2011)

3. Fredriksson, K., Mäkinen, V., Navarro, G.: Rotation and lighting invariant template matching. Inf. Comput. **205**(7), 1096–1113 (2007)
4. Yilmaz, A., Javed, O., Shah, M.: Object tracking: a survey. ACM Comput. Surv. **38**(4), 13 (2006)
5. Pluta, K., Romon, P., Kenmochi, Y., Passat, N.: Bijective digitized rigid motions on subsets of the plane. J. Math. Imaging Vis. **59**(1), 84–105 (2017)
6. Andres, E.: The quasi-shear rotation. In: Miguet, S., Montanvert, A., Ubéda, S. (eds.) DGCI 1996. LNCS, vol. 1176, pp. 307–314. Springer, Heidelberg (1996). https://doi.org/10.1007/3-540-62005-2_26
7. Carstens, H.-G., Deuber, W., Thumser, W., Koppenrade, E.: Geometrical bijections in discrete lattices. Comb. Probab. Comput. **8**(1–2), 109–129 (1999)
8. Nouvel, B., Rémila, E.: Characterization of bijective discretized rotations. In: Klette, R., Žunić, J. (eds.) IWCIA 2004. LNCS, vol. 3322, pp. 248–259. Springer, Heidelberg (2004). https://doi.org/10.1007/978-3-540-30503-3_19
9. Nouvel, B., Rémila, É.: Incremental and transitive discrete rotations. In: Reulke, R., Eckardt, U., Flach, B., Knauer, U., Polthier, K. (eds.) IWCIA 2006. LNCS, vol. 4040, pp. 199–213. Springer, Heidelberg (2006). https://doi.org/10.1007/11774938_16
10. Roussillon, T., Coeurjolly, D.: Characterization of bijective discretized rotations by gaussian integers, Technical report, LIRIS UMR CNRS 5205, January 2016
11. Pluta, K., Romon, P., Kenmochi, Y., Passat, N.: Bijective rigid motions of the 2D cartesian grid. In: Normand, N., Guédon, J., Autrusseau, F. (eds.) DGCI 2016. LNCS, vol. 9647, pp. 359–371. Springer, Cham (2016). https://doi.org/10.1007/978-3-319-32360-2_28
12. Pluta, K., Romon, P., Kenmochi, Y., Passat, N.: Bijectivity certification of 3D digitized rotations. In: Bac, A., Mari, J.-L. (eds.) CTIC 2016. LNCS, vol. 9667, pp. 30–41. Springer, Cham (2016). https://doi.org/10.1007/978-3-319-39441-1_4
13. Pluta, K., Roussillon, T., Coeurjolly, D., Romon, P., Kenmochi, Y., Ostromoukhov, V.: Characterization of bijective digitized rotations on the hexagonal grid, Technical report, HAL, submitted to Journal of Mathematical Imaging and Vision, June 2017. https://hal.archives-ouvertes.fr/hal-01540772
14. Andres, E.: Shear based bijective digital rotation in hexagonal grids. Submitted to Pattern recognition Letters 2018. https://hal.archives-ouvertes.fr/hal-01900148v1
15. Andres, E., Largeteau-Skapin, G., Zrour, R.: Shear based bijective digital rotation in triangular grids. Submitted to Pattern recognition Letters (2018). https://hal.archives-ouvertes.fr/hal-01900149v1
16. Reveillès, J.-P.: Calcul en nombres entiers et algorithmique, Ph.D. thesis. Université Louis Pasteur, Strasbourg, France (1991)
17. Jacob, M.-A., Andres, E.: On discrete rotations. In: Discrete Geometry for Computer Imagery, p. 161 (1995)

Digital Curvature Evolution Model for Image Segmentation

Daniel Antunes[1(✉)], Jacques-Olivier Lachaud[1], and Hugues Talbot[2]

[1] Université Savoie Mont Blanc, LAMA, UMR CNRS 5127, 73376 Chambéry, France
daniel.martins-antunes@univ-smb.fr,
jacques-olivier.lachaud@univ-savoie.fr
[2] CentraleSupelec Université Paris-Saclay, équipe Inria GALEN, Paris, France
hugues.talbot@centralesupelec.fr

Abstract. Recent works have indicated the potential of using curvature as a regularizer in image segmentation, in particular for the class of thin and elongated objects. These are ubiquitous in biomedical imaging (e.g. vascular networks), in which length regularization can sometime perform badly, as well as in texture identification. However, curvature is a second-order differential measure, and so its estimators are sensitive to noise. The straightforward extensions to Total Variation are not convex, making them a challenge to optimize. State-of-art techniques make use of a coarse approximation of curvature that limits practical applications.

We argue that curvature must instead be computed using a multigrid convergent estimator, and we propose in this paper a new digital curvature flow which mimics continuous curvature flow. We illustrate its potential as a post-processing step to a variational segmentation framework.

Keywords: Multigrid convergence · Digital estimator · Curvature · Shape optimization · Image segmentation

1 Introduction

Geometric quantities are particularly useful as regularizers, especially when some prior information about the object geometry is known. Length penalization is a general purpose regularizer and the literature is vast on models that make use of it [1,3]. However, this regularizer shows its limitations when segmenting thin and elongated objects, as it tends to return disconnected solutions. Such drawback can sometime be overcome by injecting curvature regularization [7].

One of the first successful uses of curvature in image processing is the inpainting algorithm described in [12]. The authors evaluate the elastica energy along the level lines of a simply-connected image to reconstruct its occluded parts.

This work has been partly funded by CoMeDiC ANR-15-CE40-0006 research grant.

M. Couprie et al. (Eds.): DGCI 2019, LNCS 11414, pp. 15–26, 2019.
https://doi.org/10.1007/978-3-030-14085-4_2

The non-intersection property of level lines allows the construction of an efficient dynamic programming algorithm. Nonetheless, it is still a challenging task to inject curvature in the context of image segmentation.

The state-of-art methods are difficult to optimize and not scalable [7,13,18]. In order to achieve reasonable running times, such approaches make use of coarse curvature estimations for which the approximation error is unknown. Improving the quality of the curvature estimator has an important impact on the accuracy of the results, but is computationally too costly in these methods. Recently, new multigrid convergent estimators for curvature have been proposed [5,16,17], motivating us to search for models in which they can be applied.

In this work, we investigate the use of a more suitable curvature estimator with multigrid convergent property and its application as a boundary regularizer in a digital flow minimizing its squared curvature. Our method decreases the elastica energy of the contour and its evolution is evaluated on several digital flows. Finally, we present an application of the model as a post-processing step in a segmentation framework. The code is freely available on github[1].

Outline. Section 2 reviews the concept of multigrid convergence and highlights its importance for the definition of digital estimators. Next, we describe two convergent estimators used in this paper, one for tangent and the other for curvature. They are used in the optimization model and in the definition of the digital elastica. Section 3 describes the proposed curvature evolution model along with several illustrations of digital flows. Section 4 explains how to use the evolution model as a post-processing step in an image segmentation framework. Finally, Sect. 5 discusses the results and points directions for future work.

2 Multigrid Convergent Estimators

A digital image is the result of some quantization process over an object X lying in some continuous space of dimension 2 (here). For example, the Gauss digitization of X with grid step $h > 0$ is defined as

$$D_h(X) = X \cap (h\mathbb{Z})^2.$$

Given an object X and its digitization $D_h(X)$, a digital estimator $\hat{u}$ for some geometric quantity u is intended to compute $u(X)$ by using only the digitization. This problem is not well-posed, as the same digital object could be the digitization of infinitely many objects very different from X. Therefore, a characterization of what constitutes a good estimator is necessary.

Let u be some geometric quantity of X (e.g. tangent, curvature). We wish to devise a digital estimator $\hat{u}$ for u. It is reasonable to state that $\hat{u}$ is a good estimator if $\hat{u}(D_h(X))$ converges to $u(X)$ as we refine our grid. For example, counting pixels is a convergent estimator for area (with a rescale of h^2); but counting boundary pixels (with a rescale of h) is not a convergent estimator

[1] https://www.github.com/danoan/BTools.

for perimeter. Multigrid convergence is the mathematical tool that makes this definition precise. Given any subset Z of $(h\mathbb{Z})^2$, we can represent it as a union of axis-aligned squares with edge length h centered on the point of Z. The topological boundary of this union of cubes is called *h-frontier* of Z. When $Z = D_h(X)$, we call it *h-boundary of* X and denote it by $\partial_h X$.

Definition 1 (Multigrid convergence for local geometric quantites). *A local discrete geometric estimator $\hat{u}$ of some geometric quantity u is (uniformly) multigrid convergent for the family $\mathbb{X}$ if and only if, for any $X \in \mathbb{X}$, there exists a grid step $h_X > 0$ such that the estimate $\hat{u}(D_h(X), \hat{x}, h)$ is defined for all $\hat{x} \in \partial_h X$ with $0 < h < h_X$, and for any $x \in \partial X$,*

$$\forall \hat{x} \in \partial_h X \text{ with } \|\hat{x} - x\|_\infty \leq h, \|\hat{u}(D_h(X), \hat{x}, h) - u(X, x)\| \leq \tau_X(h),$$

where $\tau_X : \mathbb{R}^+ \setminus \{0\} \to \mathbb{R}^+$ has null limit at 0. This function defines the speed of convergence of $\hat{u}$ towards u for X.

For a global geometric quantity (e.g. perimeter, area, volume), the definition remains the same, except that the mapping between ∂X and $\partial_h X$ is no longer necessary.

Multigrid convergent estimators provide a quality guaranty and should be preferred over non-multigrid convergent ones. In the next section, we describe two estimators that are important for our purpose.

2.1 Tangent and Perimeter Estimators

The literature presents several perimeter estimators that are multigrid convergent (see [4,6] for a review), but in order to define the digital elastica we need a local estimation of length and we wish that integration over these local length elements gives a multigrid convergent estimator for the perimeter.

Definition 2 (Elementary Length). *Let a digital curve C be represented as a sequence of grid vertices in a grid cell representation of digital objects (in grid with step h). Further, let $\hat{\theta}$ be a multigrid convergent estimator for tangent. The elementary length $\hat{s}(e)$ at some grid edge $e \in C$ is defined as*

$$\hat{s}(e) = h \cdot \hat{\theta}(l) \cdot or(e),$$

where $or(e)$ denotes the grid edge orientation.

The integration of the elementary length along the digital curve is a multigrid convergent estimator for perimeter if one uses the λ-MST [10] tangent estimator (see [9]).

2.2 Integral Invariant Curvature Estimator

Generally, an invariant σ is a real-valued function from some space Ω which value is unaffected by the action of some group $\mathfrak{G}$ on the elements of the domain

$$x \in \Omega, g \in \mathfrak{G}, \sigma(x) = v \longleftrightarrow \sigma(g \cdot x) = v.$$

Perimeter and curvature are examples of invariants for shapes on $\mathbb{R}^2$ with respect to the euclidean group (rigid transformations). Definition of integral area invariant and its one-to-one correspondence with curvature is proven in [11].

Definition 3 (Integral area invariant). *Let $X \subset \mathbb{R}^2$ and $B_r(p)$ the ball of radius r centered at point p. Further, let $\mathbb{1}_X(\cdot)$ be the characteristic function of X. The integral area invariant $\sigma_{X,r}(\cdot)$ is defined as*

$$\forall p \in \partial X, \quad \sigma_{X,r}(p) = \int_{B_r(p)} \mathbb{1}_X(x)dx.$$

The value $\sigma_{X,r}(p)$ is the intersection area of ball $B_r(p)$ with shape X. By locally approximating the shape at point $p \in X$, one can rewrite the intersection area $\sigma_{X,r}(p)$ in the form of the Taylor expansion [14]

$$\sigma_{X,r}(p) = \frac{\pi}{2}r^2 - \frac{\kappa(X,p)}{3}r^3 + O(r^4),$$

where $\kappa(X,p)$ is the curvature of X at point p. By isolating κ we can define a curvature estimator

$$\tilde{\kappa}(p) := \frac{3}{r^3}\left(\frac{\pi r^2}{2} - \sigma_{X,r}(p)\right), \tag{1}$$

Such an approximation is convenient as one can simply devise a multigrid convergent estimator for the area.

Definition 4. *Given a digital shape $D \subset (h\mathbb{Z})^2$, a multigrid convergent estimator for the area $\widehat{Area}(D,h)$ is defined as*

$$\widehat{Area}(D,h) := h^2\, Card(D)\,. \tag{2}$$

In [5], the authors combine the approximation (1) and digital estimator (2) to define a multigrid convergent estimator for the curvature.

Definition 5 (Integral Invariant Curvature Estimator). *Let $D \subset (h\mathbb{Z})^2$ a digital shape. The integral invariant curvature estimator is defined as*

$$\hat{\kappa}_r(D,x,h) := \frac{3}{r^3}\left(\frac{\pi r^2}{2} - \widehat{Area}\left(B_r(x) \cap D, h\right)\right).$$

This estimator is robust to noise and can be extended to estimate the mean curvature of three dimensional shapes.

3 Digital Curvature Evolution Model

Our goal is to deform a digital object in order to minimize the elastica energy along its contour. Our strategy is to define the digital elastica by using the elementary length and the integral invariant curvature estimators and minimize its underlying binary energy. However, the derived energy is of order four and difficult to optimize. Therefore we propose an indirect method to minimize it.

3.1 Ideal Global Optimization Model

We evaluate the quality of a boundary by evaluating the elastica energy along it. Let $\kappa(\cdot)$ denote the curvature function evaluated on the contour of some shape X. In continuous terms, the elastica energy is defined as

$$E(X) = \int_{\partial X} (\alpha + \beta\kappa^2) ds, \quad \text{for } \alpha \geq 0, \beta \geq 0.$$

We are going to use the digital version of the energy, using multigrid convergent estimators. The energy, in this case, is also multigrid convergent.

$$\hat{E}(D_h(X)) = \sum_{x \in \partial D_h(X)} \hat{s}(x) \left(\alpha + \beta \hat{\kappa}_r^2(D_h(X), x, h) \right), \tag{3}$$

where $\partial D_h(X)$ denotes the 4-connected boundary of $D_h(X)$. In the following we omit the grid step h to simplify expressions (or, putting it differently, we assume that X is rescaled by $1/h$ and we set $h = 1$).

A segmentation energy can be devised by including some data fidelity term g in (3), but we need to restrict the optimization domain to consistent regions. We cannot properly estimate length and curvature along anything different from a boundary. Let Ω be the digital domain and $\mathcal{T}$ the family of subsets of Ω satisfying the property

$$D \in \mathcal{T} \implies D \subset \Omega \text{ and } 4B(\partial D),$$

where $4B(\cdot)$ is the 4-connected closed boundary predicate.

For some $\gamma > 0$, the segmented region $D^\star$ is defined as

$$D^\star = \arg\min_{D \in \mathcal{T}} \sum_{x \in \partial D} \hat{s}(x) \left(\alpha + \beta \hat{\kappa}_r{}^2(D, x) \right) + \gamma \cdot g(D). \tag{4}$$

In its integer linear programming model [18], Schoenemann restricts the optimization domain by imposing a set of constraints that enforces compact sets as solutions. However, the main difficulty here is the minimization of a third order binary energy. We are going to explore an alternative strategy.

3.2 Nonzero Curvature Identification

We can use the curvature estimator to detect regions of positive curvature. Given a digital object D embedded in a domain Ω, we define its pixel boundary set $P(D)$ as

$$P(D) = \{ x \mid x \in D, |\mathcal{N}_4(x) \cap D| < 4 \},$$

where $\mathcal{N}_4(x)$ denotes the 4-adjacent neighbor set of x (without x). The following optimization regions are important in our process.

$$\begin{array}{lll} O & = P(D) & \text{Optimization region.} \\ F & = D - P(D) & \text{Trust foreground.} \\ B & = \Omega - D & \text{Trust background.} \\ A & = P(F) \cup P(B) & \text{Computation region.} \end{array}$$

Note that our definition of the optimization region guarantees that only connected solutions are produced. The computation region is defined around O for symmetric issues. We proceed by minimizing the squared curvature energy along A with respect to the optimization region O.

$$Y^\star = \underset{Y \in \{0,1\}^{|O|}}{\arg\min} \sum_{p \in A} \hat{\kappa}_r^2(p). \tag{5}$$

We expand the squared curvature estimator for a single point $p \in A$ using (1). Define constants $c_1 = (3/r^3)^2$, $c_2 = \pi r^2/2$. Hence,

$$\begin{aligned} \hat{\kappa}_r^2(p) &= c_1 \cdot \left(c_2 - \sigma_{D,r}(p) \right)^2 \\ &= c_1 \cdot \left(c_2^2 - 2c_2\sigma_{D,r}(p) + \sigma_{D,r}(p)^2 \right). \end{aligned}$$

Let $F_r(p) \subset F$ denote the intersection set between the estimating ball applied at p with the foreground region. The subset $Y_r(p) \subset Y$ is defined analogously. Substituting $\sigma_{D,r}(p) = |F_r(p)| + \sum_{y_i \in Y_r(p)} y_i$, we obtain

$$\hat{\kappa}_r^2(p) = c_1 \cdot \left(C + 2\left(|F_r(p)| - c_2\right) \cdot \sum_{y_i \in Y_r(p)} y_i + \sum_{y_i \in Y_r(p)} y_i^2 + 2 \cdot \sum_{\substack{y_i, y_j \in Y_r(p) \\ i<j}} y_i y_j \right),$$

where $C = c_2^2 - 2c_2 \cdot |F_r(p)| + |F_r(p)|^2$. By ignoring constants and multiplication factors and using the binary character of the variables, problem (5) is equivalent to

$$Y^\star = \underset{Y \in \{0,1\}^{|O|}}{\arg\min} \sum_{p \in A} \left((1/2 + |F_r(p)| - c_2) \cdot \sum_{y_i \in Y_r(p)} y_i + \sum_{\substack{y_i, y_j \in Y_r(p) \\ i<j}} y_i y_j \right). \tag{6}$$

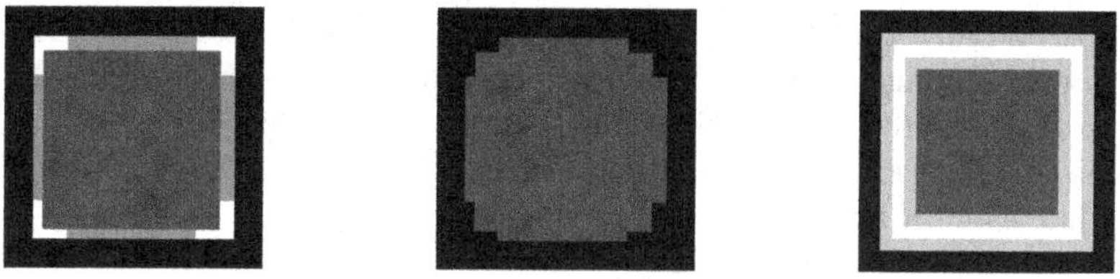

(a) (b) (c)

Fig. 1. (a): White pixels denote variables labeled as one by QPBOP, while light gray pixels variables are labeled zero; (b): Inverted labeling; (c): Regions of interest: Background (black); Foreground (dark gray); Computation (light gray); and Optimization (white) regions.

Energy (6) is non-submodular, and minimizing it is a NP-Hard problem in the general case. The QPBOP [15] method provides a partial labeling for the optimization variables with the property that all labeled variables belong to some optimal solution. However, some pixels may be left unlabeled. The optimization method is further discussed in Sect. 3.4. For $r = 3$, evaluation of the model on a digital square produces Fig. 1a.

We interpret positive curvature at some point p as a shortage of intersection points between the digital object and the estimating ball. The curvature can be reduced if the estimating ball is pulled towards the interior of the digital object, which is done by removing the highlighted pixels in Fig. 1a. In other words, the partial labeling is inverted, and unlabeled pixels remain unchanged. Points with negative curvature are detected similarly if we evaluate the model on the digital object complement.

3.3 Digital Curvature Flow

We derive the digital curvature flow by iteratively evaluating model (6) with a slight modification. We extend the computation region to take into account more level sets (ℓ) of the original object. As a practical consequence, zones of high curvature are more likely to be detected, leading to a smaller number of unlabeled pixels by QPBOP.

$$A = \bigcup_{i \leq \ell} \partial F^{-i} \cup \partial B^{-i},$$

where the $-i$ exponent means an erosion by a square of side i. Figure 1c illustrates the different regions of the optimization model.

At each flow step, the model is evaluated twice. In the second evaluation, we take care of concavities. The model is executed on $\overline{D^{+1}}$, the complement of the dilation by a square of side one, and we swap foreground and background regions. Figure 2 presents several digital curvature flows and Table 1 lists the initial and final digital elastica energy for the tested shapes.

We observe that the choice of ball radius (r) and level sets (ℓ) should take into account the image scale. For example, using a radius that is too large might lead to a disconnected intersection zone and the accuracy of the estimator is compromised. This explains the difference in flows in Fig. 2. In practice, we observe that using a ball of radius 3 is sufficient to produce good results while achieving a reasonable running time.

Table 1. Evaluation of digital elastica ($\alpha = 0$) for start and end curves of the flow. Except for the ball, all the elastica energies were decreased significantly.

	Digital elastica			
	Ball	Triangle	Square	Flower
Initial value	0.156	2.55	1.81	4.196
$\mathbf{r = 3, \ell = 3}$	0.192	0.335	0.286	0.298
$r = 5, \ell = 2$	0.156	0.556	0.423	1.477
$r = 5, \ell = 3$	0.166	0.375	0.321	0.364
$r = 5, \ell = 4$	0.207	0.508	0.311	0.174
$\mathbf{r = 5, \ell = 5}$	0.193	0.52	0.278	0.163
$\mathbf{r = 10, \ell = 10}$	0.216	1.33	0.333	0.159

3.4 Optimization Method

Let F be a function of n binary variables, i.e.

$$F(y_1, \cdots, y_n) = \sum_i F_i(y_i) + \sum_{i<j} F_{i,j}(y_i, y_j)$$

Function F is submodular if and only if the following inequality holds for each pairwise term $F_{i,j}$ [8]

$$F_{i,j}(0,0) + F_{i,j}(1,1) \leq F_{i,j}(0,1) + F_{i,j}(1,0)$$

Energy (6) is non-submodular and optimizing it is a difficult problem, which constrained us to use heuristics and approximation algorithms. The QPBO method [15] transforms the original problem in a max-flow/min-cut formulation and returns a full optimal labeling for submodular energies. For non-submodular energies the method is guaranteed to return a partial labeling with the property that the set of labeled variables is part of an optimal solution. That property is called partial optimality.

In practice, QPBO can leave many pixels unlabeled. There exist two extensions of QPBO that ameliorate this limitation: QPBOI (improve) and QPBOP (probe). The first is an approximation method that is guaranteed to not increase the energy, but we lost the property of partial optimality. The second is an exact method which is reported to label more variables than QPBO. We use QPBOP. The extended computation region also regularizes the energy and we have checked that it induces a higher number of labeled variables.

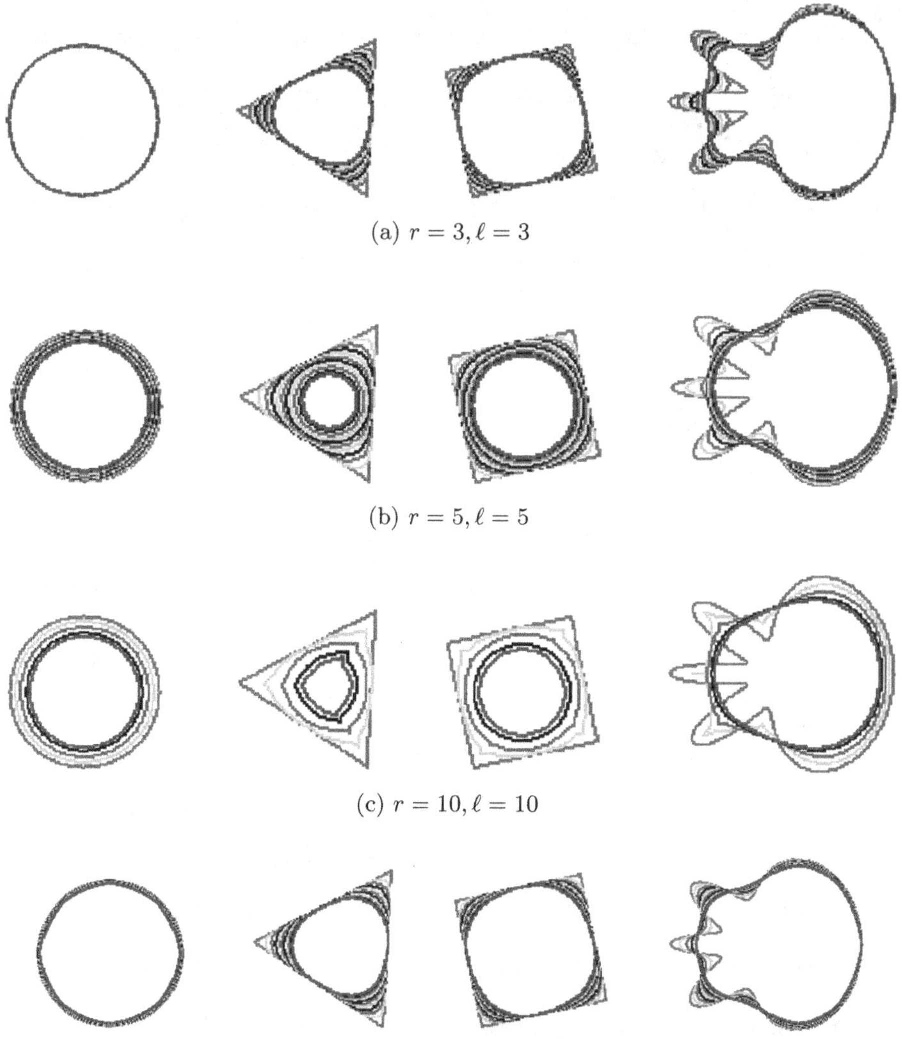

(a) $r = 3, \ell = 3$

(b) $r = 5, \ell = 5$

(c) $r = 10, \ell = 10$

(d) $r = 5, \ell = 5$ (1.5× scaling)

Fig. 2. Digital curvature flow for four different shapes. A total of 20 iterations were executed for each flow, except for (c) (7 iterations). Curves are displayed every 2 iterations. The initial curve (countour of the original shape) is in red and the end curve in blue. (Color figure online)

4 Application in Image Segmentation

The digital curvature flow can be applied as a post-processing step in an image segmentation framework. We use graph cut [2] as segmentation method and we execute the flow for n iterations. We include the graph cut data fidelity term g and standard length penalization s to the flow energy.

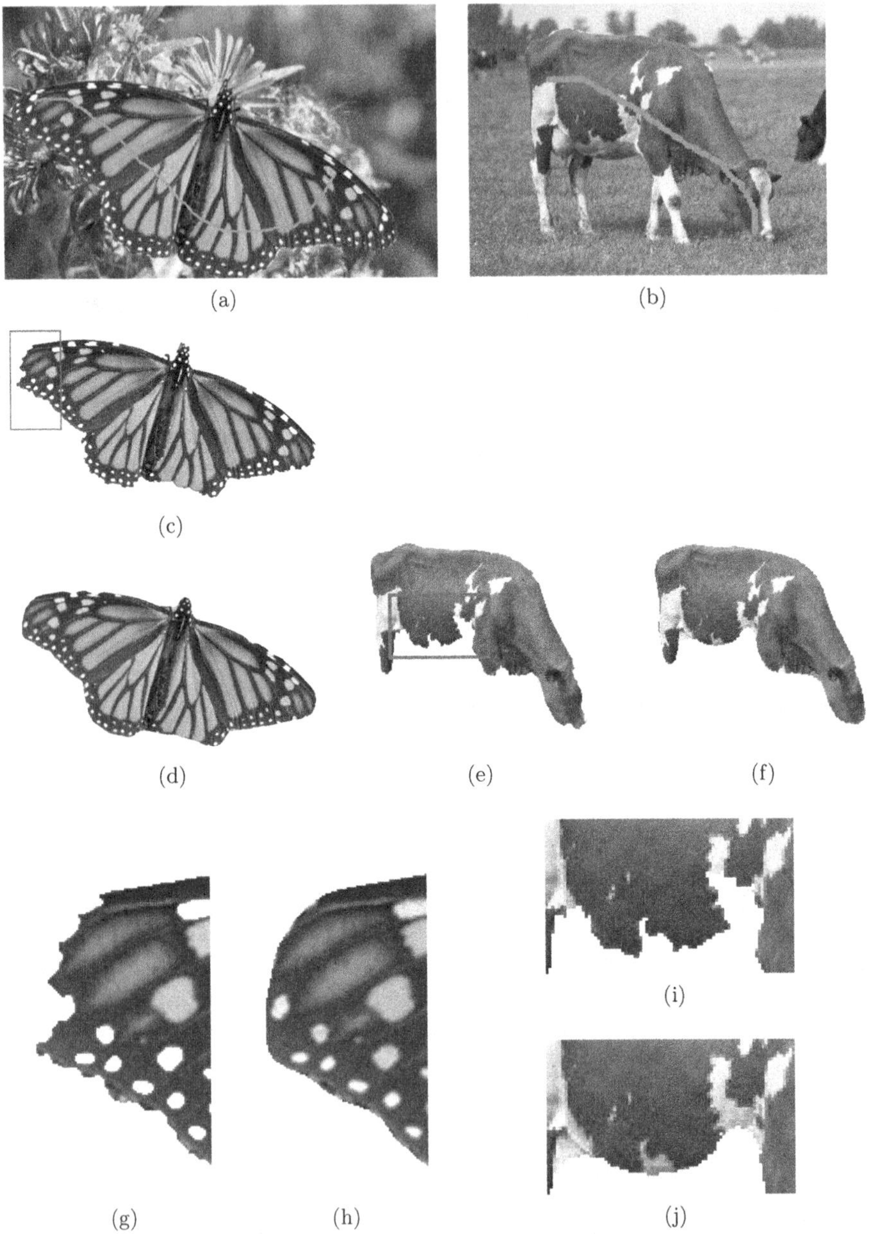

Fig. 3. Digital flow post-processing results for a total of 5 iterations ($\alpha = 0.1, \beta = 1, \gamma = 1$).

$$\min_{Y} \sum_{y \in Y} (\alpha \cdot s(y) + \gamma \cdot g(y)) + \beta \cdot \sum_{p \in A} \hat{\kappa}_r^2(p). \tag{7}$$

Let $\mathcal{N}_4(p)$ denote the four neighborhood of pixel p. Length penalization is defined as

$$s(p) = \sum_{p_k \in \mathcal{N}_4(p)} (p - p_k)^2.$$

In Fig. 3 we show some results. The flow clearly regularizes the contour of figures produced by the comparison segmentation via graph cut. In both figures, the flow is able to correct zones of high positive curvature and expand regions of low negative curvature, but without invading the background zone. Nonetheless, the flow does not expand zones of convexity. Unfortunately, as we follow a local strategy, we are unable to expand some zones that clearly belongs to the segmented object, like the cow's leg.

5 Conclusion

We have shown that the integral invariant curvature estimator can be integrated into an optimization model and can be applied together with classical penalization terms as length and data fidelity in an image processing task. We demonstrated its potential by designing a digital curvature flow that mimics continuous flow in an accurate way. Finally, we show how it can be used as a post-processing tool in an image segmentation framework.

We have some directions for future work. First, optimize the code and evaluate a runtime analysis to compare with competitor methods. We also think that we can reformulate the model in [18] using the digital estimator $\hat{\kappa_r}$.

References

1. Appleton, B., Talbot, H.: Globally optimal geodesic active contours. J. Math. Imaging Vis. **23**(1), 67–86 (2005)
2. Boykov, Y.Y., Jolly, M.P.: Interactive graph cuts for optimal boundary & region segmentation of objects in n-d images. In: Proceedings Eighth IEEE International Conference on Computer Vision, ICCV 2001, vol. 1, pp. 105–112 (2001)
3. Caselles, V., Kimmel, R., Sapiro, G.: Geodesic active contours. Int. J. Comput. Vis. **22**(1), 61–79 (1997)
4. Coeurjolly, D., Klette, R.: A comparative evaluation of length estimators of digital curves. IEEE Trans. Pattern Anal. Mach. Intell. **26**(2), 252–258 (2004)
5. Coeurjolly, D., Lachaud, J.-O., Levallois, J.: Integral based curvature estimators in digital geometry. In: Gonzalez-Diaz, R., Jimenez, M.-J., Medrano, B. (eds.) DGCI 2013. LNCS, vol. 7749, pp. 215–227. Springer, Heidelberg (2013). https://doi.org/10.1007/978-3-642-37067-0_19
6. Coeurjolly, D., Lachaud, J.O., Roussillon, T.: Multigrid convergence of discrete geometric estimators. In: Brimkov, V., Barneva, R. (eds.) Digital Geometry Algorithms. LNCVB, vol. 2, pp. 395–424. Springer, Dordrecht (2012). https://doi.org/10.1007/978-94-007-4174-4_13

7. El-Zehiry, N.Y., Grady, L.: Fast global optimization of curvature. In: 2010 IEEE Computer Society Conference on Computer Vision and Pattern Recognition, pp. 3257–3264, June 2010
8. Kolmogorov, V., Zabin, R.: What energy functions can be minimized via graph cuts? IEEE Trans. Pattern Anal. Mach. Intell. **26**(2), 147–159 (2004)
9. Lachaud, J.O.: Non-Euclidean spaces and image analysis: Riemannian and discrete deformable models, discrete topology and geometry. Ph.D. thesis, Université Sciences et Technologies - Bordeaux I, December 2006. https://tel.archives-ouvertes.fr/tel-00396332
10. Lachaud, J.O., Vialard, A., de Vieilleville, F.: Fast, accurate and convergent tangent estimation on digital contours. Image Vis. Comput. **25**(10), 1572–1587 (2007)
11. Manay, S., Hong, B.-W., Yezzi, A.J., Soatto, S.: Integral invariant signatures. In: Pajdla, T., Matas, J. (eds.) ECCV 2004. LNCS, vol. 3024, pp. 87–99. Springer, Heidelberg (2004). https://doi.org/10.1007/978-3-540-24673-2_8
12. Masnou, S., Morel, J.M.: Level lines based disocclusion. In: Proceedings 1998 International Conference on Image Processing, ICIP98 (Cat. No. 98CB36269), vol. 3, pp. 259–263, October 1998
13. Nieuwenhuis, C., Toeppe, E., Gorelick, L., Veksler, O., Boykov, Y.: Efficient squared curvature. In: 2014 IEEE Conference on Computer Vision and Pattern Recognition, pp. 4098–4105, June 2014
14. Pottmann, H., Wallner, J., Huang, Q.X., Yang, Y.L.: Integral invariants for robust geometry processing. Comput. Aided Geom. Des. **26**(1), 37–60 (2009)
15. Rother, C., Kolmogorov, V., Lempitsky, V.S., Szummer, M.: Optimizing binary MRFs via extended roof duality. In: 2007 IEEE Conference on Computer Vision and Pattern Recognition, pp. 1–8 (2007)
16. Roussillon, T., Lachaud, J.-O.: Accurate curvature estimation along digital contours with maximal digital circular arcs. In: Aggarwal, J.K., Barneva, R.P., Brimkov, V.E., Koroutchev, K.N., Korutcheva, E.R. (eds.) IWCIA 2011. LNCS, vol. 6636, pp. 43–55. Springer, Heidelberg (2011). https://doi.org/10.1007/978-3-642-21073-0_7
17. Schindele, A., Massopust, P., Forster, B.: Multigrid convergence for the MDCA curvature estimator. J. Math. Imaging Vis. **57**(3), 423–438 (2017)
18. Schoenemann, T., Kahl, F., Cremers, D.: Curvature regularity for region-based image segmentation and inpainting: a linear programming relaxation. In: 2009 IEEE 12th International Conference on Computer Vision, pp. 17–23, September 2009

Rhombic Dodecahedron Grid—Coordinate System and 3D Digital Object Definitions

Ranita Biswas[1(✉)], Gaëlle Largeteau-Skapin[2], Rita Zrour[2], and Eric Andres[2]

[1] Department of Computer Science and Engineering, Indian Institute of Technology, Roorkee, India
biswas.ranita@gmail.com
[2] University of Poitiers, Laboratory XLIM, ASALI, UMR CNRS 7252, BP 30179, 86962 Futuroscope Chasseneuil, France
{gaelle.largeteau.skapin,rita.zrour,eric.andres}@univ-poitiers.fr

Abstract. We propose a new non-orthogonal basis to express the 3D Euclidean space in terms of a regular grid. Every grid point, each represented by integer 3-coordinates, corresponds to rhombic dodecahedron centroid. Rhombic dodecahedron is a space filling polyhedron which represents the close packing of spheres in 3D space and the Voronoi structures of the face centered cubic (FCC) lattice. In order to illustrate the interest of the new coordinate system, we propose the characterization of 3D digital plane with its topological features, such as the interrelation between the thickness of the digital plane and the separability constraint we aim to obtain. A characterization of a 3D digital sphere with relevant topological features is proposed as well with the help of a 48 symmetry that comes with the new coordinate system.

Keywords: Rhombic dodecahedron · FCC grid · 3D coordinate system · Digital plane · Digital sphere

1 Introduction

The cubic grid $\mathbb{Z}^3$ is the most frequently used grid for three-dimensional images. Recently non standard, three-dimensional grids received a lot of interest with in particular applications to networks [18], image processing [6,12,19], computer vision [7] and many other fields. Among non standard grids we can cite face centered cubic (FCC), body centered cubic (BCC), honeycomb [3] and diamond grids [17]. Many works discussed coordinate system on the 3D grids such as [10] for cubic grids, [13,14] for hexagonal and triangular grids and [4,5,20] for FCC grids.

Space filled entirely by rhombic dodecahedra may be sliced by specific planes to reveal patterns of hexagons. In [9], authors proposed a generation algorithm for discrete spheres on the FCC grid using layered discrete annuli on hexagonal

M. Couprie et al. (Eds.): DGCI 2019, LNCS 11414, pp. 27–37, 2019.
https://doi.org/10.1007/978-3-030-14085-4_3

grids; the general idea is to propose digital primitive generation algorithms that are compatible with additive manufacturing techniques.

The face centered cubic grid is the densest possible packing in three dimensions [20]. The shape of the cells in an FCC grid is rhombic dodecahedron, which is a space filling polyhedron described by 12 faces, 24 edges, and 14 vertices. The FCC grid can be seen as the union of four disjoint cubic grids [4,8,15,16]. In such a rhombic dodecahedron grid system, sometimes distinction is made between the cells and not all the integer coordinates correspond to a cell. These make it difficult to use geometric transforms in those grid systems.

Our main contribution in this work is to propose a new non-orthogonal basis coordinate system that provides integer coordinates for rhombic dodecahedron centroids covering the whole set of integer points. The interest of such a system would be to facilitate work on such a grid. This will be highly beneficial considering the usage of this grid in image reconstruction and similar applications due to its correspondence with hexagonal grids in 2D, and also in additive manufacturing using spherical cells utilizing the emulation of densest sphere packing in space. The interest of the new coordinate system is illustrated in this paper with the (topological) characterization of various classes of digital planes and spheres in the FCC grid. The coordinate system offers a 48-symmetry which can be used to efficiently construct digital spheres and other symmetric objects.

Organization of the Paper. In Sect. 2, we present some preliminaries and basics as well as the Rhombic Dodecahedron Grid (RDG) and Nagy's coordinate system [4]. In Sect. 3, we detail our coordinate system and we present some of its properties. In Sect. 4, we define the digital spheres and planes in this coordinate system and explore the required thickness for different models. We conclude and present perspectives in Sect. 5.

2 Preliminaries

In this section, we recall some of the basic terminologies and definitions relevant to our problem. We put forward the general definitions of a digital plane, a digital sphere, and the different topological models already established in literature for conventional cubic grid. We also talk about the formal notion of grids and coordinate systems and present the concept and importance of designing a rhombic dodecahedron grid.

We consider a 3D space. We call integer points the points that have integer coordinates on the three axes. Let us denote d_2 as the Euclidean distance in the regular Cartesian space. The standard definitions of digital planes and digital spheres are as follows.

Definition 1 (Analytical Plane [1]). *The digital plane P corresponding to the Euclidean plane $ax + by + cz + d = 0$, with ρ the thickness of the digital plane, is the set of integer coordinate points $p(x, y, z)$ verifying the following inequalities:*

$$-\frac{\rho}{2} \leq ax + by + cz + d < \frac{\rho}{2}.$$

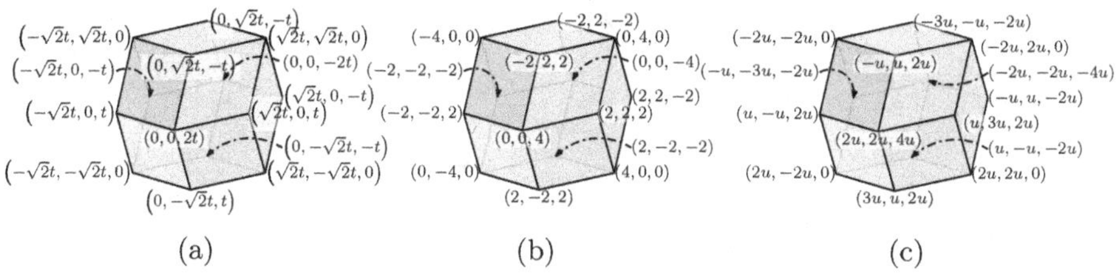

Fig. 1. (a) Unit dodecahedron in the Cartesian coordinate system, where $t = \frac{1}{\sqrt{3}}$. (b) Combinatorial coordinates for vertices in the face centered cubic grid [4]. (c) Vertex coordinates in the proposed non-orthogonal coordinate system, where $u = \frac{1}{4}$.

Definition 2 (Andres Analytical Sphere [2]). *The digital sphere S centered in c, of radius r and thickness ρ, is the set of integer coordinate points p verifying the following inequalities:*

$$(r - \frac{\rho}{2}) \leq d_2(c, p) < (r + \frac{\rho}{2}).$$

Different values of ρ in these definitions leads to different digitization models, e.g. 2-separating, 0-separating, etc. A digital object is l-separating if its inverse is not l-connected. Hence, a digital plane is 2-separating if the half-spaces specified by it are not 2-connected. Note that, 0-, 1-, and 2-connectivity here refer to the classical 26-, 18-, 6-connectivity respectively. Please consult [1,2] for more details on digital hyperplanes and hyperspheres in cubic grid.

We consider a grid based on rhombic dodecahedrons (which are space-filling). A rhombic dodecahedron has 12 face-connected neighbors and 6 strictly vertex-connected neighbors. A consistent coordinate system for the RDG will help us to utilize this grid in 3D with similar advantages as hexagonal grid gives over square grid in 2D, specially in the domain of imaging, tomography, etc. As rhombic dodecahedron space filling arrangement emulates the close sphere packing in space, RDG can be used for additive manufacturing using spherical material particle powder.

The main problem of the unit dodecahedron (see Fig. 1) is that neither its vertices nor its centroids form an integer coordinate grid. It is however simple to get integer coordinates for either centroids or vertices or both using a scale and a rotation as presented in [4]. In this paper [4], a combinatorial 3-coordinate system for cells in the face centered cubic grid is presented and some of its properties are detailed; authors made a rescaling with factor 2, in order to assign to each cell the Cartesian coordinates of its center and to have only integer coordinates for faces and edges. However, the main drawback of this system, which we want to address here, is that this coordinate system leads to two different and intricate grids, and any integer point in this space may either represent a centroid or a vertex. This will be a major cause of anomaly while doing transformation operations such as rotation, shear etc. and construction of different digital objects in this grid.

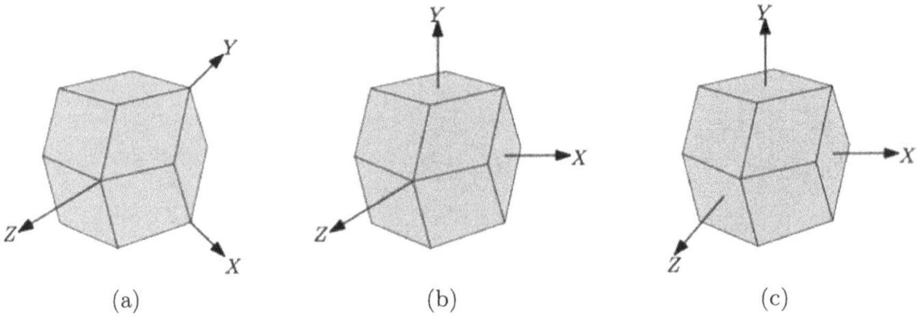

Fig. 2. Coordinate system transformation: (a) Initial arrangement of axes in Cartesian coordinate system, (b) After 45° rotation along Z-axis, (c) After required transformation to get non-orthogonal coordinate system.

3 Rhombic Dodecahedron Grid

In this section, we propose a non-orthogonal coordinate system for the rhombic dodecahedron grid (RDG). The grid is formed on the rhombic dodecahedron close packing such that each integer point corresponds to centroid of a rhombic dodecahedron and vice versa. We talk about the conversion of this new coordinate system to and from the Cartesian coordinate system and define the corresponding distance metric. We also show that there exists a 48-symmetry in RDG using this new coordinate system which can be utilized further to propose efficient algorithm for geometric object construction in this type of grid.

3.1 Non-orthogonal Coordinate System

We propose a non-orthogonal coordinate system to represent the RDG where each grid point represents the centroid of a rhombic dodecahedron cell. We transform the well-known Cartesian space to obtain non-orthogonal grid vectors. Figure 2(a) shows the initial positioning of the rhombic dodecahedron centered at the origin with respect to the coordinate axes. All three axes are orthogonal to each other and pass through vertices of the rhombic dodecahedron.

First, we give a clockwise rotation of 45° with respect to the positive Z-axis. This gives us the arrangement as shown in Fig. 2(b). Notice that, to simplify the visualization, we have kept the rhombic dodecahedron fixed while transforming the coordinate axes. In this state, X- and Y-axes passes through the center of faces of the rhombic dodecahedron and Z-axis still passes through the same vertex.

Next, we keep the X- and Y-axis same while bending the Z-axis away from the positive X- and Y-axes to make it pass through the center of the face of the rhombic dodecahedron. At the third step, a scaling may be required to make the rhombic dodecahedron cells unit in size; we consider our cells to have unit distance between opposite faces and do not show any explicit scaling step to obtain the same.

The following two lemmas states the transformation needed for a point in the new coordinate system to and from the Cartesian coordinate system. The steps for rotation and transformation of Z-axis is described in the corresponding proofs.

Lemma 1 (Cartesian to New). *The transformation of a point $p(x, y, z)$ from the Cartesian coordinate system to the new coordinate system is given by,*

$$\mathcal{N}(p) = \left(\frac{-\sqrt{2}y + z}{2}, \frac{\sqrt{2}x + z}{2}, z\right).$$

Proof. As explained using Fig. 2, the steps to get the new coordinate values for a point $p(x, y, z)$ in the Cartesian space are a clockwise rotation of 45° along positive Z-axis followed by a transformation to make the Z-axis non-orthogonal with respect to the XY-plane. The final transformation can be calculated as follows.

$$\begin{aligned}\mathcal{N}(p) &= \begin{bmatrix} \frac{1}{2} & -\frac{1}{2} & \frac{1}{2} \\ \frac{1}{2} & \frac{1}{2} & \frac{1}{2} \\ 0 & 0 & 1 \end{bmatrix} \begin{bmatrix} \cos 45^\circ & -\sin 45^\circ & 0 \\ \sin 45^\circ & \cos 45^\circ & 0 \\ 0 & 0 & 1 \end{bmatrix} (x, y, z) \\ &= \left(\frac{-\sqrt{2}y + z}{2}, \frac{\sqrt{2}x + z}{2}, z\right).\end{aligned}$$

□

Lemma 2 (New to Cartesian). *The transformation of a point $p(x, y, z)$ from the new coordinate system to the Cartesian coordinate system is given by,*

$$\mathcal{C}(p) = \left(\frac{2y - z}{\sqrt{2}}, \frac{-2x + z}{\sqrt{2}}, z\right).$$

Proof. To get the Cartesian coordinate values for a point $p(x, y, z)$ in the new coordinate space, we need to do the reverse transformation as given in Lemma 1. Therefore, the first step is to make the Z-axis orthogonal to the XY-plane and then doing a anticlockwise rotation of 45° with respect to the positive Z-axis. Following calculation can be used to obtain the same.

$$\begin{aligned}\mathcal{C}(p) &= \begin{bmatrix} \cos(-45^\circ) & -\sin(-45^\circ) & 0 \\ \sin(-45^\circ) & \cos(-45^\circ) & 0 \\ 0 & 0 & 1 \end{bmatrix} \begin{bmatrix} 1 & 1 & -1 \\ -1 & 1 & 0 \\ 0 & 0 & 1 \end{bmatrix} (x, y, z) \\ &= \left(\frac{2y - z}{\sqrt{2}}, \frac{-2x + z}{\sqrt{2}}, z\right).\end{aligned}$$

□

Lemma 3 (Euclidean Distance). *Let $p(x, y, z)$ and $p'(x', y', z')$ be two points in the new coordinate system. The Euclidean distance between p and p' is given by,*

$$d_2(p, p') = \sqrt{2((x - x')^2 + (y - y')^2 + (z - z')^2 - (x - x')(z - z') - (y - y')(z - z'))}.$$

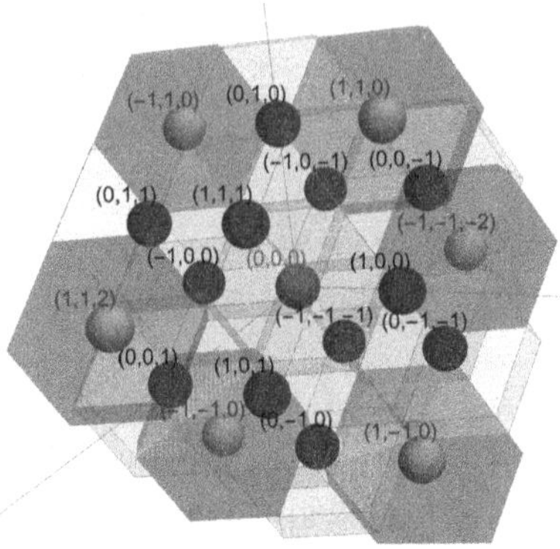

Fig. 3. The twelve face-connected neighbors and the six strictly vertex-connected neighbors in the proposed non-orthogonal coordinate system.

Proof. The proof follows directly from Lemma 2.

$$
\begin{aligned}
d_2(p,p') &= \|\mathcal{C}(p) - \mathcal{C}(p')\| \\
&= \left\| \left(\tfrac{2y-z}{\sqrt{2}}, \tfrac{-2x+z}{\sqrt{2}}, z\right) - \left(\tfrac{2y'-z'}{\sqrt{2}}, \tfrac{-2x'+z'}{\sqrt{2}}, z'\right) \right\| \\
&= \sqrt{\tfrac{1}{2}\left((2y - z - 2y' + z')^2 + (-2x + z + 2x' - z')^2 + 2(z - z')^2\right)} \\
&= \sqrt{2\left((x - x')^2 + (y - y')^2 + (z - z')^2 - (x - x')(z - z') - (y - y')(z - z')\right)}.
\end{aligned}
$$

□

We simplify the distance calculation by dropping the constant factor from the Euclidean distance as calculated in Lemma 3, and define our new distance which is to be used for further calculations.

Definition 3 (New Distance). *Let $p(x,y,z)$ and $p'(x',y',z')$ be two points in the new coordinate system. The new distance between p and p' is defined as,*

$$
\begin{aligned}
d(p,p') &= \tfrac{1}{\sqrt{2}} d_2(p,p') \\
&= \sqrt{(x - x')^2 + (y - y')^2 + (z - z')^2 - (x - x')(z - z') - (y - y')(z - z')}.
\end{aligned}
$$

A rhombic dodecahedron has 12 face-connected neighbors and 6 strictly vertex-connected neighbor as shown in Fig. 3. We make the following observations regarding the neighborhood of each rhombic dodecahedron cell.

Observation 1 (Neighbors). *Rhombic dodecahedron cell p and p' are face (respectively strictly vertex) connected neighbors iff the new distance between p and p' i.e. $d(p,p')$ is equal to 1 (respectively $\sqrt{2}$).*

Proof. From the positioning of the rhombic dodecahedron cell with respect to the non-orthogonal coordinate axes, as shown in Fig. 2(c), following are the face-connected neighbors of a cell (x,y,z): $(x+1,y,z)$, $(x-1,y,z)$, $(x,y+1,z)$,

$(x, y-1, z)$, $(x, y, z+1)$, $(x, y, z-1)$, $(x+1, y+1, z+1)$, $(x-1, y-1, z-1)$, $(x, y+1, z+1)$, $(x+1, y, z+1)$, $(x-1, y, z-1)$, $(x, y-1, z-1)$. By using the definition of $d(p, p')$, we can see that the distance between p and p' is always 1 when they are face-connected neighbors.

Similarly, we can have the strictly vertex-connected neighbors of (x, y, z), which are $(x+1, y+1, z)$, $(x+1, y-1, z)$, $(x-1, y+1, z)$, $(x-1, y-1, z)$, $(x+1, y+1, z+2)$, $(x-1, y-1, z-2)$, and the distance between two strictly vertex-connected neighbors comes out to be $\sqrt{2}$. □

3.2 48-Symmetry

The rhombic dodecahedron belongs to the octahedral symmetry group. It therefore has 24 rotational symmetries and 48 symmetries when considering transformations that combine a reflection and a rotation. Table 1 shows the 48 symmetric points corresponding to a point (x, y, z) in the non-orthogonal coordinate system. This symmetry can be utilized while constructing symmetrical geometric objects, e.g. spheres, cones, cylinders, etc., in this digital space.

Table 1. 48-symmetry in RDG using proposed non-orthogonal coordinate system.

(x, y, z)	$(y-z, y, -x+y)$	$(-x+z, y, z)$	$(y-z, y, x+y-z)$
$(-x+z, y, -x+y)$	$(x, y, x+y-z)$	$(-y+z, -x+z, z)$	$(-x, -x+z, -x+y)$
$(y, -x+z, z)$	$(-x, -x+z, -x-y+z)$	$(y, -x+z, -x+y)$	$(-y+z, -x+z, -x-y+z)$
(y, x, z)	$(x-z, x, x-y)$	$(-y+z, x, z)$	$(x-z, x, x+y-z)$
$(-y+z, x, x-y)$	$(y, x, x+y-z)$	$(x-z, y-z, -z)$	$(y, y-z, -x+y)$
$(-x, y-z, -z)$	$(y, y-z, x+y-z)$	$(-x, y-z, -x+y)$	$(x-z, y-z, x+y-z)$
$(-x+z, -y+z, z)$	$(-y, -y+z, x-y)$	$(x, -y+z, z)$	$(-y, -y+z, -x-y+z)$
$(x, -y+z, x-y)$	$(-x+z, -y+z, -x-y+z)$	$(y-z, x-z, -z)$	$(x, x-z, x-y)$
$(-y, x-z, -z)$	$(x, x-z, x+y-z)$	$(-y, x-z, x-y)$	$(y-z, x-z, x+y-z)$
$(-y, -x, -z)$	$(-x+z, -x, -x+y)$	$(y-z, -x, -z)$	$(-x+z, -x, -x-y+z)$
$(y-z, -x, -x+y)$	$(-y, -x, -x-y+z)$	$(-x, -y, -z)$	$(-y+z, -y, x-y)$
$(x-z, -y, -z)$	$(-y+z, -y, -x-y+z)$	$(x-z, -y, x-y)$	$(-x, -y, -x-y+z)$

4 Digital Objects: Definition and Topological Analysis

In this section, we define digital models of two primitive geometric objects, namely, digital sphere and digital plane. We define these using the proposed non-orthogonal coordinate space and use the defined distance to control the thickness of the objects resulting in different topological models.

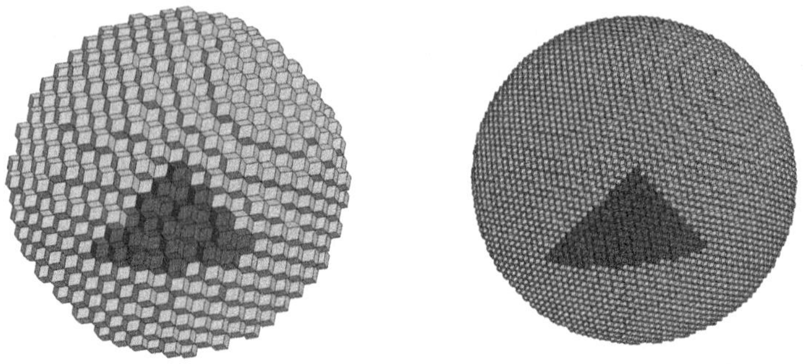

(a) (b)

Fig. 4. Digital sphere of radius 8 (a) and radius 23 (b) in the rhombic dodecahedron grid, with one of the 48th parts shown in blue, computed using the proposed non-orthogonal coordinate system.

4.1 Digital Sphere

We define a digital sphere in the proposed orthogonal coordinate space using the same thickness notion as used while defining Andres Analytical Sphere (recollected in Definition 2). However, we use our new distance measure as defined in Definition 3.

Definition 4 (Digital Sphere). *The set of Dodecahedron centroids belonging to the digitization of a Sphere centered on c of radius r and thickness ρ is given by:* $S_{(c,r)} = \{p \in \mathbb{Z}^3 | r - \frac{\rho}{2} \leq d(c,p) < r + \frac{\rho}{2}\}$

To control the topology of the resulting sphere, we use one of the two following values for the thickness.

Proposition 1 (Topology and Separation). *The thickness $\rho = 1$ provides a 2-separating sphere. The thickness $\rho = \sqrt{2}$ provides a 0-separating sphere.*

Proof. The proof for the 2-separating value comes directly from the distance between a dodecahedron and its face neighbors which is 1 (leading to a thickness 1). The distance $\sqrt{2}$ between the dodecahedron and its vertex neighbors leads to the 0-separating thickness of $\sqrt{2}$. □

The spheres presented in Fig. 4 are obtained using a brute force algorithm implemented in Mathematica that tests every integer coordinate point (i.e. dodecahedron centroid) in the bounding box with the sphere inequality using the 2-separating thickness 1. An optimization of this algorithm could be implemented using a 48th of the bounding box (see Fig. 4(a)) and build the rest of the sphere using the symmetries we have previously defined in Sect. 3.2.

4.2 Digital Plane

Similar to digital sphere, we define digital plane in the new coordinate system using the classical definition. However, note that, the real plane in consideration here is given in the new coordinate system and not in the Cartesian coordinate system.

Definition 5 (Digital Plane). *Let $ax + by + cz + d = 0$ be a plane equation in the new coordinate system. Let ρ be the thickness of the digital plane. The digitization of the plane in the dodecahedron grid is the set of dodecahedron (x, y, z) satisfying the following inequality.*

$$-\rho/2 \leq ax + by + cz + d < \rho/2. \tag{1}$$

Following we give the formulas for converting between plane equations from the new coordinate system to the Cartesian coordinate system and reverse.

Proposition 2 (New to Cartesian Plane Equation). *Let $ax_n + by_n + cz_n + d = 0$ be a plane equation in the new coordinate system. The equation of the same plane in the Cartesian coordinate system is given by:*

$$b\sqrt{2}x_c - a\sqrt{2}y_c + (a + b + 2c)z_c + 2d = 0 \tag{2}$$

Proposition 3 (Cartesian to New Plane Equation). *Let $ax_c + by_c + cz_c + d = 0$ be a plane equation in the Cartesian coordinate system. The equation of the same plane in the new coordinate system is given by:*

$$-\frac{2b}{\sqrt{2}}x_n + a\sqrt{2}y_n + (b/\sqrt{2} + c - a/\sqrt{2})z_n + d = 0 \tag{3}$$

We can control the topology of the digital plane by changing the value of ρ as mentioned in Proposition 4 below.

Proposition 4 (Topology and Separation). *The Supercover plane is obtained using:*

$$\rho_{Sup} = \max\left(|a + b|, |b - a|, |a + b + 2c|, \frac{1}{2}(|a + b| + |b - a| + |a + b + 2c|) \right). \tag{4}$$

The 0-separating (thinner than the supercover) plane is obtained using:

$$\rho_{Std} = \max\left(|a|, |b|, |c|, |a + b + c|, |a + c|, |b + c|, |a + b|, |a - b|, |a + b + 2c|\right). \tag{5}$$

The 2-separating digital plane in the dodecahedron grid is obtained for

$$\rho_N = \max\left(|a|, |b|, |c|, |a + b + c|, |a + c|, |b + c|\right). \tag{6}$$

Proof. The proof for the supercover is the same as in [11] using the following plane equation in Nagy's coordinate system:

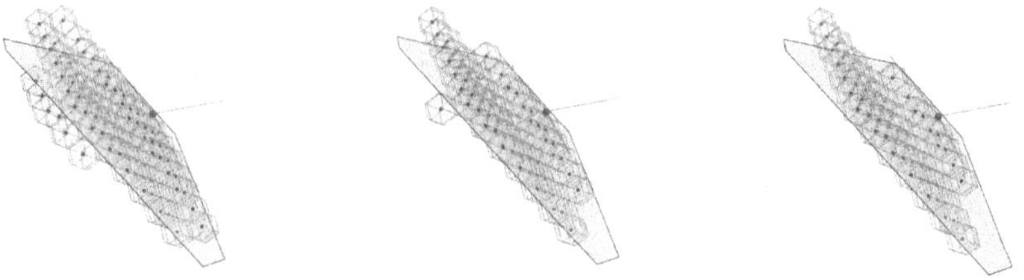

a) Supercover. b) Standard (0-separating). c) Naive (2-separating).

Fig. 5. Dodecahedron Digitizations for the plane $x + 11y + z + 20 = 0$ in the new coordinate system.

$$Ax + By + Cz + D = 0$$

where $A = \frac{a+b}{2}, B = \frac{b-a}{2}, C = \frac{a+b+2c}{2}, D = d$. It can also be proven by considering the thickness of the thickest directions we can have in the dodecahedron, i.e. the distance between opposite vertices or opposite faces as necessary.

Figure 5 presents the three different digitizations for the plane $x + 11y + z + 20 = 0$ in the dodecahedron non-orthogonal grid. Figure 5(a) shows the thickest plane, i.e. the supercover plane which uses Eq. 4 for the thickness. On Fig. 5(b), we can see the standard 0-separating plane obtained with Eq. 5 and on Fig. 5(c) we present the Naive 2-separating plane for the thickness defined by Eq. 6.

5 Conclusion

In this paper, we have defined a new coordinate system. The novelty of this system is that unlike the previous coordinate system, every single one of the integer coordinate points is the centroid of a dodecahedron. Using this system we have defined spheres (together with their 48 symmetries) and planes. Both can be topologically controlled, i.e. both are proposed with a 0- and a 2-separating forms. In future works, we want to explore the possibilities of defining lines in this coordinate system and to study the graphical transforms such as the rotation.

References

1. Andres, E., Acharya, R., Sibata, C.: Discrete analytical hyperplanes. Graph. Models Image Process. **59**(5), 302–309 (1997)
2. Andres, E., Jacob, M.-A.: The discrete analytical hyperspheres. IEEE Trans. Vis. Comput. Graph. **3**, 75–86 (1997)
3. Brimkov, V.E., Barneva, R.P.: Analytical honeycomb geometry for raster and volume graphics. Comput. J. **48**(2), 180–199 (2005)
4. Comic, L., Nagy, B.: A combinatorial 3-coordinate system for the face centered cubic grid. In: 2015 9th International Symposium on Image and Signal Processing and Analysis (ISPA), pp. 298–303 (2015)

5. Comic, L., Nagy, B.: A topological 4-coordinate system for the face centered cubic grid. Pattern Recognit. Lett. **83**, 67–74 (2016)
6. Grigoryan, A.M., Agaian, S.S.: 2D hexagonal quaternion Fourier transform in color image processing. In: Mobile Multimedia/Image Processing, Security, and Applications 2016, vol. 9869, p. 98690N. International Society for Optics and Photonics (2016)
7. He, X., Jia, W.: Hexagonal structure for intelligent vision. In: 2005 First International Conference on Information and Communication Technologies, ICICT 2005, pp. 52–64. IEEE (2005)
8. Kittel, C.: Introduction to Solid State Physics, 8th edn. Wiley, New York (2004)
9. Koshti, G., Biswas, R., Largeteau-Skapin, G., Zrour, R., Andres, E., Bhowmick, P.: Sphere construction on the FCC grid interpreted as layered hexagonal grids in 3D. In: Barneva, R.P., Brimkov, V.E., Tavares, J.M.R.S. (eds.) IWCIA 2018. LNCS, vol. 11255, pp. 82–96. Springer, Cham (2018). https://doi.org/10.1007/978-3-030-05288-1_7
10. Kovalevsky, V.A.: Geometry of Locally Finite Spaces: Computer Agreeble Topology and Algorithms for Computer Imagery. Editing House Dr. Barbel Kovalevski, Berlin (2008)
11. Linh, T.K., Imiya, A., Strand, R., Borgefors, G.: Supercover of non-square and non-cubic grids. In: Klette, R., Žunić, J. (eds.) IWCIA 2004. LNCS, vol. 3322, pp. 88–97. Springer, Heidelberg (2004). https://doi.org/10.1007/978-3-540-30503-3_7
12. Mostafa, K., Her, I.: An edge detection method for hexagonal images. Int. J. Image Process. (IJIP) **10**(4), 161 (2016)
13. Nagy, B.: Cellular topology on the triangular grid. In: Barneva, R.P., Brimkov, V.E., Aggarwal, J.K. (eds.) IWCIA 2012. LNCS, vol. 7655, pp. 143–153. Springer, Heidelberg (2012). https://doi.org/10.1007/978-3-642-34732-0_11
14. Nagy, B.: Cellular topology and topological coordinate systems on the hexagonal and on the triangular grids. Ann. Math. Artif. Intell. **75**, 117–134 (2015)
15. Nagy, B., Strand, R.: A connection between $\mathbb{Z}^n$ and generalized triangular grids. In: Advances in Visual Computing, Berlin, Heidelberg, pp. 1157–1166 (2008)
16. Nagy, B., Strand, R.: Non-traditional grids embedded in Z^n. Int. J. Shape Model. **14**, 209–228 (2008)
17. Nagy, B., Strand, R.: Neighborhood sequences in the diamond grid: algorithms with two and three neighbors. Int. J. Imaging Syst. Technol. **19**(2), 146–157 (2009)
18. Stojmenović, I.: Honeycomb networks. In: Wiedermann, J., Hájek, P. (eds.) MFCS 1995. LNCS, vol. 969, pp. 267–276. Springer, Heidelberg (1995). https://doi.org/10.1007/3-540-60246-1_133
19. Strand, R., Borgefors, G.: Resolution pyramids on the FCC and BCC grids. In: Andres, E., Damiand, G., Lienhardt, P. (eds.) DGCI 2005. LNCS, vol. 3429, pp. 68–78. Springer, Heidelberg (2005). https://doi.org/10.1007/978-3-540-31965-8_7
20. Strand, R., Nagy, B., Borgefors, G.: Digital distance functions on three-dimensional grids. Theor. Comput. Sci. **412**(15), 1350–1363 (2011)

Facet Connectedness of Arithmetic Discrete Hyperplanes with Non-Zero Shift

Eric Domenjoud[1(✉)], Bastien Laboureix[2], and Laurent Vuillon[3]

[1] Univ. de Lorraine, CNRS, Loria, 54000 Nancy, France
Eric.Domenjoud@loria.fr
[2] Ecole Normale Supérieure Cachan, 94230 Cachan, France
Bastien.Laboureix@hotmail.fr
[3] Univ. Savoie Mont Blanc, CNRS, LAMA, 73000 Chambéry, France
Laurent.Vuillon@univ-smb.fr

Abstract. We present a criterion for the arithmetic discrete hyperplane $\mathbb{P}(\boldsymbol{v}, \mu, \theta)$ to be facet connected when θ is the connecting thickness $\Omega(\boldsymbol{v}, \mu)$. We encode the shift μ in a numeration system associated with the normal vector $\boldsymbol{v}$ and we describe an incremental construction of the plane based on this encoding. We deduce a connectedness criterion and we show that when the *Fully Subtractive* algorithm applied to $\boldsymbol{v}$ has a periodic behaviour, the encodings of shifts μ for which the plane is connected may be recognised by a finite state automaton.

Keywords: Discrete hyperplane · Connectedness · Connecting thickness · Fully subtractive algorithm · Numeration system · Finite state automaton

1 Introduction

The *arithmetic discrete hyperplane* with normal vector $\boldsymbol{v} \in \mathbb{R}^d \setminus \{\boldsymbol{0}\}$, shift $\mu \in \mathbb{R}$ and thickness $\theta \in \mathbb{R}$, is the set of points in $\mathbb{Z}^d$ defined by

$$\mathbb{P}(\boldsymbol{v}, \mu, \theta) = \{\boldsymbol{x} \in \mathbb{Z}^d \mid 0 \leq \langle \boldsymbol{v}, \boldsymbol{x} \rangle + \mu < \theta\}$$

where $\langle \cdot, \cdot \rangle$ denotes the usual scalar product on $\mathbb{R}^d$ [1,9].

For κ in $\{0, \ldots, d-1\}$, two points $\boldsymbol{x}$ and $\boldsymbol{y}$ in $\mathbb{Z}^d$ are κ-neighbours if and only if $\|\boldsymbol{x}-\boldsymbol{y}\|_\infty = 1$ and $\|\boldsymbol{x}-\boldsymbol{y}\|_1 \leq d-\kappa$. In particular, $\boldsymbol{x}$ and $\boldsymbol{y}$ are $(d-1)$-neighbours if and only if $\boldsymbol{x} - \boldsymbol{y} = \pm \boldsymbol{e}_i$ for some i, where $(\boldsymbol{e}_1, \ldots, \boldsymbol{e}_d)$ is the canonical basis of $\mathbb{R}^d$. A set $S \subset \mathbb{Z}^d$ is κ-connected if and only if for all $\boldsymbol{x}$ and $\boldsymbol{y}$ in S, there exists a *path* $\boldsymbol{x} = \boldsymbol{z}_1, \ldots, \boldsymbol{z}_n = \boldsymbol{y}$ in S where $\boldsymbol{z}_i$ and $\boldsymbol{z}_{i+1}$ are κ-neighbours for all $i = 1, \ldots, n-1$.

Many works have been devoted to determining the conditions under which the hyperplane $\mathbb{P}(\boldsymbol{v}, \mu, \theta)$ is κ-connected, see [3–5,7] among others. The set of thicknesses θ for which $\mathbb{P}(\boldsymbol{v}, \mu, \theta)$ is non-empty and κ-connected is a right unbounded interval of $\mathbb{R}^+$ the lower bound of which, denoted by $\Omega_\kappa(\boldsymbol{v}, \mu)$, is called the

M. Couprie et al. (Eds.): DGCI 2019, LNCS 11414, pp. 38–50, 2019.
https://doi.org/10.1007/978-3-030-14085-4_4

κ*-connecting thickness* of $\boldsymbol{v}$ with shift μ. In this work, we consider only the $(d-1)$-connectedness which is the only case where a general algorithm is known to compute the κ-connecting thickness. We shall therefore drop the subscript κ in all denotations and speak of connectedness and connecting thickness.

By definition of $\Omega(\boldsymbol{v},\mu)$, $\mathbb{P}(\boldsymbol{v},\mu,\theta)$ is empty or disconnected for all $\theta < \Omega(\boldsymbol{v},\mu)$ and is non-empty and connected for all $\theta > \Omega(\boldsymbol{v},\mu)$. The question which arises naturally is whether $\mathbb{P}(\boldsymbol{v},\mu,\Omega(\boldsymbol{v},\mu))$ is connected or not. In [5], we showed that $\mathbb{P}(\boldsymbol{v},\mu,\Omega(\boldsymbol{v},\mu))$ is almost always disconnected, excepted when $\boldsymbol{v}$ belongs to some specific set which was proven to be Lebesgue negligible by Kraaikamp and Meester [8]. In this case, $\Omega(\boldsymbol{v},\mu) = \|\boldsymbol{v}\|_1/(d-1)$ which does not actually depend on μ and will therefore be denoted simply by $\Omega(\boldsymbol{v})$. We showed that for such vectors, $\mathbb{P}(\boldsymbol{v},\mu,\Omega(\boldsymbol{v}))$ may be connected or not, depending on the value of μ. In particular, $\mathbb{P}(\boldsymbol{v},0,\Omega(\boldsymbol{v}))$ is always connected while $\mathbb{P}(\boldsymbol{v},\Omega(\boldsymbol{v}),\Omega(\boldsymbol{v}))$ is always disconnected.

In the present work, we investigate the general case where μ is arbitrary. We first set our main definitions and recall how the connecting thickness $\Omega(\boldsymbol{v})$ may be computed by means of the *Fully Subtractive* algorithm. Then we introduce a numeration system associated with the normal vector $\boldsymbol{v}$, in which the shift μ may be encoded as an infinite sequence in $\{0,1\}^\omega$. Based on this encoding, we describe an incremental construction of $\mathbb{P}(\boldsymbol{v},\mu,\Omega(\boldsymbol{v}))$ and we show that, provided that the encoding of μ has some property called *bi-admissibility*, we actually construct the connected component of $\mathbf{0}$ in $\mathbb{P}(\boldsymbol{v},\mu,\Omega(\boldsymbol{v}))$. We first give a purely geometrical connectedness criterion based on a particular projection of a connected component of $\mathbb{P}(\boldsymbol{v},\mu,\Omega(\boldsymbol{v}))$. Then we investigate specific patterns occurring in the encoding of μ and we introduce the notion of *minimal interior patterns* which allows us to deduce a more operational connectedness criterion. Finally, we study the case where the Fully Subtractive algorithm applied to the normal vector $\boldsymbol{v}$ has a periodic behaviour. We show that in this case, the language of encodings of shifts μ for which $\mathbb{P}(\boldsymbol{v},\mu,\Omega(\boldsymbol{v}))$ is connected is a regular language which may therefore be recognised by a finite state automaton.

2 Preliminaries

The $(d-1)$-connecting thickness may be computed by means of the *Fully Subtractive* algorithm as follows. Let $\mathbf{1}$ be the vector $(1,\ldots,1)$. For each $k \in \{1,\ldots,d\}$, we define γ_k as

$$\begin{aligned} \gamma_k \;:\; & \mathbb{R}^d \to \mathbb{R}^d \\ \boldsymbol{x} \;& \mapsto \boldsymbol{x} - \langle \boldsymbol{e}_k, \boldsymbol{x}\rangle\,(\mathbf{1} - \boldsymbol{e}_k) \end{aligned}$$

Given a vector $\boldsymbol{v} \in (\mathbb{R}^+)^d \setminus \{\mathbf{0}\}$, the algorithm computes a sequence of pairs $(\boldsymbol{v}^n,\Theta^n)_{n\geq 1}$ where $(\boldsymbol{v}^1,\Theta^1) = (\boldsymbol{v},0)$. For all $n \geq 1$, if $\boldsymbol{v}^n$ has at least two components and no component is 0 then $(\boldsymbol{v}^{n+1},\Theta^{n+1}) = (\gamma_{\delta_n}(\boldsymbol{v}^n),\Theta^n + v^n_{\delta_n})$ where δ_n is the index of a minimal coordinate of $\boldsymbol{v}^n$. If some component becomes 0 in the process, it is simply *erased* and we continue with $\boldsymbol{v}' \in \mathbb{R}^{d-1}$. If $\boldsymbol{v}^n$ has only one component left, the algorithm stops. This sequence has the property that for all $n \geq 1$, $\mathbb{P}(\boldsymbol{v}, \mu, \theta)$ is non-empty and connected if and only if $\mathbb{P}(\boldsymbol{v}^n, \mu, \theta - \Theta^n)$ is non-empty and connected [4,5].

If the process terminates with $(\boldsymbol{v}^m, \Theta^m)$, which happens if and only if $\boldsymbol{v} = \lambda\, \boldsymbol{v}'$ for some integral vector $\boldsymbol{v}'$ [5], then $\Omega(\boldsymbol{v}, \mu) = \Theta^m + \lambda\left((\mu/\lambda) \bmod \gcd(v'_1, \ldots, v'_d)\right)$. Otherwise, $(\boldsymbol{v}^n, \Theta^n)$ converges to $(\boldsymbol{v}^\infty, \Theta^\infty)$ and $\Omega(\boldsymbol{v}, \mu) = \Theta^\infty + \|\boldsymbol{v}^\infty\|_\infty$ which does not depend on μ and will therefore be denoted simply by $\Omega(\boldsymbol{v})$.

The set $\mathcal{K}_d$ is defined as the set of vectors $\boldsymbol{v} \in (\mathbb{R}^+)^d \setminus \{\boldsymbol{0}\}$ such that the Fully Subtractive algorithm erases no component and $(d-1)\, \|\boldsymbol{v}^n\|_\infty \leq \|\boldsymbol{v}^n\|_1$ for all $n \geq 1$. If $\boldsymbol{v} \in \mathcal{K}_d$ then $\boldsymbol{v}^\infty = \boldsymbol{0}$ and $\Omega(\boldsymbol{v}) = \Theta^\infty = \|\boldsymbol{v}\|_1/(d-1)$. We proved in [5] that $\mathbb{P}(\boldsymbol{v}, \mu, \Omega(\boldsymbol{v}))$ may be connected only if for some $n \geq 1$, $\boldsymbol{v}^n$ belongs to some $\mathcal{K}_{d'}$, in which case, $\boldsymbol{v}^m \in \mathcal{K}_{d'}$ for all $m \geq n$. Kraaikamp and Meester proved in [8] that $\mathcal{K}_d$ is Lebesgue negligible when $d \geq 3$.

From now on, $\boldsymbol{v}$ is a vector in $\mathcal{K}_d$ with $d \geq 3$ and $\Omega = \Omega(\boldsymbol{v})$.

Given the sequence of vectors $(\boldsymbol{v}^n)_{n \geq 1}$ computed by the Fully Subtractive algorithm, we define the *directive sequence* $\Delta = (\delta_n)_{n \geq 1}$ where δ_n is the index of the minimal component of $\boldsymbol{v}^n$. This minimal component is unique since otherwise $\boldsymbol{v}^{n+1}$ would have a zero component, which is impossible by definition of $\mathcal{K}_d$. The sequence Δ is therefore well defined. For all $n \geq 1$ we set $\Delta^n = \delta_n \delta_{n+1} \ldots$ and $\Omega^n = \Omega(\boldsymbol{v}^n) = \|\boldsymbol{v}^n\|_1/(d-1)$.

To each vector $\boldsymbol{v}$ in $\mathcal{K}_d$ is associated a unique sequence Δ which has the property that each $k \in \{1, \ldots, d\}$ occurs infinitely many times in Δ. Conversely, each sequence $\Delta \in \{1, \ldots, d\}^\omega$ in which each $k \in \{1, \ldots, d\}$ occurs infinitely many times is associated with a unique direction in $\mathcal{K}_d$, which means that the vector $\boldsymbol{v}$ is determined by Δ up to a multiplicative factor [5,6].

For all $n \geq 1$, we set $\theta_n = v^n_{\delta_n} = \langle \boldsymbol{v}^n, \boldsymbol{e}_{\delta_n} \rangle$ and $\boldsymbol{T}^\Delta_n = {}^t\gamma_{\delta_1} \ldots {}^t\gamma_{\delta_{n-1}}(\boldsymbol{e}_{\delta_n})$ where ${}^t\gamma_i$ is the transpose of γ_i. We have $\theta_n = \langle \boldsymbol{v}, \boldsymbol{T}^\Delta_n \rangle = \langle \boldsymbol{v}^n, \boldsymbol{e}_{\delta_n} \rangle$ and for all $n \geq 1$, $\Omega^n = \sum_{i \geq n} \theta_i$. In particular, $\Omega = \Omega^1 = \sum_{i \geq 1} \theta_i$.

We denote by $\mathbb{Z}^\omega$ the set of sequences of integers and by $\mathbb{Z}^\star 0^\omega$ the set of sequences with finitely many non-zero terms. The *length* of $\sigma \in \mathbb{Z}^\omega$, denoted by $|\sigma|$, is the smallest index $n \geq 0$ such that $\forall m > n$, $\sigma_m = 0$. If no such index exists then $|\sigma| = +\infty$. If σ is a finite sequence then $|\sigma|$ denotes its natural length. We define the mappings ψ^Δ on $\mathbb{Z}^\star 0^\omega$ by $\psi^\Delta(\sigma) = \sum_n \sigma_n T^\Delta_n$ and φ^v on $\mathbb{Z}^\omega$ by $\varphi^v(\sigma) = \sum_n \sigma_n \theta_n$ when this sum converges. It is in particular the case when σ is bounded. Given $\sigma \in \{0,1\}^\omega$ we denote by $\widetilde{\sigma}$ the sequence obtained by replacing zeros with ones and ones with zeros in σ. We have $\varphi^v(\widetilde{\sigma}) = \Omega - \varphi^v(\sigma)$. When $\sigma \in \mathbb{Z}^\star 0^\omega$, we have $\varphi^v(\sigma) = \langle \boldsymbol{v}, \psi^\Delta(\sigma) \rangle$. These definitions are naturally extended to the case where $\sigma \in \mathbb{Z}^\star$ by $\psi^\Delta(\sigma) = \psi^\Delta(\sigma 0^\omega)$ and $\varphi^v(\sigma) = \varphi^v(\sigma 0^\omega)$.

If $\xi^{\Delta,k}_n$ is the index of the nth occurrence of k in Δ, we recall [5,6] that for all $k \in \{1, \ldots, d\}$ and for all $n \geq 1$, we have

$$\theta_1 + \cdots + \theta_{\xi^{\Delta,k}_1} = v_k \tag{1}$$

$$\theta_{\xi^{\Delta,k}_n + 1} + \cdots + \theta_{\xi^{\Delta,k}_{n+1}} = \theta_{\xi^{\Delta,k}_n} \tag{2}$$

In the sequel, we symmetrise our problem by studying the connectedness of $\overline{\mathbb{P}}(\boldsymbol{v}, \mu, \Omega) = \{\boldsymbol{x} \in \mathbb{Z}^d \mid 0 \leq \langle \boldsymbol{v}, \boldsymbol{x} \rangle + \mu \leq \Omega\}$ so that $\overline{\mathbb{P}}(\boldsymbol{v}, \Omega - \mu, \Omega) = -\overline{\mathbb{P}}(\boldsymbol{v}, \mu, \Omega)$.

We have $\overline{\mathbb{P}}(\boldsymbol{v},\mu,\Omega) \neq \mathbb{P}(\boldsymbol{v},\mu,\Omega)$ if and only if there exists $\boldsymbol{x}_0 \in \mathbb{Z}^d$ such that $\langle \boldsymbol{v}, \boldsymbol{x}_0\rangle + \mu = \Omega$. In this case, we have $\mathbb{P}(\boldsymbol{v},\mu,\Omega) = \boldsymbol{x}_0 + \mathbb{P}(\boldsymbol{v},\Omega,\Omega)$ which is disconnected when $d \geq 3$ [5]. In all other cases, $\mathbb{P}(\boldsymbol{v},\mu,\Omega)$ is connected if and only if $\overline{\mathbb{P}}(\boldsymbol{v},\mu,\Omega)$ is.

We may also reduce the study to the case where $\mu \in [0;\Omega]$. Indeed, if $q = \lfloor \mu/v_1 \rfloor$ and $\mu' = \mu - q\,v_1$, then $\mu' \in [0; v_1[\, \subset [0;\Omega]$, and $\overline{\mathbb{P}}(\boldsymbol{v},\mu,\Omega) = -q\,\boldsymbol{e}_1 + \overline{\mathbb{P}}(\boldsymbol{v},\mu',\Omega)$. Thus $\overline{\mathbb{P}}(\boldsymbol{v},\mu,\Omega)$ is a translate of $\overline{\mathbb{P}}(\boldsymbol{v},\mu',\Omega)$ by an integral vector and is connected if and only if $\overline{\mathbb{P}}(\boldsymbol{v},\mu',\Omega)$ is.

3 Δ-Numeration System

We showed in [5] that $\mathbb{P}(\boldsymbol{v},0,\Omega) = \psi^{\Delta}(\{0,1\}^{\star}0^{\omega})$, meaning that $\varphi^{\boldsymbol{v}}(\{0,1\}^{\star}0^{\omega}) = (v_1\,\mathbb{Z} + \cdots + v_d\,\mathbb{Z}) \cap [0;\Omega]$. We now extend this result.

Theorem 1. $\varphi^{\boldsymbol{v}}(\{0,1\}^{\omega}) = [0;\Omega]$

To prove this theorem, we need a technical lemma.

Lemma 2

1. $2\theta_1 < \Omega$.
2. If $d \geq 3$, $j < n$ *and* $\delta_j \neq \delta_n$, *then* $\theta_j + \sum_{i=1}^{n} \theta_i > \Omega$.
3. If $d \geq 3$ *and* $\boldsymbol{v} \in \mathcal{K}_d$ *then* $v_i + v_j > \Omega$ *for all* $i \neq j$.

Proof of Theorem 1. The inclusion $\varphi^{\boldsymbol{v}}(\{0,1\}^{\omega}) \subset [0;\Omega]$ is obvious since for all $\sigma \in \{0,1\}^{\omega}$, we have $0 \leq \varphi^{\boldsymbol{v}}(\sigma) = \sum_{n\geq 1} \sigma_n\,\theta_n \leq \sum_{n \geq 1} \theta_n = \Omega$.

Now let $x \in [0;\Omega]$, $r_1 = x$ and for all $n \geq 1$, $r_{n+1} = r_n - \sigma_n\,\theta_n$ where $\sigma_n = 1$ if $r_n \geq \theta_n$ and $\sigma_n = 0$ otherwise. Then we have $x = \sum_{k<n} \sigma_k\,\theta_k + r_n$ and $r_n \in [0;\Omega^n]$ for all n. Indeed, we obviously have $r_n \geq 0$ for all n. By hypothesis, we have $r_1 = x \in [0;\Omega^1] = [0;\Omega]$. Now, for $n \geq 1$, assume that $r_n \leq \Omega^n$ and let us show that $r_{n+1} \leq \Omega^{n+1}$. If $r_n < \theta_n$ then $\sigma_n = 0$ and $r_{n+1} = r_n < \theta_n$. By Lemma 2 applied to $\boldsymbol{v}^n$ we get $2\,\theta_n < \Omega^n$ and thus $\theta_n < \Omega^n - \theta_n = \Omega^{n+1}$. If $r_n \geq \theta_n$, then $\sigma_n = 1$ and $r_{n+1} = r_n - \theta_n \leq \Omega^n - \theta_n = \Omega^{n+1}$.

Since $\Omega^n = \|\boldsymbol{v}^n\|_1/(d-1)$ and $\lim_{n\to\infty} \boldsymbol{v}^n = \boldsymbol{0}$, we have $\lim_{n\to\infty} r_n = 0$ hence $\lim_{n\to\infty} \sum_{k<n} \sigma_k\,\theta_k = \sum_{n\geq 1} \sigma_n\,\theta_n = x$. Then we have $\sigma = (\sigma_n)_{n\geq 1} \in \{0,1\}^{\omega}$ and $x = \varphi^{\boldsymbol{v}}(\sigma) = \sum_{n\geq 1} \sigma_n\,\theta_n$. □

The sequence $(\theta_n)_{n\geq 1}$ may be seen as the basis of a numeration system in which we can encode any $\mu \in [0;\Omega]$ as a sequence $\overline{\mu} \in \{0,1\}^{\omega}$. Note that by considering a bi-infinite sequence Δ, we would define a *full* numeration system in which any non negative real number may be encoded as $\sigma.\sigma'$ with $\sigma \in \{0,1\}^{\star}$ and $\sigma' \in \{0,1\}^{\omega}$.

Definition 3 (Δ-admissible sequence)

- *A sequence* $\sigma \in \{0,1\}^{\omega}$ *is* Δ-admissible *iff there does not exist* $k \in \{1,\ldots,d\}$ *and an index* n_0 *such that* $\sigma_n = 0 \iff \delta_n = k$ *for all* $n \geq n_0$.
- *A sequence* $\sigma \in \{0,1\}^{\omega}$ *is* Δ-bi-admissible *iff* σ *and* $\widetilde{\sigma}$ *are* Δ*-admissible.*

– *If there exist* k *and* n_0 *such that* $\sigma_n = 0 \iff \delta_n = k$ *for all* $n \geq n_0$*, then* k *is unique. In this case, we say that* σ *is* Δ-non-admissible of type k.

A Δ-non-admissible sequence is analogous to the writing $1 = 0.9999\ldots$ in the classical numeration system in base 10. For instance, if $\Delta = (12213)^\omega$ then $\sigma = 11001(01101)^\omega$ is Δ-non-admissible of type 1. We have actually $\varphi^v(\sigma) = \varphi^v(1101)$. This principle is general: for all Δ-non-admissible sequence σ, a finite sequence σ' exists such that $\varphi^v(\sigma) = \varphi^v(\sigma')$.

Proposition 4. *Let* $\sigma \in \{0,1\}^\omega$ *be a* Δ*-non-admissible sequence of type* k*,* n_0 *such that* $\delta_{n_0} = k$ *and* $\sigma_n = 0 \iff \delta_n = k$ *for all* $n \geq n_0$*. Then* $\varphi^v(\sigma) = \varphi^v(\sigma_1 \ldots \sigma_{n_0-1}1)$.

Proof. For all $n \geq 0$, let $\zeta_n = \xi_n^{\Delta,k}$. Let r_0 be such that $n_0 = \zeta_{r_0}$ and for all $r \geq r_0$, let $\sigma^r = \sigma_1 \ldots \sigma_{\zeta_r-1}1$. We have $\sigma^{r_0} = \sigma_1 \ldots \sigma_{n_0-1}1$ and by Eq. 2, for all $r \geq r_0$, $\varphi^v(\sigma^r) = \varphi^v(\sigma^{r_0})$. Then

$$\varphi^v(\sigma^{r_0}) = \varphi^v(\sigma^r) = \varphi^v(\sigma_1 \ldots \sigma_{\zeta_r-1}) + \varphi^v(0^{\zeta_r-1}1)$$

and since $\lim_{r\to\infty} \varphi^v(\sigma_1 \ldots \sigma_{\zeta_r-1}) = \varphi^v(\sigma)$ and $\lim_{r\to\infty} \varphi^v(0^{\zeta_r-1}1) = \lim_{r\to\infty} \theta_{\zeta_r} = 0$, we get the result at the limit. □

If a sequence σ is not Δ-bi-admissible then, by applying this transformation to either σ or $\widetilde{\sigma}$, we get in all cases a Δ-bi-admissible sequence. Hence the corollary:

Corollary 5. *For all* $\sigma \in \{0,1\}^\omega$*, there exists a* Δ*-bi-admissible sequence* σ' *such that* $\varphi^v(\sigma') = \varphi^v(\sigma)$.

Definition 6 (Δ-normalised sequence). *A sequence* $\sigma \in \{0,1\}^\omega$ *is* Δ-normalised *if and only if it is* Δ*-admissible and it does not contain a sub-sequence* 01^k *with* $k \geq 1$ *at a position* i *such that* $\delta_i = \delta_{i+k}$.

Proposition 7. *For all* $\mu \in [0;\Omega]$*, there exists a unique* Δ*-normalised sequence* $\sigma \in \{0,1\}^\omega$ *such that* $\varphi^v(\sigma) = \mu$.

Definition 8 (Δ-normal form). *Given a sequence* $\sigma \in \mathbb{N}^\omega$ *such that* $\varphi^v(\sigma) \leq \Omega$*, the unique* Δ*-normalised sequence* $\sigma' \in \{0,1\}^\omega$ *such that* $\varphi^v(\sigma') = \varphi^v(\sigma)$ *is called the* Δ*-normal form of* σ *and is denoted by* $\sigma\Downarrow^\Delta$.

Note that given a sequence σ, there does not always exist a sequence σ' which is *both* Δ-bi-admissible *and* Δ-normalised and such that $\varphi^v(\sigma') = \varphi^v(\sigma)$. For instance, if $\Delta = (123)^\omega$ then $\sigma = (100)^\omega$ is Δ-normalised but not Δ-bi-admissible because $\widetilde{\sigma} = (011)^\omega$ is not Δ-admissible. Since the Δ-normal form is unique, there is no Δ-normalised and Δ-bi-admissible sequence σ' such that $\varphi^v(\sigma') = \varphi^v(\sigma)$.

4 Incremental Construction of $\overline{\mathbb{P}}(\boldsymbol{V}, \boldsymbol{\mu}, \boldsymbol{\Omega})$

In [5] and [6], we presented an incremental construction of $\mathbb{P}(\boldsymbol{v}, \mu, \Omega)$ when $\mu = 0$. We now extend this construction to the case where μ is arbitrary.

As said before, we may always reduce the study of the connectedness of $\overline{\mathbb{P}}(\boldsymbol{v}, \mu, \Omega)$ to the case where $\mu \in [0; \Omega]$. Then there exists a sequence $\overline{\mu} = (\mu_i)_{i \geq 1} \in \{0,1\}^\omega$ such that $\varphi^{\boldsymbol{v}}(\overline{\mu}) = \mu$. We define the sequence $(\mathcal{S}_n^{\Delta,\overline{\mu}})_{n \geq 0}$ of subsets of $\mathbb{Z}^d$ by $\mathcal{S}_0^{\Delta,\overline{\mu}} = \{\boldsymbol{0}\}$ and for all $n \geq 1$,

$$\mathcal{S}_n^{\Delta,\overline{\mu}} = \mathcal{S}_{n-1}^{\Delta,\overline{\mu}} \cup \left(\mathcal{S}_{n-1}^{\Delta,\overline{\mu}} + (1 - 2\,\mu_n)\,\boldsymbol{T}_n^{\Delta}\right) = \begin{cases} \mathcal{S}_{n-1}^{\Delta,\overline{\mu}} \cup \left(\mathcal{S}_{n-1}^{\Delta,\overline{\mu}} + \boldsymbol{T}_n^{\Delta}\right) & \text{if } \mu_n = 0 \\ \mathcal{S}_{n-1}^{\Delta,\overline{\mu}} \cup \left(\mathcal{S}_{n-1}^{\Delta,\overline{\mu}} - \boldsymbol{T}_n^{\Delta}\right) & \text{if } \mu_n = 1 \end{cases}$$

For example, let $\boldsymbol{v} = (\alpha, \alpha+\alpha^2, 1)$ where $\alpha \in [0;1]$ and $\alpha+\alpha^2+\alpha^3 = 1$. We have $\Delta = (123)^\omega$. The picture below shows $\mathcal{S}_{12}^{\Delta,\overline{\mu}}$ for $\overline{\mu} = (01)^\omega$ and $\overline{\mu} = (011001)^\omega$. We have $\varphi^{\boldsymbol{v}}((01)^\omega) = (\alpha+1)/2$ and $\varphi^{\boldsymbol{v}}((011001)^\omega) = (6\,\alpha^2 + 15\,\alpha - 3)/14$. We shall see in Sect. 6 that $\mathbb{P}(\boldsymbol{v}, \mu, \Omega)$ is connected in the first case and disconnected in the second case.

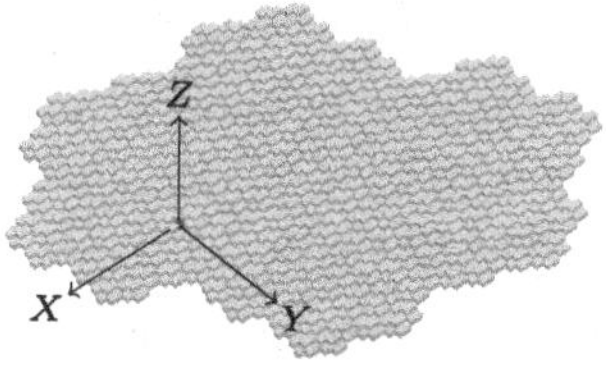

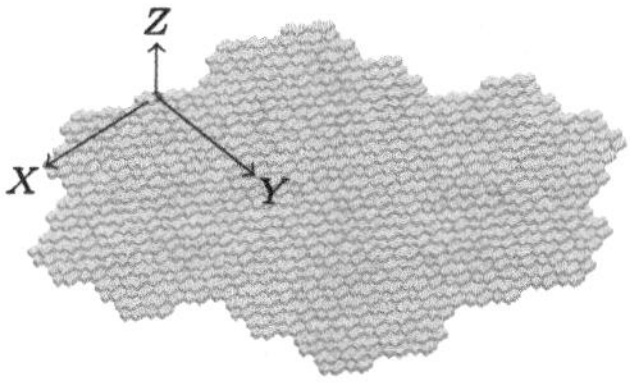

We observe that the geometry of $\mathcal{S}_n^{\Delta,\overline{\mu}}$ does not seem to depend on $\overline{\mu}$. Only the position with respect to the origin changes. Indeed, we have

$$\begin{aligned}\mathcal{S}_n^{\Delta,\overline{\mu}} &= \left\{\sum_{i \in I}(1 - 2\,\mu_i)\,\boldsymbol{T}_i^{\Delta} \mid I \subset \{1,\ldots,n\}\right\} = \left\{\sum_{i \in J}\boldsymbol{T}_i^{\Delta} \mid J \subset \{1,\ldots,n\}\right\} - \sum_{\substack{i \leq n \\ \mu_i = 1}} \boldsymbol{T}_i^{\Delta} \\ &= \mathcal{S}_n^{\Delta} - \sum_{i \leq n} \mu_i\,\boldsymbol{T}_i^{\Delta} = \mathcal{S}_n^{\Delta} - \psi^{\Delta}(\mu_1 \ldots \mu_n)\end{aligned}$$

Where $\mathcal{S}_n^{\Delta}$ stands for $\mathcal{S}_n^{\Delta,0^\omega}$. For all $\overline{\mu} \in \{0,1\}^\omega$ and all $n \geq 0$, $\mathcal{S}_n^{\Delta,\overline{\mu}}$ is a translate of $\mathcal{S}_n^{\Delta}$ by an integral vector, and therefore inherits topological properties of $\mathcal{S}_n^{\Delta}$, especially the fact that $\mathcal{S}_n^{\Delta,\overline{\mu}}$ is symmetric, connected and circuit-free [5,6].

Let $\mathcal{S}_\infty^{\Delta,\overline{\mu}}$ be the limit of the sequence $(\mathcal{S}_n^{\Delta,\overline{\mu}})_{n \geq 0}$. We have

$$\mathcal{S}_\infty^{\Delta,\overline{\mu}} = \left\{\sum_{i \in I}(1 - 2\,\mu_i)\,\boldsymbol{T}_i^{\Delta} \mid I \subset \mathbb{N} \setminus \{0\},\ |I| < \infty\right\}$$

and as an immediate corollary, we get that for all $\overline{\mu} \in \{0,1\}^\omega$, $\mathcal{S}_\infty^{\Delta,\overline{\mu}}$ is connected and circuit-free.

When $\mu = 0$, we have $\mathcal{S}_\infty^{\Delta} = \mathbb{P}(\boldsymbol{v}, 0, \Omega) = \overline{\mathbb{P}}(\boldsymbol{v}, 0, \Omega)$ [5]. The next theorem establishes the relationship between $\mathcal{S}_\infty^{\Delta,\overline{\mu}}$ and $\overline{\mathbb{P}}(\boldsymbol{v}, \varphi^{\boldsymbol{v}}(\overline{\mu}), \Omega)$.

Theorem 9. *If $d \geq 3$, $\overline{\mu} \in \{0,1\}^\omega$ is a Δ-bi-admissible sequence and $\mu = \varphi^v(\overline{\mu})$, then $\mathcal{S}_\infty^{\Delta,\overline{\mu}}$ is the connected component of $\mathbf{0}$ in $\overline{\mathbb{P}}(\boldsymbol{v}, \mu, \Omega)$.*

Proof (sketch). We observe that by hypothesis, we have $\mu \in [0;\Omega]$ and thus $\mathbf{0} \in \overline{\mathbb{P}}(\boldsymbol{v}, \mu, \Omega)$. Then we prove that any neighbour in $\overline{\mathbb{P}}(\boldsymbol{v}, \mu, \Omega)$ of a point $\boldsymbol{x} \in \mathcal{S}_\infty^{\Delta,\overline{\mu}}$ belongs to $\mathcal{S}_\infty^{\Delta,\overline{\mu}}$. Let $\Sigma^{\overline{\mu}} = (\{0,1\}^\omega - \overline{\mu}) \cap \{-1,0,1\}^\star 0^\omega$, i.e. the set of sequences $\sigma \in \{-1,0,1\}^\star 0^\omega$ such that $\sigma + \overline{\mu} \in \{0,1\}^\omega$. We have $\mathcal{S}_\infty^{\Delta,\overline{\mu}} = \psi^\Delta(\Sigma^{\overline{\mu}})$. We consider a sequence $\sigma \in \Sigma^{\overline{\mu}}$ such that $\boldsymbol{x} = \psi^\Delta(\sigma)$ and we assume $\boldsymbol{x} \pm \boldsymbol{e}_i \in \overline{\mathbb{P}}(\boldsymbol{v}, \mu, \Omega)$. Then we exhibit a sequence $\tau \in \Sigma^{\overline{\mu}}$ such that $\boldsymbol{x} \pm \boldsymbol{e}_i = \psi^\Delta(\tau)$. □

The fact that $\overline{\mu}$ is Δ-bi-admissible is crucial in the theorem above. Let us consider for instance $\Delta = (123)^\omega$ and $\overline{\mu} = (011)^\omega$. This sequence is Δ-non-admissible of type 1 since for $n \geq 1$, we have $\mu_n = 0 \iff \delta_n = 1$. We have $\varphi^v(\overline{\mu}) = \varphi^v(10^\omega)$ and $\psi^\Delta(10^\omega) = \boldsymbol{T}_1^\Delta = \boldsymbol{e}_1$ hence $\mu = \langle \boldsymbol{v}, \boldsymbol{e}_1 \rangle$. In this case, $\overline{\mathbb{P}}(\boldsymbol{v}, \mu, \Omega) = -\boldsymbol{e}_1 + \overline{\mathbb{P}}(\boldsymbol{v}, 0, \Omega)$ which is connected and is therefore equal to the connected component of $\mathbf{0}$. However, we may show that $\mathcal{S}_\infty^{\Delta,\overline{\mu}} = -\boldsymbol{e}_1 + \mathcal{C}_{\boldsymbol{e}_1}^\star$ where $\mathcal{C}_{\boldsymbol{e}_1}^\star$ is the connected component of $\boldsymbol{e}_1$ in $\overline{\mathbb{P}}(\boldsymbol{v}, 0, \Omega) \setminus \{\mathbf{0}\}$ which is disconnected. Similarly, $\overline{\mu} = (100)^\omega$ is Δ-admissible but not Δ-bi-admissible since $\widetilde{\overline{\mu}} = (011)^\omega$ which is not Δ-admissible. In this case, $\mu = \varphi^v(\overline{\mu}) = \varphi^v(01^\omega) = \Omega - \langle \boldsymbol{v}, \boldsymbol{e}_1 \rangle$ and we have $\overline{\mathbb{P}}(\boldsymbol{v}, \mu, \Omega) = \boldsymbol{e}_1 - \overline{\mathbb{P}}(\boldsymbol{v}, 0, \Omega)$ which is connected. But $\mathcal{S}_\infty^{\Delta,\overline{\mu}} = \boldsymbol{e}_1 - \mathcal{C}_{\boldsymbol{e}_1}^\star$.

Proposition 10. *If $\overline{\mu}$ is a Δ-non-admissible sequence* of type k *and $\overline{\mu}'$ is a Δ-admissible sequence such that $|\overline{\mu}'| < +\infty$ and $\varphi^v(\overline{\mu}') = \varphi^v(\overline{\mu})$, then $\mathcal{S}_\infty^{\Delta,\overline{\mu}} = \mathcal{C}_{\boldsymbol{e}_k}^\star - \psi^\Delta(\overline{\mu}')$ where $\mathcal{C}_{\boldsymbol{e}_k}^\star$ is the connected component of $\boldsymbol{e}_k$ in $\overline{\mathbb{P}}(\boldsymbol{v}, 0, \Omega) \setminus \{\mathbf{0}\}$.*

When $\overline{\mathbb{P}}(\boldsymbol{v}, \mu, \Omega)$ is disconnected, the incremental construction builds only $\mathcal{C}_{\mathbf{0}}^\mu$, the connected component of $\mathbf{0}$ in $\overline{\mathbb{P}}(\boldsymbol{v}, \mu, \Omega)$. We may build the other connected components of $\overline{\mathbb{P}}(\boldsymbol{v}, \mu, \Omega)$ as follows. Given a point $\boldsymbol{x} \in \overline{\mathbb{P}}(\boldsymbol{v}, \mu, \Omega)$, we have $\mathbf{0} \in \overline{\mathbb{P}}(\boldsymbol{v}, \mu, \Omega) - \boldsymbol{x} = \overline{\mathbb{P}}(\boldsymbol{v}, \mu + \langle \boldsymbol{v}, \boldsymbol{x} \rangle, \Omega)$. Since $\boldsymbol{x} \in \overline{\mathbb{P}}(\boldsymbol{v}, \mu, \Omega)$, we have $0 \leq \mu + \langle \boldsymbol{v}, \boldsymbol{x} \rangle \leq \Omega$ and there exists a Δ-bi-admissible encoding $\overline{\mu}'$ of $\mu' = \mu + \langle \boldsymbol{v}, \boldsymbol{x} \rangle$. Then $\mathcal{S}_\infty^{\Delta,\overline{\mu}'}$ is the connected component of $\mathbf{0}$ in $\overline{\mathbb{P}}(\boldsymbol{v}, \mu, \Omega) - \boldsymbol{x}$ which means that $\mathcal{C}_{\boldsymbol{x}}^\mu = \boldsymbol{x} + \mathcal{S}_\infty^{\Delta,\overline{\mu}'}$ is the connected component of $\boldsymbol{x}$ in $\overline{\mathbb{P}}(\boldsymbol{v}, \mu, \Omega)$. This way, we may build all the connected components provided that we know a starting point in each of them. $\mathcal{C}_{\boldsymbol{x}}^\mu$ may even be built directly by initialising the construction with $\{\boldsymbol{x}\}$ instead of $\{\mathbf{0}\}$. Note that initialising the construction with $\{\boldsymbol{x}\}$ without considering $\overline{\mu}'$ does not give the result. It would build $\boldsymbol{x} + \mathcal{C}_{\mathbf{0}}^\mu$ which is different from $\mathcal{C}_{\boldsymbol{x}}^\mu$.

For instance, let us consider a vector $\boldsymbol{v} \in \mathcal{K}_3$ the directive sequence of which is $\Delta = (1213)^\omega$, $\overline{\mu} = (1000)^\omega$ which is Δ-bi-admissible and $\mu = \varphi^v(\overline{\mu})$. Then $\overline{\mathbb{P}}(\boldsymbol{v}, \mu, \Omega)$ has two connected components. The point $\boldsymbol{x}_0 = \boldsymbol{e}_2 - \boldsymbol{e}_3$ belongs to $\overline{\mathbb{P}}(\boldsymbol{v}, \mu, \Omega) \setminus \mathcal{C}_{\mathbf{0}}^\mu$ and a Δ-bi-admissible encoding of $\mu' = \mu + \langle \boldsymbol{v}, \boldsymbol{x}_0 \rangle$ is $\overline{\mu}' = (0011)^\omega$. Figure 1 shows $\mathcal{S}_{10}^{\Delta,\overline{\mu}} \subset \mathcal{C}_{\mathbf{0}}^\mu$ and $\boldsymbol{x}_0 + \mathcal{S}_{10}^{\Delta,\overline{\mu}'} \subset \mathcal{C}_{\boldsymbol{x}_0}^\mu$.

5 Connectedness Criterion

Let $\mathcal{B}$ be the closed ball $[-\frac{1}{2}, +\frac{1}{2}]^d$, $\mathbf{1}$ the vector $(1, \ldots, 1)$ and $\Pi^{\mathbf{1}}$ the orthogonal projection on the plane $\mathbf{1}^\perp$. By a slight abuse of notation, for any set $S \subset \mathbb{R}^d$, we denote by $\mathring{\Pi}^{\mathbf{1}}(S)$ and $\partial\Pi^{\mathbf{1}}(S)$ the interior and the boundary of $\Pi^{\mathbf{1}}(S)$ in $\mathbf{1}^\perp$.

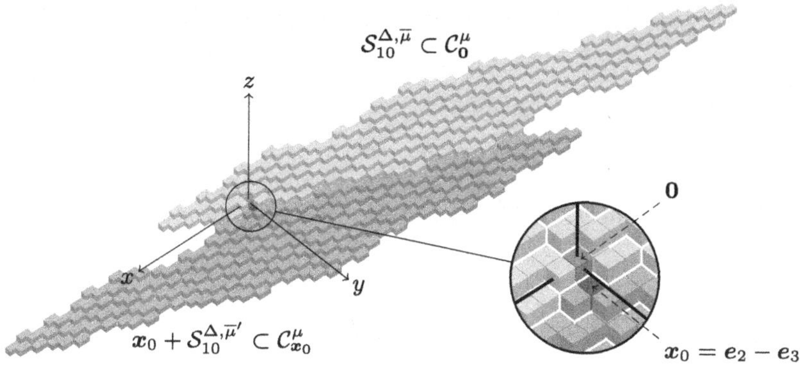

Fig. 1. The two connected components of $\overline{\mathbb{P}}(\boldsymbol{v}, \varphi^{\boldsymbol{v}}((1000)^{\omega}), \Omega)$ when $\Delta = (1213)^{\omega}$

Theorem 11. *Let $\boldsymbol{v} \in \mathcal{K}_d$, $\mu \in [0, \Omega]$ and $\mathcal{C}_{\mathbf{0}}^{\mu}$ the connected component of $\mathbf{0}$ in $\overline{\mathbb{P}}(\boldsymbol{v}, \mu, \Omega)$. Then $\overline{\mathbb{P}}(\boldsymbol{v}, \mu, \Omega)$ is connected if and only if $\partial\Pi^1(\mathcal{C}_{\mathbf{0}}^{\mu} + \mathcal{B})$ is empty.*

The proof of this theorem requires two additional lemmas.

Lemma 12. *For all $\boldsymbol{x}, \boldsymbol{y} \in \overline{\mathbb{P}}(\boldsymbol{v}, \mu, \Omega)$, we have $|\sum_i (y_i - x_i)| \leq \|\boldsymbol{y} - \boldsymbol{x}\|_{\infty}$.*

Lemma 13. *Let $\boldsymbol{x}, \boldsymbol{y} \in \overline{\mathbb{P}}(\boldsymbol{v}, \mu, \Omega)$ such that $\boldsymbol{x} \neq \boldsymbol{y}$ and $\Pi^1(\boldsymbol{x}+\mathcal{B}) \cap \mathring{\Pi}^1(\boldsymbol{y}+\mathcal{B}) \neq \emptyset$. Then $\boldsymbol{y} = \boldsymbol{x} \pm \boldsymbol{e}_i$.*

Proof of Theorem 11. We first note that $\mathbf{0} \in \overline{\mathbb{P}}(\boldsymbol{v}, \mu, \Omega)$ since $\mu \in [0, \Omega]$ and thus $\mathcal{C}_{\mathbf{0}}^{\mu} \neq \emptyset$. Since $\Omega > \|\boldsymbol{v}\|_{\infty}$, the discrete plane $\mathbb{P}(\boldsymbol{v}, \mu, \Omega)$ is $(d-1)$-separating [1] which implies $\Pi^1(\mathbb{P}(\boldsymbol{v}, \mu, \Omega) + \mathcal{B}) = \mathbf{1}^{\perp}$ and therefore $\Pi^1(\overline{\mathbb{P}}(\boldsymbol{v}, \mu, \Omega) + \mathcal{B}) = \mathbf{1}^{\perp}$. If $\overline{\mathbb{P}}(\boldsymbol{v}, \mu, \Omega)$ is connected then $\mathcal{C}_{\mathbf{0}}^{\mu} = \overline{\mathbb{P}}(\boldsymbol{v}, \mu, \Omega)$ hence $\Pi^1(\mathcal{C}_{\mathbf{0}}^{\mu} + \mathcal{B}) = \mathbf{1}^{\perp}$. Its boundary in $\mathbf{1}^{\perp}$ is therefore empty.

Assume now that $\overline{\mathbb{P}}(\boldsymbol{v}, \mu, \Omega)$ is disconnected and let $\boldsymbol{x} \in \overline{\mathbb{P}}(\boldsymbol{v}, \mu, \Omega) \setminus \mathcal{C}_{\mathbf{0}}^{\mu}$. Then $\mathcal{C}_{\mathbf{0}}^{\mu}$ contains no point of the form $\boldsymbol{x} \pm \boldsymbol{e}_i$. Indeed, if $\boldsymbol{x} \pm \boldsymbol{e}_i$ belongs to $\overline{\mathbb{P}}(\boldsymbol{v}, \mu, \Omega)$ then it is connected to $\boldsymbol{x}$ in $\overline{\mathbb{P}}(\boldsymbol{v}, \mu, \Omega)$ and cannot belong to $\mathcal{C}_{\mathbf{0}}^{\mu}$. But these points are the only $\boldsymbol{y}$ which could belong to $\overline{\mathbb{P}}(\boldsymbol{v}, \mu, \Omega)$ such that $\Pi^1(\boldsymbol{y}+\mathcal{B}) \cap \mathring{\Pi}^1(\boldsymbol{x}+\mathcal{B}) \neq \emptyset$. We deduce that $\mathring{\Pi}^1(\boldsymbol{x} + \mathcal{B})$ is disjoint from $\Pi^1(\mathcal{C}_{\mathbf{0}}^{\mu} + \mathcal{B})$ which therefore has a non-empty boundary in $\mathbf{1}^{\perp}$. □

Definition 14 (Interior and border of $\mathcal{S}_n^{\Delta,\overline{\mu}}$). *Let $\boldsymbol{x} \in \mathcal{S}_n^{\Delta,\overline{\mu}}$:*

- *$\boldsymbol{x}$ is* interior *in $\mathcal{S}_n^{\Delta,\overline{\mu}}$ if and only if $\Pi^1(\boldsymbol{x} + \mathcal{B})$ is contained in $\mathring{\Pi}^1(\mathcal{S}_n^{\Delta,\overline{\mu}} + \mathcal{B})$;*
- *$\boldsymbol{x}$ is on the* border *of $\mathcal{S}_n^{\Delta,\overline{\mu}}$ if and only if it is not interior in $\mathcal{S}_n^{\Delta,\overline{\mu}}$.*

We saw earlier that $\mathcal{S}_n^{\Delta,\overline{\mu}} = \mathcal{S}_n^{\Delta} - \psi^{\Delta}(\mu_1 \ldots \mu_n)$. Thus a point $\boldsymbol{x}$ is interior in $\mathcal{S}_n^{\Delta,\overline{\mu}}$ if and only if $\boldsymbol{x} + \psi^{\Delta}(\mu_1 \ldots \mu_n)$ is interior in $\mathcal{S}_n^{\Delta}$. Therefore we shall mainly study $\mathcal{S}_n^{\Delta}$. All results will apply immediately to $\mathcal{S}_n^{\Delta,\overline{\mu}}$ by translation.

Proposition 15. *Let $\sigma \in \{0,1\}^n$ such that $\psi^{\Delta}(\sigma)$ is interior in $\mathcal{S}_n^{\Delta}$.*

- *For all $\sigma' \in \{0,1\}^{\star}$, $\psi^{\Delta}(\sigma\sigma')$ is interior in $\mathcal{S}_{n+|\sigma'|}^{\Delta}$.*

– *For all* $\Delta_0 \in \{0,\ldots,d\}^\star$ *and all* $\sigma_0 \in \{0,1\}^{|\Delta_0|}$, $\psi^{\Delta_0\Delta}(\sigma_0\sigma)$ *is interior in* $\mathcal{S}^{\Delta_0\Delta}_{|\Delta_0|+n}$.

Again, we need two technical lemmas.

Lemma 16. *Let* $\mathcal{V}_\mathbf{0}$ *be the 0-neighbourhood of* $\mathbf{0}$ *in* $\mathbb{Z}^d$. *A point* $\boldsymbol{x} \in \mathcal{S}^\Delta_n$ *is interior in* $\mathcal{S}^\Delta_n$ *if and only if for all* $\boldsymbol{y} \in \mathcal{V}_\mathbf{0}$ *such that* $\langle \mathbf{1}, \boldsymbol{y}\rangle = 0$, *and all* i, j *such that* $i \neq j$, $y_i \leq 0$ *and* $y_j \geq 0$, $\mathcal{S}^\Delta_n$ *contains at least one point in* $\boldsymbol{x} + \{\boldsymbol{y}, \boldsymbol{y} + \boldsymbol{e}_i, \boldsymbol{y} - \boldsymbol{e}_j\}$.

Lemma 17. *Let* $\Delta = \delta_1 \ldots \delta_{n_0} \in \{1,\ldots,d\}^{n_0}$, $\Delta' \in \{1,\ldots,d\}^\omega$ *and* $\Gamma^\Delta = {}^t\gamma_{\delta_1} \ldots {}^t\gamma_{\delta_{n_0}}$ *where* ${}^t\gamma$ *is the transpose of* γ. *For all* $n \geq 0$, *we have*

$$\mathcal{S}^{\Delta\Delta'}_{n_0+n} = \mathcal{S}^\Delta_{n_0} + \Gamma^\Delta(\mathcal{S}^{\Delta'}_n)$$

The picture below shows an example of this composition.

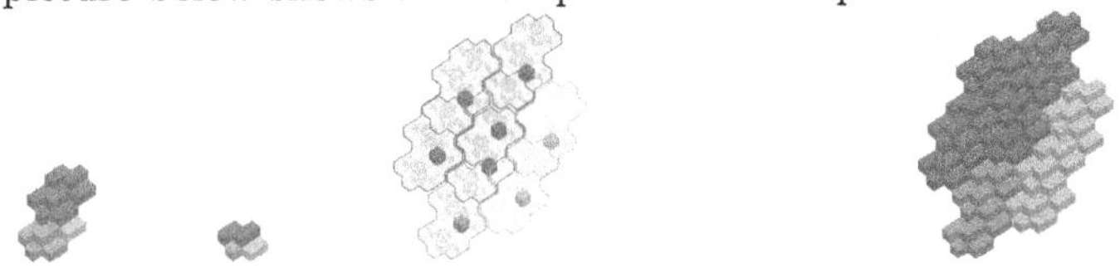

$\mathcal{S}^{12313}_5$ $\mathcal{S}^{123}_3$ $\Gamma^{12313}(\mathcal{S}^{123}_3)$ $\mathcal{S}^{12313123}_8 = \mathcal{S}^{12313}_5 + \Gamma^{12313}(\mathcal{S}^{123}_3)$

Proof of Proposition 15 *(sketch).* The first point is obvious since $\mathcal{S}^\Delta_n \subset \mathcal{S}^\Delta_m$ for all $n \leq m$.

We prove the second point for $|\Delta_0| = 1$, i.e. $\Delta_0 = \delta \in \{1,\ldots,d\}$, and $\sigma_0 = \varepsilon \in \{0,1\}$. The general result follows by induction. We consider more generally a set $\mathcal{S}$ having the property of Lemma 12. Using Lemmas 16 and 17, we determine that we have to prove that if $\boldsymbol{x}$ is interior in $\mathcal{S}$, then for all $\boldsymbol{y} \in \mathcal{V}_\mathbf{0}$ such that $\langle \mathbf{1}, \boldsymbol{y}\rangle = 0$, and all i, j such that $i \neq j$, $y_i \leq 0$ and $y_j \geq 0$, $\mathcal{S}$ contains at least one point in

$$\begin{aligned}\boldsymbol{x} + \{&\boldsymbol{y} + (\varepsilon - y_\delta)\,\boldsymbol{e}_\delta,\ \boldsymbol{y} + (\varepsilon - y_\delta - 1)\,\boldsymbol{e}_\delta,\ \boldsymbol{y} + \boldsymbol{e}_i + (\varepsilon - y_\delta - \langle \boldsymbol{e}_i, \boldsymbol{e}_\delta\rangle)\,\boldsymbol{e}_\delta,\\ &\boldsymbol{y} + \boldsymbol{e}_i + (\varepsilon - y_\delta - \langle \boldsymbol{e}_i, \boldsymbol{e}_\delta\rangle + 1)\,\boldsymbol{e}_\delta,\ \boldsymbol{y} - \boldsymbol{e}_j + (\varepsilon - y_\delta + \langle \boldsymbol{e}_j, \boldsymbol{e}_\delta\rangle - 1)\,\boldsymbol{e}_\delta,\\ &\boldsymbol{y} - \boldsymbol{e}_j + (\varepsilon - y_\delta + \langle \boldsymbol{e}_j, \boldsymbol{e}_\delta\rangle - 2)\,\boldsymbol{e}_\delta\}\end{aligned}$$

We consider 18 cases according to the respective values of ε, i, j and y_δ. □

Lemma 16 allows us to determine whether a point $\boldsymbol{x}$ is interior in $\mathcal{S}^\Delta_n$ by looking only at its 0-neighbourhood. Using this lemma we may compute all possible configurations in which $\boldsymbol{x}$ is interior. These are subsets of the 0-neighbourhood of $\boldsymbol{x}$ which satisfy the conditions of the lemma. By Lemma 12, we may restrict this neighbourhood to the points $\boldsymbol{y}$ such that $\langle \mathbf{1}, \boldsymbol{y} - \boldsymbol{x}\rangle \in \{-1, 0, 1\}$. We may also keep only the configurations which are minimal with respect to inclusion. Then $\boldsymbol{x}$ is interior in $\mathcal{S}^\Delta_n$ if and only if its 0-neighbourhood in $\mathcal{S}^\Delta_n$ contains one of these minimal configurations.

We find 64 minimal configurations in dimension 3 and 1498 in dimension 4. If we consider only configurations which appear in planes the coordinates of the normal vector of which are sorted in ascending order, we find, in dimension 3, the 21 minimal configurations shown below. Other minimal configurations are obtained by permuting the coordinates.

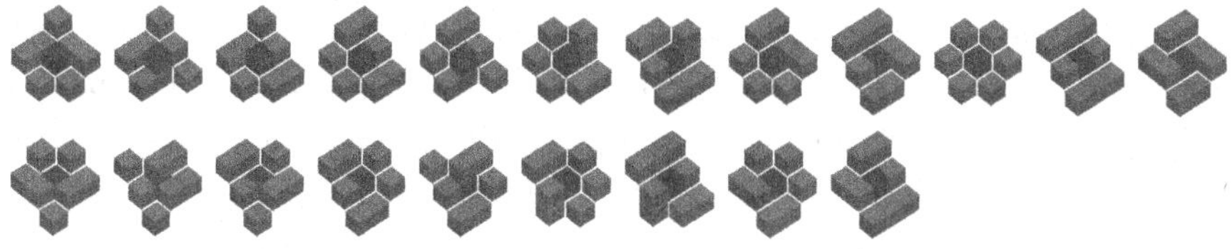

Lemma 18. *If for all n, $\mathbf{0}$ is at bounded distance of the border of $\mathcal{S}_n^{\Delta,\overline{\mu}}$, then there exists n_0 such that $\mathbf{0}$ is on the border of $\mathcal{S}_n^{\Delta^{n_0},\overline{\mu}^{n_0}}$ for all n, where $\overline{\mu}^{n_0}$ denotes the sequence $(\mu_n)_{n\geq n_0}$.*

Proof. Since $\mathbf{0}$ is at bounded distance of the border of $\mathcal{S}_n^{\Delta,\overline{\mu}}$, $\mathcal{S}_\infty^{\Delta,\overline{\mu}}$ has a non-empty border.

If $\mathbf{0}$ is on the border of $\mathcal{S}_\infty^{\Delta,\overline{\mu}}$ then we may take $n_0 = 1$. Otherwise, let m be such that $\mathbf{0}$ is interior in $\mathcal{S}_m^{\Delta,\overline{\mu}}$ or equivalently, $\psi^\Delta(\mu_1 \dots \mu_m)$ is interior in $\mathcal{S}_m^\Delta$. For all $k \geq 0$, we have $\mathcal{S}_{m+k}^\Delta = \mathcal{S}_m^{\delta_1\dots\delta_m} + \Gamma^{\delta_1\dots\delta_m}(\mathcal{S}_k^{\Delta^{m+1}})$. The distance from $\psi^{\Delta^{m+1}}(\mu_{m+1}\dots\mu_{m+k})$ to the border of $\mathcal{S}_k^{\Delta^{m+1}}$ is strictly less than the distance from $\psi^\Delta(\mu_1\dots\mu_{m+k})$ to the border of $\mathcal{S}_{m+k}^\Delta$. By repeating this operation, we find eventually n_0 such that the distance from $\psi^{\Delta^{n_0}}(\overline{\mu}^{n_0})$ to the border of $\mathcal{S}_k^{\Delta^{n_0}}$ is zero for all $k \geq 0$. This means that $\psi^{\Delta^{n_0}}(\overline{\mu}^{n_0})$ is on the border of $\mathcal{S}_k^{\Delta^{n_0}}$, or equivalently $\mathbf{0}$ is on the border of $\mathcal{S}_k^{\Delta^{n_0},\overline{\mu}^{n_0}}$. □

Proposition 19. *Let $\mu \in [0;\Omega]$ and $\overline{\mu} \in \{0,1\}^\omega$ a Δ-bi-admissible sequence such that $\varphi^{\boldsymbol{v}}(\overline{\mu}) = \mu$. Then $\overline{\mathbb{P}}(\boldsymbol{v},\mu,\Omega)$ is connected if and only if for all $m \geq 1$, there exists $n \geq m$ such that $\psi^{\Delta^m}(\mu_m \cdots \mu_n)$ is interior in $\mathcal{S}_{n-m+1}^{\Delta^m}$.*

Since each point in $\mathcal{S}_n^\Delta$ may be coded by a word in $\{0,1\}^n$, we consider the language of words which code interior points. We define the language $\mathcal{L}^\Delta$ as the set of words $\sigma \in \{0,1\}^\star$ such that $\psi^\Delta(\sigma)$ is interior in $\mathcal{S}_{|\sigma|}^\Delta$ and for all $\boldsymbol{p} \in \mathbb{Z}^d$, we define the language $\mathcal{L}_{\boldsymbol{p}}^\Delta$ as

$$\mathcal{L}_{\boldsymbol{p}}^\Delta = \{\sigma \in \{0,1\}^\star \mid \psi^\Delta(\sigma) + \boldsymbol{p} \in \mathcal{S}_{|\sigma|}^\Delta\}$$

Intuitively, $\mathcal{L}_{\boldsymbol{p}}^\Delta$ is the set of codes of points $\boldsymbol{x} \in \overline{\mathbb{P}}(\boldsymbol{v},0,\Omega)$ such that $\boldsymbol{x}$ and $\boldsymbol{x}+\boldsymbol{p}$ belong to the same $\mathcal{S}_n^\Delta$, i.e. $\boldsymbol{x} \in \bigcup_n \mathcal{S}_n^\Delta \cap (\mathcal{S}_n^\Delta - \boldsymbol{p})$.

Lemma 20. *For all $\boldsymbol{p} \in \mathbb{Z}^d$:*

1. $\mathcal{L}_{-\boldsymbol{p}}^\Delta = \widetilde{\mathcal{L}_{\boldsymbol{p}}^\Delta} = \{\widetilde{\sigma} \mid \sigma \in \mathcal{L}_{\boldsymbol{p}}^\Delta\}$;
2. $\mathcal{L}_{\boldsymbol{p}}^\Delta$ *is non-empty if and only if* $\boldsymbol{p} \in \overline{\mathbb{P}}(\boldsymbol{v},0,\Omega)$ *or* $-\boldsymbol{p} \in \overline{\mathbb{P}}(\boldsymbol{v},0,\Omega)$.

The language $\mathcal{L}^\Delta$ may then be defined from the minimal interior configurations we defined earlier. If $\mathcal{C}$ is the set of these minimal configurations, then

$$\mathcal{L}^\Delta = \bigcup_{c\in\mathcal{C}} \bigcap_{\boldsymbol{p}\in c} \mathcal{L}_{\boldsymbol{p}}^\Delta.$$

We saw that if $\psi^{\Delta}(\mu_1 \dots \mu_n)$ is interior in $\mathcal{S}_n^{\Delta}$, then $\psi^{\Delta}(\mu_1 \dots \mu_m)$ is interior in $\mathcal{S}_m^{\Delta}$ for all $m \geq n$. Then there exists a minimal index n_1 such that $\psi^{\Delta}(\mu_1 \dots \mu_{n_1})$ is interior in $\mathcal{S}_{n_1}^{\Delta}$, meaning that $\psi^{\Delta}(\mu_1 \dots \mu_{n_1-1})$ is on the border of $\mathcal{S}_{n_1-1}^{\Delta}$. Similarly, there exists a maximal index n_0 such that $\psi^{\Delta^{n_0}}(\mu_{n_0} \dots \mu_{n_1})$ is interior in $\mathcal{S}_{n_1-n_0+1}^{\Delta^{n_0}}$, which means that $\psi^{\Delta^{n_0+1}}(\mu_{n_0+1} \dots \mu_{n_1})$ is on the border of $\mathcal{S}_{n_1-n_0}^{\Delta^{n_0+1}}$. Thus the pair $(\delta_{n_0} \dots \delta_{n_1}, \mu_{n_0} \dots \mu_{n_1})$ is a *minimal interior pattern*. For all Δ' and all $\overline{\mu}'$, if there exists an index i such that $\delta'_i \dots \delta'_{i+n_1-n_0} = \delta_{n_0} \dots \delta_{n_1}$ and $\mu'_i \dots \mu'_{i+n_1-n_0} = \mu_{n_0} \dots \mu_{n_1}$, then $\psi^{\Delta'}(\mu'_1 \dots \mu'_m)$ is interior in $\mathcal{S}_m^{\Delta'}$ for all $m \geq i + n_1 - n_0$.

Definition 21 (Interior Patterns)

- *A* pattern *is an element of* $\cup_{n \geq 0}(\{1, \dots, d\}^n \times \{0,1\}^n)$.
- *A pattern* $(\delta_1 \dots \delta_n, \sigma_1 \dots \sigma_n)$ *is* interior *if and only if* $\psi^{\delta_1 \dots \delta_n}(\sigma_1 \dots \sigma_n)$ *is interior in* $\mathcal{S}_n^{\delta_1 \dots \delta_n}$.
- *An interior pattern* $(\delta_1 \dots \delta_n, \sigma_1 \dots \sigma_n)$ *is* minimal *if and only if neither* $(\delta_2 \dots \delta_n, \sigma_2 \dots \sigma_n)$ *nor* $(\delta_1 \dots \delta_{n-1}, \sigma_1 \dots \sigma_{n-1})$ *is interior.*

We set $\mathcal{L}_n^{\Delta} = \mathcal{L}^{\Delta^n}$ and

$$\mathcal{M}_n^{\Delta} = \mathcal{L}_n^{\Delta} \cap \{0,1\}\overline{\mathcal{L}_{n+1}^{\Delta}} \cap \overline{\mathcal{L}_n^{\Delta}}\{0,1\}$$

where $\overline{\mathcal{L}} = \{0,1\}^\star \setminus \mathcal{L}$.

$\mathcal{M}_n^{\Delta}$ is the set of words $\sigma \in \{0,1\}^\star$ such that $(\delta_n \dots \delta_{n+|\sigma|-1}, \sigma)$ is a minimal interior pattern. Thus the set of all minimal interior patterns which may appear in $(\Delta, \overline{\mu})$ is

$$\bigcup_{n \geq 1} \{(\delta_n \dots \delta_{n+|\sigma|-1}, \sigma) \mid \sigma \in \mathcal{M}_n^{\Delta}\}$$

and the set of all minimal interior patterns is

$$\bigcup_{\Delta \in \{1,\dots,d\}^\omega} \{(\delta_1 \dots \delta_{|\sigma|}, \sigma) \mid \sigma \in \mathcal{M}_1^{\Delta}\}$$

Theorem 22. *Let $\mu \in [0; \Omega]$ and $\overline{\mu}$ a Δ-bi-admissible encoding of μ. Then $\overline{\mathbb{P}}(\boldsymbol{v}, \mu, \Omega)$ is connected if and only if $(\Delta, \overline{\mu})$ contains infinitely many minimal interior patterns.*

6 The Periodic Case

We consider now the specific case where Δ is periodic, that is $\Delta = (\delta_1 \dots \delta_p)^\omega$ for some $p \geq 1$ and $\{\delta_1, \dots, \delta_p\} = \{1, \dots, d\}$. This means that after p application of the Fully Subtractive algorithm to the normal vector $\boldsymbol{v}$, we obtain a vector $\boldsymbol{v}'$ which is proportional to $\boldsymbol{v}$. In other words, $\boldsymbol{v}' = \gamma_{\delta_p} \cdots \gamma_{\delta_1}(\boldsymbol{v}) = \beta\, \boldsymbol{v}$ for some $\beta \in [0; 1[$. The value β is the inverse of the Pisot eigenvalue of $\gamma_{\delta_1}^{-1} \dots \gamma_{\delta_p}^{-1}$ which was proven to be Pisot in [2].

For each $\boldsymbol{p} \in \mathbb{P}(\boldsymbol{v}, \mu, \Omega)$, there exists $\pi \in \{0,1\}^\star$ such that $\psi^\Delta(\pi) = \boldsymbol{p}$. The language $\mathcal{L}_{\boldsymbol{p}}^\Delta$ is obviously recognised by the infinite automaton which contains for each word $\sigma \in \{0,1\}^\star$ a state labelled with σ and for each $\alpha \in \{0,1\}$ the transition $\sigma \xrightarrow{\alpha} \sigma\alpha$. A state is final if and only if $\varphi^{\boldsymbol{v}}(\sigma + \pi) < \Omega$ and $|(\sigma + \pi)\Downarrow^\Delta 0^\omega| \leq |\sigma|$. When Δ is periodic, this infinite automaton is actually equivalent to a finite one, which means that $\mathcal{L}_{\boldsymbol{p}}^\Delta$ is a regular language.

Theorem 23. *If Δ is periodic then $\mathcal{L}_{\boldsymbol{p}}^\Delta$ is a regular language for all $\boldsymbol{p} \in \mathbb{Z}^d$.*

Proof (sketch). We saw that $\mathcal{L}_{\boldsymbol{p}}^\Delta$ is non-empty if and only if $\boldsymbol{p} \in \mathbb{P}(\boldsymbol{v}, 0, \Omega)$ or $-\boldsymbol{p} \in \mathbb{P}(\boldsymbol{v}, 0, \Omega)$ and $\mathcal{L}_{-\boldsymbol{p}}^\Delta = \widetilde{\mathcal{L}_{\boldsymbol{p}}^\Delta}$. We have obviously $\mathcal{L}_{\boldsymbol{0}}^\Delta = \{0,1\}^\star$. It is therefore sufficient to prove the result when $\boldsymbol{p} \in \mathbb{P}(\boldsymbol{v}, 0, \Omega) \setminus \{\boldsymbol{0}\}$. A word $\sigma \in \{0,1\}^\star$ belongs to $\mathcal{L}_{\boldsymbol{p}}^\Delta$ if and only if $\psi^\Delta(\sigma + \pi)$ belongs to $\mathcal{S}_{|\sigma|}^\Delta$, i.e. if and only if $\varphi^{\boldsymbol{v}}(\sigma + \pi) < \Omega$ and $|(\sigma + \pi)\Downarrow^\Delta 0^\omega| \leq |\sigma|$.

We consider an incremental construction of the automaton in which states are labelled with triples $(\sigma, k, n) \in \{0,1\}^\star \times (\mathbb{N} \setminus \{0\}) \times (\mathbb{N} \setminus \{0\})$. The initial state is $(\pi\Downarrow^\Delta, 1, 1)$ and we add two special states: a final state $\top$ and a *trap* state $\bot$ with the transitions $\top \xrightarrow{0,1} \top$ and $\bot \xrightarrow{0,1} \bot$. For each state (σ, k, n) and each $\alpha \in \{0,1\}$ we have a transition $(\sigma, k, n) \xrightarrow{\alpha} \mathrm{simp}(\sigma + 0^{k-1}1, k+1, n)$ where 'simp' is a function which simplifies states. If $\phi^{\boldsymbol{v}^n}(\sigma + 0^{k-1}1) > \Omega^n$ then $\mathrm{simp}(\sigma + 0^{k-1}1, k+1, n) = \bot$. If $\phi^{\boldsymbol{v}^n}(\sigma + 0^{k-1}1) < \Omega^n$ and $|(\sigma + 0^{k-1}1)\Downarrow^{\Delta^n} 0^\omega| \leq k$ then $\mathrm{simp}(\sigma + 0^{k-1}1, k+1, n) = \top$. In all other cases, the simplification of (σ, k, n) consists in removing from $\sigma\Downarrow^{\Delta^n}$ a *useless* prefix of size ℓ: if $\sigma\Downarrow^{\Delta^n} = \tau_1 \ldots \tau_\ell \sigma'$, then the result of the simplification is $(\sigma', k - \ell, 1 + (n + \ell - 1) \bmod p)$. Intuitively, a prefix is *useless* if it does not actually participate in the computation of the Δ^n-normal form of $\sigma + 0^{k-1}\mu$, i.e. for all $\mu \in \{0,1\}^\star$,

$$\begin{aligned} &\phi^{\boldsymbol{v}^n}(\sigma + 0^{k-1}\mu) < \Omega^n \wedge |(\sigma + 0^{k-1}\mu)\Downarrow^{\Delta^n} 0^\omega| \leq k \\ \iff\ &\phi^{\boldsymbol{v}^{n+\ell}}(\sigma' + 0^{k-\ell-1}\mu) < \Omega^{n+\ell} \wedge |(\sigma' + 0^{k-\ell-1}\mu)\Downarrow^{\Delta^{n+\ell}} 0^\omega| \leq k - \ell \end{aligned}$$

which may be tested in finite time. We first show that k is bounded by $p + 2$. Then, from Lemma 17 we have $\mathcal{S}_{p+n}^\Delta = \mathcal{S}_p^{\delta_1 \ldots \delta_p} + \Gamma^{\delta_1 \ldots \delta_p}(\mathcal{S}_n^\Delta)$ where $\Gamma^{\delta_1 \ldots \delta_p}$ is the inverse of a Pisot operator [2]. Using this decomposition, we are able to prove that $2\mathcal{S}_m^\Delta \cap \mathcal{S}_\infty^\Delta \subset \mathcal{S}_{m+r}^\Delta$ for some fixed r. We deduce that after m steps of the construction of the automaton, $|(\sigma + \pi)\Downarrow^\Delta 0^\omega|$ has not grown more than r. Using the fact that k is bounded, we deduce that after simplification, the length of σ is also bounded. Since Δ is periodic, we may replace n with $1 + (n-1) \bmod p$ which is also bounded. Therefore, the automaton has only finitely many states. □

From this theorem, we deduce that $\mathcal{L}_n^\Delta$ and $\mathcal{M}_n^\Delta$ are regular languages for all n and may therefore be recognised by finite state automata. In particular, the words σ such that $\psi^\Delta(\sigma)$ is interior in $\mathcal{S}_{|\sigma|}^\Delta$ are recognised by a finite state automaton. Since Δ is periodic, the sequences $(\mathcal{L}_n^\Delta)_{n \geq 1}$ and $(\mathcal{M}_n^\Delta)_{n \geq 1}$ are also periodic and their periods divide the period of Δ.

For example, if $\boldsymbol{v} = (\alpha, \alpha + \alpha^2, 1)$, where $\alpha \in [0;1]$ and $\alpha + \alpha^2 + \alpha^3 = 1$, then $\Delta = (123)^\omega$ and we find for all n, $\mathcal{M}_n^\Delta = \mathcal{M} \cup \widetilde{\mathcal{M}}$ where

$$\mathcal{M} = 1111 \,|\, 0111 \,|\, 1(100)^+11 \,|\, 1(010)^+10 \,|\, 0(001)^+010$$

As announced in the example under Sect. 4, we deduce that $\overline{\mathbb{P}}(\boldsymbol{v},\mu,\Omega)$ is connected when $\mu = \varphi^{\boldsymbol{v}}((01)^\omega) = (\alpha+1)/2$ and disconnected when $\mu = \varphi^{\boldsymbol{v}}((011001)^\omega) = (6\alpha^2+15\alpha-3)/14$.

If we take $\boldsymbol{v} = (1, 2-\alpha, 2+\alpha-\alpha^2)$, where $\alpha \in [0;1]$ and $\alpha^3-\alpha^2-3\alpha+1=0$, then $\Delta = (1213)^\omega$. We find for all $n \geq 0$, $\mathcal{M}^{\Delta}_{2n+1} = \emptyset$ and $\mathcal{M}^{\Delta}_{2n+2} = \mathcal{M} \cup \widetilde{\mathcal{M}}$ where

$$\mathcal{M} = 01001 \,|\, (00\,|\,10\,|\,11)000 \,|\, (00\,|\,10\,|\,11)(0101)^{+}1 \,|\, 10(1000)^{\star}10(11\,|\,010\,|\,0011)$$

In order for $\mathbb{P}(\boldsymbol{v},\varphi^{\boldsymbol{v}}(\overline{\mu}),\Omega)$ to be disconnected, $\overline{\mu}$ must have a suffix which avoids completely the minimal interior patterns. As stated by the next theorem, this happens almost never which means that $\mathbb{P}(\boldsymbol{v},\mu,\Omega)$ is actually almost always connected.

Theorem 24. *The set* $\{\mu \in [0;\Omega] \mid \mathbb{P}(\boldsymbol{v},\mu,\omega)$ *is disconnected*$\}$ *is Lebesgue negligible.*

References

1. Andrès, E., Acharya, R., Sibata, C.: Discrete analytical hyperplanes. CVGIP: Graph. Model Image Process. **59**(5), 302–309 (1997)
2. Avila, A., Delecroix, V.: Some monoids of Pisot matrices, preprint (2015). https://arxiv.org/abs/1506.03692
3. Brimkov, V.E., Barneva, R.P.: Connectivity of discrete planes. Theor. Comput. Sci. **319**(1–3), 203–227 (2004). https://doi.org/10.1016/j.tcs.2004.02.015
4. Domenjoud, E., Jamet, D., Toutant, J.-L.: On the connecting thickness of arithmetical discrete planes. In: Brlek, S., Reutenauer, C., Provençal, X. (eds.) DGCI 2009. LNCS, vol. 5810, pp. 362–372. Springer, Heidelberg (2009). https://doi.org/10.1007/978-3-642-04397-0_31
5. Domenjoud, E., Provençal, X., Vuillon, L.: Facet connectedness of discrete hyperplanes with zero intercept: the general case. In: Barcucci, E., Frosini, A., Rinaldi, S. (eds.) DGCI 2014. LNCS, vol. 8668, pp. 1–12. Springer, Cham (2014). https://doi.org/10.1007/978-3-319-09955-2_1
6. Domenjoud, E., Vuillon, L.: Geometric palindromic closures. Uniform Distribution Theory **7**(2), 109–140 (2012). https://math.boku.ac.at/udt/vol07/no2/06DomVuillon13-12.pdf
7. Jamet, D., Toutant, J.L.: Minimal arithmetic thickness connecting discrete planes. Discrete Appl. Math. **157**(3), 500–509 (2009). https://doi.org/10.1016/j.dam.2008.05.027
8. Kraaikamp, C., Meester, R.: Ergodic properties of a dynamical system arising from percolation theory. Ergodic Theory Dyn. Syst. **15**(04), 653–661 (1995). https://doi.org/10.1017/S0143385700008592
9. Réveillès, J.P.: Géométrie discrète, calcul en nombres entiers et algorithmique. Thèse d'état, Université Louis Pasteur, Strasbourg, France (1991)

Local Turn-Boundedness: A Curvature Control for a Good Digitization

Étienne Le Quentrec(✉), Loïc Mazo, Étienne Baudrier, and Mohamed Tajine

ICube-UMR 7357, 300 Bd Sébastien Brant - CS 10413,
67412 Illkirch Cedex, France
elequentrec@unistra.fr

Abstract. This paper focuses on the classical problem of the control of information loss during the digitization step. The properties proposed in the literature rely on smoothness hypotheses that are not verified by the curves including angular points. The notion of *turn* introduced by Milnor in the article *On the Total Curvature of Knots* generalizes the notion of integral curvature to continuous curves. Thanks to the turn, we are able to define the *local turn-boundedness*. This promising property of curves do not require smoothness hypotheses and shares several properties with the par(r)-regularity, in particular well-composed digitizations. Besides, the local turn-boundedness enables to constraint spatially the continuous curve in function of its digitization.

1 Introduction

The loss of information caused by a digitization process is inevitable. Therefore a fundamental point concerns the control of this information loss. Indeed, the border of a compact connected shape S of the Euclidean plane can be arbitrarily far from the digitization of S or can oscillate around this latter border. Therefore, hypotheses on the border of the shape S, which is a Jordan curve denoted by $\mathcal{C}$, are needed. One of the most used hypothesis, called *par(r)-regularity*, was introduced by Pavlidis in [7]. It demands that any point $c \in \mathcal{C}$ has an interior osculating disk entirely included in the interior of $\mathcal{C}$ except for the point c and an exterior osculating disk entirely included in the exterior of $\mathcal{C}$ except for the point c. It has been used to prove some preserving topology properties [2,5,7] or to control the behavior of the projection from the digitized curve to the continuous curve $\mathcal{C}$ [3,4]. However, the notion can be hard to manipulate in geometry and most of the authors using this notion add the assumption that the curve $\mathcal{C}$ is of class C^2. The par(r)-regularity encompasses two ideas: the border of the shape has a curvature bounded from above and the shape has a positive minimal thickness. In this article, relaxing the assumption of a bounded curvature, we develop a new notion called *local turn-boundedness* that is defined on continuous curves, including polygons. The local turn-boundedness relies on the notion of *turn* adapted to regular curves and polygons, firstly introduced by Milnor [6] to study the geometry of knots. The main properties and definitions around

M. Couprie et al. (Eds.): DGCI 2019, LNCS 11414, pp. 51–61, 2019.
https://doi.org/10.1007/978-3-030-14085-4_5

the notion of turn are recalled in Sect. 2. The local turn-boundedness involves the Euclidean distance between any two points of a curve and the turn of this curve between these two points. The properties of locally turn-bounded curves are given and illustrated in Sect. 3.

2 Turn of a Simple Curve

Although the notion was introduced by Milnor in [6], the definitions and properties given in this section come from the book of Alexandrov and Reshetnyak [1]. As presented in Proposition 2, the turn extends to continuous curves the notion of integral curvature already defined for regular curves.

Terminology and Notations

- Let $c \in \mathbb{R}^2$ and $r \geq 0$, $B(c, r)$ is the open disk of center c and radius r.
- A *parametrized curve* is a continuous application from an interval $[a, b]$ of $\mathbb{R}$ ($a < b$) to $\mathbb{R}^2$. It is *simple* if it is injective on $[a, b)$ and *closed* if $\gamma(b) = \gamma(a)$. A *(geometric) curve* is the image of a parametrized curve. A *Jordan curve* is a simple closed curve.
- For a simple parametrized curve γ, an order is defined on the points of the associated geometric curve $\mathcal{C}$ by:

$$\gamma(\alpha) \leq_\gamma \gamma(\beta) \Leftrightarrow \alpha \leq \beta$$

  and $\leq_\gamma$ is noted $\leq$ if there is no ambiguity. A simple parametrized curve γ with such an order is called *oriented curve* and its image is called *oriented geometric curve*. A *chain* is a finite increasing sequence of points on an oriented geometric curve $\mathcal{C}$.
- A polygonal line with vertices x_0, ..., x_N is denoted by $[x_0 x_1 ... x_N]$ (if $x_N = x_0$, the polygonal line is a polygon). A polygonal line L is *inscribed* into an oriented geometric curve $\mathcal{C}$ if the vertices of the polygonal line L form a chain. For a Jordan curve, a polygonal line L is *inscribed* if all its vertices but the last form a chain and its first and last vertice are equal.
- Let N be a positive integer and $x_0, x_1, ..., x_N$ points of $\mathbb{R}^2$. The polygonal line $\mathrm{PL} = [x_0 x_1 \dots x_N]$ can be considered as the image of the parametrized curve $pl : [0, N] \mapsto \mathbb{R}^2$ such that $pl(t) = x_{\lfloor t \rfloor}(t - \lfloor t \rfloor) + (1 - t + \lfloor t \rfloor) x_{\lfloor t \rfloor + 1}$ where for $r \in \mathbb{R}$, $\lfloor r \rfloor$ in the integer part of the real r. In other words, for any integer i between 0 and N, if $t \in [i, i + 1)$, then $pl(t) = (t - i)x_i + (1 - t + i)x_{i+1}$, and thus $pl([i, i + 1])$ is the segment $[x_i, x_{i+1}]$ of $\mathbb{R}^2$. A polygonal line is simple if it is simple for the previous parametrization and thus a simple polygon is a Jordan curve.
- Given a curve $\mathcal{C}$ and two points a, b on $\mathcal{C}$ ($a \neq b$), we write $\mathcal{C}_a^b$ for the arc ending at a and b if $\mathcal{C}$ is not closed. If $\mathcal{C}$ is closed, $\mathcal{C}_a^b$ and $\mathcal{C}_b^a$ stand for the two arcs of $\mathcal{C}$ ending at a and b.
- The angle between two vectors $\boldsymbol{u}$ and $\boldsymbol{v}$ is denoted by $(\boldsymbol{u}, \boldsymbol{v})$ ($(\boldsymbol{u}, \boldsymbol{v}) \in \mathbb{R}/2\pi\mathbb{Z}$). The geometric angle between two vectors $\boldsymbol{u}$ and $\boldsymbol{v}$, denoted by $(\widehat{\boldsymbol{u}, \boldsymbol{v}})$, or two directed straight lines oriented by $\boldsymbol{u}$ and $\boldsymbol{v}$, is the absolute value of the reference angle taken in $(-\pi, \pi]$ between the two vectors. Thus, $(\widehat{\boldsymbol{u}, \boldsymbol{v}}) \in [0, \pi)$.

Definition 1 (Turn).

- The turn $\kappa(L)$ *of a polygonal line* $L = [x_i]_{i=0}^N$ *is defined by:*

$$\kappa(L) := \sum_{i=1}^{N-1} (\widehat{x_{i-1}x_i, x_i x_{i+1}}).$$

- The turn $\kappa(P)$ *of a polygon* $P = [x_i]_{i=0}^N$ *(where* $x_N = x_0$*) is defined by:*

$$\kappa(P) := \sum_{i=1}^{N} (\widehat{x_{i-1}x_i, x_i x_{i+1}}).$$

- The turn $\kappa(\mathcal{C})$ *of a simple curve* $\mathcal{C}$ *is the upper bound of the turn of its inscribed polygonal lines*
- The turn $\kappa(\mathcal{C})$ *of a Jordan curve* $\mathcal{C}$ *is the upper bound of the turn of its inscribed polygons.*

It should be noticed that the turn does not depend on the orientation of the curve. Indeed, it is well-known that $(\boldsymbol{u}, \boldsymbol{v}) = (-\boldsymbol{u}, -\boldsymbol{v}) = -(-\boldsymbol{v}, -\boldsymbol{u})$. Thus $\kappa\left([x_i]_{i=0}^N\right) = \kappa\left([x_i]_{i=N}^0\right)$.

Furthermore, since the turn of a polygon is equal to the upper bound of the turn of the polygonal lines inscribed in it (cf Corollary p. 119 [1]), the turn of the polygon seen as a closed curve is equal to the turn of the polygon. Hence the notation κ is well defined (Fig. 1).

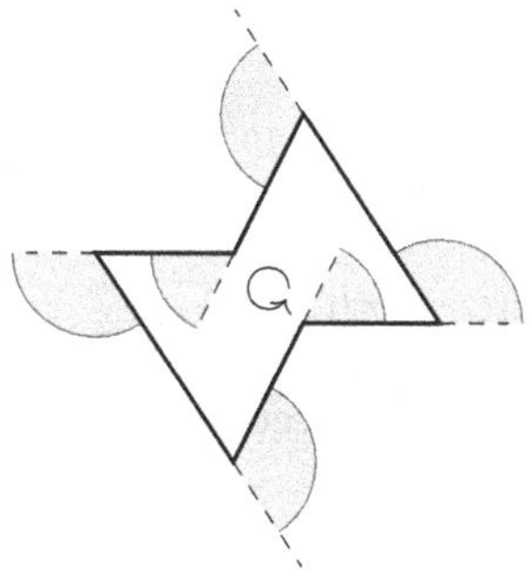

Fig. 1. The turn of the polygon is the sum of the green angles. (Color figure online)

In the same way that we estimate the length of a curve, the following proposition makes it possible to calculate the turn thanks to multiscale samplings. Given a curve $\mathcal{C}$, we denote by $\mathcal{L}(\mathcal{C})$ the length of $\mathcal{C}$.

Proposition 1 (Convergence of the length and turn of a sequence of polygonal lines [1], p. 23, 30, 121, 122). *Let* $\mathcal{C}$ *be a simple curve and* $(L_m)_{m\in\mathbb{N}}$ *a sequence of polygonal lines inscribed in* $\mathcal{C}$ *and with same endpoints* *as $\mathcal{C}$. If $\lim_{m \to +\infty} \lambda_m = 0$, where λ_m is the maximal Euclidean distance between two consecutive vertices of L_m, then*

$$\lim_{m \to +\infty} \mathcal{L}(L_m) = \mathcal{L}(\mathcal{C})$$

and

$$\lim_{m \to +\infty} \kappa(L_m) = \kappa(\mathcal{C}).$$

Moreover, if $\kappa(\mathcal{C})$ is finite, then $\mathcal{L}(\mathcal{C})$ is also finite, that is $\mathcal{C}$ is rectifiable.

In Proposition 1, if we assume that the sequence (L_m) is increasing (L_m is inscribed in L_{m+1}), then the sequences $(\mathcal{L}(L_m))$ and $(\kappa(L_m))$ are both increasing [1, Lemma 5.1.1].

Proposition 2 (Turn for regular curves [1], p. 133). *Let $\gamma\colon [0,\ell] \to \mathbb{R}^2$ be a parametrization by arc length of a simple curve $\mathcal{C}$. Assume that γ is of class C^2 and denote the curvature at the point $\gamma(s)$ by $k(s)$. Then,*

$$\kappa(\gamma) = \int_0^\ell k(s)\mathrm{d}s.$$

For regular curves, therefore, the turn corresponds to the integral of the curvature (with respect to an arc-length parametrization).

The following theorem gives a lower bound of the turn for closed curves.

Theorem 1 (Fenchel's Theorem, [1] Theorem 5.1.5 p. 125). *For any Jordan curve $\mathcal{C}$, $\kappa(\mathcal{C}) \geq 2\pi$. Moreover $\kappa(\mathcal{C}) = 2\pi$ if and only if the interior of $\mathcal{C}$ is convex.*

Proposition 3. *Let $\mathcal{C}$ be a oriented simple curve from a to b. Let $\mathcal{C}_a^{x_2}$ and $\mathcal{C}_{x_1}^b$ be two arcs of $\mathcal{C}$ that overlap with $a < x_1 < x_2 < b$. Then,*

$$\kappa(\mathcal{C}) \leq \kappa(\mathcal{C}_a^{x_2}) + \kappa(\mathcal{C}_{x_1}^b).$$

Proof. Let (L_m) be a sequence of polygonal lines $[l_{m,i}]_{i=0}^{N_m}$ inscribed in $\mathcal{C}$ such that

$$l_{m,0} = a, \quad l_{m,N_m} = b,$$

$$\|l_{m,i+1} - l_{m,i}\|_2 \leq \frac{1}{m},$$

and for each m there exists j such that

$$x_1 < l_{m,j} < x_2.$$

For each m, let us denote by l_{m,i_2} the largest $l_{m,i}$ such that $l_{m,i}$ lies in $\mathcal{C}_a^{x_2}$ and l_{m,i_1} the smallest $l_{m,i}$ that lies in $\mathcal{C}_{x_1}^b$. Then,

$$l_{m,i_1} \leq l_{m,j} \leq l_{m,i_2}.$$

Therefore, from Definition 1, we derive that

$$\kappa([l_m]_{i=0}^{N_m}) \leq \kappa([l_m]_{i=0}^{i_2}) + \kappa([l_m]_{i=i_1}^{N_m}).$$

Then, by Proposition 1,

$$\kappa(\mathcal{C}) \leq \kappa(\mathcal{C}_a^{x_2}) + \kappa(\mathcal{C}_{x_1}^b).$$

□

Without the hypothesis of an overlapping non reduced to a singleton, Proposition 3 is false in presence of angular points as it is illustrated in Fig. 2.

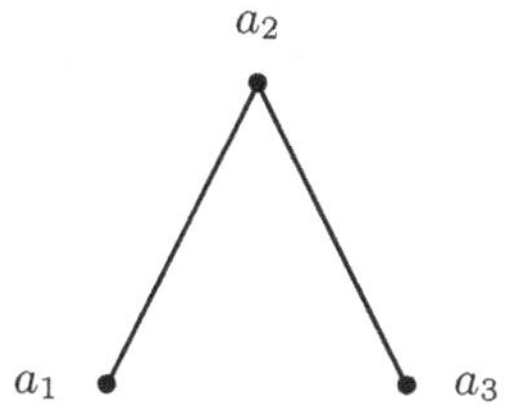

Fig. 2. Counterexample: the turn of the arcs $\mathcal{C}_{a_1}^{a_2}$ and $\mathcal{C}_{a_2}^{a_3}$ are zero but the turn of the arc $\mathcal{C}_{a_1}^{a_3}$ is nonzero. This is why the arcs are required to be overlapping in Proposition 3.

Remark 1. The turn is stable under homothetic maps. Indeed, obviously, the turn is invariant by any conformal map, in particular by the homotheties.

3 Locally Turn-Bounded Curves

Definition 2. *Let $\theta \geq 0$. The θ-turn step $\sigma(\theta)$ of a Jordan curve $\mathcal{C}$ is the infimum of the (Euclidean) distances between two points a and b in $\mathcal{C}$ such that the turns of $\mathcal{C}$ between a and b are both greater than θ:*

$$\sigma(\theta) = \inf\{\|a - b\|_2 \mid \kappa(\mathcal{C}_a^b) > \theta \wedge \kappa(\mathcal{C}_b^a) > \theta\}. \tag{1}$$

Let $\theta \geq 0$, $\delta \geq 0$. A Jordan curve $\mathcal{C}$ is locally turn-bounded *with parameters (θ, δ) if $\delta \leq \sigma(\theta)$.*

The Fig. 3 illustrates the definition of the turn step with different curves.

On a locally turn-bounded curve with parameters (θ, δ) the smallest turn of the curve $\mathcal{C}$ between two points of $\mathcal{C}$ a and b such that $\|b - a\|_2 < \delta$ is less than or equal to θ.

The *turn step function* $\theta \mapsto \sigma(\theta)$ is increasing. Indeed, the value of the θ-turn step is an infimum of a set that decreases (for the inclusion order) in function of θ. If the turn of $\mathcal{C}$ is finite, there exists a value $\theta_{\max}$ above which this set is empty and the $\theta_{\max}$-turn step is then infinite.

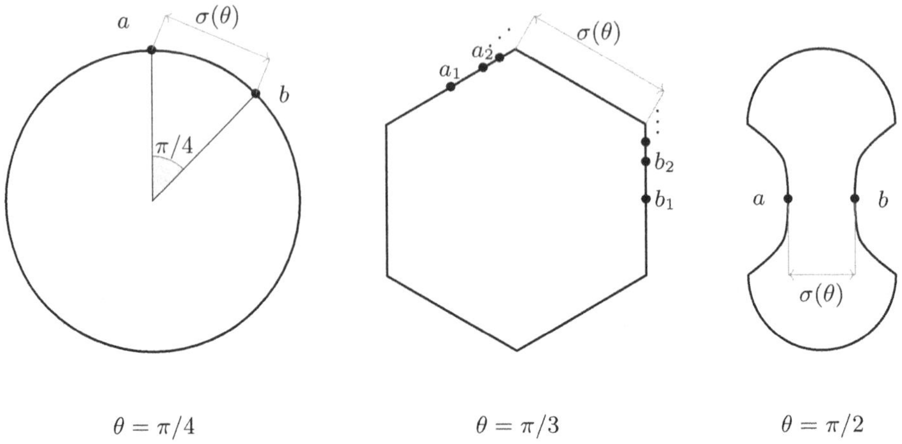

Fig. 3. For the chosen value of θ, the corresponding θ-turn-step $\sigma(\theta)$.

Examples

- The 0-turn step of a Jordan curve is 0.
- The π-turn step of a convex Jordan curve is $+\infty$ (see Theorem 1).
- The θ-turn step of a circle with radius r is $2r\sin(\theta/2)$ if $\theta \leq \pi$ and is infinite for $\theta > \pi$.
- The θ-turn step of a polygon having vertices with interior angles strictly less than $\pi - \theta$ is zero. Indeed, the turn of an arc joining two points arbitrary close, located before and after such a vertex is greater than θ.
- The turn between two points on a polygonal curve is a finite sum of geometric angles, then the turn step function is a step function.

Remark 2. Local turn-boundedness is scale invariant: let $\mathcal{C}$ be a locally turn-bounded Jordan curve with parameters (θ, δ). Then, the curve $k\mathcal{C}$, $k>0$, is locally turn-bounded with parameters $(\theta, k\delta)$. It is a direct consequence of Remark 1.

The next proposition makes it possible to localize a locally turn-bounded curve from a sufficiently tight sampling. Figure 4 illustrates the proposition.

Proposition 4. *Let $\mathcal{C}$ be a simple curve locally turn-bounded with parameters $(\theta \in (0, \pi), \delta)$. Let $a < b$ be two points on $\mathcal{C}$ such that $\|a - b\|_2 < \delta$. Then, one of the arc of $\mathcal{C}$ between a and b is included in the union of the two truncated closed disks where the line segment $[a, b]$ is seen from an angle greater than or equal to $\pi - \theta$.*

Proof. Since the Euclidean distance between a and b is less than the θ-turn step, the turn of one of the arcs of $\mathcal{C}$ between a and b is less than or equal to θ. Denote this arc by $\mathcal{C}_0$. Let c be a point on $\mathcal{C}_0$. By definition, the turn of the polygonal line $[a, c, b]$ is less than or equal to the turn of $\mathcal{C}_0$ Then the geometric angle $(\widehat{a-c, b-c})$ is greater than or equal to $\pi - \theta$. We conclude the proof by invoking the inscribed angle theorem. □

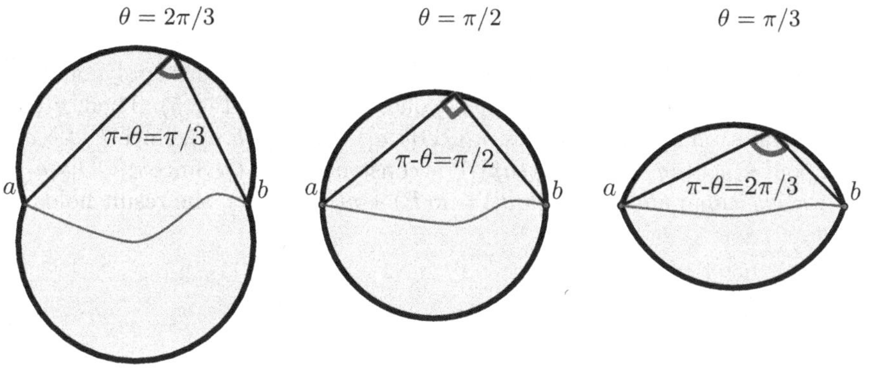

Fig. 4. Illustration of Proposition 4 for three values of the parameter θ: $\pi/3$, $\pi/2$, $2\pi/3$. Given two points $a, b \in \mathcal{C}$ such that $\|a - b\|_2 < \sigma(\theta)$, then one of the arcs of $\mathcal{C}$ between a and b belongs to the grey area.

Locally turn-bounded curves for angles $\theta \leq \pi/2$ are locally connected subsets of the Euclidean plane.

Proposition 5. *Let $\mathcal{C}$ be a Jordan curve locally turn-bounded with parameters $(\theta \in (0, \pi/2], \delta)$ and $a \in \mathcal{C}$. Then the intersection of $\mathcal{C}$ with the open disk $B(a, \epsilon)$ is path-connected whenever $\epsilon \leq \delta$.*

Proof. Let $b \in \mathcal{C} \cap B(a, \epsilon)$. Then, by the very definition of δ, the turn of one of the arc of $\mathcal{C}$ between a and b is less than or equal to θ. So, from Proposition 4 and for $\theta \leq \pi/2$, this arc is included in the disk with diameter $[a, b]$ which is itself included in $B(a, \epsilon)$. Hence, $\mathcal{C} \cap B(a, \epsilon)$ is path-connected. □

For $\pi/2 < \theta < \pi$, in particular for polygons with acute angles, Proposition 5 does not hold (the intersection of the curve with a ball near an acute angle may have two connected components).

Proposition 6. *A locally turn-bounded Jordan curve with parameters $(\theta \in (0, \pi/2], \delta)$ has a finite turn and is thus rectifiable.*

Proof. Let $\mathcal{C}$ be a locally turn-bounded curve with parameters (θ, δ). The open balls $B(a, \delta/2)$, $a \in \mathcal{C}$, cover the compact set $\mathcal{C}$. Then, there exists a finite subset of $\mathcal{C}$, $\{a_0, \dots, a_m\}$ such that $\bigcup_{i=0}^{m} B(a_i, \delta/2)$ covers $\mathcal{C}$. By Proposition 5, for each i, $\mathcal{C} \cap B(a_i, \delta/2)$ is an arc of $\mathcal{C}$.

Since the balls are open and thus overlaps, by Proposition 3 $\kappa(\mathcal{C}) \leq \sum_{i=0}^{m} \kappa(\mathcal{C} \cap B(a_i, \delta/2))$. Besides, by hypothesis, $\kappa(\mathcal{C} \cap B(a, \delta/2)) \leq \theta$. Therefore, $\kappa(\mathcal{C}) \leq (m+1)\theta$. □

The next technical lemma is used in the proofs of Propositions 7 and 8.

Lemma 1. *Let $\mathcal{C}$ be a curve with endpoints a, b such that the line segment $[a, b]$ does not intersect the curve $\mathcal{C}$. Let P be a polygonal line from a to b lying in the interior of the Jordan curve $\mathcal{C} \cup [a, b]$ and such that $P \cup [a, b]$ is convex. Then $\kappa(\mathcal{C}) \geq \kappa(P)$.*

Proof. Let c be any point in (a, b) and $Q = [aq_1 \dots q_m b]$ the polygonal line obtained by projecting $P = [ap_1 \dots p_m b]$ on $\mathcal{C}$ from c (see Fig. 5). Then, $\kappa(\mathcal{C}) \geq \kappa(Q)$ by definition of $\kappa(\mathcal{C})$. Besides, $\kappa(Q \cup [b, a]) \geq \kappa(P \cup [b, a])$, since $P \cup [b, a]$ is convex, and $\kappa([q_m baq_1]) \leq \kappa([p_m bap_1])$ by construction of Q. Since $\kappa(Q \cup [b, a]) = \kappa(Q) + \kappa([q_m baq_1])$ and $\kappa(P \cup [b, a]) = \kappa(P) + \kappa([p_m bap_1])$, the result holds. □

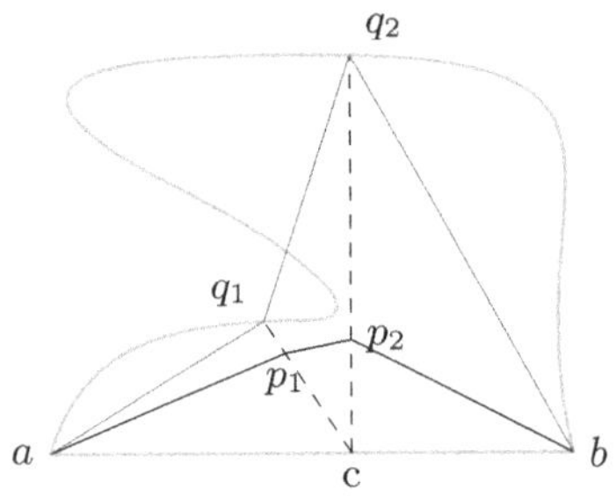

Fig. 5. Thick, blue: the curve $\mathcal{C}$ and the line segment $[a, b]$. Black: the polygonal line $P = [ap_1p_2b]$. Black, dashed: the projection of p_1 and p_2 on $\mathcal{C}$ yields the points q_1 and q_2. Red: the polygonal line $Q = [aq_1q_2b]$. (Color figure online)

The digitization process may digitize several continuous curves in a single digitization. The information on the turn makes it possible to select the continuous curves staying close to the digitization points as illustrated in Fig. 6.

Proposition 7 (Hausdorff distance between the curve and a pixel). *Assuming an n-regular tiling of the plane with edge length h ($n \in \{3, 4, 6\}$), let $\mathcal{C}$ be a locally turn-bounded Jordan curve with parameters ($\theta < 2\pi/n, \delta > h\sqrt{n-2}$). Let T be a tile crossed by $\mathcal{C}$ and a, b be respectively the infimum and the supremum of $\mathcal{C} \cap T$ ($\mathcal{C}$ is ordered by some parametrization). Then, the Hausdorff distance between T and one of the segments of $\mathcal{C}$ bounded by a and b is less than $h(1 - \cos(\theta))/2\sin(\theta)$.*

Proof. Let $\mathcal{C}_1$, $\mathcal{C}_2$ be the arcs of $\mathcal{C}$ bounded by a and b. We assume $\kappa(\mathcal{C}_1) \leq \kappa(\mathcal{C}_2)$. As the diameter of T is $\sqrt{n-2}h$, by the hypothesis $\delta \geq \sqrt{n-2}h$, $\kappa(\mathcal{C}_1) \leq \theta$. If the arc $\mathcal{C}_1$ leaves T, it intersects the border of T on a point c. By the definition of a and b there exists $d \neq c$ such that d is on the border of T. The point d belongs to the same edge as c. Indeed, if it wasn't the case, there would be a polygonal line included in the border of T containing at least one vertex and separating an arc segment of $\mathcal{C}_1$ from the line segment $[c, d]$. Then, by Lemma 1, the turn of $\mathcal{C}_1$ would be greater than or equal to $2\pi/n > \theta$. Hence, c and d belongs to the same edge. Therefore, by Proposition 4, we derive that $\mathcal{C}_1$ lies in the union of T with the interior of n truncated circles whose Hausdorff distance to T is $h(1 - \sin\theta)/2\cos\theta$. □

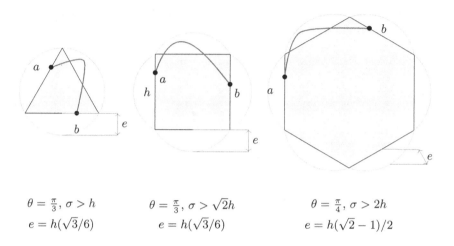

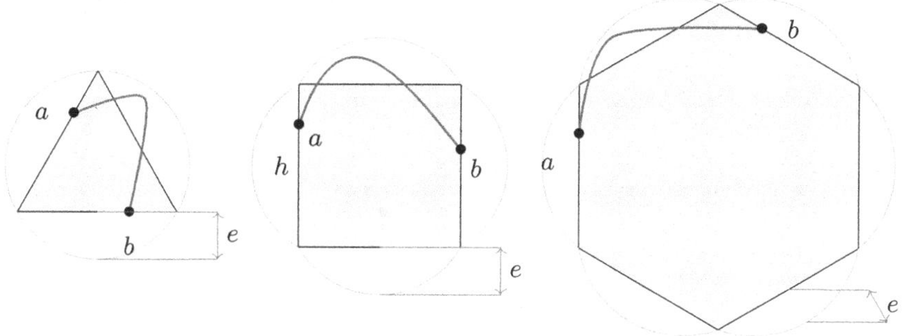

$\theta = \frac{\pi}{3}$, $\sigma > h$ $e = h(\sqrt{3}/6)$

$\theta = \frac{\pi}{3}$, $\sigma > \sqrt{2}h$ $e = h(\sqrt{3}/6)$

$\theta = \frac{\pi}{4}$, $\sigma > 2h$ $e = h(\sqrt{2} - 1)/2$

Fig. 6. Gray: a tile T with edge length h. Blue, thick: a locally turn-bounded curve arc passing through T. Red: the union of n circle arcs from which the edges of T are viewed from an angle $2\pi - \theta$. The Hausdorff distance between the circle arcs and the pixel T is e. (Color figure online)

The next proposition makes a link between the well-composedness of the digitized object and a local turn-boundedness hypothesis. Let us first define the Gauss digitization and the well-composedness. Let $h > 0$ be a sampling grid step, the *Gauss digitization of a shape* K is defined as $K \cap (h\mathbb{Z})^2$. By abuse of language, given a Jordan curve $\mathcal{C}$ which is the border of the compact shape K, we define its *Gauss digitization* —we write $\mathrm{Dig}_h(\mathcal{C})$— as the border of the union of the squares $p \oplus [-h/2, h/2] \times [-h/2, h/2]$ where $\oplus$ denotes the Minkowski sum and $p \in K \cap (h\mathbb{Z})^2$. The Gauss digitization of $\mathcal{C}$ is *well-composed* if it is a Jordan curve.

Proposition 8. *Let $\mathcal{C}$ be a locally turn-bounded Jordan curve with parameters $(\theta \in (0, \pi/2], \delta)$ where δ is less than the diameter of $\mathcal{C}$. Then, the Gauss digitization of $\mathcal{C}$ for a grid step $h < \delta/\sqrt{2}$ is almost surely a disjoint union of Jordan curves (the digitization of the shape is well-composed). More precisely, given a Jordan curve $\mathcal{C}$, the set of translations of the grid for which the digitization is not well-composed (in bijection with a subset of the Euclidean plane) is of Lebesgue measure zero.*

Proof. The proof is made by contradiction. So, let a be a double point on $\mathrm{Dig}_h(\mathcal{C})$ traveled counterclockwise. Since the interior points of $(h\mathbb{Z})^2$ are on the left of the discrete curve, there is only one configuration modulo rotations and symmetries depicted in Fig. 7. Furthermore, we can assume almost surely $\mathcal{C} \cap (h\mathbb{Z})^2 = \emptyset$. Indeed, the set of translations for which a given point $x \in (h\mathbb{Z})^2$ lies on the translation of $\mathcal{C}$ is $\{x - p \mid p \in \mathcal{C}\}$. In other words, it is a translation of $-\mathcal{C}$ (in the translation parameter space). Then, since $\mathcal{C}$ is rectifiable, this set is one-dimensional. Therefore, the set S of translations for which there exists a point of $(h\mathbb{Z})^2$ on the translation of $\mathcal{C}$ is a countable union of one-dimensional sets. Thus, the Lebesgue measure of S is zero.

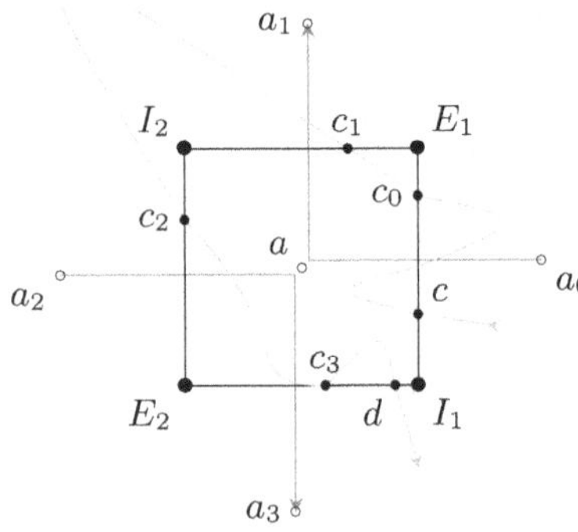

Fig. 7. Blue: the curve $\mathcal{C}$. Red: the line segments $[a_0a]$, $[aa_1]$, $[a_2a]$, $[aa_3]$ are edges of $\mathrm{Dig}_h(\mathcal{C})$. Black: the points E_1, E_2, I_1, I_2 are in $(h\mathbb{Z})^2$, E_1, E_2 are exterior to $\mathcal{C}$ while I_1, I_2 are interior to $\mathcal{C}$. (Color figure online)

With the notation of the figure, we claim that on each edge of the square $T = [I_1, E_1, I_2, E_2, I_1]$ lies a point of the curve $\mathcal{C}$. Indeed, each edge links an interior point and an exterior point. Furthermore, thanks to the assumptions that $\mathcal{C} \cap (h\mathbb{Z})^2 = \emptyset$, these four points actually lie on the open edges of T. Let b, d be respectively the infimum and the supremum of $\mathcal{C} \cap T$ and $[bc_0c_1c_2c_3d]$ be the inscribed polygon formed by these six points ($\mathcal{C}$ is ordered by some parametrization). We denote by $\mathcal{C}_b^d$ the oriented arc of $\mathcal{C}$ from b to d. On the one hand, the arc $\mathcal{C}_b^d$ contains the four points c_0, c_1, c_2 and c_3. Thus, whatever the geometric configuration ($[c_0c_1c_2c_3]$ simple or not), the turn of $\mathcal{C}_b^d$ is greater than $\pi/2$. Nevertheless, since $\delta > h\sqrt{2}$, the points b, c_0, c_1, c_2, c_3, d are included in the ball $B(a, \delta)$. Thus, from Proposition 5, the arc $\mathcal{C}_b^d$ is included in the ball $B(a, \delta)$. On the other hand, since $\kappa(\mathcal{C}_b^d) \geq \pi/2$ and $\delta > h\sqrt{2}$, the turn of $\mathcal{C} \setminus \mathcal{C}_b^d$ is less than θ and a fortiori less than $\pi/2$. Then, from Proposition 4, $\mathcal{C} \setminus \mathcal{C}_b^d$ is included in the disk with diameter $[b, d]$ which is itself included in the disk $B(a, \delta)$. Thereby, the whole curve $\mathcal{C}$ is included in $B(a, \delta)$ which contradicts the assumption on its diameter. □

We began the proof of Proposition 8 without any restriction on the localization of the curve (omitting "almost surely"). But there are many special cases to consider and it can not be given as part of a short article.

4 Conclusion

In this article, the notion of local turn boundedness which adapted to both regular curves and polygons having large enough interior angles, has been developed to have control on curves without smoothness assumption. Under this assumption on a curve, it has been shown that the domain where an arc of this curve is lying is bounded. Then the well-composedness of the curve Gauss-digitization has been deduced almost surely. A bound on the Hausdorff distance between one pixel and an arc crossing it has also been exhibited.

The local turn boundedness also needs to be related to others notions used in discrete geometry for similar purposes. To this end, we will prove in an upcoming article that a par(r)-regular curve of class $\mathcal{C}^1$ is locally turn-bounded with parameters $(\theta, 2r\sin(\theta/2))$ where θ is any angle in $[0, \pi)$. Besides, we expect that these results will make it possible to uniquely associate the vertices of a Gauss digitization to points on the continuous curve without smoothness assumption.

References

1. Alexandrov, A., Reshetnyak, Y.G.: General Theory of Irregular Curves. Kluwer Academic Pulishers, Dordrecht (1989)
2. Gross, A.D., Latecki, L.J.: Digitizations preserving topological and differential geometric properties. Comput. Vis. Image Underst. **62**, 370–381 (1995)
3. Lachaud, J.O.: Espaces non-euclidiens et analyse d'image: modèles déformables riemanniens et discrets, topologie et géométrie discrète (2006)
4. Lachaud, J.O., Thibert, B.: Properties of gauss digitized shapes and digital surfaces integration. J. Math. Imaging Vis. **54**(2), 162–180 (2014)
5. Latecki, L.J., Conrad, C., Gross, A.: Preserving topology by a digitization process. J. Math. Imaging Vis. **8**, 131–159 (1998)
6. Milnor, J.W.: On the total curvature of knots. Ann. Math. Second Ser. **52**, 248–257 (1950)
7. Pavlidis, T.: Algorithms for Graphics and Image Processing. Springer, Heidelberg (1982). https://doi.org/10.1007/978-3-642-93208-3

Distance Transform Based on Weight Sequences

Benedek Nagy[1](✉), Robin Strand[2], and Nicolas Normand[3]

[1] Eastern Mediterranean University,
Famagusta, North Cyprus, via Mersin-10, Turkey
nbenedek.inf@gmail.com

[2] Centre for Image Analysis, Division of Visual Information and Interaction,
Uppsala University, Uppsala, Sweden
Robin.Strand@it.uu.se

[3] Université de Nantes, LS2N UMR CNRS 6004, Nantes, France
Nicolas.Normand@polytech.univ-nantes.fr

Abstract. There is a continuous effort to develop the theory and methods for computing digital distance functions, and to lower the rotational dependency of distance functions. Working on the digital space, e.g., on the square grid, *digital* distance functions are defined by minimal cost-paths, which can be processed (back-tracked etc.) without any errors or approximations. Recently, digital distance functions defined by weight sequences, which is a concept allowing multiple types of weighted steps combined with neighborhood sequences, were developed. With appropriate weight sequences, the distance between points on the perimeter of a square and the center of the square (i.e., for squares of a given size the weight sequence can be easily computed) are exactly the Euclidean distance for these distances based on weight sequences. However, distances based on weight sequences may not fulfill the triangular inequality. In this paper, continuing the research, we provide a sufficient condition for weight sequences to provide metric distance. Further, we present an algorithm to compute the distance transform based on these distances. Optimization results are also shown for the approximation of the Euclidean distance inside the given square.

Keywords: Digital distances · Weight sequences · Distance transforms · Neighborhood sequences · Chamfer distances · Combined distances · Approximation of the Euclidean distance

1 Introduction

In the digital geometry setting, it is very natural to define distance functions by minimal cost paths. Recently a new digital (i.e., path based) distance function has been investigated which provides the perfect Euclidean distance (from the Origin) for all the points having the same fixed number of steps in the corresponding shortest path from the Origin [11,12]. This new distance function,

M. Couprie et al. (Eds.): DGCI 2019, LNCS 11414, pp. 62–74, 2019.
https://doi.org/10.1007/978-3-030-14085-4_6

which is called distance by weight sequence, is obtained as a kind of mixture of the two well known digital distance approaches, namely of the distances based on neighborhood sequences [5,8,10,19] and of the chamfer distances, also known as weighted distances [1].

Discrete and digital distances may have some non-usual properties, e.g., they may not be metric. A function that is positive definite, symmetric and for which the triangular inequality holds is called a metric. The concept of metricity is an important property of distance functions. In our previous paper [12] we presented a very restricted sufficient condition for a weight sequence distance to be a metric. In Sect. 3 in this paper, we present a less restrictive condition for metricity.

Digital distances are well applicable in digital image processing. Given a distance function, a *distance transform* is a transform where each element in a set is assigned the distance to the closest (as given by the distance function) element in a complementary set. The result of a distance transform is called a distance map. This tool is often used in image processing and computer graphics [6]. In Sect. 4, we present a distance transform algorithm, which utilizes an auxiliary data structure to keep track of the number of steps of the minimal cost path at each point.

One of the 'goodness' measures of digital distances used in practice is related to their rotational dependency and how well they approximate the Euclidean distance. The most basic path-based distance functions, such as the city block and chessboard distances, have very high rotational dependency. In our previous paper [12] (which builds on [17]), we presented a formula for weight sequences that gives the exact Euclidean distance values on the border of a square, for arbitrary size of the square. This property is independent of the order of the weights in the weight sequence, and permutation optimization of the weights by a genetic algorithm is presented in Sect. 5.

2 Definitions, Preliminaries

In this section, first, we recall some basic concepts and fix our notations.

We denote the set of integers by $\mathbb{Z}$ and the set of non-negative integers by $\mathbb{N}$. Consequently the points of the digital two dimensional (square) grid are represented by $\mathbb{Z}^2$. The set of positive real numbers is denoted by $\mathbb{R}^+$.

In this paper, we consider grid points with integer coordinates. Of course, in image processing, each grid point is associated with a picture element, pixel. In a city block (resp. chessboard) distance, distinct points with unit difference in at most one (resp. two) of the coordinates have unit distance. Here, we use the notion of 1- and 2-neighbors in the following sense: Two grid points $P_1 = (x_1, y_1), P_2 = (x_2, y_2) \in \mathbb{Z}^2$ are ρ-neighbors, $\rho \in \{1, 2\}$, if

$$\begin{aligned} |x_1 - x_2| + |y_1 - y_2| &\leq \rho \text{ and} \\ \max\{|x_1 - x_2|, |y_1 - y_2|\} &= 1. \end{aligned} \tag{1}$$

The points are *strict* ρ-neighbors if the equality in (1) is attained. Two points are *adjacent* if they are ρ-neighbors for some $\rho \in \{1, 2\}$.

We also recall the notion of path. A *path* in a grid is a sequence of adjacent grid points. A path $\Pi = (P_0, P_1, \ldots, P_n)$ is a path of n steps where for all $i \in \{1, 2, \ldots, n\}$, P_{i-1} and P_i are adjacent. The path Π connects P_0 and P_n. We may also say that the path starts from P_0 and arrives to P_n, i.e., they are its start point and endpoint, respectively.

In this paper, we use the general concept of weight (or chamfer) sequences (from [11,12]), instead of neighborhood sequences [5,8–10] or chamfer distances [1]. This general description is as follows:

Definition 1 (weight-sequence distance). *Let $m \in \mathbb{N}$, $m \geq 0$ be the number of the used weights, and let $S = \{1, \infty\} \cup \{w_3, \ldots, w_m\}$ be the weight set including* 1*, the sign ∞ and the used weights ($w_i \in \mathbb{R}$, $w_i > 1$ for all $i \in \{3, \ldots, m\}$).*

A weight sequence is $W = (c(i))_{i=1}^{\infty}$, where $c(i) \in S$, for all $i \in \mathbb{N}$.

Let $P, Q \in \mathbb{Z}^2$, then the weight of the path $\Pi = (P = P_0, P_1, \ldots, P_m = Q)$ is the sum of the weights of its steps, where the weight of the j-th step is specified as
$$\begin{cases} c(j), & \text{if the } j\text{-th step is a step to a strict 2-neighbor;} \\ 1, & \text{otherwise.} \end{cases}$$

The W-distance $d(P, Q; W)$ of P and Q, is then, defined as the weight of the minimal weighted path connecting P and Q.

In fact, all city block paths are valid for $d(P, Q; W)$ and have the same weight as in the city block distance. This implies that $d(P, Q; W)$ is upper-bounded by the city block distance. When the ∞ sign used in W for some step i ($c(i) = \infty$), it denotes an arbitrary large weight that prevents paths with a strict 2-neighbor at step i to be minimal. When computing a W-distance one can vary the features of neighborhood sequences and chamfer distances: the cost of a move to a 1-neighbor is 1 in every step, and the cost of a move to a strict 2-neighbor (diagonal step) is given by the actual element $c(j)$ of the weight sequence.

In [12] it was shown that for any weight sequence W, if Π is a minimal W-path between $P, Q \in \mathbb{Z}^2$, then it does not contain any steps to strict 2-neighbors by any weight $w_i > 2$.

The way the set S is defined, m different neighborhood relations (possible steps by various weights) are allowed in our paths:

- a traditional 1-step is a step between 1-neighbors with unit weights, the sign ∞ denotes these steps in W (practically, strict 2-steps are not allowed);
- a traditional 2-step is a step between 2-neighbors with unit weights, they are denoted by value 1 in W;

and if $m > 2$, then, by the used weights $w_3, \ldots, w_m$ the further steps are as follows:

- weighted 2-steps: the steps between 1-neighbors with unit weights, and between strict 2-neighbors with weight w_k (where $3 \leq k \leq m$) with $c(i) = w_k$ for some i in W.

Note that the weight sequence W can contain m values, "weights", according to a predefined set S, i.e. $W = (c(i))_{i=1}^{\infty}$ where $c(i) \in S$.

When written in a formal way, the sum of the weights along the path is:

$$\sum_{i=1}^{n} \delta_i, \text{ where } \delta_i = \begin{cases} c(i), & \text{if } P_{i-1} \text{ and } P_i \text{ are strict 2-neighbors;} \\ 1, & \text{otherwise.} \end{cases}$$

When the weight sequence W is fixed, the term W-path is used for paths having finite cost as defined by the weight sequence W.

We recall that greedy algorithms cannot be used to provide shortest paths. If a smaller weight appears after a larger weight in W, it may be needed in the shortest path instead of the previous larger one, depending on both the weight sequence and on the difference of the coordinate values of the points.

Now we recall the formula for computing the distance between any two grid points. The formula is used for finding optimal parameters in Sect. 5. Before the theorem we define a technical notation which will also be helpful later on.

Definition 2. *Let a weight sequence $W = (c(i))_{i=1}^{\infty}$ (based on a weight set $S = \{1, \infty, w_3, \ldots, w_m\}$) be given. Let $m, n \in \mathbb{N}$ such that $n \geq m$. Then $I(n, m)$ contains the indices of the smallest m weight values among the first n elements of W, i.e., among $(c(1), \ldots, c(n))$.*

Theorem 1. *Let the weight sequence $W = (c(i))_{i=1}^{\infty}$ (based on a weight set $S = \{1, \infty, w_3, \ldots, w_m\}$) and the point $P(x, y) \in \mathbb{Z}^2$, where $x \geq y \geq 0$, be given. Then the W-distance of P from the Origin $\mathbf{0}$ is given by*

$$d(\mathbf{0}, (x, y); W) = \min_{f \in \{0..y\}} \left\{ x + y - 2f + \sum_{i \in I(x+y-f, f)} c(i) \right\}. \tag{2}$$

Since the roles of the x- and y-coordinates are similar, and our distance function is translation invariant, one can compute the W-distance of any pair of points of $\mathbb{Z}^2$ by the previous formula.

Remark 1. As a special case, the weight sequence $W = (1)_{i=1}^{\infty}$ defines the chess-board distance, since steps to 2-neighbors are always allowed with unit cost. The weight sequence $W = (2)_{i=1}^{\infty}$ defines the city block distance, since, in this case, in any path one needs to pay the weight of two city block steps for a diagonal steps if the path contains any diagonal steps. Equivalently, minimal cost paths can be obtained using only steps to 1-neighbors.

3 Metricity Properties

One important property of digital distances is that of metricity. A metric is a distance function which satisfies the metric properties: positive definiteness, symmetry, and the triangular inequality. Formally:

A distance function d is a metric if it satisfies the following three properties:

- $d(P,Q) \geq 0$ for any point pair P,Q, moreover $d(P,Q) = 0$ if and only if $P = Q$. (positive definiteness)
- $d(P,Q) = d(Q,P)$ for any point pair P,Q. (symmetry)
- $d(P,Q) + d(Q,R) \geq d(P,R)$ for any three points P,Q,R. (triangular inequality)

It is easy to find a weight sequence such that the distance generated by the sequence does not satisfy the metric conditions. The first two properties, namely, the positive definiteness and the symmetry are always satisfied for distances defined by weight sequences on the square grid (it is clear by the definitions and the formula 2 shown in the previous section). However, the triangular inequality is problematic in some cases.

In [12] it was proven that if the weight sequence W contains the weights in a non-increasing order, then the generated distance function is a metric. However, with this restriction the number of the possible weight sequences is very limited. Now, we show a more general sufficient condition for a weight sequence to define a metric. First, we define the shifted sequences of a weight sequence (in a kind of analogous way as similar concept was defined and used for the neighborhood sequences in, e.g., [8]).

Definition 3. *The k-shifted sequence of a weight sequence $W = (c(i))_{i=1}^{\infty}$ is the sequence that obtained from W by starting from its k-th element, i.e., $W^k = (c(i))_{i=k}^{\infty}$.*

It is easy to see that $W = W^1$, while for $k > 1$, W^k is obtained from W by erasing its first $k-1$ elements.

Theorem 2. *Let the weight sequence $W = (c(i))_{i=1}^{\infty}$ be given. If W has the property that for every $n,m,k \in \mathbb{N}$ $(m \leq n)$*

$$\sum_{i \in I(n,m)} c(i) \geq \sum_{i \in I^k(n,m)} c(i) \tag{3}$$

(where, according to Definition 2, $I(n,m)$ and $I^k(n,m)$ are the sets containing the indices of the m smallest weights/elements among the first n elements of W and W^k, respectively, thus $|I(n,m)| = |I^k(n,m)| = m$ is their cardinality), then W defines a metric.

Proof. It is clear that all W-distances are symmetric and positive definite, thus we work only with the triangular inequality. Let us assume, by contrary, that there are three points $P,Q,R \in \mathbb{Z}^2$ such that $d(P,Q;W) + d(Q,R;W) < d(P,R;W)$. Let the path $\Pi_{P,Q}$ be the path from P to Q with the least weight, i.e., the path which is defining the distance $d(P,Q;W)$. Further, let $\Pi_{Q,R}$ be the path from Q to R that defines the distance $d(Q,R;W)$. Then let us consider the path $\Pi_{P,R}$ from P to R which is exactly the concatenation of the path $\Pi_{P,Q}$ and the steps of the path $\Pi_{Q,R}$. We notice that since the weights may not be the same in the prefix of W and in its subsequence used from Q to R, it is generally, not true that the obtained path $\Pi_{P,R}$ has the sum of the weight $d(P,Q;W) + d(Q,R;W)$. However, if the condition of the theorem is fulfilled, let us choose the numbers n, m and k as follows. Let n be the number of the steps of $\Pi_{Q,R}$. Set m as the number of its diagonal steps, finally put k as 1 more than the number of the steps of the path $\Pi_{P,Q}$. Considering the steps of $\Pi_{Q,R}$, in the left hand side of the inequality $d(P,Q;W)+d(Q,R;W) < d(P,R;W)$ the W-distances are used, while, considering the right hand side, we have a W-path, such that its first part (between P and Q) has exactly the weight $d(P,Q;W)$, and its second part (from Q to R) has the weight that is based on the weight sequence W^k. Then, we state that, by the condition $\sum_{i \in I(n,m)} c(i) \geq \sum_{i \in I^k(n,m)} c(i)$, there is a path $\Pi'_{Q,R}$ such that its W^k weight is not more than $d(Q,R;W)$. However, in this case, concatenating $\Pi_{P,Q}$ and $\Pi'_{Q,R}$ we have a W-path from P to Q such that its weight is not more than $d(P,Q;W) + d(Q,R;W)$. That is contradicting to our indirect assumption, thus, the triangular inequality must hold, if the condition of the theorem is fulfilled by the sequence W.

What is remained to show, that if $\sum_{i \in I(n,m)} c(i) \geq \sum_{i \in I^k(n,m)} c(i)$, then there is a path $\Pi'_{Q,R}$ such that its W^k weight is not more than $d(Q,R;W)$. Let us assume, by contrary, that $d(Q,R;W^k) > d(Q,R;W)$, i.e., there exist no such path $\Pi'_{Q,R}$. Then applying the formula for the distance, actually, $d(Q,R;W) = n - m + \sum_{i \in I(n,m)} c(i)$. By the condition of the theorem there is also a W^k-path with n steps between Q and R having m diagonal steps with cost $n - m + \sum_{i \in I^k(n,m)} c(i)$. However, since $\sum_{i \in I(n,m)} c(i) \geq \sum_{i \in I^k(n,m)} c(i)$, the cost of this path is not more than $d(Q,R;W)$. Therefore, $d(P,Q;W^k)$ cannot be larger than $d(P,Q;W)$, path $\Pi'_{Q,R}$ must exist with the desired property. □

The following examples illustrate metricity of distance functions based on some specific weight sequences.

Example 1. Let the weight sequence contain additional weights $\{1.5, 1.8\}$ and consider the periodic weight sequence $W = (1.8, 1.5, 1.8, 1.5...)$. One can easily see that W fulfills the condition of Theorem 2, thus it defines a metric.

Example 2. Let the weight set be $\{1.5, 1.55, 1.6, 1.75\}$. Further, let $W = (1.75, 1.5, 1.6, 1.5, 1.55, 1.5, 1.5, 1.5, ...)$ with $c(i) = 1.5$ for all $i > 6$. Again, it is easy to check that the triangular inequality holds and that the given W-distance is a metric.

Notice that in both of our examples the weights in the sequence are not ordered in non increasing way, thus the condition presented in [12] cannot be applied to prove the metricity of the corresponding distances.

Finally, to show that the triangular inequality may fail in some cases, let us consider the next example:

Example 3. Let $W = (1.2, 1.5, 1, ...)$. Then the W-distances of the points $(0,0)$, $(1,1)$ and $(2,2)$ are computed:

$$\begin{aligned} &d((0,0),(1,1);W) = 1.2 \\ &d((1,1),(2,2);W) = 1.2 \\ &d((0,0),(2,2);W) = 1.2 + 1.5 = 2.7. \end{aligned}$$

Therefore, the triangular inequality fails for this W-distance.

4 Distance Transform (DT)

The DT is a mapping from the image domain, a subset of the grid, to the range of the distance function. In a DT, each object grid point is assigned the distance from its closest background grid point. A modified version of a wave-front propagation algorithm can be used.

Now the formal definition of an image is given.

Definition 4. *The image domain is a finite subset of $\mathbb{Z}^2$ denoted by $\mathcal{I}$. We call the function $F : \mathcal{I} \longrightarrow \mathcal{R}_d$ an image, where $\mathcal{R}_d$ is the range of the distance function d.*

An *object* is a subset $X \subset \mathcal{I}$ and the *background* is $\overline{X} = \mathcal{I} \setminus X$. We assume that $X, \overline{X} \neq \emptyset$. We denote the distance map for path-based distances with DM_d, where the subscript d indicates what distance function is used.

Definition 5. *The* distance map DM_d *generated by the distance function $d(\cdot,\cdot;W)$ of an object $X \subset \mathcal{I}$ is the mapping*

$$\begin{aligned} DM_d : \mathcal{I} \rightarrow \mathcal{R}_d \text{ defined by} \\ P \mapsto d\left(\overline{X}, P; W\right), \text{ where} \\ d\left(\overline{X}, P; W\right) = \min_{Q \in \overline{X}} \left\{ d\left(Q, P; W\right) \right\}. \end{aligned}$$

In the case of W-distances with two weights, a minor modification of the Dijkstra algorithm (and with the same time complexity) can be used, see [18] and Theorem 4.1 in [15]. However, for multiple weights, this is not necessarily true. For W-distances, the used weights are also important and they are determined by the *number of steps* of the minimal cost-path (not by the cost). Therefore, in the general case of multiple weights presented here, we need to store this value also when propagating distance information. We define the auxiliary transform DM_s that holds the number of steps of the minimal cost path at each point, see Algorithm 1.

Note that we need to store not only the best distance values at the points, but the best values that are computed by various number of steps. Therefore, for each point P a set $S(P)$ of pairs of values of the form (DM_d, DM_s) are stored, i.e., the weight DM_d of the shortest path among the ones containing exactly DM_s number of steps, with pairwise different DM_s. After the run of the algorithm for each point the minimal DM_d gives the result.

Algorithm 1 shows how the distance map can be computed based on an extended optimal search (Dijkstra algorithm) and using the data structure described above. At the initialization the object points are the only points that are reached and it is done by 0 steps and with 0 cost. Then the border points of the object are listed in an increasing order by the minimal cost path already known for them. Actually every point in the priority queue has 0 cost, but the queue will be updated by involving other points to where paths are already found. The **while** loop chooses (one of the) point(s) with minimal cost from the queue since it is sure that we have the minimal cost path to this point already. Then in the loop the data of all adjacent points of the chosen point are updated by computing the cost of the new paths through the chosen point (having last step from the chosen point to the actual adjacent point). Therefore, the algorithm holds the optimal distance attained at each point (as the usual algorithm), but this is done *for each* path length from the shortest ones till the ones obtained by only steps to 1-neighbors. So, if there are paths of different lengths ending up at the same point, distance information for each of the different path lengths are stored.

The algorithm may look more complex than similar algorithms for the other distances and this is due to the fact that greedy shortest path algorithm does not work in the general case. However, by using the weights in non-decreasing order, the greedy shortest paths algorithms work, and thus, for every point Q of the grid one need to remember only for the value obtained with the path length when the point Q is pushed into the queue.

5 Approximation of the Euclidean Distance

In this section, we apply a genetic algorithm to permute a weight sequence that gives the exact Euclidean distance values on the border of a square. The comparison to the Euclidean distance is carried out with absolute and relative mean and maximum values of the absolute difference between the Euclidean and weight sequence distance.

The following Lemma, which is from our previous paper [12], will be used for defining the fitness function in the genetic algorithm optimization.

Algorithm 1. Computing DM for W-distances given by a weight sequence W.

```
Input: W and an object X ⊂ Z^2.
Output: The distance map DM_d.
Initialization: Let S(P) ← {(0,0)} for grid points P ∈ X̄. Let
DM_d(P) = min{DM_d | (DM_d, DM_s) ∈ S(P)}. Initialize L, a priority queue of
points P sorted by increasing DM_d(P), with the set of grid points P ∈ X̄
adjacent to X.
while L is not empty do
    Pop P from L;
    foreach Q ∈ I: Q, P are strict 2-neighbors do
        foreach pair (DM_d, DM_s) ∈ S(P) do
            if c(DM_s + 1) ≤ 2 then
                if there is an element (DM'_d, DM_s + 1) ∈ S(Q) then
                    if DM'_d > DM_d + c(DM_s + 1) then
                        Replace (DM'_d, DM_s + 1) by
                        (DM_d + c(DM_s + 1), DM_s + 1) in S(Q)
                else
                    Add (DM_d + c(DM_s + 1), DM_s + 1) to S(Q)
        if DM_d(Q) > min{DM'_d | (DM'_d, DM'_s) ∈ S(Q)} then
            Let DM_d(Q) = min{DM'_d | (DM'_d, DM'_s) ∈ S(Q)}
            Insert Q in the priority queue L ordered by increasing DM_d
    foreach Q ∈ I: Q, P are 1-neighbors do
        foreach pair (DM_d, DM_s) ∈ S(P) do
            if there is an element (DM'_d, DM_s + 1) ∈ S(Q) then
                if DM'_d > DM_d + 1 then
                    Replace (DM'_d, DM_s + 1) by (DM_d + 1, DM_s + 1) in S(Q)
            else
                Add (DM_d + 1, DM_s + 1) to S(Q)
        if DM_d(Q) > min{DM'_d | (DM'_d, DM'_s) ∈ S(Q)} then
            Let DM_d(Q) = min{DM'_d | (DM'_d, DM'_s) ∈ S(Q)}
            Insert Q in the priority queue L ordered by increasing DM_d
```

Lemma 1. *If the weight sequence W is non-decreasing and all elements in W are smaller than or equal to 2, then the distance value in Eq. (2) is given by*

$$d(\mathbf{0},(x,y);W) = x - y + \sum_{i=1}^{y} c(i)$$

The following Theorem, from [12], states that the W-distance with r different weights (as given in the Theorem) can give the exact Euclidean distance values on the border points of a square of side length $2r-1$.

Theorem 3. *Given an integer $x > 0$, the Euclidean distance values from $(0,0)$ to each point of the set $\{(x,y) \in Z^2, 0 \le y \le x\}$ is given without errors by the* *weight sequence* $W = (c(i))_{i=1}^{\infty}$ *with* $c(i) = \left(1 + \sqrt{x^2 + i^2} - \sqrt{x^2 + (i-1)^2}\right)$ *for* $1 \leq i \leq x$.

Note that the distance values on the border of the square of side $2r - 1$ are independent of the order of the weights r weights in the sequence given in Theorem 3. In [12], we presented a greedy algorithm for ordering the weights in the weight sequence. Here, we extend the optimization by a global optimization, where the optimal permutation of the r weights are sought in order to minimize

- the absolute mean error,

$$\frac{2}{r \cdot (r+1)} \sum_{r \geq x \geq y \geq 0} |d_E(x,y) - d(x,y;W)| \tag{4}$$

- the relative mean error

$$\frac{2}{r \cdot (r+1)} \sum_{r \geq x \geq y \geq 0} \frac{|d_E(x,y) - d(x,y;W)|}{d_E(x,y)} \tag{5}$$

- the absolute maximum error,

$$\max_{r \geq x \geq y \geq 0} |d_E(x,y) - d(x,y;W)| \tag{6}$$

- the relative maximum error

$$\max_{r \geq x \geq y \geq 0} \frac{|d_E(x,y) - d(x,y;W)|}{d_E(x,y)} \tag{7}$$

In the experiments, a weight sequence of length 40 was calculated by Theorem 3. The weight sequence is $c(i) = \left(1 + \sqrt{40^2 + i^2} - \sqrt{40^2 + (i-1)^2}\right)$, $i = 1..40$. The weights were permuted by a genetic algorithm [7] with the fitness functions (4), (5), (6), and (7). The fitness functions are efficiently computed by Lemma 1. The *crossover* is a flip of a section (with random start and end point) of the weight sequence, the *mutation* swaps to random elements of the weight sequence. The optimization is run with a population size of 60 and the maximum number of generations was set to 1000.

The optimal permutations are found below and illustrations of the difference between the so-obtained distance functions and the Euclidean distance are found in Fig. 1.

The obtained optimal permutations are (the indices of the weights $c(i)$ are shown for brevity):

- the absolute mean error (4)
  36, 1, 32, 16, 22, 11, 23, 35, 7, 25, 10, 26, 12, 40, 4, 30, 18, 27, 5, 37, 14, 19, 38, 3, 28, 13, 24, 9, 29, 15, 39, 2, 31, 17, 33, 8, 20, 34, 6, 21
- the relative mean error (5)
  19, 7, 37, 28, 2, 32, 22, 16, 10, 31, 14, 27, 4, 39, 21, 12, 26, 6, 40, 15, 30, 3, 35, 17, 20, 8, 34, 13, 24, 29, 1, 36, 18, 11, 38, 9, 25, 23, 5, 33

- the absolute maximum error (6)
  14, 37, 18, 6, 32, 26, 1, 21, 38, 16, 25, 3, 30, 27, 11, 15, 29, 8, 40, 10, 35, 12, 28, 17, 5, 24, 36, 2, 20, 23, 33, 4, 19, 39, 7, 34, 13, 31, 9, 22
- the relative maximum error (7)
  19, 13, 36, 4, 40, 26, 1, 16, 21, 14, 27, 8, 25, 7, 34, 35, 15, 23, 29, 2, 31, 3, 18, 38, 20, 10, 12, 33, 24, 9, 37, 11, 28, 39, 5, 22, 30, 6, 32, 17

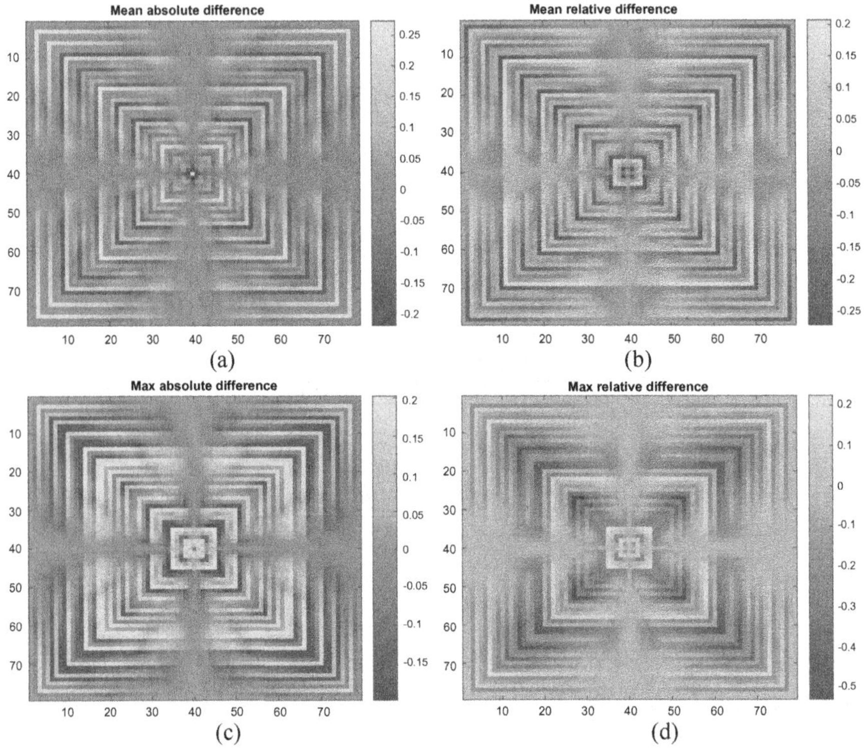

Fig. 1. Illustration of the point-wise difference between the Euclidean distance and the weight sequence distance with a permutation of the weights obtained by a genetic algorithm. The fitness functions used are Absolute mean difference (a), Relative mean difference (b), Absolute max difference (c), and Relative max difference (d). All distances are computed from a single central source point.

6 Conclusions and Discussion

This paper extends the work presented in [12], and further develops the framework for distances based on weight sequences, that is digital path-based distance functions defined as minimal cost paths on the square grid.

By the results in this paper, a digital and completely error-free distance function with very low rotational dependency, together with an algorithm that can be used to compute distance transforms, Algorithm 1, are presented. Efficient algorithms for computing distance transforms are necessary for real-world applications, where computation time often is a bottle-neck. Due to the very low rotational dependency the proposed digital distance function can be used instead of, for example, the Euclidean distance transform. The Euclidean distance transform can be computed efficiently by state-of-the-art methods, but Euclidean distance transform computation has some drawbacks: (i) the vector-propagation method inevitably produces errors for some pixel configurations [4,13], (ii) fast-marching methods are based on coarse approximations of the Eikonal equation, and thus give only approximate distance values [14], (iii) separable algorithms are not suited for constrained distance transform computation [2,3]. See also [15,16].

This paper presents a less restrictive condition for metricity, compared to what was presented in [12]. This less restrictive condition will be important when further developing the theory for these distance functions.

In addition, this paper presents an optimization procedure based on a genetic algorithm. Note that the genetic algorithm can be run with several different parameters, and the ones used in the manuscript are standard parameters with slight fine-tuning based on experimental evaluation. The optimization is not deterministic and there are no guarantees that the obtained local optimum is a global one, but the process was reasonably stable when executed several times. The low rotational dependency obtained by the optimization procedure is key in many real-world problems.

In our future work, we will analyze Algorithm 1 in depth, including both proof of correctness and space and time complexity. We also leave it as an open problem whether the condition of Theorem 2 is also a necessary condition to define metric distances.

The proposed distance function has great potential in many image processing algorithms where distance transforms with low rotational dependency is required.

References

1. Borgefors, G.: Distance transformations in digital images. Comput. Vis. Graph. Image Process. **34**, 344–371 (1986). https://doi.org/10.1016/S0734-189X(86)80047-0
2. Breu, H., Kirkpatrick, D., Werman, M.: Linear time Euclidean distance transform algorithms. IEEE Trans. Pattern Anal. Mach. Intell. **17**(5), 529–533 (1995). https://doi.org/10.1109/34.391389
3. Coeurjolly, D., Vacavant, A.: Separable distance transformation and its applications. In: Brimkov, V., Barneva, R. (eds.) Digital Geometry Algorithms. Theoretical Foundations and Applications to Computational Imaging. LNCVB, vol. 2, pp. 189–214. Springer, Dordrecht (2012). https://doi.org/10.1007/978-94-007-4174-4_7
4. Danielsson, P.E.: Euclidean distance mapping. Comput. Graph. Image Process. **14**(3), 227–248 (2008). https://doi.org/10.1016/0146-664X(80)90054-4

5. Das, P.P., Chakrabarti, P.P.: Distance functions in digital geometry. Inf. Sci. **42**, 113–136 (1987). https://doi.org/10.1016/0020-0255(87)90019-3
6. Fabbri, R., Costa, L.D.F., Torelli, J.C., Bruno, O.M.: 2D Euclidean distance transform algorithms: a comparative survey. ACM Comput. Surv. **40**(1), 1–44 (2008). https://doi.org/10.1145/1322432.1322434
7. Goldberg, D.E.: Genetic Algorithms in Search, Optimization & Machine Learning. Addison-Wesley, Boston (1989)
8. Nagy, B.: Distance functions based on neighbourhood sequences. Publ. Math. Debrecen **63**(3), 483–493 (2003)
9. Nagy, B.: Metric and non-metric distances on $\mathbb{Z}^n$ by generalized neighbourhood sequences. In: IEEE Proceedings of 4th International Symposium on Image and Signal Processing and Analysis (ISPA 2005), Zagreb, Croatia, pp. 215–220 (2005). https://doi.org/10.1109/ISPA.2005.195412
10. Nagy, B.: Distance with generalized neighbourhood sequences in nD and ∞D. Discrete Appl. Math. **156**(12), 2344–2351 (2008). https://doi.org/10.1016/j.dam.2007.10.017
11. Nagy, B., Strand, R., Normand, N.: A weight sequence distance function. In: Hendriks, C.L.L., Borgefors, G., Strand, R. (eds.) ISMM 2013. LNCS, vol. 7883, pp. 292–301. Springer, Heidelberg (2013). https://doi.org/10.1007/978-3-642-38294-9_25
12. Nagy, B., Strand, R., Normand, N.: Distance functions based on multiple types of weighted steps combined with neighborhood sequences. J. Math. Imaging Vis. **60**(8), 1209–1219 (2018). https://doi.org/10.1007/s1085
13. Ragnemalm, I.: The Euclidean distance transform in arbitrary dimensions. Pattern Recogn. Lett. **14**(11), 883–888 (1993). https://doi.org/10.1016/0167-8655(93)90152-4
14. Sethian, J.A.: Level Set Methods and Fast Marching Methods: Evolving Interfaces in Computational Geometry, Fluid Mechanics, Computer Vision, and Materials Science. Cambridge University Press, Cambridge (1999)
15. Strand, R.: Distance functions and image processing on point-lattices: with focus on the 3D face- and body-centered cubic grids. Ph.D. thesis, Uppsala University, Sweden (2008). http://urn.kb.se/resolve?urn=urn:nbn:se:uu:diva-9312
16. Strand, R.: Weighted distances based on neighbourhood sequences. Pattern Recogn. Lett. **28**(15), 2029–2036 (2007). https://doi.org/10.1016/j.patrec.2007.05.016
17. Strand, R., Nagy, B.: A weighted neighborhood sequence distance function with three local steps. In: IEEE Proceedings of 8th International Symposium on Image and Signal Processing and Analysis (ISPA 2011), Dubrovnik, Croatia, pp. 564–568 (2011)
18. Strand, R., Normand, N.: Distance transform computation for digital distance functions. Theor. Comput. Sci. **448**, 80–93 (2012). https://doi.org/10.1016/j.tcs.2012.05.010
19. Yamashita, M., Ibaraki, T.: Distances defined by neighborhood sequences. Pattern Recogn. **19**(3), 237–246 (1986). https://doi.org/10.1016/0031-3203(86)90014-2

Stochastic Distance Transform

Johan Öfverstedt[1(✉)], Joakim Lindblad[1,2], and Nataša Sladoje[1,2]

[1] Centre for Image Analysis, Uppsala University, Uppsala, Sweden
{johan.ofverstedt,natasa.sladoje}@it.uu.se, joakim@cb.uu.se
[2] Mathematical Institute of the Serbian Academy of Sciences and Arts, Belgrade, Serbia

Abstract. The distance transform (DT) and its many variations are ubiquitous tools for image processing and analysis. In many imaging scenarios, the images of interest are corrupted by noise. This has a strong negative impact on the accuracy of the DT, which is highly sensitive to spurious noise points. In this study, we consider images represented as discrete random sets and observe statistics of DT computed on such representations. We, thus, define a *stochastic distance transform* (SDT), which has an adjustable robustness to noise. Both a stochastic Monte Carlo method and a deterministic method for computing the SDT are proposed and compared. Through a series of empirical tests, we demonstrate that the SDT is effective not only in improving the accuracy of the computed distances in the presence of noise, but also in improving the performance of template matching and watershed segmentation of partially overlapping objects, which are examples of typical applications where DTs are utilized.

Keywords: Distance transform · Stochastic · Robustness to noise · Random sets · Monte Carlo · Template matching · Watershed segmentation

1 Introduction

Distance transforms (DTs) have been, since introduced to image analysis in 1966, [18], a standard tool with applications in, among others, the context of similarity measure computation [13], image registration and template matching [13], segmentation [4], and skeletonization (by computing the centres of maximal balls) of binary images [19].

The properties of DT have been explored extensively, and several aspects of their performance have been improved by a sequence of important studies: their optimization for efficient approximation of Euclidean DT by local computations [5], fast algorithms for the exact Euclidean DT [16], DT with sub-pixel precision [10,14], and extension of DT to grey-scale and fuzzy images [11,20]. Exact algorithms eliminate approximation errors, and sub-pixel precision methods can reduce the inaccuracy of distances introduced by digitization of objects. However, neither provide a solution to the challenge which spurious points and

M. Couprie et al. (Eds.): DGCI 2019, LNCS 11414, pp. 75–86, 2019.
https://doi.org/10.1007/978-3-030-14085-4_7

structures pose: a single noise-point can be sufficient to heavily corrupt the DT, and negatively affect performance of all the analysis methods relying on it.

In this study we combine prior theoretical work related to discrete random sets (DRS) with the approach to observe distributions over distances, and we propose a Stochastic Distance Transform (SDT). Rather than increasing precision, SDT introduces a gradual (adjustable) insensitivity to noise, acting to yield a smoothed DT with less weight attributed to sparse points. A stochastic Monte Carlo-based method and a deterministic method, which has similarities to robust distances developed in [6], are proposed for computing the SDT. A similar idea is explored in [17], where a robust method for estimating distributions of Hausdorff set distances between sets of points, based on random removal of the points in the observed sets, is proposed. In that work, the authors utilize DT only as a tool for estimation of the Hausdorff set distance by computing weighted distance histograms based on user-provided point-wise reliability coefficients, without exploring how these random sets can increase the robustness and accuracy of the DT itself. We fill this gap, in this study, and define and evaluate the SDT on an illustrative subset of scenarios where DTs are used.

We show that the proposed method is more accurate than the standard DT in the presence of noise, and that it can increase the performance of several common applications of the DT, such as template matching and watershed segmentation.

2 Background

2.1 Discrete Random Sets

Discrete random sets (DRS) [9,15] are random variables taking values as subsets of some discrete reference space. DRS theory provides a theoretical foundation and offers suitable tools for the modeling and analysis of shapes in images, allowing exploration of both their structural and statistical characteristics. Representing (binary) image objects as (finite) random subsets on an image domain (bounded on $\mathbb{Z}^n$) facilitates their structural analysis in presence of noise.

The coverage probability function of a DRS Y is defined such that, for each element x of a reference set X, it expresses the probability that Y contains x,

$$p_x(Y) = \mathbb{P}(x \in Y). \tag{1}$$

2.2 Distances

Definitions of distances between points and sets, or sets and sets, commonly build on definitions of a distance measure between points. The Euclidean point-to-point distance d_E is a natural and common choice.

Given a point-to-point distance d, the standard point-to-set distance between a point x and a set X is defined as

$$d(x, X) = \inf_{y \in X} d(x, y). \tag{2}$$

The (unsigned, external) distance transform (DT) with respect to a foreground set (image object) X evaluated on the image domain I, is

$$\mathtt{DT}[X](x) = \min_{y \in X} d(x, y), \qquad \text{for} \quad x \in I. \tag{3}$$

Due to the separability of the (squared) Euclidean distance, the DT with d_E as underlying point-to-point distance can be computed exactly, in time linear to the number of pixels [16], which enables its efficient use in practical applications.

Point-to-set distances can also be extended to DRS [2] and hence generate probability distributions of distance values. Given a set of realizations of a DRS $Y_1, Y_2, Y_3 \ldots Y_n$, an empirical mean (or some other statistics) of the distances $d(x, Y_1), \ldots, d(x, Y_n)$ can be computed, [12], to estimate a distance of a given point x to the DRS.

3 Stochastic Distance Transform

In this section we propose a novel type of noise-resistant DT, defined for (ordinary, non-random) sets and points. The transform builds on the theoretical framework of random sets, and distributions of distances from points to random sets, to achieve high robustness to noise.

Let $R(X, c)$ denote a DRS on a reference set X, where probability of inclusion/exclusion of each element is i.i.d., that is, independent from the inclusion/exclusion of all other elements, and identically distributed, with constant coverage probability c, i.e., let $p_x(R(X, c)) = c$, $\forall\, x \in X$.

Definition 1. *Given an image domain I, a foreground set (image object) $X \subseteq I$, uncertainty factor $\rho \in [0, 1]$, a maximal distance $d_{\mathtt{MAX}} \in \mathbb{R}_+$, and a point-to-set distance d, the (unsigned, external) Stochastic Distance Transform (SDT) is*

$$\mathtt{SDT}_\rho\, [X]\, (x) = \mathbb{E}(\min\, [d(x, R(X, 1 - \rho)), d_{\mathtt{MAX}}]), \quad \text{for } x \in I. \tag{4}$$

For $\rho \in (0, 1]$, there is a non-zero probability that all points from X are excluded in some realization of $R(X, 1 - \rho)$. Since $d(x, \emptyset) = \infty$ this leads to the expectation value $\mathbb{E}(d(x, X)) = \infty$, for all X and x. Special care has, therefore, to be taken of the case of empty sets, to ensure that the SDT is well defined. We propose one possible solution by introduction of a parameter, $d_{\mathtt{MAX}}$, a finite maximum distance which saturates the underlying point-to-set distance. This ensures that $\mathtt{SDT}_\rho\, [X]\, (x)$ is finite-valued and well-defined for all $\rho \in [0, 1]$, X and x.

Tuning of ρ depends on the amount of noise and artifacts in the images of interest, and is either performed by heuristics, or by optimisation of an application specific evaluation metric. $d_{\mathtt{MAX}}$ is typically set to the diameter of the domain.

3.1 Monte Carlo Method

An estimate of $\mathtt{SDT}_{\rho}\left[X\right](x)$ can be obtained by a Monte Carlo method, denoted MC-SDT, by drawing N random samples (sets) from $R(X, 1-\rho)$, computing the corresponding point-to-set distances, typically using a fast DT algorithm, and then computing their empirical mean:

$$\mathtt{MC\text{-}SDT}_{\rho,N}\left[X\right](x) = \frac{1}{N}\sum_{i=1}^{N} \min\left[d(x, R(X, 1-\rho)_i), d_{\mathtt{MAX}}\right]. \tag{5}$$

Here $R(X, 1-\rho)_i$ denotes realization i of random set $R(X, 1-\rho)$, which can be sampled by one independent Bernouilli trial per element in X.

3.2 Deterministic Method

The $\mathtt{SDT}_{\rho}\left[X\right](x)$ can be modeled similarly to a geometric distribution, where each trial has a corresponding distance. The nearest point to x in X will be present and selected with probability $1-\rho$; the second nearest point to x in X will be present and selected with probability $\rho(1-\rho)$, and hence the i-th nearest point in X will be present and selected with probability $\rho^{i-1}(1-\rho)$, given that such a point exists.

Let $d_{(i)}(x, X)$ denote a generalization of the point-to-set distance (2), which defines the distance between the point x and the set X to be the distance between x and its i-th nearest point in X, where $d_{(i)}(x, X) = \infty$, for $i > |X|$. Now, a deterministic formulation of SDT, denoted DET-SDT can be given as:

$$\mathtt{DET\text{-}SDT}_{\rho,k}\left[X\right](x) = \rho^k d_{\mathtt{MAX}} + \sum_{i=1}^{k} \rho^{i-1}(1-\rho)\min\left[d_{(i)}(x, X), d_{\mathtt{MAX}}\right], \tag{6}$$

where k denotes the number of considered nearest points, and $0^0 = 1$. This deterministic formulation is similar to a weighted version of the k-distance [6], which is defined as the arithmetic mean of the (squared) distances to the k-nearest neighbours (k-NN).

The DET-SDT method given in (6) is exactly equal to the SDT if $k = |X|$, i.e. all points in X are considered. In practice it tends to be impractical to consider all the points in X (especially considering the exponentially diminishing contribution of each additional point) and we may instead choose to capture a sufficiently large fraction m (for the application at hand) of the probability mass, such as $m = 0.999$. Given such a value $m \in (0, 1)$ and $\rho > 0$, using the cumulative distribution function (CDF) of the geometric distribution, we can solve for an integer $\kappa_{\rho,m}$ of minimally required nearest points which guarantee that at least m of the total probability mass is captured,

$$\kappa_{\rho,m} = \left\lceil \frac{log(1-m)}{log(\rho)} \right\rceil. \tag{7}$$

Table 1 shows the required number of points to consider for various m and ρ.

Table 1. Minimally required number of nearest points $\kappa_{\rho,m}$ to consider for various combinations of probability mass m and uncertainty factor ρ.

m	ρ											
	0.1	0.2	0.3	0.4	0.5	0.6	0.7	0.8	0.9	0.95	0.975	0.99
0.95	2	2	3	4	5	6	9	14	29	59	119	299
0.99	2	3	4	6	7	10	13	21	44	90	182	459
0.999	3	5	6	8	10	14	20	31	66	135	273	688

There are many algorithms in the literature for finding the k-NN among a set of points, with corresponding distances. For regularly spaced grids, there are efficient algorithms for computing the k-NN utilizing the properties of the grid to achieve an improved computational complexity [7]. For other scenarios, e.g. for point-clouds, algorithms based on the efficient kd-tree data-structure [3] can be used to compute the k-NN efficiently. Once the k-NN (with distances) has been found, the closed-form expression (6) can be computed directly. The best algorithm and data-structures for computing the k-NN for (6) is highly situation-dependent, and a trade-off must be found between factors such as: (i) execution time; (ii) memory usage; (iii) utilization of the image domain structure.

4 Performance Analysis

In this section we evaluate the utility and performance of the proposed method in three main ways: (i) Measuring the distance accuracy in the presence of noise; (ii) Measuring the effect of the SDT on robustness of a template matching framework, when the proposed method is inserted into a set-to-set measure based on spatial/shape information; (iii) Observing the difference in quality of the segmentation obtained by replacing the standard DT with the SDT in the context of the classical watershed segmentation framework.

If not stated otherwise in experimental setups, $N = 400$ realizations are used for the MC-SDT, and at least $m = 99.9\%$ of the probability mass is used for the DET-SDT. The parameter k is determined by this m and the used ρ, according to (7), and as illustrated in Table 1.

4.1 Distance Transform Accuracy in the Presence of Noise

The accuracy of the distances computed by the standard DT can deteriorate heavily with just a single noise-point in a background region. In this section, the accuracy of the SDT is compared to the accuracy of DT, in the presence of added noise.

Experimental Setup: We consider two test images, one containing a solid letter A, and the other containing a letter X constructed as a sparse point-cloud in the regular grid, both corrupted by random noise-points added with probability

$p = 0.001$. We compute both the MC-SDT and DET-SDT using $\rho = 0.75$. Different computed DTs, in noise-free and noisy conditions, are presented in Fig. 1 for qualitative assessment. The evaluation metric used is Average Absolute Distance Error (AADE) in the computed distance map, over all pixels, averaged over 100 repetitions with different noise realizations.

Results: Quantitative results are presented in Table 2. The stochastic methods exhibit substantially higher, and more consistent (in terms of std. dev.), accuracy than the standard DT in the presence of noise.

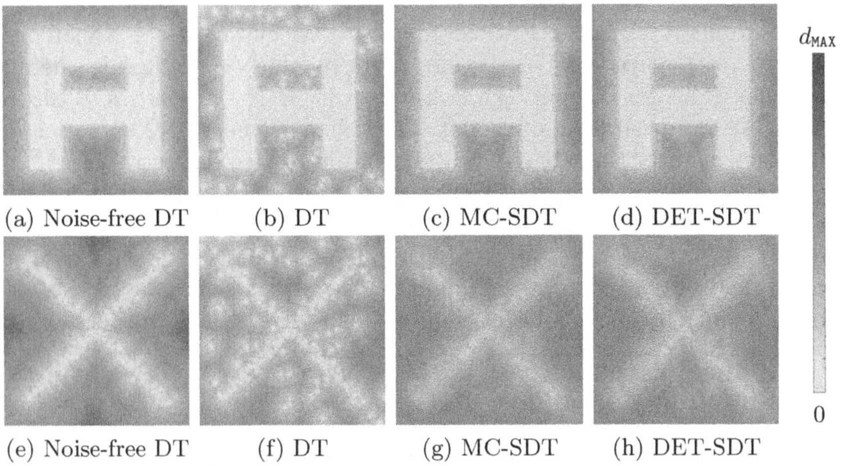

Fig. 1. (a, e) Noise-free DT of the test images; (b, f) DT of the same images, after they are corrupted by noise; (c, d, g, h) SDT applied on the noisy images.

Table 2. Average absolute distance error (AADE ±SD) for the experiments illustrated in Fig. 1. Lower is better. Bold marks the best result for each image.

	DT	MC-SDT	DET-SDT
Image A	5.41 ±0.55	**2.39** ±0.29	**2.42** ±0.27
Image X	14.31 ±0.87	**8.02** ±0.75	**8.02** ±0.74

4.2 Template Matching

Template matching of (binary) images is a process of locating a particular region/object in the image by finding a location where a given template "fits best", i.e., where a distance between the template and the image is minimal (or where a similarity measure is maximal). In the search, we consider all possible translations of the observed template by vectors with integer coordinates, such that the template is completely included in the image. We minimize the bidirectional (asymmetric, being suitable for template matching) distance [13] based on *Sum of Minimal Distances* (SMD) [8], defined as

$$d_{\rightarrow}(A,B) = \sum_{a \in A} d(a,B) + \sum_{a \in \bar{A}} d(a,\bar{B}), \tag{8}$$

where $\bar{A}$ and $\bar{B}$ denote the complement sets of A and B, respectively. This distance measure has been shown to have a number of appealing properties, such as a smooth distance field subject to translation, rotation and affine transformations. One drawback that has been observed is that this distance is quite noise-sensitive in the sense that a few spurious points can create shallow local minima in its distance landscape. As a consequence, both local search (where the search stops upon finding a local minimum) or global search may result in many false detections and must be pruned in post-processing. This part of the study aims to investigate if noise-sensitivity of template matching with $d_{\rightarrow}$ can be reduced if SDT is used in computation of $d_{\rightarrow}$ instead of the (ordinary) DT.

Experimental Setup: We consider the well-known **Cameraman** (grey-scale) image. We corrupt it with additive Gaussian noise ($\sigma = 0.1$), Fig. 2(a) and then threshold at intensity 0.5 into a binary image, Fig. 2(b). A binary template is extracted from the noise-free original, by thresholding at the same intensity level, Fig. 2(c). Within the evaluation framework we compute the distance $d_{\rightarrow}(T,X)$ between the template T and the image X, for every position in the image where the template is completely included in the image. In this computation, we use SDT_{ρ}, with $\rho \in \{0, 0.025, 0.05, \ldots 0.975, 0.99\}$, as the underlying DT for $d_{\rightarrow}$. The position where global minimum of $d_{\rightarrow}$ is reached is recorded to evaluate if the correct location is recovered. The number of minima (NoM) is also computed for the distance field, as well as the catchment basin (CB) of the global minimum. The CB is the set of all image points which would, if used as initialization for a local search (using 8-neighbourhood steps), provide convergence to the global minimum. The evaluation metrics are averaged over 50 noise realizations for each considered ρ. Since the uncertainty factor $\rho = 0$ corresponds to the standard DT, this evaluation includes comparison of performance of $d_{\rightarrow}$ using standard DT and with using the here proposed SDT.

Results: The results are presented in Fig. 2(d–g). Figures 2(d, e) show colored labelling (on a single realization) of the NoM and their corresponding CB, for standard DT, and for DET-SDT. Significantly decreased NoM, and visibly larger CB corresponding to the correct template position (red cross) characterize DET-SDT. The plots Fig. 2(f, g) show the NoM and the size of a CB of the global minimum (in a percentage of the number of pixels in the image), as a function of the uncertainty factor ρ. The results clearly show that the evaluation metrics improve in a stable and gradual way with increasing ρ. The NoM exhibits a linearly decreasing trend, where the CB size initially exhibits a linearly increasing trend until $\rho > 0.7$, where a super-linear increase is observed. The know global minimum (correct match) is successfully recovered in all tests.

(a) Noisy Input Image (b) Binary Image (c) Binary Template (d) DT: CB Labelling (e) DET-SDT: CB Labelling

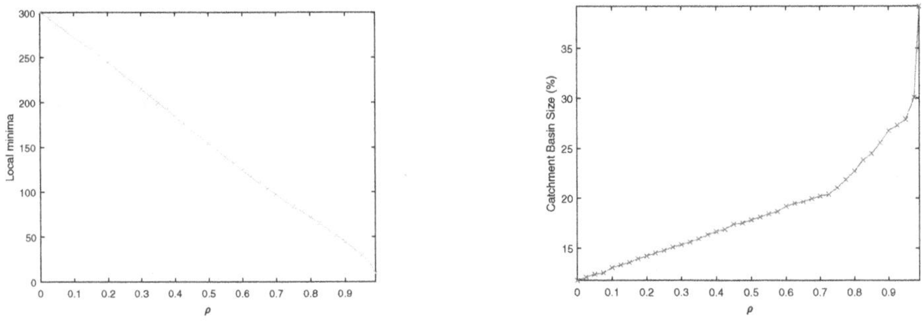

(f) Number of minima (NoM) (g) Global Minimum CB Size

Fig. 2. Template matching on binarized versions (b) of a test image (a) with additive Gaussian noise ($\sigma = 0.1$), and a noise-free template (c). The labelled catchment basins (CB) for each local minimum with standard DT (d), and $\texttt{SDT}_{0.99}$ (e). NoM (f) and size of CB (g) for different values of ρ used in the DET-SDT. The MC-SDT exhibits very similar performance in this experiment. (Color figure online)

4.3 Watershed Segmentation

The watershed transform [4] is a transformation which partitions a grey-scale image into regions associated with the local minima of the image (or a number of defined seed points). Intuitively, the graph of the grey-level image is flooded with water coming out from the seed points (minima) and filling the corresponding basins. Where the basins of different seed points meet, ridge-lines mark a delineation of the different objects. One common approach for shape based watershed segmentation is to use the negative of the DT as the grey-scale image, and its minima (maxima in the original DT) as seed-points.

It is important to note that the watershed segmentation approach used here utilizes stochasticity in a very different way than the stochastic watershed segmentation [1] method, which randomly places seed points and yields a PDF of the ridge-lines which separate the objects in the image. The method presented here employs stochasticity to remove spurious optima, with the aim of achieving robustness to noise and preventing oversegmentation.

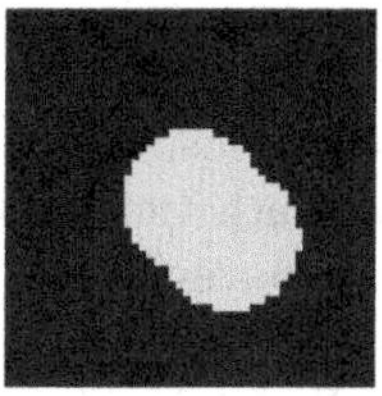

(a) DT

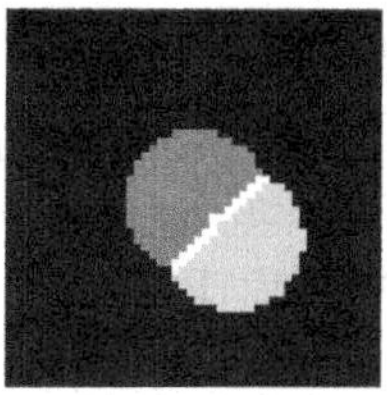

(b) MC-SDT

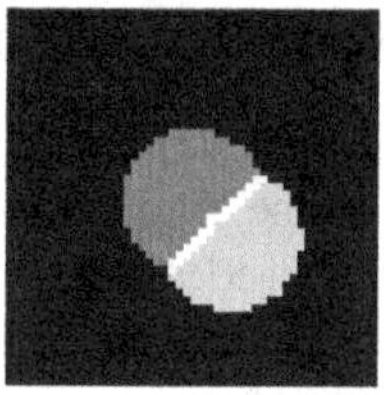

(c) DET-SDT

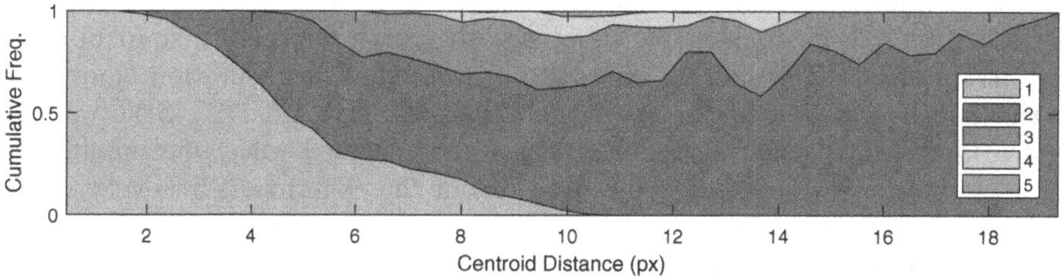

(d) DT

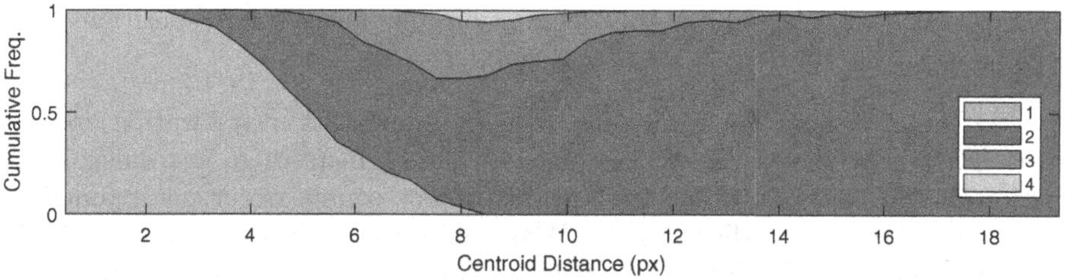

(e) MC-SDT

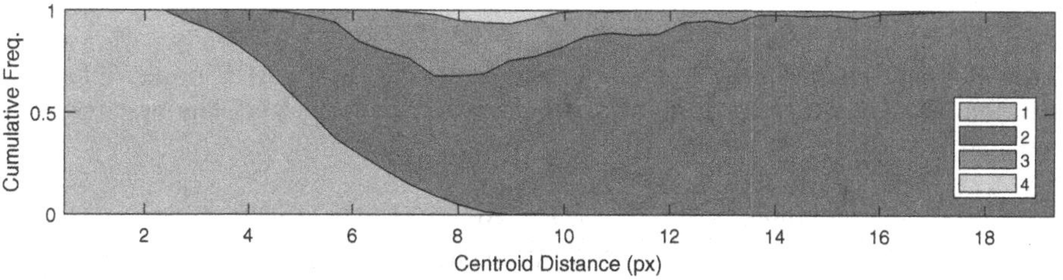

(f) DET-SDT

Fig. 3. (a–c) An example of disks segmentation by watershed algorithm based on different distance transforms: DT, MC-SDT, and DET-SDT. Segmentation based on the standard DT resulted in a single object, while segmentations with both SDT approaches successfully separate the two objects. (d–f) Quantitative evaluation of the performance on disk separation. The frequencies of appearance of the different segment counts are presented as colored areas, for increasing distance between the centers of the disks. The improvement brought by SDT, over DT, is indicated by absence of 5-segment results, and larger number of 2-segment results (indicated by the larger corresponding area), than for DT.

Separation of a Pair of Discretized Disks

Experimental Setup: To evaluate the proposed method in a scenario with no additive noise, but merely digitization artifacts/noise, a synthetic benchmark was constructed: two equisized disks are positioned with random sub-pixel placements, so that they have some overlap, and then digitized (by Gauss centre point digitisation) on a regular grid into a binary image. Watershed segmentation is applied on the negated internal DT to separate the created object into two components. Figure 3(a) shows the created binary object, where (b) an (c) present examples of segmentation (separation of the two disks). The radii of the disks are chosen to be $r = 3\pi$ pixels, to create reasonably sized objects with irrational radii. Following digitization, watershed segmentation is applied to the distance map resulting from the DT, MC-SDT and DET-SDT on the binary image. The resulting segmentations are analyzed w.r.t. the number of segmented objects, observing 200 repetitions of the experiment for every value of the distance between the centres of the disk within the range $[0.05r, 2r]$, with a step-size of $0.05r$. Considering that the disks can, in the continuous case, always be segmented into two components, we assume that 2 is the correct number of objects to result from the performed watershed segmentation. An uncertainty factor $\rho = 0.75$ is selected based on tuning on a smaller set of repetitions and steps.

Results: Figure 3 shows the results of the disks segmentation (separation) experiment. Each plot shows the performance of the segmentation in terms of the fraction of the trials at which the various object counts occur, as a function of distance between the disk centres. We observe that the SDT-based methods perform similarly, and substantially better than the standard DT. Table 3 shows the Area Under the Curve (AUC) of the detection frequency corresponding to 2 segments.

Table 3. Disk separation by watershed segmentation (Fig. 3): AUC measure for the detection frequency corresponding to 2 segments. Higher is better. The best result is presented in bold.

Method	DT	MC-SDT	DET-SDT
AUC: (2 segments)	0.563	0.661	**0.663**

Watershed Segmentation, a Realistic Example

Experimental Setup: To evaluate the performance of the watershed segmentation when used with the proposed SDT method in a realistic setting, we observe the well known image **Pears.png**, to which we apply additive Gaussian noise with $\sigma = 0.1$, Fig. 4(a). We binarize the image (by thresholding), Fig. 4(b), and segment it using watershed method with both DT and SDT.

Parameter values are: $\rho = 0.95$, $d_{\mathtt{MAX}} = 256$, binarization threshold set to 0.35 (manually selected on the noise-less image).

Results: The segmentation results are evaluated subjectively. We find that the segmentations shown in Fig. 4(d, e), which rely on SDT, clearly indicate advantage of the here presented approach, compared to classic DT which leads to heavy oversegmentation, caused by high sensitivity to noise.

(a) Input Image (b) Binary Image

(c) DT: Overlay (d) MC-SDT: Overlay (e) DET-SDT: Overlay

Fig. 4. (a) An image corrupted by a moderate amount of Gaussian noise ($\sigma = 0.1$); (b) Binarization of (a) by thresholding; (c–e) The labellings obtained by watershed segmentation with the standard DT, MC-SDT, and DET-SDT, respectively, overlayed on the image. Using the classic DT yields a highly over-segmented image while both variations of the SDT yield segmentations that are largely unaffected by the noise.

5 Conclusion

In this study we have proposed a novel type of distance transform, the Stochastic Distance Transform. SDT is based on probability distributions of distances to image objects represented as Discrete Random Sets. The main advantage of the SDT over the classic DT is its adjustable robustness to noise, allowing to choose parameters controlling a level of sensitivity according to the application at hand. The proposed method's utility and favorable properties are observed both through various synthetic tests and an illustrative natural example.

Future work includes an extended study of the theoretical properties of the proposed method, investigating (possibly adaptive) methods for reducing the biases of the resulting distance values, extending the empirical evaluation, and exploring further potential applications.

Acknowledgements. This work is supported by VINNOVA, MedTech4Health grants 2016-02329 and 2017-02447 and the Ministry of Education, Science, and Techn. Development of the Republic of Serbia (proj. ON174008 and III44006).

References

1. Angulo, J., Jeulin, D.: Stochastic watershed segmentation. In: Proceedings of the 8th International Symposium on Mathematical Morphology, pp. 265–276 (2007)
2. Baddeley, A., Molchanov, I.: Averaging of random sets based on their distance functions. J. Math. Imaging Vis. **8**(1), 79–92 (1998)
3. Bentley, J.L.: Multidimensional binary search trees used for associative searching. Commun. ACM **18**(9), 509–517 (1975)
4. Beucher, S.: Use of watersheds in contour detection. In: Proceedings of the International Workshop on Image Processing, CCETT (1979)
5. Borgefors, G.: Distance transformations in digital images. Comput. Vis. Graph. Image Process. **34**(3), 344–371 (1986)
6. Cuel, L., Lachaud, J.O., Mérigot, Q., Thibert, B.: Robust geometry estimation using the generalized voronoi covariance measure. SIAM J. Imaging Sci. **8**(2), 1293–1314 (2015)
7. Cuisenaire, O., Macq, B.: Fast k-NN classification with an optimal k-distance transformation algorithm. In: 2000 10th European Signal Processing Conference, pp. 1–4. IEEE (2000)
8. Eiter, T., Mannila, H.: Distance measures for point sets and their computation. Acta Informatica **34**(2), 109–133 (1997)
9. Goutsias, J.: Morphological analysis of discrete random shapes. J. Math. Imaging Vis. **2**(2–3), 193–215 (1992)
10. Gustavson, S., Strand, R.: Anti-aliased Euclidean distance transform. Pattern Recogn. Lett. **32**(2), 252–257 (2011)
11. Levi, G., Montanari, U.: A grey-weighted skeleton. Inf. Control **17**(1), 62–91 (1970)
12. Lewis, T., Owens, R., Baddeley, A.: Averaging feature maps. Pattern Recogn. **32**(9), 1615–1630 (1999)
13. Lindblad, J., Curic, V., Sladoje, N.: On set distances and their application to image registration. In: Proceedings of 6th International Symposium on Image and Signal Processing and Analysis, ISPA 2009, pp. 449–454. IEEE (2009)
14. Lindblad, J., Sladoje, N.: Exact linear time Euclidean distance transforms of grid line sampled shapes. In: Benediktsson, J.A., Chanussot, J., Najman, L., Talbot, H. (eds.) ISMM 2015. LNCS, vol. 9082, pp. 645–656. Springer, Cham (2015). https://doi.org/10.1007/978-3-319-18720-4_54
15. Matheron, G.: Random Sets and Integral Geometry. Wiley, New York (1975)
16. Maurer, C.R., Qi, R., Raghavan, V.: A linear time algorithm for computing exact Euclidean distance transforms of binary images in arbitrary dimensions. IEEE Trans. Pattern Anal. Mach. Intell. **25**(2), 265–270 (2003)
17. Moreno, R., Koppal, S., De Muinck, E.: Robust estimation of distance between sets of points. Pattern Recogn. Lett. **34**(16), 2192–2198 (2013)
18. Rosenfeld, A., Pfaltz, J.L.: Sequential operations in digital picture processing. J. ACM (JACM) **13**(4), 471–494 (1966)
19. Saha, P.K., Borgefors, G., di Baja, G.S.: A survey on skeletonization algorithms and their applications. Pattern Recogn. Lett. **76**, 3–12 (2016)
20. Saha, P.K., Wehrli, F.W., Gomberg, B.R.: Fuzzy distance transform: theory, algorithms, and applications. Comput. Visi. Image Underst. **86**(3), 171–190 (2002)

Discrete Topology

Filtration Simplification for Persistent Homology via Edge Contraction

Tamal K. Dey and Ryan Slechta(✉)

The Ohio State University, Columbus, OH 43210, USA
{dey.8,slechta.3}@osu.edu

Abstract. Persistent homology is a popular data analysis technique that is used to capture the changing topology of a filtration associated with some simplicial complex K. These topological changes are summarized in persistence diagrams. We propose two contraction operators which when applied to K and its associated filtration, bound the perturbation in the persistence diagrams. The first assumes that the underlying space of K is a 2-manifold and ensures that simplices are paired with the same simplices in the contracted complex as they are in the original. The second is for arbitrary d-complexes, and bounds the bottleneck distance between the initial and contracted p-dimensional persistence diagrams. This is accomplished by defining interleaving maps between persistence modules which arise from chain maps defined over the filtrations. In addition, we show how the second operator can efficiently compose with itself across multiple contractions. The paper concludes with experiments demonstrating the second operator's utility on manifolds and a brief discussion of future directions for research.

Keywords: Persistent homology · Edge contraction · Topological data analysis

1 Introduction

Edge contraction is a fundamental operation which has been famously explored by the graphics and computational geometry communities when developing tools for mesh simplification [6,12,16] and by mathematicians when developing graph minor theory [17]. However, comparatively little work has been done to incorporate edge contraction as a tool for topological data analysis. Edge contraction has been used to compute persistent homology for simplicial maps [7] and to simplify discrete Morse vector fields [9,13], but no work has been done to develop a *persistence-aware* contraction operator. *Persistent homology* is based on the observation that adding a simplex to a simplicial complex either creates or destroys a homology class [11]. Hence, the lifetime, or *persistence*, of a class through a filtered simplicial complex can be defined as the difference in the birth time and death time of the class. This permits a pairing of simplices, where σ is paired with τ if τ destroys the homology class created by σ. A summary of the

M. Couprie et al. (Eds.): DGCI 2019, LNCS 11414, pp. 89–100, 2019.
https://doi.org/10.1007/978-3-030-14085-4_8

births and deaths of homology classes is given in a *persistence diagram*. We give further details in Sect. 2.

In this paper, we aim to develop contraction operators which when applied to a filtered simplicial complex, simplify the cell structure while also controlling perturbations in the persistence diagrams associated with the complexes. We develop two such operators: one for 2-manifolds which maintains the same pairing in the contracted complex as the original, and one for arbitrary d-complexes which bounds the bottleneck distance between the persistence diagrams of the original and contracted filtrations. In addition, we show how our operator for d-complexes composes with itself to bound perturbation across multiple contractions. We provide an implementation of the operator which controls bottleneck distance and demonstrate its utility on manifolds.

This version excludes proofs in the interest of brevity. The full version can be found on arXiv.

2 Preliminaries

Throughout this paper, we will use K to refer to a finite simplicial complex of arbitrary dimension, unless otherwise specified. We assume that K is equipped with a height function

$$h : K \to \mathbb{R} \tag{1}$$

such that if σ is a face of τ, $h(\sigma) \leq h(\tau)$. This is equivalent to assuming that K is *filtered*. That is, K is equipped with a sequence of subcomplexes $\{K_a\}_{a \in A}$ where $K_a \subset K_{a'}$ if $a < a'$, $|A|$ finite. In addition, for some $a \in A$, $K_a = K$. A filtration induces a height function on K that respects the face poset, where the height of any particular simplex is the first index at which it occurs. Similarly, h induces a filtration in a canonical way.

The filtration $\{K_a\}_{a \in A}$ gives a natural partial order on the simplices of K. It induces a total order $\prec$ by giving precedence to lower dimensional simplices and arbitrarily breaking ties within each dimension. For simplices $\sigma, \tau \in K$, we write $\sigma < \tau$ if σ is a face of τ, and define $\sigma > \tau$ as expected. Similarly, we write $\sigma <_1 \tau$ if σ is a facet of τ. If $\sigma \neq \tau$ are of the same dimension and there exists a ρ where $\rho <_1 \sigma$ and $\rho <_1 \tau$, then we say that σ is *incident* to τ (and vice-versa).

2.1 Edge Contraction

For a filtered simplicial complex K, we model contracting edge $\{u, v\} \in K$ as a simplicial map

$$\xi_{\{\mathrm{u,v}\}} : K \to \Delta \tag{2}$$

where Δ is the maximal simplicial complex on the vertex set of K. We often denote $\xi_{\{\mathrm{u,v}\}}(K)$ as K'. A height function $h' : K' \to \mathbb{R}$ is induced on K' by defining $h'(\sigma) = \min\{h(\tau) \mid \tau \in {\xi_{\{\mathrm{u,v}\}}}^{-1}(\sigma)\}$. Equivalently, the filtration $\{\xi_{\{\mathrm{u,v}\}}(K_a)\}_{a \in A}$ is induced on K'. The total order $\prec$ also induces a total order on K' where if multiple simplices map to the same simplex, then the image takes the first position in the total order of its preimages. We will abuse notation and allow $\prec$ to refer to both the total order on K and K'.

In this paper, when contracting $\{u, v\}$, we always assume that $u \prec v$. Following [7], those simplices σ for which $\{u, v\} < \sigma$ are called *vanishing simplices.* If σ is the face of some vanishing simplex and contains exactly one of u or v as a face, then σ is a *mirrored simplex.*

Remark 1. Mirrored simplices come in pairs. If σ is defined by the n vertices $\{x_1, x_2, \ldots, x_{n-1}, u\}$, then the *mirror* of σ, denoted $m(\sigma)$, is the simplex defined by $\{x_1, x_2, \ldots, x_{n-1}, v\}$.

If σ is either mirrored or vanishing, then σ is a *local* simplex. If a simplex σ contains a mirrored simplex as a facet, but is not a local simplex, then σ is said to be an *adjacent simplex.* Figure 1 details the various types of simplices in a 2-complex. For nonlocal simplices, $\xi_{\{\mathrm{u,v}\}}(\sigma) = \sigma$ if $v \notin \sigma$, and $\xi_{\{\mathrm{u,v}\}}(\sigma) = (\sigma \setminus \{v\}) \cup \{u\}$ otherwise. If σ is a mirrored simplex, and $u < \sigma$, then $\xi_{\{\mathrm{u,v}\}}(\sigma) = \xi_{\{\mathrm{u,v}\}}(m(\sigma)) = \sigma$. Otherwise, $\xi_{\{\mathrm{u,v}\}}(\sigma) = \xi_{\{\mathrm{u,v}\}}(m(\sigma)) = m(\sigma)$. If τ is vanishing, then it has mirrored facet σ. We let $\xi_{\{\mathrm{u,v}\}}(\sigma)$ determine $\xi_{\{\mathrm{u,v}\}}(\tau)$.

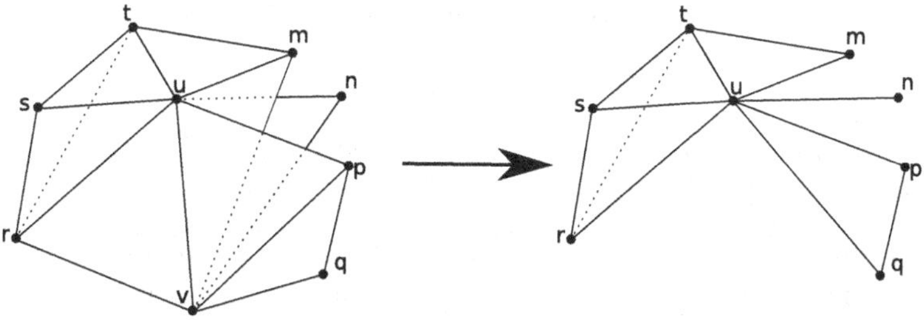

Fig. 1. A simplicial 2-complex before and after contracting edge $\{u, v\}$ where $u \prec v$. In the pre-contracted complex, edge $\{u, v\}$ and triangles $\{r, u, v\}$, $\{p, u, v\}$, $\{m, u, v\}$, and $\{n, u, v\}$ are vanishing. Vertex pair (u, v) and edge pairs $(\{r, u\}, \{r, v\})$, $(\{p, u\}, \{p, v\})$, $(\{m, u\}, \{m, v\})$, and $(\{n, u\}, \{n, v\})$ are mirrored edges. Triangles $\{r, s, u\}$, $\{m, t, u\}$, $\{p, q, v\}$, $\{r, t, u\}$ and edges $\{s, u\}$, $\{t, u\}$, $\{p, v\}$ are adjacent simplices. All other simplices are nonlocal simplices.

2.2 Persistence Modules and Filtrations

Let $\{K_a\}_{a \in A}$ denote some filtration for K. Note that if $a < b$, there is a natural map from the chains of K_a to K_b. The chain maps define a map between the p-dimensional homology groups $H_p(K_a) \to H_p(K_b)$ (consult [10] for details). These homology groups, together with all of the induced maps between them, give a *persistence module.* We use the definition given by Chazal *et al.* [4].

Definition 1. *Let R be a commutative ring with unity, and A a subset of $\mathbb{R}$. A persistence module M_A is a family $\{F_\alpha\}_{\alpha \in A}$ of R-modules indexed by the elements of A, together with a family $\{f_\alpha^{\alpha'} : F_\alpha \to F_{\alpha'}\}_{\alpha \leq \alpha' \in A}$ of homomorphisms such that, $\forall \alpha \leq \alpha' \leq \alpha'' \in A, f_\alpha^{\alpha''} = f_{\alpha'}^{\alpha''} \circ f_\alpha^{\alpha'}$.*

In this paper, we will only consider when R is $\mathbb{Z}_2$. As mentioned earlier, $\{K_a\}_{a \in A}$ gives a persistence module

$$M_A : H_p(K_{j_1}) \xrightarrow{f_{j_1,j_2}} H_p(K_{j_2}) \xrightarrow{f_{j_2,j_3}} H_p(K_{j_3}) \xrightarrow{f_{j_3,j_4}} \cdots \xrightarrow{f_{j_{n-1},j_n}} H_p(K_{j_n})$$

where additional maps are given by composition. Note that there is a second persistence module given by the filtration $\{\xi_{\{\mathrm{u,v}\}}(K_a)\}_{a \in A}$. For convenience, we will define $K'_a := \xi_{\{\mathrm{u,v}\}}(K_a)$. This gives us a module for the contracted complex:

$$M'_A : H_k(K'_{j_1}) \xrightarrow{f'_{j_1,j_2}} H_k(K'_{j_2}) \xrightarrow{f'_{j_2,j_3}} H_k(K'_{j_3}) \xrightarrow{f'_{j_3,j_4}} \cdots \xrightarrow{f'_{j_{n-1},j_n}} H_k(K'_{j_n}).$$

2.3 Persistence Diagrams

A *persistence diagram* captures the birth and deaths of homology classes in a corresponding persistence module. We let $\mu_p^{i,j}$ be the number of p-dimensional homology classes which are born in K_i and die in K_j. This gives a formalization of a persistence diagram [5].

Definition 2. *A persistence diagram* $\mathrm{Dgm}_p(f)$ *of a filtration induced by f is a multi-subset of the extended real plane, such that each point $(i,j), j > i$ has multiplicity $\mu_p^{i,j}$, and points (i,i) have infinite multiplicity.*

Persistence diagrams are often plotted as in Fig. 2. Note that a persistence diagram can equivalently be thought to capture the changes in a persistence module, so for persistence module M we often use the notation $\mathrm{Dgm}_p(M)$. In addition, we can define a distance between persistence diagrams.

Definition 3. *For persistence diagrams* $\mathrm{Dgm}_p(f), \mathrm{Dgm}_p(g)$*, let B denote the set of all bijections γ :* $\mathrm{Dgm}_p(f) \to \mathrm{Dgm}_p(g)$*. The bottleneck distance d_b is defined as*

$$d_b(\mathrm{Dgm}_p(f), \mathrm{Dgm}_p(g)) = \inf_{\gamma \in B} \sup_{x \in \mathrm{Dgm}(f)} d_\infty(x, \gamma(x))$$

where d_∞ denotes the infinity norm.

We encourage the reader to consult [10] for a more thorough treatment of persistence diagrams and bottleneck distance. Developing a contraction operator which bounds the perturbation in the bottleneck distance between the persistence diagrams of the original and contracted filtrations is one of the goals of this paper. However, this distance is quite cumbersome. Chazal et al. showed that bottleneck distance could be directly related to the persistence module corresponding to a filtration [4]. This requires a notion of similarity between persistence modules. We use the definition given in [4].

Definition 4. *Two persistence modules* $M_{\mathbb{R}}$ *and* $M'_{\mathbb{R}}$ *are strongly* ϵ*-interleaved if there exist two families of homomorphisms* $\{\phi_\alpha : M_\alpha \to M'_{\alpha+\epsilon}\}_{\alpha\in\mathbb{R}}$ *and* $\{\psi_\alpha : M_\alpha \to M'_{\alpha+\epsilon}\}_{\alpha\in\mathbb{R}}$ *such that the diagrams of Eq. 3 commute* $\forall \alpha \leq \alpha' \in \mathbb{R}$.

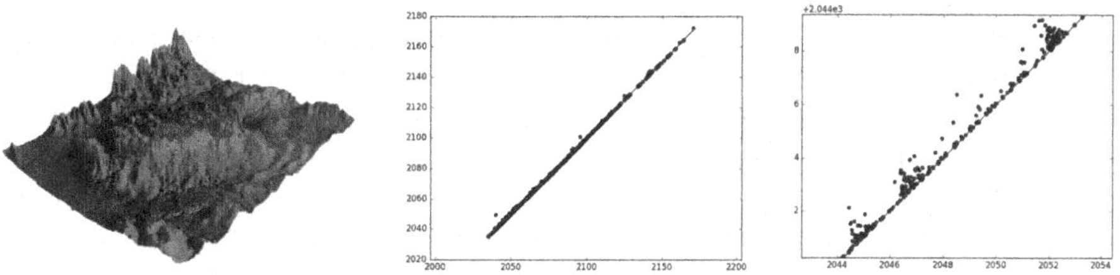

Fig. 2. A triangulated terrain (left), the 0-dimensional persistence diagram corresponding to the terrain (middle), and a closer view of the same diagram (right). The height function was extended to the entire complex such that the value at a simplex is the maximum height value of its constituent vertices.

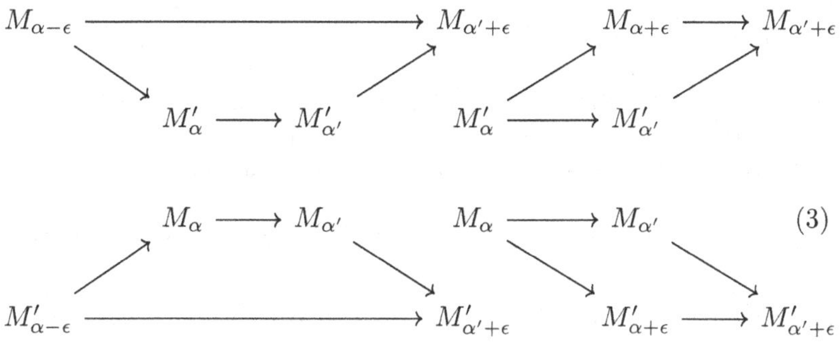

(3)

Chazal *et al.* proved the following theorem.

Theorem 1. *Let* $M_{\mathbb{R}}$ *and* $M'_{\mathbb{R}}$ *be tame persistence modules. If* $M_{\mathbb{R}}$ *and* $M'_{\mathbb{R}}$ *are strongly* ϵ*-interleaved, then* $d_b(\mathrm{Dgm}_p(M_{\mathbb{R}}), \mathrm{Dgm}_p(M'_{\mathbb{R}})) \leq \epsilon$.

As K is assumed to be finite, all persistence modules we consider will be tame. In addition, ϵ-interleavings induce a pseudometric on the space of persistence modules called *interleaving distance* [15]. In particular, this permits application of the triangle inequality.

3 Preserving Pairings

We now move to developing a contraction operator for 2-manifolds such that if σ is paired with τ, and if σ is nonlocal or a mirror that precedes its partner under $\prec$, then $\xi_{\{u,v\}}(\sigma)$ is paired with $\xi_{\{u,v\}}(\tau)$.

3.1 Pairings for Manifolds

In this section, we assume that the underlying space (of the geometric realization) of K is a 2-manifold. In particular, we assume that K is without boundary, but a slight modification works for manifolds with boundary. Hence, we can assume that each edge is the facet of exactly two triangles. Attali *et al.* observed that the persistence pairings of such complexes can be computed in near-linear time in the number of edges [1]. This is done by considering two graphs induced by the simplicial complex: the vertex graph K_v and the triangle graph K_t. To avoid confusion, we refer to edges and vertices in the vertex and triangle graphs as *arcs* and *nodes*, respectively. The vertex graph is induced in the obvious way, and the triangle graph is its dual. Note that arcs in K_t also correspond to edges in K. In both graphs, arc e is weighted with value $h(e')$, where e' is the edge corresponding to e. Persistence partners are computed by using Kruskal's algorithm to compute a minimum spanning tree on K_v and a maximum spanning tree on K_t. When an arc e is introduced to a spanning forest on K_v, it connects two trees rooted at nodes v_1, v_2. Assuming the vertex corresponding to v_1 occurs prior to that corresponding to v_2 under $\prec$, we define $\mathrm{MinV}(e) = v_1$ and $\mathrm{MaxV}(e) = v_2$. Following the introduction of e, $\mathrm{MinV}(e)$ becomes the root of the combined tree and the edge corresponding to e is paired with the vertex corresponding to $\mathrm{MaxV}(e)$. For arcs e in K_t, $\mathrm{MaxT}(e)$ and $\mathrm{MinT}(e)$ are defined analogously, except upon the introduction of e, $\mathrm{MaxT}(e)$ is the root of the new tree, and the edge corresponding to e is paired with the triangle corresponding to $\mathrm{MinT}(e)$.

Note that some simplices remain unpaired following this algorithm. We let $P(K)$ denote the set of *pairs* of simplices that results from the aforementioned algorithm. We aim to develop a contraction operator such that

$$P(K') = \{(\xi_{\{\mathrm{u,v}\}}(\sigma), \xi_{\{\mathrm{u,v}\}}(\tau)) \mid (\sigma, \tau) \in P(K) \wedge \xi_{\{\mathrm{u,v}\}}(\sigma) \neq \xi_{\{\mathrm{u,v}\}}(\tau)\}. \quad (4)$$

3.2 A Persistence-Pair Preserving Condition

We now present sufficient conditions for contracting an edge that maintains the persistence pairing. First, we assume that e satisfies the *link condition*, which ensures that the complex remains a 2-manifold following contraction [6]. For simplex σ, we define $\mathrm{Cl}(\sigma) = \{\tau \mid \tau < \sigma\}$, $\mathrm{Star}(\sigma) = \{\tau \mid \sigma < \tau\}$, and $\mathrm{Lk}(\sigma) = \mathrm{Cl}(\mathrm{Star}(\sigma)) \setminus \mathrm{Star}(\mathrm{Cl}(\sigma))$. All of these operations extend to sets of simplices in the natural way.

Definition 5 (Link Condition [6]). *An edge $e = \{u, v\}$ satisfies the link condition if $\mathrm{Lk}(e) = \mathrm{Lk}(u) \cap \mathrm{Lk}(v)$. We say e is contractible if it satisfies the link condition.*

Requiring the link condition implies that there are two sets of mirrored edges relative to $\{u, v\}$. We label the two incident triangles to $\{u, v\}$ as t_1 and t_2, and their respective constituent mirrored edges as e_1, e_1' and e_2, e_2'. We will assume without loss of generality that $e_1 \prec e_1'$ and $e_2 \prec e_2'$.

Definition 6. *An edge* $e = \{u, v\}$ *is admissible if* e *satisfies the Link Condition, and* e *is paired with* v*,* t_1 *is paired with* e'_1*, and* t_2 *is paired with* e'_2*.*

This definition, together with the pairing algorithm for 2-manifolds, gives the following results.

Theorem 2. *If* $\{u, v\}$ *is admissible, and* $e \in K$*,* $e \neq \{u, v\}, e'_1, e'_2$ *is paired with vertex* r*, then* $\xi_{\{u,v\}}(e)$ *is paired with* $\xi_{\{u,v\}}(r)$*.*

Theorem 3. *If* $\{u, v\}$ *is admissible, and* $e \in K$*,* $e \neq \{u, v\}, e'_1, e'_2$ *is paired with triangle* r*, then* $\xi_{\{u,v\}}(e)$ *is paired with* $\xi_{\{u,v\}}(r)$*.*

Corollary 1. *If* $\{u, v\}$ *is admissible, then* $K'_p = \{(\xi_{\{u,v\}}(\sigma), \xi_{\{u,v\}}(\tau)) | (\sigma, \tau) \in K_p \wedge \xi_{\{u,v\}}(\sigma) \neq \xi_{\{u,v\}}(\tau)\}$*.*

These contraction conditions are expandable, particularly for the vertex/edge pairing. For example, it is somewhat easy to extend our conditions to permit $\{u, v\}$ to be paired with u. Expanding the triangle conditions is significantly more difficult. It will be interesting to develop necessary and sufficient conditions for this problem.

4 Stable Contraction

Let $\mathrm{Dgm}_p(M_A)$ and $\mathrm{Dgm}_p(M'_A)$ denote the p-dimensional persistence diagrams corresponding to the persistence modules M_A and M'_A, where M'_A is obtained by contracting a single edge $\{u, v\}$. If $d_B(\mathrm{Dgm}_p(M_A), \mathrm{Dgm}_p(M'_A)) \leq \epsilon$, then the contraction map $\xi_{\{u,v\}}$ is said to be (p, ϵ)-stable. In this section we develop such a contraction operator. We will always assume that edge $\{u, v\}$ meets a relaxation of the link condition. In addition, we let h and h' denote height functions on K and K'.

Due to Theorem 1, it is sufficient to develop a contraction operator which bounds the interleaving distance between M_A and M'_A. Note that any persistence module defined over some index set $A \subset \mathbb{R}$ can be extended to a persistence module over $\mathbb{R}$ in a canonical way. Hence, we now refer to $M_{\mathbb{R}}$ and $M'_{\mathbb{R}}$ and will establish maps such that they are strongly interleaved. We let ξ_{j_i} denote the restriction of $\xi_{\{u,v\}}$ to the subcomplex K_{j_i}. Now, we define the maps $\xi^*_{j_i}$: $H_p(K_{j_i}) \to H_p(K'_{j_i})$ in the following diagram.

$$\begin{array}{ccccccccc}
\cdots & \xrightarrow{f_{j_0,j_1}} & H_p(K_{j_1}) & \xrightarrow{f_{j_1,j_2}} & H_p(K_{j_2}) & \xrightarrow{f_{j_2,j_3}} & H_p(K_{j_3}) & \xrightarrow{f_{j_3,j_4}} & \cdots \\
 & & \downarrow \xi^*_{j_1} & & \downarrow \xi^*_{j_2} & & \downarrow \xi^*_{j_3} & & \\
\cdots & \xrightarrow{f'_{j_0,j_1}} & H_k(K'_{j_1}) & \xrightarrow{f'_{j_1,j_2}} & H_p(K'_{j_2}) & \xrightarrow{f'_{j_2,j_3}} & H_p(K'_{j_3}) & \xrightarrow{f'_{j_3,j_4}} & \cdots
\end{array}$$

We extend ξ_{j_i} to the canonical chain map $\xi_{j_i,\#}$: $C_p(K_{j_i}) \to C_p(K'_{j_i})$. This map induces a map between homology groups. Let $\gamma \in H_p(K_{j_i})$ and let $\sum_I \sigma_i$ be a representative cycle of γ. Then we define $\xi^*_{j_i}$: $H_p(K_{j_i}) \to H_p(K'_{j_i})$ where $\xi^*_{j_i}(\gamma)$ is determined by $\xi_{j_i,\#}(\sum_I(\sigma_i))$.

Lemma 1. *If $\sum_I \sigma_i$, $\sigma_i \in K$ is a p-cycle in sublevel set K_{j_i}, then $\xi_{j_i,\#}(\sum_I \sigma_i)$ is also a p-cycle at sublevel set K'_{j_i}.*

Theorem 4. *If $\sum_I \sigma_i$, $\sigma_i \in K$ is a boundary at sublevel set at K_{j_i}, then $\xi_{j_i,\#}(\sum_I \sigma_i)$ is also a boundary at sublevel set K'_{j_i}.*

The next theorem follows immediately from the previous two results.

Theorem 5. *The maps ξ^*_j are well defined.*

For all $\epsilon > 0$, ξ^*_j extends to a map $\xi_j^{j+\epsilon,*}$ by composing $f'_{j,j+\epsilon}$ with ξ^*_j.

It is now necessary to define maps from the contracted persistence module to the original module. To do so, we will require $\{u, v\}$ to be (p, ϵ)-admissible for some fixed ϵ. First, we define the p-Link Condition [8].

Definition 7. *Edge $\{u, v\}$ satisfies the p-Link Condition in K if and only if either (i) $p \leq 0$, or (ii) $p > 0$ and every $(p-1)$-simplex $\sigma \in \mathrm{Lk}(u) \cap \mathrm{Lk}(v)$ is also in $\mathrm{Lk}(\{uv\})$.*

Requiring that $\{u, v\}$ satisfies the p- and $(p-1)$-link conditions guarantees that $H_p(K) \cong H_p(\xi_{\{\mathrm{u,v}\}}(K))$. This permits a definition for (p, ϵ)-admissible.

Definition 8. *Edge $\{u, v\}$ is (p, ϵ)-admissible if*

1. *$\{u, v\}$ satisfies the p- and $(p-1)$-link conditions and,*
2. *For each pair of p-mirrors $\sigma_1 \prec \sigma_2$ with shared vanishing cofacet τ, $|h(\sigma_2) - h(\tau)| \leq \epsilon$ and,*
3. *if $p > 0$ then for each pair of $p-1$-mirrors $\sigma_1 \prec \sigma_2$ with shared vanishing cofacet τ, $|h(\sigma_2) - h(\tau)| \leq \epsilon$.*

Intuitively, the first requirement guarantees that contraction does not destroy any cycles with infinite persistence. The second requirement ensures commutativity when contracting $\{u, v\}$ destroys homological classes in some subcomplex. The third requirement does the same when contracting $\{u, v\}$ creates cycles in a subcomplex. Clearly, contraction cannot create a new 0-dimensional class, so the third requirement does not apply in this case.

For (p, ϵ)-admissible edge $\{u, v\}$, we now define chain maps $\psi^b_a : K'_a \to C_p(K_b)$ $\forall a \in \mathbb{R}$, where $b = a + \epsilon$ and $C_p(K_b)$ is the group of p-chains over K_b. Of particular importance is determining the image under ψ^b_a of those simplices σ which are the image of mirrored simplices under $\xi_{\{\mathrm{u,v}\}}$. If $\sigma = \xi_{\{\mathrm{u,v}\}}(\tau) = \xi_{\{\mathrm{u,v}\}}(m(\tau))$, $\tau \prec m(\tau)$, then we let $\psi^b_a(\sigma) = \tau + \sum_I \sigma_i$, where σ_i are a subset of those vanishing $\dim(\sigma)$-simplices relative to $\{u, v\}$ which are incident to τ. In particular, we only include those σ_i where, for the shared mirrored facet η, $m(\eta) \prec \eta$. Similarly, if σ is the image of an adjacent simplex which contains η as a facet, η a mirror where $m(\eta) \prec \eta$, then $\psi^b_a(\sigma) = {\xi_{\{\mathrm{u,v}\}}}^{-1}(\sigma) + \tau$, where τ is the vanishing cofacet of η. If σ is nonlocal, then $\psi^b_a(\sigma) = \sigma$. The map ψ^b_a now extends linearly to a chain map $\psi^b_{a,\#} : C_p(K'_a) \to C_p(K_b)$. Note that while $\psi^b_{a,\#}$ does not commute with the boundary map, it does still have the requisite properties to induce homomorphisms.

Theorem 6. *Let $\sum_I \sigma_i$ be a cycle in K'_a. Then $\psi^b_{a,\#}(\sum_I \sigma_i)$ is a cycle in K_b.*

The next theorem follows from essentially the same parity argument as Theorem 6.

Theorem 7. *Let $\sum_J \tau_j$ be a boundary in K'_a. Then $\psi^b_{a,\#}(\sum_J \tau_j)$ is a boundary in K_b.*

Hence, $\psi^b_{a,\#}$ induces a map between homology groups which we denote $\psi^{b,*}_a$. We have already shown that ξ^*_a extends to $\xi^{b,*}_a$, $b = a + \epsilon$, for all $a \in \mathbb{R}$. All that remains is to show that the maps $\xi^{b,*}_a$ and $\psi^{b,*}_a$ commute with the homomorphisms given by the persistence modules as in Eq. 3.

Theorem 8. *If $\{u, v\}$ is (p, ϵ)-admissible, then the maps $\psi^{b,*}_a$, $\xi^{b,*}_a$, $f_{a,b}$ and $f'_{a,b}$ commute as in Eq. 3.*

Corollary 2. *If $\{u, v\}$ satisfies the link condition and is (p, ϵ)-admissible, then $\xi_{\{u,v\}}$ is (p, ϵ)-stable.*

4.1 Multiple Contractions

The existence of (p, ϵ)-stable contraction gives rise to the question of bounding interleaving distance across multiple contractions. Consider (p, ϵ)-admissible edges, $\{u_1, v_1\}$, $\{u_2, v_2\}$, $\{u_3, v_3\}, \ldots, \{u_n, v_n\}$. Specifically, $\{u_2, v_2\}$ is (p, ϵ)-admissible in the complex $\xi_{\{u_1,v_1\}}(K)$, $\{u_3, v_3\}$ is (p, ϵ)-admissible in the complex $\xi_{\{u_2,v_2\}} \circ \xi_{\{u_1,v_1\}}(K)$, and so on. Contracting these edges sequentially gives a sequence of p-dimensional persistence modules $M_0, M_1, \ldots, M_n$. Naively applying the triangle inequality implies that $d_b(\mathrm{Dgm}_p(M_0), \mathrm{Dgm}_p(M_n)) \leq n\epsilon$. We aim to find conditions under which $d_b(\mathrm{Dgm}_p(M_0), \mathrm{Dgm}_p(M_n)) \leq \epsilon$. In this section, we will use $K^{(i)}$ to denote the complex that results from contracting the first i edges, and will use $f^{(i)}$ to refer to the inclusion homomorphisms in M_i.

For each (p, ϵ)-admissible edge $e \in K$, we define a *p-window* $W_p(e) \subset \mathbb{R}$. If $0 < p < \dim(K)$, let r be the $p-1$ mirror relative to e with minimal height value, and s be the $p+1$ simplex with maximal height value. Then $W_p(e) = [f(r), f(s)]$. If $p = 0$, and u is the vertex with minimal height value, then $W_p(e) = [f(u), f(e)]$. Let z denote the maximum vanishing p-simplex relative to $\{u, v\}$. If $p = \dim(K)$, then $W_p(e) = [f(r), f(z)]$.

Definition 9. *The sequence of (p, ϵ)-admissible edges $\{u_1, v_1\} \in K$, $\{u_2, v_2\} \in K', \ldots, \{u_n, v_n\} \in K^{(n-1)}$ is (p, ϵ)-compatible if the intervals $W_p(\{u_1, v_1\}), \ldots, W_p(\{u_n, v_n\})$ are disjoint.*

This definition permits us to bound the interleaving distance between the initial and final persistence modules. Downward arrows are now given by the composition of contraction maps. Each given $\{u_i, v_i\}$ has a different collection of ψ^b_a, so we use the notation $\psi^b_{i,a} : H_p(K^{(i)}_a) \to H_p(K^{(i-1)}_b)$ to distinguish them. Note that if $W_p(\{u_i, v_i\}) = [r, s]$, then $H_p(K^{(i-1)}_j) \cong H_p(K^{(i)}_j)$ provided

$j \notin [r, s)$. It is easy to see that careful composition of $\psi_{i,a}^{b}$ with these isomorphisms defines a new collection of upward maps $\Psi_a^b : H(K_a^{(n)}) \to H(K_b)$, $b = a + \epsilon$. Similarly, through composition of contraction and inclusion maps, we get $\Xi_a^b : H_p(K_a) \to H_p(K_b^{(n)})$. It is clear that both collections of maps inherit the boundary to boundary and cycle to cycle properties from the single-contraction case. Hence, the following theorem follows immediately.

Theorem 9. *Let* $\{u_1, v_1\} \in K$, $\{u_2, v_2\} \in K', \ldots, \{u_n, v_n\} \in K^{(n-1)}$ *be* (p, ϵ)-*compatible edges. The maps* Ψ_a^b, Ξ_a^b, f, *and* $f^{(n)}$ *commute as in Eq. 3.*

5 Experiments

In this section, we implement our contraction operator and demonstrate its utility on some manifolds. Note that the notion of (p, ϵ)-compatibility lends itself to an elementary scheduling problem. At each stage, we aim to contract as many edges as possible while ensuring that the bottleneck distance between the previous and resulting persistence diagrams remains $\leq \epsilon$. Hence, by the triangle inequality, the bottleneck distance between persistence diagrams for the initial and contracted complexes is $\leq m\epsilon$, where m is the number of stages. In this section, we will contract sets of edges which are $(1, 0.5)$-compatible and $(1, 5.0)$-compatible at each stage.

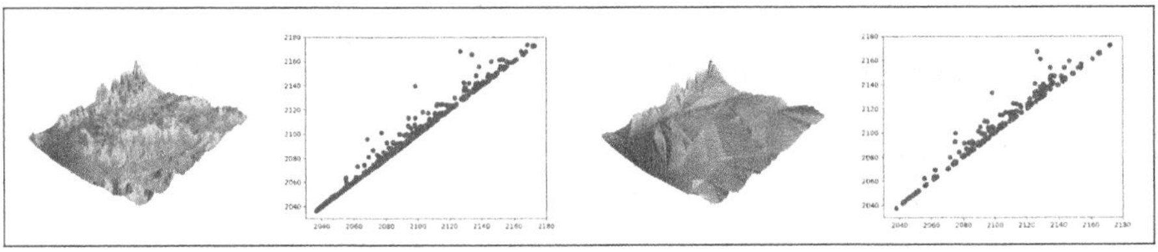

Fig. 3. A triangulated terrain (far left), the 1-dimensional persistence diagram corresponding to the terrain (center left), the complex after approximately 450,000 edge contractions resulting in a 99% reduction in the number of simplices (center right), the 1-dimensional persistence diagram corresponding to the contracted complex (right). The bottleneck distance between the two persistence diagrams is 6.540.

We consider four terrains, obtained from the National Elevation Dataset, and 3 models, obtained from the Aim@Shape repository. The terrains are from near Aspen, Colorado; Minneapolis, Minnesota; Los Alamos, New Mexico; and Columbus, Ohio. A height function is defined over the models' vertex set by a surface curvature approximation. For each manifold, we find a maximal set of $(1, \epsilon)$-compatible edges, contract them all, and repeat the process until there are no remaining $(1, \epsilon)$-admissible edges. An example of the contraction process and persistence diagrams can be seen in Fig. 3. In Tables 1 and 2 we show the resulting data for $\epsilon = 0.5$ and $\epsilon = 5.0$, respectively.

Table 1. Contraction data for seven datasets when contracting $(1, 0.5)$-compatible edges. Iterations corresponds to the number of sets of compatible edges that were contracted. Despite many iterations, note that the bottleneck distance between the 1-dimensional persistence diagrams remains comparatively close to $\epsilon = 0.5$.

Dataset	Init. Simps.	Contractions	Its.	Rem. Simps.	% Red.	d_b	Time(s)
Columbus	2,728,353	450,835	6,092	23,343	99.14	0.601	347
Los alamos	2,728,353	449,181	7,840	33,267	98.78	1.370	490
Minneapolis	10,940,401	1,818,560	66,627	29,041	99.73	1.106	19,508
Aspen	10,940,401	1,806,640	6,231	100,561	99.08	1.115	1,592
Filigree	3,086,440	438,925	85,116	452,890	85.32	0.916	20,963
Eros	2,859,566	475,401	157,636	7,160	99.75	0.489	10,054
Statue	14,994	2,470	945	174	98.84	0.074	.111

In general, the bottleneck distance is much less than $m\epsilon$. This raises an issue of the tightness of our multiple contraction scheme, which we leave to future work.

Table 2. Contraction data for seven datasets when contracting $(1, 5.0)$-compatible edges. As in the $(1, 0.5)$ case, the bottleneck distance between the 1-dimensional persistence diagrams remains comparatively close to $\epsilon = 5.0$.

Dataset	Init. Simps.	Contractions	Its.	Rem. Simps.	% Red.	d_b	Time(s)
Columbus	2,728,353	452,914	6,123	10,869	99.60	2.097	350
Los Alamos	2,728,353	452,710	7,350	12,093	99.56	6.540	394
Minneapolis	10,940,401	1,819,786	66,981	21,685	99.80	4.832	19,127
Aspen	10,940,401	1,818,593	5,664	28,843	99.74	8.570	1,232
Filigree	3,086,440	466,395	90,009	288,070	90.67	9.051	15,690
Eros	2,859,566	476,573	158,426	128	99.99	4.296	9,751
Statue	14,994	2,470	945	174	98.84	0.074	.093

6 Conclusion

We conclude with a short discussion on directions for future research. It appears that the conditions for (p, ϵ)-compatible edges are very conservative, and could possibly be expanded to permit further contraction. In addition, there are a variety of other operations that could be applied to simplicial complexes which may be persistence-sensitive. Based on the theory of strong homotopy [2], Boissonat *et al.* have recently invented a quick way to compute persistent homology by simplifying the input complex with strong collapses [3]. A natural problem is to develop an elementary collapse operator that controls the perturbation in the persistence diagrams. Other such operations worthy of investigation include vertex removal, or arbitrary vertex identification in a CW-complex.

Acknowledgments. The authors would like to thank the National Elevation Dataset for their terrain data, the Aim@Shape repository for the models, and the Hera project for their bottleneck distance code [14]. This work was supported by NSF grants CCF-1740761, DMS-1547357 and CCF-1839252.

References

1. Attali, D., Glisse, M., Morozov, D., Hornus, S., Lazarus, F.: Persistence-sensitive simplification of functions on surfaces in linear time. In: TopoInVis Workshop (2009)
2. Barmak, J.A., Minian, E.G.: Strong homotopy types, nerves and collapses. Discrete Comput. Geom. **47**(2), 301–328 (2012)
3. Boissonnat, J., Pritam, S., Pareek, D.: Strong collapse for persistence. In: 26th Annual European Symposium on Algorithms, ESA 2018, pp. 67:1–67:13 (2018)
4. Chazal, F., Cohen-Steiner, D., Glisse, M., Guibas, L.J., Oudot, S.Y.: Proximity of persistence modules and their diagrams. In: Proceedings of the Twenty-Fifth Annual Symposium on Computational Geometry, SCG 2009, pp. 237–246 (2009)
5. Cohen-Steiner, D., Edelsbrunner, H., Harer, J.: Stability of persistence diagrams. In: Proceedings of the Twenty-First Annual Symposium on Computational Geometry, SCG 2005, pp. 263–271 (2005)
6. Dey, T.K., Edelsbrunner, H., Guha, S., Nekhayev, D.V.: Topology preserving edge contraction. Publ. Inst. Math. (Beograd) (NS) **66**(80), 23–45 (1999)
7. Dey, T.K., Fan, F., Wang, Y.: Computing topological persistence for simplicial maps. In: Proceedings of the Thirtieth Annual Symposium on Computational Geometry, SOCG 2014, pp. 345:345–345:354 (2014)
8. Dey, T.K., Hirani, A.N., Krishnamoorthy, B., Smith, G.: Edge contractions and simplicial homology. arXiv e-prints arXiv:1304.0664, April 2013
9. Dey, T.K., Slechta, R.: Edge contraction in persistence-generated discrete Morse vector fields. Comput. Graph. **74**, 33–43 (2018)
10. Edelsbrunner, H., Harer, J.: Computational Topology: An Introduction. American Mathematical Society, Providence (2010)
11. Edelsbrunner, H., Letscher, D., Zomorodian, A.: Topological persistence and simplification. Discrete Comput. Geom. **28**(4), 511–533 (2002)
12. Hoppe, H., DeRose, T., Duchamp, T., McDonald, J., Stuetzle, W.: Mesh optimization. In: Proceedings of the 20th Annual Conference on Computer Graphics and Interactive Techniques, SIGGRAPH 1993, pp. 19–26 (1993)
13. Iuricich, F., De Floriani, L.: Hierarchical forman triangulation: a multiscale model for scalar field analysis. Comput. Graph. **66**, 113–123 (2017)
14. Kerber, M., Morozov, D., Nigmetov, A.: Geometry helps to compare persistence diagrams. J. Exp. Algorithmics **22**, 1.4:1–1.4:20 (2017)
15. Lesnick, M.: The theory of the interleaving distance on multidimensional persistence modules. Found. Comput. Math. **15**(3), 613–650 (2015)
16. Lindstrom, P., Turk, G.: Fast and memory efficient polygonal simplification. In: Proceedings Visualization 1998 (Cat. No. 98CB36276), pp. 279–286 (1998)
17. Robertson, N., Seymour, P.: Graph minors. I. Excluding a forest. J. Comb. Theory Ser. B **35**(1), 39–61 (1983)

One More Step Towards Well-Composedness of Cell Complexes over nD Pictures

Nicolas Boutry[1], Rocio Gonzalez-Diaz[2], and Maria-Jose Jimenez[2](✉)

[1] EPITA Research and Development Laboratory (LRDE), Le Kremlin-Bicêtre, France
nicolas.boutry@lrde.epita.efr

[2] Departamento Matematica Aplicada I, Universidad de Sevilla, Campus Reina Mercedes, 41012 Sevilla, Spain
{rogodi,majiro}@us.es

Abstract. An nD pure regular cell complex K is weakly well-composed (wWC) if, for each vertex v of K, the set of n-cells incident to v is face-connected. In previous work we proved that if an nD picture I is digitally well composed (DWC) then the cubical complex $Q(I)$ associated to I is wWC. If I is not DWC, we proposed a combinatorial algorithm to "locally repair" $Q(I)$ obtaining an nD pure simplicial complex $P_S(I)$ homotopy equivalent to $Q(I)$ which is always wWC. In this paper we give a combinatorial procedure to compute a simplicial complex $P_S(\bar{I})$ which decomposes the complement space of $|P_S(I)|$ and prove that $P_S(\bar{I})$ is also wWC. This paper means one more step on the way to our ultimate goal: to prove that the nD repaired complex is continuously well-composed (CWC), that is, the boundary of its continuous analog is an $(n-1)$-manifold.

1 Introduction

Ensuring that the cellular decomposition K of an object is continuously well-composed (CWC) allows to later work in a more efficient way. For example in the 3-dimensional (3D) setting, it allows to work with surface parameterizations [9]. Besides, K being CWC has also topological computation benefits in the sense that the homology of the object can be deduced from the homology of its boundary [7,8]. Moreover, if the boundary of the object is not a manifold, then it is well-known that the multigrid convergence of some geometrical estimators is slower (see, for example, [14]).

The authors have been working for several years in providing algorithms to "locally repair" the cubical complex $Q(I)$ canonically associated to a given picture I, in order to obtain a cell complex homotopic to $Q(I)$ and CWC.

A first attempt to solve the problem in the 3-dimensional setting (3D) was presented in [10]. In [11–13], the authors provided a satisfactory solution in the 3D case to locally repair $Q(I)$ following a purely combinatorial procedure, obtaining a polyhedral $P(I)$ that is CWC.

Authors' names listed in alphabetical order.

M. Couprie et al. (Eds.): DGCI 2019, LNCS 11414, pp. 101–114, 2019.
https://doi.org/10.1007/978-3-030-14085-4_9

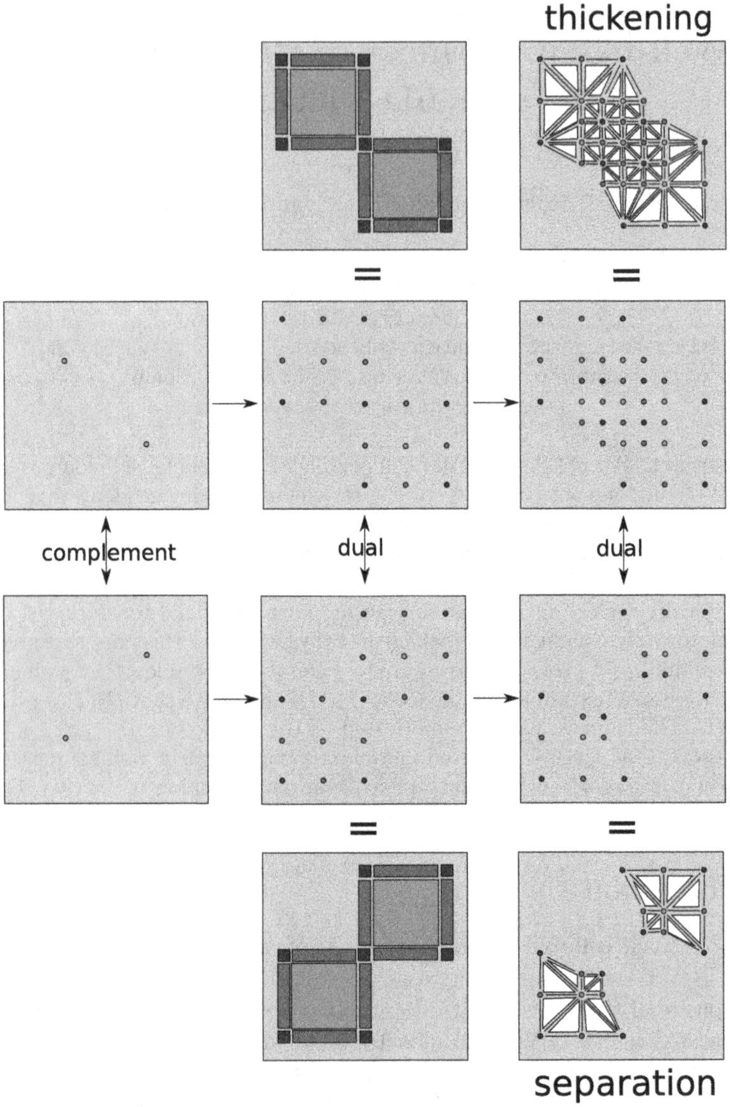

Fig. 1. $P_S(I)$ versus $P_S(\bar{I})$. In the upper half, starting from an image I made of two (green) points (on the left), we compute its associated cubical complex $Q(I)$ (in the middle column). When a critical configuration is detected in I, a reparation procedure is applied on $Q(I)$, leading to a well-composed simplicial complex $P_S(I)$ (on the right). In the lower half, we compute the complement $\bar{I}$ of the image I which leads to the dual cubical complex $Q(\bar{I})$ (in the middle). Now, the reparation procedure (which is described in this paper) "separates" the complex $Q(\bar{I})$ to yield a new simplicial complex $P_S(\bar{I})$ (on the right), which is also weakly well-composed. Notice that the picture of $\bar{I}$ (and hence, $Q(\bar{I})$ and $P_S(\bar{I})$) has been restricted to a two-by-two pixels context and the infinite set of pixels around them are missed. (Color figure online)

In [3], we extend to any dimension (nD) the previous method: Given an nD picture I, we detect non-well-composed configurations (which cause $Q(I)$ not to be CWC) and apply on those configurations a repairing process (see the upside part of Fig. 1 for a sketch of the procedure) consisting in "thickening" the picture representation with some new cells in order to remove the "pinches". The resulting computed complex $P_S(I)$ is a simplicial complex homotopic to $Q(I)$. In that paper, as a first step towards continuous well-composedness, we proved that $P_S(I)$ is weakly well-composed (wWC), what means that there always exists a face-connected path in $P_S(I)$ of n-simplices incident to a common vertex v, joining any two n-simplices incident to v. Some other recent works following the idea of the reparation method are [1] (using the inter-simplex Chebyshev distance), [5] (for face centered cubic grids) and [6] (for body centered cubic grids). A tutorial revisiting the different concepts on well-composedness depending on the nature of the object of study is [2].

Proving that $P_S(I)$ is CWC in any dimension is not an easy task and is still an open problem. CWCness is *self-dual*, that is, the boundary of a compact subset X of $\mathbb{R}^n$ is a manifold iff the boundary of its complement $X^c := \mathbb{R}^n \setminus X$ is a manifold too. Hence, proving that $P_S(I)$ and its *dual* $P_S(\bar{I})$, where $\bar{I}$ is the complement of I (see the lower half of Fig. 1), are both wWC, shows that the type of well-composedness of $P_S(I)$ we are proving here is also self-dual and then stronger than simply weak well-composedness. Consequently, in this paper we go one step towards the final goal, by proving that $P_S(\bar{I})$ is also wWC. The paper is organized as follows: Sect. 2 is devoted to recall the main concepts needed to understand the rest of the paper and we sketch the method developed in [3] to compute $P_S(I)$. In Sect. 3 we introduce the complement of an nD picture denoted by $\bar{I}$ and its canonically associated cubical complex $Q(\bar{I})$. In Sect. 4, we explain the procedure to locally repair $Q(\bar{I})$ to obtain the wWC simplicial complex $P_S(\bar{I})$, satisfying that its continuous analog is the closure of the complement space of $P_S(I)$. This procedure is a modification of the one given in [3] to compute $P_S(I)$. Section 5 is devoted to conclusions and future works.

2 Background

Let $n \geq 2$ be an integer and $\mathbb{Z}^n$ the set of points with integer coordinates in nD space $\mathbb{R}^n$. An *nD binary image* is a pair $I = (\mathbb{Z}^n, F_I)$ where F_I is a finite subset of $\mathbb{Z}^n$ called *foreground* of I. If $I = (\mathbb{Z}^n, F_I)$, with $F_I \subset 4\mathbb{Z}^n$, then we say that I is an *nD picture*. This is only a technical issue: we can encode the cubical complex associated to I in $2\mathbb{Z}^n$, but we need the foreground F_I to be included in $4\mathbb{Z}^n$ so that we have "extra" space to encode the repairing cells that will be added to "thicken" the non-manifold parts of such cubical complex.

For two integers $k \leq k'$, let $[\![k, k']\!]$ denote the set $\{k, k+1, \ldots, k'-1, k'\}$ and $\mathbb{B} = \{e^1, \ldots, e^n\}$ the canonical basis of $\mathbb{Z}^n$.

Given a point $z \in 4\mathbb{Z}^n$ and a family of vectors $\mathcal{F} = \{f^1, \dots, f^k\} \subseteq \mathbb{B}$, the *block of dimension* k associated to the couple $(z, \mathcal{F})$ is the set defined as:

$$B(z,\mathcal{F}) = \left\{ z + \sum_{i \in [\![1,k]\!]} \lambda_i \, f^i \; : \; \lambda_i \in \{0,4\}, \forall i \in [\![1,k]\!] \right\}.$$

This way, a 0-block is a point, a 1-block is an edge, a 2-block is a square, and so on. A subset $B \subset 4\mathbb{Z}^n$ is called a *block* if there exists a couple $(z, \mathcal{F}) \in 4\mathbb{Z}^n \times \mathcal{P}(\mathbb{B})$[1] such that $B = B(z, \mathcal{F})$. We will denote the set of blocks of $4\mathbb{Z}^n$ by $\mathcal{B}(4\mathbb{Z}^n)$.

Two points p, q belonging to a block $B \in \mathcal{B}(4\mathbb{Z}^n)$ are said to be *antagonists* in B if their distance equals the maximum distance using the L^1-norm[2] between two points in B:

$$\|p - q\|_1 = \max \{\|r - s\|_1 \; ; \; r, s \in B\}.$$

The antagonist of a point p in a block $B \in \mathcal{B}(4\mathbb{Z}^n)$ containing p exists and is unique. It is denoted by $\text{antag}_B(p)$. Note that when two points $(x_1, \dots, x_n)$ and $(y_1, \dots, y_n)$ are antagonists in a block of dimension $k \in [\![0, n]\!]$, then $|x_i - y_i| = 4$ for $i \in \{i_1, \dots, i_k\} \subseteq [\![1, n]\!]$ and $x_i = y_i$ otherwise.

Now, let $I = (\mathbb{Z}^n, F_I)$ be an nD picture and $B \in \mathcal{B}(4\mathbb{Z}^n)$ a block of dimension $k \in [\![2, n]\!]$. We say that I contains a *critical configuration* (CC) in the block B if $F_I \cap B = \{p, p'\}$ or $F_I \cap B = B \setminus \{p, p'\}$, with p, p' being two antagonists in B.

DWCness. The nD picture $I = (\mathbb{Z}^n, F_I)$ is said to be *digitally well-composed* (DWC) if for any block B of dimension $k \in [\![2, n]\!]$, the set $F_I \cap B$ is not a critical configuration. Additionally, $\bar{I} = (\mathbb{Z}^n, 4\mathbb{Z}^n \setminus F_I)$ is DWC iff I is DWC (we call this property *self-duality* [4] of DWCness).

We say that two elements x, y of $4\mathbb{Z}^n$ are $2n$-*neighbors* in $4\mathbb{Z}^n$ when there exists some $i \in [\![1, n]\!]$ verifying that $x_i = y_i \pm 4$ and for any $j \in [\![1, n]\!] \setminus \{i\}$, $x_j = y_j$. Also, we call $2n$-*path* joining p, q in $S \subseteq 4\mathbb{Z}^n$ a sequence $(p^0 = p, p^1, \dots, p^k = q)$ of elements of S which verifies for any $i \in [\![0, k-1]\!]$ that p^i and p^{i+1} are $2n$-neighbours in $4\mathbb{Z}^n$.

Proposition 1 ([3]). *If* $I = (\mathbb{Z}^n, F_I)$ *is DWC then, for any block* $B \in \mathcal{B}(4\mathbb{Z}^n)$ *and for any two points* p, q *in* $F_I \cap B$*, there exists a* $2n$*-path in* $F_I \cap B$ *joining* p *and* q.

We use the term *nD cell complex* K to refer to a pure regular cell complex of rank n embedded in $\mathbb{R}^n$. The underlying space (i.e., the union of the n-cells as subspaces of $\mathbb{R}^n$) will be denoted by $|K|$. Regular cell complexes have particularly nice properties, for example, their homology is effectively computable (see [15, p. 243]).

---

[1] The expression $\mathcal{P}(\mathbb{B})$ represents the set of all the subsets of $\mathbb{B}$.
[2] The L^1-norm of a vector $\alpha = (x_1, \dots, x_n)$ is $||\alpha||_1 = \sum_{i \in [\![1,n]\!]} |x_i|$.

The *boundary surface* ∂K of an nD cell complex K is the (n-1)D cell complex composed by the $(n-1)$-cells of K that are faces of exactly one n-cell, together with all their faces.

The *cone (join)* on a simplicial complex K with vertex v, denoted by $v * K$ is the simplicial complex whose simplices have the form $\langle v_0, \ldots, v_\ell, v\rangle$ (where $\langle v_0, \ldots, v_\ell\rangle$ is a simplex of K spanned by the set of points $\{v_0, \ldots, v_\ell\}$), along with all the faces of such simplices.

CWCness. An nD cell complex K is said to be *continuously well-composed* (CWC) if $|\partial K|$ is an $(n-1)$-manifold, that is, each point of $|\partial K|$ has a neighborhood homeomorphic to $\mathbb{R}^{n-1}$ into $|\partial K|$.

Let $\mathcal{S}$ be a set of ℓ-cells of K, $\ell \in [\![1, n]\!]$. We say that two ℓ-cells σ and σ' are *face-connected in* $\mathcal{S}$ if there exists a path $\pi(\sigma, \sigma') = (\sigma_1 = \sigma, \sigma_2 \ldots, \sigma_{m-1}, \sigma_m = \sigma')$ of ℓ-cells of $\mathcal{S}$ such that for any $i \in [\![1, m-1]\!]$, σ_i and σ_{i+1} share exactly one $(\ell-1)$-cell of K. The set $\mathcal{S}$ is *face-connected* if any two ℓ-cells σ and σ' in $\mathcal{S}$ are face-connected in $\mathcal{S}$.

wWCness. An nD cell complex K is *weakly well-composed (wWC)* if for any 0-cell μ in K, the set of n-cells incident to μ, denoted by $\mathcal{A}_K^{(n)}(\mu)$, is face-connected.

A *canonical size-4 n-cube* is a size-4 n-cube centered at a point in $\mathbb{Z}^n$ having their $(n-1)$-faces parallel to the coordinate hyperplanes.

Given an nD picture I, the nD cell complex whose n-cells are the canonical size-4 n-cubes centered at each point in F_I is denoted by $Q(I)$.

We say that $J = (\mathbb{Z}^n, F_J)$ *encodes* $Q(I)$ if F_J is the set of barycenters of the cells in $Q(I)$[3]. We say that $p \in F_J$ *encodes* $\sigma \in Q(I)$ if p is the barycenter of σ. In that case, we denote σ as $\sigma_{Q(I)}(p)$.

Notation 1. *Let $N, M \in \mathbb{Z}$ such that $0 \leq N < M$. Let $p = (x_1, \ldots, x_n) \in \mathbb{Z}^n$. Then $N_M(p)$ denotes the set of indices $\{i \in [\![1, n]\!] \; : \; x_i \equiv N \text{ mod } M\}$.*

Remark 1. $2\mathbb{Z}^n$ can be decomposed into the disjoint sets

$$\mathcal{E}_\ell := \{p \in 2\mathbb{Z}^n : \text{card}(0_4(p)) \text{ is } \ell\}$$

where card(S) is the number of elements in the set S. For example $\mathcal{E}_n = 4\mathbb{Z}^n$ and $\mathcal{E}_0 = (2\mathbb{Z} \setminus 4\mathbb{Z})^n$. Besides, $\mathbb{Z}^n \setminus 2\mathbb{Z}^n$ can be decomposed into the disjoint sets:

$$\mathcal{O}_\ell := \{p \in \mathbb{Z}^n \setminus 2\mathbb{Z}^n \; : \text{card}(0_2(p)) \text{ is } \ell\}.$$

Notice that ℓ-cells of the cubical complex $Q(I)$ are encoded by points in $\mathcal{E}_\ell$. Intuitively, sets $\mathcal{O}_\ell$ will serve as support to build new cells, in the repairing process, around critical configurations.

Proposition 2 ([3]). *The set of points of F_J encoding the faces of a cell $\sigma_{Q(I)}(p)$ is:*

$$\mathcal{D}_{F_J}(p) := \mathcal{D}^+_{F_J}(p) \setminus \{p\} \quad \textit{where } \mathcal{D}^+_{F_J}(p) = \left\{ p + \sum_{j \in 0_4(p)} \lambda_j \, e^j \; : \; \lambda_j \in \{0, \pm 2\} \right\}.$$

---

[3] Observe that $F_J \subset 2\mathbb{Z}^n$.

The set of points encoding the i-faces of $\sigma_{Q(I)}(p)$ will be denoted by $\mathcal{D}^i_{F_J}(p)$ and is, in fact, $\mathcal{D}_{F_J}(p) \cap \mathcal{E}_i$. In particular, the set of points encoding the vertices of $\sigma_{Q(I)}(p)$ is

$$\mathcal{D}^0_{F_J}(p) = \left\{ p + \sum_{j \in 0_4(p)} \lambda_j \, e^j \; : \; \lambda_j \in \{\pm 2\} \right\}.$$

Proposition 3 ([3]). *The set of points of F_J encoding the cells incident to a cell $\sigma_{Q(I)}(p)$ is:*

$$\mathcal{A}_{F_J}(p) := \mathcal{A}^+_{F_J}(p) \setminus \{p\} \textit{ where } \mathcal{A}^+_{F_J}(p) = \left\{ p + \sum_{j \in 2_4(p)} \lambda_j \, e^j : \lambda_j \in \{0, \pm 2\} \right\} \cap F_J.$$

The set of points encoding the i-cells incident to $\sigma_{Q(I)}(p)$ will be denoted by $\mathcal{A}^i_{F_J}(p)$. In particular,

$$\mathcal{A}^n_{F_J}(p) := F_J \cap \left\{ p + \sum_{j \in 2_4(p)} \lambda_j \, e^j \; : \; \lambda_j \in \{\pm 2\} \right\}.$$

Proposition 4 ([3]). *If $I = (\mathbb{Z}^n, F_I)$ is DWC, then $Q(I)$ is wWC.*

Cell Complexes over nD Pictures. A *cell complex over an nD picture I* is an nD cell complex, denoted by $K(I)$, such that there exists a deformation retraction from $K(I)$ onto $Q(I)$.

In [3] we detailed a procedure to "locally repair" the cubical complex $Q(I)$ when $I = (\mathbb{Z}^n, F_I)$ is not DWC to obtain a simplicial complex $P_S(I)$ such that $|P_S(I)|$ is homotopy equivalent to $|Q(I)|$ and $P_S(I)$ is a wWC simplicial complex. The principal steps are summarized below:

First, Procedure 1 from [3] is used to compute the set R of (critical) points of F_J representing the critical cells in $Q(I)$. Basically, we look for critical configurations and for the cells incident to vertices involved in such critical configurations. These cells are the candidates to be repaired in order to obtain the wWC simplicial complex $P_S(I)$. We reproduce below the procedure so that the paper is self-contained.

Once the critical points are computed, extra points are added around them, defining the set F_L that will contain all the vertices of the new construction $P_S(I)$:

$$F_L := F_J \cup \bigcup_{p \in R} S(p),$$

where $S(p) := \left\{ p + \sum_{j \in 2_4(p)} \lambda_j \, e^j \; : \; \lambda_j \in \{0, \pm 1\} \right\}$. These points are classified into sets $\mathcal{C}_\ell$, for $\ell \in [\![0, n]\!]$, defined as follows:

$$\mathcal{C}_n := (\mathcal{E}_n \cap F_L) \cup R \text{ and } \mathcal{C}_\ell := ((\mathcal{E}_\ell \setminus R) \cup \mathcal{O}_\ell) \cap F_L \; \text{ for } \ell \in [\![0, n-1]\!],$$

where $\mathcal{E}_\ell$ and $\mathcal{O}_\ell$ were defined in Remark 1.

Procedure 1. Obtaining the critical points R in F_J (Procedure 1 in [3]).

Input: $I = (\mathbb{Z}^n, F_I)$ and $J = (\mathbb{Z}^n, F_J)$.
Output: The set R of critical points in F_J.
$V := \emptyset;\ R := \emptyset;$
for $B \in \mathcal{B}(4\mathbb{Z}^n)$ *of dimension* $k \in [\![2, n]\!]$ *and* $p \in B$ **do**
  $p' := \text{antag}_B(p);$
  if $(F_I \cap B = \{p, p'\}$ *or* $B \setminus F_I = \{p, p'\})$ **then**
    $V := V \cup \mathcal{D}^0_{F_J}\left(\frac{p+p'}{2}\right)$
  end
end
for $q \in F_J$ *such that* $\mathcal{D}^0_{F_J}(q) \cap V \neq \emptyset$ **do**
  $R := R \cup \{q\}$
end

Finally, F_L is the set of vertices used to compute $P_S(I)$ successively applying the cone join operation $(*)$. For this construction, new sets $\mathcal{D}_{F_L}(p)$, for each $p \in F_L$, are defined.

Definition 1 ([3]). *For $p \in F_L$, define the set $\mathcal{D}_{F_L}(p) := \mathcal{D}^+_{F_L}(p) \setminus \{p\}$ where:*

- *If $p \in \mathcal{C}_0$ then $\mathcal{D}^+_{F_L}(p) := \{p\}$;*
- *If $p \in \mathcal{E}_\ell \setminus R$, for $\ell \in [\![1, n]\!]$, then $\mathcal{D}^+_{F_L}(p) := \mathcal{D}^+_{F_J}(p)$;*
- *If $p \in \mathcal{E}_\ell \cap R$, for $\ell \in [\![1, n]\!]$, then*

$$\mathcal{D}^+_{F_L}(p) := S(p) \sqcup (\mathcal{D}_{F_J}(p) \setminus R) \sqcup \bigsqcup_{r \in \mathcal{D}_{F_J}(p) \cap R} (S(r) \cap \mathcal{N}(p));$$

- *If $p \in \mathcal{O}_\ell$, for $\ell \in [\![1, n-1]\!]$, then $\exists!\ q \in R$ s.t. $p \in S(q)$. We have:*

$$\mathcal{D}^+_{F_L}(p) := (S(q) \cap \mathcal{N}^+(p)) \sqcup (\mathcal{D}_{F_J}(q) \setminus R) \sqcup \bigsqcup_{r \in \mathcal{D}_{F_J}(q) \cap R} (S(r) \cap \mathcal{N}(p));$$

Where $\mathcal{N}^+(p) := \left\{p + \sum_{j \in 0_2(p)} \lambda_j e^j\ :\ \lambda_j \in \{0, \pm 1\}\right\}$ *and* $\mathcal{N}(p) := \mathcal{N}^+(p) \setminus \{p\}$.

Procedure 2. Obtaining the simplicial complex $P_S(I)$ (Procedure 6 of [3]).

Input: The set F_L.
Output: The simplicial complex $P_S(I)$.
$P_S(I) := \{\langle p \rangle : p \in \mathcal{C}_0\};$
for $\ell \in [\![1, n]\!]$ **do**
  for $p \in \mathcal{C}_\ell$ **do**
    let $K_{\mathcal{D}_{F_L}}(p)$ be the set of simplices whose vertices lie in $\mathcal{D}_{F_L}(p)$;
    $P_S(I) := P_S(I) \cup (p * K_{\mathcal{D}_{F_L}}(p))$
  end
end

Proposition 5 ([3]). *The simplicial complex $P_S(I)$ is always wWC.*

3 The Complement of an nD Digital Picture I

The *complement* of an nD picture $I = (\mathbb{Z}^n, F_I)$ is the pair $\bar{I} = (\mathbb{Z}^n, 4\mathbb{Z}^n \setminus F_I)$. Observe that $\bar{I}$ is not an nD picture (as defined in this paper) since $4\mathbb{Z}^n \setminus F_I$ is not finite.

The nD cell complex whose n-cells are the canonical size-4 n-cubes centered at each point in $4\mathbb{Z}^n \setminus F_I$ is denoted by $Q(\bar{I})$.

DWCness is *self-dual* in the sense that an nD picture $I = (\mathbb{Z}^n, F_I)$ is DWC iff its complement $\bar{I} = (\mathbb{Z}^n, 4\mathbb{Z}^n \setminus F_I)$ is DWC. The following proposition is a property of DWCness (see Theorem 1 in [4]) adapted to nD pictures.

Proposition 6 ([4]). *When $I = (\mathbb{Z}^n, F_I)$ is a DWC nD picture, then for any block $B \in \mathcal{B}(4\mathbb{Z}^n)$, and for any $p, q \in B \setminus F_I$ which are antagonists in B, there exists a $2n$-path joining p and q in $B \setminus F_I$.*

Then we obtain the following result.

Proposition 7. *If $I = (\mathbb{Z}^n, F_I)$ is DWC then, for any block $B \in \mathcal{B}(4\mathbb{Z}^n)$ and for any two points p, q in $B \setminus F_I$, there exists a $2n$-path in $B \setminus F_I$ from p to q.*

Proof. Similarly to the proof of Proposition 5 in [3], let $B \in \mathcal{B}(4\mathbb{Z}^n)$ be a block such that $B \setminus F_I$ has at least two points. For any two points $p, q \in B \setminus F_I$, there exists a block $B' \subseteq B$ such that $q = \text{antag}_{B'}(p)$. Then by Proposition 6, there exists a $2n$-path joining p and q in $B' \setminus F_I \subseteq B \setminus F_I$. □

Proposition 8. *Let $J' = (\mathbb{Z}^n, F'_J)$ encoding $Q(\bar{I})$. Then:*

- $F_J \cup F'_J = 2\mathbb{Z}^n$
- *$(\mathbb{Z}^n, F_J \cap F'_J)$ encodes $\partial Q(I)$.*

Proof. First, by construction, $F_J \cup F'_J \subset 2\mathbb{Z}^n$. Let us assume that a point $p \in 2\mathbb{Z}^n$ belongs to $\mathcal{E}_\ell$. Consider $p' = p + \sum_{j \in 2_4(p)} 2e^j$. Then, either $p' \in F_I$ or $p' \in 4\mathbb{Z}^n \setminus F_I$, and hence, p encodes either, an ℓ-face of an n-cube of $Q(I)$ or an ℓ-face of an n-cube of $Q(\bar{I})$. That is, $p \in F_J \cup F'_J$.

Let $p = (x_1, \ldots, x_n) \in \mathcal{E}_{n-1}$ such that p encodes an $(n-1)$-cell of $\partial Q(I)$. Then, without loss of generality, we can assume that $x_i \equiv 0 \bmod 4$ for $i \in [\![1, n-1]\!]$ and $x_n \equiv 2 \bmod 4$. Consider the two points in $4\mathbb{Z}^n$ $q = p + 2e^n$ and $q' = p - 2e^n$. Necessarily, one of them lies in F_I and the other one in $4\mathbb{Z}^n \setminus F_I$, so $p \in F_J \cap F'_J$. If p encodes an ℓ-cell of $\partial Q(I)$, then, p is a face of an $(n-1)$-cell of $\partial Q(I)$ and the previous argument applies to conclude that $p \in F_J \cap F'_J$. Conversely, if $p \in F_J \cap F'_J$ (assume that $p \in \mathcal{E}_\ell$), then p encodes an ℓ-face of an $n-1$-cell in $Q(I)$ which is a face of both an n-cube in $Q(I)$ and an n-cube in $Q(\bar{I})$ (otherwise, it could not happen that $p \in F_J \cap F'_J$). Then, p encodes a cell in $\partial Q(I)$. Notice that the same argument proves that p encodes a cell in $\partial Q(\bar{I})$. □

Proposition 9. *If I is DWC then $Q(\bar{I})$ is wWC.*

Proof. When I is DWC, $\bar{I}$ is DWC by definition of DWCness. Then $Q(\bar{I})$ is wWC by Proposition 4. □

4 The Simplicial Complex $P_S(\bar{I})$

Similar to the work made in [3], we now detail a procedure to "locally repair" the cubical complex $Q(\bar{I})$ when $I = (\mathbb{Z}^n, F_I)$ is not DWC obtaining a simplicial complex $P_S(\bar{I})$ which is always wWC.

First, Procedure 1 is used to compute the set of points R' representing the critical cells in $Q(\bar{I})$, considering as input $I = (\mathbb{Z}^n, F_I)$ and $J' = (\mathbb{Z}^n, F'_J)$.

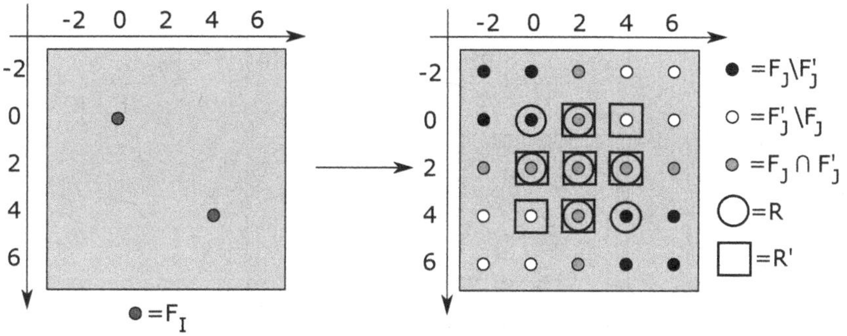

Fig. 2. The difference between R and R'.

Remark 2. In general, $R \cap R' \neq \emptyset, R, R'$ (see Fig. 2).

Second, once the critical points R' are computed, the points of F'_J which do not belong to R' are preserved, the points of R' are replaced by a set of points using an operator $S'(.)$ introduced below. This part of the process is the one that differs from that in [3].

- If $p \in R' \setminus R$ (that is, $\sigma_{Q(\bar{I})}(p)$ is not in $\partial Q(I)$) then $S'(p) := S(p)$.
- If $p \in R' \cap R$ (that is, $\sigma_{Q(\bar{I})}(p)$ is in $\partial Q(I)$) then

$$S'(p) := \left\{ p + \sum_{j \in 2_4(p)} \lambda_j e^j \; : \; \lambda_j \in \{0, \pm 1\} \text{ and } p + \sum_{j \in 2_4(p)} 2\lambda_j e^j \in R' \setminus R \right\}.$$

Figure 3 shows small examples of sets $S'(p)$. Then, the set F'_L is defined as:

$$F'_L := (F'_J \setminus R') \cup \bigcup_{p \in R'} S'(p)$$

Observe that not all the points in F'_J belong to F'_L, contrary to what happened with F_L.

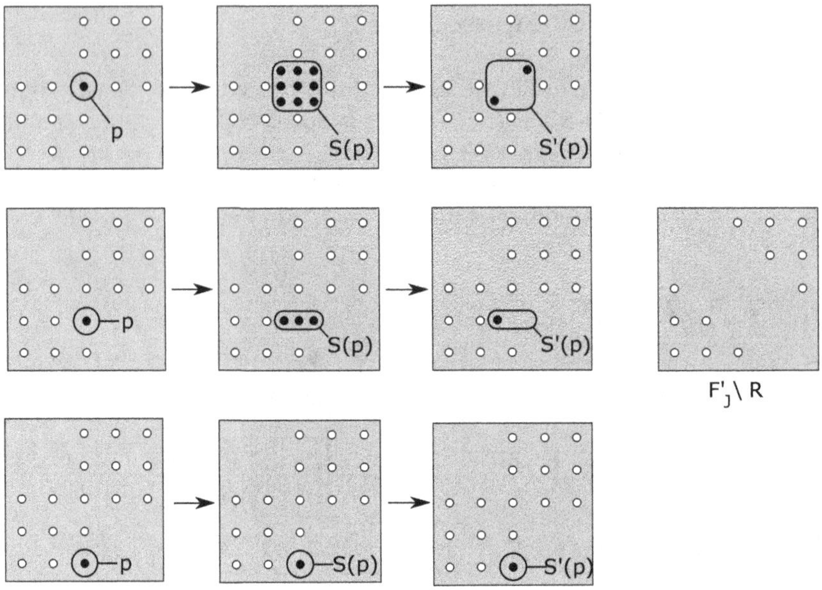

Fig. 3. Computation of $S'(p)$. From F'_J depicted in the upper left part of the figure by the black and white points, we deduce $S(p)$, and then $S'(p)$ which is the subset of $S(p)$ satisfying that they can be written $p + \sum_{j \in 2_4(p)} 2\lambda_j\, e^j \in F'_J \setminus R$, with $\lambda_j \in \{0, \pm 1\}$.

Remark 3. As it happened with F_L in [3], the set F'_L can be decomposed into the disjoint sets:

$$\mathcal{C}'_n := (\mathcal{E}_n \cap F'_L) \cup (R' \setminus R)$$
$$\mathcal{C}'_\ell := ((\mathcal{E}_\ell \setminus R') \cup \mathcal{O}_\ell) \cap F'_L \text{ for } \ell \in [\![0, n-1]\!].$$

Third, to compute $P_S(\bar{I})$, we cannot directly use Procedure 6 of [3] since F'_L has an infinite number of points. Now, the idea is to compute $P_S(\bar{I})$ locally around points in $\mathcal{C}'_n$:

- For each p in $(\mathcal{E}_n \cap F'_L) \setminus R'$, apply Procedure 2 with $\mathcal{D}^+_{F'_L}(p)$[4] as input. This process will produce a simplicial subdivision of the canonical 4-size n-cube.
- For the whole set $R' \setminus R$, apply Procedure 2 with

$$\bigcup_{p \in R' \setminus R} \mathcal{D}^+_{F'_L}(p)$$

as input (a finite set), what will produce the rest of the simplices in $P_S(\bar{I})$.

A complete procedure used to compute the simplicial complex $P_S(\bar{I})$ over an nD picture I which consists in two pixels sharing a vertex, is showed in Fig. 4.

[4] Definition of $\mathcal{D}^+_{F'_L}(p)$ can be obtained from Definition 1 replacing F_L by F'_L and R by R'.

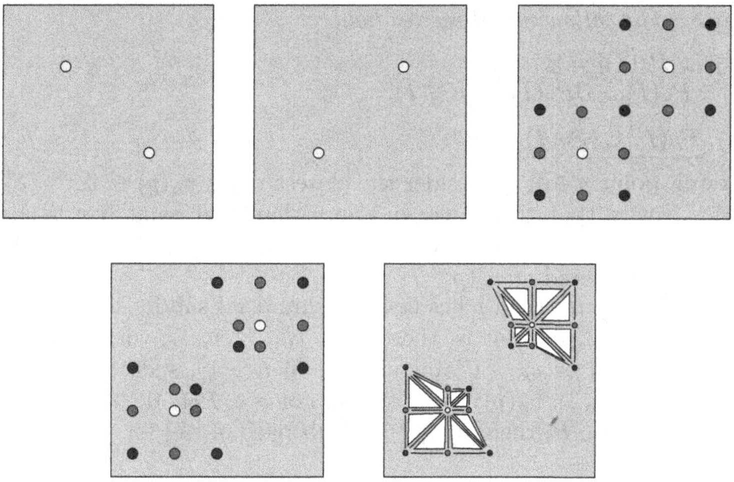

Fig. 4. Computation of $P_S(\bar{I})$: In the raster scan order, we start from I, we deduce $\bar{I}$, we compute F'_J encoding the cells of $Q(\bar{I})$, we construct the set F'_L and finally obtain the simplicial complex $P_S(\bar{I})$ built on the set of vertices F'_L.

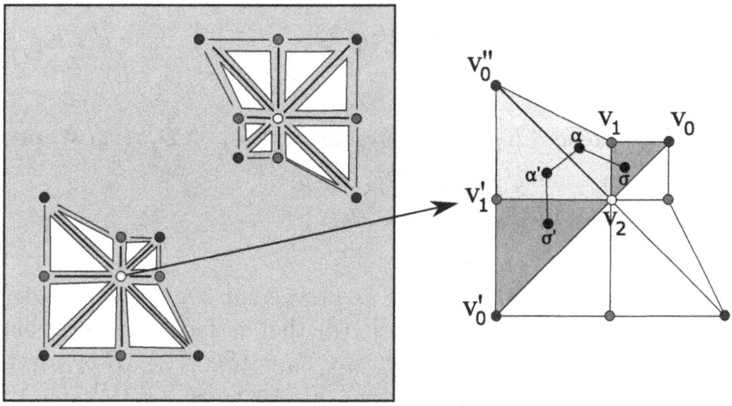

Fig. 5. How to compute the path joining two cells, $\sigma := \langle v_0, v_1, v_2 \rangle$ and $\sigma' := \langle v'_0, v'_1, v_2 \rangle$, both incident to $\langle v_2 \rangle$. Since we are in the case $i \neq i'$, we compute two new cells α and α' which "join" σ and σ': the final path joining σ and σ' is $\pi = \langle \sigma, \alpha, \alpha', \sigma' \rangle$.

Theorem 2. *$P_S(\bar{I})$ is wWC.*

Proof. The proof is the same as the one provided for Theorem 44 in [3] where we proved that for any 0-cell μ in $P_S(I)$, the set of n-cells incident to μ, denoted by $\mathcal{A}^{(n)}_{P_S(I)}(\mu)$, is face-connected. The same happens for any 0-cell μ' in $P_S(\bar{I})$. See an example in Fig. 5. □

Theorem 3. *The following properties hold:*

(a) $|P_S(I)| \cup |P_S(\bar{I})| = \mathbb{R}^n$.
(b) $P_S(I) \cap P_S(\bar{I}) = \partial P_S(I) = \partial P_S(\bar{I})$.

Proof. (a) $|P_S(I)| \cup |P_S(\bar{I})| = \mathbb{R}^n$:

For a given point $p \in \mathbb{R}^n$, consider the closest point $v_n(p) \in 4\mathbb{Z}^n = \mathcal{E}_n$ under L^1-norm. Notice that there may be more than one point if p is shared by several n-cubes; in that case, take any of them. Then p lies inside the 4-size n-cube $c_{v_n(p)}$ centered at $v_n(p)$.

- If $v_n(p) \notin R \cup R'$, then p lies inside a simplicial subdivision of the n-cube encoded by $v_n(p)$ (that is, there exist $v_0, \dots, v_{n-1}$, with $v_k \in \mathcal{C}_k$ for all $k \in [\![0, n-1]\!]$ or $v_k \in \mathcal{C}'_k$ for all $k \in [\![0, n-1]\!]$, such that p lies inside $\sigma = \langle v_0, \dots, v_{n-1}, v_n(p) \rangle$ for $\sigma \in P_S(I)$ or $\sigma \in P_S(\bar{I})$).
- Else, if $v_n(p) \in R$, then $|c_{v_n(p)}|$ is a subspace of $|X|$ for

$$X = \bigcup_{q \in \mathcal{C}_n \cap \mathcal{D}^+_{F_J}(v_n(p))} q * K_{\mathcal{D}_{F_L}}(q).$$

Then p lies in $|q * K_{\mathcal{D}_{F_J}}(q)|$ for some $q \in \mathcal{C}_n \cap \mathcal{D}^+_{F_J}(v_n(p))$.
- Else, $v_n(p) \in R' \setminus R$. Then $|c_{v_n(p)}|$ is a subspace of $|X'|$ for

$$X' = \bigcup_{q \in \mathcal{C}_n \cap \mathcal{D}^+_{F'_J}(v_n(p))} q * K_{\mathcal{D}_{F_L}}(q) \cup \bigcup_{q' \in \mathcal{C'}_n \cap \mathcal{D}^+_{F'_J}(v_n(p))} q' * K_{\mathcal{D}_{F'_L}}(q').$$

Then p lies in $|q * K_{\mathcal{D}_{F_L}}(q)|$ for some $q \in \mathcal{C}_n \cap \mathcal{D}^+_{F'_J}(v_n(p))$ or in $|q' * K_{\mathcal{D}_{F'_L}}(q')|$ for some $q' \in \mathcal{C'}_n \cap \mathcal{D}^+_{F'_J}(v_n(p))$.
Therefore, p lies in $|P_S(I)|$ or in $|P_S(\bar{I})|$.

(b) $P_S(I) \cap P_S(\bar{I}) = \partial P_S(I) = \partial P_S(\bar{I})$:
Let σ be a simplex in $\partial P_S(I)$. Let us prove that $\sigma \in P_S(\bar{I})$. By definition, σ belongs to an $(n-1)$-cell μ of $P_S(I)$ that is face of exactly one n-cell $\mu' = \langle v_0, \dots, v_n \rangle$ of $P_S(I)$. We have: $v_n \in \mathcal{C}_n = (\mathcal{E}_n \cap F_L) \cup R$ by construction and $\mu = \langle v_0, \dots, v_{n-1} \rangle$, with $v_{n-1} \in \mathcal{C}_{n-1}$, otherwise $\mu \notin \partial P_S(I)$.
When $v_n \in \mathcal{E}_n \cap F_L$, then v_n is the barycenter of a 4-size n-cube of $P_S(I)$ and there exists $v'_n \in \mathcal{E}_n \setminus F_L$ that is the barycenter of a 4-size n-cube of $P_S(\bar{I})$ such that $\langle v_0, \dots, v_{n-1}, v'_n \rangle$ is an n-cell of $P_S(\bar{I})$.
When $v_n \in R$ then $v_{n-1} \in S(v_n)$. Since $\mu \in \partial P_S(I)$ then $v_{n-1} \in S'(v_n) \cap \mathcal{C}'_{n-1}$. Therefore, for $\lambda_j \in \{0, \pm 1\}$,

$$v_{n-1} = v_n + \sum_{j \in 2_4(v_n)} \lambda_j e^j \text{ and then } v'_n = v_n + \sum_{j \in 2_4(v_n)} 2\lambda_j e^j \in R' \setminus R.$$

Then μ is an $(n-1)$-face of $\langle v_0, \dots, v_{n-1}, v'_n \rangle \in P_S(\bar{I})$, so $\mu \in P_S(\bar{I})$ and, hence $\sigma \in P_S(\bar{I})$.
The proof for $\partial P_S(\bar{I})$ is analogous.
If $\sigma \in P_S(I) \cap P_S(\bar{I})$, then all the vertices of σ lie necessarily in $\partial P_S(I) = \partial P_S(\bar{I})$. □

5 Conclusions

In this paper we provide a method to locally repair the cubical complex canonically associated to the complement of a given nD picture I obtaining a simplicial complex $P_S(\bar{I})$. We prove that $P_S(\bar{I})$ decomposes the complement space of the continuous analog of $P_S(I)$ and is weakly well-composed. As future works, we plan to study the combinatorial structure of $\partial P_S(I) = \partial P_S(\bar{I})$ and prove that it is a combinatorial $(n-1)$-manifold.

Acknowledgments. This research has been partially supported by MINECO, FEDER/UE under grant MTM2015-67072-P and Instituto de Matematicas de la Universidad de Sevilla (IMUS).

References

1. Bhunre, P.K., Bhowmick, P., Mukherjee, J.: On efficient computation of inter-simplex Chebyshev distance for voxelization of 2-manifold surface. Inf. Sci. (2018)
2. Boutry, N., Géraud, T., Najman, L.: A tutorial on well-composedness. J. Math. Imaging Vis. **60**(3), 443–478 (2018)
3. Boutry, N., Gonzalez-Diaz, R., Jimenez, M.J.: Weakly well-composed cell complexes over nD pictures. Inf. Sci. (2018)
4. Boutry, N., Géraud, T., Najman, L.: How to make nD functions digitally well-composed in a self-dual way. In: Benediktsson, J.A., Chanussot, J., Najman, L., Talbot, H. (eds.) ISMM 2015. LNCS, vol. 9082, pp. 561–572. Springer, Cham (2015). https://doi.org/10.1007/978-3-319-18720-4_47
5. Comic, L., Nagy, B.: A topological 4-coordinate system for the face centered cubic grid. Pattern Recognit. Lett. **83**, 67–74 (2016)
6. Čomić, L., Magillo, P.: Repairing 3D binary images using the BCC grid with a 4-valued combinatorial coordinate system. Inf. Sci. (2018)
7. Dey, T.K., Li, K.: Persistence-based handle and tunnel loops computation revisited for speed up. Comput. Graph. **33**(3), 351–358 (2009)
8. Dey, T.K., Li, K., Sun, J.: On computing handle and tunnel loops. In: IEEE International Conference on Cyberworlds, 357–366. IEEE (2007)
9. Floater, M.S., Hormann, K.: Surface parameterization: a tutorial and survey. In: Dodgson, N.A., Floater, M.S., Sabin, M.A. (eds.) Advances in Multiresolution for Geometric Modelling, pp. 157–186. Springer, Heidelberg (2005). https://doi.org/10.1007/3-540-26808-1_9
10. Gonzalez-Diaz, R., Jimenez, M.-J., Medrano, B.: Well-composed cell complexes. In: Debled-Rennesson, I., Domenjoud, E., Kerautret, B., Even, P. (eds.) DGCI 2011. LNCS, vol. 6607, pp. 153–162. Springer, Heidelberg (2011). https://doi.org/10.1007/978-3-642-19867-0_13
11. Gonzalez-Diaz, R., Jimenez, M.J., Medrano, B.: 3D well-composed polyhedral complexes. Discrete Appl. Math. **183**, 59–77 (2015)
12. Gonzalez-Diaz, R., Jimenez, M.-J., Medrano, B.: Encoding specific 3D polyhedral complexes using 3D binary images. In: Normand, N., Guédon, J., Autrusseau, F. (eds.) DGCI 2016. LNCS, vol. 9647, pp. 268–281. Springer, Cham (2016). https://doi.org/10.1007/978-3-319-32360-2_21

13. Gonzalez-Diaz, R., Jimenez, M.J., Medrano, B.: Efficiently storing well-composed polyhedral complexes computed over 3D binary images. J. Math. Imaging Vis. **59**(1), 106–122 (2017)
14. Lachaud, J.-O., Thibert, B.: Properties of Gauss digitized shapes and digital surface integration. J. Math. Imaging Vis. **54**(2), 162–180 (2016)
15. Massey, W.S.: A Basic Course in Algebraic Topology. Springer, New York (1991)

On the Space Between Critical Points

Walter G. Kropatsch[1(✉)], Rocio M. Casablanca[2], Darshan Batavia[1], and Rocio Gonzalez-Diaz[2]

[1] Pattern Recognition and Image Processing Group 193/03, TU Wien, Vienna, Austria
{krw,darshan}@prip.tuwien.ac.at

[2] Applied Math I, University of Seville, Seville, Spain
{rociomc,rogodi}@us.es

Abstract. The vertices of the neighborhood graph of a digital picture P can be interpolated to form a 2-manifold M with critical points (maxima, minima, saddles), slopes and plateaus being the ones recognized by local binary patterns (LBPs). Neighborhood graph produces a cell decomposition of M: each 0-cell is a vertex in the neighborhood graph, each 1-cell is an edge in the neighborhood graph and, if P is well-composed, each 2-cell is a slope region in M in the sense that every pair of s in the region can be connected by a monotonically increasing or decreasing path. In our previous research, we produced superpixel hierarchies (combinatorial graph pyramids) that are multiresolution segmentations of the given picture. Critical points of P are preserved along the pyramid. Each level of the pyramid produces a slope complex which is a cell decomposition of M preserving critical points of P and such that each 2-cell is a slope region. Slope complexes in different levels of the pyramid are always homeomorphic. Our aim in this research is to explore the configuration at the top level of the pyramid which consists of a slope complex with vertices being only the critical points of P. We also study the number of slope regions on the top.

1 Introduction

Local binary patterns (LBPs) can efficiently recognize local configurations of digital images as the critical points (maxima, minima, saddles) as well as slopes and plateaus. Of the wide range of successful applications of LBPs the classification of textures, segmentation and feature extraction have shown the great potential of these simple features (see, for example, [11]). Adjacent LBPs are redundant through the inequalities shared by the two adjacent windows. This redundancy applies to all edges of monotonic paths along which all bits of the LBP are same. This fact has been used in our previous research [2,3] to produce a hierarchy of successively smaller graphs (combinatorial graph pyramid) such that all critical points are preserved wile providing a multiresolution segmentations of an image. The main differences from most of previous multiresolution structures are that they preserve the critical points and they preserve textures and high frequency

M. Couprie et al. (Eds.): DGCI 2019, LNCS 11414, pp. 115–126, 2019.
https://doi.org/10.1007/978-3-030-14085-4_10

image features allowing image reconstructions with only a small percentage of regions.

At the top of this pyramid we expect only to have critical points and their adjacencies. Since our original graphs are planar and since our reduction algorithms preserve the embedding of the graphs in the plane we automatically generate also the dual graphs built by the regions between the critical points.

In [4], Edelsbrunner et al. presented algorithms for constructing a hierarchy of increasingly coarse Morse-Smale complexes that decompose a piecewise linear 2-manifold M. Morse-Smale regions are defined in the continuous setting and consist in the set of points of M that lie in integral lines with common origin and destination. In that paper, they simplify Morse-Smale complexes by canceling pairs of critical points in order of increasing persistence. Each region of the Morse-Smale complex is a quadrangle with vertices of index 0, 1, 2, 1, in this order around the region. The boundary is possibly glued to itself along vertices and arcs.

In this paper, we define the concept of slope complex which can be seen as an extension of the concept of Morse-Smale complex of Edelsbrunner et al [4]. We compute a combinatorial graph pyramid by merging slope regions only if the resulting region is also a slope. At the top of the pyramid, only slope regions bounded by critical points should remain. We study the possible configurations of slope regions at the top of the pyramid and give bounds for the number of that regions on the top.

The paper is organized as follows. Section 2 is devoted to recall the main concepts needed to understand the rest of the paper. In Sect. 3, we study the configuration of slope complexes which are cell decompositions with 2-cells being slope regions. Section 4 is the main section of the paper. In that section we study the structure (in terms of the number of 2-cells and configurations of the 2-cells) of the cell decomposition with vertices being only critical points. Last section is devoted to conclusions and future work.

2 Preliminaries

Given a grayscale digital picture $P = (\mathbb{Z}^2, F_P, g)$, being F_P a finite subset of $\mathbb{Z}^2$, the intensity of a pixel p of F_P, denoted by $g(p)$, is expressed within a given range between a minimum and a maximum. Without loss of generality we suppose that the range is $[0, 255]$. The Local Binary Pattern $LBP(P)$ [10] is again a grayscale digital picture which represents the texture element at each pixel in P. This is currently the most frequently used texture descriptor[1] with outstanding results in applications ranging from object detection to segmentation and classification [11].

The vertices of the adjacency graph of a digital picture P can be interpolated to form a 2-manifold M with critical points (maxima, minima, saddles), slopes and plateaus being the ones that are recognized by LBPs.

[1] See LBP'2014 Workshop on Computer Vision With Local Binary Pattern Variants: https://sites.google.com/site/lbp2014ws/.

In [2,3], we proposed an equivalent LBP encoding which transfers the code from the pixels to the neighbor relations thus only using one bit per edge of the adjacency graph. That LBP encoding is the one that we follow in this paper.

Below, we provide the background needed to understand the rest of the paper.

Primal and Dual Graphs. A *neighborhood graph*, denoted by G, encodes the 4-adjacency of pixels of P. A vertex v of G is associated to each pixel p of P and it is labeled with the gray value of p, i.e., $g(v) := g(p)$. Vertices of neighboring pixels are connected by edges of G.

The *dual* graph of G is denoted by $\overline{G}$. The vertices of $\overline{G}$ are associated with the faces of G and the edges of $\overline{G}$ corresponds to the border between the two faces in G. In simple words there is there is a one-to-one correspondence between the edges of G and $\overline{G}$ so as the faces of G and vertices of $\overline{G}$ as mentioned in [9, Sect. 4.6].

Thus the operation *edge contraction* in G is equivalent to *edge removal* in $\overline{G}$. Conversely, *edge removal* in G is equivalent to *edge contraction* in $\overline{G}$. The graphs G and $\bar{G}$ are planar if they represent a 2D decomposition of M into regions. In the original picture P, a vertex v in G is a *border vertex* if the number of edges incident to it, denoted by $deg(v)$ is 2 or 3. Otherwise ($deg(v) = 4$), it is an *inner vertex*. A *border edge* connects two border vertices. Otherwise, it is an *inner edge* as mentioned in [9, Sect. 1.3].

Monotonically Increasing (Decreasing) Path. A *path* in G is a finite sequence of vertices $v_1, v_2 \ldots, v_n$ and edges (v_1, v_2), (v_2, v_3), $\ldots$, (v_{n-1}, v_n). A path in G is *monotonically increasing (resp. decreasing)* if $g(v_i) \leq g(v_{i+1})$ (resp. $g(v_i) \geq g(v_{i+1})$), for $1 \leq i < n$. Two monotonically increasing (resp. decreasing) paths can be concatenated if both have the *same orientation*. Observe that if $g(v_i) = g(v_{i+1})$, for $1 \leq i < n$, then the path is a level curve. For simplicity we will use *monotonic paths* irrespective whether it is increasing or decreasing.

LBP Codes. The adjacent vertices of a vertex v in H are compared with v to obtain the LBP code of v. Given a vertex v in H, follow these adjacent vertices along a circle (for example, clockwise orientation). Whenever the gray value of v is greater than the gray value of the adjacent vertex w, i.e., when $g(w) < g(v)$, write LBP value 0. Otherwise (i.e., when $g(w) > g(w)$), write LBP value 1. Observe that the case $g(w) = g(v)$ is not considered when computing LBP codes.

If vertices of G are points (pixels) in F_P, then a direction can be associated to an edge (u, v) in G between vertices $v \neq u$ when $g(u) > g(v)$:

$$\boxed{(u, v) \text{ has direction } u \to v \textbf{ iff } g(u) > g(v).}$$

In this way LBP codes are moved from the vertices to the orientation of edges. As a result the LBP-codes are independent of the degree of the vertices. In this particular case, the one-to-one correspondence between edges in G and edges in $\bar{G}$ can be geometrically obtained by rotating 90^0 (for example clockwise orientation). Then, we can associate a direction to an edge in $\bar{G}$ if the corresponding edge in G also has a direction.

Graph Pyramids. A *(dual) irregular graph pyramid* [6,7] is a stack of successively reduced planar graphs $\mathcal{P} = \{(G_0, \bar{G}_0), \ldots, (G_n, \bar{G}_n)\}$. Each level $(G_k, \bar{G}_k)$, $0 < k \leq n$ is obtained by first contracting edges in selected contraction kernels of G_{k-1}, if the corresponding regions should be merged, and then removing edges in G_{k-1} to simplify redundant structure (such as regions completely bordered by a self-loop or by two parallel edges).

Merging Plateaus. In order to characterize each vertex in the neighborhood graph in terms of its LBP code, we first contract edges in G (i.e., merge regions in $\bar{G}$) having the same gray value. Therefore, after merging plateaus, we obtain a new neighborhood graph denoted by H satisfying that $g(v) \neq g(w)$, for any two adjacent vertices v, w in H. Plateau regions may contain *holes*. They can degenerate when the width of the region shrinks to 0. The degenerate case is a *level curve*. We will contract plateau regions to a single vertex. Every hole in a plateau region generates a self-loop surrounding the hole after contraction.

Critical Vertices. A vertex v of H is a *local maximum* if its LBP code is composed just by 0s. A *local minimum* produces an LBP code only with 1s. The LBP code describes a *slope vertex* if there are exactly two transitions from 0s to 1s or 1s to 0s in the code, when traversed circularly. More transitions identify a *saddle vertex*. In the particular case of v being a border vertex, if the two border edges incident to v produce equal LBP values, it counts as an extra transition (in order to later get monotonic paths along the boundary of the slope complex, see next section). Self-loops are not taken into account when computing the LBP code. Since LBP codes can also be described by the orientation of their incident edges, we can also associate an LBP code to each vertex in $\bar{H}$. Besides, observe that self-loops in H are never oriented whereas self-loops in $\bar{H}$ can be oriented. In the case of a self-loop incident to a vertex $\bar{v}$ in H is oriented, it counts twice when computing the LBP code of $\bar{v}$.

Remark 1. Vertices in $\bar{H}$ can only be slopes and saddles vertices.

Non Well-Composed Configurations. A gray scale digital picture is called *well-composed* [8] if it does not contain the following *non-well composed configuration* of pixels (modulo reflection and 90-degree rotation). Look at the drawings on the left of Fig. 1 where $g(a) < g(b)$, $g(a) < g(c)$, $g(d) < g(b)$ and $g(d) < g(c)$, being a, b, c and d vertices in G.

In order to avoid the non-well composed configurations (see [2,3]), we insert a new *dummy* vertex v adjacent to vertices a, b, c, d in G creating new four regions that correspond to four slope vertices in $\bar{G}$. The new vertex v with new gray value $g(v)$ reflects the relation between a, b, c, d:

$$\begin{aligned} g(v) = &(g(a) + g(b) + g(c) + g(d))/2 \\ &-(\max\{g(a), g(b), g(c), g(d)\} + \min\{g(a), g(b), g(c), g(d)\})/2. \end{aligned}$$

From now on, *to repair* a non-well composed configuration means that such vertex v has been inserted.

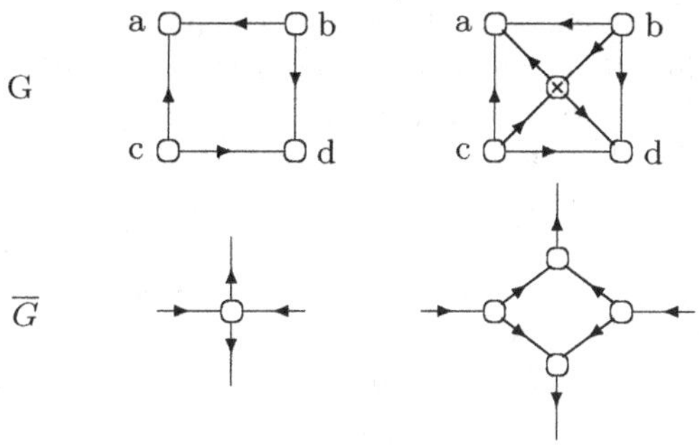

Fig. 1. Repairing non-well composed configurations.

Remark 2. After repairing all non-well composed configurations, we obtain a new neighborhood graph J for which all the vertices in $\bar{J}$ are slope vertices.

3 Slope Complexes

This section is devoted to introduce the most important concept of our paper: slope complexes. Roughly, a slope complex is a cell complex formed by slope regions. We provide a different definition of a slope region which helps us explore the top of the pyramid.

We can naturally associate an *abstract cell complex* $K[G]$ to a neighborhood graph denoted by G, where the 0-cells of $K[G]$ are the vertices of G, the 1-cells of $K[G]$ are the edges (or loops) of G, and the 2-cells of $K[G]$ are the regions of M (which correspond to vertices of $\bar{G}$). The border of a 2-cell is the set of edges (1-cells) delimiting such 2-cell. The *boundary of the cell complex* $K[G]$, denoted by $\partial K[G]$ is the subcomplex of $K[G]$ formed by the border vertices and the border edges of G. See [5] for more details.

Definition 1. *A 2-cell in $K[G]$ is a* slope region *if the corresponding vertex $\bar{v}$ in $\bar{G}$ is a* slope vertex. *Besides, $K[G]$ is a slope complex if all its 2-cells are slope regions.*

After merging plateau regions, we obtain a new cell complex $K[H]$ which is homeomorphic to $K[G]$ (see [1]). The following result provides us an important characterization for slope regions.

Proposition 1. *A 2-cell in $K[H]$ is a slope region if and only if there exist two vertices $u \neq v$ in $K[H]$ such that the border of the 2-cell is composed exactly by two monotonic paths between u and v with the same orientation.*

Proof. Let r be a 2-cell in $K[H]$ and let u, v be two vertices of $K[H]$. Assume first that r is a slope region in $K[H]$. Then, the corresponding vertex $\bar{v}$ in $\bar{H}$ is a slope vertex. It means that there are exactly one transition from 0s to 1s and one from 1s to 0s in the LBP code for $\bar{v}$, when traversed circularly.

Then, the corresponding edges in H can be grouped in two disjoint sets, say E_1 and E_2, where all the edges in each set have the same orientation. These two sets correspond to two monotonic paths. Hence, there exist two different vertices u and v in H which are connected by two monotonic paths. These two paths have the same orientation. By contradiction, suppose that one path is monotonically increasing and the other path is monotonically decreasing. Then, it holds $g(v_i) < g(v_{i+1})$, for $1 \leq i \leq m$, for the first path and $g(v_i) > g(v_{i+1}))$, for $1 \leq i \leq n$, for the second path. Then, the union of both paths generates a directed cycle where $g(u) \leq g(v)$ and also $g(u) \geq g(v)$, which is a contradiction because H contains no plateaus.

Suppose now that the border of r is composed exactly by two monotonic paths between two different vertices u and v with the same orientation. Then, the corresponding vertex $\bar{v}$ in $\bar{H}$ has $deg(\bar{v})$ (here, with loops counted twice) equal to the number of edges that are in both paths. Moreover, computing the LBP code of $\bar{v}$ we add bit 1 for each edge that corresponds to one path and 0 for each edge that correspond to the other. Hence, $\bar{v}$ is a slope vertex and r is a slope region. □

After repairing non-well composed configurations in H, we obtain a new cell complex $K[J]$ which is homeomorphic[2] to $K[H]$. Let us highlight an important property for this level of the pyramid.

Proposition 2. *The cell complex $K[J]$ is a slope complex.*

Proof. Let J be the new neighborhood graph obtained by repairing all the non-well composed configurations in H. Then, by Remark 2, every vertex in $\bar{J}$ is a slope vertex, so every 2-cell in $K[J]$ is a slope region, and therefore, $K[J]$ is a slope complex. □

In order to study the decomposition of a slope complex, we introduce the concept of minimal slope regions.

Definition 2. *A minimal slope region is a slope region whose border has a length of three edges.*

We differentiate between *simple* and *non-simple* minimal slope regions. A simple minimal slope region is bounded by three edges which does not contain self-loop (see Fig. 2a). A non-simple minimal slope region contains a self-loop (see Fig. 2b and c) but the dual face also has degree three.

Figure 2c exchanges the inside and the outside of the self-loop. To follow the boundary of the non-simple slope region showed in Fig. 2b, cross the edge AC twice while in Fig. 2c, the edge is split and forms a double edge. If the inside of the self-loop is empty, it is not a slope region because it is contracted in the construction of the graph pyramid when it appears. Consequently, the double edge does not surround any substructure and can be collapsed to a single edge. Hence the self-loop must be filled with a non-empty subgraph "?" (Fig. 2c).

[2] A repair (Fig. 1) inserts one vertex four edges into $K[H]$ and leaves the remaining $K[H]$ unchanged.

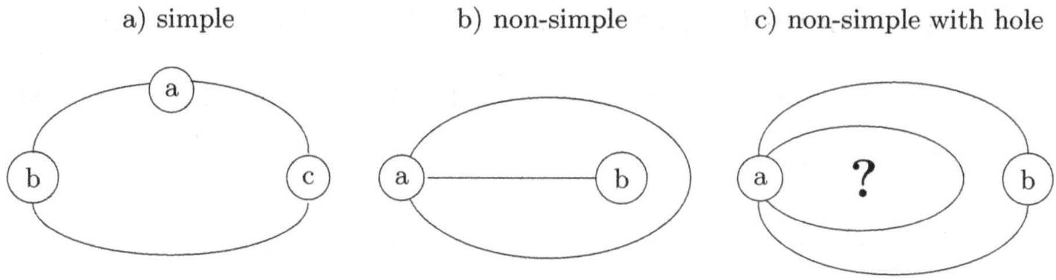

Fig. 2. Types of minimal slope regions

We are interested in finding tight bounds on the number of slope regions in a slope complex. We finish this section with an upper bound on the number of slope regions in a slope complex.

Denote by $|F|$ the number of slope regions (faces), by $|E|$ the number of edges, and by $|V|$ the number of vertices of $K[G]$. Denote by B and I, the number of border edges and inner edges of $\partial K[G]$, respectively. Notice that $E = B \cup I$.

Since $K[G]$ is a cellular decomposition of a given 2-manifold M (with boundary), it is possible to use the Euler characteristic formula in [9, Theorem 4.2.7] (without the outer-face): $|V| - |E| + |F| = 1$.

The next result shows that the number of slope regions in a slope complex depends on the number of vertices and the length of the boundary of such slope complex.

Proposition 3. *Let $K[G]$ be a slope complex where all the slope regions are minimal. Then $|F| = 2|V| - |B| - 2$.*

Proof. Let $K[G]$ be a slope complex where all the slope regions are minimal. Notice that every inner edge bounds two minimal slope regions and every border edge bounds only one. Since $E = B \cup I$ and every slope region is bounded by three edges, then $2|I|+|B| = 3|F|$. This allows us to express the number of edges depending on the number of edges on the boundary and the number of faces: $2|I| + |B| = 3|F|$; $2|I| + 2|B| = 3|F| + |B|$; $2(|I| + |B|) = 3|F| + |B|$; $|E| = (|B| + 3|F|)/2$. Finally, the Euler-equation yields $|V| - (|B| + 3|F|)/2 + |F| = 1$, from which we derive the number of slope regions $|F| = 2|V| - |B| - 2$. □

To see an example where $|F| = 2|V| - |B| - 2$, it is sufficient to consider a neighborhood graph (denoted by G) where all the vertices in $\bar{G}$ are saddles. Then, after repairing all the non-well composed configurations in G, all the vertices in $\bar{G}$ are slope vertices and the corresponding regions in G are minimal slope regions.

As an immediate consequence, next result follows.

Corollary 1. *Let $K[G]$ be a slope complex. Then $|F| \leq 2|V| - |B| - 2$, with equality if and only if all the slope regions are minimal.*

4 Critical Slope Complex

Our goal is to describe the slope complex on the top of the pyramid. As said in the introduction each cell of the complex should only contain critical vertices in its border.

Definition 3. *A critical slope region is a slope region with only critical vertices in its border.*

Definition 4. *Given a neighborhood graph, denoted by G, a slope complex homeomorphic to $K[G]$ with only critical slope regions, is called critical slope complex.*

The following result gives an upper bound on the number of 2-cells in a given critical slope complex.

Proposition 4. *A critical slope complex with $|V|$ critical vertices, $|W|$ of them being in the boundary of the complex, has at most $2|V| - |W| - 2$ critical slope regions.*

Proof. The result is a direct consequence of Corollary 1. Observe that since all the vertices in a critical slope complex are critical, the number of critical vertices in the boundary is equal to the number of edges in the boundary of the complex. □

4.1 On the Number of 2-cells in a Critical Slope Complex

Now, let us study the number of 2-cells in a critical slope complex, depending on the number of edges in the border of each 2-cell.

Proposition 5. *A critical slope complex with $|V|$ critical vertices, $|W|$ of them being in the boundary of the complex, has at most $2(|V| - 1) - |W|$ critical slope regions.*

Proof. Let F denote the number of 2-cells of the critical slope complex (excluding the outer-face), F_d the number of 2-cells with d edges in their border and $|E|$ the number of edges. Clearly $F = \sum_{d\geq 3} F_d$ (since special cases of $d \in \{1, 2\}$ are removed in the construction of the pyramid). Therefore, $2|E| - |W| = \sum_{d\geq 3} dF_d$. Every inner edge $E \setminus W$ bounds two different faces and the edges W on the boundary is part of only one face. Expressing the number of edges $|E|$ of the critical slope complex in terms of the edges on the boundary and the faces yields $|E| = (|W| + \sum_{d\geq 3} dF_d)/2$. Combining it with the Euler characteristic equation we get

$$1 = |V| - (|W| + \textstyle\sum_{d\geq 3} dF_d)/2 + \sum_{d\geq 3} F_d$$
$$2 = 2|V| - |W| - \textstyle\sum_{d\geq 3} dF_d + 2\sum_{d\geq 3} F_d = 2|V| - |W| - \sum_{d\geq 3}(d-2)F_d$$

and then $\sum_{d\geq 3}(d-2)F_d = 2(|V| - 1) - |W|$. □

In other words, we can now calculate the number of faces depending on the total number of critical vertices of the critical slope complex and number of critical vertices of its boundary with the choice of the number of edges in the border of the 2-cells.

For example, assume that we have only critical slope regions with 3 or 4 edges in their border. Then $F_3 + 2F_4 = 2(|V| - 1) - |W|$. So the total number of critical slope regions becomes $F_3 + F_4 = 2(|V| - 1) - |W| - F_4$. That means that the number of critical slope regions is reduced by F_4 with respect to its upper bound. That motivates the study of slope configurations with respect to the number of edges in their border since this would reduce the total number of critical slope regions necessary to cover the complete area.

4.2 Critical Vertices in the Border of a Critical Slope Region

In this subsection we explore how the critical vertices are distributed along the border of a critical slope region.

Proposition 6. *A critical slope region cannot contain more than one minimum or more than one maximum in its border.*

Proof. Let us proceed by contradiction, assume that there are two maxima in the border of a slope region. Then, the corresponding vertex of such region in $\bar{G}$ is a saddle since there are at least four transitions in its LBP code. □

As an immediate consequence of the previous result, next we describe all the possible critical slope regions. Other different configurations are not possible because they would produce no slope vertices in the dual graph, which contradicts the slope region definition. First, we focus on the critical slope regions which are minimal. The smallest critical slope region is a minimal slope region with only critical vertices. The configurations comes from to the minimal slope regions depicted in Fig. 2.

A simple minimal slope region can be formed by two saddles and one extremum. The alternating sequence of orientation of the edges attached to a saddle would allow a minimal slope region with three saddles. However in this case the triangle would form an oriented cycle which contradicts the transitivity of the inequalities (if $A > B$ and $B > C$ then $A > C$). Then, the only possibility for a minimal slope region with only critical vertices is to contain all three types of critical points (Fig. 3a).

For non-simple minimal slope regions, there are two options: Fig. 2b or c. In Fig. 2b, it is a critical slope region if the inner vertex B in Fig. 2b is an extremum and the vertex A is either the opposite extremum or a saddle. In Fig. 2c, the inside of the self-loop cannot be empty, then the vertex B have to be an extremum and the vertex A the opposite extremum or a saddle (see examples in Fig. 3b and c).

Now we describe the critical slope regions which are not minimal, that is, those which have more than three edges in their border. By Proposition 6, the only possibilities are the generalized triangle and generalized quadrangle (see Fig. 4).

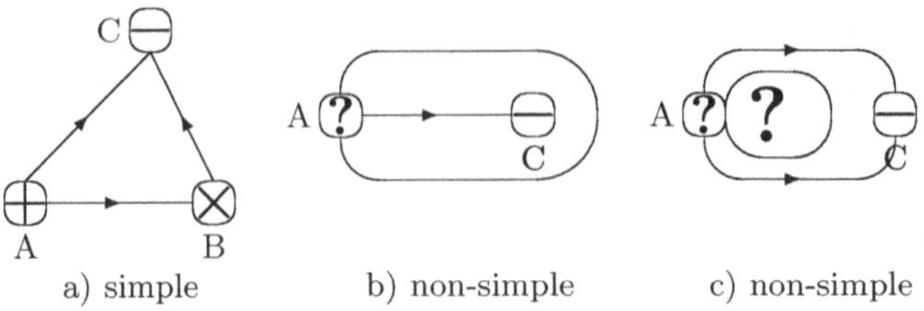

Fig. 3. Types of minimal slope regions with critical vertices in their border

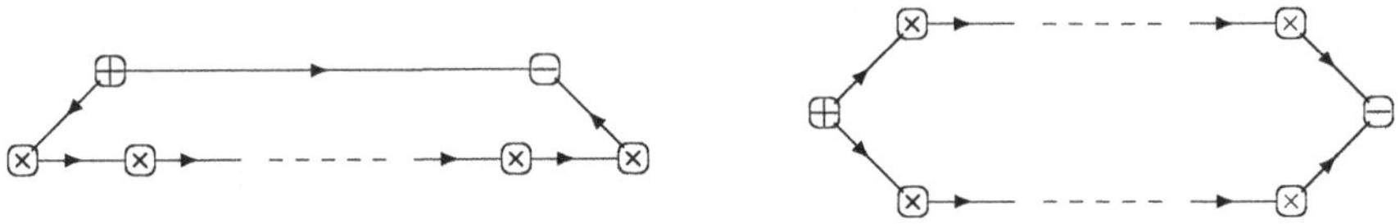

Fig. 4. Critical slope regions which are not minimal.

4.3 Configurations of Critical Slope Regions

We explore now possible configurations of critical slope regions which can compose a critical slope complex.

First, we combine only simple critical slope regions, it means to combine simple triangles. Then, the possible configurations are those a saddle is surrounded by an even and alternating sequences of maxima and minima, or a maximum (minimum) is surrounded by alternating sequences of minima (maxima) and saddles (see Fig. 5). Notice that by combining critical slope regions with minimum number of edges produce the maximum number of 2-cells in the critical slope complex.

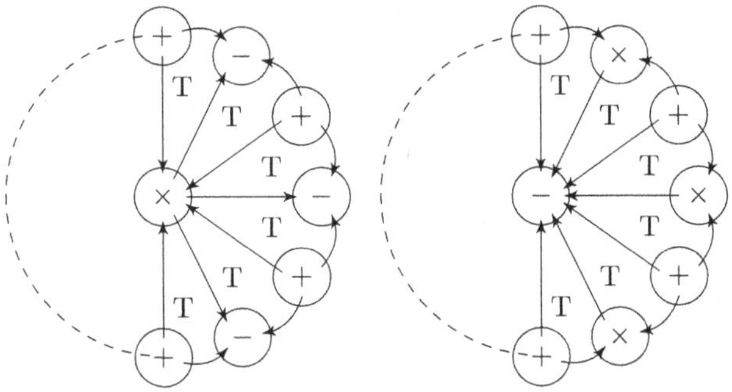

Fig. 5. Combination of simple critical slope regions, denoted by T.

Now let us study possible configurations that can be obtained by combining non-simple critical slope regions shown in Fig. 2b and c.

1. Combination of self-loop inside a double edge anchored at an extremum: Region S_+ in Fig. 6a contains on the outside a self-loop attached to a maximum. This self-loop is required in order to match the outside configuration. Hence it can only be a self-loop anchored at a maximum, another region S_+. The case of S_- in Fig. 6b is symmetric to S_+. This combination contains on the outside an alternating sequence of maximum and minimum. Again we have only a single primitive configuration that can encapsulate S_- and the result is an outer boundary with alternating minima and maxima.

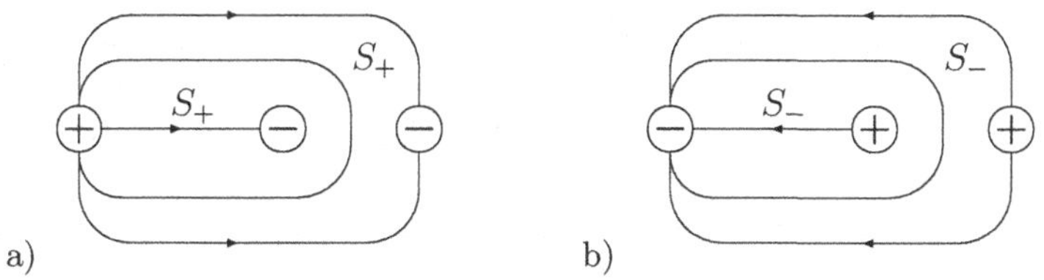

Fig. 6. Combination of non-simple critical slope regions (two maxima and two minima).

2. In Fig. 7, the self-loops attached to saddles show a simple $(S_{\times+}, S_{\times-})$ and a more complex configuration involving two more simple critical slope regions. The simple configurations can be combined similar to the self-loops at extrema:

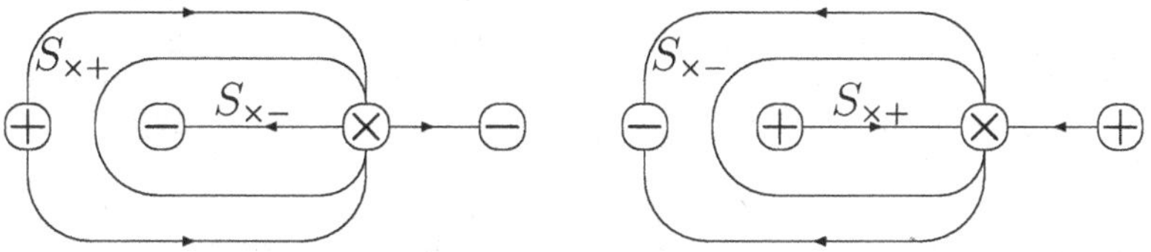

Fig. 7. Combination of non-simple critical slope regions (a saddle with maximum $S_{\times+}$ and minimum $S_{\times-}$).

In this case $S_{\times+}$ and $S_{\times-}$ complement each other to yield as outside a double edge between the saddle and the extremum. The pending edge must be complemented by the opposite extremum. It can be completed by a cycle around $\otimes$ similar to the simple triangles T in Fig. 5.
The complex configurations $(S_{\times+}, T, T), (S_{\times-}, T, T)$ have on the outside a self-loop attached to the saddle. The self-loop of $S_{\times+}$ can be encapsulated into a $S_{\times-}$-configuration and the self-loop $S_{\times-}$ into a $S_{\times+}$- configuration. In both cases the outside is a double edge connecting two different extrema. Hence it can be combined easily with any of the other configurations with alternating extrema on the outside.

3. Alternatively, any of the configurations bounded by a double edge can be recursively embedded into a self-loop generating one more of the above primitive self-loop configurations. The completion towards the outside is then analogously as in the primitive configurations.

5 Conclusion and Future Work

In this paper, our starting point is the neighborhood graph associated to a grayscale digital picture $P = (\mathbb{Z}^2, F_P, g)$ with critical vertices computed from the LBP code of F_P. These critical vertices coincide with the ones of the 2-manifold obtained by interpolating the points of F_P. We study the structure (in terms of the number of 2-cells and configurations of the 2-cells) of the critical cell decomposition K of M in the sense that the vertices in the border of each 2-cell in K are critical.

As future work we plan to show that the minimum number of slope regions in a slope complex is the one of the critical slope complex and that critical slope complexes are in some sense unique with respect to the initial configuration of the critical points of P. In this sense, critical slope complexes can be seen as a cell-complex representation of the image P.

References

1. Gonzalez-Diaz, R., Ion, A., Iglesias-Ham, M., Kropatsch, W.G.: Invariant representative cocycles of cohomology generators using irregular graph pyramids. Comput. Vis. Image Underst. **115**(7), 1011–1022 (2011)
2. Cerman, M., Gonzalez-Diaz, R., Kropatsch, W.: LBP and irregular graph pyramids. In: Azzopardi, G., Petkov, N. (eds.) CAIP 2015. LNCS, vol. 9257, pp. 687–699. Springer, Cham (2015). https://doi.org/10.1007/978-3-319-23117-4_59
3. Cerman, M., Janusch, I., Gonzalez-Diaz, R., Kropatsch, W.G.: Topology-based image segmentation using LBP pyramids. Mach. Vis. Appl. **27**(8), 1161–1174 (2016)
4. Edelsbrunner, H., Harer, J., Zomorodian, A.: Hierarchical morse - smale complexes for piecewise linear 2-manifolds. Discrete Comput. Geom. **30**(1), 87–107 (2003)
5. Hatcher, A.: Algebraic Topology. Cambridge University Press, Cambridge (2002)
6. Kropatsch, W.G.: Building irregular pyramids by dual-graph contraction. IEEE Proceedings on Vision Image and Signal Processing, vol. 142, no. 6, pp. 366–374 (1995)
7. Kropatsch, W.G., Haxhimusa, Y., Pizlo, Z., Langs, G.: Vision pyramids that do not grow too high. Pattern Recognit. Lett. **26**(3), 319–337 (2005)
8. Latecki, L., Eckhardt, U., Rosenfeld, A.: Well-composed sets. Comput. Vis. Image Underst. **61**, 70–83 (1995)
9. Reinhard, D.: Graph Theory. Graduate Texts in Maths. Springer, Heidelberg (1997)
10. Ojala, T., Pietikainen, M., Harwood, D.: A comparative study of texture measures with classification based on featured distributions. Pattern Recognit. **29**(1), 51–59 (1996)
11. Pietikäinen, M., Hadid, A., Zhao, G., Ahonen, T.: Computer Vision Using Local Binary Patterns, vol. 40. Springer, London (2011). https://doi.org/10.1007/978-0-85729-748-8

Rigid Motions in the Cubic Grid: A Discussion on Topological Issues

Nicolas Passat[1](✉), Yukiko Kenmochi[2], Phuc Ngo[3], and Kacper Pluta[4]

[1] Université de Reims Champagne-Ardenne, CReSTIC, Reims, France
nicolas.passat@univ-reims.fr
[2] Université Paris-Est, LIGM, CNRS - ENPC - ESIEE Paris - UPEM, Champs-sur-Marne, France
yukiko.kenmochi@esiee.fr
[3] Université de Lorraine, LORIA, UMR 7503, 54506 Vandoeuvre-lès-Nancy, France
[4] Technion – Israel Institute of Technology, Haifa, Israel

Abstract. Rigid motions on 2D digital images were recently investigated with the purpose of preserving geometric and topological properties. From the application point of view, such properties are crucial in image processing tasks, for instance image registration. The known ideas behind preserving geometry and topology rely on connections between the 2D continuous and 2D digital geometries that were established via various notions of regularity on digital and continuous sets. We start by recalling these results; then we discuss the difficulties that arise when extending them from $\mathbb{Z}^2$ to $\mathbb{Z}^3$. On the one hand, we aim to provide a discussion on strategies that prove to be successful in $\mathbb{Z}^2$ and remain valid in $\mathbb{Z}^3$; on the other hand, we explain why certain strategies cannot be extended to the 3D framework of digitized rigid motions. We also emphasize the relationships that may exist between specific concepts initially proposed in $\mathbb{Z}^2$. Overall, our objective is to initiate an investigation about the most promising approaches for extending the 2D results to higher dimensions.

1 Introduction

Geometric transformations are often involved in 2D and 3D digital image processing such as image registration [26]. Among them, rigid motions, i.e. translations, rotations and their composition, are fundamental ones. When a rigid motion is applied to a digital image, we need to digitize the result in order to map back each point onto the Cartesian grid. In such a point-wise model of rigid motions in $\mathbb{Z}^n$, this final digitization step induces discontinuities in the transformation space. A direct consequence is the loss of geometric and topological invariance during rigid motions in $\mathbb{Z}^n$, as shown in Fig. 1, by contrast with rigid motions in $\mathbb{R}^n$ where geometry and topology are preserved.

This work was funded by the French *Agence Nationale de la Recherche*, grant agreements ANR-15-CE40-0006 (CoMeDiC, https://lama.univ-savoie.fr/comedic) and ANR-15-CE23-0009 (MAIA, http://recherche.imt-atlantique.fr/maia); and by the French *Programme d'Investissements d'Avenir* (LabEx Bézout, ANR-10-LABX-58).

M. Couprie et al. (Eds.): DGCI 2019, LNCS 11414, pp. 127–140, 2019.
https://doi.org/10.1007/978-3-030-14085-4_11

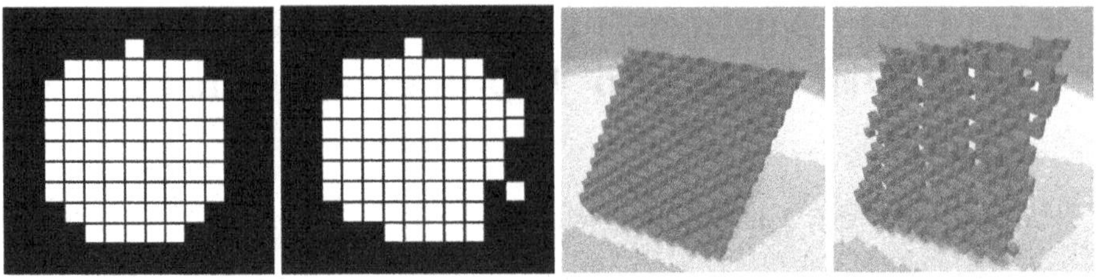

(a) (b)

Fig. 1. Digital images (left) and their images under digitized rigid motions (right). (a) A digital disk. (b) A digital plane. These sets preserve neither topology nor geometry.

These topological issues have been studied in $\mathbb{Z}^2$. In particular, a class of digital images that preserve their topological properties during rigid motions—called regular images—was identified, as well as the "regularization" process based on an up-sampling strategy. This regularization approach allows creating regular images out of non-regular ones [18]. However, this up-sampling strategy in $\mathbb{Z}^2$ cannot be directly extended to $\mathbb{Z}^3$, leading to topological open problems in 3D digitized rigid motions [1]. In this paper, we investigate this topic.

In particular, we consider an alternative approach to regularity, based on quasi-r-regular polygons, which are used as intermediate continuous models of digital shapes for their rigid motions in $\mathbb{Z}^2$ [16]. This approach, which relies on a mixed discrete–continuous paradigm (see [9] for related works), relies on three steps: polygonizing the boundary of a given digital set; applying a rigid motion on the polygon; and digitizing the transformed polygon. Topological issues may occur during the last step. In this context, the class of quasi-r-regular polygons provides guarantees on topological preservation between the polygons and their digitized analog. It should be mentioned that quasi-r-regularity is related to the classical notions of r-regularity [20] and r-half-regularity [25] for continuous sets with smooth and polygonal boundaries, respectively. The main advantage of this approach is its possible extension to 3D [17], by contrast to the notion of regularity.

Our first contribution is a link between the two concepts of regularity and quasi-r-regularity, in 2D. We also show that such a link does not exist in 3D. This difference explains why a straightforward extension of image regularity to 3D does not preserve topology under the point-wise rigid motions in $\mathbb{Z}^3$. This fact emphasizes that quasi-r-regular polyhedra may be a key concept for topology-preserving 3D rigid motion on $\mathbb{Z}^3$. Then, a question arises: which polyhedrization method(s) we can guarantee to generate quasi-r-regular polyhedra from 3D binary images? Our first investigations show that polyhedral isosurfaces generated by the marching cubes method [12]—mostly used for 3D digital images—do not fulfill quasi-r-regularity requirements in $\mathbb{R}^3$, whereas, its 2D analog, namely the marching squares method allows one to generate quasi-r-regular polygons in $\mathbb{R}^2$.

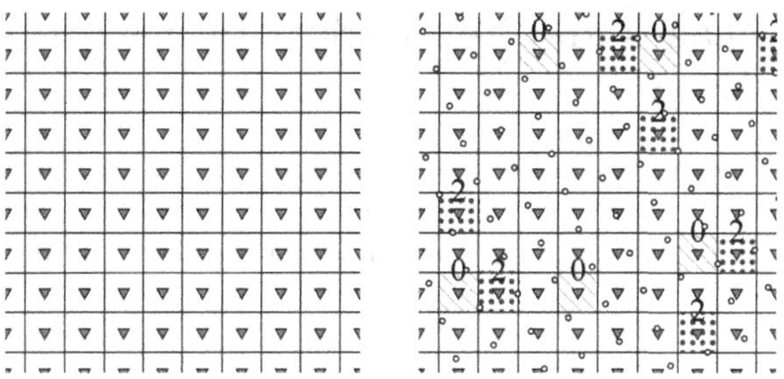

(a) (b)

Fig. 2. (a) Points of $\mathbb{Z}^2$ (triangles), initially located at the center of unit squares of the Cartesian tiling of the Euclidean space. (b) After a rigid motion $\mathfrak{T}$ of these points (circles), some of the unit square cells contain no point (non-surjectivity, green cells) or two points (non-injectivity, red cells); the transformation $\mathcal{T}$ is no longer bijective. (Color figure online)

2 Rigid Motions on $\mathbb{Z}^n$

Let us consider a bounded, closed, connected subset X of the Euclidean space $\mathbb{R}^n$, $n \geq 2$. A rigid motion on $\mathbb{R}^n$ is defined by a mapping

$$\left| \begin{array}{lll} \mathfrak{T} : \mathbb{R}^n & \to & \mathbb{R}^n \\ \quad \mathrm{x} & \mapsto & R\mathrm{x} + \mathrm{t} \end{array} \right. \tag{1}$$

where R is a rotation matrix and $\mathrm{t} \in \mathbb{R}^n$ is a translation vector. Such bijective transformation $\mathfrak{T}$ is isometric and orientation-preserving, so that $\mathfrak{T}(X)$ has the same shape as X i.e., both its geometry and topology are preserved.

If we simply apply a rigid motion $\mathfrak{T}$, such as defined in Eq. (1), to the discrete set $\mathbb{Z}^n$, we generally have $\mathfrak{T}(\mathbb{Z}^n) \nsubseteq \mathbb{Z}^n$. Then, in order to map back onto $\mathbb{Z}^n$, we need a digitization operator

$$\left| \begin{array}{lll} \mathfrak{D} : \mathbb{R}^n & \to & \mathbb{Z}^n \\ \quad (x_1, \ldots, x_n) & \mapsto & \left(\lfloor x_1 + \frac{1}{2} \rfloor, \ldots, \lfloor x_n + \frac{1}{2} \rfloor\right) \end{array} \right. \tag{2}$$

where $\lfloor s \rfloor$ denotes the greatest integer lower than s. A discrete version of $\mathfrak{T}$ is then obtained by

$$\mathcal{T} = \mathfrak{D} \circ \mathfrak{T}_{|\mathbb{Z}^n} \tag{3}$$

so that the point-wise rigid motion of a finite subset X on $\mathbb{Z}^n$ is given by $\mathcal{T}(\mathsf{X})$. Due to the behavior of $\mathfrak{D}$ that maps $\mathbb{R}^n$ onto $\mathbb{Z}^n$, digitized rigid motions are, most of the time, non-bijective (see Fig. 2). Besides, they guarantee neither topology nor geometry preservation of X (see Fig. 1).

3 Regular Images and Topological Invariance Under Rigid Motions

The above problems were studied in [18] and led to the notion of regularity in $\mathbb{Z}^2$ defined in the frameworks of digital topology [7] and well-composed images [11] (see Sect. 3.2). Unfortunately, this notion of regularity is inadequate in $\mathbb{Z}^3$ (Sect. 3.3).

3.1 Digital Topology and Well-Composed Images

Digital topology [7] provides a simple framework for handling the topology of binary images in $\mathbb{Z}^n$. It is shown in [14] that it is compliant with other discrete models (e.g. Khalimsky grids [5] and cubical complexes [8]) but also with continuous notions of topology [15].

Practically, digital topology relies on adjacency relations: two distinct points $\mathsf{p}, \mathsf{q} \in \mathbb{Z}^2$ are k-adjacent if $\|\mathsf{p} - \mathsf{q}\|_\ell \leq 1$ with $k = 2n$ (resp. $3^n - 1$) when $\ell = 1$ (resp. ∞). In the case of $\mathbb{Z}^2$ (resp. $\mathbb{Z}^3$), we retrieve the well-known 4- and 8-adjacency (resp. 6- and 26-adjacency) relations. If two points p, q are k-adjacent, we note $\mathsf{p} \frown_k \mathsf{q}$.

From the reflexive–transitive closure of the k-adjacency relation on a finite subset $\mathsf{X} \subset \mathbb{Z}^n$, we derive the k-connectivity relation on X. It is an equivalence relation whose equivalence classes are called the k-connected components of X. Due to paradoxes related to the discrete version of the Jordan theorem [13], some dual adjacencies are used for X and its complement $\overline{\mathsf{X}} = \mathbb{Z}^n \setminus \mathsf{X}$, namely the $(k, \overline{k})$-adjacencies [22], where $(k, \overline{k}) = (2n, 3^n - 1)$ or $(3^n - 1, 2n)$.

The notion of well-composedness [11] was then introduced to characterize some digital sets X whose structure intrinsically avoids the topological issues of the Jordan theorem in $\mathbb{Z}^2$ and further in higher dimensions.

Definition 1 (Well-composed sets [11]). *We say that a set* $\mathsf{X} \subset \mathbb{Z}^2$ *is weakly well-composed if any* 8*-connected component of* X *is also a* 4*-connected component. We say that* X *is well-composed if both* X *and* $\overline{\mathsf{X}}$ *are weakly well-composed.*

The notion of well-composedness on sets is trivially extended to binary images: an image $I : \mathbb{Z}^2 \to \{0, 1\}$, defined by the finite set $\mathsf{X} = I^{-1}(\{1\}) = \{\mathsf{p} \in \mathbb{Z}^2 \mid I(\mathsf{p}) = 1\}$ is well-composed when X is well-composed.

This definition implies that the boundary[1] of X is a set of 1-manifolds whenever X is well-composed. In particular, a definition of well-composedness in $\mathbb{Z}^n$, $n \geq 3$, is based on this $(n-1)$-manifoldness characterization. This discussion is out of the scope of this paper; the interested reader is referred to [4,10] for more details.

[1] Here, the notion of boundary is related to the continuous embedding of X into the Euclidean space $\mathbb{R}^2$. More precisely, we associate each point $\mathsf{p} \in \mathsf{X}$ with the closed unit square i.e., a Voronoi cell or a pixel centered in p. The union of these squares forms a polygon P in $\mathbb{R}^2$, and we consider the boundary of this polygon. The way of passing from $\mathsf{X} \subset \mathbb{Z}^2$ to $P \subset \mathbb{R}^2$ will be called "polygonization" and more extensively discussed in Sect. 4.

3.2 Topological Invariance Under Rigid Motions in $\mathbb{Z}^2$

Given a binary image I, a rigid motion $\mathcal{T} : \mathbb{Z}^2 \to \mathbb{Z}^2$, and the transformed image[2] $I_\mathcal{T}$ obtained from I and $\mathcal{T}$, a frequent question in image analysis is: *"does $\mathcal{T}$ preserve the topology between I and $I_\mathcal{T}$?"*. It is generally answered by observing the topological invariants of these images.

Among the simplest topological invariants are the Euler-Poincaré characteristics and the Betti numbers. However, these are too weak to accurately model the notion of "topology preservation" between digital images [13]. Therefore, it is necessary to consider stronger topological invariants, e.g. the (digital) fundamental group [6], the homotopy-type (considered via various notions of simple points [2,3] or simple sets [19]), or the adjacency tree [21]. The adjacency tree was considered in [18] and allowed to develop comprehensive proofs of the topology preservation properties. Indeed, in the 2D case, this topology preservation is equivalent to the preservation of the homotopy-type [23], that is the most commonly used topological invariant in 2D image processing.

Definition 2 (Topological invariance [18]**).** *Let I be a binary and well-composed image. We say that I is topologically invariant if any transformed image $I_\mathcal{T}$ has an adjacency-tree isomorphic to that of I.*

In [18], a new notion of regularity was then introduced for 2D images.

Definition 3 (Regular image [18]**).** *Let $I : \mathbb{Z}^2 \to \{0,1\}$ be a non-singular*[3]*, well-composed image. Let $v \in \{0,1\}$. We say that I is v-regular if for any $\mathsf{p},\mathsf{q} \in I^{-1}(\{v\})$, we have*

$$(\mathsf{p} \frown_4 \mathsf{q}) \Rightarrow (\exists \boxplus \subseteq I^{-1}(\{v\}), \mathsf{p},\mathsf{q} \in \boxplus) \tag{4}$$

where $\boxplus = \{x, x+1\} \times \{y, y+1\} \subset \mathbb{Z}^2$. We say that I is regular if it is both 0- and 1-regular.

The regularity—which strengthens the notion of well-composedness—provides sufficient conditions for topological invariance under rigid motions.

Theorem 1 ([18]**).** *An image $I : \mathbb{Z}^2 \to \{0,1\}$ is topologically invariant if it is regular.*

3.3 Topological Alterations Under Rigid Motions on $\mathbb{Z}^3$

In $\mathbb{Z}^2$, Definition 3 describes a regular set X (resp. its complement $\overline{\mathsf{X}}$) as a cover of 2×2 squares that must locally intersect everywhere. Intuitively, the extension of

[2] In practice, we consider the backward transformation model such that $\mathcal{T} = \mathfrak{D} \circ (\mathfrak{T}^{-1})_{|\mathbb{Z}^2}$ rather than Eq. (3), so that $\mathcal{T}$ is surjective. This means that the transformed image $I_\mathcal{T} = I \circ \mathfrak{D} \circ (\mathfrak{T}^{-1})_{|\mathbb{Z}^2} = I \circ \mathcal{T}$ has no point with either no or double/conflicted values.

[3] An image I is singular if $\exists \mathsf{p} \in \mathbb{Z}^2, \forall \mathsf{q} \in \mathbb{Z}^2, (\mathsf{q} \frown_4 \mathsf{p}) \Rightarrow (I(\mathsf{p}) \neq I(\mathsf{q}))$.

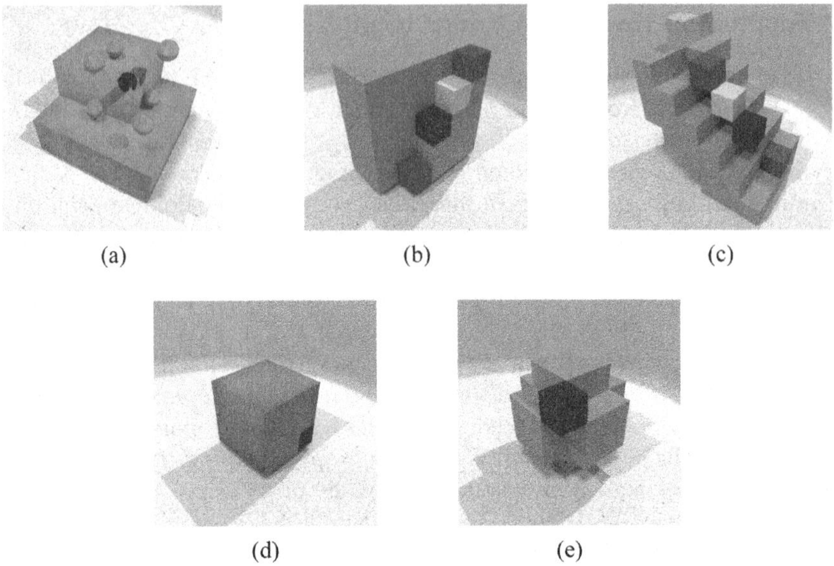

Fig. 3. (a) A sample of a regular set $\mathsf{X} \subset \mathbb{Z}^3$, illustrated as its voxel polyhedron $P \subset \mathbb{R}^3$ (in green); X is 6-connected. Let us consider a rigid motion $\mathfrak{T}^{-1}$ of $\mathbb{Z}^3$, a part of which (transformation of a point and its 6-adjacent neighborhood) is illustrated by blue dots. The central point (in dark blue) lies in P, whereas all of its 6-adjacent points (in light blue) do not. With such a transformation, $[\mathfrak{D}\circ(\mathfrak{T}^{-1})]^{-1}(\mathsf{X})$ is not 6-connected anymore. (b,c) A counterexample to the topology-preservation of a part of a 3D regular image under rigid motion: (b) a part of a regular set, which is 6-connected; (c) the transformed set, obtained after applying a rigid motion, which is no longer 6-connected; see, e.g. the blue voxel, which has all of its 6-adjacent points in the background. (d,e) Another counterexample to the topology-preservation of a 3D regular image under rigid motion: (d) a regular set, which is 6-connected; (e) the transformed set, which is no longer 6-connected (see the blue voxel). (Color figure online)

this definition to $\mathbb{Z}^3$ would consist of considering a cover of $2 \times 2 \times 2$ cubes, that would also locally overlap everywhere. One may expect that a regular image in $\mathbb{Z}^3$ would also be topologically invariant. However, this is false, in general.

Indeed, for any regular image I containing a connected set X (composed of at least one $2 \times 2 \times 2$ cube), we can find an ad hoc transformation $\mathcal{T}$ and a point $\mathsf{p} \in \mathbb{Z}^3$ such that $I_{\mathcal{T}}(\mathsf{p}) = 1$ whereas for any $\mathsf{q} \frown_6 \mathsf{p}$ we have $I_{\mathcal{T}}(\mathsf{q}) = 0$. An example of such a case is illustrated in Fig. 3(a). This provides us with counterexamples to the putative extension of Theorem 1 to the 3D case. For an illustration, we refer the reader to Fig. 3(b, c).

This implies that the notion of regularity reaches its limit of validity[4] in $\mathbb{Z}^2$. Alternative approaches are then required to handle the case of topology preservation under rigid motions in higher dimensions.

4 Quasi-regular Polytopes and Their Digitization

As discussed previously, the topological properties of digital sets in $\mathbb{Z}^n$ may be altered by rigid motions. This is due, in particular, to the process (Eq. (3)) that aims to map back the transformed result from $\mathbb{R}^n$ to $\mathbb{Z}^n$. In practice, this issue is the same as the problem of digitization encountered for defining the digital analog of a continuous set.

Recently, the notion of quasi-r-regularity [16] was introduced together with an algorithmic scheme in order to perform rigid motions on digital sets in $\mathbb{Z}^2$. The scheme relies on the use of intermediate modeling of a 2D digital set as a piecewise affine subset of $\mathbb{R}^2$, namely a polygon. The rigid motion transforms this polygon, and a result in $\mathbb{Z}^2$ is then retrieved by a final digitization of the transformed polygon. The polygon in $\mathbb{R}^2$ and its digitized analog in $\mathbb{Z}^2$ have the same topology if the polygon is quasi-r-regular.

Then, the use of an intermediate continuous model allows one to avoid the alterations induced by the standard pointwise definition of rigid motions that led to the difficulties identified in the case of regularity (see Sect. 3.2).

We recall the definitions of quasi-r-regularity that was initially defined in $\mathbb{R}^2$ [16] and then extended to $\mathbb{R}^3$ [17]. These definitions and associated results are developed in the case of simply connected (i.e. connected, without tunnels/holes) digital sets.

Definition 4 (Quasi-r-regularity [16,17]**).** *Let $X \subset \mathbb{R}^n$ $(n = 2,3)$ be a bounded, simply connected set. We say that X is quasi-r-regular with margin $r' - r$ (with $r' \geq r > 0$) if it satisfies the following four properties:*

- *$X \ominus B_r$ is non-empty and connected,*
- *$\overline{X} \ominus B_r$ is connected,*
- *$X \subseteq X \ominus B_r \oplus B_{r'}$,*
- *$\overline{X} \subseteq \overline{X} \ominus B_r \oplus B_{r'}$,*

with $\oplus, \ominus$ the standard dilation and erosion operators and $B_r, B'_r \subset \mathbb{R}^n$ the closed balls of radius r and r', respectively.

The Gauss digitization of a quasi-r-regular set $X \subset \mathbb{R}^2$ (namely $X \cap \mathbb{Z}^2$) is a well-composed set that remains simply connected, thus preserving its topological properties from $\mathbb{R}^2$ to $\mathbb{Z}^2$.

[4] Beyond the limitations of regularity, the notion of topological invariance of Definition 2 is also insufficient in $\mathbb{Z}^3$. Indeed, the adjacency tree cannot model topological patterns that appear in $\mathbb{Z}^3$, such as the tunnels: e.g. a sphere and a torus have isomorphic adjacency trees. Considering stronger topological invariants, e.g. homotopy type or fundamental group, becomes mandatory.

Proposition 1 (**[16]**). *If $X \subset \mathbb{R}^2$ is quasi-1-regular with margin $\sqrt{2}-1$, then $\mathsf{X} = X \cap \mathbb{Z}^2$ and $\overline{\mathsf{X}} = \overline{X} \cap \mathbb{Z}^2$ are both 4-connected. In particular, $\mathsf{X} \subset \mathbb{Z}^2$ is well-composed.*

In [17], a similar result[5] was obtained for convex quasi-r-regular sets of $\mathbb{R}^3$. This result is extended hereafter to any quasi-r-regular sets of $\mathbb{R}^3$.

Proposition 2 (Extended from [17]). *If $X \subset \mathbb{R}^3$ is quasi-1-regular with margin $\frac{2}{\sqrt{3}} - 1$, then $\mathsf{X} = X \cap \mathbb{Z}^3$ and $\overline{\mathsf{X}} = \overline{X} \cap \mathbb{Z}^3$ are both 6-connected.*

Proof. We only prove the 6-connectedness of X; the same reasoning holds for $\overline{\mathsf{X}}$. Let us first prove that $(X \circ B_1) \cap \mathbb{Z}^3$ is 6-connected. Let p and q be two distinct points of $(X \circ B_1) \cap \mathbb{Z}^3$. Let B_1^{p} and B_1^{q} be two balls of radius 1, included in $X \circ B_1$ and such that $\mathsf{p} \in B_1^{\mathsf{p}}$ and $\mathsf{q} \in B_1^{\mathsf{q}}$ (such balls exist, from the definition of opening). Let b_{p} and b_{q} be the centers of B_1^{p} and B_1^{q}, respectively. We have $b_{\mathsf{p}}, b_{\mathsf{q}} \in X \ominus B_1$, from the definition of erosion. Since $X \ominus B_1$ is connected in $\mathbb{R}^3$, there exists a continuous path Π from b_{p} to b_{q} in $X \ominus B_1$. Note that for any ball B_1, we always have $B_1 \cap \mathbb{Z}^3$ non-empty and 6-connected; in particular it contains at least two points of $\mathbb{Z}^3$. For a value $\varepsilon > 0$ small enough, two balls B_1 and B_1' with centres distant of ε are such that $B_1 \cap B_1' \cap \mathbb{Z}^3 \neq \emptyset$. As a consequence, the union $\bigcup_{b \in \Pi} B_1(b) \cap \mathbb{Z}^3$ (with $B_1(b)$ the ball of center b) is a 6-connected set of $\mathbb{Z}^3$, and p, q are then connected in $(X \circ B_1) \cap \mathbb{Z}^3$. Our purpose is then to prove that any integer point p in $X \setminus (X \circ B_1)$ is 6-adjacent to a point of $(X \circ B_1) \cap \mathbb{Z}^3$. Let $\mathsf{p} \in X \setminus (X \circ B_1)$ be such point. From Definition 4, we have $\mathsf{p} \in X \subseteq X \ominus B_1 \oplus B_{\frac{2}{\sqrt{3}}}$. Then, from the definition of dilation, there exists $b \in X \ominus B_1$ such that b is the center of a ball $B_{\frac{2}{\sqrt{3}}}(b)$ of radius $\frac{2}{\sqrt{3}}$, and p is a point within this ball. The distance between b and p is lower than $\frac{2}{\sqrt{3}}$. As b is a point of $X \ominus B_1$, it is also the center of a ball $B_1(b)$ of radius 1 included in $X \circ B_1$. From the definition of adjacency, any point q being 6-adjacent to p belongs to the sphere $S_1(\mathsf{p})$ of radius 1 and center p. Let us consider the intersection D between $S_1(\mathsf{p})$ and $B_1(b)$. This set D is a spherical dome, namely a part of the sphere $S_1(\mathsf{p})$ with a circular boundary C. This set C also corresponds to the intersection of $S_1(\mathsf{p})$ and the 2D plane orthogonal to the line $(b\mathsf{p})$ and intersecting the segment $[b\mathsf{p}]$ at an equal distance lower than $\frac{1}{\sqrt{3}}$ from both b and p. Then, the radius of this circle C is greater than $\sqrt{1^2 - (\frac{1}{\sqrt{3}})^2} = \frac{\sqrt{6}}{3}$. In particular, C encompasses an equilateral triangle of edge length $\sqrt{2}$. As a consequence, the spherical dome D of $S_1(\mathsf{p})$ bounded by this circle always contains at least one point $\mathsf{q} \frown_6 \mathsf{p}$. As such point q lies in $(X \circ B_1) \cap \mathbb{Z}^3$, it follows that X is 6-connected. □

Remark 1. The value r' (see Definition 4), required to define the margin $r' - 1$ for quasi-1-regular sets, is $\sqrt{2} = \frac{2}{\sqrt{2}}$ in $\mathbb{Z}^2$ and $\frac{2}{\sqrt{3}}$ in $\mathbb{Z}^3$. The above proof allows

[5] Erratum: In [17], the proposed proof contains an error. The 6-connectedness of $(X \circ B_1) \cap \mathbb{Z}^3$ is claimed by using [24, Theorem 16] with an invalid value. We correct this error in the proposed proof of Proposition 2, that does no longer rely on [24, Theorem 16].

us to understand that in $\mathbb{Z}^n$, the required value r' is $\frac{2}{\sqrt{n}}$. Indeed, the crucial part of the proof is to ensure that a point p of $X \setminus (X \circ B_1) \cap \mathbb{Z}^n$ remains $2n$-adjacent to points of $(X \circ B_1) \cap \mathbb{Z}^n$. To this end, let us consider a $(n-1)$-simplex of $\mathbb{R}^n$, whose vertices are spatially organized as the n points induced by the orthonormal basis of $\mathbb{R}^n$. This $(n-1)$-simplex must be encompassed by the $(n-2)$-sphere C that is the intersection between the $(n-1)$-sphere $S_1(\mathsf{p})$ of center p and radius 1 and the hyperplane orthogonal to the segment $[b\mathsf{p}]$ and equidistant to p and a point $b \in X \ominus B_1$ defined as the centre of a ball $B_{\frac{2}{\sqrt{n}}}(b)$ that contains p. Note that p is, by construction, at a distance r' from b. The distance between the barycentre of this $(n-1)$-simplex and its vertices, i.e. the radius of the $(n-1)$-sphere C is $\sqrt{1-\frac{1}{n}}$. Since each point of C is at a distance 1 of p while the center of C is on the segment $[b\mathsf{p}]$ at a distance $\frac{r'}{2}$ from p, it follows that the distance between p and b, namely r', is $\frac{2}{\sqrt{n}}$.

The following property is a direct corollary of this remark and it provides a dimensional limit of validity for the notion of quasi-1-regularity.

Property 1. In $\mathbb{R}^4$, the value r' required for quasi-1-regularity is $\frac{2}{\sqrt{4}} = 1$. Then, the margin $r' - r$ is equal to 0, in other words, no margin is permitted. The notion of quasi-r-regularity then becomes similar to that of r-regularity [20]. In particular, only smooth sets of $\mathbb{R}^4$ can be quasi-1-regular.

5 Links Between Regularity and Quasi-*r*-Regularity: The Cubic Polygonal Model

We now investigate the links between the notions of regularity [18] (Sect. 3) and quasi-r-regularity [16,17] (Sect. 4). Note that we still focus on simply connected sets.

5.1 2D Case

The paradigm of quasi-r-regularity for rigid motion of digital sets $\mathsf{X} \subset \mathbb{Z}^2$ acts in three steps. First, a polygon $P \subset \mathbb{R}^2$ is defined as a continuous representation of X. Note that there exist various ways of carrying out polygonization. The main constraint is the coherence between the polygon P and its digital analog X. In particular, it is important to satisfy $P \cap \mathbb{Z}^2 = \mathsf{X}$. Second, the polygon P is transformed by $\mathfrak{T}$ (Eq. (1)). In other words, we build a new polygon $P_{\mathfrak{T}} = \mathfrak{T}(P) = \{\mathfrak{T}(x) \mid x \in P\}$. Third, the transformed polygon $P_{\mathfrak{T}} \subset \mathbb{R}^2$ is digitized to map the result back onto $\mathbb{Z}^2$. To this end, we use the Gauss digitization model, i.e. we set $\mathsf{X}_{\mathcal{T}} = P_{\mathfrak{T}} \cap \mathbb{Z}^2 = \mathfrak{T}(P) \cap \mathbb{Z}^2$.

When considering the notion of regularity, at the first sight, the paradigm for rigid motion of digital sets $\mathsf{X} \subset \mathbb{Z}^2$ may appear as a different one. It is however the same. In order to compute the transformed object $\mathsf{X}_{\mathcal{T}}$ from X, we do not use the forward transformation model, but the backward one. More precisely, we define $\mathsf{X}_\mathcal{T}$ as $[\mathfrak{D} \circ (\mathfrak{T}^{-1})]^{-1}(\mathsf{X})$, that is $\mathsf{X}_\mathcal{T} = \{x \in \mathbb{Z}^2 \mid \mathfrak{D} \circ (\mathfrak{T}^{-1})(x) \in \mathsf{X}\}$. But this formula is equivalent[6] to $\mathsf{X}_\mathcal{T} = \mathfrak{T}(P_\square(\mathsf{X})) \cap \mathbb{Z}^2$ where $P_\square(\mathsf{X}) \subset \mathbb{R}^2$ is the polygon defined as $P_\square(\mathsf{X}) = \mathsf{X} \oplus \square$, with $\square \subset \mathbb{R}^2$ the closed, unit square centered on $(0, 0)$. In other words, we implicitly apply the three-step polygonization-based algorithm involved in the context of quasi-r-regularity, with a specific kind of polygonization, called cubic polygonization. This polygonization associates a digital set $\mathsf{X} \subset \mathbb{Z}^2$ with its set of pixels, i.e. Voronoi cells in $\mathbb{R}^2$. In particular, it is plain that such polygonization satisfies $P_\square(\mathsf{X}) \cap \mathbb{Z}^2 = \mathsf{X}$.

The question which now arises is to determine whether a *regular* digital set X leads to a *quasi-*1*-regular* polygon $P_\square(\mathsf{X})$. The answer is negative; this emphasizes the fact that quasi-1-regularity is a sufficient, yet non-necessary condition for topology preservation.

Property 2. The regularity of a simply connected set $\mathsf{X} \subset \mathbb{Z}^2$ does not imply the quasi-1-regularity of $P_\square(\mathsf{X})$.

To prove this property, it is sufficient to exhibit a counterexample. A simple one is the object $\mathsf{X} \subset \mathbb{Z}^2$ defined as the union of two 2×2 patterns $\boxplus$ intersecting in exactly one point; for instance, $\mathsf{X} = \{(0, 0), (1, 0), (0, 1), (1, 1), (2, 1), (1, 2), (2, 2)\}$. This set is obviously regular, but the associated polygon $P_\square(\mathsf{X})$ is not quasi-1-regular. Indeed, we have $P_\square(\mathsf{X}) \ominus B_1 = \{(\frac{1}{2}, \frac{1}{2}), (\frac{3}{2}, \frac{3}{2})\}$, which is composed of two points in $\mathbb{R}^2$ and is non-connected.

In [18] the strategy for building a regular set from a well-composed set X was to up-sample X, i.e. to define a new set $\mathsf{X}_2 \in \mathbb{Z}^2$ such that $(x, y) \in \mathsf{X} \Leftrightarrow (2x, 2y) + \{0, 1\} \times \{0, 1\} \subseteq \mathsf{X}_2$. By applying the same strategy on a regular set X, we can build an up-sampled set $\mathsf{X}_2 \subset \mathbb{Z}^2$ which is still regular, but also quasi-1-regular. In other words, with the cubic polygonization model, regularity implies quasi-r-regularity—up to an up-sampling realized by doubling the resolution of the Cartesian grid.

Proposition 3. *If a simply connected set $\mathsf{X} \subset \mathbb{Z}^2$ is regular, then $\mathsf{X}_2 \subset \mathbb{Z}^2$ is regular and $P_\square(\mathsf{X}_2) \subset \mathbb{R}^2$ is quasi-*1*-regular with margin $\sqrt{2} - 1$.*

Proof. The regularity of X_2 is obvious. We show the non-vacuity and connectedness of $P_\square(\mathsf{X}_2) \ominus B_1$ and the fact that $P_\square(\mathsf{X}_2) \subseteq P_\square(\mathsf{X}_2) \ominus B_1 \oplus B_{\sqrt{2}}$; the same reasoning holds for $\overline{P_\square(\mathsf{X}_2)}$. Since X is regular, it is defined as $\mathsf{X} = S \oplus \{0, 1\}^2$ where $S = \{(x, y) \in \mathsf{X} \mid (x, y) + \{0, 1\}^2 \subset \mathsf{X}\}$. By definition, we have $\mathsf{X}_2 = \bigcup_{(x,y) \in S} (2x, 2y) + \{0, 1, 2, 3\}^2$. Let $(x, y) \in S$. We have $((2x, 2y) + \{0, 1, 2, 3\}^2) \oplus \square = [2x - \frac{1}{2}, 2x + \frac{7}{2}] \times [2y - \frac{1}{2}, 2y + \frac{7}{2}]$, and then $((2x, 2y) + \{0, 1, 2, 3\}^2) \oplus \square \ominus B_1 = [2x + \frac{1}{2}, 2x + \frac{5}{2}] \times [2y + \frac{1}{2}, 2y + \frac{5}{2}]$. Then we have $P_\square(\mathsf{X}_2) = \bigcup_{(x,y) \in S} (2x, 2y) + \{0, 1, 2, 3\}^2 \oplus \square = \bigcup_{(x,y) \in S} [2x - \frac{1}{2}, 2x + \frac{7}{2}] \times [2y - \frac{1}{2}, 2y + \frac{7}{2}]$ on the one hand, and $P_\square(\mathsf{X}_2) \ominus B_1 = \mathsf{X}_2 \oplus \square \ominus B_1 \supseteq \bigcup_{(x,y) \in S} [2x + \frac{1}{2}, 2x + \frac{5}{2}] \times [2y + \frac{1}{2}, 2y + \frac{5}{2}]$, on the other hand. Due to the specific square structure of X_2 and its regularity, we have $P_\square(\mathsf{X}_2) \ominus \square_2 = \bigcup_{(x,y) \in S} [2x - \frac{1}{2}, 2x + \frac{7}{2}] \times [2y - \frac{1}{2}, 2y + \frac{7}{2}] \ominus \square_2 =$

[6] This equivalence, presented here in the 2D case, holds for any dimension $n \geq 2$, with $\square$ being the unit n-cube.

$\bigcup_{(x,y)\in S}[2x+\frac{1}{2}, 2x+\frac{5}{2}] \times [2y+\frac{1}{2}, 2y+\frac{5}{2}]$, where $\square_2$ is the square of edge size 2 centered on $(0,0)$. In particular, we have $P_{\square}(\mathsf{X}_2)\ominus\square_2 \subseteq P_{\square}(\mathsf{X}_2)\ominus B_1$, and we note $R = (P_{\square}(\mathsf{X}_2) \ominus B_1) \setminus (P_{\square}(\mathsf{X}_2) \ominus \square_2)$ the residue between both. The non-vacuity of $P_{\square}(\mathsf{X}_2) \ominus B_1$ directly follows from the non-vacuity of $P_{\square}(\mathsf{X}_2) \ominus \square_2$. Up to a translation of $(-\frac{1}{2}, -\frac{1}{2})$ and a scaling of factor $\frac{1}{2}$, the set $P_{\square}(\mathsf{X}_2) \ominus \square_2$ is equal to $\bigcup_{(x,y)\in S}[x, x+1] \times [y, y+1]$. The existence of a continuous path between two points of $\bigcup_{(x,y)\in S}[x, x+1] \times [y, y+1]$ is equivalent to the existence of a 4-path between two points of $\bigcup_{(x,y)\in S}\{x, x+1\}\times\{y, y+1\} = \bigcup_{(x,y)\in S}(x,y)+\{0,1\}^2 = \mathsf{X}$. Since X is 4-connected in $\mathbb{Z}^2$, it follows that $P_{\square}(\mathsf{X}_2) \ominus \square_2$ is connected in $\mathbb{R}^2$. The residue R is composed of connected components of $\mathbb{R}^2$ (namely "triangular" shapes formed by two edges of length 1 adjacent to the border of $P_{\square}(\mathsf{X}_2) \ominus \square_2$ and a third concave, edge defined as a the quadrant of a circle of radius 1); the connectedness of $P_{\square}(\mathsf{X}_2) \ominus B_1 = (P_{\square}(\mathsf{X}_2) \ominus \square_2) \cup R$ then follows from that of $P_{\square}(\mathsf{X}_2) \ominus \square_2$. We have $P_{\square}(\mathsf{X}_2) = P_{\square}(\mathsf{X}_2) \ominus \square_2 \oplus \square_2$. Since $B_1 \subset \square_2 \subset B_{\sqrt{2}}$, the decreasingness of erosion and increasingness of dilation lead to $P_{\square}(\mathsf{X}_2) \subseteq P_{\square}(\mathsf{X}_2) \ominus B_1 \oplus B_{\sqrt{2}}$. □

5.2 3D Case

As already observed in Sect. 3.3, the extension of the notion of regularity to $\mathbb{Z}^3$ leads to 3D regular sets that may not be topologically invariant (see Fig. 3 for examples). Actually, we have an even stronger result, regularity in 3D never leads to quasi-r-regularity.

Property 3. Let $\mathsf{X} \subset \mathbb{Z}^3$ be a simply connected, regular set. Let $P_{\square}(\mathsf{X}) = \mathsf{X} \oplus \square$, with $\square \subset \mathbb{R}^3$ the closed, unit cube centered at $(0,0,0)$. Then $P_{\square}(\mathsf{X})$ is never quasi-1-regular with margin $\frac{2}{\sqrt{3}} - 1$.

To prove this property, it is sufficient to observe that for any salient vertex v of $P_{\square}(\mathsf{X})$ (such vertex exists, as X is finite and $P_{\square}(\mathsf{X})$ is then bounded), the distance between v and $P_{\square}(\mathsf{X}) \ominus B_1$ is $\sqrt{3} > \frac{2}{\sqrt{3}}$. Then, v does not belong to $P_{\square}(\mathsf{X}) \ominus B_1 \oplus B_{\frac{2}{\sqrt{3}}}$.

In particular, the up-sampling strategy proposed in the 2D case is useless in 3D. Indeed, in 2D this up-sampling allowed us to tackle connectedness issues in $P_{\square}(\mathsf{X}) \ominus B_1$, whereas in 3D connectedness issues occur in the complement part $P_{\square}(\mathsf{X}) \setminus (P_{\square}(\mathsf{X}) \ominus B_1)$, and the size of this residue is not impacted by increasing the resolution of the Cartesian grid.

6 Links Between Regularity and Quasi-r-Regularity: The Marching Squares/Cubes Polygonal Model

6.1 2D Case

The above cubic polygonization is the model implicitly considered when applying a pointwise rigid motion with the backward transformation model, and with the nearest neighbor digitization operator $\mathfrak{D}$ (Eq. (2)). This trivial polygonization is directly mapped on the pixel structure of the image, leading to poor modeling of the shape of the underlying continuous set.

There exist numerous ways of polygonizing a digital set. Here, we investigate the marching squares (MS) model, which is probably the simplest polygonization, except from the cubic one. In our case, the considered images are binary and regular. Then, the MS polygonization is the same as the cubic polygonization, except in the 2×2 configurations $(x, y) + \{0, 1\}^2$ where one point (for instance (x, y)) belongs to X (resp. $\overline{\mathsf{X}}$) while the other three belong to $\overline{\mathsf{X}}$ (resp. X). In that case, the edge of the MS polygon, noted $P_{\Diamond}(\mathsf{X})$, associated to X is a segment between the points $(x, y+\frac{1}{2}), (x+\frac{1}{2}, y)$ (whereas the cubic polygon $P_{\Box}(\mathsf{X})$ would locally have edges/segments between the points $(x, y+\frac{1}{2}), (x+\frac{1}{2}, y+\frac{1}{2})$ and $(x+\frac{1}{2}, y+\frac{1}{2}), (x+\frac{1}{2}, y)$).

In the case of a regular object $\mathsf{X} \subset \mathbb{Z}^2$, the MS polygonization can be formalized as follows. We have $\mathsf{X} = S \oplus \{0, 1\}^2$, where $S = \{(x, y) \in \mathsf{X} \mid (x, y) + \{0, 1\}^2 \subset \mathsf{X}\}$. By setting $S' = (\frac{1}{2}, \frac{1}{2}) + S$, this rewrites as $\mathsf{X} = \bigcup_{(x,y) \in S'} (x, y) + \{-\frac{1}{2}, \frac{1}{2}\}^2$. In other words, S' is the set of barycenters of the 2×2 square subsets of points forming X. Let C be the octagon centered on $(0, 0)$, formed by the two edges $[(-\frac{1}{2}, 1), (\frac{1}{2}, 1)]$ and $[(\frac{1}{2}, 1), (1, \frac{1}{2})]$, and the other six edges obtained by rotation of center $(0, 0)$ and angles $k.\pi/2$, $k = 1, 2, 3$, of these two edges. (Note that the distance between $(0, 0)$ and the four edges induced by $[(-\frac{1}{2}, 1), (\frac{1}{2}, 1)]$ (resp. $[(\frac{1}{2}, 1), (1, \frac{1}{2})]$) is 1 (resp. $\frac{3}{2\sqrt{2}} > 1$)). Let $G = \bigcup\{[p, q] \subset \mathbb{R}^2 \mid p, q \in S' \wedge 0 < \|p - q\|_2 \leq \sqrt{2}\}$. In other words, G is the set of the continuous straight segments linking the points of S' that are either 4- or 8-adjacent in the grid $\mathbb{Z}^2 + (\frac{1}{2}, \frac{1}{2})$.

Property 4. Let X be a simply connected set. If X is regular, then $P_{\Diamond}(\mathsf{X}) = G \oplus C$

Despite its simplicity, this MS polygonization model is sufficient for linking the notions of regularity and quasi-r-regularity.

Proposition 4. *Let $\mathsf{X} \subset \mathbb{Z}^2$ be a simply connected set. If X is regular, then $P_{\Diamond}(\mathsf{X}) \subset \mathbb{R}^2$ is quasi-1-regular with margin $\sqrt{2} - 1$.*

Proof. We show the non-vacuity and connectedness of $P_{\Diamond}(\mathsf{X}) \ominus B_1$ and the fact that $P_{\Diamond}(\mathsf{X}) \subseteq P_{\Diamond}(\mathsf{X}) \ominus B_1 \oplus B_{\sqrt{2}}$; the same reasoning holds for $\overline{P_{\Diamond}(\mathsf{X})}$. We have $B_1 \subset C$. Then, it leads to $S' \subset G \subseteq G \oplus C \ominus B_1 = P_{\Diamond}(\mathsf{X}) \ominus B_1$. The non-vacuity of $P_{\Diamond}(\mathsf{X}) \ominus B_1$ derives from that of S' and S. From the definition of regularity, it is plain that G is a connected subset of $\mathbb{R}^2$. Since C is convex, $G \oplus C$ is also a connected subset of $\mathbb{R}^2$. But as B_1 is convex and $B_1 \subset C$, $G \oplus C \ominus B_1 = P_{\Diamond}(\mathsf{X}) \ominus B_1$ is also connected in $\mathbb{R}^2$. We have $C \subset B_{\sqrt{2}}$, then $G \oplus C \subseteq G \oplus B_{\sqrt{2}}$. But we also have $G \subseteq P_{\Diamond}(\mathsf{X}) \ominus B_1$, then $G \oplus B_{\sqrt{2}} \subseteq P_{\Diamond}(\mathsf{X}) \ominus B_1 \oplus B_{\sqrt{2}}$. It follows that $P_{\Diamond}(\mathsf{X}) = G \oplus C \subseteq P_{\Diamond}(\mathsf{X}) \ominus B_1 \oplus B_{\sqrt{2}}$. □

6.2 3D Case

Similarly to the case of the cubic polyhedrization, we observe that the standard marching cubes (MC) method [12], namely the 3D version of the marching squares, also fails to generate a quasi-r-regular polyhedron from a regular set.

Property 5. Let $\mathsf{X} \subset \mathbb{Z}^3$ be a simply connected, regular set. Let $P_{\Diamond}(\mathsf{X})$ be the polyhedron generated from X by MC polyhedrization. Then $P_{\Diamond}(\mathsf{X})$ is never quasi-1-regular with margin $\frac{2}{\sqrt{3}} - 1$.

To prove this property, it is sufficient to observe that there exists a vertex v on a convex part of $P_{\Diamond}(\mathsf{X})$ (such vertex exists, as X is finite and $P_{\Diamond}(\mathsf{X})$ is then bounded), such that the distance between v and $P_{\Diamond}(\mathsf{X}) \ominus B_1$ is $\frac{\sqrt{6}}{2} > \frac{2}{\sqrt{3}}$. Thus, v does not belong to $P_{\Diamond}(\mathsf{X}) \ominus B_1 \oplus B_{\frac{2}{\sqrt{3}}}$.

7 Conclusion

In this article, we observed that the notion of quasi-r-regularity allows one to define polygons/polyhedra that preserve their topology under digitization in $\mathbb{Z}^n$ for $n = 2, 3$ (this property is no longer valid in $\mathbb{Z}^n$, $n \geq 4$). As a consequence, building a quasi-r-regular polygon/polyhedron from a digital set in $\mathbb{Z}^2$ or $\mathbb{Z}^3$ for handling topology-preserving rigid motions is relevant. In this context, we established that two simple polygonization models (cubic model and marching squares) can link the notions of regularity and quasi-r-regularity in $\mathbb{Z}^2$. However, in 3D, the corresponding models fail to generate quasi-r-regular polyhedra in $\mathbb{R}^3$. Our further works will consist of investigating other kinds of polyhedrization devoted to 3D regular images.

References

1. Bazin, P.L., Ellingsen, L.M., Pham, D.L.: Digital homeomorphisms in deformable registration. In: Karssemeijer, N., Lelieveldt, B. (eds.) Information Processing in Medical Imaging, pp. 211–222. Springer, Berlin Heidelberg (2007). https://doi.org/10.1007/978-3-540-73273-0_18
2. Bertrand, G.: On P-simple points. Comptes Rendus de l'Académie des Sciences, Série Mathématique **I**(321), 1077–1084 (1995)
3. Bertrand, G., Malandain, G.: A new characterization of three-dimensional simple points. Pattern Recognit. Lett. **15**(2), 169–175 (1994)
4. Boutry, N., Géraud, T., Najman, L.: A tutorial on well-composedness. J. Math. Imaging Vis. **60**(3), 443–478 (2018)
5. Khalimsky, E.: Topological structures in computer science. J. Appl. Math. Simul. **1**(1), 25–40 (1987)
6. Kong, T.Y.: A digital fundamental group. Comput. Graph. **13**(2), 159–166 (1989)
7. Kong, T.Y., Rosenfeld, A.: Digital topology: introduction and survey. Comput. Vis. Graph. Image Process. **48**(3), 357–393 (1989)
8. Kovalevsky, V.A.: Finite topology as applied to image analysis. Comput. Vis. Graph. Image Process. **46**(2), 141–161 (1989)
9. Largeteau-Skapin, G., Andres, E.: Discrete-Euclidean operations. Discret. Appl. Math. **157**(3), 510–523 (2009)
10. Latecki, L.J.: 3D well-composed pictures. Graph. Model Image Process. **59**(3), 164–172 (1997)
11. Latecki, L.J., Eckhardt, U., Rosenfeld, A.: Well-composed sets. Comput. Vis. Image Underst. **5**(1), 70–83 (1995)

12. Lorensen, W.E., Cline, H.E.: Marching cubes: a high-resolution 3D surface construction algorithm. ACM SIGGRAPH Comput. Graph. **21**(4), 163–169 (1987)
13. Maunder, C.R.F.: Algebraic Topology. Dover (1996)
14. Mazo, L., Passat, N., Couprie, M., Ronse, C.: Digital imaging: a unified topological framework. J. Math. Imaging Vis. **44**(1), 19–37 (2012)
15. Mazo, L., Passat, N., Couprie, M., Ronse, C.: Paths, homotopy and reduction in digital images. Acta Appl. Math. **113**(2), 167–193 (2011)
16. Ngo, P., Passat, N., Kenmochi, Y., Debled-Rennesson, I.: Geometric preservation of 2D digital objects under rigid motions. J. Math. Imaging Vis. **61**(2), 204–223 (2019). https://doi.org/10.1007/s10851-018-0842-9
17. Ngo, P., Passat, N., Kenmochi, Y., Debled-Rennesson, I.: Convexity invariance of voxel objects under rigid motions. In: International Conference on Pattern Recognition, pp. 1157–1162 (2018)
18. Ngo, P., Passat, N., Kenmochi, Y., Talbot, H.: Topology-preserving rigid transformation of 2D digital images. IEEE Trans. Image Process. **23**(2), 885–897 (2014)
19. Passat, N., Mazo, L.: An introduction to simple sets. Pattern Recognit. Lett. **30**(15), 1366–1377 (2009)
20. Pavlidis, T.: Algorithms for Graphics and Image Processing. Springer/Computer Science Press, Berlin/Rockville (1982)
21. Rosenfeld, A.: Adjacency in digital pictures. Inf. Control **26**(1), 24–33 (1974)
22. Rosenfeld, A.: Digital topology. Am. Math. Mon. **86**(8), 621–630 (1979)
23. Rosenfeld, A., Kong, T.Y., Nakamura, A.: Topology-preserving deformations of two-valued digital pictures. Graph. Models Image Process. **60**(1), 24–34 (1998)
24. Stelldinger, P., Latecki, L.J., Siqueira, M.: Topological equivalence between a 3D object and the reconstruction of its digital image. IEEE Trans. Pattern Anal. Mach. Intell. **29**(1), 126–140 (2007)
25. Stelldinger, P., Terzic, K.: Digitization of non-regular shapes in arbitrary dimensions. Image Vis. Comput. **26**(10), 1338–1346 (2008)
26. Zitová, B., Flusser, J.: Image registration methods: a survey. Image Vis. Comput. **21**(11), 977–1000 (2003)

Graph-Based Models, Analysis and Segmentation

A New Entropy for Hypergraphs

Isabelle Bloch[1(✉)] and Alain Bretto[2]

[1] LTCI, Télécom ParisTech, Université Paris-Saclay, Paris, France
isabelle.bloch@telecom-paristech.fr
[2] NormandieUnicaen, GREYC CNRS-UMR 6072, Caen, France
alain.bretto@info.unicaen.fr

Abstract. This paper introduces a new definition of entropy for hypergraphs. It takes into account the fine structure of a hypergraph by considering its partial hypergraphs, leading to an entropy vector. This allows for more precision in the description of the underlying complexity of the hypergraph. Properties of the proposed definitions are analyzed.

Keywords: Hypergraphs · Entropy · Algebraic and graphical structures

1 Introduction

The concept of entropy has been introduced to measure the amount of information contained or delivered by an information source. Its purpose is to model the combinatorial possibilities of the different states of a given set. Naturally, these different states are modeled by a probability distribution of their occurrences. This makes it possible to have a "relatively simple" (in terms of summarization) vision of the content of the information of the objects under study. It is therefore primarily a probabilistic theory of information. After its introduction, this notion has been developed in different parts of sciences, such as dynamic systems, computer science, stochastic theory, among others. Introducing entropy into information theory is credited to Shannon [15]. This notion has been adapted to quantum physics by von Neumann using the spectrum of the density matrix [13].

More recently this notion has been adapted to the Laplacian of a graph, thus showing the utility of this concept in graph theory and its application in particular in image analysis [7,10–12].

Hypergraphs (generalizing the notion of graphs with higher arity of edges) are useful to model real data where relationships between different data items have to be taken into account. This includes images, where relations between pixels or regions can involve more than two elements (e.g. proximity, parallelism...). A combinatorial object such as a hypergraph can be very complicated, and the introduction of an entropy measure on such a structure is a relevant way to assess this complexity. This measure can be used in feature selection methods such as in [19] or for other tasks such as compression or similarity assessment.

M. Couprie et al. (Eds.): DGCI 2019, LNCS 11414, pp. 143–154, 2019.
https://doi.org/10.1007/978-3-030-14085-4_12

Usual approaches rely on the Laplacian of hypergraphs, in a similar way as for graphs. In previous work [6], we defined similarities between hypergraphs based on mathematical morphology on hypergraphs [5] and valuations in lattices [4,16], from which partitions and entropies can be defined, e.g. up to a morphological transformation, to introduce additional filtering and gain in robustness. In all these approaches, the entropy is defined as a single number, considering the hypergraph globally.

In this paper, we propose to define the entropy of a hypergraph as a vector, by using substructures - namely partial hypergraphs. An entropy vector associated with n random variables is a kind of generalization of Shannon's entropy, which links domains such as geometry, combinatorics and information theory [9]. In this paper we introduce a new concept of entropy called *entropy vector* associated with a hypergraph. It is based on the incidence matrix $I(H)$ associated with the hypergraph H, more precisely on $I(H)I(H)^t$, as well as all the main sub-matrices of $I(H)I(H)^t$. The defined vector conveys much more detailed information than the information measured by conventional entropy.

In Sect. 2, basic definitions on hypergraphs are recalled, and the notations used in this paper are introduced. In Sect. 3, the usual notion of entropy of a hypergraph is given, for which we show a few properties that are useful in this paper. The main contribution of this paper is introduced in Sect. 4, where a definition of entropy vector of a hypergraph is proposed, and in the next sections, where properties are analyzed. A lattice structure is introduced in Sect. 5. Since complexity can be a concern, a fast approximate calculus method is proposed in Sect. 6. Finally links with the Zeta function are suggested in Sect. 7.

2 Preliminaries

Let us first recall some useful notions on hypergraphs [3,8]. A hypergraph H denoted by $H = (V, E = \{e_i, i = 1...m\})$ is defined as a pair of a finite set V (vertices) and a finite family $\{e_i, i = 1...m\}$ of hyperedges. In this paper we consider only simple hypergraphs (i.e. without repeated hyperedges), and E is hence a set. Hyerpedges can be considered equivalently as subsets of vertices or as a relation between vertices of V. The first interpretation is considered here, and we will note $x \in e_i$ the fact that a vertex $x \in V$ belongs to the hyperedge e_i. In this paper, it is further assumed that E is non-empty and that each hyperedge is non-empty (i.e. $\forall e_i \in E, e_i \neq \emptyset$). We denote by $r(H)$ the rank of H, defined as $r(H) = \max_{e \in E} |e|$.

In order to study the fine structure of an hypergraph, the following notions are important in this paper:

- A *partial hypergraph* H' of H generated by $J \subseteq \{1...m\}$ is a hypergraph $(V', \{e_j, j \in J\})$, where $\cup_{j \in J} e_j \subseteq V' \subseteq V$. In the sequel we will suppose that $V' = V$. This will be denoted by $H' \leq H$.
- Given a subset $V' \subseteq V$, a *subhypergraph* H' of H is the partial hypergraph $H' = (V', \{e_i, e_i \in E \mid e_i \subseteq V'\})$.

- The *induced subhypergraph* H' of H with $V' \subseteq V$ is the hypergraph defined as $H' = (V', E')$ with $E' = \{e' = \{e \cap V'\} \mid e \in E \text{ and } e \cap V' \neq \emptyset\}$. Note that if $V' = V$ and hypergraphs are considered without empty hyperedges, then $H' = H$.
- A hypergraph $H = (V, E)$ is *isomorphic* to a hypergraph $H' = (V', E')$ ($H \simeq H'$), if there exist a bijection $f : V \to V'$ and a bijection $\pi : \{1...m\} \to \{1...m'\}$, where $m = |E|$ and $m' = |E'|$, which induce a bijection $g : E \to E'$ (i.e. $g(e) = \{f(x) \mid x \in e\}$) such that $g(e_i) = e'_{\pi(i)}$ for all $e_i \in E$ and $e'_{\pi(i)} \in E'$. The mapping f is then called *isomorphism* of hypergraphs.

Finally, let us introduce some notations on symmetric matrices. Let $A \in \mathcal{M}_{n\times n}(\mathbb{R})$, $A = ((a_{i,j}))_{i,j\in\{1...n\}}$ be a symmetric matrix on $\mathbb{R}$. Let $\alpha = \{\alpha_1, \ldots \alpha_t,\} \subseteq \{1, \ldots n\}$. We denote by $A(\alpha; \alpha)$ the submatrix of A generated by keeping only the rows and columns of A indexed by α, i.e.

$$A(\alpha; \alpha) = ((b_{i,j}))_{i,j\in\{1...t\}}, \; b_{i,j} = a_{\alpha_i,\alpha_j}, \; \alpha_i, \alpha_j \in \alpha.$$

Such a matrix is called *principal submatrix* of A. If $\alpha = \{1, 2, \ldots, k\}$, $k \leq n$, $A(\alpha; \alpha)$ is called *leading principal submatrix* of A.

3 Entropy of a Hypergraph

Let (H, V) be a simple hypergraph with $|V| = n$, $|E| = m$. Let $I(H) = ((a_{i,j}))_{(i,j)\in\{1...m\}\times\{1...n\}}$ be the incidence matrix of H: $a_{i,j} = 1$, if $x_j \in e_i$ and $a_{i,j} = 0$ otherwise. We consider the matrix $m \times m$ defined as:

$$L(H) = I(H)I(H)^t = ((|e_i \cap e_j|))_{i,j\in\{1...m\}}.$$

This matrix is positive semi-definite. Its eigenvalues $\lambda_i(L(H))$, $i = 1...m$, are positive and can be ordered as follows:

$$0 \leq \lambda_1(L(H)) \leq \lambda_2(L(H)) \leq \ldots \leq \lambda_m(L(H)).$$

Lemma 1. *The trace of the matrix $L(H)$ is the sum of the degrees $d(x)$ of the vertices x of H (i.e. the number of hyperedges that contain x):*

$$\mathrm{Tr}(L(H)) = \sum_{x\in V} d(x).$$

Proof. $\mathrm{Tr}(L(H)) = \sum_{i=1}^{m} \lambda_i = \sum_{e\in E} |e| = \sum_{x\in V} d(x)$.

□

Let us define:

$$\mu_i = \frac{\lambda_i(L(H))}{\sum_{i=1}^{m} \lambda_i(L(H))} = \frac{\lambda_i(L(H))}{\mathrm{Tr}(L(H))}.$$

The μ_i are the eigenvalues of the normalized matrix

$$\mathcal{L}(H) = \frac{1}{\mathrm{Tr}(L(H))} L(H)$$

which is also positive semi-definite.

Classically, the *Shannon entropy* of the hypergraph H (see e.g. [8]) is defined as:

$$S(H) = -\sum_{i=1}^{m} \mu_i \log_2(\mu_i).$$

Note that other forms of entropy exist, such as the *Renyi entropy* for instance, which is defined for a hypergraph H as:

$$R_s(H) = \frac{1}{1-s} \ln(\sum_{i=1}^{m} \mu_i^s), \; s \geq 0.$$

In this paper we mostly consider Shannon entropy, except in Sect. 7.

Proposition 1. *Let $H = (V, E)$ be a simple hypergraph without isolated vertex, without empty hyperedge, and with $|V| = n$ and $|E| = m$ ($m > 0$). We have the two following properties:*

(a) *$S(H) = 0$ if and only if $|E| = 1$,*
(b) *$S(H) = \log_2(n) - \log_2(r(H)) = \log_2(m)$, where $r(H)$ (the rank of H) is a constant equal to $\frac{n}{m}$, if and only if H is uniform (i.e. $\forall e \in E$, $|e| = r(H)$) and the intersection of any two distinct hyperedges is empty (i.e. for all e, e' in E such that $e \neq e'$, $|e \cap e'| = 0$).*

Proof. **(a)** Assume that $|E| = 1$. Then $L(H)$ is reduced to a scalar value, which is non-zero since $e \neq \emptyset$, and the unique normalized eigenvalue is $\mu = 1$. Hence $S(H) = -1 \log_2(1) = 0$.

Conversely, suppose that $S(H) = 0$. Since $E \neq \emptyset$ and H does not contain any empty hyperedge, then $S(H) = 0$ implies that $\forall i, \mu_i = 0$ or $\mu_i = 1$. Since hyperegdes are not empty, $\mu_i = 0$ is not possible, and since $\sum_i \mu_i = 1$, there is a unique eigenvalue, equal to 1. Hence $|E| = 1$.

(b) Assume now that H is uniform, with $|e| = r(H)$ for each hyperedge e, and that for all $e, e' \in E$ such that $e \neq e'$, $|e \cap e'| = 0$. Note that in this case $L(H)$ is diagonal and so $\lambda_i = |e_i| = r(H)$. Since $e \cap e' = \emptyset$, $d(x) = 1$ for each vertex x and from Lemma 1, $\text{Tr}(L(H)) = \sum_{x \in V} d(x) = n$. This corresponds to a situation where vertices are uniformly distributed among the hyperedges, i.e. $r(H) = \frac{n}{m}$. Therefore we have:

$$\begin{aligned} S(H) &= -\sum_{i=1}^{m} \mu_i \log_2(\mu_i) \\ &= -\sum_{e \in E} \frac{|e|}{\text{Tr}(L(H))} \log_2(\frac{|e|}{\text{Tr}(L(H))}) \\ &= -\frac{mr(H)}{n} \log_2(\frac{r(H)}{n}) \\ &= -\log_2(\frac{r(H)}{n}) = \log_2(n) - \log_2(r(H)) = \log_2(m). \end{aligned}$$

Conversely suppose that $S(H) = \log_2(n) - \log_2(r(H)) = \log_2(m)$. This implies $n = mr(H)$.

Moreover from Lemma 1, from $|e| \leq r(H)$ for all e by definition of $r(H)$, and since $\sum_{x \in V} d(x) \geq n$ (no isolated vertex), we have:

$$n \leq \sum_{e \in E} |e| = \sum_{x \in V} d(x) \leq mr(H) = n.$$

It follows that $\sum_{x \in V} d(x) = n = \mathrm{Tr}(L(H))$.

It also follows that $\sum_{e \in E} |e| = n = mr(H)$. Since $E \neq \emptyset$ and there is no empty hyperedge, we can derive that $|e| = r(H)$ for all $e \in E$. This means that H is uniform, and since $n = mr(H)$, for all $e, e' \in E$ such that $e \neq e'$, we have $|e \cap e'| = 0$. Note that in this case the matrix $L(H)$ is diagonal, and for all $i \in \{1, 2, \ldots m\}$, $\mu_i = \frac{\lambda_i}{\mathrm{Tr}(L(H))} = \frac{r(H)}{n}$.

□

This proposition shows that the entropy is closely related with the parameters of the hypergraph. Note that Case **b** implicitly assumes that $\frac{n}{m}$ is an integer. Moreover, Case **a** is a consequence of Case **b**.

A straightforward extension of this result deals with hypergraphs that may contain isolated vertices. A similar result holds by replacing n by n', the number of non-isolated vertices (i.e. that belong to at least one hyperedge).

4 Entropy Vector Associated with a Hypergraph

Now, we built on the classical definition of hypergraph entropy to propose a new entropy, defined as a vector, based on a finer analysis of the structure of the hypergraph by considering all its partial hypergraphs.

Definition 1 (Entropy vector). *Let $H = (V, E)$ be a hypergraph. For $i \leq m$ ($m = |E|$), let*

$$SE_i(H) = \{S(H_i) \mid H_i = (V, E_i), H_i \leq H, |E_i| = i\}$$

be the set of entropy values of all partial hypergraphs of H whose set of hyperedges has cardinality i, arranged in increasing order.

The entropy vector *of the hypergraph H is then the vector:*

$$SE(H) = (SE_1(H), SE_2(H), \ldots SE_m(H))$$

with $2^m - 1$ coordinates.

Note that if $H' \leq H$ then it is easy to see that the matrix $L(H')$ is a principal submatrix of $L(H)$.

From Proposition 1, the vector $S(H)$ begins with at least m values equal to 0.

Example 1. Let us illustrate this definition on a very simple example, illustrated in Fig. 1, for a hypergraph H with three hyperedges.

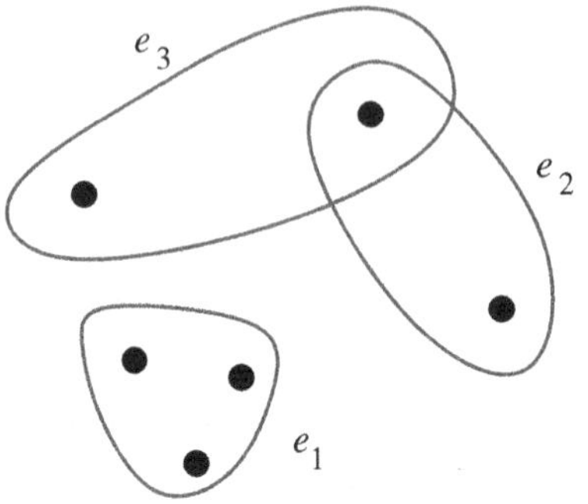

Fig. 1. A simple example of hypergraph, with three hyperedges (indicated by blue lines): e_1 contains three vertices, e_2 and e_3 both contain two vertices, with one common vertex. (Color figure online)

Let us compute the entropy vector:

- For SE_1, there are three partial hypergraphs containing one hyperedge (e_1, e_2 and e_3, respectively), and $SE_1 = (0, 0, 0)$.
- For SE_2, there are three partial hypergraphs containing two hyperedges. For (e_1, e_2), the matrix L is equal to $\begin{pmatrix} 3 & 0 \\ 0 & 2 \end{pmatrix}$, with eigenvalues 2 and 3, and the corresponding entropy is equal to $s_1 = -\frac{2}{5}\log_2\frac{2}{5} - \frac{3}{5}\log_2\frac{3}{5} \simeq 0.29$. The same reasoning applies for (e_1, e_3). For (e_2, e_3), the matrix L is equal to $\begin{pmatrix} 2 & 1 \\ 1 & 2 \end{pmatrix}$, with eigenvalues 1 and 3, and the corresponding entropy is equal to $s_2 = -\frac{1}{4}\log_2\frac{1}{4} - \frac{3}{4}\log_2\frac{3}{4} \simeq 0.24$. Then $SE_2 = (s_2, s_1, s_1) \simeq (0.24, 0.29, 0.29)$.
- For SE_3, there is one partial hypergraph containing three hyperedges, i.e. H. The matrix L is equal to $\begin{pmatrix} 3 & 0 & 0 \\ 0 & 2 & 1 \\ 0 & 1 & 2 \end{pmatrix}$, with eigenvalues 1, 3 and 3, and the corresponding entropy is equal to $s_3 = -\frac{1}{7}\log_2\frac{1}{7} - 2\frac{3}{7}\log_2\frac{3}{7} \simeq 0.44$. Hence $SE_3 = (s_3) \simeq (0.44)$.
- Finally, the entropy vector is

$$SE(H) = (0, 0, 0, s_2, s_1, s_1, s_3) \simeq (0, 0, 0, 0.24, 0.29, 0.29, 0.44).$$

Proposition 2. *Let $H = (V, E)$ and $H' = (V', E')$ be two isomorphic hypergraphs. Then there is a permutation σ such that*

$$SE(H) = (SE_1(H), SE_2(H), SE_3(H) \ldots, SE_m(H)) =$$
$$SE(H') = \left(SE'_{\sigma(1)}(H), SE'_{\sigma(2)}(H), SE'_{\sigma(3)}(H) \ldots, SE'_{\sigma(m)}(H)\right)$$

Proof. Let $I(H)$ and $I(H')$ be the incidence matrices of H and H', respectively. Since H and H' are isomorphic there are two permutation matrices P and Q such that

$$I(H') = PI(H)Q^t.$$

Consequently

$$I(H')^t = QI(H)^t P^t$$

and

$$I(H')I(H')^t = PI(H)Q^t QI(H)^t P^t.$$

Therefore we obtain

$$L(H') = PL(H)P^t \text{ and } \mathcal{L}(H') = P\mathcal{L}(H)P^t.$$

It is well known that if $A = PBP^t$ (with $P^{-1} = P^t$) then A and B have the same eigenvalues. The matrix P represents the isomorphism and gives rise to the permutation σ (the permutation on vertices induces a permutation on hyperedges). The isomorphism guarantees that the two hypergraphs have the same structure. Hence their sets of partial hypergraphs are in one-to-one correspondence and each partial hypergraph of H is isomorphic to a partial hypergraph of H'. The result follows.

□

5 Partial Ordering and Lattice Structure

In this section we further analyze the properties of partial ordering on hypergraphs and on vector entropy, which result in lattice structures.

We first recall results from [5].

Definition 2. *Let $\mathcal{H}$ be the set of isomorphism classes of hypergraphs. A partial order $\leq_f$ on $\mathcal{H}$ is defined as: $\forall H', H \in \mathcal{H}$, $H' \leq_f H$ if $\exists V' \subseteq V$ such that H' is isomorphic (by f) to the subhypergraph of H induced by V'.*

It is clear that $\leq_f$ is a partial order relation. We denote by $=_f$ the corresponding equality, and by $<_f$ the corresponding strict ordering.

Hereafter, we will denote a class by a representative hypergraph H in this class.

Proposition 3 ([5]). *The structure $(\mathcal{H}, \leq_f)$ is a complete lattice. The supremum of any $H_1 = (V_1, E_1), H_2 = (V_2, E_2)$ is $\sup\{H_1, H_2\} = H_1 \vee H_2 = (V_1 \cup V_2, E_1 \cup E_2)$, and the infimum $\inf\{H_1, H_2\}$ is the maximum common induced subhypergraph (and their extensions to any family).*

Let us now move to the vector entropy. Let $\mathcal{V}$ be a set of vectors. A partial order is defined on $\mathcal{V}$ as follows. For $x = (x_1, x_2, \ldots x_k) \in \mathcal{V}$ and $y = (y_1, y_2, \ldots y_t) \in \mathcal{V}$, $x \leq y$ if $\forall i \leq \min(k, t)$, $x_i \leq y_i$. Note that this relation is equivalent to the usual Pareto ordering, if the shortest length vector (say x, i.e. $k \leq t$) is completed by $t - k$ components equal to 0. The set $\mathcal{V}$ endowed with this partial ordering is called an *ordered vector set*.

Proposition 4. *The set* $SE_{\mathcal{H}} = \{SE(H) \mid H \in \mathcal{H}\}$ *is an ordered vector set.*

Note that while each $SE_i(H)$ has its values increasingly ordered, this is not the case for $SE(H)$. Proposition 4 means that a partial ordering can be defined on $SE_{\mathcal{H}}$, as defined above.

Proposition 5. *The set* $SE_{\mathcal{H}}$ *endowed with the partial ordering* $\leq$ *is a lattice.*

Proof. It is clear that the partial ordering $\leq$ (Pareto-like ordering) gives rise to a supremum (least upper-bound) and an infimum (greatest lower bound) for any finite family of $SE(H)$. The supremum is computed as the component-wise maximum and the infimum as the component-wise minimum. It is bounded by 0, obtained for $H = (V, E = \{e\})$ from Proposition 1, so any infinite family has also an infimum.

□

Proposition 6. *Let* $H = (V, E)$ *and* $H' = (V', E')$*. If* $H' \leq_f H$*, then* $SE(H') \leq SE(H)$*.*

Proof. Let $H = (V, E)$ and $H' = (V', E')$ such that $H' \leq_f H$, with $|E| = m$ and $|E'| = m'$. Two cases arise:

1. If $H' =_f H$ then, by Proposition 2 there is a permutation σ such that $SE_\sigma(H) = SE(H)$.
2. If $H' <_f H$ then by definition there is an induced subhypergraph $H'' = (V'', E'')$ of H which is isomorphic to H'. From Proposition 2, $\mathcal{L}(H')$ and $\mathcal{L}(H'')$ have the same eigenvalues, and moreover, since f is an isomorphism between H' and H'', there is a permutation σ (which comes from f) such that
$$\forall i, 1 \leq i \leq m', \; SE_{\sigma(i)}(H') \leq SE_i(H'').$$
Since $H' <_f H$ then $m' < m$, hence $2^{m'} - 1 < 2^m - 1$. Hence, by adding $2^m - 1 - (2^{m'} - 1)$ components equal to 0 in $(SE_{\sigma(1)}(H), SE_{\sigma(2)}(H), \ldots, SE_{\sigma(m')}(H))$ at a good place we obtain a vector with $2^m - 1$ components such that
$$SE(H') \leq SE(H).$$

□

Let $H = (V, E)$ and $H' = (V', E')$ be two hypergraphs such that $V = \{x_1, x_2, \ldots x_n\}$, $V' = \{x'_1, x'_2, \ldots x'_{n'}\}$, $E = \{e_1, e_2, \ldots e_m\}$, $E' = \{e'_1, e'_2, \ldots e'_{m'}\}$. Let $I(H) = ((a_{i,j}))_{(i,j) \in \{1 \ldots m\} \times \{1 \ldots n\}}$ and $I(H') = ((a'_{i,j}))_{(i,j) \in \{1 \ldots m'\} \times \{1 \ldots n'\}}$. Let $H'' = H \cup H' = (V \cup V', E \cup E')$. Its incidence matrix is built as follows: add rows and columns with 0's in $I(H)$ (respectively $I(H')$) for vertices and hyperedges present in H' and not in H (respectively present in H and not in H'), so as to get two matrices of size $m'' \times n''$ whose generic terms are still denoted $a_{i,j}$ and $a'_{i,j}$. Then the matrix $I(H'')$ has m'' rows corresponding to $E \cup E'$,

n'' columns corresponding to $V \cup V'$, and coefficients $a''_{i,j} = \max(a_{i,j}, a'_{i,j})$, for $1 \le i \le m'', 1 \le j \le n''$.

We define $H \cap H' = (V \cap V', E \cap E')$ in the same way by suppressing rows and columns in $I(H)$ and $I(H')$ corresponding to non-common vertices and edges, and by replacing max by min to define the coefficients of the resulting incidence matrix.

Clearly both $L(H \cup H')$ and $L(H \cap H')$ are well defined.

6 An Algorithm for Approximate Calculus of the Entropy Vector

When the size of the hypergraph increases, the entropy vector may become very costly. Suggestions to reduce the complexity would be to disregard partial hypergraphs with very few hyperedges (bringing reduced information), or in contrast with near to m hyperedges (and hence less relevant). Another way would be to order the hyperedges in order of increasing cardinality and to take the leading principal matrices.

In this section, we propose approximations that alleviate two main drawbacks to calculate the entropy vectors.

The first drawback is the calculation of the eigenvalues. Since entropy is a mean value of the information, we can take an approximate value. Recall that $\log(x) = \log(1 + (x - 1)) \simeq x - 1$ for $-1 \le x - 1 \le 1$. Since $\mu_i \in [0, 1]$, and hence $\mu_i - 1 \in [-1, 0]$, $S(H) = -\sum_{i=1}^{m} \mu_i \log_2(\mu_i)$ can be approximated by:

$$\begin{aligned} -\sum_{i=1}^{m} \mu_i(1 - \mu_i) &= \mathrm{Tr}(\mathcal{L}(H)) - \mathrm{Tr}(\mathcal{L}(H)^2) \\ &= 1 - \frac{1}{\left(\sum_{e \in E} |e|\right)^2} \left(\sum_{e \in E} |e|^2 + \sum_{e \in E} \sum_{e' \in A_e} |e \cap e'|^2 \right) \quad (1) \end{aligned}$$

where $A_e = \{e' \in E \mid e' \ne e \text{ and } e \cap e' \ne \emptyset\}$. This expression is easy to compute.

The second drawback is that we have $2^m - 1$ principal matrices. We can take only the leading principal matrices, hence, we have to manage $m - 1$ matrices and we get an entropy vector with $m - 1$ components. In this case the order of the hyperedges in the matrix $\mathcal{L}(H)$ is important: if we permute two hyperedges in $\mathcal{L}(H)$, for instance the first one with the last one, we may change $SE(H)$. This comes from Eq. 1. Consequently we have to find an appropriate order on hyperedges. We sort hyperedges when building $\mathcal{L}(H)$ so that e is before e' (denoted by $e \preceq e'$) if:

- $|e| < |e'|$, or
- $|e| = |e'|$ and $\sum_{a \in A_e} |e \cap a| \le \sum_{a' \in A_{e'}} |e' \cap a'|$.

The case where $|e| = |e'|$ and $\sum_{a \in A_e} |e \cap a| = \sum_{a' \in A_{e'}} |e' \cap a'|$ is not important because it does not change Eq. 1, hence we can choose to put either e before e' or e' before e. Relation $\preceq$ is a pre-order (both reflexive and symmetric relation). After pre-ordering $\mathcal{L}(H)$, the vector $SE(H)$ is canonically associated to this matrix.

7 Zeta Function and Entropy

In this section, we suggest some links between the Zeta function and the notion of hypergraph entropy.

The Zeta spectral function plays an important role in areas where linear operators are often present [17], that is in most fields of physics. It has also been introduced on graphs and in image analysis [11,18]. Define now the *spectral Zeta function* of a hypergraph $H = (V, E)$, with $|E| = m$, as:

$$\zeta_H(s) = \mathrm{Tr}(\mathcal{L}(H)^{-s}) = \sum_{i=1,\mu_i\neq 0}^{m} \mu_i^{-s}$$

In a similar way as for entropy, we can define the *spectral Zeta vector function.*

It is easy to show that the derivative of ζ with respect to s can be expressed as:

$$\zeta_H^{'}(s) = \frac{\zeta_H(s)}{ds} = -\sum_{i=1}^{m} \mu_i^{-s} \ln(\mu_i).$$

Consequently we have:

$$\zeta_H^{'}(-1) = -\sum_{i=1}^{m} \mu_i \ln(\mu_i) = \ln(2)S(H),$$

$$\zeta_H^{'}(0) = -\sum_{i=1}^{m} \ln(\mu_i) = -\ln(\prod_{i=1}^{m} \mu_i) = -\ln(\det(\mathcal{L}(H))).$$

The following result also holds.

Proposition 7. *Let $H = (V, E)$ be a simple hyperpgraph, with $|E| = m$, and strictly positive eigenvalues of $\mathcal{L}(H)$. The Zeta function is related to the Renyi entropy by the following equation:*

$$\forall s \geq 0, \; \zeta_H(-s) = e^{(1-s)R_s(H)}$$

Proof. We have, for $s \geq 0$:

$$(1-s)R_s(H) = \ln(\sum_{i=1}^{m} \mu_i^s)$$

and

$$e^{(1-s)R_s(H)} = \sum_{i=1}^{m} \mu_i^s = \zeta_{\mathcal{H}}(-s)$$

□

These relations show how to relate spectro-analytic theory to information theory. This is of great importance, especially in dynamic networks seen as a series of hypergraphs. Indeed the notion of dynamical Zeta functions are an important tool to analyze chaotic dynamical systems [2,14] and should be effective in quantifying and tracking the evolution of entropy vectors, by applying the results in this section to the entropy values of all partial hypergraphs.

8 Conclusion

In this paper, a new measure of entropy for hypergraphs was proposed, defined as a vector. It generalizes Shannon entropy, representing a richer information on the complexity of a hypergraph, taking into account all its partial hypergraphs, hence its sub-structures. Algebraic properties were proved, as well as some links with the Zeta function.

Future work aims at exploring further the proposed notion, including additional properties and applications, in particular for image description. For instance, as suggested in [6], a hypergraph could be built from an image considering pixels or regions resulting from an over-segmentation as vertices, and hyperedges could be defined from relations (e.g. neighborhood, grey-levels or colors, spatial relations between regions...). Looking at the sub-structures of the hypergraph would then provide a description of the image complexity which would be finer and more precise than a global description.

For concrete applications to be tractable, the computation cost may be an issue for even moderately large hypergraphs, since the size of the entropy vector grows exponentially with the number of hyperedges. We suggested a few ways to address this issue in Sect. 6. Other could be developed, for instance by choosing randomly partial hypergraphs, or by looking at specific patterns in the hypergraph.

Another extension of hypergraph entropy to sequences of entropy values was proposed in [1]. The approach we proposed here is different since all the partial hypergraphs are considered, instead of centroid expansion subgraphs as in this earlier work. In future work it would be also interesting to compare both approaches based on their respective properties, as well as on concrete examples.

Another direction of research consists in extending the proposed notion to other forms of entropy (such as Renyi entropy), to weighted hyperedges (for instance using the distance between vertices), to weighted terms in the matrix $L(H)$ (e.g. using a distance between two hyperedges e_i and e_j to weight $|e_i \cap e_j|$), or more generally to attributed hypergraphs.

References

1. Bai, L., Escolano, F., Hancock, E.R.: Depth-based hypergraph complexity traces from directed line graphs. Pattern Recogn. **54**, 229–240 (2016)
2. Baladi, V.: Dynamical zeta functions. In: Branner, B., Hjorth, P. (eds.) Real and Complex Dynamical Systems. NATO ASI Series (Series C: Mathematical and Physical Sciences), vol. 464, pp. 1–26. Springer, Dordrecht (1995). https://doi.org/10.1007/978-94-015-8439-5_1
3. Berge, C.: Hypergraphs. Elsevier Science Publisher, The Netherlands (1989)
4. Birkho, G.: Lattice Theory, vol. 25, 3rd edn. American Mathematical Society, Providence (1979)
5. Bloch, I., Bretto, A.: Mathematical morphology on hypergraphs, application to similarity and positive kernel. Comput. Vis. Image Underst. **117**(4), 342–354 (2013)

6. Bloch, I., Bretto, A., Leborgne, A.: Robust similarity between hypergraphs based on valuations and mathematical morphology operators. Discrete Appl. Math. **183**, 2–19 (2015)
7. Braunstein, S.L., Ghosh, S., Severini, S.: The Laplacian of a graph as a density matrix: a basic combinatorial approach to separability of mixed states. Ann. Comb. **10**(3), 291–317 (2006)
8. Bretto, A.: Hypergraph Theory: An Introduction. Mathematical Engineering. Springer, New York (2013). https://doi.org/10.1007/978-3-319-00080-0
9. Cover, T.M., Thomas, J.A.: Elements of Information Theory. Wiley, Hoboken (2012)
10. Dehmer, M., Mowshowitz, A.: A history of graph entropy measures. Inf. Sci. **181**(1), 57–78 (2011)
11. Han, L., Escolano, F., Hancock, E.R., Wilson, R.C.: Graph characterizations from von Neumann entropy. Pattern Recogn. Lett. **33**(15), 1958–1967 (2012)
12. Lin, H., Zhou, B.: On the von Neumann entropy of a graph. Discrete Appl. Math. **247**, 448–455 (2018)
13. von Neumann, J.: Mathematische Grundlagen der Quantenmechanik, vol. 38. Springer, New york (1955)
14. Shanker, O.: Graph zeta function and dimension of complex network. Mod. Phys. Lett. B **21**(11), 639–644 (2007)
15. Shannon, C.E.: A mathematical theory of communication. Bell Syst. Tech. J. **27**(3), 379–423 (1948)
16. Simovici, D.: Betweenness, metrics and entropies in lattices. In: 38th IEEE International Symposium on Multiple Valued Logic, ISMVL 2008, pp. 26–31 (2008)
17. Voros, A.: Spectral zeta functions. Adv. Stud. Pure Math. **21**, 327–358 (1992)
18. Xiao, B., Hancock, E.R., Wilson, R.C.: Graph characteristics from the heat kernel trace. Pattern Recogn. **42**(11), 2589–2606 (2009)
19. Zhang, Z., Hancock, E.: Hypergraph based information-theoretic feature selection. Pattern Recogn. Lett. **33**, 1991–1999 (2012)

Graph-Based Segmentation with Local Band Constraints

Caio de Moraes Braz[1], Paulo A. V. Miranda[1(✉)], Krzysztof Chris Ciesielski[2], and Fábio A. M. Cappabianco[3]

[1] Institute of Mathematics and Statistics, University of São Paulo, São Paulo, SP 05508-090, Brazil
{caiobraz,pmiranda}@ime.usp.br
[2] Department of Mathematics, West Virginia University, Morgantown, USA
Krzysztof.Ciesielski@mail.wvu.edu
[3] Instituto de Ciência e Tecnologia, Universidade Federal de São Paulo, São José dos Campos, SP, Brazil
cappabianco@unifesp.br

Abstract. Shape constraints are potentially useful high-level priors for object segmentation, allowing the customization of the segmentation to a given target object. In this work, we present a novel shape constraint, named Local Band constraint (LB), for the generalized graph-cut framework, which in its limit case is strongly related to the Boundary Band constraint, preventing the generated segmentation to be irregular in relation to the level sets of a given reference cost map or template of shapes. The LB constraint is embedded in the graph construction with additional arcs defined by a translation-variant adjacency relation, making it easy to combine with other high-level constraints. The LB constraint demonstrates competitive results as compared to Geodesic Star Convexity, Boundary Band, and Hedgehog Shape Prior in Oriented Image Foresting Transform (OIFT) for various scenarios involving natural and medical images, with reduced sensibility to seed positioning.

Keywords: Boundary Band constraint · Hedgehog Shape Prior · Image Foresting Transform · Graph-cut segmentation

1 Introduction

Image segmentation is one of the most fundamental and challenging problems in image processing and computer vision. In many scenarios, the high-level, application-domain specific knowledge of the user is often required in the segmentation process because of the presence of heterogeneous backgrounds, objects

Thanks to CNPq (308985/2015-0, 486988/2013-9, FINEP 1266/13), FAPESP (2014/12236-1, 2016/21591-5), Coordenação de Aperfeiçoamento de Pessoal de Nível Superior - Brasil (CAPES) - Finance Code 001, and NAP eScience - PRP - USP for funding and Dr. J. K. Udupa (MIPG-UPENN) for the medical images.

M. Couprie et al. (Eds.): DGCI 2019, LNCS 11414, pp. 155–166, 2019.
https://doi.org/10.1007/978-3-030-14085-4_13

with ill-defined borders, field inhomogeneity, noise, artifacts, partial volume effects, and their interplay [19]. It may be thought of as consisting of two related processes – object recognition and delineation [10]. Recognition is the task of determining an object's approximate whereabouts in the image. Delineation completes segmentation by defining the exact spatial extent of that object. In this work, we are interested in solving the delineation problem by fast methods to efficiently deal with large amounts of data, but which must also be versatile enough to support the inclusion of high-level constraints from prior object knowledge.

The segmentation problem can be interpreted as a graph partition problem subject to hard constraints, such as seed pixels selected in the image domain for object recognition, by modelling neighborhood relations of picture elements from digital images. Examples of seed-based methods are watershed [8], random walks [12], fuzzy connectedness [5], graph cuts (GC) [2], grow cut [18], minimum barrier distance [7], and image foresting transform (IFT) [6,9]. Some methods, including the min-cut/max-flow algorithm, can provide global optimal solutions according to a graph-cut measure in graphs and can be described in a unified manner according to a common framework, which we refer to as Generalized GC (GGC) [4].

Oriented Image Foresting Transform (OIFT) [24] and *Oriented Relative Fuzzy Connectedness* (ORFC) [1] are extensions of some GGC methods for directed weighted graphs, which have lower computational complexity compared to the min-cut/max-flow algorithm [2]. OIFT is a flexible method, which has been extended to support the processing of global object properties, such as connectedness [20,21], shape constraints [23,25], boundary polarity [1,22], and hierarchical constraints [16]. These high-level priors are potentially useful for object segmentation, allowing the customization of the segmentation to a given target object. Shape constraints can be used to eliminate undesirable intricate forms, improving the segmentation of objects with more regular contour. Some shape constraints demand more sophisticated algorithms, such as the *Boundary Band constraint* (BB) [25]. The OIFT with the BB constraint allows the segmentation to follow a pre-established template of shapes, with variances within a range of permitted deformations around an arbitrary scale, while other approaches handle scale inefficiently based on brute force, by computing the graph cut for each level of a gaussian pyramid [11].

In this work, we propose a novel shape constraint, named *Local Band constraint* (LB), to be used for object segmentation in the Generalized GC framework and which, in its limit case, is strongly related to the Boundary Band constraint [25]. The LB constraint demonstrates competitive results with higher accuracy when compared to BB, Hedgehog [14,15], and Geodesic Star Convexity [13] in various scenarios. It can also be easily combined with other high-level priors already supported by OIFT, considerably advancing the targeted segmentation [17].

The next section gives the required background on image graphs and GGC. The proposed Local Band constraint is presented in Sect. 3. In Sect. 4, we experimentally evaluate LB, comparing it to previous graph-based works on shape constraints. Our conclusions are stated in Sect. 5.

2 Background

An image can be interpreted as a directed graph (digraph) $G = (\mathcal{I}, \mathcal{A})$ whose nodes/vertices are the image pixels in its image domain $\mathcal{I} \subset \mathbb{Z}^n$ and whose arcs/edges, elements of $\mathcal{A}$, are the ordered pixel pairs (s, t) of vertices that are adjacent, that is, spatially close (e.g., 4-neighborhood, or 8-neighborhood, in the case of 2D images). We write $t \in \mathcal{A}(s)$ or $(s, t) \in \mathcal{A}$ to indicate that t is adjacent to s. We will usually assume also that our image graph G is edge-weighted, that is, that each arc $(s, t) \in \mathcal{A}$ has a fixed weight $w(s, t) \in [-\infty, \infty]$ (often $w(s, t) = \|\mathcal{I}(t) - \mathcal{I}(s)\|$ for an image with values given by $\mathcal{I}(t)$). An edge weighted digraph will be denoted as $G = (\mathcal{I}, \mathcal{A}, w)$. A digraph G is symmetric if, for all $(s, t) \in \mathcal{A}$, the pair (t, s) is also an arc of G. Note that in symmetric graphs we can have $w(s, t) \neq w(t, s)$. In this work, all considered graphs are symmetric and connected.

Image segmentation can be formulated as a graph partition problem subject to hard constraints. In the case of binary segmentation (object/background), we consider two non-empty disjoint seed sets $\mathcal{S}_1$ and $\mathcal{S}_0$ containing pixels selected inside the object $\mathcal{O}$ and in its exterior, respectively. A label, $L(t) = 1$ for all $t \in \mathcal{S}_1$ and $L(t) = 0$ for all $t \in \mathcal{S}_0$, is propagated to all unlabeled pixels during the execution of seed-based segmentation algorithms, see e.g. [24]. For a label map $L \colon \mathcal{I} \to \{0, 1\}$ an object identified with it is defined as $\mathcal{O} := \{t \in \mathcal{I} \colon L(t) = 1\}$.

In the case of directed weighted graphs, there are two important classes of energy formulations within the Generalized GC framework: the Max-Min[1] and Min-Sum optimizers [4]. OIFT and ORFC algorithms are Max-Min optimizers while the min-cut/max-flow algorithm is a Min-Sum optimizer. The resulting segmentation by OIFT gives a global optimum solution by maximizing the following graph-cut measure

$$\varepsilon_{\min}(L) = \min\{w(s, t) \colon (s, t) \in \mathcal{A} \ \&\ L(s) > L(t)\} \qquad (1)$$

subject to the seed constraints [24].

The segmentation L by OIFT can be computed by Algorithm 1, which comes from [22]. This algorithm, which can also be adapted for multi-object segmentation by computing a related variant in a hierarchical layered digraph [16], will be a part of our new algorithm. Note that in line 11 of Algorithm 1, the weight $w(t, s)$ of the anti-parallel arc (t, s) is used (rather than that of chosen $(s, t) \in \mathcal{A}$). That is why a symmetric digraph is required.

For the results presented in this work, the most important property of the OIFT algorithm is the following result, see [24, Theorem 3]:

Proposition 1 [Miranda, Mansilla 2014]. *Let $G = (\mathcal{I}, \mathcal{A}, w)$ be a symmetric edge weighted image digraph. Let L be a segmentation returned by* Algorithm 1

[1] Min-Max optimizer is a dual equivalent problem.

Algorithm 1. Segmentation by OIFT Algorithm

Input: Symmetric edge weighted image digraph $(\mathcal{I}, \mathcal{A}, w)$ and non-empty disjoint seed sets $\mathcal{S}_1$ and $\mathcal{S}_0$.
Output: The label map $L\colon \mathcal{I} \to \{0,1\}$.
Auxiliary: Priority queue Q, variable tmp, and an array of status $S\colon \mathcal{I} \to \{0,1\}$, where $S(t) = 1$ for processed nodes and $S(t) = 0$ for unprocessed nodes. The value $V(t)$ represents a potential penalty that a change of $L(t)$ would contribute to $\varepsilon_{\min}(L)$.

For each $t \in \mathcal{I}$, **do**
- *Set* $S(t) \leftarrow 0$.
- **If** $t \in \mathcal{S}_0$, **then** $V(t) \leftarrow -\infty$, $L(t) \leftarrow 0$ *and insert* t *in* Q.
- **Else If** $t \in \mathcal{S}_1$, **then** $V(t) \leftarrow -\infty$, $L(t) \leftarrow 1$ *and insert* t *in* Q.
- **Else** $V(t) \leftarrow \infty$.

While $Q \neq \emptyset$ **do**
- *Remove* s *from* Q *such that* $V(s)$ *is minimum.*
- *Set* $S(s) \leftarrow 1$.
- **For each** $(s,t) \in \mathcal{A}$ *such that* $S(t) = 0$ **do**
  - **If** $L(s) = 1$, **then** $tmp \leftarrow w(s,t)$.
  - **Else If** $L(s) = 0$, **then** $tmp \leftarrow w(t,s)$.
  - **If** $tmp < V(t)$, **then**
    - *Set* $V(t) \leftarrow tmp$ *and* $L(t) \leftarrow L(s)$.
    - **If** $t \notin Q$, **then** *insert* t *in* Q.

Return L.

applied to G and non-empty disjoint seed sets $\mathcal{S}_1$ and $\mathcal{S}_0$. Then L satisfies the seed constraints and maximizes the energy $\varepsilon_{\min}$, given by (1), among all segmentations satisfying these constraints.

3 The Local Band Constraint

Let $C\colon \mathcal{I} \to [0,\infty)$ be a fixed vertex cost function associated with an image digraph $G = (\mathcal{I}, \mathcal{A})$. Usually $C(t)$ is defined as a minimum of all possible path cost functions for the paths from $\mathcal{S}_1$ to t. The path cost can be its geodesic length, as used in Geodesic Star Convexity, but other path costs are also useful. It can also be based on templates of shapes discussed in [3], which will be considered for evaluation in Sect. 4.

To relate Local Band constraint to Boundary Band constraint introduced in [25], we first introduce the following notion of *Local Boundary Band constraint*, LBB. In this definition the symbol $\|\cdot\|$ denotes the standard Euclidean L_2 norm on $\mathcal{I} \subset \mathbb{Z}^2$. The *boundary* of an object $\mathcal{O}$ is defined as

$$\mathrm{bd}(\mathcal{O}) = \{t \in \mathcal{O}\colon \exists s \in \mathcal{A}(t) \text{ such that } s \notin \mathcal{O}\}.$$

Definition 1 (Local Boundary Band constraint (LBB)**).** *For $\Delta, R > 0$ and a cost map $C\colon \mathcal{I} \to [0,\infty)$, a pixel $t \in \mathcal{O}$ is LBB_Δ^R (satisfies Local Boundary*

Band Constraint with band size Δ *and parameter* R*) provided* $C(t) < C(s) + \Delta$ *for all* $s \in \text{bd}(\mathcal{O})$ *such that* $\|s - t\| \leq R$*. An object* $\mathcal{O}$ *is* LBB^R_Δ *provided every* $t \in \mathcal{O}$ *is* LBB^R_Δ*.*

Definition 2 (Boundary Band constraint (BB)). *For* $\Delta > 0$*, an object* $\mathcal{O}$ *is* BB_Δ *(satisfies Boundary Band constraint with band size* Δ*) provided it is* LBB^∞_Δ*, that is, when* $C(t) < C(s) + \Delta$ *for all* $t \in \mathcal{O}$ *and* $s \in \text{bd}(\mathcal{O})$*. As a consequence,* $\text{bd}(\mathcal{O})$ *is contained in the band* $\{s \in \mathcal{I}: C(s) \in (m - \Delta, m]\}$*, where* $m = \max\{C(t): t \in \mathcal{O}\}$*. In particular,* $|C(s) - C(t)| < \Delta$ *for all* $s, t \in \text{bd}(\mathcal{O})$*. Consequently, this regularizes the shape of* $\text{bd}(\mathcal{O})$*, see* [25].

The idea of BB is to establish a maximum possible variation of the cost C between the boundary points $\text{bd}(\mathcal{O})$ of the object $\mathcal{O}$ to be segmented. This is expected to prevent the generated segmentation to be irregular in relation to the C-level sets [25]. During the OIFT computation subject to BB, the band changes its reference level set, allowing a better adaptation to the image content, while its width is kept fixed (Fig. 1). Note that this bears some resemblance to narrow band level set [26] and to the regional context of a level line used in [27].

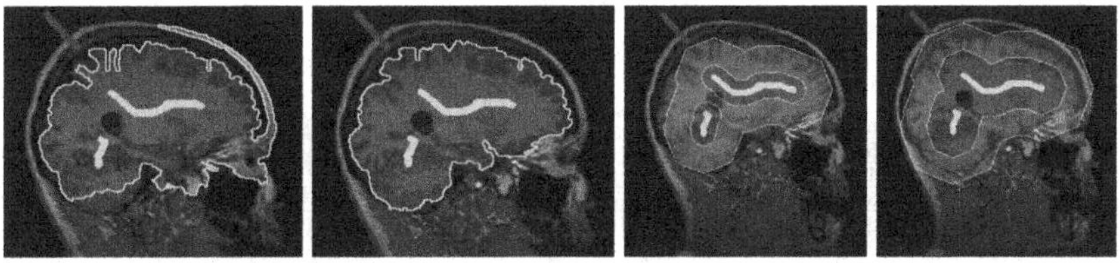

(a) (b) (c) (d)

Fig. 1. Brain segmentation example in MRI exam. (a–b) Segmentation results by OIFT without and with the BB constraint, respectively. (c–d) The BB fixed size band evolves from the seeds, adapting to the image contents. Note that the segmentation boundary achieved in (b) resides within the band area in (d).

In BB, however, local changes in a part of the object can generate constraint violations in any other part of its boundary, usually resulting in greater sensitivity to the initialization of the cost map C and to the positioning of internal seeds, while in LBB its consistency checks are limited locally, leading to a more flexible solution. Clearly, every BB_Δ object is LBB^R_Δ, but the converse is not true. However, for every C and Δ, there exists an $R \in (0, \infty)$ such that the property LBB^R_Δ implies BB_Δ (this certainly holds for any $R \geq \max\{\|s - t\|: s, t \in \mathcal{I}\}$). Thus, BB_Δ can be considered as a limit, as $R \to \infty$, of LBB^R_Δ.

In order to facilitate the implementation, we consider an approximate alternative definition, named the *Local Band constraint* (LB), in order to avoid the continuous analysis of the dynamic set of boundary pixels inside the disks of radius R at runtime, but keeping the main idea of locally restricting the band effects. This effort resulted in the following similar definition.

Definition 3 (Local Band constraint (LB)). *For $\varDelta, R > 0$ and a cost map $C\colon \mathcal{I} \to [0, \infty)$, a pixel $t \in \mathcal{O}$ is $\mathrm{LB}^R_\varDelta$ (satisfies Local Band constraint with band size $\varDelta$ and parameter R) provided $C(t) < C(s) + \varDelta$ for all $s \in \mathcal{I} \setminus \mathcal{O}$ such that $\|s - t\| \le R$. An object $\mathcal{O}$ is $\mathrm{LB}^R_\varDelta$ provided every $t \in \mathcal{O}$ is $\mathrm{LB}^R_\varDelta$.*

In other words, if $\mathcal{O}$ is $\mathrm{LB}^R_\varDelta$, then for any pair of pixels s and t such that $\|s - t\| \le R$ and $C(t) - C(s) \ge \varDelta$, we have that $t \in \mathcal{O}$ implies $s \in \mathcal{O}$. Note that neither of the statements "$\mathcal{O}$ is $\mathrm{LB}^R_\varDelta$" and "$\mathcal{O}$ is $\mathrm{LBB}^R_\varDelta$" implies the other. Nevertheless, they are closely related (Fig. 2), as shown by the following result.

Proposition 2. *Let $r = \max_{(s,t)\in\mathcal{A}} \|s - t\|$ and $\delta = \max_{(s,t)\in\mathcal{A}} |C(t) - C(s)|$. If $\varDelta, R > 0$ and $\mathcal{O}$ is $\mathrm{LB}^{R+r}_\varDelta$, then $\mathcal{O}$ is $\mathrm{LBB}^R_{\varDelta+\delta}$.*

Proof. Choose a $t \in \mathcal{O}$. Then $C(t) < C(s) + \varDelta$ for all $s \in \mathcal{I}\setminus\mathcal{O}$ such that $\|s - t\| \le R + r$. We need to show that t is $\mathrm{LBB}^R_{\varDelta+\delta}$, that is, that $C(t) < C(u) + \varDelta + \delta$ for all $u \in \mathrm{bd}(\mathcal{O})$ such that $\|u - t\| \le R$. So, take such u. Then, there is an $s \in \mathcal{I}\setminus\mathcal{O}$ with $(u, s) \in \mathcal{A}$. Notice that $\|s - t\| \le \|s - u\| + \|u - t\| \le r + R$. Using this and the definition of δ, we get $C(t) < C(s) + \varDelta \le C(u) + \varDelta + |C(s) - C(u)| \le C(u) + \varDelta + \delta$, as needed.

Since usually numbers δ and r are small, so should be the difference between the objects with properties $\mathrm{LB}^R_\varDelta$, $\mathrm{LB}^{R+r}_\varDelta$, $\mathrm{LBB}^R_{\varDelta+\delta}$, or $\mathrm{LBB}^R_\varDelta$ and, for large R, each approximates $\mathrm{BB}_\varDelta$.

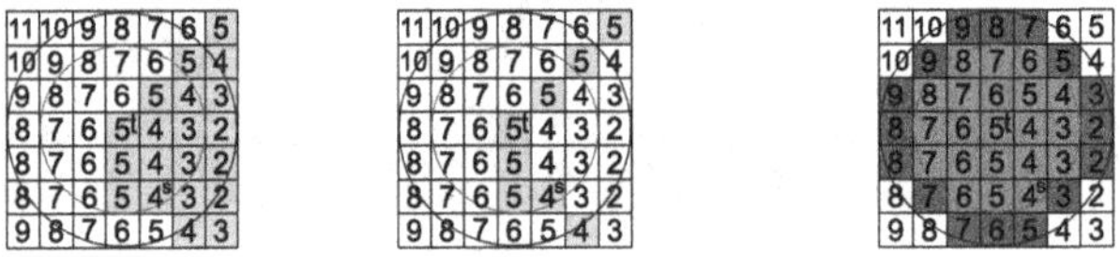

(a) $\mathcal{O}$ (in yellow) (b) $\mathrm{bd}(\mathcal{O})$ (in yellow) (c) Disks of R and $R + r$.

Fig. 2. Example of Proposition 2, where "t is $\mathrm{LB}^{R+r}_\varDelta$" and "$t$ is $\mathrm{LBB}^R_{\varDelta+\delta}$" for $R = 2.5$, $r = 1.0$, $\varDelta = 1$ and $\delta = 1$. (a) $\mathcal{O}$, (b) $\mathrm{bd}(\mathcal{O})$, and (c) the disks of radii R and $R + r$. (Color figure online)

The LB constraint can be implemented, as proposed in Algorithm 2 for OIFT, by considering a modified graph G' with the LB constraint embedded on its arcs. In general, the worst cost should be ∞ for Min-Sum optimizers and $-\infty$ for Max-Min optimizers. In order to maintain a symmetric graph, we also create anti-parallel arcs with the best cutting cost (zero for Min-Sum and ∞ for Max-Min optimizers) if they do not exist (line 5 in Algorithm 2). Note that in G' the set of displacement vectors $D(s) = \{t - s : t \in \mathcal{A}'(s)\}$ varies for different positions of s, leading therefore to a translation-variant adjacency relation.

Theorem 3. *Let $G = (\mathcal{I}, \mathcal{A}, w)$ be a symmetric edge weighted image digraph with $w\colon \mathcal{A} \to \mathbb{R}$. Let L be a segmentation returned by* Algorithm 2 *applied to G, non-empty disjoint seed sets $\mathcal{S}_1$ and $\mathcal{S}_0$, cost map $C\colon \mathcal{I} \to [0, \infty)$, and parameters $R > 0$ and $\varDelta > 0$. Assume that $\mathcal{S}_1$ and $\mathcal{S}_0$ are $\mathrm{LB}^R_\varDelta$-consistent, that is, that*

($\star$) there exists a labeling satisfying seeds and $\mathrm{LB}^R_\varDelta$ constraints.

Algorithm 2. SEGMENTATION BY OIFT SUBJECT TO THE LB CONSTRAINT

INPUT: Symmetric edge weighted image digraph $G = (\mathcal{I}, \mathcal{A}, w)$, non-empty disjoint seed sets $\mathcal{S}_1$ and $\mathcal{S}_0$, cost map $C\colon \mathcal{I} \to [0, \infty)$, and parameters $R > 0$ and $\Delta > 0$.
OUTPUT: The label map $L\colon \mathcal{I} \to \{0, 1\}$.
AUXILIARY: Edge weighted digraph $G' = (\mathcal{I}, \mathcal{A}', w')$ with $\mathcal{A} \subset \mathcal{A}'$.

1. *Set* $\mathcal{A}' \leftarrow \mathcal{A}$ *and* $w' \leftarrow w$.
2. **For each** $(s,t) \in \{(p,q) \in \mathcal{I} \times \mathcal{I} : \|p-q\| \leq R \ \&\ C(p) \geq C(q) + \Delta\}$ **do**
3.   **If** $(s,t) \notin \mathcal{A}'$ **then** *Set* $\mathcal{A}' \leftarrow \mathcal{A}' \cup \{(s,t)\}$ *and define* $w'(s,t) := -\infty$.
4.   **Else** *Redefine* $w'(s,t) := -\infty$.
5.   **If** $(t,s) \notin \mathcal{A}'$ **then** *Set* $\mathcal{A}' \leftarrow \mathcal{A}' \cup \{(t,s)\}$ *and define* $w'(t,s) := \infty$.
6. *Compute, by Algorithm 1,* $L\colon \mathcal{I} \to \{0,1\}$ *for* G' *and seed sets* $\mathcal{S}_1$ *and* $\mathcal{S}_0$.
7. *Return* L.

Then L *satisfies seeds and* LB_{Δ}^{R} *constraints and maximizes the energy* $\varepsilon_{\min}$, *given by (1) w.r.t.* G, *among all segmentations satisfying these constraints.*

Proof. In this proof $\varepsilon_{\min}^{G}$ and $\varepsilon_{\min}^{G'}$ denote the energy $\varepsilon_{\min}$ with respect to G and G', respectively. Let $\mathcal{L} := \{(p,q) \in \mathcal{I} \times \mathcal{I}\colon 0 < \|p-q\| \leq R \ \&\ C(p) \geq C(q) + \Delta\}$ and $\mathcal{M} := \{(t,s)\colon (s,t) \in \mathcal{L}\} \setminus \mathcal{A}$. It is easy to see that after the execution of lines 1-5 we have $\mathcal{A}' = \mathcal{A} \cup \mathcal{L} \cup \mathcal{M}$ and

$$w'(s,t) = \begin{cases} -\infty & \text{for } (s,t) \in \mathcal{L}, \\ \infty & \text{for } (s,t) \in \mathcal{M}, \\ w(s,t) & \text{otherwise, that is for } (s,t) \in \mathcal{A} \setminus \mathcal{L}. \end{cases}$$

Also, by Proposition 1, after the execution of line 6 the labeling L satisfies the seed constraints and maximizes the energy $\varepsilon_{\min}^{G'}$ among all segmentations satisfying seeds constraints. We need to show that L satisfies also LB_{Δ}^{R} constraints an that it maximizes $\varepsilon_{\min}^{G}$ among all segmentations satisfying these constraints.

To see this, let $L'\colon \mathcal{I} \to \{0,1\}$ be an arbitrary labeling satisfying seeds and LB_{Δ}^{R} constraints. It exists by $(\star)$. Then, by the definition of LB_{Δ}^{R} constraints, the set $T' := \{(p,q) \in \mathcal{A}'\colon L'(p) > L'(q)\}$ is disjoint with $\mathcal{L}$. In particular,

$$\varepsilon_{\min}^{G'}(L) \geq \varepsilon_{\min}^{G'}(L') = \min\{w'(s,t)\colon (s,t) \in \mathcal{A}' \ \&\ L'(s) > L'(t)\} > -\infty.$$

Hence

$$\varepsilon_{\min}^{G'}(L) = \min\{w'(s,t)\colon (s,t) \in \mathcal{A}' \ \&\ L(s) > L(t)\} > -\infty,$$

so that the set $T := \{(p,q) \in \mathcal{A}'\colon L(p) > L(q)\}$ must be also disjoint with $\mathcal{L}$. This means that L satisfies LB_{Δ}^{R} constraints. To finish the proof we need to show that $\varepsilon_{\min}^{G}(L) \geq \varepsilon_{\min}^{G}(L')$. For this notice first that

$$\varepsilon_{\min}^{G'}(L') = \varepsilon_{\min}^{G}(L'). \tag{2}$$

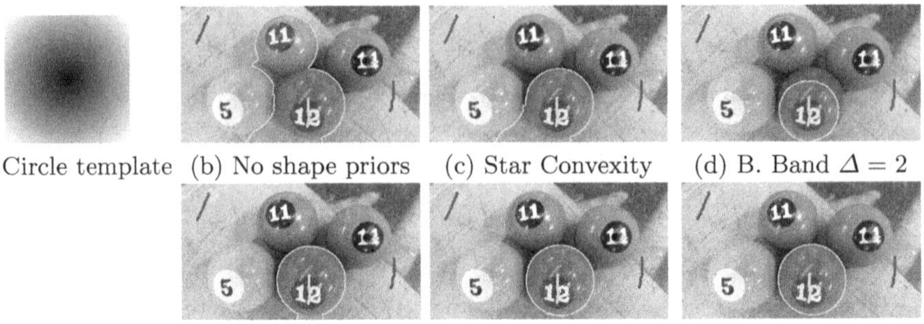

(a) Circle template (b) No shape priors (c) Star Convexity (d) B. Band $\Delta = 2$

(e) B. Band $\Delta = 40$ (f) Hedgehog $\theta = 45°$ (g) Local Band $\Delta = 2$

Fig. 3. Pool ball OIFT segmentation with a circle template in a 600×338 image.

Indeed, $T' \cup T$ is disjoint with $\mathcal{L}$, so $(s,t) \in \mathcal{A}'$ & $L'(s) > L'(t)$ implies that $(s,t) \in (\mathcal{A} \setminus \mathcal{L}) \cup \mathcal{M}$. Thus, since $w' = w$ on $\mathcal{A} \setminus \mathcal{L}$ and $w' = \infty$ on $\mathcal{M}$,

$$\begin{aligned}\varepsilon_{\min}^{G'}(L') &= \min\{w'(s,t) \colon (s,t) \in \mathcal{A}' \;\&\; L'(s) > L'(t)\} \\ &= \min\left(\{w'(s,t) \colon (s,t) \in \mathcal{A} \setminus \mathcal{L} \;\&\; L'(s) > L'(t)\} \cup \{\infty\}\right) \\ &= \min\left(\{w(s,t) \colon (s,t) \in \mathcal{A} \;\&\; L'(s) > L'(t)\} \cup \{\infty\}\right) = \varepsilon_{\min}^{G}(L'),\end{aligned}$$

as needed. Finally, using (2) for L and L', we obtain

$$\varepsilon_{\min}^{G}(L) = \varepsilon_{\min}^{G'}(L) \geq \varepsilon_{\min}^{G'}(L') = \varepsilon_{\min}^{G}(L'),$$

finishing the proof.

4 Experimental Results

In this section we compare LB with shape constraints commonly employed in graph-based segmentation: Geodesic Star Convexity [13], Boundary Band [25], and Hedgehog Shape Prior [14,15]. We opted to compare them using Max-Min optimizers, because BB is not yet supported by Min-Sum optimizers [25].

From the IFT [9] perspective, when the cost map C is the geodesic length from $\mathcal{S}_1$ in $G = (\mathcal{I}, \mathcal{A})$, the previous constraints are based on different attributes of a previously computed minimal forest in G rooted at $\mathcal{S}_1$: Geodesic Star Convexity uses the predecessor map [23], BB and LB constraints exploit the cost map directly, and Hedgehog uses the gradient of the cost map as vector field.

Figure 3 shows the segmentation results by OIFT using different methods and a circle template, as reference cost map, centered on the center of mass of the internal seeds. The BB constraint fails to give good results compared to Local Band and Hedgehog, due to its greater sensitivity to the template positioning. Figure 4 shows some results of a tile segmentation using a square template. In order to measure the sensitivity of the most promising methods for different seed positioning, in Fig. 5 we show the accuracy curves using internal seeds in

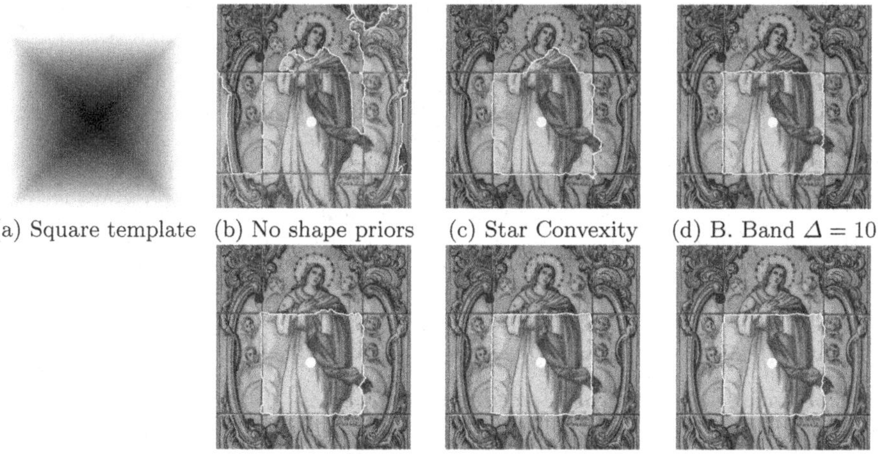

(a) Square template (b) No shape priors (c) Star Convexity (d) B. Band $\varDelta = 10$

(e) Hedgehog $\theta = 45°$ (f) Local Band $\varDelta = 1$ (g) Local Band $\varDelta = 2$

Fig. 4. Wall tile segmentation by OIFT with a square template in a 576×881 image.

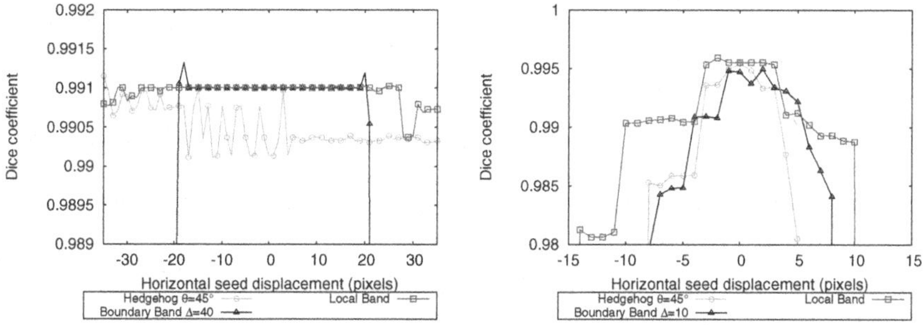

(a) pool ball (b) wall tile

Fig. 5. The accuracy curves for different horizontal displacements of the internal seeds.

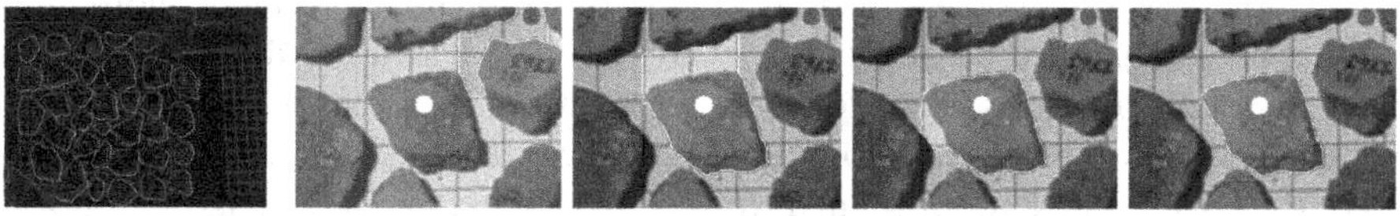

(a) Sobel gradient (b) No priors (c) B. Band (d) Hedgehog (e) Local Band

Fig. 6. Archaeological fragment segmentation.

a circular brush of radius 3 pixels with horizontal displacements relative to the object's center. Note that, in both cases, LB ($R = 3.5$ and $\varDelta = 2$) has the most accurate and slightly more stable results, giving almost perfect results for 41.4% and 10.3% of the maximum possible horizontal shift in the pool ball (radius 84 pixels) and wall tile (radius 145 pixels), respectively.

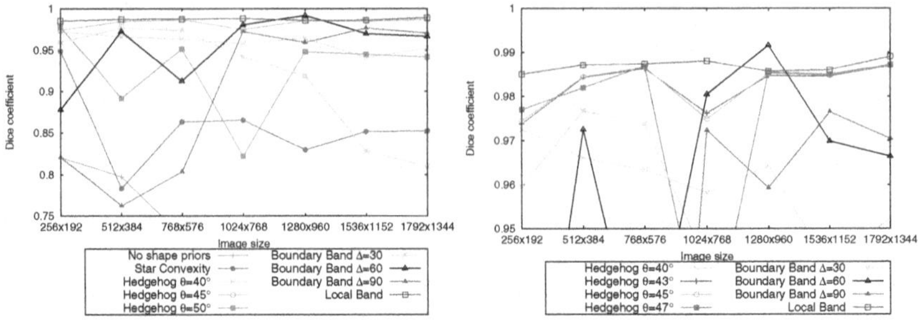

(a) (b)

Fig. 7. (a) The mean accuracy values to segment the archaeological fragments for different image resolutions. (b) Zoomed results (accuracy $\geq 95\%$).

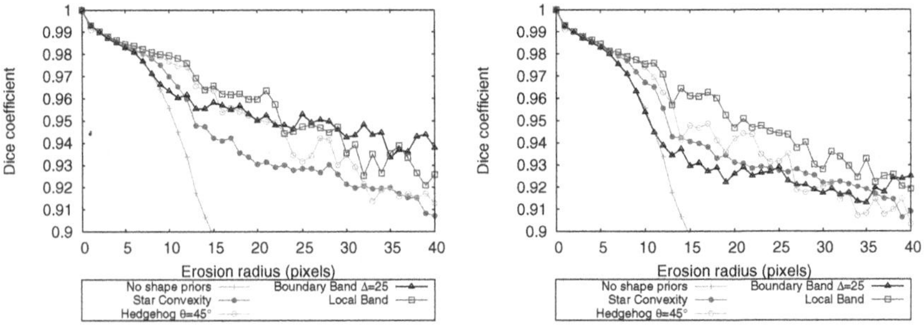

(a) $\mathcal{S}_1$ centralized (b) $\mathcal{S}_1$ shifted to the left

Fig. 8. The mean accuracy curves to segment the liver for seed sets obtained by erosion.

We also tested their robustness in relation to different image resolutions by quantitative experiments, to segment archaeological fragments in seven different resolutions with the geodesic cost. In order to make the experiment more challenging, the simple arc weight $w(s,t) = G(s) + G(t)$ was used, disregarding any prior color information, where $G(t)$ denotes the magnitude of Sobel gradient, such that we have several false boundaries (Fig. 6). Figure 7 shows the mean values of the Dice coefficient for segmenting ten fragments for each image resolution, totalizing 70 executions for each method. The overall best results were obtained by LB using $R = 3.5$ and $\Delta = 2$. Hedgehog for different θ values and the same radius presented unstable results (Fig. 6d). Further increasing its radius is not recommended, since it drastically increases the computational cost.

Finally, we conducted experiments with the geodesic cost to segment the liver in medical images of 40 slices of thoracic CT studies of size 512×512, using regular weights $w(s,t) = \|\mathcal{I}(t) - \mathcal{I}(s)\|$ and seed sets progressively obtained by eroding the ground truth and its background with twice the radius size. Although this scenario is apparently advantageous for the BB constraint, in view of the well-distributed and centralized seeds, LB ($R = 3.5$ and $\Delta = 2$) demonstrated good results with the highest accuracy for a large part of the curve (Fig. 8a). We repeated the experiments, but now with the internal seeds shifted by 5 pixels to the left (25% of the maximum possible displacement in the central part of the curves) whenever possible. In this new scenario, the results clearly show that LB is more robust than BB in relation to seed positioning (Fig. 8b).

5 Conclusion

We have proposed the Local Band shape constraint, for the Generalized GC framework, which in its limit case (i.e., $R \to \infty$) is strongly related to Boundary Band constraint and is less sensitive to the seed/template positioning. To the best of our knowledge, we are also the first to report OIFT with the Hedgehog shape prior. As future work, we intend to test LB in 3D medical applications.

References

1. Bejar, H.H., Miranda, P.A.: Oriented relative fuzzy connectedness: theory, algorithms, and its applications in hybrid image segmentation methods. EURASIP J. Image Video Process. **2015**(1), 21 (2015)
2. Boykov, Y., Funka-Lea, G.: Graph cuts and efficient N-D image segmentation. Int. J. Comput. Vis. **70**(2), 109–131 (2006)
3. Braz, C.D.M.: Segmentação de imagens pela transformada imagem-floresta com faixa de restrição geodésica. Master's thesis, Instituto de Matemática e Estatística, Universidade de São Paulo, São Paulo, Brasil (2016)
4. Ciesielski, K., Udupa, J., Falcão, A., Miranda, P.: A unifying graph-cut image segmentation framework: algorithms it encompasses and equivalences among them. In: Proceedings of SPIE on Medical Imaging: Image Processing, vol. 8314 (2012)
5. Ciesielski, K., Udupa, J., Saha, P., Zhuge, Y.: Iterative relative fuzzy connectedness for multiple objects with multiple seeds. Comput. Vis. Image Underst. **107**(3), 160–182 (2007)
6. Ciesielski, K.C., Falcão, A.X., Miranda, P.A.V.: Path-value functions for which Dijkstra's algorithm returns optimal mapping. J. Math. Imaging Vis. **60**(7), 1025–1036 (2018)
7. Ciesielski, K.C., Strand, R., Malmberg, F., Saha, P.K.: Efficient algorithm for finding the exact minimum barrier distance. Comput. Vis. Image Underst. **123**, 53–64 (2014)
8. Cousty, J., Bertrand, G., Najman, L., Couprie, M.: Watershed cuts: thinnings, shortest path forests, and topological watersheds. Trans. Pattern Anal. Mach. Intell. **32**, 925–939 (2010)
9. Falcão, A., Stolfi, J., Lotufo, R.: The image foresting transform: theory, algorithms, and applications. IEEE TPAMI **26**(1), 19–29 (2004)
10. Falcão, A., Udupa, J., Samarasekera, S., Sharma, S., Hirsch, B., Lotufo, R.: User-steered image segmentation paradigms: Live-wire and live-lane. Graph. Mod. Image Process. **60**, 233–260 (1998)
11. Freedman, D., Zhang, T.: Interactive graph cut based segmentation with shape priors. In: IEEE Computer Society Conference on 2005 Computer Vision and Pattern Recognition, CVPR 2005, vol. 1, pp. 755–762. IEEE (2005)

12. Grady, L.: Random walks for image segmentation. IEEE Trans. Pattern Anal. Mach. Intell. **28**(11), 1768–1783 (2006)
13. Gulshan, V., Rother, C., Criminisi, A., Blake, A., Zisserman, A.: Geodesic star convexity for interactive image segmentation. In: Proceedings of Computer Vision and Pattern Recognition, pp. 3129–3136 (2010)
14. Isack, H., Veksler, O., Sonka, M., Boykov, Y.: Hedgehog shape priors for multi-object segmentation. In: 2016 IEEE Conference on Computer Vision and Pattern Recognition (CVPR), pp. 2434–2442, June 2016
15. Isack, H.N., Boykov, Y., Veksler, O.: A-expansion for multiple "hedgehog" shapes. CoRR abs/1602.01006 (2016). http://arxiv.org/abs/1602.01006
16. Leon, L.M.C., Miranda, P.A.V.D.: Multi-object segmentation by hierarchical layered oriented image foresting transform. In: 2017 30th SIBGRAPI Conference on Graphics, Patterns and Images (SIBGRAPI), pp. 79–86, October 2017
17. Lézoray, O., Grady, L.: Image Processing and Analysis with Graphs: Theory and Practice. CRC Press, California (2012)
18. Li, X., Chen, J., Fan, H.: Interactive image segmentation based on grow cut of two scale graphs. In: Zhang, W., Yang, X., Xu, Z., An, P., Liu, Q., Lu, Y. (eds.) Advances on Digital Television and Wireless Multimedia Communications, pp. 90–95. Springer, Berlin Heidelberg, Berlin, Heidelberg (2012). https://doi.org/10.1007/978-3-642-34595-1_13
19. Madabhushi, A., Udupa, J.: Interplay between intensity standardization and inhomogeneity correction in MR image processing. IEEE Trans. Med. Imaging **24**(5), 561–576 (2005)
20. Mansilla, L.A.C., Miranda, P.A.V.: Oriented image foresting transform segmentation: Connectivity constraints with adjustable width. In: 29th SIBGRAPI Conference on Graphics, Patterns and Images, pp. 289–296, October 2016
21. Mansilla, L.A.C., Miranda, P.A.V., Cappabianco, F.A.M.: Oriented image foresting transform segmentation with connectivity constraints. In: 2016 IEEE International Conference on Image Processing (ICIP), pp. 2554–2558, September 2016
22. Mansilla, L., Miranda, P.: Image segmentation by oriented image foresting transform: handling ties and colored images. In: 18th International Conference on Digital Signal Processing, pp. 1–6. Greece July 2013
23. Mansilla, L.A.C., Miranda, P.A.V.: Image segmentation by oriented image foresting transform with geodesic star convexity. In: Wilson, R., Hancock, E., Bors, A., Smith, W. (eds.) CAIP 2013. LNCS, vol. 8047, pp. 572–579. Springer, Heidelberg (2013). https://doi.org/10.1007/978-3-642-40261-6_69
24. Miranda, P., Mansilla, L.: Oriented image foresting transform segmentation by seed competition. IEEE Trans. Image Process. **23**(1), 389–398 (2014)
25. de Moraes Braz, C., Miranda, P.: Image segmentation by image foresting transform with geodesic band constraints. In: IEEE International Conference on Image Processing (ICIP) 2014, pp. 4333–4337, October 2014
26. Sethian, J.: A fast marching level set method for monotonically advancing fronts. Proc. Nat. Acad. Sci. USA **93**(4), 1591–5 (1996)
27. Xu, Y., Géraud, T., Najman, L.: Context-based energy estimator: application to object segmentation on the tree of shapes. In: 2012 19th IEEE International Conference on Image Processing, pp. 1577–1580, September 2012

A Study of Observation Scales Based on Felzenswalb-Huttenlocher Dissimilarity Measure for Hierarchical Segmentation

Edward Cayllahua-Cahuina[1,2](✉), Jean Cousty[1], Silvio Guimarães[3], Yukiko Kenmochi[1], Guillermo Cámara-Chávez[4], and Arnaldo de Albuquerque Araújo[1,2]

[1] Université Paris-Est, LIGM, CNRS - ENPC - ESIEE Paris - UPEM, Noisy-le-Grand, France
ecayllahua1@gmail.com

[2] Computer Science Department, Universidade Federal de Minas Gerais, Belo Horizonte, Brazil

[3] PUC Minas - ICEI - DCC - VIPLAB, Belo Horizonte, Brazil

[4] Computer Science Department, Universidade Federal de Ouro Preto, Ouro Preto, Brazil

Abstract. Hierarchical image segmentation provides a region-oriented scale-space, *i.e.*, a set of image segmentations at different detail levels in which the segmentations at finer levels are nested with respect to those at coarser levels. Guimarães *et al.* proposed a hierarchical graph based image segmentation (HGB) method based on the Felzenszwalb-Huttenlocher dissimilarity. This HGB method computes, for each edge of a graph, the minimum scale in a hierarchy at which two regions linked by this edge should merge according to the dissimilarity. In order to generalize this method, we first propose an algorithm to compute the intervals which contain all the observation scales at which the associated regions should merge. Then, following the current trend in mathematical morphology to study criteria which are not increasing on a hierarchy, we present various strategies to select a significant observation scale in these intervals. We use the BSDS dataset to assess our observation scale selection methods. The experiments show that some of these strategies lead to better segmentation results than the ones obtained with the original HGB method.

1 Introduction

Hierarchical image segmentation provides a multi-scale approach to image analysis. Hierarchical image analysis was pioneered by [11] and has received a lot of attention since then, as attested by the popularity of [1]. Mathematical morphology has been used in hierarchical image analysis with, *e.g.*, hierarchical watersheds [3,12], binary partition trees [14], quasi-flat zones hierarchies [10],

M. Couprie et al. (Eds.): DGCI 2019, LNCS 11414, pp. 167–179, 2019.
https://doi.org/10.1007/978-3-030-14085-4_14

and tree-based shape spaces [17]. Other methods for hierarchical image analysis consider regular and irregular pyramids [9], scale-set theory [7], multiscale combinatorial grouping [13] and series of optimization problems [16].

A hierarchical image segmentation is a series of image segmentations at different detail levels where the segmentations at higher detail levels are produced by merging regions from segmentations at finer detail levels. Consequently, the regions at finer detail levels are nested in regions at coarser levels. The level of a segmentation in a hierarchy is also called an *observation scale*. In [8], Guimarães *et al.* proposed a hierarchical graph based image segmentation (HGB) method based on the Felzenszwalb-Huttenlocher dissimilarity measure. The HGB method computes, for each edge of a graph, the minimum observation scale in a hierarchy at which two regions linked by this edge should merge according to the dissimilarity.

In this article, we provide a formal definition of the criterion which is implicitly used in the HGB method. Then, we show that this criterion is not increasing with respect to the observation scales. An important consequence of this observation is that selecting the minimum observation scale for which the criterion holds true, as done with the original HGB method, is not the unique strategy that makes sense with respect to practical needs. Hence, following a recent trend of mathematical morphology (see, *e.g.*, [17]) to study non-increasing criteria on a hierarchy, we investigate scale selection strategies, leading to new variations of the original HGB method. The proposed methods are assessed with the evaluation framework of [1]. The assessment shows that some of the proposed variations significantly outperform the original HGB method (see illustration in Fig. 1).

Fig. 1. Saliency maps resulting from the HGB method using the original observation scale (middle) and from one of our proposed observation scale (right).

Section 2 presents the basic notions for the HGB method. Section 3 discusses the non-increasing property of the criterion used by the HGB method and introduces an algorithm to find all the scales (associated to a given edge) which satisfy the criterion. Section 4 presents a series of strategies to select different observation scales based on this non-increasing criterion and Sect. 5 provide the comparative analysis between the proposed strategies and the original HGB method.

2 Hierarchical Graph-Based Image Segmentation

This section aims at explaining the method of hierarchical graph-based image segmentation (HGB) [8]. We first give a series of necessary notions such as quasi-flat zones hierarchies [10], and then describe the HGB method.

2.1 Basic Notions

Hierarchies. Given a finite set V, a *partition* of V is a set $\mathbf{P}$ of nonempty disjoint subsets of V whose union is V. Any element of $\mathbf{P}$ is called a *region* of $\mathbf{P}$. Given two partitions $\mathbf{P}$ and $\mathbf{P}'$ of V, $\mathbf{P}'$ is said to be a refinement of $\mathbf{P}$, denoted by $\mathbf{P}' \preceq \mathbf{P}$, if any region of $\mathbf{P}'$ is included in a region of $\mathbf{P}$. A hierarchy on V is a sequence $\mathcal{H} = (\mathbf{P}_0, \ldots, \mathbf{P}_\ell)$ of partitions of V, such that $\mathbf{P}_{i-1} \preceq \mathbf{P}_i$, for any $i \in \{1, \ldots, \ell\}$.

Graph and Connected-Component Partition. A *graph* is a pair $G = (V, E)$ where V is a finite set and E is a subset of $\{\{x, y\} \subseteq V | x \neq y\}$. Each element of V is called a *vertex* of G, and each element of E is called an *edge* of G. A subgraph of G is a graph (V', E') such that $V' \subseteq V$ and $E' \subseteq E$. If X is a graph, its vertex and edge sets are denoted by $V(X)$ and $E(X)$, respectively.

If two vertices of a graph G are joined by an edge, we say that they are adjacent. From the reflexive–transitive closure of this adjacency relation on a finite set $V(G)$, we derive the connectivity relation on $V(X)$. It is an equivalence relation, whose equivalence classes are called connected components of G. We denote by $\mathbf{C}(G)$ the set of all connected components of G. Note that $\mathbf{C}(G)$ is a partition of $V(G)$, called the *connected-component partition induced by G*.

Quasi-flat Zone Hierarchies. Given a graph $G = (V, E)$, let w be a map from E into the set $\mathbb{R}$ of real numbers. For any edge u of G, the value $w(u)$ is called the weight of u (for w), and the pair (G, w) is called an *edge-weighted graph*. We now make from an edge-weighted graph a series of connected-component partitions, which constitutes a hierarchy. Such a hierarchy is called a quasi-flat zone hierarchy of (G, w) and the quasi-flat zone hierarchy transform is a bijection between the hierarchies and a subset of the edge weighted graphs called the saliency maps [4]. Hence, any edge-weighted graph induces a quasi-flat zone hierarchy and any hierarchy $\mathcal{H}$ can be represented by an edge-weighted graph whose quasi-flat zone hierarchy is precisely $\mathcal{H}$ [4]. This bijection allows us to handle quasi-flat zone hierarchies through edge-weighted graphs.

Given an edge-weighted graph (G, w), let X be a subgraph of G and let λ be a value of $\mathbb{R}$. The *λ-level edge set* of X for w is defined by $w_\lambda(X) = \{u \in E(X) \mid w(u) < \lambda\}$, and the *$\lambda$-level graph* of X for w is defined as the subgraph $w_\lambda^V(X)$ of X such that $w_\lambda^V(X) = (V(X), w_\lambda(X))$. Then, the connected-component partition $\mathbf{C}(w_\lambda^V(X))$ induced by $w_\lambda^V(X)$ is called the *λ-level partition* of X for w.

As we consider only finite graphs and hierarchies, the set of considered level values is reduced to a finite subset of $\mathbb{R}$ that is denoted by $\mathbb{E}$ in the remaining parts of this article. In order to browse the values of this set and to round real values to values of $\mathbb{E}$, we define, for any $\lambda \in \mathbb{R}$: $\mathrm{p}_{\mathbb{E}}(\lambda) = \max\{\mu \in \mathbb{E}\cup\{-\infty\} \mid \mu < \lambda\}$, $\mathrm{n}_{\mathbb{E}}(\lambda) = \min\{\mu \in \mathbb{E}\cup\{\infty\} \mid \mu > \lambda\}$ and $\hat{\mathrm{n}}_{\mathbb{E}}(\lambda) = \min\{\mu \in \mathbb{E}\cup\{\infty\} \mid \mu \geq \lambda\}$.

Let (G, w) be an edge-weighted graph and let X be a subgraph of G. The sequence of all λ-level partitions of X for w, ordered by increasing value of λ, is a hierarchy, defined by $\mathcal{QFZ}(X, w) = (\mathbf{C}(w_\lambda^V(X)) \mid \lambda \in \mathbb{E} \cup \{\infty\})$, and called the *quasi-flat zone hierarchy of X for w*. Let $\mathcal{H}$ be the quasi-flat zone hierarchy of G for w. Given a vertex x of G and a value λ in $\mathbb{E}$, the region that contains x in the λ-level partition of the graph G is denoted by $\mathcal{H}_x^\lambda$.

Let us consider a minimum spanning tree T of (G, w). It has been shown in [4] that $\mathcal{QFZ}(T, w)$ of T for w is the same as $\mathcal{QFZ}(G, w)$ of G for w. This indicates that the quasi-flat zone hierarchy for G can be handled by its minimum spanning tree.

2.2 Hierarchical Graph-Based Segmentation Method

In this article, we consider that the input is the edge-weighted graph (G, w) representing an image, where the pixels correspond to the vertices of G and the edges link adjacent pixels. The weight of each edge is given by a dissimilarity measure between the linked pixels such as the absolute difference of intensity between them.

Before explaining the HGB method, we first describe the following observation scale dissimilarity [8], which is required by the method and whose idea originates from the region merging criterion proposed in [6].

Observation Scale Dissimilarity. Let R_1 and R_2 be two adjacent regions, the dissimilarity measure compares the so-called inter-component and within-component differences [6]. The *inter-component difference* between R_1 and R_2 is defined by $\Delta_{inter}(R_1, R_2) = \min\{w(\{x, y\}) \,|x \in R_1, y \in R_2, \{x, y\} \in E(T)\}$, while the *within-component difference* of a region R is defined by $\Delta_{intra}(R) = \max\{w(\{x, y\}) \,|x, y \in R, \{x, y\} \in E(T)\}$. It leads to the *observation scale of R_1 relative to R_2*, defined by $S_{R_2}(R_1) = (\Delta_{inter}(R_1, R_2) - \Delta_{intra}(R_1)) \,|R_1|$, where $|R_1|$ is the cardinality of R_1. Then, a symmetric metric between R_1 and R_2, called the *observation scale dissimilarity between R_1 and R_2*, is defined by

$$D(R_1, R_2) = \max\{S_{R_2}(R_1), S_{R_1}(R_2)\}. \qquad (1)$$

This dissimilarity is used to determine if two regions should be merged or not at a certain observation scale in the following.

HGB Method. The HGB method [8] is presented in Method 1. The input is a graph G representing an image with its associated weight function w, where the minimum spanning tree T of G is taken. Given (T, w), the HGB method computes a new weight function f which leads to a new hierarchy $\mathcal{H} = \mathcal{QFZ}(T, f)$.

Method 1. HGB method

Input : A minimum spanning tree T of an edge-weighted graph (G, w)
Output: A hierarchy $\mathcal{H} = \mathcal{QFZ}(T, f)$

1 **for** *each* $u \in E(T)$ **do** $f(u) := \max\{\lambda \in \mathbb{E}\}$;
2 **for** *each* $u \in E(T)$ *in non-decreasing order for* w **do**
3 $\quad \mathcal{H} := \mathcal{QFZ}(T, f)$;
4 $\quad f(u) := \mathrm{p}_{\mathbb{E}}\left(\lambda^{\star}_{\mathcal{H}}(u)\right)$;
5 **end**
6 $\mathcal{H} := \mathcal{QFZ}(T, f)$;

The resulting hierarchy $\mathcal{H}$ is considered as the hierarchical image segmentations of the initial image. Thus, the core of the method is the generation of the weight function f for T.

After initializing all values of f to infinity (see Line 1), we compute an observation scale value $f(u)$ for each edge $u \in E(T)$ in non-decreasing order with respect to the original weight w (see Line 2). Note that each iteration in the loop requires updating the hierarchy $\mathcal{H} = \mathcal{QFZ}(T, f)$ (see Line 3). An efficient algorithm for the hierarchy update can be found in [2]. Once $\mathcal{H}$ is updated, the value $\lambda^{\star}_{\mathcal{H}}(u)$ of a finite subset $\mathbb{E}$ of $\mathbb{R}$ is obtained by

$$\lambda^{\star}_{\mathcal{H}}(\{x, y\}) = \min\left\{\lambda \in \mathbb{E} \mid D\left(\mathcal{H}^{\lambda}_{x}, \mathcal{H}^{\lambda}_{y}\right) \leq \lambda\right\}. \tag{2}$$

We first consider the regions $\mathcal{H}^{\lambda}_{x}$ and $\mathcal{H}^{\lambda}_{y}$ at a level λ. Using the dissimilarity measure D, we check if $D\left(\mathcal{H}^{\lambda}_{x}, \mathcal{H}^{\lambda}_{y}\right) \leq \lambda$. Equation (2) states that the observation scale $\lambda^{\star}_{\mathcal{H}}(\{x, y\})$ is the minimum value λ for which this assertion holds.

3 Observation Scale Intervals

3.1 Non-increasing Observation Criterion

In this section, we provide a formal definition of the observation criterion which is involved in Eq. (2). Then, we discuss its non-increasing behaviour opening the doors towards new strategies to select interesting observation scale values based on Felzenszwalb-Huttenlocher dissimilarity measure as used in Method 1.

In the remaining part of this section, we consider that $\mathcal{H}$ is any hierarchy and that $u = \{x, y\}$ is any edge of T.

Let λ be any element in $\mathbb{E}$. We say that λ is a *positive observation scale (for* $(\mathcal{H}, u)$*)* whenever $D(\mathcal{H}^{\lambda}_{x}, \mathcal{H}^{\lambda}_{y}) \leq \lambda$. We denote by $\mathcal{T}$ the Boolean criterion such that $\mathcal{T}(\lambda)$ is true if and only if λ is a positive observation scale. The criterion $\mathcal{T}$ is called *the observation criterion.* Dually, if λ is not a positive observation scale, then we say that λ is a *negative observation scale (for* $(\mathcal{H}, u)$*)*. If λ is a negative observation scale, then we have $D(\mathcal{H}^{\lambda}_{x}, \mathcal{H}^{\lambda}_{y}) > \lambda$.

Observe that the value $\lambda^{\star}_{\mathcal{H}}(x, y)$ defined in Eq. (2) is simply the lowest element of $\mathbb{E}$ such that $\mathcal{T}(\lambda)$ is true. Dually, we denote by $\overline{\lambda}^{\star}_{\mathcal{H}}(x, y)$, the largest negative observation scale.

Intuitively, a positive observation scale corresponds to a level of the hierarchy $\mathcal{H}$ for which the two regions linked by u should be merged according to the observation criterion $\mathcal{T}$ which is based on the dissimilarity measure D. On the other hand, a negative observation scale corresponds to a level of the hierarchy for which the two associated regions should remain disjoint. A desirable property would be that the observation criterion $\mathcal{T}$ be increasing with respect to scales, a Boolean criterion $\mathcal{T}$ being increasing whenever, for any scale value $\lambda \in \mathbb{E}$, $\mathcal{T}(\lambda)$ holds true implies that $\mathcal{T}(\lambda')$ holds true for any scale λ' greater than λ. Indeed, in such desirable case, any level in $\mathbb{E}$ greater than $\lambda^{\star}_{\mathcal{H}}(x, y)$ would be a positive observation scale, whereas any level not greater than $\lambda^{\star}_{\mathcal{H}}(x, y)$ would be a negative scale. In other words, we would have $\lambda^{\star}_{\mathcal{H}}(x, y) = \mathrm{n}_{\mathbb{E}}\left(\overline{\lambda}^{\star}_{\mathcal{H}}(x, y)\right)$. Hence, it would be easily argued that the observation scale of the edge u must be set to $\lambda^{\star}_{\mathcal{H}}(x, y)$. However, in general, the criterion $\mathcal{T}$ is not increasing (see a counterexample in Fig. 2) and we have $\lambda^{\star}_{\mathcal{H}}(x, y) < \mathrm{n}_{\mathbb{E}}\left(\overline{\lambda}^{\star}_{\mathcal{H}}(x, y)\right)$. Therefore, it is interesting to investigate strategies which can be adopted to select a significant observation scale between $\lambda^{\star}_{\mathcal{H}}(x, y)$ and $\overline{\lambda}^{\star}_{\mathcal{H}}(x, y)$ (see in Fig. 3 a graphical illustration of different situations which may occur). In other words, the criterion $\mathcal{T}$ is transformed into an increasing criterion $\mathcal{T}'$.

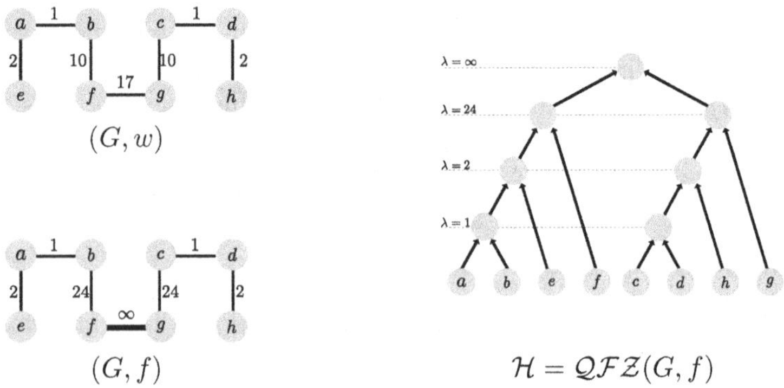

Fig. 2. Counterexample for the increasing property of the observation criterion $\mathcal{T}$: for the edge $u = \{x, y\}$ with $x = f$, $y = g$, we have $D(\mathcal{H}^1_x, \mathcal{H}^1_y) = 17$, $D(\mathcal{H}^{23}_x, \mathcal{H}^{23}_y) = 17$, and $D(\mathcal{H}^{25}_x, \mathcal{H}^{25}_y) = 28$. Hence, we have $\mathcal{T}(1) = \mathit{false}$, $\mathcal{T}(23) = \mathit{true}$, and $\mathcal{T}(25) = \mathit{false}$, which proves that $\mathcal{T}$ is not increasing.

In the framework of mathematical morphology, non-increasing regional attributes/criteria are known to be useful but difficult to handle. Several rules or strategies to handle non-increasing criteria have been considered in the context of connected filters. Among them, one may cite the min- and max-rules [15] or the

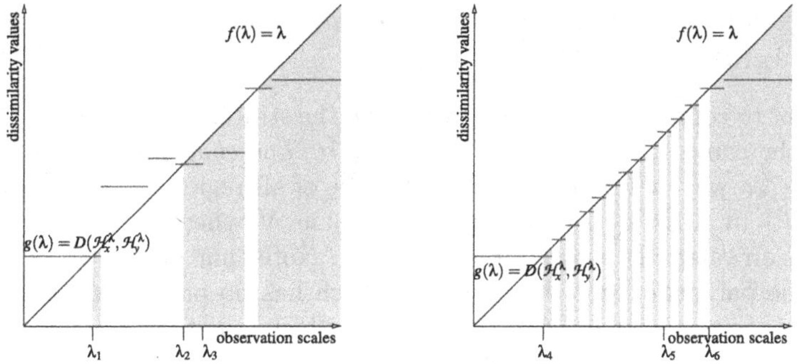

Fig. 3. Illustration of possible observation scale selection strategies. The positive observation intervals are represented in gray. On the left, the min-, the lower α-length and the lower p-rank selection strategies select the scales λ_1, λ_2 and λ_3, respectively (for a value of α which is a little larger than the leftmost gray interval and for $p = 0.3$), whereas, on the right, the max-, the upper α-length and the upper p-rank selection strategies select the scales λ_4, λ_5 and λ_6, respectively.

Viterbi [15] and the shape-space filtering [17] strategies. Note that the strategy adopted in Eq. (2) corresponds to the min-rule and that the strategy consisting of selecting $\overline{\lambda^\star_{\mathcal{H}}}(x, y)$ corresponds to the max-rule. Our main goal in this article is to investigate other strategies to efficiently handle the non-increasing observation criterion $\mathcal{T}$ in the context of hierarchical segmentation and edge-observation scale selection based on the Felzenszwalb-Huttenlocher region dissimilarity measure. Before presenting our proposed selection strategies, we first define positive and negative observation intervals together with an algorithm to compute them.

3.2 Algorithm for Computing Observation Intervals

Let λ_1 and λ_2 be any two real numbers in $\mathbb{E} \cup \{-\infty\}$ such that $\lambda_1 < \lambda_2$. We denote by $\rrbracket\lambda_1, \lambda_2\rrbracket_{\mathbb{E}}$ the subset of $\mathbb{E}$ that contains every element of $\mathbb{E}$ that is both greater than λ_1 and not greater than λ_2: $\rrbracket\lambda_1, \lambda_2\rrbracket_{\mathbb{E}} = \{\lambda \in \mathbb{E} \mid \lambda_1 < \lambda \leq \lambda_2\}$. We say that a subset I of $\mathbb{E}$ is an *open-closed interval of* $\mathbb{E}$, or simply an *interval*, if there exist two real values λ_1 and λ_2 in $\mathbb{E}$ such that I is equal to $\rrbracket\lambda_1, \lambda_2\rrbracket_{\mathbb{E}}$.

Definition 1 (observation interval). *Let $\mathcal{H}$ be any hierarchy, let u be any edge in $E(T)$, and let I be an interval. We say that I is a* positive observation interval (resp. a negative observation interval) *for $(\mathcal{H}, u)$ if the two following statements hold true:*

1. *any element in I is a positive (resp. negative) observation scale for $(\mathcal{H}, u)$; and*
2. *I is maximal among all intervals for which statement (1) holds true,* i.e., *any interval which is a proper superset of I contains a negative (resp. positive) observation scale for $(\mathcal{H}, u)$.*

The set of all positive (resp. negative) observation intervals is denoted by $\Lambda_{\mathcal{H}}(u)$ *(resp. by* $\overline{\Lambda}_{\mathcal{H}}(u)$*).*

In order to compute $\Lambda_{\mathcal{H}}(\{x, y\})$, we follow the strategy presented in [2], which relies on the *component tree* of the hierarchy $\mathcal{H}$. The *component tree of* $\mathcal{H}$ is the pair $\mathcal{T}_{\mathcal{H}} = (\mathcal{N}, parent)$ such that $\mathcal{N}$ is the set of all regions of $\mathcal{H}$ and such that a region R_1 in $\mathcal{N}$ is a *parent* of a region R_2 in $\mathcal{N}$ whenever R_1 is a minimal (for inclusion relation) proper superset of R_2. Note that every region in $\mathcal{N}$ has exactly one parent, except the region V which has no parent and is called the *root* of the component tree of $\mathcal{H}$. Any region which is not the parent of another one is called a *leaf* of the tree. It can be observed that any singleton of V is a leaf of $\mathcal{T}_{\mathcal{H}}$ and that conversely any leaf of $\mathcal{T}_{\mathcal{H}}$ is a singleton of V. The *level* of a region R in $\mathcal{H}$ is the highest index of a partition that contains R in $\mathcal{H}$. Then, the proposed algorithm, whose precise description is given in Algorithm 1, browses in increasing order the levels of the regions containing x and y until finding a value λ such that $D(\mathcal{H}_x^\lambda, \mathcal{H}_y^\lambda) \leq \lambda$. This value is then $\lambda_{\mathcal{H}}^\star(x, y)$ defined by Eq. (2). This value is also the lower bound of the first positive observation interval. If we keep browsing the levels of the regions containing x and y in this tree, as long as $D(\mathcal{H}_x^\lambda, \mathcal{H}_y^\lambda) \leq \lambda$, we can identify the upper bound of this first positive observation interval. We can further continue to browse the levels of the regions containing x and y in the tree in order to identify all positive observation intervals. Therefore, at the end of the execution, we can return the set $\Lambda_{\mathcal{H}}(\{x, y\})$ of all positive observation intervals. From the set $\Lambda_{\mathcal{H}}(\{x, y\})$, we can obtain by duality the set $\overline{\Lambda}_{\mathcal{H}}(\{x, y\})$ of all negative observation intervals.

The time complexity of Algorithm 1 depends linearly on the number of regions in the branches of the component tree of $\mathcal{H}$ containing x and y since it consists of browsing all these regions from the leaves to the root. In the worst case, at every level of the hierarchy the region containing x is merged with a singleton region. Hence, as there are $|V|$ vertices in G, in this case, the branch of x contains $|V|$ regions. Thus, the worst-case time complexity of Algorithm 1 is $O(|V|)$. However, in many practical cases, the component tree of $\mathcal{H}$ is well balanced and each region of $\mathcal{H}$ results from the merging of two regions of (approximately) the same size. Then, if the tree is balanced, the branch of x contains $O(\log_2(|V|))$ nodes and the time-complexity of Algorithm 1 reduces to $O(\log_2(|V|))$.

4 Selecting Observation Scales

Let K be any subset of $\mathbb{E}$. We consider the two following selection rules from K to set the value of $f(u)$ in Method 1:

$$\text{min-rule: } f(u) := \min\{k \in K\}; \text{ and}$$
$$\text{max-rule: } f(u) := \max\{k \in K\}.$$

When K is the set of the positive observation scales, the result obtained with the min-rule is called the *min-selection strategy*. Note that the results obtained with the min-selection strategy correspond exactly to the results obtained with the method presented in [2,8], as described by Eq. (2). In this article, we also consider the *max-selection strategy*, that is the result obtained with the max-rule when K is the set of the negative observation scales. Furthermore, we also apply these rules to filtered sets of positive and of negative observation scales. The motivation for introducing these strategies is to regularise the observation criterion with respect to scales in order to cope with situations such as the ones shown in Fig. 3. As filterings, we investigate the well-known rank and area filters.

Algorithm 1. Positive observation intervals

Input : The component tree $\mathcal{T} = (\mathcal{N}, parent)$ of a hierarchy $\mathcal{H}$, an edge $u = \{x, y\}$ of T, an array *level* that stores the level of every region of $\mathcal{H}$

Output: A set containing all elements of $\Lambda_{\mathcal{H}}(\{x, y\})$

$C_x := \{x\}$; $C_y := \{y\}$; $\Lambda_{\mathcal{H}}(\{x, y\}) = \{\}$;
$\lambda := \min(level[C_x], level[C_y])$; $\lambda_{prev} := -\infty$;
do
  while $D(C_x, C_y) > \lambda$ **do**
    $\lambda_{prev} := \lambda$;
    $\lambda := \min(level[parent[C_x]], level[parent[C_y]])$;
    if $level[parent[C_x]] = \lambda$ **then** $C_x := parent[C_x]$;
    if $level[parent[C_y]] = \lambda$ **then** $C_y := parent[C_y]$;
  end
  $\lambda_{lower} := \max(\mathrm{n}_{\mathbb{E}}(\lambda_{prev}), \hat{\mathrm{n}}_{\mathbb{E}}(D(\mathcal{H}_x^{\lambda}, \mathcal{H}_y^{\lambda})))$;
  while $D(C_x, C_y) \leq \lambda$ ***and*** $(parent[C_x] \neq root$ ***and*** $parent[C_y] \neq root)$ **do**
    $\lambda_{prev} := \lambda$;
    $\lambda := \min(level[parent[C_x]], level[parent[C_y]])$;
    if $level[parent[C_x]] = \lambda$ **then** $C_x := parent[C_x]$;
    if $level[parent[C_y]] = \lambda$ **then** $C_y := parent[C_y]$;
  end
  $\lambda_{upper} := \max(\mathrm{n}_{\mathbb{E}}(\lambda_{prev}), \hat{\mathrm{n}}_{\mathbb{E}}(\lambda_{prev}))$;
  $\Lambda_{\mathcal{H}}(\{x, y\}).add(]\!]\lambda_{lower}, \lambda_{upper}]\!])$;
while $parent[C_x] \neq root$ ***and*** $parent[C_y] \neq root$;

Let us first provide a precise definition of the rank filters that we apply to the positive and to the negative observation scales. The intuitive idea of the selection strategies based on these filters is to remove a lower percentile of the positive observation scales, which is then considered as non significant, before applying the min-rule and to remove an upper percentile of the negative observation scales before applying the max-rule.

Let K be any subset of $\mathbb{E}$ with n elements. Let k be any positive integer less than n. We denote by $\text{rank}_{k/n}(K)$ the element e of K such that there are exactly k distinct elements in K which are less than e. Let p be any real value between 0 and 1, we set $\text{rank}_p(K) = \text{rank}_{\lfloor p.n \rfloor/n}(K)$, where $\lfloor p.n \rfloor$ is the largest integer which is not greater than the real value $p.n$.

Let p be any real value between 0 and 1. A *p-positive observation scale for* $(\mathcal{H}, u)$ is any positive observation scale for $(\mathcal{H}, u)$ that is greater than $\text{rank}_p(K)$ where K is the set of all positive observation scales not greater than $\overline{\lambda}^{\star}_{\mathcal{H}}(x, y)$. A *p-negative observation scale for* $(\mathcal{H}, u)$ is any negative observation scale for $(\mathcal{H}, u)$ that is less than $\text{rank}_{1-p}(K)$ where K is the set of all negative observation scales not greater than $\lambda^{\star}_{\mathcal{H}}(x, y)$. The min-rule from the set of all p-positive observation scales is called the *lower p-rank selection strategy* while the max-rule from the set of all p-negative observation scales is called the *upper p-rank selection strategy.*

Let us now describe the selection strategies obtained by applying an area filter on the positive and on the negative observations scales before applying the min- and max-rules. Let $\mathcal{H}$ be any hierarchy, let $\{x, y\}$ be any edge of G and let α be any positive integer. We set $\Lambda^{\alpha}_{\mathcal{H}}(\{x, y\}) = \{]\!]\lambda_1, \lambda_2]\!]_{\mathbb{E}} \in \Lambda_{\mathcal{H}}(\{x, y\}) \mid \lambda_2 - \lambda_1 \geq \alpha\}$ and $\overline{\Lambda}^{\alpha}_{\mathcal{H}}(\{x, y\}) = \{]\!]\lambda_1, \lambda_2]\!]_{\mathbb{E}} \in \overline{\Lambda}_{\mathcal{H}}(\{x, y\}) \mid \lambda_2 - \lambda_1 \geq \alpha\}$. The min-rule from the set $\cup \Lambda^{\alpha}_{\mathcal{H}}(\{x, y\})$ and the max-rule from the set $\cup \overline{\Lambda}^{\alpha}_{\mathcal{H}}(\{x, y\})$ are called the *lower α-length selection strategy* and the *upper α-length selection strategy*, respectively.

The six selection strategies introduced in this section are illustrated in Fig. 3.

5 Experiments

In this section we aim to compare the segmentation results obtained from the original HGB method against the segmentations obtained by our strategies. To this end, we use the Berkeley Segmentation Dataset (BSDS) and associated evaluation framework [1] for our experiments. This dataset consists of 500 natural images of size 321×481 pixels. In order to perform a quantitative analysis, we use the F-measures defined from the precision-recall for regions F_r. The segmentation is perfect when $F_r = 1$ and totally different from the ground-truth when $F_r = 0$. Each image is represented by a graph where the vertices are the pixels, where the edges are given by the 4-adjacency relation, and where the weight of any edge is given by the Euclidean distance between the colors of the pixels linked by this edge. We compute for each image a set of segmentations at different scales. From each pair made of an image segmentation and the associated ground truth, we obtain one F-measure value. Then, we keep the best F_r-measure obtained for each image of the database. Alternatively, we can also keep the F_r-measure for a constant scale over the database, such that the constant scale is chosen to maximize the average F_r-measure of the overall database. They are called optimal image scale (OIS) and optimal database scale (ODS) respectively.

In Table 1, we see the average F_r scores for ODS and OIS on the BSDS dataset. As we can observe, we obtain much better segmentation results from the selection strategies that use max-rule over the selection strategies using min-rule. Furthermore, among the selection strategies that use max-rule, the *upper p-rank* selection shows a slight improvement over the max selection. We also show in Fig. 4 the distribution of the best F_r scores for our strategies. In Fig. 1, we can see a qualitative comparison between the saliency maps resulting from the HGB method using the min selection strategy over our *upper p-rank* strategy which shows a significant improvement.

Table 1. Average F_r scores for BSDS dataset. In the table, avg., param. and med. stands for average, parameter, and median, respectively.

Strategy	Param.	ODS		OIS	
		avg.	med.	avg.	med.
Min	-	0.463	**0.453**	**0.570**	0.555
Lower p-rank	0.005	0.432	0.431	0.551	0.546
	0.01	0.432	0.431	0.551	0.544
	0.05	0.431	0.427	0.552	0.547
	0.1	0.431	0.426	0.543	0.531
Lower α−length	10	**0.465**	0.450	0.563	**0.564**
	100	0.439	0.430	0.552	0.537
	500	0.420	0.416	0.546	0.536

Strategy	Param.	ODS		OIS	
		avg.	med.	avg.	med.
Max	-	0.547	0.543	0.638	0.642
Upper p-rank	0.005	0.552	**0.553**	0.646	**0.649**
	0.01	**0.553**	0.541	**0.647**	0.648
	0.05	0.553	0.541	0.643	0.637
	0.1	0.548	0.541	0.641	0.637
Upper α−length	10	0.548	0.545	0.638	0.640
	100	0.547	0.545	0.638	0.638
	500	0.546	0.543	0.640	0.643

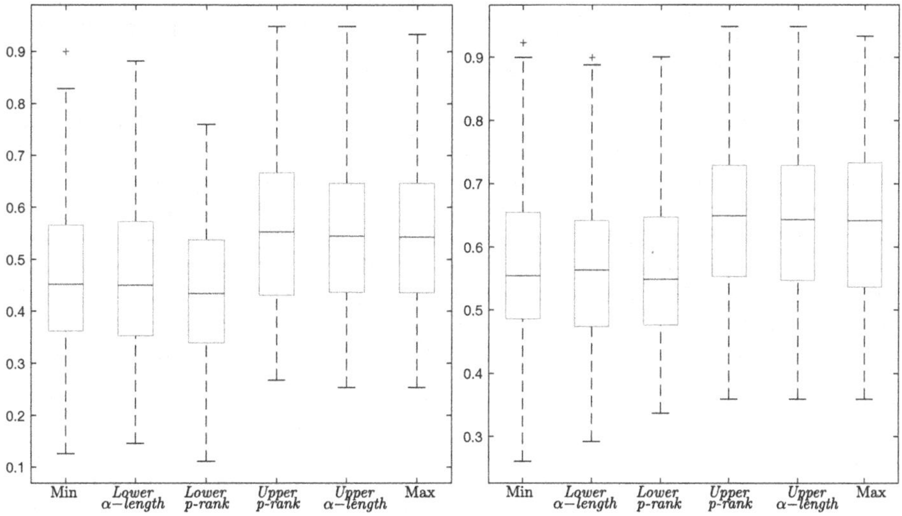

Fig. 4. Distribution F_r scores for ODS and OIS on the BSDS dataset, respectively.

Fig. 5. Saliency maps resulting from the HGB method with upper p-rank selection strategy using as edge weights Euclidean distance (middle) and the structured edge detector from [5] (right).

6 Conclusions

In this article, we study the HGB method with the aim of proposing new strategies for selecting an observation scale that can lead to better segmentation results. To this end, we propose an algorithm that computes all the scales for which the Felzenswalb-Huttenlocher dissimilarity measure indicates that the regions should merge. Dually, we are able to obtain based on the min- and max-rule selection with filtering techniques the negative intervals. Then, we propose several strategies to select scales at both positive and negative intervals. We validate the performance of our strategies on the BSDS dataset. The best performance was achieved by our *upper p-rank* strategy (see Table 1).

As future work, we plan to use other gradients to weight the edges of the graph. This along with our proposed strategies can lead to better segmentation results, as we can observe in Fig. 5.

Acknowledgements. The research leading to these results has received funding from the French Agence Nationale de la Recherche, grant ANR-15-CE40-0006 (CoMeDiC), the Brazilian Federal Agency of Support and Evaluation of Postgraduate Education (program CAPES/PVE: grant 064965/2014-01), Brazilian Federal Agency of Research (CNPq/Universal 421521/2016-3 and CNPq/PQ 307062/2016-3), Fundo de Amparo Pesquisa do Estado de Minas Gerais (FAPEMIG/PPM 00006-16), the Peruvian agency Consejo Nacional de Ciencia, Tecnológica CONCYTEC (contract N 101-2016-. FONDECYT-DE). The first author would like to thank Brazilian agencies CNPq and CAPES and Peruvian agency CONCYTEC for the financial support during his thesis.

References

1. Arbelaez, P., Maire, M., Fowlkes, C., Malik, J.: Contour detection and hierarchical image segmentation. TPAMI **33**(5), 898–916 (2011)
2. Cahuina, E.J.Y.C., Cousty, J., Kenmochi, Y., Araújo, D.A., Cámara-Chávez, G.: Algorithms for hierarchical segmentation based on the Felzenszwalb-Huttenlocher dissimilarity. In: ICPRAI. pp. 1–6 (2018)

3. Cousty, J., Najman, L.: Incremental algorithm for hierarchical minimum spanning forests and saliency of watershed cuts. In: Soille, P., Pesaresi, M., Ouzounis, G.K. (eds.) ISMM 2011. LNCS, vol. 6671, pp. 272–283. Springer, Heidelberg (2011). https://doi.org/10.1007/978-3-642-21569-8_24
4. Cousty, J., Najman, L., Kenmochi, Y., Guimarães, S.: Hierarchical segmentations with graphs: quasi-flat zones, minimum spanning trees, and saliency maps. JMIV **60**(4), 479–502 (2018)
5. Dollár, P.: Zitnick: fast edge detection using structured forests. TPAMI **37**(8), 1558–1570 (2015)
6. Felzenszwalb, P.F., Huttenlocher, D.P.: Efficient graph-based image segmentation. IJCV **59**(2), 167–181 (2004)
7. Guigues, L., Cocquerez, J.P., Le Men, H.: Scale-sets image analysis. IJCV **68**(3), 289–317 (2006)
8. Guimarães, S., Kenmochi, Y., Cousty, J., Patrocinio Jr., Z., Najman, L.: Hierarchizing graph-based image segmentation algorithms relying on region dissimilarity - the case of the Felzenszwalb-Huttenlocher method. MMTA **2**(1), 55–75 (2017)
9. Haxhimusa, Y., Ion, A., Kropatsch, W.G.: Irregular pyramid segmentations with stochastic graph decimation strategies. In: Martínez-Trinidad, J.F., Carrasco Ochoa, J.A., Kittler, J. (eds.) CIARP 2006. LNCS, vol. 4225, pp. 277–286. Springer, Heidelberg (2006). https://doi.org/10.1007/11892755_28
10. Meyer, F., Maragos, P.: Morphological scale-space representation with levelings. In: Nielsen, M., Johansen, P., Olsen, O.F., Weickert, J. (eds.) Scale-Space 1999. LNCS, vol. 1682, pp. 187–198. Springer, Heidelberg (1999). https://doi.org/10.1007/3-540-48236-9_17
11. Pavlidis, T.: Structural Pattern Recognition. Springer Series in Electronics and Photonics, vol. 1. Springer, Heidelberg (1977)
12. Perret, B., Cousty, J., Guimarães, S.J., Maia, D.S.: Evaluation of hierarchical watersheds. IEEE Trans. Image Process. **27**(4), 1676–1688 (2018)
13. Pont-Tuset, J., Arbeláez, P., Barron, J.T., Marques, F., Malik, J.: Multiscale combinatorial grouping for image segmentation and object proposal generation. IEEE TPAMI **39**(1), 128–140 (2017)
14. Salembier, P., Garrido, L.: Binary partition tree as an efficient representation for image processing, segmentation, and information retrieval. TIP **9**(4), 561–576 (2000)
15. Salembier, P., Oliveras, A., Garrido, L.: Antiextensive connected operators for image and sequence processing. TIP **7**(4), 555–570 (1998)
16. Syu, J.H., Wang, S.J., Wang, L.: Hierarchical image segmentation based on iterative contraction and merging. TIP **26**(5), 2246–2260 (2017)
17. Xu, Y., Géraud, T., Najman, L.: Connected filtering on tree-based shape-spaces. TPAMI **38**(6), 1126–1140 (2016)

The Role of Optimum Connectivity in Image Segmentation: Can the Algorithm Learn Object Information During the Process?

Alexandre Falcão(✉) and Jordão Bragantini(✉)

University of Campinas, Campinas, SP 13083-852, Brazil
afalcao@ic.unicamp.br, jordao.bragantini@gmail.com

Abstract. Image segmentation is one of the most investigated research topics in Computer Vision and yet presents challenges due to the difficulty of modeling all possible appearances of objects in images. In this sense, it is important to investigate methods that can learn object information before and during delineation. This paper addresses the problem by exploiting optimum connectivity between image elements (pixels and superpixels) in the image domain and feature space to improve segmentation. The study uses the Image Foresting Transform (**IFT**) framework to explain and implement all methods and describes some recent advances related to superpixel and object delineation. It provides a guideline to learn prior object information from the target image only based on seed pixels, superpixel clustering, and classification, evaluates the impact of using object information in several connectivity-based delineation methods using the segmentation by a deep neural network as baseline, and shows the potential of a new paradigm, namely *Dynamic Trees*, to learn object information from the target image only during delineation.

Keywords: Superpixel and object segmentation · Seed-based delineation algorithms · Image Foresting Transform

1 Introduction

The definition of the exact spatial extent of relevant image segments for shape and texture analysis, namely *segmentation*, is certainly one of the most challenging problems in Computer Vision and Image Processing. Such segments may be *superpixels*—connected regions in which the pixels share common texture properties—or *objects*—connected components with characteristic shapes.

Superpixel segmentation aims at considerably reducing the number of image elements for analysis, such that the objects of interest can still be represented by the union of their superpixels (Fig. 1a). The problem requires an accurate

FAPESP 2014/12236-1, 2018/08951-8, and CNPq 302970/2014-2.

M. Couprie et al. (Eds.): DGCI 2019, LNCS 11414, pp. 180–194, 2019.
https://doi.org/10.1007/978-3-030-14085-4_15

(a) (b) (c)

Fig. 1. (a) Superpixel segmentation by recursive spanning forests [21]. Object segmentation by (b) live markers [36] with internal (yellow) and external (red) seeds, some created from live-wire segments [20]. (c) Deep Extreme Cut (**DEC**) [25] with the four extreme points (magenta) needed for object localization. Arrows indicate three delineation errors in (c). (Color figure online)

delineation algorithm from seed pixels with one seed per superpixel. The first set of seed pixels *localizes* the superpixels and a few iterations of superpixel delineation and seed recomputation refine segmentation [1,5,9,21]. Superpixels are then expected to increase the efficiency of higher level operations without compromising their effectiveness. One example is object segmentation using a superpixel graph [32].

Localization and delineation are also crucial tasks for object segmentation. Localization requires to specify the whereabouts of the object in the image—e.g., by providing a bounding box around the object [34], selecting points on its boundary [20], or drawing scribbles inside and outside the object [15,16, 32]. Humans can more easily locate the object than machines. From effective localization, a delineation algorithm can define the spatial extent of the object in the image (Fig. 1b). Machines are more precise, not subject to fatigue, and so better than humans for object delineation.

A single iteration of localization and delineation does not usually solve object segmentation, which often requires multiple user interventions. Prior object information—i.e., texture [10,24,25] and shape [18,26,31,39] models built from a training set with interactively segmented images—have been proposed to automate the segmentation process. Without connectivity-based delineation algorithms, these models may force their prior knowledge ignoring the appearance of the object in the image. For example, after a user indicates four extreme points to localize the object, a texture model based on deep learning [25] can provide an impressive object approximation (Fig. 1c). However, the problem is not solved in a single iteration and the method does not allow further user intervention. This suggests the combination of texture and shape models with connectivity-based delineation algorithms [10,31], methods that can automatically adapt the model to the target image [26], and methods that can learn object information during delineation [8].

In this paper, we demonstrate the role of optimum connectivity in image segmentation through the challenge of learning object information from the target image only. The strategy aims at improving object delineation with minimum user intervention. The *Image Foresting Transform* (**IFT**) is the framework [19] of choice since it can unify several methods used for delineation [5,8,21,32,36], clustering [33], and classification [2,3,30], by exploiting connectivity between a seed set and the remaining image elements. We combine some of these methods to learn object information before and during segmentation, and evaluate their impact on object delineation.

Section 2 presents the **IFT** framework with examples of adjacency relations and connectivity functions for superpixel and object segmentation. Sections 3 and 4 cover some recent advances in **IFT**-based superpixel and object segmentation, by calling attention to dynamic trees [8]—a new paradigm that learns object information during the delineation process. Section 4.3 then explains how to use superpixel segmentation, clustering, and classification to learn object information from the target image before object delineation. Section 5 evaluates the results of incorporating object information in connectivity-based delineation algorithms. Finally, Sect. 6 states conclusion and suggests research directions related to image segmentation.

2 Image Foresting Transform (IFT)

In the **IFT** framework [19], an image is interpreted as a graph in which the nodes may be pixels, superpixels, or any other element from the image domain, each one being a point in some n-dimensional feature space. The arcs of the graph must satisfy some *adjacency relation*, which may be defined in the image domain and/or feature space. The **IFT** uses Dijkstra's algorithm, with more general *connectivity functions* and multiple sources [12], to transform the image graph into an optimum-path forest rooted at the minima of a resulting cost map (e.g., a seed set for delineation), then reducing the image operator to a local processing of its attributes (optimum paths, their costs, and root labels).

2.1 Images as Graphs

Let $\hat{I} = (D_I, \mathbf{I})$ be a 2D natural image in which $\mathbf{I}(p) \in \mathbb{R}^3$ represents the CIELab color components of each pixel $p \in D_I \subset \mathcal{Z}^2$. Image $\hat{I}$ may be interpreted as a weighted graph $(\mathcal{N}, \mathcal{A}, w)$ in several ways such that the nodes $s, t \in \mathcal{N}$ may be pixels or superpixels, the arcs $(s,t) \in \mathcal{A}$ may connect nodes in the image domain and/or feature space, depending on the *adjacency relation* $\mathcal{A} \subset \mathcal{N} \times \mathcal{N}$, and the arc weights $w(s,t)$ may be defined from image properties and object information.

Let Π be the set of all possible paths in $(\mathcal{N}, \mathcal{A}, w)$. A path $\pi_{r \to s}$ is a sequence $\langle r = t_1, t_2, \ldots, t_n = s\rangle$ of adjacent nodes $(t_i, t_{i+1}) \in \mathcal{A}$, $i = 1, 2, \ldots, n-1$, being π_s any path with terminus s, $\pi_s = \langle s \rangle$ a *trivial* path, and $\pi_s \cdot \langle s, t\rangle$ the concatenation of π_s and $\langle s, t\rangle$ with the joining instances of s merged into one. For a suitable *connectivity function* $f\colon \Pi \to \mathbb{R}$, the **IFT** algorithm on $(\mathcal{N}, \mathcal{A}, w)$ can output an *optimum-path forest* P—i.e., a spanning forest (acyclic map) that assigns to all nodes $t \in \mathcal{N}$ a predecessor $P(t) = s \in \mathcal{N}$ in the optimum path with terminus t or a marker $P(t) = nil \notin \mathcal{N}$ when t is a *root* of the map—by minimizing a path-cost map V (also called connectivity map).

$$V(t) = \min_{\forall \pi_t \in \Pi} \{f(\pi_t)\}. \tag{1}$$

The map V can also be maximized depending on the connectivity function and the sufficient conditions for optimal cost mapping have been revised recently [12]. Even when these conditions are not satisfied, the algorithm can output a spanning forest useful for superpixel [5] and object [28] segmentation, being the minimization criterion a graph-cut measure in the latter.

2.2 Adjacency Relations

Let $\mathcal{K}(s,k)$ be the set of the k-closest nodes $t \in \mathcal{N}$ of s in some feature space (e.g., the CIELab color space) according to the Euclidean metric $\|.,.\|$. Examples of adjacency relations used in this work are given below.

$$\mathcal{A}_1 = \{s,t \in \mathcal{N} \mid \|s,t\| \leq d\}, \tag{2}$$
$$\mathcal{A}_2 = \{s,t \in \mathcal{N} \mid t \in \mathcal{K}(s,k), 1 \leq k \ll |\mathcal{N}|\}, \tag{3}$$
$$\mathcal{A}_3 = \{s,t \in \mathcal{N} \mid s \neq t\}, \tag{4}$$

where $\mathcal{A}_1$ with $d = 1$ and $\mathcal{N} = D_I$ (4-neighbor graph) is used for superpixel and object delineation (Sects. 3 and 4), $\mathcal{A}_2$ (k-nearest neighbor graph) is used for clustering of superpixels (Sect. 4.3), and $\mathcal{A}_3$ (complete graph) is used to create an *object membership map* O based on pixel classification from a small training subset $\mathcal{N} \subset D_I$ (Sect. 4.3).

2.3 Algorithm and Connectivity Functions

The **IFT** algorithm starts with all nodes $s \in \mathcal{N}$ being trivial paths $\langle s \rangle$ with values $f(\langle s \rangle) = V_o(s) \geq f(\pi_s)$ for all $\pi_s \in \Pi$. This step sets $V(s) \leftarrow V_o(s)$ and $P(s) \leftarrow nil$, and inserts all nodes in a priority queue Q. At each iteration, a node s with minimum cost $V(s)$ is removed from Q and it offers the extended path $\pi_s \cdot \langle s,t \rangle$ to its adjacent nodes in Q. Whenever $V(t) > f(\pi_s \cdot \langle s,t \rangle)$, the node s conquers the node t by substituting its current path π_t by $\pi_s \cdot \langle s,t \rangle$. This step updates $V(t) \leftarrow f(\pi_s \cdot \langle s,t \rangle)$ and $P(t) \leftarrow s$, and the process stops after $|\mathcal{N}|$ iterations. The roots of the forest can be forced to be a *seed set* $\mathcal{S} \subset \mathcal{N}$, when $V_o(s)$ is defined as

$$V_o(s) = \begin{cases} 0 & \text{if } s \in \mathcal{S}, \\ +\infty & \text{otherwise.} \end{cases} \tag{5}$$

Otherwise, the root set $\mathcal{R}$ derives from the minima (flat zones) of V (a subset of the flat zones of V_0). It is also possible to guarantee a single root per minimum if the nodes are inserted in Q with costs $f(\langle s \rangle) = V_o(s) + \delta$, where

$0 < \delta = \min_{\forall(s,t)\in\mathcal{A}}\{f(\pi_s \cdot \langle s,t\rangle) - f(\pi_s)\}$ for monotonically incremental path-cost functions [19]. By that, when a node s is removed from Q with $P(s) = nil$, it has never been conquered by another node and so it is immediately identified as a root in $\mathcal{R}$ with cost updated to $V_o(s)$. This is equivalent to define

$$f(\langle s\rangle) = \begin{cases} V_o(s) & \text{if } s \in \mathcal{R}, \\ V_o(s) + \delta & \text{otherwise.} \end{cases} \tag{6}$$

Examples of connectivity functions for non-trivial paths are given below. They usually use Eq. 5 for trivial paths, but Sect. 4.3 shows one example of f_1 with trivial-path cost given by Eq. 6.

$$f_1(\pi_s \cdot \langle s,t\rangle) = \max\{f_1(\pi_s), w(s,t)\}, \tag{7}$$

$$f_2(\pi_s \cdot \langle s,t\rangle) = w(s,t), \tag{8}$$

$$f_3(\pi_{r\to s} \cdot \langle s,t\rangle) = f_3(\pi_{r\to s}) + aw^b(r,t) + \|s,t\|, \tag{9}$$

where r is the root of s, $a > 0$ and $b \geq 1$. Clearly, the success of the image operations strongly depends on the *arc-weight function* w. Next sections illustrate different choices of w and connectivity functions for superpixel and object segmentation, clustering, and classification.

3 Superpixel Segmentation

Superpixel segmentation has evolved as an important topic with several applications, such as object detection, tracking, and segmentation. A superpixel is a connected image region in which pixels present similar texture properties. However, some methods cannot guarantee connected superpixels during delineation [1,9], except by post-processing. In [40], the authors propose a framework, named *Iterative Spanning Forest* (**ISF**), for the generation of connected superpixels by choice of four independent components: (a) a seed sampling strategy, (b) an adjacency relation, (c) a connectivity function, and (d) a seed recomputation procedure. They present and evaluate different choices for those components. Similar to some popular approaches, for a desired number of superpixels, the **ISF** algorithm starts by finding an initial seed set $\mathcal{S}$ that represents meaningful locations for those superpixels. Subsequently, the **IFT** algorithm followed by seed recomputation is executed multiple times to improve the seed set $\mathcal{S}$ and refine superpixel delineation. Connected superpixels require the adjacency relation in Eq. 2 with $d = 1$ for 2D and 3D images (4- and 6-neighbor graphs, respectively). Equations 5 and 9 represent one example of connectivity function in which $\|s,t\|$ contributes to obtain compact superpixels, $w^b(r,t) = \|\mathbf{I}(r), \mathbf{I}(t)\|^b$ increases the boundary adherence for higher values of b (e.g., $b = 12$), and a (e.g., $a \in \{0.08, 0.1, 0.2, 0.5\}$) controls the compromise between boundary adherence and compactness of the superpixels. In **ISF**, each connected superpixel is a spanning tree rooted at one seed pixel $r \in \mathcal{S}$.

In [5], the authors present the first **ISF**-based method and the authors in [38] propose a connectivity function that increases the boundary adherence of superpixels to the boundary of a given object mask created by automated segmentation. The method can be used to transform any object segmentation into an optimum-path forest rooted at superpixel seeds for correction by differential **IFTs** [17]. By adapting the connectivity function in [38], the authors in [6] incorporate object information in superpixel delineation such that **ISF** can considerably reduce the number of superpixels while the object representation is preserved. The method is called *Object-based **ISF*** (**OISF**). **ISF** can also be executed recursively on superpixel graphs to create a hierarchy of segmentations [21] for distinct applications. This hierarchy is explored in [21] to introduce a geodesic seed sampling strategy, which can considerably improve superpixel delineation.

At each execution of the **IFT** algorithm in the **ISF**-based methods, the seed set partially changes by preserving the most stable seeds from previous executions, eliminating the unstable ones, and adding new seeds for replacement. For the sake of efficiency, this requires the update of the spanning forest in a differential way. However, the differential **IFT** algorithm [17] does not apply for non-monotonically incremental connectivity functions [19]. A solution for root-based connectivity functions (e.g., Eq. 9) has been presented in [14].

In this work, the superpixel representation reduces the number of image elements from 481×321 pixels (the resolution of the images in the GrabCut dataset [34]) to about 800 superpixels. We then use the method in [33] to find groups in the superpixel set much faster and with higher efficiency than using the pixel set (Sect. 4.3). For pixel sets, this algorithm is usually applied to a considerably smaller subset of randomly selected pixels and then the group labels are propagated to the remaining pixels. In our case, the algorithm can use the entire image content as represented by the superpixel set.

4 Object Segmentation

In the **IFT** framework, object localization is usually solved by selecting seeds on the object's boundary [20,27,36] or inside and outside the object [11,13, 19,28,29,32], being some approaches hybrid (e.g., Fig. 1b [36]). Region-based approaches can be naturally extended to 3D images and so more easily combined with shape models for automated segmentation [26,31]. In this section, our focus is on region-based object delineation algorithms from a seed set $\mathcal{S} = \mathcal{S}_o \cup \mathcal{S}_b$ with internal ($\mathcal{S}_o$) and external ($\mathcal{S}_b$) pixels. These algorithms run on 4-neighbor graphs and output two sets in a label map L, the object set $\mathcal{O} = \{p \in D_I, L(p) = 1\}$ and the background set $\mathcal{B} = \{p \in D_I, L(p) = 0\}$, such that $\mathcal{O} \cup \mathcal{B} = D_I$ and $\mathcal{O} \cap \mathcal{B} = \emptyset$. The partition defines a graph cut $\mathcal{C} = \{\forall\ (p, q) \in \mathcal{A} \mid p \in \mathcal{O}, q \in \mathcal{B}\}$. We are interested in algorithms that define connected sets $\mathcal{O}$ and $\mathcal{B}$, with $\mathcal{S}_o \subset \mathcal{O}$ and $\mathcal{S}_b \subset \mathcal{B}$, and optimize one of the two graph-cut measures below [13].

$$\mathbf{GC_{max}} = \max_{\forall \mathcal{C} \subset \mathcal{A}} \left\{ \min_{\forall (p,q) \in \mathcal{C}} \{w(p,q)\} \right\}, \tag{10}$$

$$\mathbf{GC_{sum}} = \min_{\forall \mathcal{C} \subset \mathcal{A}} \left\{ \sum_{\forall (p,q) \in \mathcal{C}} w(p,q) \right\}. \tag{11}$$

For $\mathbf{GC_{sum}}$, we use the min-cut/max-flow algorithm [7] and the **IFT** algorithm [12] is used for $\mathbf{GC_{max}}$. Note that, $w(p,q)$ must be higher on the object's boundary than elsewhere for $\mathbf{GC_{max}}$ and lower on the object's boundary than elsewhere for $\mathbf{GC_{sum}}$. In order to avoid small cuts in $\mathbf{GC_{sum}}$, we must adopt an exponent $\beta > 1$ when computing $w(p,q)$. Note that, for $\beta \gg 1$, $\mathbf{GC_{sum}}$ should provide similar results to $\mathbf{GC_{max}}$ [29].

$\mathbf{GC_{sum}}$ extends the image graph by adding binary arc weights, with values either 0 or $+\infty$, between pixels and the source (object) and sink (background) seeds in $\mathcal{S}$, while the **IFT** algorithm uses Eqs. 5 and 7 to satisfy the constraints. In the **IFT** algorithm, the seeds in $\mathcal{S}_o$ and $\mathcal{S}_b$ compete among themselves such that the object $\mathcal{O}$ is defined in P by the optimum-path trees rooted in $\mathcal{S}_o$.

4.1 Localization by a User Robot

In order to avoid user subjectivity in selecting seeds, we use a geodesic user robot [22], which uses the ground-truth segmentation to simulate the user knowledge about the object. For the first iteration of object delineation, the robot selects one marker (i.e., a pixel and its four neighbors) at the geodesic center of each, object and background. At each subsequent iteration, the largest error components of both, object and background, are identified and the robot selects one marker at the geodesic center of each one. The markers must be at least two pixels from the object's boundary and the process proceeds for 15 iterations, but it can halt earlier when the error components are small to contain markers. As it occurs in practice, distinct methods require seeds in different locations to correct segmentation. In [8], we use benchmarks of seed sets, but the methods are compared for a single execution only.

4.2 Arc-Weight Assignment

We consider the following arc-weight assignment functions.

$$w_1(p,q) = \|\mathbf{I}(p), \mathbf{I}(q)\|, \tag{12}$$

$$w_2(p,q) = (1-\alpha) w_1(p,q) + \alpha |O(p) - O(q)|, \tag{13}$$

$$w_3(p,q) = \left(1 - \frac{w_1(p,q)}{dI_{max}}\right)^{\beta}, \tag{14}$$

$$w_4(p,q) = \left[(1-\alpha)\left(1 - \frac{w_1(p,q)}{dI_{max}}\right) + \alpha\left(1 - \frac{|O(p) - O(q)|}{dO_{max}}\right)\right]^{\beta}, \tag{15}$$

$$w_5(p,q) = \|\mu_{R(p)}, \mathbf{I}(q)\|, \tag{16}$$

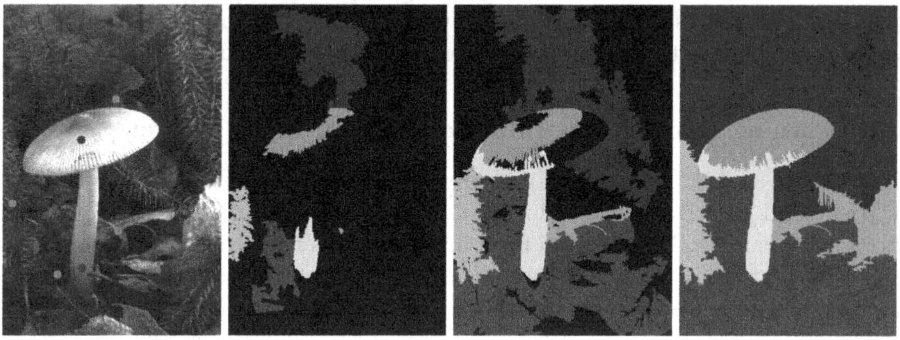

(a) (b) (c) (d)

Fig. 2. (a) Original image with object (blue) and background (red) markers. (b-c-d) Dynamic tree growth during segmentation. Each color is a different optimum-path tree. (Color figure online)

$$w_6(p,q) = (1-\alpha)w_5(p,q) + \alpha|O(p) - O(q)|, \tag{17}$$
$$w_7(p,q) = w_5(p,q) + \|\mathbf{I}(p), \mathbf{I}(q)\|, \tag{18}$$
$$w_8(p,q) = (1-\alpha)w_7(p,q) + \alpha|O(p) - O(q)|, \tag{19}$$
$$w_9(p,q) = \min_{\forall r \in \mathcal{S} | L(r)=L(p)} \{\|\mu_r, \mathbf{I}(q)\|\}, \tag{20}$$
$$w_{10}(p,q) = (1-\alpha)w_9(p,q) + \alpha|O(p) - O(q)|, \tag{21}$$
$$w_{11}(p,q) = w_9(p,q) + \|\mathbf{I}(p), \mathbf{I}(q)\|, \tag{22}$$
$$w_{12}(p,q) = (1-\alpha)w_{11}(p,q) + \alpha|O(p) - O(q)|, \tag{23}$$

where O is an object map (Sect. 4.3), $dI_{max} = \max_{\forall (p,q) \in \mathcal{A}_1}\{\|\mathbf{I}(p), \mathbf{I}(q)\|\}$, $dO_{max} = \max_{\forall (p,q) \in A_1}\{|O(p) - O(q)|\}$, $L(p) \in \{0, 1\}$ indicates which set p belongs to, either $\mathcal{O}$ or $\mathcal{B}$, by the time $w(p,q)$ is estimated, α controls the balance between object and image information (e.g., $\alpha = 0.25$), β avoids small cuts in $\mathbf{GC_{sum}}$ (e.g., $\beta = 70$), and $\mu_{R(p)}$ is the mean CIELab color of the optimum-path tree that contains p with root $R(p) \in \mathcal{S}$ at the moment the arc weight $w(p,q)$ is computed. In the **IFT** algorithm, at this moment the optimum-path trees have already been assigned to either $\mathcal{O}$ or $\mathcal{B}$. Note that, Eqs. 12, 13, 16–23 are used for $\mathbf{GC_{max}}$ while Eqs. 14 and 15 are used for $\mathbf{GC_{sum}}$. A new paradigm, named *dynamic trees* [8], is defined by Eqs. 16–23. In this paradigm, object information is extracted from the growing optimum-path trees to estimate the arc weights during the process. Figure 2 illustrates the growing process of those trees for three distant iterations. One can see that, as they grow, it is possible to obtain shape and texture information. We have evaluated two simple measures so far—the mean color μ_{Rp} of the tree that contains p (Eqs. 16–19) and, for the object/background forest that contains p, the mean color μ_r of the tree closest to the color of p (Eqs. 20–23). Although we are addressing binary segmentation, $\mathbf{GC_{max}}$ is extensive to multiple object segmentation with no additional cost.

4.3 Learning Prior Object Information from Input Markers

Equations 16 and 20 use only object information as extracted and represented by $\mu_{R(p)}$ and μ_r, respectively, during delineation. In addition to that, object information can also be extracted from input markers by executing another **IFT** prior to object delineation. The process takes as input the markers $\mathcal{S} = \mathcal{S}_o \cup \mathcal{S}_b$ on the target image and outputs an *object map* O, which is used for arc-weight assignment in Eqs. 13, 15, 17, 19, 21, and 23. Other knowledge about objects, such as a hierarchy among them [23] and border polarity [20,28], may also be used.

An object map can be created from markers inside and outside the object when those markers are interpreted as training sets for a pattern classifier [29]. By defining the graph $(\mathcal{N}, \mathcal{A}, w)$ such that $\mathcal{N} \subset \mathcal{S}$ (*some representative seeds*) in $\mathcal{S}_o$ and $\mathcal{S}_b$, $\mathcal{A}$ is given by Eq. 4, and $w = \|\mathbf{I}(t), \mathbf{I}(s)\|$, the **IFT** algorithm for path-cost function f_1 (Eqs. 5 and 7) can output two optimum-path forests, one rooted in $\mathcal{S}_o$ and the other rooted in $\mathcal{S}_b$, with the paths first computed in $\mathcal{S}$ and then propagated to the remaining pixels in $D_I \setminus \mathcal{S}$ [30]. The object map O is defined as $O(p) = \frac{V^{(b)}(p)}{V^{(o)}(p)+V^{(b)}(p)}$ from the costs $V^{(o)}(p)$ and $V^{(b)}(p)$ of the optimum paths that reach each pixel $p \in D_I$ from its roots in $\mathcal{S}_o$ and $\mathcal{S}_b$, respectively [29]. However, object and background markers selected in regions with similar properties can impair the object map. The problem is solved in [35] by a method that eliminates seeds with similar properties and distinct labels that fall in a same cluster of the color space. Figure 3 illustrates the process using markers drawn by real users from the database in [4]. Intelligent seed selection selects groups (Fig. 3a) with a minimum percentage (e.g., 80%) of seeds from the same label (Fig. 3b). Without seed selection, the object map (above) is impaired.

For clustering, we use a graph $(\mathcal{N}, \mathcal{A}, w)$ such that $\mathcal{N}$ is a superpixel set with 800 superpixels [21], $\mathcal{A}$ is given by Eq. 3, and $w = \|\mathbf{I}(t), \mathbf{I}(s)\|$, being $\mathbf{I}(s)$ the mean CIELab color of superpixel s. The **IFT** algorithm for path-cost function f_1 (Eqs. 6 and 7) can output one optimum-path tree (cluster) rooted at each minimum of $V_o(s)$, being $V_o(s)$ the complement of a probability density function (**pdf**) [33]. The parameter k in $\mathcal{K}(s,k)$ is found by optimization in $[1, k_{max}]$ (e.g., $k_{max} = 24$) as the one that produces a minimum normalized cut in the graph [33]. This way each dome of the estimated pdf becomes one cluster.

5 Experimental Results

In order to evaluate object delineation using $\mathbf{GC_{max}}$ and $\mathbf{GC_{sum}}$, we first randomly divided the GrabCut dataset [34] in two parts: 50% of the images to optimize the parameters of the methods and 50% of the images to evaluate the

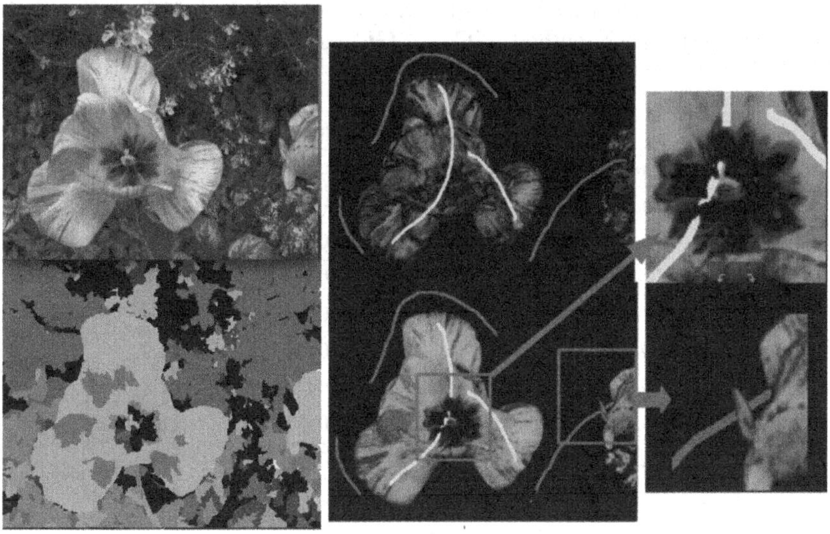

(a) (b)

Fig. 3. (a) Original image (top) and clusters of superpixels (bottom). (b) Object map O without (top) and with (bottom and zoomed regions on the right) intelligent seed selection.

methods. The number of superpixels was found in [100, 1000], by keeping $a = 0.5$ and $b = 12$ fixed in Eq. 9 [40], using two hierarchy levels [21], being the first level with three times more superpixels, and classifying as part of the object the superpixels with at least 80% of the pixels inside the ground-truth object mask. The number 800 was the minimum to obtain a segmentation accuracy (Dice coefficient) above 98%. The clustering parameter k_{max} [33] was found in [8 – 80] and the purity of the clusters for seed selection was found in [50% – 95%] by using the real user markers from [4] to generate an object map and then measuring the accuracy of segmentation based on the binary classification of the map at 50%. The best values were $k_{max} = 24$ and purity equal to 80%. Afterwards, α and β in Eqs. 12–23 were searched in [0.0 – 0.95] and [50 – 150], respectively, based on the accuracy of delineation. The best values were $\alpha = 0.25$ for w_2, $\alpha = 0.1$ for w_4, w_{10} and w_{12}, $\alpha = 0.2$ for w_6 and w_8, which indicates that the color of the pixels are more important than the object map, and $\beta = 70$ for $\mathbf{GC_{sum}}$. We have also included Deep Extreme Cut (**DEC**) [25] as baseline and optimized the threshold on its probability map. We found the best threshold at 85% in [50%, 95%] based on the accuracy of segmentation.

Table 1 presents the experimental results on the test images when the geodesic user robot is used for at most 15 iterations. In general, methods based on $\mathbf{GC_{max}}$ with dynamic arc-weight assignment—i.e., Dynamic Trees (**DT**) [8]—

Table 1. Mean dice coefficient, standard deviation, and number of iterations required for convergence using each method, $\mathbf{GC_{max}}$, $\mathbf{GC_{max}}$**-DT** (Dynamic Trees), $\mathbf{GC_{sum}}$, and **DEC** (Deep Extreme Cut).

Arc-weight	Mean accuracy (dice)	Standard dev.	Iterations
GC_{max}-DT w_{10}	0.970	0.0237	11.3
GC_{max}-DT w_8	0.968	0.0295	11.8
GC_{max}-DT w_6	0.966	0.0364	12.0
GC_{max} w_2	0.961	0.0421	12.6
GC_{max}-DT w_7	0.961	0.0491	11.9
GC_{max}-DT w_9	0.959	0.0498	10.9
GC_{max}-DT w_5	0.958	0.0458	11.8
GC_{max}-DT w_{12}	0.958	0.0878	11.4
GC_{max}-DT w_{11}	0.957	0.0599	11.5
DEC	0.942	0.0814	4.00
GC_{max} w_1	0.933	0.0854	13.8
GC_{sum} w_4	0.932	0.0845	13.3
GC_{sum} w_3	0.918	0.1080	13.6

provide the best results, not only in accuracy but also in the number of iterations, which can be related to less user intervention. **DEC** reduces user intervention to four extreme points (4 iterations), which are inserted by a robot based on the ground truth, but it does not allow user corrections. The object map plays an important role to improve the results, making $\mathbf{GC_{max}}$ with w_2 more competitive. The good performance of $\mathbf{GC_{max}}$**-DT** with w_9 and w_5 indicates that Dynamic Trees can learn object information from the tree and forest that contain p at the moment $w(p, q)$ is estimated. On the other hand, the local distance $\|\mathbf{I}(p), \mathbf{I}(q)\|$ seems to have a negative influence on $\mathbf{GC_{max}}$**-DT**, when using the forest that contains p (w_9 and w_{11}), but the same is not observed when using the tree that contains p (w_5 and w_7). The object map can also improve $\mathbf{GC_{sum}}$, but it is clear that the small size of the markers placed by the robot affects more $\mathbf{GC_{sum}}$ than $\mathbf{GC_{max}}$ [13,29]. Given that **DEC** needs training from several segmented images, these results are a strong indication that optimum connectivity is crucial not only for delineation but also to extract object information from target images with no prior training.

Finally, a better idea about these numbers can be obtained in Fig. 4 when comparing $\mathbf{GC_{max}}$**-DT** with w_{10} and w_9 against **DEC** and $\mathbf{GC_{sum}}$ with w_3.

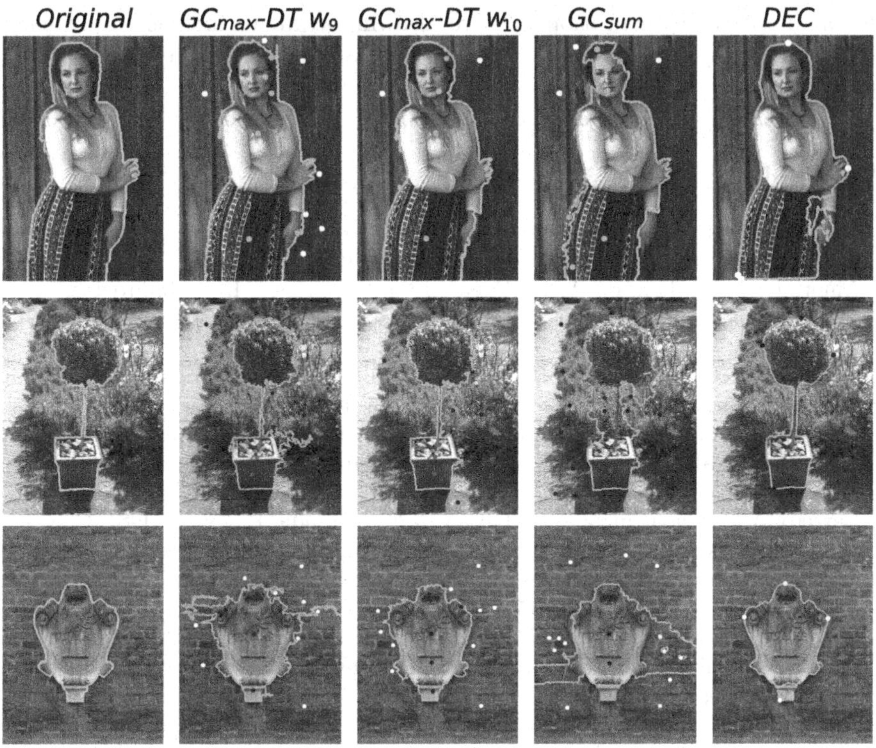

Fig. 4. From left to right, original images with ground-truth border (magenta), segmentation results with their respective methods written above, object (extreme points) and background markers with enhanced size to improve visualization. (Color figure online)

6 Conclusion and Research Directions

We described the **IFT** framework[1], some of its recent results, and applications to delineation, clustering, and classification. We also provided a guideline to extract object information from the target image before and during delineation. Experimental results showed significant improvements in delineation when object information is used, calling attention to a new paradigm—Dynamic Trees ($\mathbf{GC_{max}}$-**DT**) [8].

As research directions, one can directly extend $\mathbf{GC_{max}}$-**DT** to 3D images and multiple objects, learn other object properties from dynamic trees, and integrate it with shape [26] and texture models [24] for automated segmentation. When it fails, methods that transform the result into an optimum-path forest [38] for user correction by differential **IFTs** [14, 19] are needed. Interactive machine learning

[1] Segmentation framework binaries available at https://github.com/JoOkuma/PyIFT.

and collaborative segmentation [37] are also important to increase intelligence and reliability in segmentation, creating large annotated image databases for research.

References

1. Achanta, R., et al.: SLIC superpixels compared to state-of-the-art superpixel methods. IEEE Trans. Pattern Anal. Mach. Intell. **34**(11), 2274–2282 (2012)
2. Amorim, W.P., Falcão, A.X., Papa, J.P., Carvalho, M.H.: Improving semi-supervised learning through optimum connectivity. Pattern Recogn. **60**, 72–85 (2016)
3. Amorim, W.P., Falcão, A.X., Papa, J.P.: Multi-label semi-supervised classification through optimum-path forest. Inf. Sci. **465**, 86–104 (2018)
4. Andrade, F., Carrera, E.V.: Supervised evaluation of seed-based interactive image segmentation algorithms. In: Symposium on Signal Processing, Images and Computer Vision, pp. 1–7 (2015)
5. Barreto, A.E., Chowdhury, A.S., Falcao, A.X., Miranda, P.A.V.: IFT-SLIC: a general framework for superpixel generation based on simple linear iterative clustering and image foresting transform. In: Conference on Graphics, Patterns and Images (SIBGRAPI), pp. 337–344 (2015)
6. Belém, F., Guimarães, S., Falcão, A.X.: Superpixel segmentation by object-based iterative spanning forest. In: 23rd Iberoamerican Congress on Pattern Recognition (2018, to appear)
7. Boykov, Y., Kolmogorov, V.: An experimental comparison of min-cut/max-flow algorithms for energy minimization in vision. IEEE Trans. Pattern Anal. Mach. Intell. **26**(9), 1124–1137 (2004)
8. Bragantini, J., Martins, S.B., Castelo-Fernandez, C., Falcão, A.X.: Graph-based image segmentation using dynamic trees. In: 23rd Iberoamerican Congress on Pattern Recognition (2018, to appear)
9. Chen, J., Li, Z., Huang, B.: Linear spectral clustering superpixel. IEEE Trans. Image Process. **26**(7), 3317–3330 (2017)
10. Chen, L.C., Papandreou, G., Kokkinos, I., Murphy, K., Yuille, A.L.: DeepLab: semantic image segmentation with deep convolutional nets, atrous convolution, and fully connected CRFs. IEEE Trans. Pattern Anal. Mach. Intell. **40**(4), 834–848 (2018)
11. Ciesielski, K.C., et al.: Joint graph cut and relative fuzzy connectedness image segmentation algorithm. Med. Image Anal. **17**(8), 1046–1057 (2013)
12. Ciesielski, K.C., et al.: Path-value functions for which Dijkstra's algorithm returns optimal mapping. J. Math. Imaging Vis. **60**(7), 1–12 (2018)
13. Ciesielski, K., Udupa, J., Falcão, A., Miranda, P.: Fuzzy connectedness image segmentation in graph cut formulation: a linear-time algorithm and a comparative analysis. J. Math. Imaging Vis. **44**(3), 375–398 (2012)
14. Condori, M.A.T., Cappabianco, F.A.M., Falcão, A.X., Miranda, P.A.V.: Extending the differential image foresting transform to root-based path-cost functions with application to superpixel segmentation. In: Conference on Graphics, Patterns and Images (SIBGRAPI), pp. 7–14 (2017)
15. Couprie, C., Grady, L., Najman, L., Talbot, H.: Power watershed: a unifying graph-based optimization framework. IEEE Trans. Pattern Anal. Mach. Intell. **33**(7), 1384–1399 (2011)

16. Cousty, J., Bertrand, G., Najman, L., Couprie, M.: Watershed cuts: minimum spanning forests and the drop of water principle. IEEE Trans. Pattern Anal. Mach. Intell. **31**(8), 1362–1374 (2009)
17. Falcão, A.X., Bergo, F.P.G.: Interactive volume segmentation with differential image foresting transforms. IEEE Trans. Med. Imaging **23**(9), 1100–1108 (2004)
18. Falcão, A.X., Spina, T.V., Martins, S.B., Phellan, R.: Medical image segmentation using object shape models. In: Invited Lecture in Proceedings of the ECCOMAS Thematic (VipIMAGE), pp. 9–15. CRC Press (2015)
19. Falcão, A.X., Stolfi, J., de Lotufo, R.A.: The image foresting transform: theory, algorithms, and applications. IEEE Trans. Pattern Anal. Mach. Intell. **26**(1), 19–29 (2004)
20. Falcão, A.X., Udupa, J.K., Samarasekera, S., Sharma, S., Hirsch, B.E., Lotufo, R.A.: User-steered image segmentation paradigms: live wire and live lane. Graph. Models Image Process. **60**(4), 233–260 (1998)
21. Galvão, F.L., Falcão, A.X., Chowdhury, A.S.: RISF: recursive iterative spanning forest for superpixel segmentation. In: Conference on Graphics, Patterns and Images (SIBGRAPI). IEEE Xplore (2018, to appear)
22. Gulshan, V., et al.: Geodesic star convexity for interactive image segmentation. In: IEEE Conference on Computer Vision and Pattern Recognition, pp. 3129–3136 (2010)
23. Leon, L.M.C., Miranda, P.A.V.: Multi-object segmentation by hierarchical layered oriented image foresting transform. In: Conference on Graphics, Patterns and Images (SIBGRAPI), pp. 79–86 (2017)
24. Long, J., Shelhamer, E., Darrell, T.: Fully convolutional networks for semantic segmentation. In: IEEE Conference on Computer Vision and Pattern Recognition, pp. 3431–3440 (2015)
25. Maninis, K.K., et al.: Deep extreme cut: from extreme points to object segmentation. In: IEEE Conference on Computer Vision and Pattern Recognition (2018)
26. Martins, S.B., Spina, T.V., Yasuda, C., Falcão, A.X.: A multi-object statistical atlas adaptive for deformable registration errors in anomalous medical image segmentation. In: SPIE on Medical Imaging: Image Processing, p. 101332G(2017)
27. Miranda, P.A.V., Falcão, A.X., Spina, T.V.: Riverbed: a novel user-steered image segmentation method based on optimum boundary tracking. IEEE Trans. Image Process. **21**(6), 3042–3052 (2012)
28. Miranda, P.A.V., Mansilla, L.A.C.: Oriented image foresting transform segmentation by seed competition. IEEE Trans. Image Process. **23**(1), 389–398 (2014)
29. Miranda, P.A., Falcão, A.X.: Links between image segmentation based on optimum-path forest and minimum cut in graph. J. Math. Imaging Vis. **35**(2), 128–142 (2009)
30. Papa, J., Falcão, A., de Albuquerque, V., Tavares, J.: Efficient supervised optimum-path forest classification for large datasets. Pattern Recogn. **45**(1), 512–520 (2012)
31. Phellan, R., Falcão, A.X., Udupa, J.K.: Medical image segmentation via atlases and fuzzy object models: improving efficacy through optimum object search and fewer models. Med. Phys. **43**(1), 401–410 (2016)
32. Rauber, P.E., Falcão, A.X., Spina, T.V., Rezende, P.J.: Interactive segmentation by image foresting transform on superpixel graphs. In: Conference on Graphics, Patterns and Images (SIBGRAPI), pp. 131–138 (2013)
33. Rocha, L., Cappabianco, F., Falcão, A.: Data clustering as an optimum-path forest problem with applications in image analysis. Int. J. Imaging Syst. Technol. **19**(2), 50–68 (2009)

34. Rother, C., Kolmogorov, V., Blake, A.: GrabCut: interactive foreground extraction using iterated graph cuts. ACM Trans. Graph. **23**, 309–314 (2004)
35. Spina, T.V., Miranda, P.A.V., Falcão, A.X.: Intelligent understanding of user interaction in image segmentation. Int. J. Pattern Recogn. Artif. Intell. **26**(02), 1265001 (2012)
36. Spina, T.V., Miranda, P.A.V., Falcão, A.: Hybrid approaches for interactive image segmentation using the live markers paradigm. IEEE Trans. Image Process. **23**(12), 5756–5769 (2014)
37. Spina, T.V., Stegmaier, J., Falcão, A.X., Meyerowitz, E., Cunha, A.: SEGMENT3D: a web-based application for collaborative segmentation of 3D images used in the shoot apical meristem. In: International Symposium on Biomedical Imaging (ISBI), pp. 391–395 (2018)
38. Tavares, A.C.M., Miranda, P.A.V., Spina, T.V., Falcão, A.X.: A supervoxel-based solution to resume segmentation for interactive correction by differential image-foresting transforms. In: Angulo, J., Velasco-Forero, S., Meyer, F. (eds.) ISMM 2017. LNCS, vol. 10225, pp. 107–118. Springer, Cham (2017). https://doi.org/10.1007/978-3-319-57240-6_9
39. Udupa, J.K., et al.: Body-wide hierarchical fuzzy modeling, recognition, and delineation of anatomy in medical images. Med. Image Anal. **18**(5), 752–771 (2014)
40. Vargas-Muñoz, J.E., et al.: An iterative spanning forest framework for superpixel segmentation. IEEE Trans. Image Process. (2019, to appear)

On the Degree Sequence of 3-Uniform Hypergraph: A New Sufficient Condition

Andrea Frosini[1(✉)] and Christophe Picouleau[2]

[1] Dipartimento di Matematica e Informatica, Università di Firenze, Viale Morgagni 65, 50134 Florence, Italy
andrea.frosini@unifi.it
[2] CEDRIC, CNAM, 292 rue St-Martin, 75141 Paris Cedex 03, France
christophe.picouleau@cnam.fr

Abstract. The study of the degree sequences of h-uniform hypergraphs, say h-sequences, was a longstanding open problem in the case of $h > 2$, until very recently where its decision version was proved to be NP-complete. Formally, the decision version of this problem is: Given $\pi = (d_1, d_2, \ldots, d_n)$ a non increasing sequence of positive integers, is π the degree sequence of a h-uniform simple hypergraph?

Now, assuming $P \neq NP$, we know that such an effective characterization cannot exist even for the case of 3-uniform hypergraphs.

However, several necessary or sufficient conditions can be found in the literature; here, relying on a result of S. Behrens et al., we present a sufficient condition for the 3-graphicality of a degree sequence and a polynomial time algorithm that realizes one of the associated 3-uniform hypergraphs, if it exists. Both the results are obtained by borrowing some mathematical tools from discrete tomography, a quite recent research area involving discrete mathematics, discrete geometry and combinatorics.

Keywords: h-uniform hypergraph · Hypergraph degree sequence · Discrete tomography · Reconstruction problem

1 Introduction

In order to model complex systems with one-to-one interactions, one among the most versatile and used mathematical structure is that of graph so that the elements of the systems are represented by nodes, and their mutual interactions by edges. A wide interest in graph theory started in the half of 20th century, and few years later, took shape the idea of generalizing the interactions' possibility to more than two elements of the systems. So, edges naturally evolved into hyperedges, regarded as subsets of nodes, and the notion of graph changed into hypergraph, accordingly.

In addition to the natural generalization of graphs, hypergraphs found their own relevance in different research areas, ranging from the most theoretical ones

M. Couprie et al. (Eds.): DGCI 2019, LNCS 11414, pp. 195–205, 2019.
https://doi.org/10.1007/978-3-030-14085-4_16

such as Geometry, Algebra and Number Theory, to more applicative as Optimization, Physics, Chemistry, etc.

The seminal book by Berge [5] will give to the reader the formal definitions and vocabulary, some results with their proofs, and more about applications of hypergraphs.

Most of the times what is required is to infer some statistics and characteristics of the modelled system from partial and sometimes inaccurate information about it. A typical situation is when the data concern only the number of interactions that involves each node, say the degree of a node, without detailing the subjects of those interactions; this case is referred to as *degree based* reconstruction problem and it involves various subproblems concerning the reconstruction of a hypergraph from a given degree sequence π, counting the number of different hypergraphs having a given π, possibly reconstructing all of them, and also sampling a typical element among them.

In this paper, we concentrate on the first of these problems related to a specific subclass of hypergraphs, i.e. the h-uniform simple ones, with $h = 3$. This choice is motivated by the fact that degree sequences for $h = 2$, i.e. simple graphs, have been studied by many authors, including the celebrated work of Erdös and Gallai [17], which effectively characterizes them. From their result, a P-time algorithm was designed to reconstruct the adjacency matrix of a graph having degree sequence π (if it exists). On the other hand, the case $h \geq 3$ remained open till nowadays: in 2018, Deza et al. in [15] proved its NP-completeness, i.e., they showed that for any fixed integer $h \geq 3$ it is NP-complete to decide if a sequence of positive integers can be the degree sequence of a h-uniform hypergraph.

So, it acquires relevance to restrict the set of intractable degree sequences and find fast reconstruction algorithms for those remaining: in [18], h-uniform regular and almost regular hypergraphs are considered and the related degree sequences have been characterized and efficiently reconstructed. Successively, Behrens et al. in [2] propose a sufficient condition for a degree sequence to be h-graphic; unfortunately the characterization gives no information about the associated h-uniform hypergraphs. Finally, in [7], an efficient generalization of the algorithm in [18] fills this gap. Our studies aim to push further the efficient extension of the condition of [2] in case of 3-uniform hypergraphs.

We give the definitions and the results useful for our study in the next section. Then in the Sect. 3 we will give our new sufficient condition and the related polynomial time algorithm that given a sequence of integers π satisfying it, builds the incidence matrix of a 3-uniform hypergraph that realizes π, if such hypergraph exists.

2 Definitions and State of the Art

Borrowing the notation from [5], we define *hypergraph* to be the couple $G = (Vert, \mathcal{E})$ such that $Vert = \{v_1, \ldots, v_n\}$ is a ground set of vertices and $\mathcal{E} \subset 2^{|Vert|} \setminus \emptyset$ is the set of hyperedges. We choose to consider *simple* hypergraphs only, i.e. such that $e \not\subseteq e'$ for any pair e, e' of $\mathcal{E}$, and we admit isolated points as vertices, so $\bigcup \mathcal{E} \subseteq Vert$, see Fig. 1. The *degree* of a vertex $v \in Vert$ is the number of hyperedges $e \in \mathcal{E}$ such that $v \in e$, and the *degree sequence* of a hypergraph, also called graphic sequence, is the list of its vertex degrees, usually written in nonincreasing order, as $\pi = (d_1, d_2, \ldots, d_n), d_1 \geq d_2 \geq \cdots \geq d_n$. In the sequel, it will be useful to indicate by $\sigma(\pi)$ the sum of the elements of π, and π^- the sequence π with the first element removed, i.e. $\pi^- = (d_2, \ldots, d_n)$.

A hypergraph is h-uniform if $|e| = h$ for all hyperedge $e \in \mathcal{E}$. In Fig. 1 two 3-uniform hypergraphs, i.e., (a) and (c), and a 2-uniform one, (b), are depicted; the last one turns out to be a simple graph.

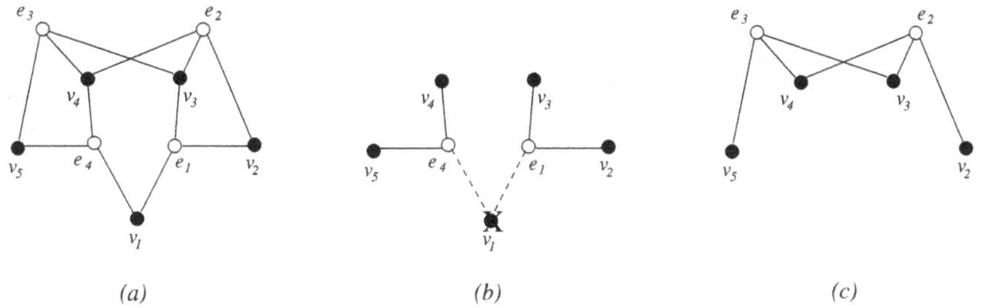

Fig. 1. (a): a 3-uniform hypergraph with vertices $V = \{v_1, v_2, v_3, v_4, v_5\}$ and hyperedges $\mathcal{E} = \{e_1, e_2, e_3, e_4\}$. Its degree sequence (arranged in non increasing order) is $\pi = (3, 3, 2, 2, 2)$; (b): the link hypergraph of the decomposition of the hypergraph in (a) w.r.t. the removed vertex v_1, according to Theorem 2. The link hypergraph is 2-uniform, i.e. it is a simple graph; (c): the residual 3-uniform hypergraph of the same decomposition.

The problem of the combinatorial and algorithmically efficient characterization of the degree sequences of h-uniform hypergraphs, say h-graphic sequences, has been one of the most relevant in the theory of hypergraphs: the case of simple graphs, i.e. when $h = 2$, was solved in 1960 by Erdös and Gallai in the following milestone theorem (see [4]).

Theorem 1 (Erdös, Gallai). *A sequence $\pi = (d_1, d_2, \ldots, d_n)$ where $d_1 \geq d_2 \geq \cdots \geq d_n$ is (2-)graphic if and only if $\sigma(\pi)$ is even and*

$$\sum_{i=1}^{k} d_i \leq k(k-1) + \sum_{i=k+1}^{n} \min\{k, d_i\}, 1 \leq k \leq n.$$

Concerning the general case of h-graphical sequences, we recall the following (non efficient) result from [14]:

Theorem 2 (Dewdney). *Let $\pi = (d_1, \ldots, d_n)$ be a non-increasing sequence of non-negative integers. π is h-graphic if and only if there exists a non-increasing sequence $\pi' = (d'_2, \ldots, d'_n)$ of non-negative integers such that*

1. *π' is $(h-1)$-graphic,*
2. *$\sum_{i=2}^{n} d'_i = (h-1)d_1$, and*
3. *$\pi'' = (d_2 - d'_2, \ldots, d_n - d'_n)$ is h-graphic.*

The underlying idea of the theorem rests on the possibility of splitting a h-uniform hypergraph G into two parts: for each vertex v, the first one consists of the hypergraph obtained from G after deleting all the hyperedges not containing v, and then removing, from all the remaining hyperedges, the vertex v; this hypergraph is identified in the literature with $L_G(v)$, say the *link* of v, and its degree sequence the *link sequence* of v. The second hypergraph G_v^-, say the *residual* of v, is obtained from G after removing all hyperedges containing v. It is clear that G can be obtained from $L_G(v)$ and G_v^-; furthermore one can notice that $L_G(v)$ is $(h-1)$-uniform, while G_v^- preserves the h-uniformity. Such a decomposition can be recursively carried on till reaching trivial hypergraphs.

Relying on this result, the authors of [2] provided a sufficient conditions for the h-graphicality of a degree sequence:

Theorem 3 (Behrens et al.). *Let π be a non-increasing sequence of length n with maximum entry Δ and t entries that are at least $\Delta - 1$. If h divides $\sigma(\pi)$ and*

$$\binom{t-1}{h-1} \geq \Delta \tag{1}$$

then π is h-graphic.

Unfortunately, this theorem does not furnish an efficient way to construct a h-uniform realization of the sequence π.

It is worth mentioning that very recently, in [15], it has been proved the non polynomiality of the reconstruction of a 3-uniform hypergraph that realizes a given degree sequence, spreading the result to each $h \geq 3$. So, it has acquired relevance the study of sets of degree sequences whose h-graphicality can be certified in polynomial time and the definition of a strategy to construct their related hypergraphs.

Following the direction, in these last years the result of Behrens et al. has been investigated from a different perspective, as an inverse problem in the discrete environment. The required mathematical tools come from *discrete tomography* that is wide research area whose aim, among many others, is that of reconstructing (or at least retrieve information about) unknown binary matrices regarded as homogeneous finite sets of points, from projections, i.e. measurements of the number of elements lying on each line intersecting the set and having a given direction.

One can refer to the books of Herman and Kuba [21,22] for basics on the theory, algorithms and applications of discrete tomography.

Coming back to our context, the problem of the characterization of the degree sequence $(d_1, d_2, \ldots, d_n)$ of an h-uniform (simple) hypergraph G asks whether there is a binary matrix A with projections $H = (h, h, \ldots, h)$ and $V = (d_1, d_2, \ldots, d_n)$ and having distinct rows, i.e., A is the adjacency matrix of G where rows and columns correspond to hyperedges and vertices, respectively. Ryser's Theorem [24] answers the question for generic hypergraphs, since it admits the presence of equal rows in the reconstructed matrix, so, as mentioned in [5], the reconstruction of a multi-hypergraph (parallel hyper edges are authorized) from a given degree sequence can be efficiently done. In [7], the Theorem 3 has been considered and translated into an inverse reconstruction problem, gaining efficiency to the sufficient condition it introduces. In the next section, we rely on this result to efficiently generalize the sufficient condition it proposes to 3-uniform hypergraphs.

3 A Sufficient Condition for the 3-Graphicality

We recall the notion of *dominance order* defined by Brylawski in [6]: let $\pi = (d_1, \ldots, d_n)$ and $\pi' = (d'_1, \ldots, d'_n)$ be two integer sequences such that $\sigma(\pi) = \sigma(\pi')$, we define

$$\pi \leq_d \pi' \quad \text{if and only if} \quad \sum_{i=1}^{k} d_i \geq \sum_{i=1}^{k} d'_i$$

for each $1 \leq k \leq n$.

Lemma 1. *Let $\pi = (d_1, \ldots, d_n)$ and $\pi' = (d'_1, \ldots, d'_n)$ be two sequences such that $\pi \leq_d \pi'$. If π is graphic, then also π' is.*

Proof. The graphic characterization in Theorem 1 related to π states that, for each $1 \leq k \leq n$, $\sum_{i=1}^{k} d_i \leq k(k-1) + \sum_{i=k+1}^{n} min\{k, d_i\}$. Since $\pi \leq_d \pi'$, we have

$$\sum_{i=1}^{k} d'_i \leq \sum_{i=1}^{k} d_i \leq k(k-1) + \sum_{i=k+1}^{n} min\{k, d_i\} \leq k(k-1) + \sum_{i=k+1}^{n} min\{k, d'_i\},$$

the last inequality holding since $\sigma(\pi) = \sigma(\pi')$, so π' is also graphic. □

The following lemma states the trivial property that in a n-nodes (simple and loopless) graph, each node can be edge connected to the remaining $n-1$ nodes at most:

Lemma 2. *Let $\pi = (d_1, \ldots, d_n)$ be a non increasing graphic sequence. It holds that $d_1 \leq n-1$.*

The following definitions and some notations introduce the main result of this section: let $\pi = (d_1, \ldots, d_n)$ be a nonincreasing integer sequence. We define the *S cut (sequence)* of π to be the sequence $\pi' = (d'_1, \ldots, d'_n)$ such that $\sigma(\pi') = S$, and

$$d'_i = \begin{cases} d_i - c - 1 & \text{if } i \leq k' \\ d_i - c & \text{if } k' < i \leq k \\ 0 & \text{otherwise} \end{cases}$$

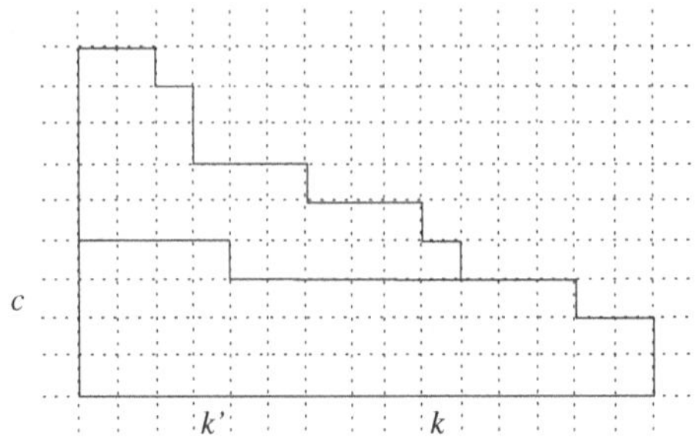

Fig. 2. The Ferrer diagram of the sequence $\pi = (9, 9, 8, 6, 6, 6, 5, 5, 5, 4, 3, 3, 3, 2, 2)$. The cut sequence of sum 29 is $\pi' = Cut(\pi, 29) = (5, 5, 4, 2, 3, 3, 2, 2, 2, 1, 0, 0, 0, 0, 0)$. The height $c = 3$ of the cut and the indexes k' and k are also highlighted. Observe that π' may loose the nonincreasing property.

with $0 \leq k' < k \leq n$ and $0 \leq c < d_1$. We indicate $\pi' = Cut(\pi, S)$ and we refer to c as the *height* of π'; Fig. 2 helps in visualizing the definition.

Furthermore, let A and B be two $m \times n$ and $m' \times n'$ matrices, respectively; we introduce the following standard operators: if $m = m'$, then the *horizontal concatenation* of A and B, write $A \oplus B$, is the matrix obtained by orderly concatenating the columns of A to those of B. Similarly, if $n = n'$, we define the *vertical concatenation* of A and B, write $A \ominus B$, to be the ordered concatenation of the rows of A and B.

Theorem 4. *Let $\pi = (d_1, \ldots, d_n)$ be a nonincreasing sequence such that $\sigma(\pi)$ is a multiple of 3. If the cut sequence $\pi' = Cut(\pi^-, 2d_1)$ is graphic, then π is 3-graphic.*

Proof. We proceed by first constructing a 3-uniform realization G' of the sequence (d_1, π'), with $\pi' = Cut(\pi^-, 2d_1) = (d'_2, \ldots, d'_n)$: the adjacency matrix of G' has dimension $d_1 \times n$ since $\sigma(d_1, \pi') = 3d_1$, so its first column contains elements 1 only. By hypothesis, π' is graphic, so we fill the remaining $n - 1$ columns by one of its realizations, obtaining the desired adjacency matrix of G'. Since its horizontal projections have common value 3, then G' is 3-uniform (see Examples 1 and 2).

Now, we focus on the residual (in the sense of Theorem 2, and as shown in Fig. 2) sequence $\pi'' = (\pi - (d_1, \pi'))^- = (d''_2, \ldots, d''_n)$ that we consider arranged in non increasing order by the permutation α, if needed. By construction, it holds that $\sigma(\pi'')$ is multiple of 3. If π'' is the zero sequence, then G' itself is a 3-uniform realization of π. On the other hand, let $Cut(\pi^-, 2d_1)$ have eight $c < d_1$; some cases according to the value of d'_2 arise:

(*i*) $d'_2 \leq 1$: the maximum element of π' is 1, and, by definition of cut sequence, π'' has an initial almost regular sequence of length greater than or equal to

d_1. Since $d'_2 \leq d_1$, then π'' satisfies the conditions in Theorem 3. The result in [7] that relies back to [18], provides in P-time a 3-uniform realization G'' of π'';

(*ii*) $d_1 - 1 \leq d'_2 \leq d_1$: π'' has maximum element 2 at most, and consequently it has an immediate 3-uniform realization G'';

(*iii*) $\frac{d_1}{2} \leq d'_2 < d_1 - 1$: the sequence $\pi' = (d'_2, \ldots, d_k, 0 \ldots, 0)$ has $k - 1 \geq d'_2 + 1$ elements different from zero, by Lemma 2, with k ranging from 4 to n (if $k = 3$, then $d'_2 = 1$, as in case (*i*) above). The residual sequence can be written as

$$\pi'' = (\underbrace{c+1, \ldots, c+1}_{t \text{ times}}, c, \ldots, c, d_{k+1}, \ldots, d_n) \text{ possibly with } t = 0.$$

Since $c \leq d'_2 < k$, it follows that π'' satisfies the hypothesis of Theorem 3 and so it admits a 3-uniform realization in P-time as in [7] (see Example 1);

(*iv*) $1 < d'_2 < \frac{d_1}{2}$: again the sequence $\pi' = (d'_2, \ldots, d'_k, 0 \ldots, 0)$ has $k - 1 \geq d'_2 + 1$ elements different from 0, by Lemma 2. We observe that the maximum number of edges, i.e. the maximum value of d_1, of a graph with $d'_2 + 1$ nodes is $\frac{d'_2(d'_2+1)}{2}$, provided by the complete graph that we indicate as $K_{d'_2+1}$, and this same upper bound also holds for $d_2 (\leq d_1)$.
If the upper bounds realize for d_1 and d_2, with the minimum k, i.e. $k = d'_2 + 2$, then π'' has an initial sequence of $d'_2 + 1$ elements having common value

$$c = d_2 - d'_2 = \frac{d'_2(d'_2 + 1)}{2} - d'_2 = \frac{(d'_2 - 1)d'_2}{2},$$

that is the number of edges of the complete graph with $(d'_2 - 1)$ nodes. So, the sequence $\pi''' = Cut(\pi'', 2d''_2)$ is again graphic and π'' satisfies the conditions of this theorem, allowing the described decomposition to apply recursively till reaching the constant sequence. A 3-uniform realization G'' of π'' can be computed in P-time. If the upper bounds on d_1 and d_2 with the minimum value of k do not realize, then the sequence π'' that originates is obviously greater in the dominance order, so π''' is graphic a fortiori (see Example 2).

In all the cases, the final 3-uniform realization of π can be recursively obtained as $G = G' \ominus ([0] \oplus \alpha(G'')^{-1})$, where [0] stands for the zero column, after observing that G' has no common edges with G''. □

To make clearer the reconstruction process just described, we provide two examples of cases (*iii*) and (*iv*), being (*i*) and (*ii*) easiest subcases.

Example 1. *Let us consider the sequence* $\pi = (10, 9, 7, 5, 4, 4, 4, 3, 1, 1)$, *and compute*

$$\pi' = Cut(\pi^-, 20) = (6, 4, 3, 2, 2, 2, 1, 0, 0) \text{ and } \pi'' = (3, 3, 2, 2, 2, 2, 2, 1, 1).$$

Since the sequence π' *is graphic by Theorem 1, then* π'' *satisfies the hypothesis of Theorem 3, i.e.* $3 \leq \binom{6}{2}$, *and so one of its realizations* G'' *can be computed in polynomial time.*

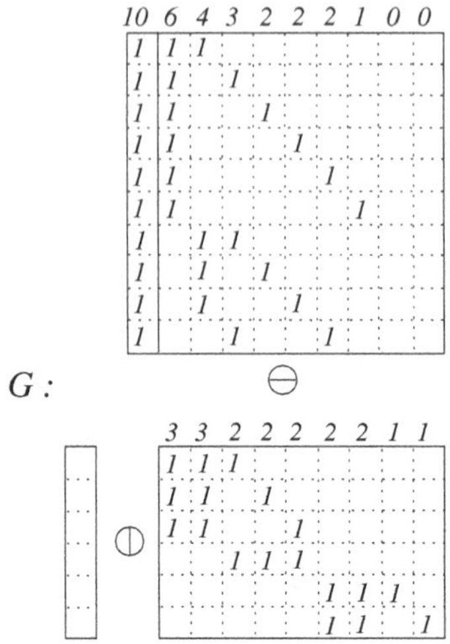

Fig. 3. A 3-uniform hypergraph G whose degree sequence is $\pi = (10, 9, 7, 5, 4, 4, 4, 3, 1, 1)$. The hypergraph G is obtained as the composition of G' and G'', i.e., the 3-uniform realizations of the related sequences π' and π'', respectively.

The final realization G of π is obtained by composing the realizations G' and G'' of π' and π'', respectively, as indicated in the proof of Theorem 4, case (iii), and depicted in Fig. 3.

Example 2. *In order to clarify the proof of Theorem 4, case (iv), let us consider the sequence $\pi = (10, 10, 10, 10, 10, 10)$, and compute*

$$\pi' = Cut(\pi, 20)^{-} = (4, 4, 4, 4, 4) \text{ and } \pi'' = (6, 6, 6, 6, 6).$$

It holds $d'_2(= 4) < \frac{d_1}{2}$, the sequence π' is graphic and its realization is the complete graph K_5.

The sequence π'' still satisfies the hypothesis of Theorem 4, since $\pi''' = Cut((\pi'')^{-}, 12) = (3, 3, 3, 3)$ is again the degree sequence of K_4.

The final realization G of π is obtained recursively by composing the 3-uniform realizations of K_5, K_4, K_3 and K_2 as in Fig. 4.

Finally, let us consider the sequence $\pi_1 = (10, 9, 9, 8, 8, 8, 5, 3)$, and verify that $\pi'_1 = (4, 4, 3, 4, 4, 1, 0)$ is graphic. The related $\pi''_1 = (5, 5, 5, 4, 4, 4, 3)$ is greater than π'' in the dominance order and consequently it admits a fortiori a 3-uniform realization as stated in Theorem 4, case (iv) (see the hypergraph G_1 in Fig. 4).

Corollary 1. *Let $\pi = (d_1, \ldots, d_n)$ be a nonincreasing integer sequence satisfying the hypothesis of Theorem 4. The reconstruction of a 3-uniform hypergraph consistent with π can be performed in P-time.*

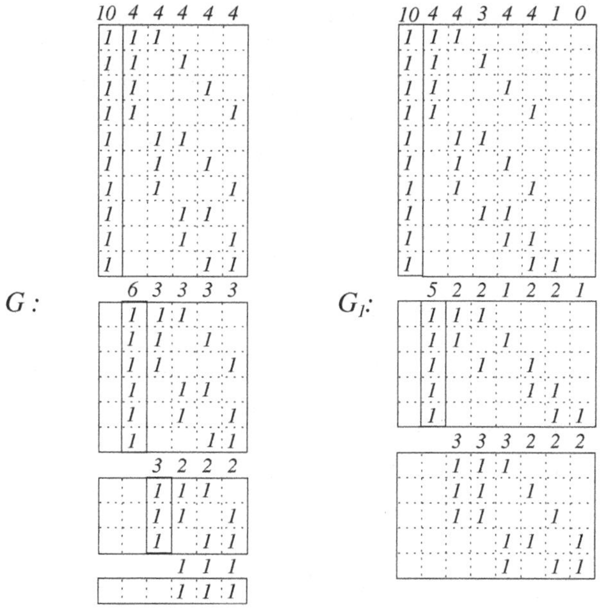

Fig. 4. Two 3-uniform hypergraphs G and G_1 related to case (iv) of Theorem 4. The complete graphs K_5, K_4, K_3, and K_2 that are part of G. The 3-uniform hypergraph G_1 is still in case (iv) and it is a realization of π_1.

In the proof of Theorem 4 we defined the algorithm to reconstruct the adjacency matrix of a 3-uniform hypergraph consistent with π. It is easy to check that all the steps can be performed in P-time (with respect to the dimensions of the matrix) and they are recursively called polynomially many times.

4 Conclusion and Open Problems

Our study gave a new sufficient condition for a sequence π of integers to be the degree sequence of a 3-uniform hypergraph, that can be efficiently checked. Furthermore, we also defined a polynomial time algorithm to reconstruct the adjacency matrix of a 3-uniform hypergraph realizing π if such hypergraph exists.

The defined condition and the algorithm are tuned for 3-uniform hypergraphs. An open problem is to find similar conditions for h-uniform hypergraphs, with $h \geq 4$.

One of the most important feature in discrete tomography is the study of the instances of a reconstruction problem admitting a unique realization. So, another interesting open problem is to characterize such instances in the case of the reconstruction of uniform hypergraphs. On the opposite side, another challenge is to count the number of realizations of a given graphic sequence. The interested reader can find some hints in this direction both in the last part of [2], and in [23].

Coming to an end, we recalled that the characterization of the degree sequences of h-uniform hypergraphs, with $h \geq 3$, is an NP-hard problem. So, under the assumption that $P \neq NP$, there is no hope to find a *good* characterization of them, but it would be of great interest to find a compact *nice looking* one in order to design algorithms for real-life applications.

References

1. Barcucci, E., Brocchi, S., Frosini, A.: Solving the two color problem: an heuristic algorithm. In: Aggarwal, J.K., Barneva, R.P., Brimkov, V.E., Koroutchev, K.N., Korutcheva, E.R. (eds.) IWCIA 2011. LNCS, vol. 6636, pp. 298–310. Springer, Heidelberg (2011). https://doi.org/10.1007/978-3-642-21073-0_27
2. Behrens, S., et al.: New results on degree sequences of uniform hypergraphs. Electron. J. Comb. **20**(4), P14 (2013)
3. Bentz, C., Costa, M.-C., Picouleau, C., Ries, B., de Werra, D.: Degree-constrained edge partitioning in graphs arising from discrete tomography. J. Graph Algorithms Appl. **13**(2), 99–118 (2009)
4. Berge, C.: Graphs. Gauthier-Villars, Paris (1983)
5. Berge, C.: Hypergraphs. North Holland, Amsterdam (1989)
6. Brylawski, T.: The lattice of integer partitions. Discrete Math. **6**, 210–219 (1973)
7. Brlek, S., Frosini, A.: A tomographical interpretation of a sufficient condition on h-graphical sequences. In: Normand, N., Guédon, J., Autrusseau, F. (eds.) DGCI 2016. LNCS, vol. 9647, pp. 95–104. Springer, Cham (2016). https://doi.org/10.1007/978-3-319-32360-2_7
8. Brocchi, S., Frosini, A., Rinaldi, S.: A reconstruction algorithm for a subclass of instances of the 2-color problem. Theor. Comput. Sci. **412**, 4795–4804 (2011)
9. Brocchi, S., Frosini, A., Rinaldi, S.: Solving some instances of the 2-color problem. In: Brlek, S., Reutenauer, C., Provençal, X. (eds.) DGCI 2009. LNCS, vol. 5810, pp. 505–516. Springer, Heidelberg (2009). https://doi.org/10.1007/978-3-642-04397-0_43
10. Brocchi, S., Frosini, A., Rinaldi, S.: The 1-color problem and the Brylawski model. In: Brlek, S., Reutenauer, C., Provençal, X. (eds.) DGCI 2009. LNCS, vol. 5810, pp. 530–538. Springer, Heidelberg (2009). https://doi.org/10.1007/978-3-642-04397-0_45
11. Colbourne, C.J., Kocay, W.L., Stinson, D.R.: Some NP-complete problems for hypergraph degree sequences. Discrete Appl. Math. **14**, 239–254 (1986)
12. Costa, M.-C., de Werra, D., Picouleau, C., Schindl, D.: A solvable case of image reconstruction in discrete tomography. Discrete Appl. Math. **148**(3), 240–245 (2005)
13. Costa, M.-C., de Werra, D., Picouleau, C.: Using graphs for some discrete tomography problems. Discrete Appl. Math. **154**(1), 35–46 (2006)
14. Dewdney, A.: Degree sequences in complexes and hypergraphs. Proc. Am. Math. Soc. **53**(2), 535–540 (1975)
15. Deza, A., Levin, A., Meesum, S.M., Onn, S.: Optimization over degree sequences. SIAM J. Discrete Math. **32**(3), 2067–2079 (2018)
16. Dürr, C., Guíñez, F., Matamala, M.: Reconstructing 3-colored grids from horizontal and vertical projections is NP-hard: a solution to the 2-atom problem in discrete tomography. SIAM J. Discrete Math. **26**(1), 330–352 (2012)

17. Erdös, P., Gallai, T.: On the maximal number of vertices representing the edges of a graph. MTA Mat. Kut. Int. Közl. **6**, 337–357 (1961)
18. Frosini, A., Picouleau, C., Rinaldi, S.: On the degree sequences of uniform hypergraphs. In: Gonzalez-Diaz, R., Jimenez, M.-J., Medrano, B. (eds.) DGCI 2013. LNCS, vol. 7749, pp. 300–310. Springer, Heidelberg (2013). https://doi.org/10.1007/978-3-642-37067-0_26
19. Gale, D.: A theorem on flows in networks. Pacific J. Math. **7**, 1073–1082 (1957)
20. Guíñez, F., Matamala, M., Thomassé, S.: Realizing disjoint degree sequences of span at most two: a tractable discrete tomography problem. Discrete Appl. Math. **159**(1), 23–30 (2011)
21. Herman, G.T., Kuba, A.: Discrete Tomography: Foundations Algorithms and Applications. Birkhauser, Boston (1999)
22. Herman, G.T., Kuba, A.: Advances in Discrete Tomography and Its Applications. Birkhauser, Boston (2007)
23. Kocay, W., Li, P.C.: On 3-hypergraphs with equal degree sequences. Ars Combinatoria **82**, 145–157 (2006)
24. Ryser, H.J.: Combinatorial properties of matrices of zeros and ones. Can. J. Math. **9**, 371–377 (1957)
25. Sierksma, G., Hoogeveen, H.: Seven criteria for integer sequences being graphic. J. Graph Theory **15**(2), 223–231 (1991)
26. Tripathi, A., Tyagi, H.: A simple criterion on degree sequences of graphs. Discrete Appl. Math. **156**(18), 3513–3517 (2008)

Optimization of Max-Norm Objective Functions in Image Processing and Computer Vision

Filip Malmberg[1(✉)], Krzysztof Chris Ciesielski[2,3], and Robin Strand[1]

[1] Centre for Image Analysis, Department of Information Technology, Uppsala University, Uppsala, Sweden
{filip.malmberg,robin.strand}@it.uu.se

[2] Department of Mathematics, West Virginia University, Morgantown, WV 26506-6310, USA
Krzysztof.Ciesielski@mail.wvu.edu

[3] Department of Radiology, MIPG, University of Pennsylvania, Philadelphia, PA 19104, USA

Abstract. Many fundamental problems in image processing and computer vision, such as image filtering, segmentation, registration, and stereo vision, can naturally be formulated as optimization problems.

We consider binary labeling problems where the objective function is defined as the max-norm over a set of variables. It is well known that for a limited subclass of such problems, globally optimal solutions can be found via watershed cuts, i.e., cuts by optimum spanning forests. Here, we propose a new algorithm for optimizing a broader class of such problems. We prove that the proposed algorithm returns a globally optimal labeling, provided that the objective function satisfies certain given conditions, analogous to the submodularity conditions encountered in min-cut/max-flow optimization. The proposed method is highly efficient, with quasi-linear computational complexity.

Keywords: Energy minimization · Pixel labeling · Minimum cut · Submodularity

1 Introduction

Many fundamental problems in image processing and computer vision, such as image filtering, segmentation, registration, and stereo vision, can naturally be formulated as optimization problems. Often, these optimization problems can be described as *labeling* problems, in which we wish to assign to each image element (pixel, or vertex of an associated graph) $v \in V$ an element $\ell(v)$ from some finite, K-element, set of labels, usually $\{0, \ldots, K-1\}$. The interpretation of these labels depends on the optimization problem at hand. In image segmentation, the labels might indicate object categories. In registration and stereo disparity

M. Couprie et al. (Eds.): DGCI 2019, LNCS 11414, pp. 206–218, 2019.
https://doi.org/10.1007/978-3-030-14085-4_17

problems the labels represent correspondences between images, and in image reconstruction and filtering the labels represent intensities in the filtered image.

In this paper, we seek binary label assignments $\ell\colon V \to \{0,1\}$ that minimizes a given objective (energy) function E_∞ of the form

$$E_\infty(\ell) := \max\{\max_{s\in V} \phi_s(\ell(s)), \max_{\{s,t\}\in\mathcal{E}} \phi_{st}(\ell(s),\ell(t))\}. \quad (1)$$

In this equation, the image domain is identified with an undirected[1] graph $\mathcal{G} = (V, \mathcal{E})$, where the set V of vertices of the graph is the set of all pixels of the image, while $\mathcal{E}$ is the set of all its edges, that is, of pairs $\{s,t\}$ of vertices/pixels that are *adjacent* according to some given adjacency relation.

At first glance, the restriction to binary labeling may appear very limiting. Many successful methods for multi-label optimization, however, rely on iteratively minimizing binary labeling problems via *move-making* strategies [1]. Thus, the ability to find optimal solutions for problems with two labels has high relevance also for the multi-label case.

The functions $\phi_s(\cdot)$ are referred to as *unary* terms. Each unary term depends only on the label $\ell(s)$ assigned to the pixel s, and they are generally used to indicate the preference of an individual pixel to be assigned each particular label, typically based on some prior information.

The functions $\phi_{st}(\cdot,\cdot)$ are referred to as *pairwise* or *binary* terms. Each such function depends on the labels assigned to two pixels simultaneously, and thus introduces a dependency between the labels of different pixels. Typically, this dependency between pixels is used to express that the desired solution should have some degree of smoothness, or regularity.

Our main contribution is an algorithm for solving labeling problems of the form described above. Specifically, the algorithm is guaranteed to produce a labeling that is globally optimal with respect to the energy function E_∞, under the condition that all pairwise terms ϕ_{st} satisfy the condition

$$\max\{\phi_{st}(0,0), \phi_{st}(1,1)\} \leq \max\{\phi_{st}(1,0), \phi_{st}(0,1)\}. \quad (2)$$

The proposed algorithm is very efficient, with an asymptotic time complexity bound by the time required to sort $\mathcal{O}(|V| + |\mathcal{E}|)$ values.[2]

1.1 Background and Related Work

In their seminal work, Kolmogorov and Zabih [5] considered binary labeling problems where the objective function has the form

$$E_1(\ell) := \sum_{s\in V} \phi_s(\ell(s)) + \sum_{\{s,t\}\in\mathcal{E}} \phi_{st}(\ell(s),\ell(t)), \quad (3)$$

[1] The energy formula (1) must be expressed in terms of undirected edges. But the algorithm can be used for directed graphs as well.

[2] Here, $|\cdot|$ denotes set cardinality.

and showed that a globally optimal solution can be computed by solving a max-flow/min-cut problem on a suitably constructed graph under the condition that all pairwise terms ϕ_{st} are *submodular*. A pairwise term ϕ_{st} is said to be submodular if

$$\phi_{st}(0,0) + \phi_{st}(1,1) \leq \phi_{st}(0,1) + \phi_{st}(1,0). \tag{4}$$

Looking at the objective functions E_1 and E_∞, we can view them both as consisting of two parts:

- A *local* error measure, in our case defined by the unary and pairwise terms.
- A *global* error measure, aggregating the local errors into a final score.

In the case of E_1, the global error measure is obtained by summing all the local error measures, and in the case of E_∞ the global error measure is taken to be the maximum of the local error measures. If we assume for a moment that all the local error measurements are non-negative, then E_1 can be seen as measuring the L_1-norm of a vector[3] containing all local errors. Similarly, E_∞ can be interpreted as measuring the L_∞ or max norm of the vector. The L_1 and L_∞ norms are both special cases of L_p norms, and in this sense we can view both E_1 and E_∞ as special cases of a more general objective function

$$E_p(\ell) := \left(\sum_{s \in V} \phi_s^p(\ell_s) + \sum_{\{s,t\} \in \mathcal{E}} \phi_{st}^p(\ell_s, \ell_t) \right)^{1/p}, \tag{5}$$

where $\phi_s^p(\cdot) = (\phi_s(\cdot))^p$ and $\phi_{st}^p(\cdot,\cdot) = (\phi_{st}(\cdot,\cdot))^p$. The value p can be seen as a parameter controlling the balance between minimizing the overall cost versus minimizing the magnitude of the individual terms. For $p = 1$, the optimal labeling may contain arbitrarily large individual terms as long as the sum of the terms is small. As p increases, a larger penalty is assigned to solutions containing large individual terms. In the limit as p approaches infinity, E_p approaches E_∞ and the penalty assigned to a solution is determined by the largest individual term only.

Labeling problems with objective functions of the form E_p can be solved using minimal graph cuts, provided that all pairwise terms ϕ_{st}^p are submodular [6]. As shown by Malmberg and Strand [6], the submodularity of ϕ_{st}^p is guaranteed for any $p \geq 1$ if ϕ_{st} is submodular and satisfies (2).

Additionally, it turns out that in some problem instances the limit case E_∞ can be optimized for directly, using efficient greedy algorithms. For example, consider a labeling problem with objective function E_∞, where all pairwise terms satisfy the restriction $\phi_{st}(1,0) = \phi_{st}(0,1)$ and $\phi_{st}(0,0) = \phi_{st}(1,1) = 0$. In this simplified case, a globally optimal solution can be found by computing a cut by optimum spanning forest (MSF cut, or watershed cut) on a suitably constructed graph [2–4].This interesting result has a high practical value, since the computation time for finding an MSF cut is substantially lower than the

[3] This vector is identified with the function ϕ_ℓ defined in the next section.

computation time for solving a min-cut/max-flow problem, asymptotically as well as in practice [3]. An interesting question is therefore whether it is possible to use similar greedy techniques to optimize the objective function E_∞, beyond the special case outlined above. The results presented in this paper answers this question affirmatively.

We observe that our proposed algorithm has some structural similarity to Kruskal's algorithm for computing minimum spanning trees, and in this sense the algorithm can be seen as a generalization of the MSF/Watershed cut approach [4].

2 Preliminaries

The exposition of our proposed algorithm relies on the notion of unary and binary solution *atoms*, which we introduce in this section. Informally, a unary atom represents one possible label configuration for a single vertex, and a binary atom represent a possible label configuration for a pair of adjacent vertices. Thus, for a binary labeling problem, there are two atoms associated with every vertex and four atoms for every edge.

Formally, we let $\mathcal{V} = \{\{v\} : v \in V\}$, put $\mathcal{D} = \mathcal{V} \cup \mathcal{E}$, and let $\mathcal{A}$ be the family of all binary maps from $D \in \mathcal{D}$ into $\{0, 1\}$. An atom, in this notation, is an element of $\mathcal{A}$. If we identify, as it is common, maps with their graphs, then each unary atom associated with a vertex $s \in V$ has form $\{(s, i)\}$, with $i \in \{0, 1\}$. Similarly, each binary atom associated with an edge $\{s, t\} \in \mathcal{E}$ has the form $\{(s, i), (t, j)\}$, with $i, j \in \{0, 1\}$.

Notice, that the maps ϕ_s and ϕ_{st} used for the unary and binary terms in (1) can be combined to form a single function $\Phi\colon \mathcal{A} \to [0, \infty)$ defined, for every $A \in \mathcal{A}$, as

$$\Phi(A) := \begin{cases} \phi_s(i) & \text{for } A = \{(s,i)\}, \\ \phi_{s,t}(i,j) & \text{for } A = \{(s,i),(t,j)\}. \end{cases}$$

For a given labeling ℓ, we define $\phi_\ell\colon \mathcal{D} \to [0, \infty)$, for every $D \in \mathcal{D}$, as

$$\phi_\ell(D) := \Phi(\ell \restriction D) = \begin{cases} \phi_s(\ell(s)) & \text{for } D = \{s\} \in \mathcal{V}, \\ \phi_{s,t}(\ell(s), \ell(t)) & \text{for } D = \{s,t\} \in \mathcal{E}, \end{cases}$$

where $\ell \restriction D$ is the restriction of ℓ to D. With this notation, we may write the objective function E_∞ as

$$E_\infty(\ell) = \|\phi_\ell\|_\infty = \max_{D \in \mathcal{D}} \phi_\ell(D). \tag{6}$$

2.1 Global and Local Consistency, Incompatible Atoms

Conceptually, the proposed algorithm works as follows: starting from the set of all possible unary and binary atoms, the algorithm iteratively removes one atom at a time until the remaining atoms define a unique labeling. A key issue in this process is to ensure that, at all steps of the algorithm, at least one labeling can be constructed from the set of remaining atoms.

Let ℓ be a binary labeling. We define $\mathcal{A}(\ell)$, the *atoms for* ℓ, as the family

$$\mathcal{A}(\ell) = \{\ell \restriction D \colon D \in \mathcal{D}\}.$$

Notice that ℓ can be easily recovered from $\mathcal{A}(\ell)$ as its union: $\ell = \bigcup \mathcal{A}(\ell)$.

Let $\mathcal{A}' \subset \mathcal{A}$ be a set of atoms. We say that $\mathcal{A}'$ is *globally consistent* if there exists at least one labeling ℓ such that $\mathcal{A}(\ell) \subseteq \mathcal{A}'$.

In general, determining whether a given set of atoms is globally consistent is difficult. Therefore we also introduce a seemingly weaker property of local consistency, which will be used in Sect. 4 to establish the correctness of our proposed algorithm. A set of atoms $\mathcal{A}'$ is said to be *locally consistent* if, for every vertex $s \in V$ and edge $\{s,t\} \in \mathcal{E}$ there are $i, j \in \{0,1\}$ such that the atoms $\{(s,i)\}$ and $\{(s,i),(t,j)\}$ both belong to $\mathcal{A}'$.

Furthermore we introduce the notion of an incompatible atom, which will be needed for the exposition of the proposed algorithm. For a given set of $\mathcal{A}'$, we say that an atom $A \in \mathcal{A}'$ is *incompatible* (w.r.t. $\mathcal{A}'$) if either

1. A is a unary atom so that $A = \{(v,i)\}$ for some vertex v, and there exists some edge $\{v,w\}$ adjacent to v such that $\mathcal{A}'$ contains neither $\{(v,i),(w,0)\}$ nor $\{(v,i),(w,1)\}$; or
2. A is a binary atom so that $A = \{(v,i),(w,j)\}$ for some edge $\{v,w\}$, and at least one of $\{(v,i)\}$ and $\{(w,j)\}$ is not in $\mathcal{A}'$.

Note that a locally consistent set of atoms may still contain incompatible atoms.

3 Proposed Algorithm

In this section, we introduce the proposed algorithm for finding a binary label assignment $\ell\colon V \to \{0,1\}$ that globally minimizes the objective function E_∞ given by (1), under the condition that all pairwise terms in the objective function satisfy (2). Informally, the general outline of the proposed algorithm is as follows:

- Start with a set S consisting of all possible atoms.
- For each atom A, in order of decreasing cost $\Phi(A)$:
  - If A is still in S, and is not the only remaining atom for that vertex/edge, remove A from S.
  - After the removal of A, S may contain incompatible atoms. Iteratively remove incompatible atoms until S contains no more incompatible atoms.

Before we formalize this algorithm, we introduce a specific preordering relation $\succ$ on the atoms $\mathcal{A}$. For $A_0, A_1 \in \mathcal{A}$ we will write $A_0 \succ A_1$ if either $\Phi(A_0) > \Phi(A_1)$, or else $\Phi(A_0) = \Phi(A_1)$ and A_1 is a binary atom of the form $\{(s,i),(t,i)\}$ (equal labeling) while A_0 is not in this form.

With these preliminaries in place, we are now ready to introduce the proposed algorithm, for which pseudocode is given in Algorithm 1.

Algorithm 1. Labeling Algorithm

Data: A graph $\mathcal{G} = (V, \mathcal{E})$ and associated $\Phi\colon \mathcal{A} \to [0, \infty)$ generating energy E_∞
Result: A labeling $\ell\colon V \to \{0, 1\}$ minimizing energy E_∞
Additional Structure: An array A of buckets of atoms, indexed by $\mathcal{D} = \mathcal{V} \cup \mathcal{E}$; a list H of atoms; a queue K of vertices/edges such that every vertex in K precedes any edge.

1 **foreach** vertex/edge $D \in \mathcal{D}$ **do** insert all D-atoms to $\mathsf{A}[D]$
2 create a list H of all atoms $\mathcal{A}$ such that A_0 precedes A_1 in $\mathcal{A}$ whenever $A_0 \succ A_1$
3 **while** $\mathsf{H} \neq \emptyset$ **do**
4   **remove** the first atom A **from** H
5   **if** $D \in \mathcal{D}$ is a vertex/edge of A and $\mathsf{A}[D]$ has more than one element **then**
6   **remove** A **from** $\mathsf{A}[D]$ and insert D to (previously empty) K
7   **while** $\mathsf{K} \neq \emptyset$ **do**
8   **remove** a vertex/edge C **from** K
9   **foreach** edge/vertex D adjacent to C **do**
10   **remove** from $\mathsf{A}[D]$ and H all A incompatible with $\bigcup_{D' \in \mathcal{D}} \mathsf{A}[D']$
11   **if** any atom was removed **from** $\mathsf{A}[D]$ and H in line 10 **then**
12   insert to K any vertex/edge C' adjacent to D: to its top, when C' is a vertex and its bottom when C' is an edge

13 return $\ell = \bigcup_{D \in \mathcal{D}} \mathsf{A}[D]$

4 Analysis of the Algorithm

In this section, we analyze the computational complexity of the proposed algorithm and prove that it is guaranteed to return a globally optimal solution to the labeling problem given in the introduction.

4.1 Computational Complexity

We now analyze the asymptotic computational complexity of Algorithm 1. First, let $\eta := |A| = 2|V| + 4|\mathcal{E}|$. In image processing applications the graph $\mathcal{G}$ is commonly sparse, in the sense that $\mathcal{O}(|V|) = \mathcal{O}(|\mathcal{E}|)$. In this case, we have $\mathcal{O}(\eta) = \mathcal{O}(|V|)$.

Creating the list H requires us to sort all atoms in A. The sorting can be performed in $\mathcal{O}(\eta \log \eta)$ time. In some cases, e.g., if all unary and binary terms are integer valued, the sorting may be possible to perform in $\mathcal{O}(\eta)$ time using, e.g., *radix* or *bucket* sort.

We make the reasonable assumption that the following operations can all be performed in $\mathcal{O}(1)$ time:

- Remove an atom from H.
- Remove an atom from $\mathsf{A}(D)$.
- Remove or insert elements in K.
- Given an atom, find its corresponding edge or vertex.

- Given a vertex, find all edges incident at that vertex.
- Given an edge, find the vertices spanned by the edge.

The combined number of the executions of the main loop, lines 3–12, and of the internal loop, lines 7–12, equals to $|\mathcal{A}|$, that is, $\mathcal{O}(\eta)$. This is so, since any insertion of an atom into K requires its prior removal from the list H. If the assumptions above are satisfied, it is easily seen that only $\mathcal{O}(1)$ operations are needed between consecutive removals of an atom from H. Therefore, the amortized cost of the execution of the main loop is $\mathcal{O}(\eta)$.

Thus, the total computational cost of the algorithm is bound by the time required to sort $\mathcal{O}(\eta)$ elements, i.e., at most $\mathcal{O}(\eta \log \eta)$.

4.2 Example Demonstrating the Need for Condition (2)

In Sect. 4.3 we will prove that if all binary terms satisfy (2), then this is a sufficient condition for the proposed algorithm to return an optimal labeling. In this section, we first give an example showing that when this condition is violated, the algorithm may indeed fail to produce a labeling.

Let $\mathcal{G}$ be a complete graph with three vertices $V = \{a, b, c\}$. Define ϕ for every $i \in \{0, 1\}$ via:

(i) $\phi(\langle a, i\rangle, \langle b, i\rangle) = \phi(\langle a, i\rangle, \langle c, i\rangle) = 1$;
(ii) $\phi(\langle a, i\rangle, \langle b, 1-i\rangle) = \phi(\langle a, i\rangle, \langle c, 1-i\rangle) = 9$;
(iii) $\phi(\langle b, i\rangle, \langle c, i\rangle) = 5$;
(iv) $\phi(\langle b, i\rangle, \langle c, 1-i\rangle) = 2$;
(v) $\phi(\langle v, i\rangle) = 0$ for all $v \in V$.

Consider the following possible steps in the execution of Algorithm 1.

Steps 1–4: Remove 4 atoms with $\phi(\langle a, i\rangle, \langle b, 1-i\rangle) = \phi(\langle a, i\rangle, \langle c, 1-i\rangle) = 9$.
Steps 5–6: Remove 2 atoms with $\phi(\langle b, 0\rangle, \langle c, 0\rangle) = \phi(\langle b, 1\rangle, \langle c, 1\rangle) = 5$.

Then, after these 6 steps, the system is inconsistent, since pairs $\langle a, b\rangle$ and $\langle a, c\rangle$ must have the same labels, while $\langle b, c\rangle$ must have different labels. So, this execution of Algorithm 1 fails to produce a valid labeling.

To motivate the introduction of the specific preordering relation $\succ$ for H, consider modifying the labeling problem given above so that

(iv) $\phi(\langle b, i\rangle, \langle c, 1-i\rangle) = 5$.

If we only require H to be ordered by decreasing Φ, steps 1–6 still represent a possible execution of Algorithm 1. By instead requiring H to be ordered according to $\succ$, we force the atoms $\{(b, 0), (c, 1)\}$ and $\{(b, 1), (c, 0)\}$ to be removed from H before $\{(b, 0), (c, 0)\}$ and $\{(b, 1), (c, 1)\}$, and the algorithm will in this case return a valid labeling.

4.3 Proof of Correctness

Theorem 1. *If all binary terms of the map $\Phi\colon \mathcal{D} \to [0,\infty)$ associated with graph $\mathcal{G} = (V, \mathcal{E})$ satisfy the condition (2), then ℓ returned by Algorithm 1 is indeed a labeling of V minimizing the objective function E_∞.*

Proof. Let $\mathsf{n} := |V| + 3|\mathcal{E}|$, the number of removals of an atom from A. For every $D \in \mathcal{D}$ and $k \in \{0, \ldots, \mathsf{n}\}$ let $\mathsf{A}_k[D]$ be equal to the value of $\mathsf{A}[D]$ directly after the k-th removal of some atom(s) from A, which can happen only as a result of execution of either line 6 or of line 10. (For $k = 0$ we mean, directly after the execution of line 2.) Let $\mathcal{A}_k = \bigcup_{D\in\mathcal{D}} \mathsf{A}_k[D]$.

Let $1 = k_1 < \cdots < k_m$ be the list of all values of $k \in \{1, \ldots, \mathsf{n}\}$ such that $\mathcal{A}_k$ is a proper refinement of $\mathcal{A}_{k-1}$ resulting from the execution of line 6. Note that it is conceivable that the numbers k_j and k_{j+1} are consecutive—this happens when the execution of loop 8–12 directly after the execution of line 5 has been used to create $\mathcal{A}_{k_j}$ resulted in removal of no atoms from $\mathcal{A}_{k_j}$.

The key element of this proof is to show, by induction, that the following properties hold for every $k \leq \mathsf{n}$.

(P0) For every edge $D = \{v, w\}$, if $\mathsf{A}_k[D]$ is missing either $\{(v,0),(w,0)\}$ or $\{(v,1),(w,1)\}$, then it must be also missing $\{(v,1),(w,0)\}$ or $\{(v,0),(w,1)\}$.
(P1) $\mathsf{A}_k[D]$ contains at least one atom for every $D \in \mathcal{D}$.
(P2) $\mathcal{A}_k$ is locally consistent.
(P3) $\mathcal{A}_k$ has no incompatible atoms directly before any execution of line 4.

It is enough to prove that if for some $\kappa \leq \mathsf{n}$ these properties hold for every $k < \kappa$, then they also hold for κ. Clearly, these properties hold immediately after the execution of line 2, that is, for $\kappa = 0$. So, we can assume that $\kappa > 0$. We need to show that (P0)–(P3) are preserved by each operation of the algorithm. More specifically, by the execution of lines 6 or 10, since the status of each of these properties can change only when an atom is removed from A during their execution.

Proof of (P0): Fix an edge $D = \{v, w\}$ and assume that (P0) holds for this D and all $k < \kappa$. Now, if $\mathsf{A}_{\kappa-1}[D]$ has less than 4 elements, then by the inductive assumption it must be already missing either $\{(v,1),(w,0)\}$ or $\{(v,0),(w,1)\}$, and so the same will be true for $\mathsf{A}_\kappa[D]$, as needed. So, assume that $\mathsf{A}_{\kappa-1}[D]$ has still all 4 elements. This means, that these 4 elements are still present in H and, by (2) and the choice of the ordering of H, the atoms $\{(v,1),(w,0)\}$ and $\{(v,0),(w,1)\}$ must precede in H any of the atoms $\{(v,0),(w,0)\}$ and $\{(v,1),(w,1)\}$. In particular, if $\kappa = k_j$ for some j, then $\mathsf{A}_\kappa[D]$ is obtained as a result of execution of line 3 and the ordering of H ensures that $\mathsf{A}_\kappa[D]$ still satisfies (P0). So, assume that $\kappa = k_j$ for no j; that is, that $\mathsf{A}_\kappa[D]$ is obtained from $\mathsf{A}_{\kappa-1}[D]$ by the execution of line 10. Since one of the atoms from $\mathsf{A}_{\kappa-1}[D]$ was removed as a result of this execution, for one of vertices of D, say v, the bucket $\mathsf{A}_{\kappa-1}[\{v\}]$ must be missing one of its atoms, say $\{(v,i)\}$. But this means that $\mathsf{A}_{\kappa-1}[D]$ must have been missing both $\{(v,i),(w,0)\}$ and $\{(v,i),(w,1)\}$, so indeed $\mathsf{A}_\kappa[D]$ satisfies (P0).

Proof of (P1)-(P3): This will be proved by the simultaneous induction on κ.

(P1) must be preserved by the execution of line 10, by the inductive assumption (P2) that $\mathcal{A}_{\kappa-1}$ is locally consistent. It also cannot be destroyed by the execution of line 6, since this is prevented by the condition of line 5. Thus, $\mathsf{A}_\kappa[D]$ still has the property (P1).

To see (P3) we can assume that $\kappa = k_j$ for some $j > 0$. Clearly (P3) holds for $k = k_{j-1}$. Thus, we need only to show that removal of an atom A in line 6 and consecutive execution of loop 7–12 preserves (P3). Indeed, the potential incompatibility can occur only in relation of the vertices associated with the atoms removed from $\bigcup_{D\in\mathcal{D}} \mathsf{A}[D]$. However, each time such an atom is removed, all adjacent atoms are inserted into the queue K and the execution of the loop 7–12 does not end until all such potential incompatibilities are taken care off.

The proof of the preservation of (P2) is more involved. Let j be the largest such that $k_j \le \kappa$. First notice that if $\kappa = k_j$, then (P2) holds. Indeed, by the inductive assumptions (P2) and (P3), $\mathcal{A}_{\kappa-1}$ is locally consistent and has no incompatible atoms. Since $\mathcal{A}_\kappa \neq \mathcal{A}_{\kappa-1}$, the bucket $\mathsf{A}[D]$ must have contained two or more atoms prior to the removal of A in line 6. Since $\mathcal{A}_{\kappa-1}$ did not contain any incompatible atoms, $\mathcal{A}_\kappa = \mathcal{A}_{\kappa-1} \setminus \{A\}$ must remain locally consistent. So, we can assume that $\mu := \kappa - k_j$ is non-zero. We will examine families $\mathcal{A}_{k_j}, \mathcal{A}_{k_j+1}, \dots, \mathcal{A}_{k_j+\mu} = \mathcal{A}_\kappa$.

Let $A = A_0, \dots, A_\mu$ be the order in which the atoms were removed from K during of this time execution of loop 8–12. Also, let $x_0, \dots, x_\mu$ be the vertices/edges associated with the atoms $A_0, \dots, A_\mu$, respectively. We will show, by induction on $\nu \le \mu$, the following property (I_ν), which in particular imply that $\mathcal{A}_{k_j+\nu}$ is locally consistent.

To state (I_ν) first notice that if an atom for a vertex v is among $x_0, \dots, x_{\nu-1}$, then $\mathcal{A}_{k_j+\nu}$ must contain precisely one of two atoms $\{(v,0)\}$ and $\{(v,1)\}$. (Must contain at least one, by (P1). It cannot contain both, since this would mean that no v-atom was removed so far and hence $A_{k_j+\nu}$ could not have been removed from $\mathcal{A}_{k_j+\nu-1}$.) In particular, this means that there is an $i_v \in \{0,1\}$ for which $\mathcal{A}_{k_j+\nu}$ already ensures that the final value of $\ell(v)$ is i_v. This means, that $\mathsf{A}_{k_j+\nu}[\{v\}] = \{\{(v,i_v)\}\}$.

We will prove, by induction on $\nu \le \mu$, that

(I_ν) $\mathcal{A}_{k_j+\nu}$ is locally consistent and if vertices v and w are among $x_0, \dots, x_\nu$, then $i_v = i_w$.

Of course, this will finish the proof of (P2).

Clearly, (I_0) holds, as we already shown that $\mathcal{A}_{k_j}$ is locally consistent, and the other condition is satisfied in void. So, fix $\nu \in \{1, \dots, \mu\}$ such that (I_ξ) holds for all $\xi < \nu$. We will show that (I_ν) holds as well.

For this, assume first that x_ν is an edge $\{v,w\}$. We need to show only that $\mathcal{A}_{k_j+\nu}$ remains locally consistent, the other part of (I_ν) being ensured in this case by $(I_{\nu-1})$. Since $x_\nu = \{v,w\}$, there must exist a $j < \nu$ such that x_j is a vertex and $x_j \in \{v, w\}$. For simplicity we assume that $x_j = v$ and that $i_v = 0$, the other cases being similar.

We need to show that $\mathcal{A}_{k_j+\nu}$, obtained from $\mathcal{A}_{k_j+\nu-1}$ by removing from it atoms $\{(v,1),(w,0)\}$ and $\{(v,1),(w,1)\}$, cannot be locally inconsistent.

Note that such removal from locally consistent $\mathcal{A}_{k_j+\nu-1}$ can potentially influence local consistency of $\mathcal{A}_{k_j+\nu}$ only of $\{v,w\}$ with respect to the vertices v and w. However, since $\mathsf{A}_{k_j+\nu-1}[\{v\}] = \{\{(v,0)\}\}$, this is also equal to $\mathsf{A}_{k_j+\nu}[\{v\}]$. Also, both $\mathcal{A}_{k_j+\nu-1}$ and $\mathcal{A}_{k_j+\nu}$ must contain either $\{(v,0),(w,0)\}$ or $\{(v,0),(w,1)\}$. Hence, $\mathcal{A}_{k_j+\nu}$ it cannot have local inconsistency of $\{v,w\}$ with v. Therefore, we must show only that $\mathcal{A}_{k_j+\nu}$ contains no local inconsistency between $\{v,w\}$ and w.

To see this, first notice that there will be no such inconsistency when

$$\mathsf{A}_{k_j-1}[\{w\}] \subsetneq \{\{(w,0)\},\{(w,1)\}\}. \tag{7}$$

Indeed, then $\mathsf{A}_{k_j-1}[\{w\}] = \{\{(w,i)\}\}$ for some $i \in \{0,1\}$ and, by the property (P3), $\mathcal{A}_{k_j-1} \supset \mathcal{A}_{k_j+\mu}$ cannot contain atom $\{(v,0),(w,1-i)\}$. Hence $\mathcal{A}_{k_j+\mu}$ must contain $\{(v,0),(w,i)\}$ and local consistency is preserved.

To finish the argument consider the following three cases.

$\mathsf{A}_{k_j+\nu}[\{w\}] = \{\{(w,0)\},\{(w,1)\}\}$: Then $\mathcal{A}_{k_j+\nu}$ is indeed locally consistent, since it contains either $\{(v,0),(w,0)\}$ or $\{(v,0),(w,1)\}$.

$\mathsf{A}_{k_j+\nu}[\{w\}] = \{\{(w,1)\}\}$: Then also $\mathsf{A}_{k_j+\nu-1}[\{w\}] = \{\{(w,1)\}\}$ and w cannot be among $x_0,\ldots,x_{\nu-1}$, since this would contradict the second part of $(I_{\nu-1})$. In particular, (7) holds and so local consistency is preserved.

$\mathsf{A}_{k_j+\nu}[\{w\}] = \{\{(w,0)\}\}$: We can assume that (7) does not hold. Then there exists $p \in \{0,\ldots,\nu-1\}$ such that $x_j = w$. Therefore, $\mathcal{A}_{k_j+p} \supset \mathcal{A}_{k_j+\nu}$ cannot contain $\{(v,0),(w,1)\}$. So, $\mathcal{A}_{k_j+\nu}$ must contain $\{(v,0),(w,0)\}$ and local consistency is preserved.

Before we proceed further, notice that for every $\nu \le \mu$,

(J_ν) for every vertex v there is at most one edge $D = \{v,w\}$ such that $\mathsf{A}_{k_j+\nu}[\{v\}]$ contains an atom incompatible with all atoms in $\mathsf{A}_{k_j+\nu}[D]$.

Indeed, by (P3), this clearly holds for $\nu = 0$. Also, if x_ν is an edge, than the ordering conditions we imposed on the queue K ensure that the atoms of no other edge can be added to K and subsequently modified, before each vertex (adjacent to x_ν) that can have incompatible atoms with that for x_ν is added to K and subsequently modified, so that the potential incompatibilities are removed.

Finally, consider the case when x_ν is a vertex v. Then we must have had $\mathsf{A}_{k_j+\nu-1}[\{v\}] = \{\{(v,0)\},\{(v,1)\}\}$. Also, there exists a $p \in \{0,\ldots,\nu-1\}$ such that x_p is an edge $D = \{v,w\}$ and $\mathsf{A}_{k_j+p}[D] \subsetneq \mathsf{A}_{k_j+p-1}[D]$. Moreover, by (J_ν), such p is unique. Therefore, $\mathcal{A}_{k_j+\nu}$ must be locally consistent, since the only potential local inconsistency in $\mathcal{A}_{k_j+\nu}$ could be between v and $\{v,w\}$. But our choice of $\mathsf{A}_{k_j+\nu}[\{v\}] \subset \mathsf{A}_{k_j+\nu-1}[\{v\}] = \{\{(v,0)\},\{(v,1)\}\}$ ensures that such inconsistency cannot occur.

Notice also that the second part of (I_ν) holds as well. Indeed, this is satisfied in void when there is no vertex among $x_0, \dots, x_{\nu-1}$. So, assume that such vertex exists. Then, w, the second vertex of the above chosen edge $x_p = D = \{v, w\}$, must be among such $x_0, \dots, x_{\nu-1}$. Indeed, if $p = 0$ then we must have $\nu = 2$ and $x_1 = w$. Since $i_w = 0$, we must have $\mathsf{A}_{k_j}[D] \subset \{\{(v,0),(w,0)\},\{(v,1),(w,0)\}\}$. Also, since $\mathsf{A}_{k_j+2}[\{v\}] \subsetneq \mathsf{A}_{k_j+1}[\{v\}]$, the bucket $\mathsf{A}_{k_j+1}[D] = \mathsf{A}_{k_j}[D]$ must contain precisely only one of the atoms $\{(v,0),(w,0)\}$ or $\{(v,1),(w,0)\}$. However, $\mathsf{A}_{k_j}[D]$ cannot be equal to $\{\{(v,1),(w,0)\}\}$, since, by (P0), this would mean that $\mathsf{A}_{k_j-1}[D] = \{\{(v,0),(w,1)\},\{(v,1),(w,1)\}\}$. But this contradicts (P3). So, $\mathsf{A}_{k_j+1}[D] = \{\{(v,0),(w,0)\}\}$, and indeed $i_v = 0$.

Finally, assume that $p > 0$. Then $w = x_q$ for some $q \in \{0, \dots, p-1\}$ and so $\mathsf{A}_{k_j+q}[\{w\}] = \{\{(w,0)\}\}$. Thus, $\mathsf{A}_{k_j+p}[D] \subset \{\{(v,0),(w,0)\},\{(v,1),(w,0)\}\}$ and $\mathsf{A}_{k_j+\nu-1}[D]$ must contain precisely one of these atoms to ensure that the inclusion $\mathsf{A}_{k_j+\nu}[\{v\}] \subsetneq \mathsf{A}_{k_j+\nu-1}[\{v\}]$ holds. We need to show that the equality $\mathsf{A}_{k_j+p}[D] = \{\{(v,1),(w,0)\}\}$ is impossible. Indeed, this would imply that $\mathsf{A}_{k_j+q-1}[D] \subset \{\{(v,1),(w,0)\},\{(v,0),(w,1)\},\{(v,1),(w,1)\}\}$ and using property (P0), that $\mathsf{A}_{k_j+q-1}[D] \subset \{\{(v,1),(w,0)\},\{(v,1),(w,1)\}\}$. But this means that $\mathcal{A}_{k_j+q-1}$ already decided the value of $\lambda(v)$ as 1. Since the value of $\lambda(w)$ was previously decided, the reasoning as for (J_ν) shows that v should appear already in $x_0, \dots, x_q$, while $q < \nu$ contradicts this. This finishes the proof of (P1)-(P3).

To finish the proof of Theorem 1, we still need to argue for two facts. First notice that the algorithm does not stop until all buckets $\mathsf{A}_\mathsf{n}[D]$, $D \in \mathcal{D}$, have precisely one element. Thus, since $\mathcal{A}_\mathsf{n}$ is locally consistent $\ell = \bigcup_{D\in\mathcal{D}} \mathsf{A}[D]$ is indeed a function from V into $\{0,1\}$.

Finally we prove that ℓ indeed minimizes energy E_∞. For this, first notice that at any time of the execution of the algorithm, any atom in H is also in $\bigcup_{D\in\mathcal{D}} \mathsf{A}[D]$. Indeed, these sets are equal immediately after the initialization and we remove from $\bigcup_{D\in\mathcal{D}} \mathsf{A}[D]$ only those atoms, that have been already removed from H. Now, let $L\colon V \to \{0,1\}$ be a labeling minimizing E_∞. We claim, that the following property holds any time during the execution of the algorithm:

(P) if $\Phi(A') > E_\infty(L)$ for some $A' \in \bigcup_{D\in\mathcal{D}} \mathsf{A}[D]$, then $\mathcal{A}[L] \subset \bigcup_{D\in\mathcal{D}} \mathsf{A}[D]$.

Indeed, it certainly holds immediately after the initialization. This cannot be changed during the execution of line 6 when the assumption is satisfied, since then A considered there has just been removed from $\mathsf{H} \supset \bigcup_{D\in\mathcal{D}} \mathsf{A}[D]$ and

$$\Phi(A) \geq \max_{H\in\mathsf{H}} \Phi(H) \geq \max_{H\in\bigcup_{D\in\mathcal{D}} \mathsf{A}[D]} \Phi(H) \geq \Phi(A') > E_\infty(L) = \max_{H\in\mathcal{A}[L]} \Phi(H),$$

so $A \notin \mathcal{A}[L]$. Also, (P) is not affected by an execution of line 10, since the inclusion $\mathcal{A}[L] \subset \bigcup_{D\in\mathcal{D}} \mathsf{A}[D]$ is not affected by it: no atom in $\mathcal{A}[L]$ is incompatible with $\mathcal{A}[L]$ so also with $\bigcup_{D\in\mathcal{D}} \mathsf{A}[D]$. This concludes the proof of (P).

Now, by the property (P), after the termination of the main loop, we have either $\mathcal{A}[L] \subset \bigcup_{D\in\mathcal{D}} \mathsf{A}[D]$, in which case $\ell = L$ have minimal E_∞ energy, or else

$$E_\infty(L) \geq \max_{H\in\bigcup_{D\in\mathcal{D}} \mathsf{A}[D]} \Phi(H) = \max_{H\in\mathsf{H}} \mathcal{A}[\ell] = E_\infty(\ell)$$

once again ensuring optimality of ℓ.

5 Conclusions

We have presented an efficient algorithm for finding optimal solutions of binary labeling problem with objective functions of the form E_∞, related to the L_∞ norm. We have showed that the algorithm is guaranteed to find a globally optimal labeling, under the condition that all binary terms satisfy the condition (2).

As we observed in Sect. 1.1, there is an interesting similarity between our results and the work by Boykov et al. [1] and Kolmogorov and Zabih [5] on optimizing binary labeling problem via minimal graph cuts. Specifically, the condition (2) strongly resembles the submodularity condition required for the minimal graph cut approach to be applicable. We note that the condition (2) also appeared in a different context in the recent work by Malmberg and Strand [6], regarding minimization of L_p norm objective functions by minimal graph cuts. This connection should be investigated further to determine if the similarity is only superficial, or the result of a deeper connection between these problems.

As discussed in Sect. 1.1, special cases of the max norm labeling problems considered here can be solved by computing a MSF/watershed cut on a suitably constructed graph. The set of problems that are solvable by this approach appears to be a strict subset of the problems solvable by our proposed algorithm. We note, however, that the extensions to the MSF-cut concept proposed by Malmberg et al. [7] and Wolf et al. [8] can be used to solve a subclass of max norm optimization problems where the binary terms do not satisfy (2). Thus, an interesting direction for future work is to determine the precise class of max-norm problems that can be solved via efficient greedy algorithms.

References

1. Boykov, Y., Veksler, O., Zabih, R.: Fast approximate energy minimization via graph cuts. IEEE Trans. Pattern Anal. Mach. Intell. **23**(11), 1222–1239 (2001)
2. Ciesielski, K.C., Udupa, J.K., Falcão, A.X., Miranda, P.A.: Fuzzy connectedness image segmentation in graph cut formulation: a linear-time algorithm and a comparative analysis. J. Math. Imaging Vis. **44**(3), 375–398 (2012)
3. Couprie, C., Grady, L., Najman, L., Talbot, H.: Power watershed: a unifying graph-based optimization framework. IEEE Trans. Pattern Anal. Mach. Intell. **33**(7), 1384–1399 (2011)
4. Cousty, J., Bertrand, G., Najman, L., Couprie, M.: Watershed cuts: minimum spanning forests and the drop of water principle. IEEE Trans. Pattern Anal. Mach. Intell. **31**(8), 1362–1374 (2009)
5. Kolmogorov, V., Zabih, R.: What energy functions can be minimized via graph cuts? IEEE Trans. Pattern Anal. Mach. Intell. **26**(2), 147–159 (2004)
6. Malmberg, F., Strand, R.: When can l_p-norm objective functions be minimized via graph cuts? In: Barneva, R.P., Brimkov, V.E., Tavares, J.M.R.S. (eds.) IWCIA 2018. LNCS, vol. 11255, pp. 112–117. Springer, Cham (2018). https://doi.org/10.1007/978-3-030-05288-1_9

7. Malmberg, F., Strand, R., Nyström, I.: Generalized hard constraints for graph segmentation. In: Heyden, A., Kahl, F. (eds.) SCIA 2011. LNCS, vol. 6688, pp. 36–47. Springer, Heidelberg (2011). https://doi.org/10.1007/978-3-642-21227-7_4
8. Wolf, S., Pape, C., Bailoni, A., Rahaman, N., Kreshuk, A., Köthe, U., Hamprecht, F.A.: The mutex watershed: efficient, parameter-free image partitioning. In: Ferrari, V., Hebert, M., Sminchisescu, C., Weiss, Y. (eds.) ECCV 2018. LNCS, vol. 11208, pp. 571–587. Springer, Cham (2018). https://doi.org/10.1007/978-3-030-01225-0_34

Common Object Discovery as Local Search for Maximum Weight Cliques in a Global Object Similarity Graph

Cong Rao[1(✉)], Yi Fan[2,3], Kaile Su[3], and Longin Jan Latecki[1]

[1] Temple University, Philadelphia, USA
{cong.rao,latecki}@temple.edu
[2] Guangxi Key Lab of Trusted Software, Guilin University of Electronic Technology, Guilin 541004, China
[3] Griffith University, Brisbane, Australia
{yi.fan4,k.su}@griffithuni.edu.au

Abstract. In this paper, we consider the task of discovering the common objects in images. Initially, object candidates are generated in each image and an undirected weighted graph is constructed over all the candidates. Each candidate serves as a node in the graph while the weight of the edge describes the similarity between the corresponding pair of candidates. The problem is then expressed as a search for the Maximum Weight Clique (MWC) in this graph. The MWC corresponds to a set of object candidates sharing maximal mutual similarity, and each node in the MWC represents a discovered common object across the images. Since the problem of finding the MWC is NP-hard, most research of the MWC problem focuses on developing various heuristics for finding good cliques within a reasonable time limit. We utilize a recently very popular class of heuristics called local search methods. They search for the MWC directly in the discrete domain of the solution space. The proposed approach is evaluated on the PASCAL VOC image dataset and the YouTube-Objects video dataset, and it demonstrates superior performance over recent state-of-the-art approaches.

Keywords: Common Object Discovery · Visual Similarity · Maximum Weight Clique · Local search algorithm

1 Introduction

For an undirected weighted graph $G = (V, E)$, where V is the set of vertices and E is the set of edges, a clique C is a subset of vertices in V in which each pair of vertices is connected by an edge in E. The Maximum Weight Clique (MWC) problem is to find a clique C which maximizes

$$w(C) = \sum_{v_i \in C} w_V(v_i) + \sum_{v_i, v_j \in C} w_E(v_i, v_j), \qquad (1)$$

M. Couprie et al. (Eds.): DGCI 2019, LNCS 11414, pp. 219–233, 2019.
https://doi.org/10.1007/978-3-030-14085-4_18

where $w_V : V \rightarrow \mathbb{R}$ and $w_E : E \rightarrow \mathbb{R}$ are the weight functions for the vertices and edges respectively. Successfully solving the MWC problem leads to various applications in practice.

In this paper, we focus on the task of common object discovery, which aims at discovering the objects of the same class in an image collection. Co-localizing objects in unconstrained environment is challenging. For images in the real-world applications, such as those in the PASCAL datasets [8,9], the objects of the same class may look very different due to viewpoint, occlusion, deformation, illumination, *etc.* Also, there could be considerable diversities within certain object class such as human beings, for their differences in gender, age, costume, hair style or skin color. Besides, there could be multiple common objects in the given set of images, thus the definition of "common" may be ambiguous. In addition, the efficiency of the involved method is very significant in time sensitive applications such as object co-localization in large collections of images or video streams.

To achieve robust and efficient object co-localization, we formulate the task as a Maximum Weight Clique (MWC) problem. It aims at finding a group of objects that are most similar to each other, which corresponds to a MWC in the associated graph. The nodes in the graph correspond to the object candidates generated from the given image collection, while the weight on an edge indicates how similar two given candidates are. We can discover a set of common objects by finding the MWC in the associated graph. Each node in the MWC is a discovered common object across the images. The main idea of the paper is illustrated in Fig. 1.

Fig. 1. Given a set of object candidates generated from an image collection (left), our goal is to find common objects by searching for the maximum weight clique in the associated graph. Each node in the clique (right) corresponds to a discovered common object.

The main contributions of this work are as follows. (1) We address the task of object co-localization as a well-defined MWC problem in the associated graph. It provides a practical and general solution for research and applications related to the MWC problem. (2) We develop a hashing based mechanism to detect the revisiting of the local optimum in the local search based MWC solver [35]. It can alleviate the cycling issue in the optimization process. (3) The Region Proposal Network (RPN) [25] is applied for efficiently generating the object candidates. The candidates are then re-ranked to improve the robustness against the background noise. (4) A Triplet Network (TN) is learned to obtain the feature embeddings of the object candidates, so as to construct a reliable affinity measure between the candidates. (5) The performance is evaluated on the PASCAL VOC 2007 image dataset [8] and the YouTube-Objects video dataset [16]. Superior performance is obtained compared to recent state-of-the-art methods.

2 Related Works

The problem of common object discovery has been investigated extensively in the past few years. Papazoglou *et al.* [22] view the task as a foreground object mining problem, where Optical Flow is used to estimate the object motion and the Gaussian Mixture model is utilized to capture the appearance of the foreground and background. Cho *et al.* [6] tackle the problem using a part-based region matching method, where a probabilistic Hough transform is used to evaluate the quality of each candidate correspondence. Joulin *et al.* [15] extend the method in [6] to co-localize objects in video frames, and a Frank-Wolfe algorithm is used to optimize the proposed quadratic programming problem. Zhang *et al.* [37] apply a part-based object detector and a motion aware region detector to generate object candidates. The problem is then formulated as a joint assignment problem and the solution is refined by inferring shape likelihoods afterwards. Kwak *et al.* [17] also focus on the problem of localizing dominant objects in videos, where an iterative process of detection and tracking is applied. Li *et al.* [18] devise an entropy-based objective function to learn a common object detector, and they address the task with a Conditional Random Field (CRF) model. Wei *et al.* [36] perform Principal Component Analysis (PCA) on the convolutional feature maps of all the images, and locate the most correlated regions across the images. Wang *et al.* [32] use segmentations produced by Fully Convolutional Networks (FCN) as object candidates. Then they discover common objects by solving a N-Partite Graph Matching problem.

Many of these methods explicitly or implicitly employ graph based models to interpret the task of object co-localization. Similarly in this paper, an undirected weighted graph is first constructed over the given set of images, modeling the visual affinities between the object candidates. We find the common objects as the Maximum Weight Clique (MWC) in this graph, where each node in the clique corresponds to a detected common object across the images. The MWC problem is NP-hard and it is difficult to obtain a global optimal solution. Generally, there are two types of algorithms to solve the MWC problem: the exact methods such as [12,14] and the heuristic methods such as [2,10,21,24,35]. Most existing works on the MWC problem focused on the heuristic approaches due to their efficiency in space and time. In this paper, our optimization algorithm adopts a simple variant of Tabu Search (TS) heuristic to discover the MWC, and it has several features: (1) it considers the local circumstance of a vertex in each step; (2) it takes only an auxiliary Boolean array for implementation; (3) it requires no extra parameters besides the time limit.

3 Problem Formulation

Given a set of N images $\mathcal{I} = \{I_1, I_2, \ldots, I_N\}$, we generate a set of object candidates from all images $\mathcal{B} = \{\boldsymbol{b} \mid \boldsymbol{b} \in \mathcal{P}(I),\ I \in \mathcal{I}\}$, where $\mathcal{P}(I)$ is the set of object candidates extracted from image I and $\boldsymbol{b}$ is a bounding box of that object candidate. Suppose n_i object candidates are extracted from image I_i, then $|\mathcal{B}| = \sum_{i=1}^{N} n_i$ candidates will be generated from the image collection $\mathcal{I}$ in total. We denote $n = |\mathcal{B}|$ in the remainder of the paper. Let $o(\boldsymbol{b}_i)$ be the score of some bounding box $\boldsymbol{b}_i$ containing the common object, and let $s(\boldsymbol{b}_i, \boldsymbol{b}_j)$ represent the similarity between two object candidates in $\boldsymbol{b}_i$ and $\boldsymbol{b}_j$, then the task of object co-localization can be formulated as finding an optimal subset $\mathcal{B}^* \subset \mathcal{B}$ such that

$$w(\mathcal{B}^*) = \sum_{\boldsymbol{b}_i \in \mathcal{B}^*} o(\boldsymbol{b}_i) + \sum_{\boldsymbol{b}_i, \boldsymbol{b}_j \in \mathcal{B}^*, \boldsymbol{b}_i \neq \boldsymbol{b}_j} s(\boldsymbol{b}_i, \boldsymbol{b}_j) \tag{2}$$

is maximized, with the constraint that at most one object candidate can be selected from each image. For the reason explained in Sect. 4.2, we set $o(\boldsymbol{b}_i) = 0$ for all $\boldsymbol{b}_i$, which means it is a Maximum Edge Weight Clique problem. However, the proposed MWC solver can optimize problems with both vertex and edge weights.

Further, we assign a label $x_i \in \{0, 1\}$ to each object candidate $\boldsymbol{b}_i$, where $x_i = 1$ means that the object candidate $\boldsymbol{b}_i$ is selected in the subset $\mathcal{B}^*$. Thus, an indicator vector $\boldsymbol{x} \in \{0, 1\}^n$ is used to identify the common objects discovered in $\mathcal{B}$. Besides, an affinity matrix $A \in \mathbb{R}^{n \times n}$ is constructed, where

$$A_{ii} = o(\boldsymbol{b}_i),\ \forall \boldsymbol{b}_i \in \mathcal{B}, \quad \text{and} \quad A_{ij} = s(\boldsymbol{b}_i,\ \boldsymbol{b}_j),\ \forall \boldsymbol{b}_i, \boldsymbol{b}_j \in \mathcal{B}. \tag{3}$$

Here we assume the similarity metric $s(\boldsymbol{b}_i,\ \boldsymbol{b}_j)$ is symmetric and non-negative, namely $A_{ij} = A_{ji} \geq 0$. On the other hand, we remove the edge between object candidates $\boldsymbol{b}_i$ and $\boldsymbol{b}_j$ if they are present in the same image, hence they cannot be simultaneously selected in $\mathcal{B}^*$. Then the problem in (2) can be expressed as finding an optimal indicator vector $\boldsymbol{x} \in \{0, 1\}^n$, such that $\boldsymbol{x}^T A \boldsymbol{x}$ is maximized. Hence the selected nodes in $\mathcal{B}^*$ correspond to a MWC in the constructed graph, and they represent the discovered set of common objects. To summarize, the overall objective function of the MWC problem can be written in the matrix form as

$$\boldsymbol{x}^* = \underset{\boldsymbol{x}}{\text{argmax}}\ \ \boldsymbol{x}^T A \boldsymbol{x}, \quad \text{s.t.}\ \ \boldsymbol{x} \in \{0, 1\}^n. \tag{4}$$

To this end, the task of object co-localization is formulated as a Maximum Weight Clique (MWC) problem as described in (1).

4 Graph Construction

4.1 Object Candidates Generation

The nodes in the associated graph correspond to the object candidates in all the images. We expect those candidates to cover as many foreground objects as possible. Meanwhile, the total number of candidates will also influence the search space for the MWC. Therefore, our first priority is to find a proper method to extract the object candidates. The Region Proposal Networks (RPN) [25] is used in our approach to generate rectangular object candidates from each image. We use the raw RPN proposals in the intermediate stage and apply Non-Maximum Suppression (NMS) [28] to remove redundant boxes. We choose the top-K scoring proposals from each image to construct the associated graph for computational efficiency. We consider two different proposal scoring measures. The first one is commonly used and is based on RPN objectness score of each object candidate. RPN also generates a vector of class likelihoods for each object candidate, and we propose to re-rank the object candidates according to the entropy of the class distribution. Since the entropy is a measure of uncertainty, it serves a similar purpose as the objectness score but tend to be more accurate in this setting. Hence we can re-rank the raw RPN proposals according to the entropy, and select the top-K scoring boxes with low uncertainty as object candidates in each image.

4.2 Common Objectness Score

For object co-localization, the underlying class of the common object is unknown in advance. Thus, the score $o(\boldsymbol{b})$ of some object $\boldsymbol{b}$ being the common one is difficult to estimate. A possible way is to set $o(\boldsymbol{b})$ as the objectness score of $\boldsymbol{b}$. But this can be problematic when $\boldsymbol{b}$ indeed contains an object but not the common one. Thus, it may lead to unexpected results if the objectness score is directly used, as observed in [31]. Therefore, we set the contribution of the score to the objective function (2) to zero, *i.e.*,

$$A_{ii} = o(\boldsymbol{b}_i) = 0, \forall \boldsymbol{b}_i \in \mathcal{B}. \tag{5}$$

In the case of object co-localization, it means we focus on the MWC problem with edge weight only. However, as shown in Sect. 5, the proposed MWC problem solver is generic and can be applied to other tasks where both vertex weights and edge weights are present.

4.3 Object Representation and Similarity

The edge weights in the associated graph represent visual similarity between the selected object candidates. Thus, we need an accurate way to represent the object candidates and evaluate their similarities. In this paper, we employ the Triplet Network framework [13] to learn the deep feature embeddings of the object candidates. Suppose a pre-trained Convolutional Neural Network (CNN) is selected to extract the deep features $f(\boldsymbol{b}; \boldsymbol{w})$ for each object candidate $\boldsymbol{b} \in \mathcal{B}$, where $\boldsymbol{w}$ is the set of parameters of the CNN. In this framework, a set of triplets is then constructed for fine-tuning the parameters $\boldsymbol{w}$. Each triplet consists of a reference object $\boldsymbol{b}_r$, a positive object $\boldsymbol{b}_p$ and a negative object $\boldsymbol{b}_n$. Namely, $\boldsymbol{b}_r$ and $\boldsymbol{b}_p$ represent a pair of similar objects, while $\boldsymbol{b}_r$ and $\boldsymbol{b}_n$ are a pair of dissimilar objects. Two objects are viewed as similar if they belong to the same category and otherwise dissimilar. Then, the hinge loss of a triplet is defined as

$$l(\boldsymbol{b}_r, \boldsymbol{b}_p, \boldsymbol{b}_n) = \max\{0, \lambda + s(\boldsymbol{b}_r, \boldsymbol{b}_n) - s(\boldsymbol{b}_r, \boldsymbol{b}_p)\}, \quad (6)$$

where λ is a margin threshold controlling how different $s(\boldsymbol{b}_r, \boldsymbol{b}_n)$ and $s(\boldsymbol{b}_r, \boldsymbol{b}_p)$ should be. The goal of the Triplet Network learning is to find a set of optimal parameters $\boldsymbol{w}$, such that the sum of the hinge loss of all triplets

$$L(\mathcal{T}) = \sum_{(\boldsymbol{b}_r, \boldsymbol{b}_p, \boldsymbol{b}_n) \in \mathcal{T}} l(\boldsymbol{b}_r, \boldsymbol{b}_p, \boldsymbol{b}_n) \quad (7)$$

is minimized over a training set of triplets $\mathcal{T}$. Namely, in the specified metric space, the learning process makes similar objects closer to each other, while dissimilar objects are pushed away. In the triplet hinge loss $l(\boldsymbol{b}_r, \boldsymbol{b}_p, \boldsymbol{b}_n)$, frequently used similarity metrics include dot-product (the linear kernel) and the Euclidean distance. But the output ranges of these metrics are not bounded, and this may invalidate the margin threshold λ in the loss function, as observed in [5]. In addition, more complex metrics can be also used here, such as the polynomial kernel and the Gaussian kernel (the RBF kernel). But there are a few more parameters in these kernel functions and they have to be chosen wisely. For simplicity, we define $s(\boldsymbol{b}_i, \boldsymbol{b}_j)$ as the cosine similarity between two CNN feature vectors $f(\boldsymbol{b}_i; \boldsymbol{w})$ and $f(\boldsymbol{b}_j; \boldsymbol{w})$, namely

$$A_{ij} = s(\boldsymbol{b}_i, \boldsymbol{b}_j) = \frac{f(\boldsymbol{b}_i; \boldsymbol{w})^T f(\boldsymbol{b}_j; \boldsymbol{w})}{\|f(\boldsymbol{b}_i; \boldsymbol{w})\| \|f(\boldsymbol{b}_j; \boldsymbol{w})\|}, \quad (8)$$

since it is already neatly bounded and parameter free. The parameters $\boldsymbol{w}$ in the overall loss function (7) can be updated via the standard Stochastic Gradient Descent (SGD) method. An intuitive description of our triplet network can be found in Fig. 2.

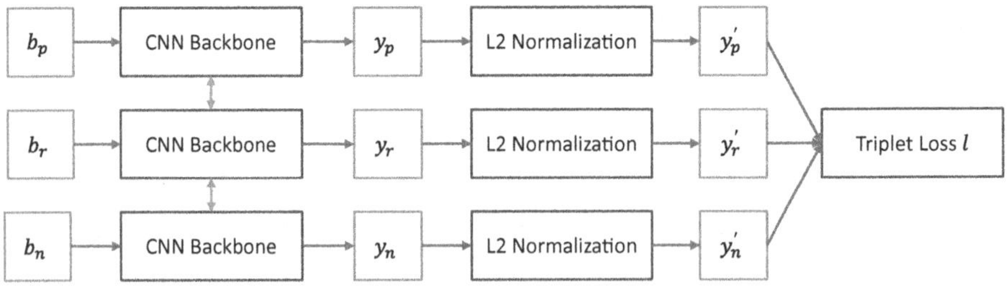

Fig. 2. The architecture of our triplet network. The weights in the CNN backbones are shared in the three branches. The goal is to learn a feature embedding such that similar objects are closer to each other in the metric space while dissimilar objects are pushed away.

5   The MWC Problem Solver

We use a local search based method to solve the MWC problem (4). The local search usually moves from one clique to another until it reaches the cutoff, then the best clique found is kept as the solution. The pipeline of our MWC solver is summarized in Algorithm 1.

Algorithm 1. Our MWC Problem Solver

Input : An undirected weighted graph $G = (V, E)$ and a time limit t
Output: A clique C^* with the maximum clique weight
1 $C^* \leftarrow C \leftarrow \emptyset$; $lastStepImproved \leftarrow true$;
2 $step \leftarrow 1$; $confChange(v) \leftarrow 1, \forall v \in V$;
3 **while** *elapsed time* $< t$ **do**
4 $\quad$ **if** $C = \emptyset$ **then** add a random vertex into C;
5 $\quad$ $v \leftarrow \text{argmax}_v \; score(v, C), v \in S_{add}(C)$, s.t. $confChange(v) = 1$;
6 $\quad$ $(u, u') \leftarrow \text{argmax}_{(u,u')} \; score(u, u', C), (u, u') \in S_{swap}(C)$, s.t.
   $\quad$ $confChange(u') = 1$;
7 $\quad$ **if** $v \neq null$ **then**
8 $\quad\quad$ **if** $(u, u') = $ *(null, null)* or $score(v) > score(u, u')$ **then**
9 $\quad\quad\quad$ $C \leftarrow C \cup \{v\}$;
10 $\quad\quad$ **else**
11 $\quad\quad\quad$ $C \leftarrow C\backslash\{u\} \cup \{u'\}$;
12 $\quad\quad$ $lastStepImproved \leftarrow true$;
13 $\quad$ **else**
14 $\quad\quad$ **if** $(u, u') = (null, null)$ *or* $score(u, u') < 0$ **then**
15 $\quad\quad\quad$ **if** $lastStepImproved = $ true **then**
16 $\quad\quad\quad\quad$ **if** $w(C) > w(C^*)$ **then** $C^* \leftarrow C$;
17 $\quad\quad\quad\quad$ **if** $hash(C)$ *is already marked* **then**
18 $\quad\quad\quad\quad\quad$ Drop all the vertices in C;
19 $\quad\quad\quad\quad\quad$ **continue**;
20 $\quad\quad\quad\quad$ Label $hash(C)$ as marked;
21 $\quad\quad\quad$ $lastStepImproved \leftarrow false$;
22 $\quad\quad$ **else**
23 $\quad\quad\quad$ $lastStepImproved \leftarrow true$;
24 $\quad\quad$ $v' \leftarrow \text{argmax}_{v'} \; score(v', C), v' \in C$;
25 $\quad\quad$ **if** $(u, u') = $ *(null, null)* or $score(v', C) > score(u, u', C)$ **then**
26 $\quad\quad\quad$ $C \leftarrow C\backslash\{v'\}$;
27 $\quad\quad$ **else**
28 $\quad\quad\quad$ $C \leftarrow C\backslash\{u\} \cup \{u'\}$;
29 $\quad$ Apply the Strong Configuration Checking (SCC) strategy;
30 $\quad$ $step$++;
31 **return** C^*;

Compared to RSL and RRWL in [11], our algorithm starts from a random single-vertex clique, while they start with a random maximal clique. This is particularly useful when the run-time is restricted. Besides, while RSL and RRWL restart when a solution is revisited in the so-called *first growing step*, our algorithm simply restarts when a local optimum is revisited. In this way, our solver spends less time on searching the local area that has been visited intensively.

5.1 Detecting Revisiting via a Hash Table

In recent methods, the local search typically moves in a deterministic way, *i.e.*, no randomness exists in this process. Thus, a sequence of steps from a previously visited local optimum would be simply repeated, and it may not improve the best clique found so far. Hence, we improve this kind of methods by introducing a cycle elimination based restart strategy, where a hash table is used to approximately detect the revisiting of a local optimum. Given a candidate solution $\mathcal{B}_c^*$ and a prime number p, we define the hash value of $\mathcal{B}_c^*$ as

$$hash(\mathcal{B}_c^*) = (\sum_{\boldsymbol{b}_i \in \mathcal{B}_c^*} 2^i) \mod p, \tag{9}$$

where $i \in \{1, 2, \ldots, n\}$ is the index of $\boldsymbol{b}_i$ in the entire object candidate set $\mathcal{B}$. If p is large enough, the chance of collision is negligible. The parameter p can be set according to the memory capacity of the machine. In the proposed algorithm, the revisiting of a local optimum is detected by checking whether the respective hash entry has been visited. If the local optimum was not visited before, the local search continues. Otherwise, the solver will be restarted and try to look for a better solution.

5.2 Scoring Functions and Candidate Nodes

Given an undirected weighted graph $G = (V, E)$, we describe our approach to finding the MWC in Algorithm 1. To begin with, we first introduce some notations used in our algorithm. In the local search for the MWC, the `add` operation adds a new node to the current clique C. The `drop` operation drops an existing node from the current clique C. The `swap` operation swaps two nodes from inside and outside the current clique C. Each operation returns a new clique as the current solution, which maximizes the gain of the clique weight. Suppose $w(C)$ is the weight of a clique C defined in Eq. (1), then for the `add` and `drop` operation, the gain of adding and dropping a node v is computed as

$$score(v, C) = \begin{cases} w(C \cup \{v\}) - w(C) & \text{if } v \notin C; \\ w(C \backslash \{v\}) - w(C) & \text{if } v \in C. \end{cases} \tag{10}$$

For `swap` operation, the gain of clique weight when swapping two nodes (u, v) is

$$score(u, v, C) = w(C \backslash \{u\} \cup \{v\}) - w(C), u \in C, v \notin C, (u, v) \notin E. \tag{11}$$

We denote the set of neighbors of a vertex v as $\mathcal{N}(v) = \{u|(u,v) \in E\}$. To ensure that the local search always maintains a clique, we define two operand sets. Firstly for a clique C, we define the set of candidate nodes for the `add` operation as

$$S_{add}(C) = \begin{cases} \{v|v \notin C, v \in \mathcal{N}(u), \forall u \in C\} & \text{if } |C| > 0; \\ \emptyset, \text{ otherwise.} \end{cases} \tag{12}$$

Secondly, the set of candidate node pairs for the `swap` operation is defined as

$$S_{swap}(C) = \begin{cases} \{(u,v)|u \in C, v \notin C, (u,v) \notin E, v \in \mathcal{N}(w), \forall w \in C \backslash \{u\}\} \text{ if } |C| > 1; \\ \emptyset, \text{ otherwise.} \end{cases} \tag{13}$$

To maximize the gain of clique weight in each step, the `add` operation adds a node v^* to the current clique C such that $v^* = \text{argmax}_v\, score(v, C), v \in S_{add}(C)$. The `drop` operation drops a node $v^* = \text{argmax}_v\, score(v, C), v \in C$ from the current clique C. The `swap` operation swaps two nodes (u^*, v^*) such that $(u^*, v^*) = \text{argmax}_{(u,v)}\, score(u, v, C), (u, v) \in S_{swap}(C)$.

5.3 The Strong Configuration Checking Strategy

We apply the Strong Configuration Checking (SCC) strategy [35] to avoid revisiting a solution too early. The main idea of the SCC strategy works as follows. After a vertex v is dropped or swapped from a clique C, it can be added or swapped back into C only if one of its neighbors is added into C. Suppose $confChange(v)$ is an indicator function of node v, where $confChange(v) = 1$ means v is allowed to be added or swapped into the candidate solution and $confChange(v) = 0$ means v is forbidden to be added or swapped into the candidate solution, then the SCC strategy specifies the following rules:

1. Initially $confChange(v)$ is set to 1 for each vertex v;
2. When v is added, $confChange(u)$ is set to 1 for all $u \in \mathcal{N}(v)$;
3. When v is dropped, $confChange(v)$ is set to 0;
4. When $(u, v) \in S_{swap}(C)$ are swapped, $confChange(u)$ is set to 0.

6 Experiments

To evaluate the performance of our method in comparison to other approaches, experiments are conducted on the PASCAL VOC 2007 image dataset [8] and the YouTube-Objects video dataset [16]. The standard PASCAL criterion Intersection over Union (IoU) is adopted for evaluation. Namely, a predicted bounding box $\boldsymbol{b}^p$ is correct if $IoU(\boldsymbol{b}^p, \boldsymbol{b}^{gt}) = \frac{area(\boldsymbol{b}^p \cap \boldsymbol{b}^{gt})}{area(\boldsymbol{b}^p \cup \boldsymbol{b}^{gt})} > 0.5$, where $\boldsymbol{b}^{gt}$ is a ground-truth annotation of the bounding box. Finally, the percentage of images with correct object localization (CorLoc) [18] is used as the evaluation protocol. Our method is denoted as LSMWC for local search MWC solver.

6.1 Implementation Details

Our experiments are carried out on a desktop machine with two Intel(R) Core(TM) i7 CPUs (2.80 GHz) and 64 GB memory. A GeForce GTX Titan X GPU is used for training and testing related deep neural networks. The proposed MWC solver is implemented in C/C++. The deep learning framework Caffe and MatConvNet are utilized as carriers for building the Region Proposal Network and the Triplet Network. The pipeline of the system is organized in MATLAB with some utilities written as MEX files, due to the efficiency for high level data management and visualization. The default parameters are used to learn RPN and generate the object candidates. A threshold of 0.5 is used for the NMS process to remove redundant object proposals. The best $K = 20$ object candidates are selected in each image. We set $\lambda = 0.25$ in the hinge loss (6) of a triplet. The prime number p in the hash function (9) is set to $10^9 + 7$, thus the hash table consumes around 1 GB memory. The RPN and Triplet Network in our method are built upon the VGG-f model [30] as well as the VGG-16 model [4]. Compared to the VGG-16 model, the structure of the VGG-f model is much simpler thus more computationally efficient. The VGG-f and VGG-16 models are pre-trained on the ImageNet dataset [29] and fine-tuned on the Microsoft COCO dataset [19]. All parameters are fixed the same in the experiments unless explicitly stated otherwise.

6.2 Experiments on the PASCAL07 Dataset

The PASCAL VOC 2007 dataset [8] is used to evaluate the performance of object co-localization in images. The dataset is split as a training-validation set and a test set, each with about 5,000 images in 20 classes. We follow [15] to construct a collection of images for object co-localization from the training-validation set and denote it as PASCAL07. This is fine in our framework, since our RPN and Triplet networks are not trained on this dataset but on the ImageNet and COCO datesets as stated in Sect. 6.1.

We first compare the co-localization accuracy of different MWC problem solvers on the PASCAL07 dataset in Table 1. The graph instances of these MWC problems are constructed based on the VGG-16 model. Since the PASCAL07 dataset has images from 20 different classes, we construct 20 different graphs, one graph for each image class. For the experiments on the PASCAL07 dataset, the average number of nodes in the constructed graphs is 6081.67, and the average number of edges is 2.16×10^7. The average density of the graphs is 0.9962. Different solvers are evaluated on exactly the same MWC problem instances constructed by our co-localization framework. As randomized processes may exist in different methods, the reported accuracy is taken as the average over 10 runs with different seeds for the random number generator.

For the method [20], it solves the MWC problem in the relaxed continuous domain and a modified Frank-Wolfe algorithm is proposed to attack the problem. Similar to our approach, the solver TBMA [1] also solves the MWC problem directly in the discrete domain. Compared to their solver, our solver will restart if a local optimum is revisited, while TBMA will restart if the solution quality has not been improved for a specified number of steps. As for the solver LSCC [35], originally it is dedicated to solve the MWC problems where the edge weights are absent. Namely, the add, swap or drop operations change the weight of a clique considering related vertex weights only. Here we modify it so that the edge weights are taken into account in these operations. Two versions of the LSCC solver in the original paper are evaluated, and they serve as the baseline results in our experiment. The experiments justify our choice of the MWC problem solver, which improves the accuracy of object co-localization.

Table 1. Co-localization CorLoc (%) of different MWC solvers on the PASCAL07 dataset.

Method	Aero	Bike	Bird	Boat	Bottle	Bus	Car	Cat	Chair	Cow	Table	Dog	Horse	Motor	Person	Plant	Sheep	Sofa	Train	TV	Avg
MC [20]	63.0	45.7	56.1	51.9	14.3	47.8	71.9	61.1	36.2	**72.3**	**46.0**	57.2	61.3	79.6	62.3	34.3	69.8	39.3	59.4	9.8	52.0
LSCC [35]	61.3	68.3	61.2	48.1	16.8	67.7	76.9	58.5	41.8	**72.3**	24.5	63.9	68.3	75.5	69.0	28.6	**76.0**	47.2	62.1	68.4	57.8
[35] + BMS	63.9	68.3	60.9	50.3	49.2	66.7	76.9	59.3	41.1	**72.3**	23.0	**65.1**	68.3	77.1	**69.8**	27.8	**76.0**	45.9	62.8	68.4	59.7
TBMA [1]	62.2	**69.5**	**62.1**	52.5	18.4	**71.5**	78.5	61.1	**49.2**	70.9	30.0	62.0	69.7	80.8	66.0	**49.4**	70.8	50.2	63.6	68.0	60.3
LSMWC	**64.7**	58.4	60.3	**54.1**	**52.0**	71.0	**79.2**	**63.8**	43.1	71.6	40.5	64.4	**72.1**	**84.9**	69.5	45.3	75.0	**51.1**	**66.3**	**71.1**	**62.9**

The co-localization accuracy of different object candidate generation and feature embedding methods on the PASCAL07 dataset is compared in Table 2. Different CNN models are used to extract the object candidate features, then the cosine similarity is applied on these deep neural network features. It shows that re-ranking the object proposals in each image according to the entropy of the class distribution of each object proposal leads to significantly better results than directly using the RPN objectness score of each proposal for ranking. With the involvement of the Triplet Network learning framework, the co-localization performance improves further. The experiments validate that the performance of the object co-localization is benefited from the proper choice of the object candidate generation and feature embedding scheme.

Table 2. Co-localization CorLoc (%) of different strategies on the PASCAL07 dataset.

CNN backbone	RPN objectness	Object proposal re-ranking	Triplet loss fine-tuning
Pre-trained VGG-f model	31.2	53.8	59.3
Pre-trained VGG-16 model	33.1	56.1	62.9

The co-localization accuracy of different object co-localization methods on the PASCAL07 dataset is reported in Table 3. The results of the compared methods are directly taken from the corresponding literature. Among these methods using deep CNN features as visual descriptors [3,7,18,26,27,34,36], our method demonstrates superior results over recent state-of-the-art methods. The experiments confirm the effectiveness of the proposed object co-localization framework.

Table 3. Co-localization CorLoc (%) of different methods on the PASCAL07 dataset.

Method	Aero	Bike	Bird	Boat	Bottle	Bus	Car	Cat	Chair	Cow	Table	Dog	Horse	Motor	Person	Plant	Sheep	Sofa	Train	TV	Avg
Joulin *et al* [15]	32.8	17.3	20.9	18.2	4.5	26.9	32.7	41.0	5.8	29.1	34.5	31.6	26.1	40.4	17.9	11.8	25.0	27.5	35.6	12.1	24.6
Cho *et al* [6]	50.3	42.8	30.0	18.5	4.0	62.3	64.5	42.5	8.6	49.0	12.2	44.0	64.1	57.2	15.3	9.4	30.9	34.0	61.6	31.5	36.6
Li *et al* [18]	73.1	45.0	43.4	27.7	6.8	53.3	58.3	45.0	6.2	48.0	14.3	47.3	69.4	66.8	24.3	12.8	51.5	25.5	65.2	16.8	40.0
Wang *et al* [34]	37.7	58.8	39.0	4.7	4.0	48.4	70.0	63.7	9.0	54.2	33.3	37.4	61.6	57.6	30.1	31.7	32.4	52.8	49.0	27.8	40.2
Bilen *et al* [3]	66.4	59.3	42.7	20.4	21.3	63.4	74.3	59.6	21.1	58.2	14.0	38.5	49.5	60.0	19.8	39.2	41.7	30.1	50.2	44.1	43.7
Ren *et al* [26]	79.2	56.9	46.0	12.2	15.7	58.4	71.4	48.6	7.2	69.9	16.7	47.4	44.2	75.5	41.2	39.6	47.4	32.2	49.8	18.6	43.9
Wei *et al* [36]	67.3	63.3	61.3	22.7	8.5	64.8	57.0	**80.5**	9.4	49.0	22.5	72.6	73.8	69.0	7.2	15.0	35.3	**54.7**	75.0	29.4	46.9
Wang *et al* [33]	**80.1**	63.9	51.5	4.9	21.0	55.7	74.2	43.5	26.2	53.4	16.3	56.7	58.3	69.5	14.1	38.3	58.8	47.2	49.1	60.9	48.5
Cinbis *et al* [7]	67.1	66.1	49.8	34.5	23.3	68.9	**83.5**	44.1	27.7	71.8	**49.0**	48.0	65.2	79.3	37.4	42.9	65.2	51.9	62.8	46.2	54.2
Rochan *et al* [27]	78.5	63.3	**66.3**	**56.3**	19.6	**82.2**	74.7	69.1	22.4	72.3	31.0	62.9	**74.9**	78.3	48.6	29.3	64.5	36.2	**75.8**	69.5	58.8
LSMWC (VGG-f)	59.7	**67.1**	60.3	46.4	51.2	68.8	75.9	57.9	40.4	**77.3**	21.5	**64.6**	65.2	74.7	67.3	41.6	**77.1**	48.0	60.9	60.9	59.3
LSMWC (VGG-16)	64.7	58.4	60.3	54.1	**52.0**	71.0	79.2	63.8	**43.1**	71.6	40.5	64.4	72.1	**84.9**	**69.5**	**45.3**	75.0	51.1	66.3	**71.1**	**62.9**

Table 4. Co-localization CorLoc (%) of different methods on the YouTube-Objects dataset.

Method	Aeroplane	Bird	Boat	Car	Cat	Cow	Dog	Horse	Motorbike	Train	Average
Prest *et al.* [23]	51.7	17.5	34.4	34.7	22.3	17.9	13.5	26.7	41.2	25.0	28.5
Joulin *et al.* [15]	25.1	31.2	27.8	38.5	41.2	28.4	33.9	35.6	23.1	25.0	30.9
Papazoglou *et al.* [22]	65.4	67.3	38.9	65.2	46.3	40.2	65.3	48.4	39.0	25.0	50.1
Zhang *et al.* [37]	**75.8**	60.8	43.7	71.1	46.5	54.6	55.5	54.9	42.4	35.8	54.1
Rochan *et al.* [27]	56.0	30.1	39.6	**85.7**	24.7	**87.8**	55.6	60.2	61.8	51.7	55.3
LSMWC (VGG-f)	48.5	**74.4**	52.8	61.6	**59.4**	69.3	**71.4**	68.5	73.6	43.0	62.3
LSMWC (VGG-16)	44.3	68.6	**56.7**	63.5	50.0	70.7	71.2	**75.9**	**73.8**	**55.5**	**63.0**

6.3 Experiments on the YouTube-Objects Dataset

The YouTube-Objects dataset [16] is used for object co-localization in videos. The dataset contains videos collected from YouTube with 10 object classes. There are about 570,000 frames with 1,407 annotations in the first version of the dataset [23]. According to our knowledge, it is the largest available video dataset with bounding-box annotations on multiple classes. The individual video frames after decompression are used in our experiments to avoid possible confusion when applying different video decoders. We only perform object co-localization on video frames with ground-truth annotations, following the practice in [15]. No additional spatial-temporal information is utilized in our method. The Youtube-Objects dataset comes with the test videos divided in 10 classes according to which dominant object is mostly present in them. Hence we construct 10 different graphs for this dataset. The co-localization accuracy of different methods on the YouTube-Objects dataset are summarized in Table 4. Among all the methods, [27,37] also utilize deep networks for visual representation. The experiments justify that the proposed object co-localization framework is also very effective for mining common objects in videos.

7 Conclusion

In this paper, we present a novel framework to address the problem of object co-localization. It provides a practical and general solution for research and applications related to the MWC problem. Besides, deep learning based methods are utilized to localize the candidates of the common objects and describe their visual characteristics. This makes it possible to better discriminate the inter-class similarities and identify the intra-class variations. Finally, a cycle elimination based restart strategy is proposed to guide the local search for the MWC. It successfully resolves the cycling issue in the optimization process. The experimental results on the object co-localization tasks demonstrate that our MWC solver is particularly suitable for graphs with high density. The proposed method shows significant improvements over several strong baselines.

Acknowledgements. This work was supported in part by NSF grants IIS-1814745 and IIS-1302164. Yi Fan is supported by the Natural Science Foundation of China (Nos. U1711263, U1811264, 61463044, 61572234, 61672441, and 61603152) and the Shenzhen Basic Research Program (No. JCYJ20170818141325209). We also acknowledge donations of Titan X GPU cards by NVIDIA corporation.

References

1. Adamczewski, K., Suh, Y., Mu Lee, K.: Discrete tabu search for graph matching. In: ICCV, pp. 109–117 (2015)
2. Alidaee, B., Glover, F., Kochenberger, G., Wang, H.: Solving the maximum edge weight clique problem via unconstrained quadratic programming. Eur. J. Oper. Res. **181**(2), 592–597 (2007)
3. Bilen, H., Pedersoli, M., Tuytelaars, T.: Weakly supervised object detection with convex clustering. In: CVPR, pp. 1081–1089 (2015)
4. Chatfield, K., Simonyan, K., Vedaldi, A., Zisserman, A.: Return of the devil in the details: delving deep into convolutional nets. In: BMVC (2014)
5. Chen, W., Chen, X., Zhang, J., Huang, K.: Beyond triplet loss: a deep quadruplet network for person re-identification. arXiv (2017)
6. Cho, M., Kwak, S., Schmid, C., Ponce, J.: Unsupervised object discovery and localization in the wild: Part-based matching with bottom-up region proposals. In: CVPR, pp. 1201–1210 (2015)
7. Cinbis, R.G., Verbeek, J., Schmid, C.: Weakly supervised object localization with multi-fold multiple instance learning. TPAMI **39**, 189–203 (2016)
8. Everingham, M., Van Gool, L., Williams, C.K.I., Winn, J., Zisserman, A.: The PASCAL Visual Object Classes Challenge 2007 (VOC2007) Results (2007)
9. Everingham, M., Van Gool, L., Williams, C.K.I., Winn, J., Zisserman, A.: The PASCAL Visual Object Classes Challenge 2012 (VOC2012) Results (2012)
10. Fan, Y., Li, C., Ma, Z., Wen, L., Sattar, A., Su, K.: Local search for maximum vertex weight clique on large sparse graphs with efficient data structures. In: Kang, B.H., Bai, Q. (eds.) AI 2016. LNCS (LNAI), vol. 9992, pp. 255–267. Springer, Cham (2016). https://doi.org/10.1007/978-3-319-50127-7_21
11. Fan, Y., Li, N., Li, C., Ma, Z., Latecki, L.J., Su, K.: Restart and random walk in local search for maximum vertex weight cliques. In: IJCAI (2017)

12. Gouveia, L., Martins, P.: Solving the maximum edge-weight clique problem in sparse graphs with compact formulations. EURO J. Comput. Optim. **3**(1), 1–30 (2015)
13. Hoffer, E., Ailon, N.: Deep metric learning using triplet network. In: Feragen, A., Pelillo, M., Loog, M. (eds.) SIMBAD 2015. LNCS, vol. 9370, pp. 84–92. Springer, Cham (2015). https://doi.org/10.1007/978-3-319-24261-3_7
14. Hosseinian, S., Fontes, D.B., Butenko, S.: A nonconvex quadratic optimization approach to the maximum edge weight clique problem. J. Glob. Optim. **72**, 1–22 (2018)
15. Joulin, A., Tang, K., Fei-Fei, L.: Efficient image and video co-localization with Frank-Wolfe algorithm. In: Fleet, D., Pajdla, T., Schiele, B., Tuytelaars, T. (eds.) ECCV 2014. LNCS, vol. 8694, pp. 253–268. Springer, Cham (2014). https://doi.org/10.1007/978-3-319-10599-4_17
16. Kalogeiton, V., Ferrari, V., Schmid, C.: Analysing domain shift factors between videos and images for object detection. TPAMI **38**(11), 2327–2334 (2016)
17. Kwak, S., Cho, M., Laptev, I., Ponce, J., Schmid, C.: Unsupervised object discovery and tracking in video collections. In: ICCV, pp. 3173–3181 (2015)
18. Li, Y., Liu, L., Shen, C., van den Hengel, A.: Image co-localization by mimicking a good detector's confidence score distribution. In: Leibe, B., Matas, J., Sebe, N., Welling, M. (eds.) ECCV 2016. LNCS, vol. 9906, pp. 19–34. Springer, Cham (2016). https://doi.org/10.1007/978-3-319-46475-6_2
19. Lin, T.-Y., et al.: Microsoft COCO: common objects in context. In: Fleet, D., Pajdla, T., Schiele, B., Tuytelaars, T. (eds.) ECCV 2014. LNCS, vol. 8693, pp. 740–755. Springer, Cham (2014). https://doi.org/10.1007/978-3-319-10602-1_48
20. Ma, T., Latecki, L.J.: Maximum weight cliques with mutex constraints for video object segmentation. In: CVPR, pp. 670–677 (2012)
21. Mascia, F., Cilia, E., Brunato, M., Passerini, A.: Predicting structural and functional sites in proteins by searching for maximum-weight cliques. In: AAAI (2010)
22. Papazoglou, A., Ferrari, V.: Fast object segmentation in unconstrained video. In: ICCV, pp. 1777–1784 (2013)
23. Prest, A., Leistner, C., Civera, J., Schmid, C., Ferrari, V.: Learning object class detectors from weakly annotated video. In: CVPR, pp. 3282–3289 (2012)
24. Pullan, W.: Approximating the maximum vertex/edge weighted clique using local search. J. Heuristics **14**(2), 117–134 (2008)
25. Ren, S., He, K., Girshick, R., Sun, J.: Faster R-CNN: towards real-time object detection with region proposal networks. TPAMI **39**(6), 1137–1149 (2017)
26. Ren, W., Huang, K., Tao, D., Tan, T.: Weakly supervised large scale object localization with multiple instance learning and bag splitting. TPAMI **38**(2), 405–416 (2016)
27. Rochan, M., Wang, Y.: Weakly supervised localization of novel objects using appearance transfer. In: CVPR, pp. 4315–4324 (2015)
28. Rosten, E., Drummond, T.: Machine learning for high-speed corner detection. In: Leonardis, A., Bischof, H., Pinz, A. (eds.) ECCV 2006. LNCS, vol. 3951, pp. 430–443. Springer, Heidelberg (2006). https://doi.org/10.1007/11744023_34
29. Russakovsky, O., et al.: Imagenet large scale visual recognition challenge. IJCV **115**(3), 211–252 (2015)
30. Simonyan, K., Zisserman, A.: Very deep convolutional networks for large-scale image recognition. arXiv (2014)
31. Tang, K., Joulin, A., Li, L.J., Fei-Fei, L.: Co-localization in real-world images. In: CVPR, pp. 1464–1471 (2014)

32. Wang, C., Zhang, H., Yang, L., Cao, X., Xiong, H.: Multiple semantic matching on augmented n-partite graph for object co-segmentation. TIP **26**(12), 5825–5839 (2017)
33. Wang, C., Ren, W., Huang, K., Tan, T.: Weakly supervised object localization with latent category learning. In: Fleet, D., Pajdla, T., Schiele, B., Tuytelaars, T. (eds.) ECCV 2014. LNCS, vol. 8694, pp. 431–445. Springer, Cham (2014). https://doi.org/10.1007/978-3-319-10599-4_28
34. Wang, X., Zhu, Z., Yao, C., Bai, X.: Relaxed multiple-instance SVM with application to object discovery. In: ICCV, pp. 1224–1232 (2015)
35. Wang, Y., Cai, S., Yin, M.: Two efficient local search algorithms for maximum weight clique problem. In: AAAI, pp. 805–811 (2016)
36. Wei, X.S., et al.: Deep descriptor transforming for image co-localization. arXiv (2017)
37. Zhang, Y., Chen, X., Li, J., Wang, C., Xia, C.: Semantic object segmentation via detection in weakly labeled video. In: CVPR, pp. 3641–3649 (2015)

Reconstruction of the Crossing Type of a Point Set from the Compatible Exchange Graph of Noncrossing Spanning Trees

Marcos Oropeza[1] and Csaba D. Tóth[1,2](✉)

[1] California State University Northridge, Los Angeles, CA, USA
marcos.oropeza.774@my.csun.edu, csaba.toth@csun.edu
[2] Tufts University, Medford, MA, USA

Abstract. Let P be a set of n points in the plane in general position. The *order type* of P specifies, for every ordered triple, a positive or negative orientation; and the *x-type* (a.k.a. *crossing type*) of P specifies, for every unordered 4-tuple, whether they are in convex position. Geometric algorithms on P typically rely on primitives involving the order type or x-type (i.e., triples or 4-tuples). In this paper, we show that the x-type of P can be reconstructed from the *compatible exchange* graph $\mathcal{G}_1(P)$ of noncrossing spanning trees on P. This extends a recent result by Keller and Perles (2016), who proved that the x-type of P can be reconstructed from the *exchange graph* $\mathcal{G}_0(P)$ of noncrossing spanning trees, where $\mathcal{G}_1(P)$ is a subgraph of $\mathcal{G}_0(P)$. A crucial ingredient of our proof is a structure theorem on the maximal sets of pairwise noncrossing edges (MSNEs) between two components of a planar straight-line graph on the point set P.

1 Introduction

Let P be a set of n points in the plane in general position (i.e., no three points are collinear), and let $K(P)$ be the straight-line drawing of the complete graph on P. The crossings between the edges of $K(P)$ have a rich combinatorial structure that is not fully understood. For example, the rectilinear crossing number, $\overline{\text{cr}}(K_n)$, is the minimum number of crossings between the edges of $K(P)$ over all n-element points sets P in general position [1,2]. For a given graph G, computing $\overline{\text{cr}}(G)$ is known to be NP-hard; although approximation algorithms are available [8,11].

The *order type* of P specifies, for every ordered triple in P, a positive or negative orientation. The *crossing type* (for short, *x-type*) of P specifies, for every unordered pair of edges in $K(P)$, whether they properly cross; which in turn determines, for every unordered 4-tuple in P, whether the four points are in convex position (i.e., whether they span two crossing edges). Clearly, the order type determines the x-type of a point set, but an x-type may correspond to up to $O(n)$ distinct order types [4,18]. There are $\exp(\Theta(n \log n))$ order types

Research supported in part by the NSF award CCF-1423615.

M. Couprie et al. (Eds.): DGCI 2019, LNCS 11414, pp. 234–245, 2019.
https://doi.org/10.1007/978-3-030-14085-4_19

realized by n points in the plane [12]. Nevertheless, there are efficient algorithms to decide whether two (unlabeled) point sets have the same order type [6] or x-type [18]. The x-type also uniquely determines the *rotation system* of a point set, which specifies the cyclic order of edges of $K(P)$ incident to each point in P [16, Proposition 6]. Geometric algorithms typically rely on elementary predicates involving a constant number of points (such as orientations or convexity). It is a fundamental problem to understand the relations between these predicates [15].

In this paper, we wish to reconstruct the x-type of a point set P from relations between noncrossing subgraphs of $K(P)$. Let $\mathcal{T} = \mathcal{T}(P)$ be the set of all noncrossing spanning trees of $K(P)$. There are n^{n-2} abstract spanning trees on $n \geq 3$ labeled vertices [9]; but the number of noncrossing straight-line spanning trees on n points is in $O(141.07^n)$ [13] and $\Omega(6.75^n)$ [3,10].

Let $\mathcal{G}_0 = \mathcal{G}_0(P)$ be the graph on the vertex set $\mathcal{T}$ where a pair of trees $\{T_1, T_2\}$ forms an edge if we can obtain T_2 from T_1 by exchanging one edge for another (i.e., the symmetric difference of the edge sets of T_1 and T_2 has cardinality 2). The graph $\mathcal{G}_0$ is called the *exchange graph* of the trees in $\mathcal{T}$. It is derived from the well-known exchange property of graphic matroids.

Recently, Keller and Perles [14] proved that the x-type of P can be reconstructed from the (unlabeled) graph $\mathcal{G}_0(P)$. In particular, they showed that the set system of maximal cliques in $\mathcal{G}_0(P)$ already provides enough information to reconstruct the x-type of P.

Over the last few decades, several graphs have been introduced on the noncrossing spanning trees $\mathcal{T}$, besides the classic exchange graph $\mathcal{G}_0(P)$. A *compatible exchange* between T_1 and T_2 exchanges two edges that do not cross each other. See [17] for the hierarchy of five relations. Each of these relations defines a graph on $\mathcal{T}$. In particular, we denote by $\mathcal{G}_1 = \mathcal{G}_1(P)$ the *compatible exchange graph*. By definition, $\mathcal{G}_1(P)$ is a subgraph of $\mathcal{G}_0(P)$. Both $\mathcal{G}_0(P)$ and $\mathcal{G}_1(P)$ are known to be connected, and their diameters are bounded above by $2n - 4$ [7].

Our Results. Our main result (Theorem 3) is that the compatible exchange graph $\mathcal{G}_1(P)$ determines the exchange graph $\mathcal{G}_0(P)$, and hence the x-type of P. The key ingredient of our proof is Theorem 1, which is a structural theorem about maximal sets of pairwise noncrossing edges (for short, MSNEs) between two geometric graphs. An MSNE is a bipartite variant of a triangulation, but (unlike triangulations) MSNEs do not necessarily have the same cardinality, and they need not be connected under edge flip operations.

Organization. In Sect. 2, we distinguish two types of cliques of size 3 or higher in the compatible exchange graph $\mathcal{G}_1(P)$, and prove some of their basic properties. Since $\mathcal{G}_1(P)$ is a subgraph of $\mathcal{G}_0(P)$, many of these properties hold for the cliques of both $\mathcal{G}_0(P)$ and $\mathcal{G}_1(P)$. In Sect. 3, we characterize the *maximal* cliques of $\mathcal{G}_1(P)$, which allows us to determine the type of these cliques in many, but not all cases. Section 4 describes an algorithm for finding all edges of the exchange graph $\mathcal{G}_0(P)$ that are not present in the compatible exchange graph $\mathcal{G}_1(P)$, and hence reconstruct both $\mathcal{G}_o(P)$ and the x-type of the point set P. This is the main result of our paper. We conclude in Sect. 5 with related open problems. All omitted proofs are available in the full paper.

2 Preliminaries

Let P be a set of points in general position in the plane. A clique in $\mathcal{G}_0(P)$ can be formed in the following two ways (refer to Fig. 1):

1. Let F be a noncrossing spanning forest on P consisting of two trees A and B. Let E_{AB} be a set of all edges in $K(P)$ between A and B that do not cross any edge of A and B. Let C be a set of trees $T = (P, E)$, where $E = A \cup B \cup \{e\}$ and $e \in E_{AB}$. The intersection of the edge sets of any two trees in C is the forest F. The set C induces a clique in $\mathcal{G}_0(P)$. We call C a clique of **type 1**.
2. Let $G = (P, E_c)$ be a noncrossing connected spanning graph with n edges. Then G contains a unique cycle that we denote by U. Let C be a set of trees $T = (P, E)$, where $E = E_c \setminus \{e\}$ and e is an edge of U. The union of any two trees in the clique C is the graph G. The set C induces a clique in $\mathcal{G}_0(P)$. We call C a clique of **type 2**.

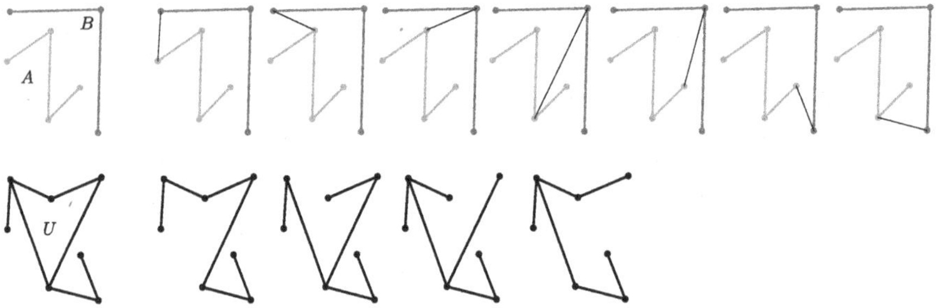

Fig. 1. Top row: A noncrossing spanning forest $F = A \cup B$; and the noncrossing spanning trees obtained by adding an edge from E_{AB}. Bottom row: A connected noncrossing spanning graph with a unique cycle U; and the noncrossing spanning trees obtained by deleting an edge from U.

It is easy to see that *every* clique of size three or higher in $\mathcal{G}_0(P)$ is of type 1 or 2. All omitted proofs are available in the full paper.

Lemma 1. *Every clique of size 3 in $\mathcal{G}_0(P)$ is either of type 1 or of type 2.*

Lemma 2. *If a clique C_1 in $\mathcal{G}_0(P)$ contains a clique C_2 of size 3, then C_1 is of the same type as C_2.*

Corollary 1. *For every $k \geq 3$, every clique of size k in $\mathcal{G}_0(P)$ is either of type 1 or of type 2.*

Remark 1. Since $\mathcal{G}_1(P)$ is a subgraph of $\mathcal{G}_0(P)$, Lemmas 1 and 2 as well as Corollary 1 hold for the cliques of $\mathcal{G}_1(P)$, as well.

Lemma 3. *If cliques C_1 and C_2 in $\mathcal{G}_0(P)$ intersect in a clique C_3 of size 3 or more, then their union induces a clique in $\mathcal{G}_0(P)$.*

Remark 2. Keller and Perles [14] show that in the exchange graph $\mathcal{G}_0(P)$, every edge T_1T_2 is part of at most two maximal cliques: at most one clique of type 1 and at most one clique of type 2. The maximal cliques of type 1 (resp., 2) in $\mathcal{G}_0$ are called *I-cliques* (resp., *U-cliques*). This observation plays a crucial role in their method for determining the types of all maximal cliques in $\mathcal{G}_0(P)$, and ultimately reconstructing the x-type of P. The compatible exchange graph $\mathcal{G}_1(P)$, however, does not have this property: An edge T_1T_2 in $\mathcal{G}_1(P)$ may be part of many *maximal* cliques of type 2. In particular, if two maximal cliques in $\mathcal{G}_1(P)$ intersect in a single edge T_1T_1, then both cliques might be of type 2.

For this reason, the proof strategy in [14] is no longer viable for compatible exchange graphs. Instead, we analyse the interactions of maximal cliques of type 2, and show how to reconstruct the maximal cliques of $\mathcal{G}_0(P)$ from the maximal cliques of $\mathcal{G}_1(P)$.

3 Properties of the Compatible Exchange Graph

Let P be set of n points in the plane in general position, and F a spanning forest on P consisting of two trees, A and B. Let E_{AB} be a maximal set of pairwise noncrossing edges (for short, MSNE) between A and B that do not cross any edges in F. Let C be a set of trees $T = (P, E)$, where $E = A \cup B \cup \{e\}$ and $e \in E_{AB}$. This set C of trees induces a clique in $\mathcal{G}_1(P)$, that we call a **T-clique**.

Lemma 4. *Every T-clique of size 3 or higher is a maximal clique in $\mathcal{G}_1(P)$.*

Lemma 5. *Every maximal clique in $\mathcal{G}_1(P)$ of size three or higher is either a U-clique (i.e., a maximal clique of type 2 in $\mathcal{G}_0$) or a T-clique.*

Definition 1 ***(Special Configurations).*** *A noncrossing straight-line graph $G = (P, E)$ is **special** if G consists of two connected components, A and B, that contain edges $a_1a_2 \in A$ and $b_1b_2 \in B$ such that the vertices (a_1, b_1, b_2, a_2) are on the boundary of* conv(P) *in this counterclockwise order, both a_1b_1 and a_2b_2 are edges of* conv(P)*, and the interior of* $\text{conv}\{a_1, b_1, b_2, a_2\}$ *does not contain any point of P (see Fig. 2(left)). Note that $T_1 = (P, A \cup B \cup \{a_1b_1\})$ and $T_2 = (P, A \cup B \cup \{a_2b_2\})$ are noncrossing spanning trees; and T_1 and T_2 are adjacent in $\mathcal{G}_1(P)$. We call the edge T_1T_2 of $\mathcal{G}_1(P)$ a **special edge**.*

Remark 3. For the pair of trees, A and B, in the configuration of a special edge, there are precisely two MSNEs between A and B: $E'_{AB} = \{a_1b_1, a_1b_2, a_2b_2\}$ and $E''_{AB} = \{a_1b_1, a_2b_1, a_2b_2\}$. The sets E'_{AB} and E''_{AB} define two T-cliques, each of size 3. The edge T_1T_2 is a part of both T-cliques.

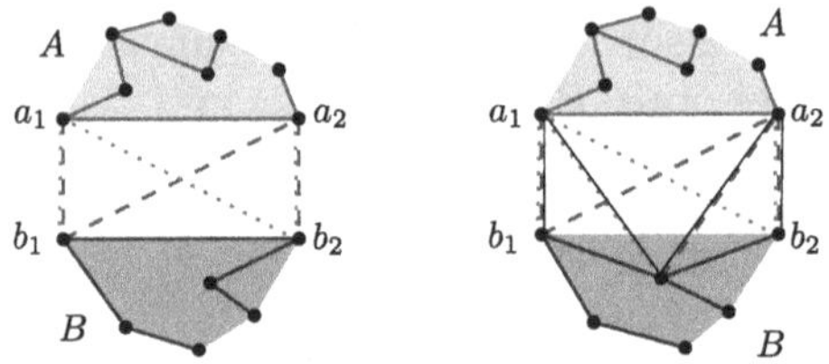

Fig. 2. Left: A special edge is T_1T_2, where $T_1 = (P, A \cup B \cup \{a_1b_1\})$ and $T_2 = (P, A \cup B \cup \{a_2b_2\})$. Right: The edge T_1T_2 is in three T-cliques: C_1 (solid edges), C_2 (dotted edges), and C_3 (dashed edges), where $C_2 \cap C_3 = \{T_1, T_2\}$.

If T_1T_2 is a special edge of the compatible exchange graph $\mathcal{G}_1$, then it is part of precisely two T-cliques. We next prove the converse: if an edge of $\mathcal{G}_1$ is part of precisely two T-cliques, then it is special (Corollary 2). We first prove a crucial property of MSNEs, which may be of independent interest, since MSNEs are considered a bipartite variant of triangulations.

Theorem 1. *Let $G = (P, E)$ be a noncrossing straight-line graph with two connected components, A and B. Then one of the following holds:*

1. *there is a unique* MSNE *between A and B,*
2. *G is special (and there are precisely two* MSNE*s between A and B, cf. Definition 1), or*
3. *for every* MSNE *E_{AB} between A and B, there exists another* MSNE *E'_{AB} such that E_{AB} and E'_{AB} share at least three edges (i.e., $|E_{AB} \cap E'_{AB}| \geq 3$).*

Proof. If there is a unique MSNE between A and B, or if G is special, then the proof is complete. We may assume that neither A nor B is a singleton (otherwise there would be a unique MSNE), consequently every MSNE between A and B has at least three edges. Assume that E'_{AB} and E''_{AB} are two distinct MSNEs between A and B. It suffices to show that there exists a MSNE E'''_{AB} such that $|E'_{AB} \cap E'''_{AB}| \geq 3$. If $|E'_{AB} \cap E''_{AB}| \geq 3$, then we can take $E'''_{AB} = E''_{AB}$. Hence we may assume that $|E'_{AB} \cap E''_{AB}| \leq 2$. We distinguish between two cases.

Case 1: A and B each have a vertex on the boundary of conv(P). Since both A and B are connected, the boundary of conv(P) contains precisely two edges between A and B. These two edges are contained in every MSNE between A and B (since convex hull edges do not cross any other edges). We may assume that $E'_{AB} \cap E''_{AB} = \{e_1, e_2\}$, where both e_1 and e_2 are convex hull edges.

We **claim** that there exist edges $e_3 \in E'_{AB} \setminus \{e_1, e_2\}$ and $e_4 \in E''_{AB} \setminus \{e_1, e_2\}$ (possibly $e_3 = e_4$) such that e_3 and e_4 do not cross each other, and they cross neither e_1 nor e_2. Assuming the claim is true, there is a set $\{e_1, e_2, e_3, e_4\}$ of at least three pairwise noncrossing edges between A and B. We can augment this set to an MSNE, denoted by E'''_{AB}, as required.

Suppose, to the contrary, that the claim is false. This implies the following:

$$\text{Every edge in } E'_{AB} \setminus \{e_1, e_2\} \text{ crosses all edges in } E''_{AB} \setminus \{e_1, e_2\}. \qquad (\star)$$

Let $e_1 = a_1b_1$, where $a_1 \in A$ and $b_1 \in B$. We claim that there is an edge in $E'_{AB} \setminus \{e_1, e_2\}$ incident to a_1 or b_1. (Similarly, there is an edge in $E''_{AB} \setminus \{e_1, e_2\}$ incident to a_1 or b_1.).

Consider $G_0 = (P, A \cup B \cup E'_{AB})$, which is a noncrossing graph on P. In a triangulation of G_0, the edge a_1b_1 is incident to some triangle $\Delta a_1b_1c'$. If $c' \in A$, then $b_1c' \in E'_{AB}$ because it connects A and B and it does not cross any edge in E'_{AB}. Otherwise $c' \in B$, and $a_1c' \in E'_{AB}$. Assume w.l.o.g. that $c' \in B$, hence $a_1c' \in E'_{AB}$. Since A is not a singleton, a_1c' cannot be a convex hull edge. Therefore $a_1c' \in E'_{AB} \setminus \{e_1, e_2\}$. By assumption ($\star$), a_1c' crosses all edges in $E''_{AB} \setminus \{e_1, e_2\}$.

Similarly, a_1b_1 is incident to some triangle $\Delta a_1b_1c''$ in the triangulation of $(P, A \cup B \cup E''_{AB})$, where a_1c'' or b_1c'' is in $E''_{AB} \setminus \{e_1, e_2\}$. Since a_1c' and a_1c'' are adjacent, they do not cross. By assumption ($\star$), $b_1c'' \in E''_{AB} \setminus \{e_1, e_2\}$, consequently $c'' \in A$. In particular $c' \neq c''$. Let d be a vertex in A such that $b_1d \in E''_{AB} \setminus \{e_1, e_2\}$ and $\angle a_1b_1d$ is minimal (Fig. 3).

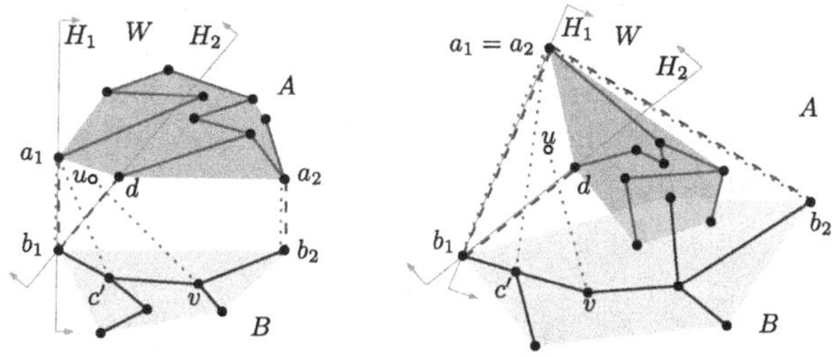

Fig. 3. The configurations described in the proof of Theorem 1, Case 1. Left: an instance where a_1, a_2, b_1, and b_2 are distinct. Right: another instance where $a_1 = a_2$.

Next we claim that every edge in $E'_{AB} \setminus \{e_1, e_2\}$ is incident to a_1. Let uv be an edge in $E'_{AB} \setminus \{e_1, e_2\}$. Suppose for the sake of contradiction that neither u nor v is equal to a_1. By assumption, uv crosses b_1d. Let H_1 be the halfplane bounded by a_1b_1 that contains all points in P, and let H_2 be the halfplane bounded by b_1d that contains a_1. Let $W = H_1 \cap H_2$ and note that it is a wedge with apex b_1. Since uv crosses b_1d, we can assume without loss of generality that $u \in H_2$. Note that $u \in H_1$ since all vertices are in H_1. We have that $u \in W$ because it is in both H_1 and H_2.

Let $P(a_1, d)$ be a unique path in A between a_1 and d. Let C_0 be the cycle $P(a_1, d) \cup (d, b_1, a_1)$. Let C be the set of points in the interior or on the boundary of C_0. We want to show that $u \in A$. Suppose, to the contrary, that $u \in B$. Then u is in the interior of C_0. That means that there exists a path $P(b_1, u)$ in B. Since C_0 has only one vertex in B, namely b_1, this path lies inside C_0. Among all vertices of B in the interior of C_0, let $u' \in B$ be a vertex u' farthest from the line b_1d (in Euclidean distance). Augment E'_{AB} to a triangulation arbitrarily. Since u' lies in the interior of C_0, it has a neighbor v' farther from the line b_1d.

By the choice of u', we have $v' \in A$, hence $u'v' \in E'_{AB}$. Since the edge $u'v'$ lies inside C_0, it does not cross b_1d, contradicting the assumption that every edge in $E'_{AB} \setminus \{e_1, e_2\}$ crosses b_1d. We have shown that every edge in $E'_{AB} \setminus \{e_1, e_2\}$ is incident to a_1. Symmetrically, every edge in $E'_{AB} \setminus \{e_1, e_2\}$ is incident to b_2.

Overall, every edge in $E'_{AB} \setminus \{e_1, e_2\}$ equals to a_1b_2. Hence $E'_{AB} = \{a_1b_2, e_1, e_2\}$. Analogously, $E''_{AB} = \{a_2b_1, e_1, e_2\}$. Since these are MSNEs between A and B, we have $a_1a_2 \in A$ and $b_1b_2 \in B$. In this case, the edge T_1T_2 is a special edge. This completes the proof in Case 1.

Case 2: A or B lies in the interior of conv(P). Assume without loss of generality that B lies in the interior of conv(P). Let $e_1 = a_1b_1$ and $e_2 = a_2b_2$ be two arbitrary edges in E'_{AB}, with $a_1, a_2 \in A$ and $b_1, b_2 \in B$. Let $D = \text{conv}(B \cup \{a_1, a_2\})$. (See Fig. 4.) Note that $B \cup \{a_1, a_2\}$ has at least three vertices that are not collinear since P is in general position. Consequently, D has at least three extremal vertices. One of these vertices, denote it by b_3, is in B. Create a triangulation of $A \cup B \cup E'_{AB}$. Since B is in the interior of the convex hull P, vertex b_3 is incident to triangles whose angles add up to 2π. Since b_3 is an extremal vertex of D, it is incident to an edge in the triangulation in the exterior of D. Let one such edge be $e_3 = b_3a'$. Since B is contained in D, $a' \notin B$, and therefore $a' \in A$. It is also clear that $b_3a' \in E'_{AB}$ as b_3a' does not cross any edge in E'_{AB} because it is from a triangulation. Note that also $b_3a' \neq e_1, e_2$ since both e_1 and e_2 are contained in D.

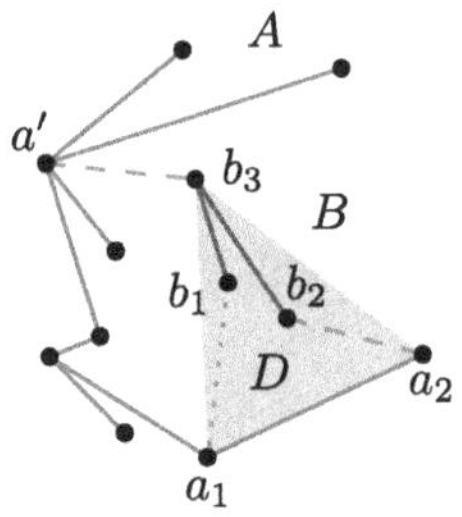

Fig. 4. The configurations described in the proof of Theorem 1, Case 2.

Similarly, there exists an $a'' \in A$ such that $b_3a'' \in E''_{AB}$, and $b_3a'' \neq e_1, e_2$. Let $e_4 = b_3a''$. Since e_3 and e_4 have a common endpoint, namely b_3, they do not cross (possibly $e_3 = e_4$). Therefore, we have a set $\{e_1, e_2, e_3, e_4\}$ of at least three pairwise noncrossing edges between A and B. We can augment this set into an MSNE, and denote it E'''_{AB}. By construction, $E'_{AB} \cap E'''_{AB} = \{e_1, e_2, e_3\}$, as required. □

Corollary 2. *If an edge T_1T_2 of $\mathcal{G}_1(P)$ is contained in two or more T-cliques, then it is either a special edge of $\mathcal{G}_1(P)$ or every T-clique containing T_1T_2 intersects some other T-clique containing T_1T_2 in at least three vertices.*

Proof. Let T_1T_2 be an edge in $\mathcal{G}_1(P)$ contained in two or more T-cliques. Let $T_1 = (P, E_1)$ and $T_2 = (P, E_2)$ such that $E_2 = E_1 - e_1 + e_2$. Let $F = (P, E_1 \cap E_2)$, which is a forest of two trees that we denote by A and B. Every T-clique containing T_1T_2 corresponds to a set of trees that contains the forest F and one edge from some MSNE E_{AB} between A and B. The T-cliques containing T_1T_2 differ in the sets E_{AB}, but share the same forest F. Theorem 1, applied for $G = F$, completes the proof: Either T_1T_2 is a special edge of $\mathcal{G}_1(P)$, or every T-clique containing T_1T_2 intersects some other T-clique containing T_1T_2 in at least three vertices. □

Cliques of Type 2. It is easy to show that $\mathcal{G}_0(P)$ and $\mathcal{G}_1(P)$ have the same maximal cliques of type 2 of size 3 or higher. Adopting terminology from [14], we call these cliques U-cliques.

Proposition 1. *If T_1T_2 is an edge of $\mathcal{G}_1(P)$, then T_1T_2 is part of a unique U-clique.*

Proof. The union of T_1 and T_2 is a connected noncrossing graph with n edges. Therefore it contains a unique cycle u. If the edge T_1T_2 belongs to any U-clique generated by a plane graph $G = (P, E_u)$ with a unique cycle u, then we have $E_1 \subseteq E_u$ and $E_2 \subseteq E_u$. Therefore, $E_u = E_1 \cup E_2$. □

Proposition 2. *Every U-clique has size at least three.*

Proof. Every cycle in a graph has at least three edges. Therefore the U-clique constructed from any cycle has size at least three. □

Propositions 1 and 2 immediately imply the following.

Corollary 3. *Every edge in $\mathcal{G}_1(P)$ is contained in at least one U-clique of size at least three.*

4 Reconstruction of the Exchange Graph

Recall that the compatible exchange graph $\mathcal{G}_1(P)$ is a subgraph of the exchange graph $\mathcal{G}_0(P)$. In this section, we explore how the maximal cliques of the two graphs are related, and then show how to find all edges of $\mathcal{G}_0(P)$ that are not present in $\mathcal{G}_1(P)$.

Lemma 6. *For every edge T_1T_2 in $\mathcal{G}_1(P)$, the maximal cliques that contain T_1T_2 satisfy precisely one of the following conditions:*

1. *T_1T_2 is contained in exactly one maximal clique, which is a U-clique.*
2. *T_1T_2 is contained in precisely two maximal cliques: one U-clique and one T-clique.*
3. *T_1T_2 is contained in precisely three maximal cliques, two of which are of size three and the third of size four. The clique of size four must be a U-clique and the other two are T-cliques. In particular, T_1T_2 is a special edge.*

4. T_1T_2 is contained in three or more maximal cliques, one of which intersects the other cliques in only T_1T_2 and, for all other cliques, there is another clique of size at least three that intersects with them.

Lemma 7. *If T_1T_2 is an edge of the exchange graph $\mathcal{G}_0(P)$, but not an edge of the compatible exchange graph $\mathcal{G}_1(P)$, then:*

1. *T_1 and T_2 are elements of two T-cliques, C_1 and C_2 respectively, that intersect in three or more vertices; or*
2. *T_1 and T_2 are elements of two T-cliques, C_1 and C_2 respectively, that each intersect some T-clique C_3 in three or more vertices; or*
3. *$T_1 = (P, A \cup B \cup \{e_1\})$ and $T_2 = (P, A \cup B \cup \{e_2\})$ where $e_1 = a_1b_2$ and $e_2 = a_2b_1$ are the two crossing diagonal edges of the special configuration (Fig. 2).*

Proof. Let T_1T_2 be an edge in $\mathcal{G}_0(P)$. Then $T_1 = (P, E_1)$ and $T_2 = (P, E_2)$ and there exist edges $e_1 \in E_1$ and $e_2 \in E_2$ such that $E_2 = E_1 - e_1 + e_2$ and $E_1 = E_2 - e_2 + e_1$. Since T_1T_2 is not an edge in the compatibility exchange graph, e_1 and e_2 cross. The edge T_1T_2 of $\mathcal{G}_0(P)$ is contained in a unique maximal clique of type 1, which is called an *I-clique* (see [14, Section 2.3]). Denote by C_0 this I-clique in $\mathcal{G}_0(P)$, and note that it corresponds to a forest F of two trees, A and B, and the set E_{AB} of all edges between A and B (which may cross each other, but do not cross edges in F). An MSNE between A and B is a T-clique contained in C_0. Let $e_1 = a_1b_1$ and $e_2 = a_2b_2$, where $a_1, a_2 \in A$ and $b_1, b_2 \in B$. Let x be the intersection point of e_1 and e_2. Let P_1 be the shortest path between a_1 and b_2 homotopic to (a_1, x, b_2) w.r.t. F (possibly, $P_1 = (a_1, b_2)$). Let P_2 be the shortest path between a_2 and b_1 homotopic to (a_2, x, b_1) w.r.t. F (possibly $P_2 = (a_2b_1)$). The paths P_1 and P_2 each contain an edge between A and B. Denote two such edges by f_1 and f_2, respectively. By construction, neither f_1 nor f_2 crosses e_1 and e_2. (See Fig. 5).

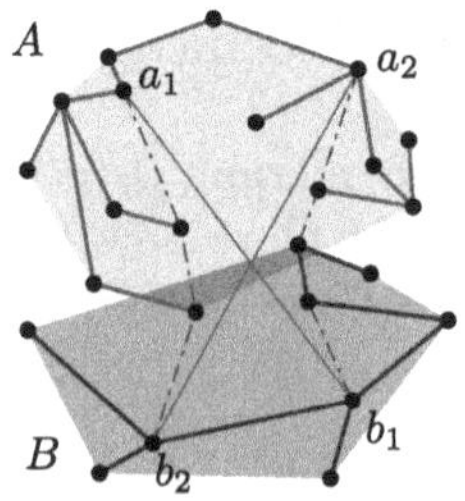

Fig. 5. The shortest paths P_1 and P_2 in the proof of Lemma 7.

Similar to the proof of Theorem 1, we distinguish between two cases based on the position of A and B relative to conv(P).

Case 1: A and B each have a vertex on the boundary of conv(P). Every MSNE contains two convex hull edges between A and B. Since e_1 and e_2 cross, they are not convex hull edges. If E'_{AB} is an MSNE that contains e_1 and E''_{AB} is an MSNE that contains e_2, then $|E'_{AB} \cap E''_{AB}| \geq 2$. We would like to find edge sets E'_{AB} and E''_{AB} such that they share at least three edges.

Let E'_{AB} be an MSNE that contains e_1, f_1, and f_2. Let E''_{AB} be an MSNE that contains e_2, f_1, and f_2. If f_1 or f_2 is not a convex hull edge, then $|E'_{AB} \cap E''_{AB}| = 3$ because their intersection is f_1, f_2, and a convex hull edge. This implies that T_1 and T_2 are elements of two T-cliques, C_1 and C_2, respectively, that intersect in three or more vertices.

Assume that both $f_1 = a_1b_1$ and $f_2 = a_2b_2$ are convex hull edges. If a_1a_2 and b_1b_2 are edges of A and B, respectively, then we are in the case that $e_1 = a_1b_2$ and $e_2 = a_2b_1$ are the two crossing diagonal edges of the special configuration.

Assume without loss of generality that a_1a_2 is not an edge in A. We want to choose a point $a_3 \in A$ that is not equal to a_1 or a_2 such that the line segment a_3x does not cross any edge in A. There is a unique path $P(a_1, a_2)$ between a_1 and a_2 in the tree A. Let R be the region enclosed by $P(a_1, a_2) \cup (a_2, x, a_1)$. Triangulate R. Let a_3 be the third vertex adjacent to a_1x. Let P_3 be the shortest path between a_3 and b_1 homotopic to (a_3, x, b_1) (possibly $P_3 = a_3b_1$). Let P_4 be the shortest path between a_3 and b_2, homotopic to (a_3, x, b_2) (possibly $P_4 = (a_3, b_2)$). These paths each contain an edge between A and B. Denote two such edges by f_3 and f_4, respectively. By construction, f_3 does not cross e_1 (but may cross e_2) and f_4 does not cross e_2 (but may cross e_1). These edges are distinct, i.e., $f_3 \neq f_4$, as they lie on opposite sides of a_1b_1 and a_2b_2.

Let $S_1 = \{e_1, f_1, f_2, f_3\}$, $S_2 = \{e_2, f_1, f_2, f_4\}$, and $S_3 = \{f_1, f_2, f_3, f_4\}$. Note that each set contains pairwise noncrossing edges. Let E'_{AB}, E''_{AB}, and E'''_{AB} be MSNEs that contain S_1, S_2, and S_3, respectively. We then have that T_1 and T_2 are elements of two T-cliques C' and C'', respectively, that each intersect some T-clique C''' in three or more vertices in $\mathcal{G}_1(P)$.

Case 2: A or B lies in the interior of conv(P). Assume without loss of generality that B lies in the interior of conv(P). Similar to the proof of Case 2 of Theorem 4, we find vertices $a' \in A$ and $b_3 \in B$ such that the edge $a'b_3$ does not cross any of the edges in A and B as well as does not cross the edges e_1, e_2, f_1, and f_2. Let E'_{AB} and E''_{AB} be MSNEs that contain $\{e_1, f_1, f_2, a'b_3\}$ and $\{e_2, f_1, f_2, a'b_3\}$, respectively. Then T_1 and T_2 are elements of two T-cliques, C' and C'' respectively, that intersect in three or more vertices. □

Theorem 2. *The compatible exchange graph $\mathcal{G}_1(P)$, for a point set P, determines the exchange graph $\mathcal{G}_0(P)$.*

Proof. Our proof is constructive. We are given $\mathcal{G}_1(P)$, which is an unlabeled graph. We can find all special edges T_1T_2 in $\mathcal{G}_1(P)$ based on the characterization in Lemma 6(3), that is, T_1T_2 is contained in precisely 3 maximal cliques, two of which are of size 3 and the third of size 4.

1. Find all maximal cliques of $\mathcal{G}_1(P)$.
2. Find all special edges T_1T_2 in $\mathcal{G}_1(P)$.
3. Put $\mathcal{H} := \mathcal{G}_1(P)$.

4. For every special edge $e = T_1T_2$ in $\mathcal{G}_1(P)$, let $T_3(e)$ and $T_4(e)$ be the 3rd vertices of the two cliques of size 3 in $\mathcal{G}_1(P)$ that contain T_1T_2; and augment $\mathcal{H}$ with the edge $T_3(e)T_4(e)$ (if it is not already present in $\mathcal{H}$).
5. While $\mathcal{H}$ contains two maximal cliques that intersect in three or more vertices, augment $\mathcal{H}$ with the edge to merge them into a single clique.
6. Return $\mathcal{H}$.

The "dual" special edges added in step 2 are in $\mathcal{G}_0(P)$ by Lemma 6. Edges added in the merge steps are in $\mathcal{G}_0(P)$ by Lemma 3. We see that the algorithm returns a subgraph of $\mathcal{G}_0(P)$. Conversely, the algorithm adds all edges in $\mathcal{G}_0(P) \setminus \mathcal{G}_1(P)$ by Lemma 7. Therefore it returns $\mathcal{G}_0(P)$, as required. □

The combination of Theorem 2 with [14] readily implies our main result.

Theorem 3. *The compatible exchange graph $\mathcal{G}_1(P)$, for a point set P in the plane in general position, determines the x-type of P.*

5 Future Directions

We have shown (Theorem 3) that the compatible exchange graph $\mathcal{G}_1(P)$ determines the exchange graph $\mathcal{G}_0(P)$, and hence the x-type of the point set P. Our results and our proof techniques raise several interesting problems. We list a few of them here.

1. Can the x-type of a point set P be reconstructed from any of the more restrictive relations, e.g., the *rotation graph* or the *edge slide graph* on $\mathcal{T}(P)$ (see [17] for precise definitions)? While all these graphs are known to be connected, the edge slide graph of $\mathcal{T}(\mathcal{P})$, for example, may have vertices of degree 1 [5], hence its clique complex is a 1-dimensional simplicial complex.
2. By Corollary 3, the compatible exchange graph $\mathcal{G}_1(P)$ is biconnected for $|P| \geq 3$ (in fact, the clique complex of $\mathcal{G}_1(P)$ is a 2-dimensional simplicial complex). However, the minimum degree is $\Omega(n)$. It remains an open problem to find tight bounds for the diameter, the minimum degree, and the vertex- and edge-connectivity of $\mathcal{G}_1(P)$ over all n-element point sets P.
3. T-cliques played a crucial role in the proof of our main result (cf. Theorem 2), but the underlying geometric structures raise intriguing open problems. Every maximal T-clique corresponds to a MSNE E_{AB} between A and B, which are two trees in a noncrossing spanning forest $F = (P, A \cup B)$. The set E_{AB} can be thought of as a bipartite analogue of straight-line triangulations on a point set P (which is a maximal set of pairwise noncrossing edges in $K(P)$). It is well known that every triangulation on P has the same number of edges, and the space of triangulations on P is connected under the so-called *edge flip* operation. For a forest $F = (P, A \cup B)$, we can define the collection $\mathcal{E}_{AB}$ of all maximal sets of pairwise noncrossing edges between A and B. Two sets in $\mathcal{E}_{AB}$ may have different cardinalities, and there is no clear operation that would generate all sets in $\mathcal{E}_{AB}$. It remains an open problem to understand the combinatorial structure of $\mathcal{E}_{AB}$.

References

1. Ábrego, B.M., Cetina, M., Fernández-Merchant, S., Leaños, J., Salazar, G.: On $\leq$ k-edges, crossings, and halving lines of geometric drawings of K_n. Discrete Comput. Geom. **48**(1), 192–215 (2012)
2. Ábrego, B.M., Fernández-Merchant, S.: The rectilinear local crossing number of K_n. J. Combin. Theory Ser. A **151**, 131–145 (2017)
3. Aichholzer, O., Hackl, T., Huemer, C., Hurtado, F., Krasser, H., Vogtenhuber, B.: On the number of plane geometric graphs. Graphs Combin. **23**(1), 67–84 (2007)
4. Aichholzer, O., Kusters, V., Mulzer, W., Pilz, A., Wettstein, M.: An optimal algorithm for reconstructing point set order types from radial orderings. Int. J. Comput. Geom. Appl. **27**(1–2), 57–84 (2017)
5. Aichholzer, O., Reinhardt, K.: A quadratic distance bound on sliding between crossing-free spanning trees. Comput. Geom. **37**(3), 155–161 (2007)
6. Aloupis, G., Iacono, J., Langerman, S., Ozkan, Ö., Wuhrer, S.: The complexity of order type isomorphism. In: Proceedings of 25th ACM-SIAM Symposium on Discrete Algorithms (SODA), pp. 405–415. SIAM (2014)
7. Avis, D., Fukuda, K.: Reverse search for enumeration. Discrete Appl. Math. **65**(1), 21–46 (1996)
8. Bald, S., Johnson, M.P., Liu, O.: Approximating the maximum rectilinear crossing number. In: Dinh, T.N., Thai, M.T. (eds.) COCOON 2016. LNCS, vol. 9797, pp. 455–467. Springer, Cham (2016). https://doi.org/10.1007/978-3-319-42634-1_37
9. Cayley, A.: A theorem on trees. Q. J. Math. **23**, 376–378 (1889)
10. Flajolet, P., Noy, M.: Analytic combinatorics of non-crossing configurations. Discrete Math. **204**(1), 203–229 (1999)
11. Fox, J., Pach, J., Suk, A.: Approximating the rectilinear crossing number. In: Hu, Y., Nöllenburg, M. (eds.) GD 2016. LNCS, vol. 9801, pp. 413–426. Springer, Cham (2016). https://doi.org/10.1007/978-3-319-50106-2_32
12. Goodman, J.E., Pollack, R.: The complexity of point configurations. Discrete Appl. Math. **31**(2), 167–180 (1991)
13. Hoffmann, M., Schulz, A., Sharir, M., Sheffer, A., Tóth, C.D., Welzl, E.: Counting plane graphs: flippability and its applications. In: Pach, J. (ed.) Thirty Essays on Geometric Graph Theory, pp. 303–325. Springer, New York (2013). https://doi.org/10.1007/978-1-4614-0110-0_16
14. Keller, C., Perles, M.A.: Reconstruction of the geometric structure of a set of points in the plane from its geometric tree graph. Discrete Comput. Geom. **55**(3), 610–637 (2016)
15. Knuth, D.E. (ed.): Axioms and Hulls. LNCS, vol. 606. Springer, Heidelberg (1992). https://doi.org/10.1007/3-540-55611-7
16. Kynčl, J.: Enumeration of simple complete topological graphs. Eur. J. Combin. **30**(7), 1676–1685 (2009)
17. Nichols, T.L., Pilz, A., Tóth, C.D., Zehmakan, A.N.: Transition operations over plane trees. In: Bender, M.A., Farach-Colton, M., Mosteiro, M.A. (eds.) LATIN 2018. LNCS, vol. 10807, pp. 835–848. Springer, Cham (2018). https://doi.org/10.1007/978-3-319-77404-6_60
18. Pilz, A., Welzl, E.: Order on order types. Discrete Comput. Geom. **59**(4), 886–922 (2018)

Mathematical Morphology

Vector-Based Morphological Operations on Polygons Using Straight Skeletons for Digital Pathology

Daniel Felipe González Obando[1,2], Jean-Christophe Olivo-Marin[1], Laurent Wendling[3], and Vannary Meas-Yedid[1(✉)]

[1] BIA, Institut Pasteur, UMR 3691, CNRS, 75015 Paris, France
vmeasyed@pasteur.fr

[2] ED EDITE, Sorbonne Université, Collège doctoral, 75005 Paris, France

[3] Université Paris Descartes, LIPADE, Sorbonne Paris Cité, 75006 Paris, France

Abstract. In this work we present an efficient implementation of vector-based mathematical morphology operators applied to simple polygons by performing wavefront propagation and computing polygon straight skeletons. In Digital Pathology (DP), the slide scanner generates important volume of images from tissues called Whole Slide Image (WSI). The main goal of the DP is to detect the biological stained structures in order to quantify the tissue pathology, such as lesions or cancerous regions. We propose the use of Adapted Straight Skeletons on polygons as an efficient technique in time and memory, to improve image segmentation and image analysis. Thanks to the use of polygons instead of bitmaps to store segmentation results, the performance of straight skeletons depends only on the polygon control points. These straight skeletons can be applied in order to perform fast morphological operations such as dilation, erosion, closing, opening, skeletonizing. When combined, these operations offer different interesting outcomes: (i) multiple disjoint-segmented shapes can be linked together to create a joint skeleton, (ii) the topological structure of segmentation can be extracted as a straight skeleton. Then, it can be used as features for structural and spatial tissue analysis.

Keywords: Polygonal morphological operations · Straight skeletons · Digital Pathology

1 Introduction

The main goal of histopathology is to assess the biological tissue samples by the examination of the tissue in order to diagnose or prognose many diseases, such as cancers, or tissue lesions. Nowadays, slide scanners allow to acquire images with a high speed and high resolution. These images are called virtual slides, but image analysis tools do not exploit the high potential content and the clinical analysis of WSI largely remains the work of human experts. The images resulting from slide scanner devices are of high dimensions (image size greater than 1 GB) containing

M. Couprie et al. (Eds.): DGCI 2019, LNCS 11414, pp. 249–261, 2019.
https://doi.org/10.1007/978-3-030-14085-4_20

up to several thousands of objects and thus, are considered as big data. Therefore these images need to be processed by fast and efficient algorithms, both in time and in memory. In [17], the authors propose a sparse coding and multi-scale dynamic sampling approach. However, this approach is not adapted to object detection, such as capillaries or early tumor areas. In order to cope with this big data issue, object extraction will be coded by polygons which appear to be an efficient structure to represent Regions of Interest (ROIs) within the tissue such as vessels, capillaries, nuclei or pathological regions (fibrosis or cancerous areas). They allow to capture information with a small amount of points in the context of WSI. In this work, we will focus on manipulating and processing area objects extracted from image segmentation of the tissue. In order to tackle this issue, polygons appear to be an efficient structure to manipulate ROIs, in the context of WSI. Indeed, polygons allow to capture region information with a relatively small amount of points.

In this study, we will focus on morphological operations on polygons and we assume that the polygons are the inputs of our algorithms. The notion of skeleton was introduced by Blum [8] as a result of the Medial Axis Transform or Symmetry Axis Transform. The research on mathematical morphology for image processing has been very active, and we can distinguish two kinds of families: raster-image based approach and vector-based approach [28]. In this paper, as the aim is to handle polygons, we will only consider the vector based approach. Among them, we find a popular approach based on the principle of the Voronoï Diagram. The "Voronoï skeleton" is computed on a polygonal representation of an object, which is the sampling of contour points, and is computed through geometry algorithms. This method allows to compute the skeleton based on the edges formed from vertices of the shape, but it does not allow to perform morphological operations such as dilations or erosions. Minkowski sums allow for computing offsets at specific distances using the sum of a circle with the polygon boundaries [29]. Although this method can compute smooth contours, it does not allow to recover the internal structure of the shape and requires recomputing the sum when multiple distances are required, which can be computationally expensive. Another category is to use straight skeletons [2]. The skeleton of a polygon is a thin version of that polygon where every point is equidistant to the polygon boundaries. Unlike skeletons based on Voronoï diagram which can contain parabolic curves [1], straight skeletons contain only straight segments. This property allows for easier and faster operations by avoiding calculus on curves, notably the possibility to compute multiple dilations and erosions at different distances in linear time, with respect to the number of vertices in the polygon. Several works have used this approach to compute straight skeletons (see Table 1). Aichholzer and Aurenhammer [3] set the base grounds by formalizing the concept of straight skeleton as a **planar straight-line graph**, as well as introducing their wavefront propagation method performed with a time complexity of $O(n^2 \log n)$, where n is the amount of vertices in the polygon. Eppstein and Erickson presented a sub-quadratic algorithm that used efficient closest-pair data structures to optimize processing time up to $O(n^{1+\epsilon} + n^{8/11+\epsilon} r^{9/11+\epsilon}) \subseteq O(n^{17/11+\epsilon})$, where r is the number of reflex vertices in the polygon and $0 < \epsilon \ll 1$ is a fixed value [13].

Table 1. Overview of existing straight skeleton algorithms: n: total number of vertices, r: number of reflex vertices.

Algorithm	Time	Memory space
Aichholzer [2]	$O(nr\log n)$	O(n)
Aichholzer [3]	$O(n^3\log n)$ (pract. $O(n\log n)$)	O(n)
CGAL [9]	$O(n^2\log n)$	$O(n^2)$
Eppstein [13]	$O(n^{1+\epsilon} + n^{8/11+\epsilon}\log n)$	ditto
Cheng	$O(n^{1+\epsilon} + n^{8/11+\epsilon}r^{9/11+\epsilon})$	O(n)
STALGO [18]	$O(n^2\log n)$ (pract. $O(n\log n)$)	O(n)

Cacciola also presented an implementation currently available in CGAL to compute straight skeletons with a complexity of $O(n^2 \log n)$ in time, and $O(n^2)$ in space [9]. More recently, Huber and Held have presented an implementation (called STALGO) by computing the straight skeleton aided by a motorcycle graph in $O(n \log n)$ in time and $O(n)$ in space [18]. We based our solution by following this idea [18], with adaptations to fulfill our needs in Digital Pathology.

2 Straight Skeletons

The straight skeleton $\mathcal{S}(P)$ of a simple polygon P is defined by a wavefront propagation process. A wavefront $\mathcal{W}_P(t)$ of P is formed by edges that are parallel to those of P moving all at unit speed to the interior of P for erosion computing (resp. to the exterior of P for dilation). $\mathcal{W}_P(t)$ is also defined by the vertices moving on the internal angular bisectors of the vertices of P. These vertices move until either of the following events happen:

- An edge *collapses* when two linked vertices join together at the same point at a given time t. From this event, a new vertex is created following the direction computed from neighboring edges of the collapsed edge.
- An edge is *split* into two new edges when a vertex forming a reflex angle (an angle larger than π on the propagation side) meets a wavefront edge. This results in two new edges with the same speed of the original edge, on each side of the splitting vertex.

This two kinds of events are repeatedly handled, up until all wavefront edges have collapsed and no new events happen. At this point, the skeleton $\mathcal{S}(P)$ of P is defined as the set of loci that are traced out by the wavefront vertices (see the magenta lines on Fig. 2b).

Motorcycle Graph. Computing straight skeletons on convex simple polygons is really straight forward since only edge collapses may happen during the wavefront propagation process. In contrast, non-convex simple polygons require

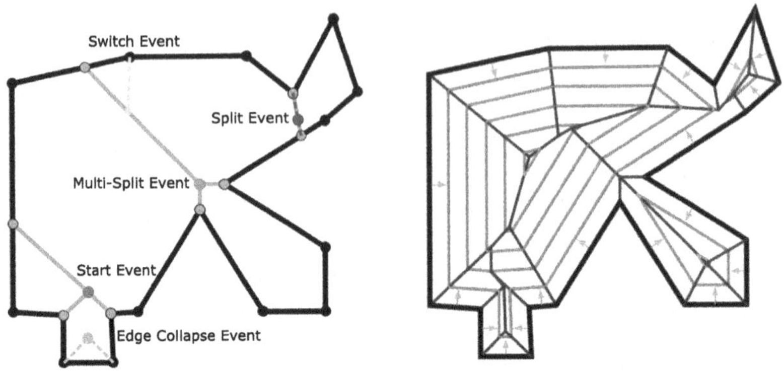

(a) Motorcycle graph and initial events (b) Wavefront propagation and straight skeleton

Fig. 1. (a) The cyan segments split the initial concave polygon into convex polygons, (b) straight skeletons are represented by blue segments (Color figure online)

detecting potential split events during the propagation. A naive algorithm to detect this kind of events could take $O(n^2 \log n)$ in time by looking all possible splitting events in the planar straight-line graph. However, using a motorcycle graph for this end comes convenient because it allows to detect all the trajectories of splitting vertices that occur during the propagation in $O(n \log n)$ time. Motorcycle graphs divide polygons into convex tessellations that guarantee no split events inside the subdivisions, thus reducing the complexity of the calculation of $\mathcal{S}(P)$ [18].

On a motorcycle graph $\mathcal{M}(P)$ a *"motorcycle"* m_i is launched for each reflex vertex p_i of the P. These motorcycles are allowed to move on the bisectors of the edges leaving a trace behind them until either they crash with another motorcycle or trace, or escape to infinity if no crash happens (see green lines on Fig. 2a). When two or more motorcycles crash at the same time, the pair of successive motorcycles forming a reflex angle is used to compute a new motorcycle. The reader is referred to [18] for more details.

Straight Skeleton Construction. Using all convex and reflex vertices from the input polygon, as well as the motorcycle crash points from $\mathcal{M}(P)$, the wavefront $\mathcal{W}_P(t)$ is initialized. From this wavefront it is possible to compute $\mathcal{S}(P)$ by following the evolution of $\mathcal{W}_P$. Vertices on $\mathcal{W}_P(t)$ move either in the direction of the bisector of the vertices incident edges, or in the inverse direction of the crashed motorcycles from their crash point. The wavefront propagation is then processed in chronological order (see Algorithm 1), handling each crash event to build $\mathcal{S}(P)$ step by step (see event examples in Fig. 1a and the wavefront evolution in Fig. 1b). Once all crash events are processed, $\mathcal{S}(P)$ is fully conformed by the trace of the wavefront vertices. With this result, multiple morphological operations are possible, presented in the following section.

3 Morphological Operations on Polygons

Although straight skeletons are easier to handle and to model, they still have a drawback. As they are built by evolving the wavefront at a constant speed, the speed of reflex vertices tends to be faster than the rest of the wavefront. This causes premature edge splitting and results in a very disturbed skeleton. To address this issue, we propose a refined straight skeleton by adding more than one vertex on reflex angles. This allows to homogenize the speed of vertices on the wavefront avoiding premature split events. As a result, the produced skeleton is smoother (see Figs. 2b and 6b, c).

3.1 Refined Straight Skeletons

To provide a more accurate method, we propose to launch two motorcycles instead of just one on reflex angles when creating the motorcycle graph (see red lines against green lines on Fig. 2a). From this, the wavefront is built in the same way as the original straight skeleton. Performing the wavefront evolution with this change allows to prevent excessive evolution on the reflex wavefront vertices (see orange lines against light green lines on Fig. 2b), which can create early polygon divisions when performing dilations and erosions.

Result: $\mathcal{S}(P)$
$\mathcal{W}_P \leftarrow initializeWavefront(P, \mathcal{M}_P)$;
$\mathcal{S}_P \leftarrow initalizeStraightSkeleton(\mathcal{W}_P)$;
Fill event priority queue Q with initial events;
while *Q is not empty* **do**
  $e \leftarrow Q.poll()$;
  if *$isEventStillValid(e)$* **then**
    $newPossibleEvents \leftarrow processEvent(e)$;
    `// This will update` $\mathcal{W}_P$ `and` $\mathcal{S}(P)$)
    $Q.addAll(newPossibleEvents)$;
  else
    continue;
  end
end

Algorithm 1. Event handling on the wavefront propagation when building the straight skeleton $\mathcal{S}(P)$.

Dilating and Eroding Polygons. With this consistent change on the wavefront construction it is now possible to obtain dilations and erosions that are closer to those made by classical bitmap methods [24]. In this case, dilations are created by recreating the wavefront $W_P(t)$ from $\mathcal{S}(P)$. For this, the edges of S are followed from the leafs of the graph S up to the desired time (distance) t. From there the dilation is created by connecting the valid edges at instant t (see Fig. 3).

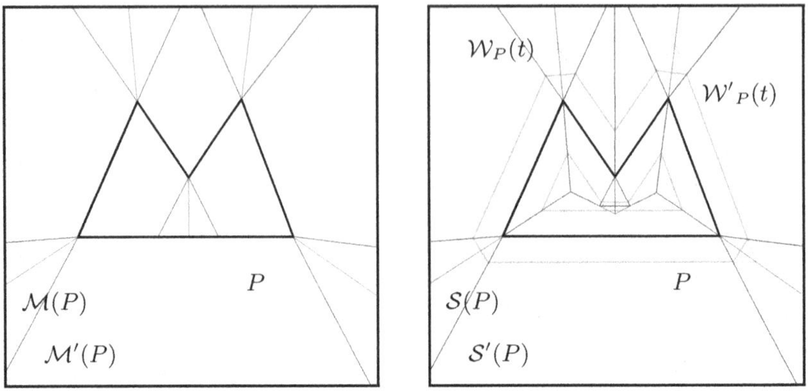

(a) Motorcycle graph (b) Straight skeleton and offset at time t

Fig. 2. Adaptation of Huber and Held's method to improve offset results. In the motorcycle graph (a), Huber's method creates only one motorcycle for each reflex vertex ($\mathcal{M}(P)$), whereas our method creates two ($\mathcal{M}'(P)$). In the straight skeleton (b), Huber's method creates only one wavefront vertex for each reflex vertex ($\mathcal{S}(P)$), producing sharp edges ($\mathcal{W}_P(t)$). Our method creates two vertices on reflex vertices ($\mathcal{S}'(P)$), accentuating sharp polygon vertices on the wavefront. These changes produce smoother offsets ($\mathcal{W}'_P(t)$). (Color figure online)

3.2 Medial Axis

Since its introduction by Blum [8], the notion of medial axis has proven to be useful for many purposes in morphological analysis. He introduced the medial axis of a shape X as the set of points $x \in X$ that have more than one nearest point on the boundary ∂X of X. To compute this medial axis several solutions have been proposed making use of discrete geometry [14,16,23], digital topology [10,22,26,27], computational geometry [4,5,21], partial differential equations [25], and level-sets [20]. In our case, we focus on the medial axis for skeletons strictly made out of straight line segments, which has not been yet formalized as far as we know.

Here we propose a medial axis based on the straight skeleton $\mathcal{S}(P)$ of a polygon P, with only one parameter for pruning. Let $\mathcal{M}(P)$ be the straight medial axis of the polygon P defined as the edges of $\mathcal{S}(P)$ filtered by the minimum allowed distance from an end-point of any skeleton edge to the shape boundaries. That is, any edge $e \in \mathcal{S}(P)$ with an end-point whose distance to the shape boundary is smaller than a given threshold will be discarded as part of the straight medial axis. This definition can cause topology issues as the connectivity of the medial axis can differ from that of the skeleton (see an example in Fig. 4a). To address this issue we perform the edge filtering using a priority on the minimal distance from the edge to the shape boundaries. This means looking both edge ends and making sure that at least one of them is of degree one. Otherwise the edge is not removed keeping the connectivity of the result (see Fig. 4b).

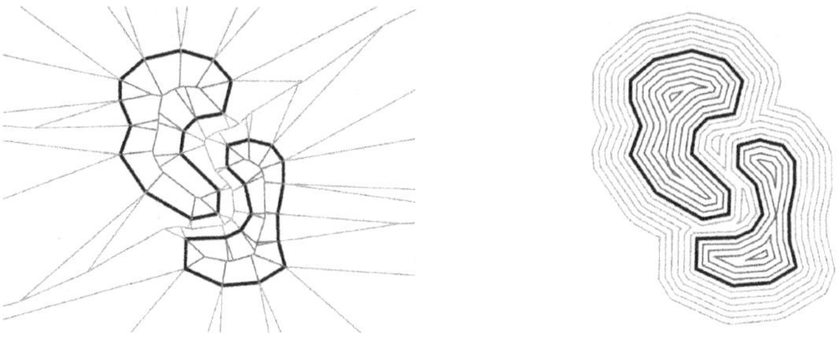

Fig. 3. Refined straight skeletons of two polygons. Inwards skeleton in red and outwards in green. Successive dilations (in magenta) and erosions (in cyan) created from refined straight skeletons. (Color figure online)

3.3 Weighted Straight Skeletons for Directional Operators

The weighted straight skeleton was first proposed in 1999 by Eppstein and Erickson [13], where the wavefront edges may move with arbitrary but fixed speeds. Since its definition some partial implementations have been proposed to address several of the issues associated to dealing with different speeds on edges. These implementations handle issues such as negative and zero edge weights [19], edge crash event ambiguities [6,7], and algorithm complexity improvements [12]. This kind of skeletons are interesting for oriented morphological operations such as oriented dilations/erosions to restrict the search space of neighbourhood based on the orientation with respect to the shape, etc. We have started the development of the weighted straight skeletons on convex polygon by taking into account two main issues (see Fig. 5):

- The collapse of an edge with two parallel neighboring edges with different weights. In this case a decision must be taken on the speed of the resulting edges joined at this point of the propagation. One option is to take the highest or lowest speed between the involved edges and apply it to them all. Another option (the one chosen for our implementation) is to let the edges following their own propagation speed and adding a zero-speed edge joining them. This way helps to keep speeds without changes and the wavefront continues to be consistent on the propagation.
- On multiple split events, special care must be taken while reconnecting edges in order to keep a consistent topology of the wavefront. If other events happen at the same time, these events must be processed before processing the splitting to avoid confusion when reconnecting edges participating on the event.

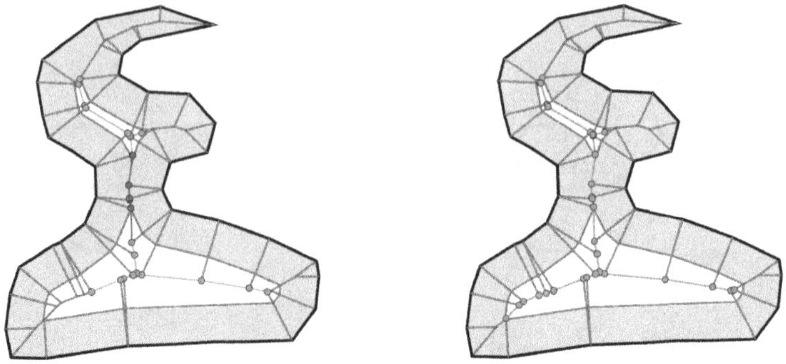

(a) axis without connectivity constraint

(b) axis with connectivity constraint

Fig. 4. A straight medial axis example. When a connectivity constraint is not applied while pruning the straight skeleton, its topology can differ from that of the original shape. (a) shows in red the edges ignored when no connectivity constraint is applied when building the medial axis (in green) with a threshold criterion expressed by the blue polygons. (b) shows the result of the connectivity-constrained straight medial axis. (Color figure online)

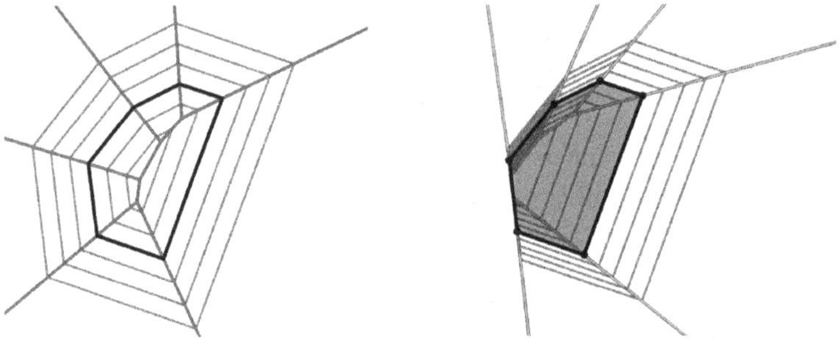

(a) Non weighted straight skeletons.

(b) Weighted straight skeletons.

Fig. 5. Comparison between the unweighted and weighted straight skeletons (in cyan) applied to the same polygon, with successive dilations (in orange). (Color figure online)

4 Results

4.1 Applications to Digital Pathology

The structures in the tissue image can be split into several segments due to several factors: the sample slice cut, the staining, the acquisition setup, etc. and can result in over-segmented objects. Figure 7 shows the lumen around glomerulus split into two regions. To help the object detection, we ran the weighted straight skeleton algorithm to perform the directional dilation in order to find the second segmented polygon thanks to this spatial context processing.

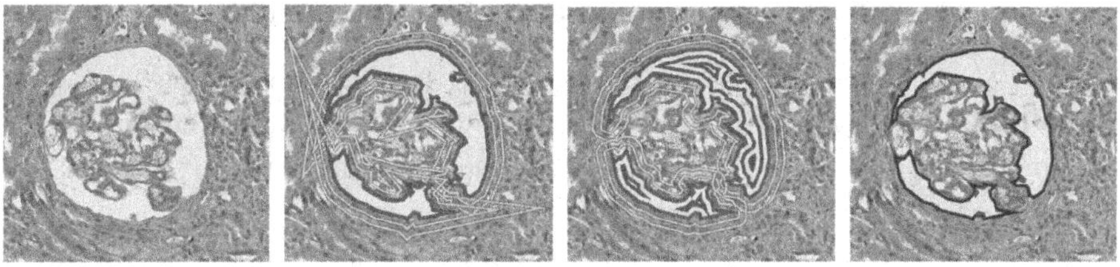

(a) (b) (c) (d)

Fig. 6. Morphological operations applied to a segmentation of a glomerulus from a stained kidney tissue (input polygons in green). (b) Dilation of polygon (yellow) using Huber's method on polygons. (c) Dilations (yellow) and erosions (red) using our method. (d) Closing (*dilation o erosion*) using our method (Blue) (Color figure online)

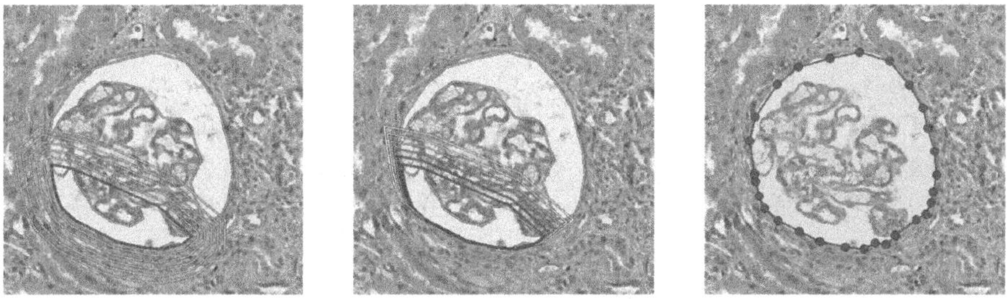

(a) (b) (c)

Fig. 7. Weighted straight skeletons applied to a glomerulus segmentation (input polygon in orange). (a) Dilations of convex polygon in green using non-weighted straight skeleton algorithm. (b) Directional dilations (green) of weighted straight skeletons. (c) Ellipse detected after fusion of the two lumen segments (Color figure online)

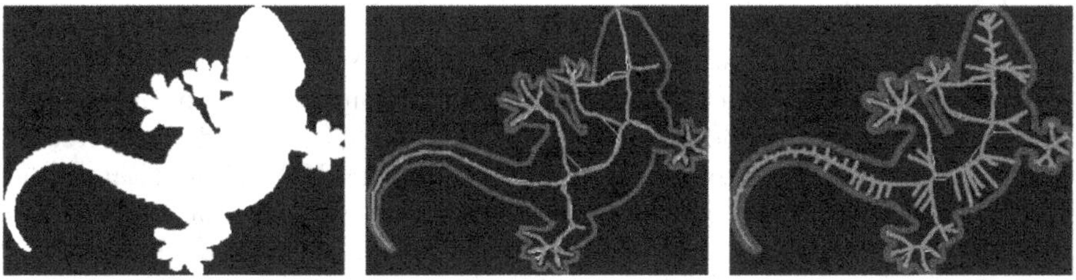

(a) (b) (c)

Fig. 8. The lizzard image used for comparison tests. In green, the polygons used for the straight skeleton algorithm and in cyan, the medial axis with a distance value set to 1. Note that the polygon is not composed by the same amount of points in (b) and (c). (Color figure online)

Table 2. Comparison of skeletonization methods on Fig. 8.

Image size W × H (px)	Points in polygon	[27] Time (ms)	[15] Time (ms)	[30] Time (ms)	Proposed method time (ms)
250 × 194	52	13	137.239	110.254	10.9
300 × 233	57	17	145.683	143.966	11.43
350 × 272	67	33	155.221	145.637	12.3
400 × 311	74	42	172.495	158.056	17.17
450 × 350	83	63	187.624	172.375	15.57
500 × 389	87	89	202.811	182.217	19.45
550 × 428	95	121	224.427	187.787	18.33
600 × 467	98	143	245.708	190.95	21.64
650 × 506	101	183	274.336	199.867	22.63
700 × 545	108	297	299.136	228.684	23.07
750 × 584	118	463	321.907	248.115	24.4
800 × 623	120	646	386.24	289.133	22.05
850 × 662	127	756	423.903	352.17	24.2
900 × 701	169	971	457.716	361.06	38.89
1000 × 779	166	1452	664.056	431.305	38.17
1500 × 1169	169	5394	1625.04	1102.85	34.33
2000 × 1559	172	10407	3528.13	2361.7	39.84
2500 × 1949	174	17504	6714.95	4563.72	39.52
3000 × 2339	181	28505	11456.6	7525.11	44.73

4.2 Evaluation

The complexity of the algorithms in time and memory can be found in Table 1. We have performed a quick performance comparison of skeletonization algorithms. Table 2 presents the comparison between 3 available and considered as fast algorithms of skeletonization: Vincent [27], Guo and Hall [15] and Zhang and Suen [30]. The algorithms have been performed on the lizzard image that shows several levels of resolution (see Fig. 8). The method based on straight skeletons is faster than the other methods especially for big shape sizes. Figure 8 shows the medial axis extracted from straight skeletons. Figure 8b and c highlight the fact that the medial axis could vary according to the number of points describing the polygon: more points will give more branches of skeletons. As we do not have any groundtruth to make a fair quantitative comparison of the skeleton accuracy, we just show some qualitative results. The white points describe the skeleton result from the method [30] in Fig. 8b and the method [15] in Fig. 8c. Figure 9 shows the medial axis of a polygon containing holes (image of letter B). We assume that a good skeleton should lay on the maximum value of the Chamfrein's distance map [8]. Our medial axis lays close to this maximum distance value.

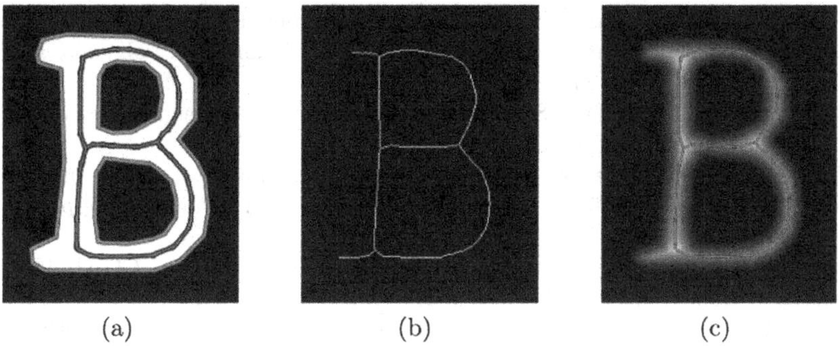

Fig. 9. Medial axis on a shape containing holes. (a) polygons of B shape and its medial axis derived from the straight skeleton algorithm. (b) medial axis transform (c) overlay of the proposed medial axis on the Chamfrein's distance map.

5 Conclusion and Perspective

Straight skeletons are thus an interesting and fast tool in the process of tissue analysis, but could also be applied as part of a characterization framework in other domains such as Geographical Data or Document Analysis. This approach can be used for map generalization: for example, river or road to line segment simplification. We want to explore other propagation modes for the weighted straight skeleton method. Other polygonal extraction methods to set the initial one are under consideration to show the robustness of the achieved representation, for instance considering blurred segments calculated from different widths of discrete curves [11].

References

1. Aggarwal, A., Guibas, L.J., Saxe, J., Shor, P.W.: A linear-time algorithm for computing the Voronoï diagram of a convex polygon. Discrete Comput. Geom. **4**(6), 591–604 (1989)
2. Aichholzer, O., Alberts, D., Aurenhammer, F., Gartner, B.: Straight skeletons of simple polygons. In: Proceedings of 4th International Symposium of LIESMARS, pp. 114–124 (1995)
3. Aichholzer, O., Aurenhammer, F.: Straight skeleton for general polygonal figures in the plane. In: Samoilenko, A.M. (ed.) Voronoï's Impact on Modern Science, vol. 2, pp. 7–21. Institute of Mathematics of the National Academy of Sciences of Ukraine (1998)
4. Attali, D., Lachaud, J.O.: Delaunay conforming iso-surface, skeleton extraction and noise removal. Comput. Geom. **19**(2), 175–189 (2001)
5. Attali, D., Montanvert, A.: Modeling noise for a better simplification of skeletons. In: Proceedings of 3rd IEEE ICIP, vol. 3, pp. 13–16, September 1996

6. Biedl, T., Held, M., Huber, S., Kaaser, D., Palfrader, P.: A simple algorithm for computing positively weighted straight skeletons of monotone polygons. Inf. Process. Lett. **115**(2), 243–247 (2015)
7. Biedl, T., Held, M., Huber, S., Kaaser, D., Palfrader, P.: Weighted straight skeletons in the plane. Comput. Geom. **48**(2), 120–133 (2015)
8. Blum, H.: A transformation for extracting new descriptors of shape. In: Wathen-Dunn, W. (ed.) Models for the Perception of Speech and Visual Form, pp. 362–380. MIT Press, Cambridge (1967)
9. Cacciola, F.: 2D straight skeleton and polygon offsetting. In: CGAL User and Reference Manual, 4.10.1 edn. CGAL Editorial Board (2017)
10. Davies, E., Plummer, A.: Thinning algorithms: a critique and a new methodology. Pattern Recognit. **14**(1), 53–63 (1981)
11. Debled-Rennesson, I., Feschet, F., Rouyer-Degli, J.: Optimal blurred segments decomposition of noisy shapes in linear time. Comput. Graph. **30**(1), 30–36 (2006)
12. Eder, G., Held, M.: Computing positively weighted straight skeletons of simple polygons based on a bisector arrangement. Inf. Process. Lett. **132**, 28–32 (2018)
13. Eppstein, D., Erickson, J.: Raising roofs, crashing cycles, and playing pool: applications of a data structure for finding pairwise interactions. Discrete Comput. Geom. **22**(4), 569–582 (1999)
14. Ge, Y., Fitzpatrick, J.M.: On the generation of skeletons from discrete euclidean distance maps. IEEE Trans. Pattern Anal. Mach. Intell. **18**, 1055–1066 (1996)
15. Guo, Z., Hall, R.W.: Parallel thinning with two-subiteration algorithms. Commun. ACM **32**(3), 359–373 (1989)
16. Hesselink, W.H., Roerdink, J.B.T.M.: Euclidean skeletons of digital image and volume data in linear time by the integer medial axis transform. IEEE Trans. Pattern Anal. Mach. Intell. **30**(12), 2204–2217 (2008)
17. Huang, C.H., Veillard, A., Roux, L., Lomenie, N., Racoceanu, D.: Time efficient sparse analysis of histopathological whole slide images. Comput. Med. Imaging Graph. **35**(7), 579–591 (2011)
18. Huber, S., Held, M.: A fast straight-skeleton algorithm based on generalized motorcycle graphs. Int. J. Comput. Geom. Appl. **22**(5), 471–498 (2012)
19. Kelly, T.: Unwritten procedural modeling with the straight skeleton. Ph.D. thesis, University of Glasgow (2013)
20. Kimmel, R., Shaked, D., Kiryati, N., Bruckstein, A.M.: Skeletonization via distance maps and level sets. Comput. Vis. Image Underst. **62**(3), 382–391 (1995)
21. Ogniewicz, R., Kübler, O.: Hierarchic Voronoï skeletons. Pattern Recognit. **28**(3), 343–359 (1995)
22. Pudney, C.: Distance-ordered homotopic thinning. Comput. Vis. Image Underst. **72**(3), 404–413 (1998)
23. Remy, E., Thiel, E.: Exact medial axis with euclidean distance. Image Vis. Comput. **23**(2), 167–175 (2005)
24. Serra, J.: Image Analysis and Mathematical Morphology. Academic Press Inc., Orlando (1983)
25. Siddiqi, K., Bouix, S., Tannenbaum, A., Zucker, S.W.: Hamilton-jacobi skeletons. Int. J. Comput. Vis. **48**(3), 215–231 (2002)
26. Talbot, H.: Euclidean skeletons and conditional bisectors. In: 1992 Visual Communications and Image Processing, pp. 862–876 (1992)
27. Vincent, L.M.: Efficient computation of various types of skeletons. In: Proceedings of SPIE, vol. 1445, pp. 1445–1445-15 (1991)

28. Vizilter, Y.V., Pyt'ev, Y.P., Chulichkov, A.I., Mestetskiy, L.M.: Morphological image analysis for computer vision applications. In: Favorskaya, M.N., Jain, L.C. (eds.) Computer Vision in Control Systems-1. ISRL, vol. 73, pp. 9–58. Springer, Cham (2015). https://doi.org/10.1007/978-3-319-10653-3_2
29. Wein, R., Baram, A., Flato, E., Fogel, E., Hemmer, M., Morr, S.: 2D minkowski sums. In: CGAL User and Reference Manual, 4.10.1 edn. CGAL Editorial Board (2017)
30. Zhang, T.Y., Suen, C.Y.: A fast parallel algorithm for thinning digital patterns. Commun. ACM **27**(3), 236–239 (1984)

Morphological Networks for Image De-raining

Ranjan Mondal[1(✉)], Pulak Purkait[2], Sanchayan Santra[1], and Bhabatosh Chanda[1]

[1] Indian Statistical Institute, Kolkata, India
{ranjan15_r,sanchayan_r,chanda}@isical.ac.in
[2] The University of Adelaide, Adelaide, Australia
pulak.isi@gmail.com

Abstract. Mathematical morphological methods have successfully been applied to filter out (emphasize or remove) different structures of an image. However, it is argued that these methods could be suitable for the task only if the type and order of the filter(s) as well as the shape and size of operator kernel are designed properly. Thus the existing filtering operators are problem (instance) specific and are designed by the domain experts. In this work we propose a morphological network that emulates classical morphological filtering consisting of a series of erosion and dilation operators with trainable structuring elements. We evaluate the proposed network for image de-raining task where the SSIM and mean absolute error (MAE) loss corresponding to predicted and ground-truth clean image is back-propagated through the network to train the structuring elements. We observe that a single morphological network can de-rain an image with any arbitrary shaped rain-droplets and achieves similar performance with the contemporary CNNs for this task with a fraction of trainable parameters (network size). The proposed morphological network (MorphoN) is not designed specifically for de-raining and can readily be applied to similar filtering/noise cleaning tasks. The source code can be found here https://github.com/ranjanZ/2D-Morphological-Network.

Keywords: Mathematical morphology · Optimization · Morphological network · Image filtering

1 Introduction

Morphological Image processing with hand-crafted filtering operators has been applied successfully to solve many problems like image segmentation ([8,16]), object shape detection and noise filtering. Due to rich mathematical foundation, image analysis by morphological operators is found to be very effective and popular. The basic building block operations are *dilation* and *erosion*, which are defined in terms of a structuring element (SE). Many problem can be solved

M. Couprie et al. (Eds.): DGCI 2019, LNCS 11414, pp. 262–275, 2019.
https://doi.org/10.1007/978-3-030-14085-4_21

(a) Input (b) CNN (#para 6M) (c) MorphoN(#para 2K)

(d) Input (e) CNN(#para 6M) (f) MorphoN(#para 2K)

Fig. 1. Some examples of the results by the proposed morphological network. Note that the same network was used to clean different types and amounts of rain—(a) input rainy image with vertical structures and (d) input rainy image with sliding structures. The proposed network emulates standard 2D morphological operations where the structuring elements are trained using the back propagation. The size of our network is drastically smaller than the conventional CNNs and capable of producing very high quality results ((c) and (f) vs (b) and (e)). More results can be found in the experiment section.

by choosing shape and size of the structuring element intelligently [14]. Finding customized/tailored size and shape of the structuring elements and also the order in which erosion and dilation operations are to be applied still remain a huge challenge. Furthermore, the design could be data dependent, *i.e.,* the expert might have to design the operator depending on the problem instances. For example, for de-raining task one needs to design different filtering operator for different rain pattern.

In this work we utilize a network architecture that consists of trainable morphological operators to de-rain the rainy images irrespective of the rain pattern.

1.1 Motivation and Contributions

The recent developments of Convolution Neural Networks (CNN) have unveiled a huge success in image processing and computer vision tasks. A number of examples could be found in [5]. To mention a few, CNNs are very popular in solving problems like object detection [10], image dehazing [6] and image segmentation [2]. A CNN consists of an input layer, an output layer, and multiple hidden layers. An image input passes through the stack of hidden layers of the trained network and produces the desired output. The hidden layers of a CNN typically consist of convolutional layers and optionally—pooling layers, fully connected layers and normalization layers. The convolutional layers apply a convolution operation with a trained mask on the input and pass the result to the next layer.

Inspired by the success of convolutional networks and similarity between the convolution and morphological dilation and erosion (both are neighbourhood operator with respect to some kernel), we propose morphological layers by replacing convolution operators by max or min operator that yields morphological networks. The network consists of a sequence of dilation-erosion operators for which the structuring elements are trained using back-propagation for a particular task, very similar to the way we train the weights of a convolutional layers of CNNs. An example of results after learning the structuring elements is displayed in Fig. 1. Note that the proposed network can also be considered as a neural network (containing stack of morphological layers) where the neurons are morphological neurons performing dilation or erosion operation.

Therefore, the contribution of the paper can be summarized as follows:

- We propose morphological networks that extends the basic concepts of morphological perceptron [3] and emulates 2D dilation and 2D erosion for gray scale image processing.
- Here we have utilized a pair of series of dilation and erosion operation along two different paths and intermediate outputs are combined to predict the final output. Ideally a number of paths could be incorporated where each path corresponds to a single compound morphological operator (*i.e.* concatenation of dilation and erosion operators with different structuring elements).
- The proposed network is evaluated for the de-raining task. We observe that a tiny morphological network (ours) is capable of producing a high quality result as that of large and complex CNNs.

The rest of the paper is organized as follows. In Sect. 1.2 we will discuss the works related to the proposed morphological network. In Sect. 2 we describe the building blocks of the learning structuring elements and define basic operations of mathematical morphology, *i.e.*, dilation-erosion in 2D. We have evaluated our algorithm on rain dataset [4] and presented the results in Sect. 3. Lastly, we will conclude the paper in Sect. 4.

1.2 Related Work

In our work we have used basic concepts of morphological perceptron. Morphological perceptron was introduced by [3] and the authors used morphological network to solve template identification problem. Later it was generalized by [11] to tackle the problem of binary classification by restricting the network to single layer architecture. The decision boundaries were considered as parallel to the axes. Later in [13] the network was extended to two layers and it was shown that the decision boundary need not be axis parallel for the classification.

To train the structuring elements of the network, researchers [9] have tried to use gradient descent by combining classical perceptron with morphological perceptron. In [1] they used linear function to apply it to regression problems. With dendritic structure of morphological neurons, Zamora *et al.* [17] replaced the `argmax` operator by `softmax` function to overcome the problem of gradient computation and used gradient descent to train the network. Similar work has been done by Ranjan *et al.* [7] where they have learned 1D structuring elements in a dense network by back-propagation. It may be noted that though the functions with max or min are not, in general, differentiable, those are piecewise differentiable. Using this property, in our work we could use gradient descent during the back-propagation to learn the 2D structuring elements for dilation and erosion operations in the network. In the next section we have defined the basic building blocks of the network, *i.e.* 2D dilation and 2D erosion by max and min operations.

2 Method

Here we first describe the 2D dilation and erosion operations which are the basic building blocks of the network, and then discuss the network architecture in detail followed by the choice of the loss.

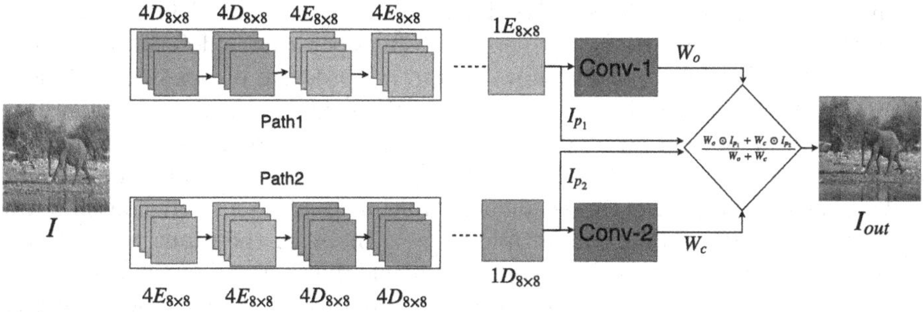

Fig. 2. In our proposed network (MorphoN) we consider two parallel branches of the network with an alternate sequence of dilation and erosion operators. The output of the branches are then combined with the weights predicted from the network to yield final output. Details can be found in the text.

2.1 Morphological Layers

The classical morphological algorithms are described using dilation and erosion operators and the morphological filtering are defined by a combination of these operators with often times utilizing different structuring elements. These operations are successfully used in many different applications in image processing. In this work, we design a morphological network to emulate those two-dimensional gray-scale morphological operations with different structuring elements.

Let I is the input gray scale image of size $m \times n$. Dilation ($\oplus$) and erosion ($\ominus$) operation on image I is defined as the following

$$(I \oplus W_d)(x,y) = \max_{i \in S_1 j \in S_2} (I(x-i, y-j) + W_d(i,j)), \tag{1}$$

$$(I \ominus W_e)(x,y) = \min_{i \in S_1 j \in S_2} (I(x+i, y+j) - W_e(i,j)). \tag{2}$$

where $W_d \in R^{a \times b}$, $W_e \in R^{a \times b}$, $S_1 = \{1, 2, .., a\}$ and $S_2 = \{1, 2, .., b\}$. W_d and W_e are dilation and erosion kernels or structuring elements. After applying dilation and erosion on an image, we call the resultant as dilation map and erosion map respectively. Note that the operations defined in (1) and (2) function very similarly as the convolutional layers. The operators perform in a windowed fashion of window size $a \times b$. Padding is incorporated to have output as same size as input. The structuring elements (W_d, W_e) are initialized randomly and, then, optimized by back-propagation.

In the following section we describe the network architecture using the Dilation and erosion operation on the image.

2.2 Morphological Network

In morphological image processing, the *Opening* and *Closing* operations are commonly used as filters to remove noise. The opening operation is defined by applying dilation on an eroded image; whereas closing operation is defined by erosion on a dilated image. For noise removal a specific ordering of opening and closing needs to be applied on the noisy image. We have considered a sequence of alternate morphological layers with dilation and erosion to implement such filters. Each layer leads to a different dilation or erosion map because there could be different trained structuring elements. Multiple dilation and erosion map are useful because there could be multiple types of noise in the input image. So a single morphological network employing multiple dilation and erosion (or effectively opening and closing) would be able to filter out various types of noise. Furthermore, as shown in Fig. 2. Here we have considered exactly two different paths of stacks of morphological layers, first starting with dilation (or closing) followed by erosion (or opening) and the second path is totally complement of the first one *i.e.*, starting with erosion (or opening).

Since it is hard to know which particular path is more effective for noise removal in a specific situation, we further yield a weight map for each path to combine them to a single output. Let W_o, W_c are the weight maps for paths starting with opening and starting with closing respectively which are of the same size as input image. We have taken `sigmoid` as activation function in the last layer so the value of each pixel in W_o and W_c are greater than zero and less than 1.0. Finally, we get the output I_{out} by the following equation.

Table 1. The architectures of paths shown in Fig. 2. $4D_{8\times8}$ denotes a layer with 4 dilations with different trainable SEs of size 8×8. $4E_{8\times8}$ is defined similarly. $2@8 \times 8$–tanh denotes 2 feature map which has been produced by convolving with kernel 8×8 followed by tanh activation

Description of the MorphoN	
Path1-Conv	$4D_{8\times8}$–$4D_{8\times8}$–$4E_{8\times8}$–$4E_{8\times8}$–$4D_{8\times8}$–$4D_{8\times8}$–$4E_{8\times8}$–$4E_{8\times8}$–$4E_{8\times8}$–$2@8 \times 8$–tanh–$3@8 \times 8$–tanh–$1@8 \times 8$–sigmoid
Path2-Conv	$4E_{8\times8}$–$4E_{8\times8}$–$4D_{8\times8}$–$4D_{8\times8}$–$4E_{8\times8}$–$4E_{8\times8}$–$4D_{8\times8}$–$4D_{8\times8}$–$4D_{8\times8}$–$2@8 \times 8$–tanh–$3@8 \times 8$–tanh–$1@8 \times 8$–sigmoid
Description of the smaller MorphoN	
Path1-Conv (small)	$1D_{8\times8}$–$1D_{8\times8}$–$1D_{8\times8}$–$1E_{8\times8}$–$1E_{8\times8}$–$1D_{8\times8}$–$1D_{8\times8}$–$1E_{8\times8}$–$1E_{8\times8}$–$1E_{8\times8}$–$2@8\times8$–tanh–$3@8\times8$–tanh–$1@8\times8$–sigmoid
Path2-Conv (small)	$1E_{8\times8}$–$1E_{8\times8}$–$1E_{8\times8}$–$1D_{8\times8}$–$1D_{8\times8}$–$1E_{8\times8}$–$1E_{8\times8}$–$1D_{8\times8}$–$1D_{8\times8}$–$1D_{8\times8}$–$2@8\times8$–tanh–$3@8\times8$–tanh–$1@8\times8$–sigmoid

$$I_{out} = \frac{W_o \odot I_{p_1} + W_c \odot I_{p_2}}{W_o + W_c} \tag{3}$$

where I_{p_1} and I_{p_2} are the outputs from path1 and path2 respectively, and $\odot$ is the pixel-wise multiplication.

2.3 Learning of Structuring Elements

As defined in Sect. 2.1, dilation and erosion consist of max and min operations respectively. The expression containing max and min are piece-wise differentiable. So, we could use back propagation algorithm to learn the structuring elements of the network as well as the weights combination. We hypothesize that SSIM [15] is a good measure which quantifies image quality degradation between two images. SSIM measure between two images x and y of same size is defined by

$$\text{SSIM}(x, y) = \frac{(2\mu_x\mu_y + c_1)(2\sigma_{xy} + c_2)}{(\mu_x^2 n + \mu_y^2 + c_1)(\sigma^2 + \sigma^2 + c_2)} \tag{4}$$

where μ_x and μ_y are the mean of the image x and y respectively and σ_x^2 and σ_y^2 are the variance of the image x and y respectively. σ_{xy} is covariance between x and y. c_1 and c_2 is constant taken as 0.0001 and 0.0009 respectively. To train the network we have used structural dissimilarity (DSSIM) as the objective function over a small patch of the output, where DSSIM is related to SSIM by the following equation,

$$\text{DSSIM}(I_{out}, I_{gt}) = \frac{1}{M} \sum_i \frac{1 - \text{SSIM}(P_{out}^i, P_{gt}^i)}{2} \tag{5}$$

where P_{out}^i and P_{gt}^i are i^{th} spatially same patch of the network predicted output image I_{out} and ground truth image I_{gt} respectively. M is the total number of such patches. In our experiment we have taken the patch size as 100 × 100. In practice, we combine DSSIM and MAE loss by the following equation.

$$Loss_{total} = \text{DSSIM}(I_{out}, I_{gt}) + \lambda \text{MAE}(I_{out}, I_{gt}) \tag{6}$$

where λ is the weighting constant between two losses and $\text{MAE}(P_{out}, P_{gt})$ is defined as follows

$$\text{MAE}(I_{out}, I_{gt}) = \frac{1}{N} \|I_{out} - I_{gt}\|_1 \tag{7}$$

where N is the number of pixels. For all the experiments we have taken $\lambda = 1$. In the next section we evaluate the proposed morphological networks.

3 Experiments

We have evaluated the proposed morphological network for image de-raining task on publicly available Rain dataset [4]. The dataset contains 1,000 clean images, and for each clean image it has 14 different rainy images with different streak orientations and sizes. Out of 1,000, 80% of the data has been considered for training, 10% of the data considered for validation and remaining 10% of the data *i.e*, 100 images have been kept for testing. Since all the images are of different sizes, we have resized them to 512×512 by bilinear interpolation (implementation bottleneck). Since the proposed morphological layers are designed for gray scale image, we have converted all the images to gray scale. An extension to color images is also possible by adding more channels in the morphological layers, but it is not exploited in the current work.

In our experiment, to evaluate each path (corresponding to a single compound morphological operation), we have also trained path1 and path2 separately. Quantitative and qualitative results on path1 and path2 is also reported. The proposed morphological layers are implemented on Keras python scripts with back-end TensorFlow library. The evaluations have been carried out in a machine, which has a Intel Xeon 16 core processor and 128 GB RAM. We have used Nvidia Titan Xp GPU for the parallel processing. In the next section we have shown qualitative and quantitative evaluation on test data.

3.1 Parameter Settings

For the initialization of the network, we have used the standard glorot uniform initializer. It draws samples from a uniform distribution within $[-l, l]$ where l is $\sqrt{(6/(f_{in} + f_{out}))}$ and f_{in} is the size of SE and fan_{out} is the number of morphological operators acting parallel. Proposed network is concatenation of dilation and erosion layers that involves max and min operators and are piece-wise differentiable with respect to SE. Therefore, standard backpropagation method can update the SEs. In this work, the Adam optimizer is used with default parameter settings ($lr = 0.001$, $\beta_1 = 0.9$, $\beta_2 = 0.999$, $\epsilon = 1e-8$).[1]

3.2 Qualitative Evaluation

We have compared our result with the results of image de-raining using a standard convolutional neural network. To be precise, U-net architecture [12] has been considered as baseline. The results of each path is also reported as baselines. In Fig. 5 we have shown the comparison of different methods with ours. We observe that the proposed morphological network is able to clean small raindrops irrespective of different inclinations or rain patterns. It is interesting to see that a separately trained network along path2 produces better results compared to the network along path1. We believe that for de-raining task the noisy pixels are usually bright pixels compared to the neighbouring ones that leads to such behaviour.

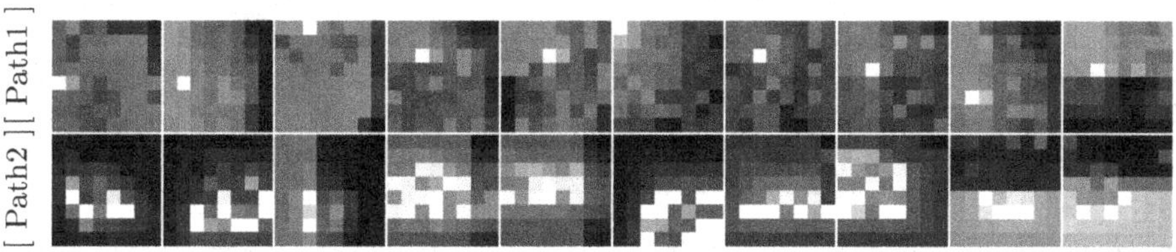

Fig. 3. We display the learned structuring elements at different layers of the small network along different paths. The structuring elements are normalized for visualization between 0 to 255. The most bright pixel is displayed by the maximum value of the structuring elements and the darker pixel by minimum value.

We have also carried out our experiments taking a single dilation/erosion map instead of taking 4 dilation/erosion and term the network as MorphoN (small). The architecture is shown in Table 1. Here We have combined path1 and path2 with Conv-1 and Conv-2 respectively in the same way as we did for the original model (with 4 copies of dilation and erosion). Using small network we are getting very similar results as the original model. In Fig. 3, we have shown the learned structuring elements in each path of dilation and erosion for the small network. We see that all the learned structuring elements are different to each other. In Fig. 4, we have also displayed layer-wise output from the MorphoN (Small) after applying erosion and dilation. Outputs of each paths are then combined by a predicted weighted combination to produce the final clear output. In Fig. 5 we have compared our results with CNN.

[1] Source code : https://github.com/ranjanZ/2D-Morphological-Network.

The proposed MorphoN and MorphoN (small) produces high quality results which are very similar to standard CNNs with a fraction of parameters (0.2% – 0.04%). There are some failure cases as shown in Fig. 6. MorphoN (small) produces blur images while removing thick rain Fig. 6*a*, *b*, *d*. As shown in Fig. 6c, the proposed MorphoN (small) is able to clear the rain but it also clears out some structures in the house including basement. However, image quality can be further improved with stacking more morphological layers and incorporating multiple paths. The architecture is not optimized for the task and the results can be further improved upon fine-tuning the network. A separate experiment is conducted to check if the proposed network is able to clean the noise of a partially degraded image. In Fig. 7, we display such synthetic examples where half of the images were degraded by rainy structure and rest were kept unaltered. A single trained MorphoN is applied on such images–clean and the noisy part simultaneously. We observe that the clean portions are unaltered while the rainy portions are cleaned.

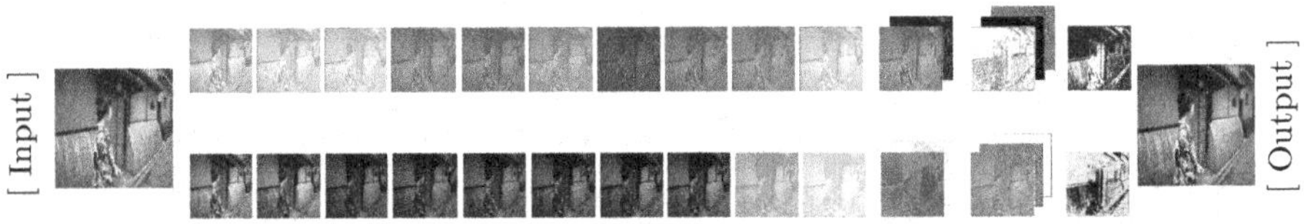

Fig. 4. Visualization of the sequential layer-wise morphological operations. The input rainy image passes through each layer of the network (essentially a morphological dilation/erosion operation with the trained structuring elements). The produced filtered images are displayed at each step. The outputs of each paths are then combined by a predicted weighted combination to produce the final output.

3.3 Quantitative Evaluation

For quantitative evaluation, we have evaluated our algorithm in terms of SSIM [15] and PSNR values. The estimated de-rain image is compared against the ground truth clean image. The methods are applied to all the images of the test dataset and the average value is reported. In Table 2, we have reported the results of different methods on test data of Rainy dataset. CNN (U-Net) archives SSIM and PNSR on an average about 0.92 and 29.12 respectively, whereas our network gives similar results, *i.e.*, 0.92 and 28.03. Our MorphoN (small) network also produces similar results to MorphoN. We have also reported the number of parameters in the Table 2. Notice that MorphoN (small) with 0.04% numbers of parameters of CNN produces similar results with CNN.

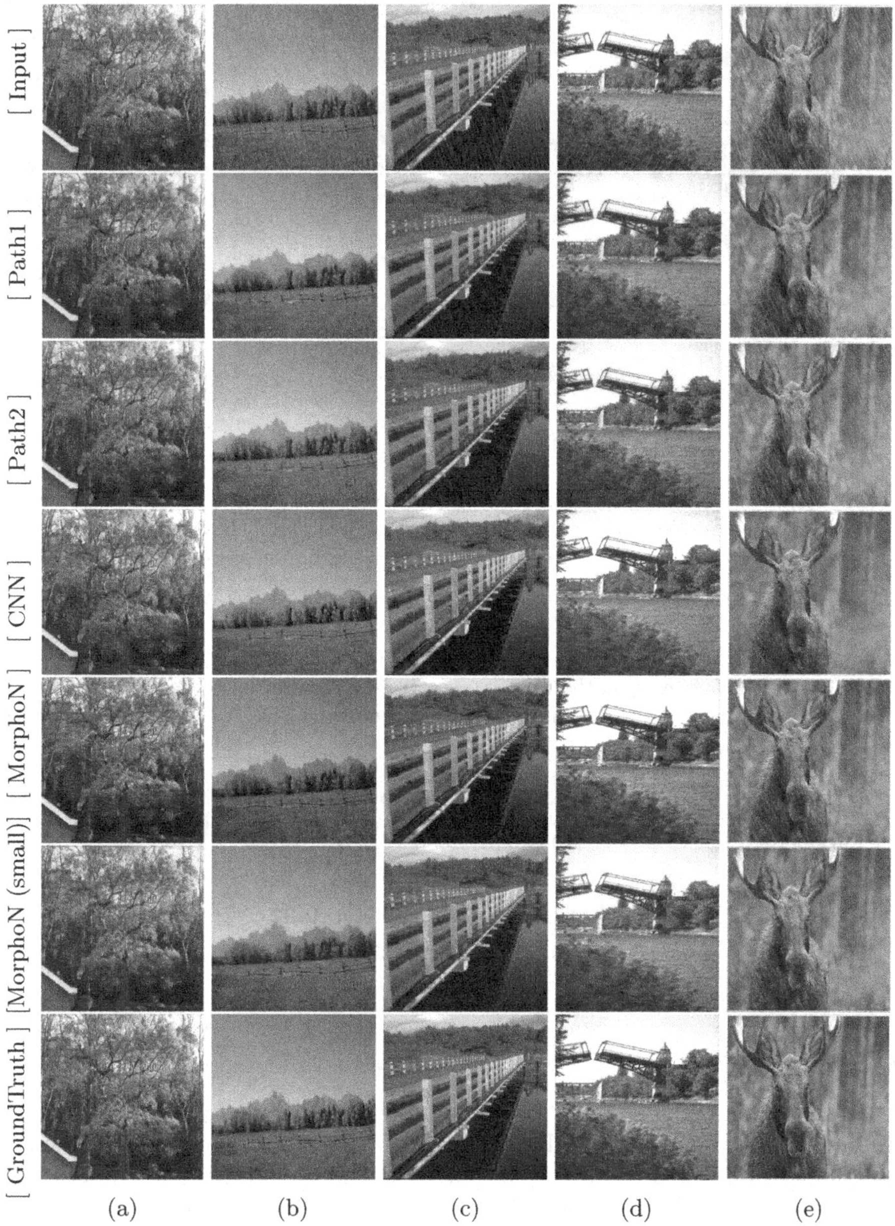

Fig. 5. Qualitative results on Rain image dataset [4]. Note that the proposed morphological network produces high quality results, very similar to standard CNNs with a fraction of parameters. Note that image quality can be further improved with stacking more morphological layers and incorporating multiple paths. The architecture is not optimized for the task and the results can be further improved upon fine-tuning the network.

Fig. 6. Failure cases of the proposed MorphoN (small). Note that most of the failure cases occur corresponding to the large rain structures and we believe with larger structuring elements ($> 8 \times 8$) and with more number of channels, better results can be produced.

Table 2. Results achieved on the rain dataset [4] by different networks. Note that with a tiny morphological network compared to a standard CNN (U-Net) [12], a similar accuracy can be achieved.

Metric	Input	CNN	Path1	Path2	MorphoN	MorphoN (small)
#Parameters	-	6,110,773	7,680	7,680	16,780	2,700
#Params w.r.t. CNN	-	100.0%	0.12%	0.12%	0.27%	**0.04%**
SSIM	0.85	**0.92**	0.87	0.90	**0.92**	0.91
PSNR	24.3	**29.12**	26.27	27.20	28.03	27.45

3.4 Real Data

A dataset with real images is collected by capturing photographs of rain at different outdoor scenes and during different times of the day. Some sample images and evaluation results are displayed in Fig. 8. Note that as the ground-truth clean images are unavailable, we only consider qualitative comparison for the evaluation. The MorphoN, trained with synthetic data, consistently produces similar or better results than the baselines.

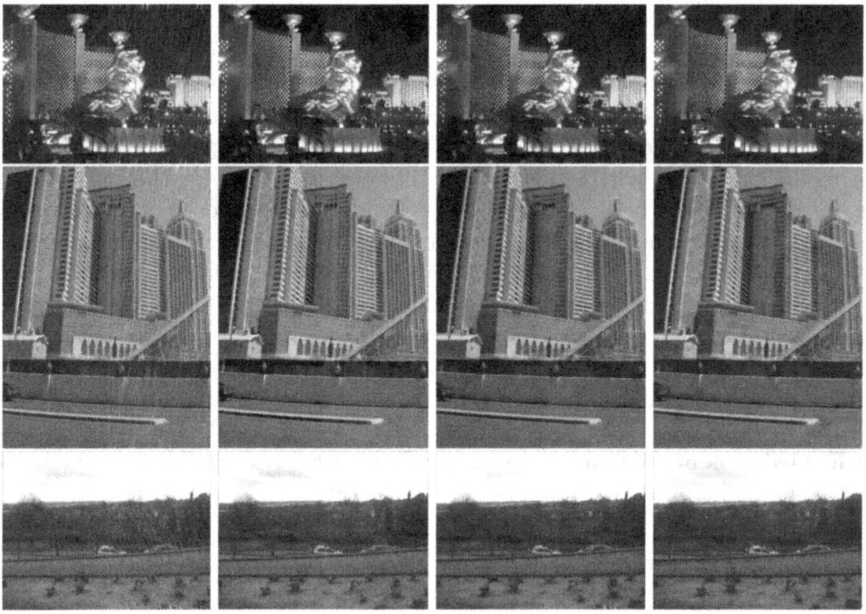

(a) Input (b) CNN (c) MorphoN (d) MorphoN (small)

Fig. 7. Results on partially degraded images. The partially degraded images are generated by creating rains synthetically on half of the image. Note that our method does not degrade the clean portion while removing the rain structures.

(a) Input (b) CNN (c) MorphoN (d) MorphoN (small)

Fig. 8. Real world examples: The proposed method and the baselines are evaluated on a number of real world examples. Our much smaller network produces results on par with the baselines.

4 Conclusion

In this work, a morphological network is proposed that emulates classical morphological filtering *i.e.*, 2D dilation and 2D erosion for gray scale image processing. The proposed network architecture consists of a pair of sequences of morphological layers with exactly two different paths where the outputs are combined to predict the final result. We evaluated the proposed network for the de-raining task and obtained very similar results with heavy-weighted convolutional neural networks. The proposed network is not tailored for de-raining and could be applied to any other filtering task. Further, this is one of the forerunner work and it opens a many directions of future research, for example, best architecture search for morphological networks. The source code is shared to encourage reproducibility and facilitate.

Acknowledgements. Initial part of the experiment has been carried out on Intel AI DevCloud. Authors want to acknowledge Intel for that.

References

1. de A. Araujo, R.: A morphological perceptron with gradient-based learning for Brazilian stock market forecasting. Neural Netw. **28**, 61–81 (2012)
2. Chen, L.C., Papandreou, G., Kokkinos, I., Murphy, K., Yuille, A.L.: Deeplab: Semantic image segmentation with deep convolutional nets, atrous convolution, and fully connected CRFs. IEEE TPAMI **40**(4), 834–848 (2018)
3. Davidson, J.L., Hummer, F.: Morphology neural networks: An introduction with applications. Circ. Syst. Sig. Process. **12**(2), 177–210 (1993)
4. Fu, X., Huang, J., Ding, X., Liao, Y., Paisley, J.: Clearing the skies: a deep network architecture for single-image rain removal. IEEE TIP **26**(6), 2944–2956 (2017)
5. Goodfellow, I., Bengio, Y., Courville, A., Bengio, Y.: Deep Learning, vol. 1. MIT press, Cambridge (2016)
6. Mondal, R., Santra, S., Chanda, B.: Image dehazing by joint estimation of transmittance and airlight using bi-directional consistency loss minimized FCN. In: CVPR Workshops, pp. 920–928 (2018)
7. Mondal, R., Santra, S., Chanda, B.: Dense morphological network: an universal function approximator. arxiv e-prints arXiv:1901.00109, January 2019
8. Perret, B., Cousty, J., Ura, J.C.R., Guimarães, S.J.F.: Evaluation of morphological hierarchies for supervised segmentation. In: Benediktsson, J.A., Chanussot, J., Najman, L., Talbot, H. (eds.) ISMM 2015. LNCS, vol. 9082, pp. 39–50. Springer, Cham (2015). https://doi.org/10.1007/978-3-319-18720-4_4
9. Pessoa, L.F.C., Maragos, P.: Neural networks with hybrid morphological/rank/linear nodes: a unifying framework with applications to handwritten character recognition. Pattern Recogn. **33**, 945–960 (2000)
10. Ren, S., He, K., Girshick, R., Sun, J.: Faster R-CNN: towards real-time object detection with region proposal networks. In: Advances in Neural Information Processing Systems, pp. 91–99 (2015)
11. Ritter, G.X., Sussner, P.: An introduction to morphological neural networks. In: ICPR, vol. 4, pp. 709–717, August 1996

12. Ronneberger, O., Fischer, P., Brox, T.: U-Net: convolutional networks for biomedical image segmentation. In: Navab, N., Hornegger, J., Wells, W.M., Frangi, A.F. (eds.) MICCAI 2015. LNCS, vol. 9351, pp. 234–241. Springer, Cham (2015). https://doi.org/10.1007/978-3-319-24574-4_28
13. Sussner, P.: Morphological perceptron learning. In: ICRA, pp. 477–482 September 1998
14. Vincent, L.: Morphological grayscale reconstruction in image analysis: applications and efficient algorithms. IEEE TIP **2**(2), 176–201 (1993)
15. Wang, Z., Bovik, A.C., Sheikh, H.R., Simoncelli, E.P.: Image quality assessment: from error visibility to structural similarity. IEEE TIP **13**(4), 600–612 (2004)
16. Wdowiak, M., Markiewicz, T., Osowski, S., Swiderska, Z., Patera, J., Kozlowski, W.: Hourglass shapes in rank grey-level hit-or-miss transform for membrane segmentation in HER2/neu images. In: Benediktsson, J.A., Chanussot, J., Najman, L., Talbot, H. (eds.) ISMM 2015. LNCS, vol. 9082, pp. 3–14. Springer, Cham (2015). https://doi.org/10.1007/978-3-319-18720-4_1
17. Zamora, E., Sossa, H.: Dendrite morphological neurons trained by stochastic gradient descent. Neurocomputing **260**, 420–431 (2017)

Minimal Component-Hypertrees

Alexandre Morimitsu[1(✉)], Wonder Alexandre Luz Alves[2], Dennis Jose Silva[1], Charles Ferreira Gobber[2], and Ronaldo Fumio Hashimoto[1(✉)]

[1] Department of Computer Science, Institute of Mathematics and Statistics, Universidade de São Paulo, São Paulo, Brazil
{alexandre.morimitsu,ronaldo}@usp.br
[2] Informatics and Knowledge Management Graduate Program, Universidade Nove de Julho, São Paulo, Brazil

Abstract. Component trees are interesting structures of nested connected components, efficiently represented by max-trees, used to implement fast algorithms in Image Processing. In these structures, connected components are constructed using a single neighborhood. In recent years, an extension of component trees, called component-hypertrees, was introduced. It consists of a sequence of component trees, generated from a sequence of increasing neighborhoods, in which their connected components are also hierarchically organized. Although this structure could be useful in applications dealing with clusters of objects, not much attention has been given to component-hypertrees. A naive implementation can be costly both in terms of time and memory. So, in this paper, we present algorithms and data structures to efficiently compute and store these structures without redundancy obtaining a minimal representation of component-hypertrees. Experimental results using our efficient algorithm show that the number of nodes is reduced by approximately 70% in comparison to a naive implementation.

Keywords: Mathematical Morphology · Component-hypertree · Component tree · Connected component · Connected operators

1 Introduction

In recent years, component trees have received increasing attention in Image Processing, particularly in Mathematical Morphology, since they can be used for many tasks such as image filtering [9] and shape recognition [11]. A component tree represents a gray-level image by storing connected components of its level sets for a given connectivity and relating them by the inclusion relation. Although multiple types of connectivity can be used, like 4 and 8-connectivity, mask-based [6] and hyper-connectivity [7], only a single tree can be extracted, and thereby limiting the number of problems that can be solved by this approach.

Recently, Passat and Naegel [8] proposed a structure which represents a gray-level image by a set of component trees built with increasing neighborhoods called component-hypertree. Due to increaseness of the neighborhoods,

M. Couprie et al. (Eds.): DGCI 2019, LNCS 11414, pp. 276–287, 2019.
https://doi.org/10.1007/978-3-030-14085-4_22

component-hypertrees relate connected components of different neighborhoods according to their inclusion. For example, in text extraction problem, a word and its letters are nodes of a component-hypertree and they are related by the inclusion relation.

In the original paper [8], Passat and Naegel provided a theoretical background for mask-based connectivities. Although the structure is explained, no explicit optimized algorithm for component-hypertrees construction is provided. Also, some theoretical explanations are given about how to perform simplification in the graph, but many repeated nodes still remain. In [3], algorithms are provided for a specific type of dilation-based connectivities. That work showed an efficient way of computing hypertrees for multiple neighborhoods but it did not focus on the problem of storing hypertrees efficiently.

Using an approach that stores the CCs of all component trees of all neighborhoods, the component-hypertree would have a prohibitive memory usage. However, most of the CCs in different trees are repeated, meaning a considerable amount of memory can be saved by smartly allocating only relevant nodes and arcs of the hypertree. So in this paper, we present the minimal component-hypertree: a data structure which efficiently stores the component-hypertree by keeping only the minimum number of nodes and arcs needed to efficiently recover all CCs. In this structure, repeated CCs obtained from different level sets and neighborhoods are discarded, meaning that each CC is stored only once.

For that, we adopt the following strategy: before explaining our data structure and algorithms, we first recall the theory behind component trees and component-hypertrees in Sect. 2. We present our main contributions in Sect. 3 where we show the theory and the algorithms used to build minimal component-hypertrees. In Sect. 4, we show some experiments that quantify the obtained memory saving using our strategy. Finally, we conclude our paper in Sect. 5.

2 Background

2.1 Images and Connectivity

In this study, we consider an image f as a mapping from a subset of a rectangular grid $\mathcal{D} \subset \mathbb{Z}^2$ to a set of gray-levels $\mathbb{K} = \{0, 1, \ldots, K-1\}$. An element $p \in \mathcal{D}$ is called pixel and the gray-level associated with p in image f is denoted by $f(p)$. In addition, a pixel $p \in \mathcal{D}$ can be a neighbor of others pixels of $\mathcal{D}$ in their surroundings. This notion of neighborhood can be defined by a structuring element (SE), that is, a subset $\mathcal{S} \subset \mathbb{Z}^2$. Thus, given a set of pixels $P \subseteq \mathcal{D}$ and a SE $\mathcal{S}$, we define $\mathcal{P} \oplus \mathcal{S}$, the *dilation* of $\mathcal{P}$ by $\mathcal{S}$ as $\mathcal{P} \oplus \mathcal{S} = \{p + s : p \in \mathcal{P}, s \in \mathcal{S} \text{ and } (p+s) \in \mathcal{D}\}$. In this sense, we say that two pixels p and q are $\mathcal{S}$-neighbors if $q \in (\{p\} \oplus \mathcal{S})$. In this paper, we only take into consideration symmetric SEs (i.e., if $s \in \mathcal{S}$, so does $-s$), so $q \in (\{p\} \oplus \mathcal{S})$ is equivalent to $p \in (\{q\} \oplus \mathcal{S})$. Given a SE $\mathcal{S}$, we define the set of neighboring pixels defined by $\mathcal{S}$, or simply *neighborhood*, as $\mathcal{A}(\mathcal{S}) = \{(p, q) : p \in \{q\} \oplus \mathcal{S}, p \in \mathcal{D}, q \in \mathcal{D}\}$. If the SE $\mathcal{S}$ is not relevant, we may simply denote a neighborhood by $\mathcal{A}$. In addition, considering a set $X \subset \mathcal{D}$ and a set of neighborhood $\mathcal{A}$, we say that two pixels $p, q \in X$ are

$\mathcal{A}$-connected if and only if there exists a sequence of pixels $(p_1, p_2, \ldots, p_m)$, with $p_j \in X$, such that all the following conditions hold: (a) $p_1 = p$; (b) $p_m = q$; (c) $\{p_j, p_{j+1}\} \in \mathcal{A}$ for all $1 \leq j < m$. *Connectedness* can be used to define *$\mathcal{A}$-connected components* ($\mathcal{A}$-CCs) or simply *connected components* (CCs). In this way, $\mathcal{A}$-CCs are defined as maximal sets of $\mathcal{A}$-connected pixels in X.

2.2 Component Trees

From an image f we define, for any $\lambda \in \mathbb{K}$, the set $X_\lambda(f) = \{p \in \mathcal{D} : f(p) \geq \lambda\}$ as the *upper level set* at value λ of the image f. Level sets are nested, i.e., $X_0(f) \supseteq X_1(f) \supseteq \ldots \supseteq X_{K-1}(f)$. By defining $CC(f, \mathcal{A}) = \{(\mathcal{C}, \lambda) : \mathcal{C} \text{ is a } \mathcal{A}\text{-CC of } X_\lambda(f), \lambda \in \mathbb{K}\}$, we can denote an order relation between these elements: given two elements $(\mathcal{C}, \alpha)$, $(\mathcal{C}', \beta) \in CC(f, \mathcal{A})$, we define that $(\mathcal{C}, \alpha) \sqsubset (\mathcal{C}', \beta) \Leftrightarrow \mathcal{C} \subseteq \mathcal{C}'$ and $\alpha > \beta$. This means these elements can be organized hierarchically in a tree structure according to this order relation. This tree is commonly referred as *component tree*. It is easy to note that, in a component tree, a single CC $\mathcal{C}$ can be represented by multiple nodes, if $\mathcal{C}$ exists in multiple level sets. In order to simplify the structure and save memory, an option is to store only one node per CC. It is easy to show that, in order to recover the complete tree, storing only the highest gray-level that generates each CC suffices. Using this property, we can create a simplified component tree without repeated CCs. This simplified structure is usually called *max-tree*. An example is shown in Fig. 1.

2.3 Algorithmic Background

There are many efficient ways of building max-trees [1]. One of the most efficient strategies consists in a generalization of the classic CC-labeling algorithm based on disjoint set structure (union-find) [4,12]. In this generalized structure, each CC has a representative pixel, and links between representative pixels give information about the inclusion relation of the CCs. In this way, each CC is composed of a representative and all other elements that point towards it. Disjoint sets can be efficiently stored using an array, which is called parent, so that parent[p] refers to the element that p is pointing to. Algorithm 1 shows the union-find algorithm.

Algorithm 1. Max-tree building based on disjoint sets.

```
procedure UNIONFIND(f, A, parent)
    for {p, q} ∈ A do
        rP ← FIND(parent, f, p);
        rQ ← FIND(parent, f, q);
        UNION(parent, rP, rQ);
    return parent;

procedure MAKESET(parent, f)
    for p ∈ D do
        parent[p] ← ∅;

procedure FIND(parent, f, p)
    if p = ∅ then
        return p;
    if f(p) ≠ f(parent[p]) then
        return p;
    else
        parent[p] ←
                FIND(parent, f, parent[p]);
        return parent[p];
```

In this algorithm, there are three basic operations: (1) MAKESET, which initializes the array with each pixel as a disjoint CC; (2) FIND, which obtains the representative of a given pixel p; and (3) UNION, which given two representatives p and q, makes the adjustments to merge CCs containing p to CCs containing q. The basic idea of the algorithm to build max-trees using disjoint sets is simply to connect all neighboring pixels by calling the UNION function when they do not belong to the same CC. Figure 1 shows a graphical example of union-find (rightmost column).

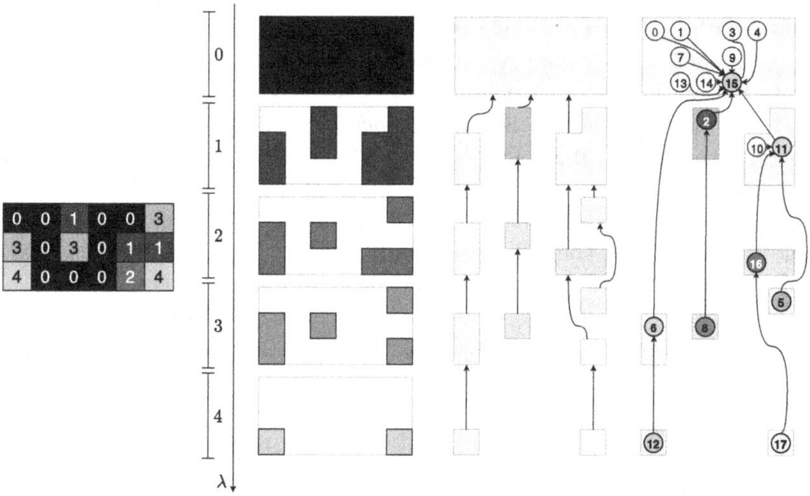

Fig. 1. From left to right: a gray-level image f; the upper level sets of f; the respective component tree with each CC as a node and arcs indicating the order relation; a graphical representation of the parent array that represents the max-tree. Each element p is represented by a circle and parent[p] is given by the arc leaving from p.

2.4 Component-Hypertrees

An extension of component trees can be obtained by considering another way of organizing the CCs of upper level sets in a hierarchical way using multiple neighborhoods. Let $\mathbb{A}$ be a sequence of neighborhood sets, i.e., $\mathbb{A} = (\mathcal{A}_1, \mathcal{A}_2, \ldots, \mathcal{A}_n)$. If these neighborhoods are increasing (i.e., $\mathcal{A}_i \subset \mathcal{A}_{i+1}$ with $1 \leq i < n$), then for any $\mathcal{A}_i$-CC $\mathcal{C}$, there exists an $\mathcal{A}_{i+1}$-CC $\mathcal{C}'$ that contains $\mathcal{C}$, that is, $\mathcal{C} \subseteq \mathcal{C}'$. An easy way of obtaining increasing neighborhoods is to consider a sequence of SEs $\mathbb{S} = (\mathcal{S}_1, \ldots, \mathcal{S}_n)$ such that $\mathcal{S}_i \subset \mathcal{S}_{i+1}, 1 \leq i < n$ (an example is provided in Fig. 2). Then, if we define $\mathcal{A}_i = \mathcal{A}(\mathcal{S}_i)$ and $\mathbb{A} = (\mathcal{A}_1, \ldots, \mathcal{A}_n)$, $\mathbb{A}$ will be a sequence of increasing neighborhoods. By combining the CCs of upper level sets and increasing neighborhoods, we can build a hierarchical graph known as *component-hypertree*. More formally, given an image f and a sequence of increasing neighborhoods $\mathbb{A}$, we define the set of $\mathbb{A}$-CCs, denoted by $N(f, \mathbb{A})$, as

$$N(f, \mathbb{A}) = \{(\mathcal{C}, \lambda, i) : \mathcal{C} \in CC(f, \mathcal{A}_i), \lambda \in \mathbb{K}, i \in \mathbb{I}\}, \tag{1}$$

where $\mathbb{I} = \{1, 2, \ldots, n\}$. Considering this set, two distinct elements $(\mathcal{C}, \alpha, i)$ and $(\mathcal{C}', \beta, j)$ of $N(f, \mathbb{A})$ are said to be nested if and only if $\mathcal{C} \subseteq \mathcal{C}', \alpha \geq \beta$ and $i \leq j$ with $(\mathcal{C}, \alpha, i) \neq (\mathcal{C}', \beta, j)$. In this case, we write $(\mathcal{C}, \alpha, i) \sqsubset (\mathcal{C}', \beta, j)$. Given two elements $\mathcal{N}, \mathcal{N}' \in N(f, \mathbb{A})$ such that $\mathcal{N} \sqsubset \mathcal{N}'$, if there is no $\mathcal{N}'' \in N(f, \mathbb{A})$ satisfying $\mathcal{N} \sqsubset \mathcal{N}'' \sqsubset \mathcal{N}'$, then we write $\mathcal{N} \prec \mathcal{N}'$.

The *hypertree* of an image f using a sequence of neighborhoods $\mathbb{A}$, denoted by $HT(f, \mathbb{A})$, is simply the directed graph $(V(f, \mathbb{A}), E(f, \mathbb{A}))$ where its vertices $V(f, \mathbb{A})$ are the $\mathbb{A}$-CCs, i.e., $V(f, \mathbb{A}) = \{v \in N(f, \mathbb{A})\}$ and its arcs $E(f, \mathbb{A})$ is defined as $E(f, \mathbb{A}) = \{(\mathcal{N}, \mathcal{N}') \in N(f, \mathbb{A}) \times N(f, \mathbb{A}) : \mathcal{N} \prec \mathcal{N}'\}$.

Each arc $e = (\mathcal{N}, \mathcal{N}')$ is named by either *parent arc* or *composite arc*. On one hand, if e is such that $\mathcal{N}$ and $\mathcal{N}'$ are nodes representing CCs from consecutive level sets but with the same neighborhood, we say that e is a *parent arc*; in addition, we also say that $\mathcal{N}$ is a *child node* of $\mathcal{N}'$, or $\mathcal{N}'$ is a *parent node* of $\mathcal{N}$. On the other hand, if $\mathcal{N}$ and $\mathcal{N}'$ have the same gray-level but consecutive neighborhood indices, we say that e is a *composite arc*; in addition, we also say that $\mathcal{N}'$ is a *composite node* of $\mathcal{N}$, or $\mathcal{N}$ being a *partial node* of $\mathcal{N}'$. An example is provided at the left side of Fig. 3, where parent and composite arcs are respectively represented by black and blue colors.

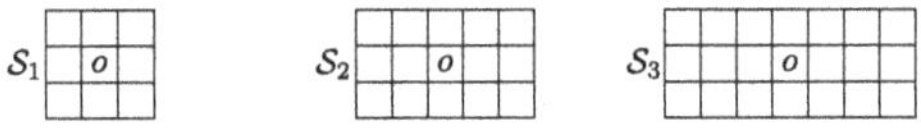

Fig. 2. A sequence $\mathbb{S} = (\mathcal{S}_1, \mathcal{S}_2, \mathcal{S}_3)$ with 3 increasing SEs.

3 Minimal Component-Hypertree

As explained, component trees can be efficiently stored using max-trees. Similarly, we want to define a minimal structure for storing hypertrees. For that, it suffices to keep in memory the CCs along with their highest threshold and their lowest adjacency index. Thus, we define the compact node of a node $\mathcal{N} = (\mathcal{C}, \lambda, i)$, denoted by $cn((\mathcal{C}, \lambda, i))$, of a hypertree $HT(f, \mathbb{A})$ as

$$\begin{aligned} &cn((\mathcal{C}, \lambda, i)) = (\mathcal{C}', \beta, \ell) \in N(f, \mathbb{A}) \text{ such that } \mathcal{C}' = \mathcal{C} \text{ and} \\ &\forall (\mathcal{C}'', \alpha, j) \in N(f, \mathbb{A}) \text{ with } \mathcal{C}'' = \mathcal{C}, \beta \geq \alpha \text{ and } \ell \leq j. \end{aligned} \tag{2}$$

In the same way, given an arc $(\mathcal{N}, \mathcal{N}') \in E(f, \mathbb{A})$, we define its *compact arc* as

$$ca((\mathcal{N}, \mathcal{N}')) = (cn(\mathcal{N}), cn(\mathcal{N}')). \tag{3}$$

So, from the above definitions, we can derive the *compact representation* of a hypertree $HT(f, \mathbb{A})$ as the directed graph $(CN(f, \mathbb{A}), CE(f, \mathbb{A}))$ where

$$\begin{aligned} &CN(f, \mathbb{A}) = \{cn(\mathcal{N}) : \mathcal{N} \in N(f, \mathbb{A})\} \text{ and} \\ &CE(f, \mathbb{A}) = \{ca((\mathcal{N}, \mathcal{N}')) : (\mathcal{N}, \mathcal{N}') \in E(f, \mathbb{A})\} \end{aligned} \tag{4}$$

are, respectively, the sets of compact nodes and compact arcs. Figure 3 shows an example of a hypertree (left) and its compact representation (right).

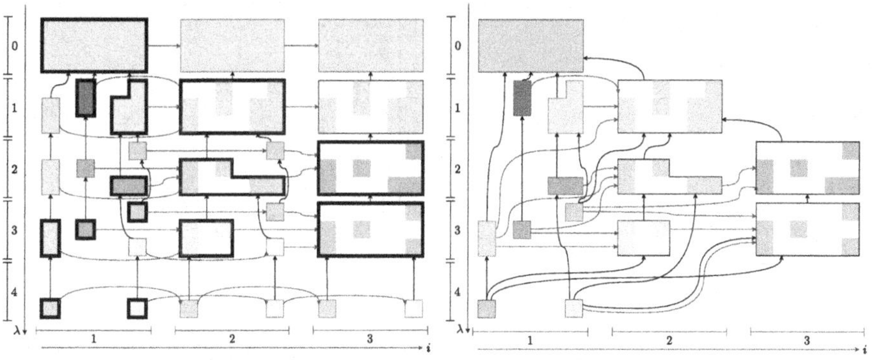

Fig. 3. Left: the complete hypertree of f from Fig. 1 using $\mathbb{A} = (\mathcal{A}(\mathcal{S}_1), \mathcal{A}(\mathcal{S}_2), \mathcal{A}(\mathcal{S}_3))$, where $\mathcal{S}_i, i = \{1, 2, 3\}$ is in Fig. 2. Each shape represents a node, and the colored squares represent their respective CCs. Parent arcs are colored black, while composite arcs are blue. Compact nodes are drawn with thick border. Right: the simplified hypertree with only compact arcs and compact nodes. (Color figure online)

From the compact representation, we still can eliminate some redundant arcs (see an example in Fig. 4). We say that a compact arc $e = (\mathcal{N}, \mathcal{N}') \in CE(f, \mathbb{A})$ is a *redundant arc* if and only if there exists a directed path $\pi = (\mathcal{N} = \mathcal{N}_1, \mathcal{N}_2, \ldots, \mathcal{N}_m = \mathcal{N}')$ in the compact hypertree such that

1. $\mathcal{N}_1 = \mathcal{N} = (\mathcal{C}, \alpha, i_1)$;
2. $\mathcal{N}_m = \mathcal{N}' = (\mathcal{C}', \beta, i_m)$;
3. For any k, with $1 \leq k < m$, $\mathcal{N}_k = (\mathcal{C}_k, \gamma_k, i_k)$, $i_k \leq i_{k+1}$.

We say that these arcs are redundant because, in terms of inclusion relation of CCs, they give the same information of the alternative path π. In addition, the above Condition (3) enforces that any path does not go "backward", i.e., we cannot have a path π that goes from a node with bigger neighborhood to a node with smaller neighborhood.

If a compact arc is not redundant, it is called a *minimal arc*. Let $MA(f, \mathbb{A})$ be the set of all minimal arcs of a compact hypertree built from an image f and using all neighborhoods in $\mathbb{A}$. Then, we call the *minimal component-hypertree* of f using $\mathbb{A}$, the directed graph $(CN(f, \mathbb{A}), MA(f, \mathbb{A}))$. Since all arcs that are not present in the minimal hypertree are redundant, we can easily see that this minimal representation preserves all the inclusion relation of CCs of the hypertree. An example of a minimal component-hypertree is shown in Fig. 4.

3.1 Data Structure

Since a hypertree is a DAG (which could be neither a tree nor a forest), only an array is not enough to store it. In this way, we explicitly allocate nodes and arcs in order to keep hypertrees in memory. Besides that, since nodes from trees with indices higher than 1 are composed of cluster of nodes from the first tree,

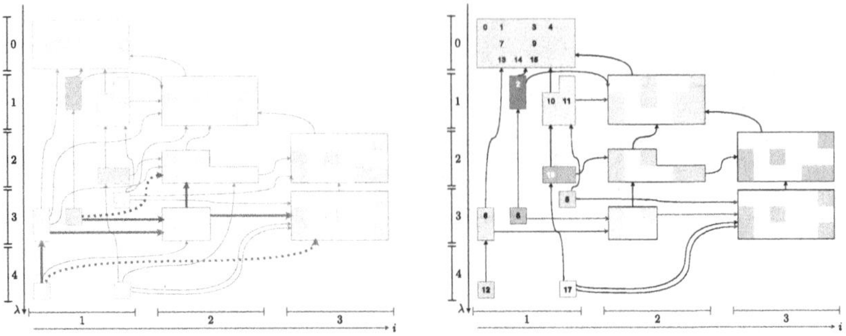

Fig. 4. Left: two examples of redundant arcs in the compact hypertree. For each color (red, green), the dotted arcs are redundant since they are equivalent to the paths with the same color. Right: the minimal component-hypertree i.e., the compact hypertree shown in the left without redundant arcs. Number inside nodes indicate the pixels stored in each node. (Color figure online)

only pixels of the first tree need to be stored. This strategy saves a considerable amount of memory, since each pixel is stored only once in the entire hypertree.

In terms of the data stored in each allocated node $\mathcal{N}$ of a hypertree, we keep the following information: (a) $rep(\mathcal{N})$ is its representative pixel; (b) $level(\mathcal{N})$ is the (highest) gray-level that generates its CC; (c) $index(\mathcal{N})$ is the (smallest) neighborhood index that generates its CC; (d) $pixels(\mathcal{N})$ is the set of pixels used to reconstruct its CC; (e) $par(\mathcal{N})$ is its set of parent nodes; (f) $child(\mathcal{N})$ is its set of children nodes; (g) $comp(\mathcal{N})$ is its set of composite nodes and (h) $part(\mathcal{N})$ is its set of partial nodes. An allocated node will be initialized as a triple $N = (rep(\mathcal{N}), level(\mathcal{N}), index(\mathcal{N}))$, with other fields empty.

3.2 Algorithm Template

The naive way of building hypertrees is to compute each max-tree from $\mathcal{A}_1$ to $\mathcal{A}_n$ by allocating its nodes without checking for repetition and then linking the corresponding nodes between two consecutive trees. In terms of memory usage, this is not efficient even if the repeated nodes are removed later, since it is needed to first allocate the whole hypertree in memory (which is a procedure that we want to keep away).

To avoid this problem, we build up the minimal hypertree by using a variant of the unordered version of the UNION procedure. The original algorithm was first presented in [13]. It has the property that it only changes a parent relation of a node if it really needs, i.e., a new neighborhood makes two originally disjoint CCs become connected. By keeping track of these changes, we can predict when new nodes need to be allocated. So our minimal hypertree construction strategy will follow the template of Algorithm 2.

In this sense, our template can be divided into two phases: one that updates the parent array (at Line 10) and other that updates the graph itself (at Lines

Algorithm 2. Minimal hypertree construction template. Underlined parameters are modified during the function call.

1: **procedure** BUILDHYPERTREE(f, $\mathbb{A} = (\mathcal{A}_1, \ldots, \mathcal{A}_n)$)
2:   $nodes \leftarrow \emptyset$;
3:   **for** $p \in \mathcal{D}$ **do**
4:     parent[p] $\leftarrow \emptyset$;
5:     compactNode[p] $\leftarrow \emptyset$;
6:   **for** $1 \leq i \leq n$ **do**
7:     $updateNode \leftarrow \emptyset$;
8:     $updateArc \leftarrow \emptyset$;
9:     **for** $(p, q) \in \mathcal{A}_i$ **do**
10:       UNION(parent, f, p, q, i, $False$, $updateNode$, $updateArc$);
11:     ALLOCATENODES($nodes$, compactNode, $updateNode$, parent, f, i);
12:     UPDATENEWARCS(compactNode, $updateArc$, parent);

11 and 12). After Line 10, we have the parent array representing the max-tree of f using $\mathcal{A}_i$, and it is used to allocate new nodes and update inclusion arcs.

The time complexity of this algorithm highly depends on the number of neighbors in each step i (at Line 9). We do not focus on the choice of the sequence of neighborhoods in this paper, but the usage of richer connectivities, such as mask-based connectivities [6,8,13] and dilation-based connectivities [3,10], can considerably reduce the number of neighbors we have to process, especially because all elements of $\mathcal{A}_{i-1}$ also belong to $\mathcal{A}_i$.

3.3 Detection of New Nodes Needed to Be Allocated

A key property needed for efficient node allocation is to quickly detect the emergence of new nodes in the parent array for the next component tree. For that, we need to change the UNION (called at Line 10) to detect these new nodes.

Given a pair of neighboring pixels (p, q), the UNION procedure makes the necessary changes so that the parent array reflects the next updated component tree with p and q as neighbors. In a nutshell, these changes consist of merging all disjoint CCs containing either p or q. This means that these changes are limited to the branches linking p to r and q to r, where r is the first common ancestor of p and q in the parent array.

In particular, the CCs represented in the branch from p to r are all CCs containing p that do not contain q and, likewise, the CCs from q to r do not contain p. So, the changes we need to do in the UNION procedure is to mark elements of the array when a change in the parent array occurs. Let $\pi_p = (p = p_1, p_2, \ldots, p_j = r)$ and $\pi_q = (q = q_1, q_2, \ldots, q_k = r)$ be the paths in the parent array, respectively, from p to r and from q to r. If a pixel q' of π_q becomes the new parent of p' from π_p (since they become connected in the adjacency $\mathcal{A}_i$), then we have a change in the parent array so that p' has its parent changed to q' and all elements from q' to r will have at least a new pixel p' included, and will produce new CCs. Analogously, the same property holds if a pixel from π_p

becomes a parent of an element of π_q. Once a change happens, we mark all arcs and nodes until reaching the common ancestor. These marks are used later to allocate nodes and update the inclusion relation of new nodes (Lines 11 and 12).

An example is provided in Fig. 5. In the left, the original state of the union-find representation is shown. Suppose p_8 and p_{16} become neighbors (p_j refers to the pixel in the figure with label j). Notice that p_{15} is their first common ancestor and all changes are limited to the branches from p_8 to p_{15} and p_{16} to p_{15}. When parent of p_8 is updated to p_{16}, all elements from p_{16} up to p_{15} now have the new pixel p_8. Likewise, parent of pixel p_{11} is now pixel p_2, and this triggers the creation of a new node. This change makes p_{11} not be a representative anymore.

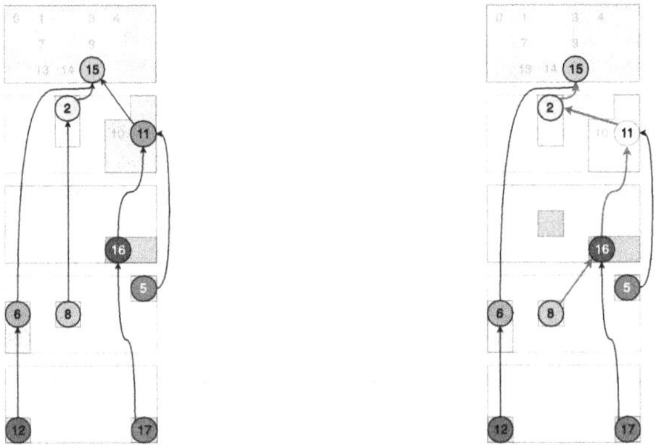

Fig. 5. Left: representation of the parent array from Fig. 1, with representatives highlighted. Right: updating the array when pixels p_8 and p_{16} are neighbors. In Algorithm 3, the red arcs will be added in the set *updateArc* and nodes p_{16}, p_{11} and p_2 will be added to the set *updateNode*, since they all have a new descendant (p_8). Node p_{15} is not added because it is the first common ancestor of p_8 and p_{16}.

In this way, we obtain a version of UNION procedure that detects when new nodes need to be created by marking changes in the parent array (see Algorithm 3), where marked elements are added to the set *updateNode*. Likewise, to create new arcs, whenever a node is created, the path leading that node to the common ancestor is added to the set *updateArc*.

3.4 Graph Update: Node Allocation and Arc Addition

Node allocation procedure is shown in Algorithm 4. For each marked pixel (i.e., the one that is in *updateNode*) which is representative, we allocate a node. We still need to deal with some arcs from different component trees, since it is possible to have arcs linking nodes with same representative. If the newly allocated node has one partial node (the node from the previous tree with the same representative), then we add an arc linking them. In addition, we update compactNode to always point to the lastly allocated node for each representative (i.e. the last and smallest compact node that contains the representative). Arc addition is

Algorithm 3. Modified union-find to mark changes in the parent array.

1: **procedure** UNION(parent, f, p, q, i, *changed*, *updateNode*, *updateArc*)
2:   **if** $f(p) < f(q)$ **then**   ▷ we consider $f(\emptyset) = -1$
3:     UNION(parent, f, q, p, i, *changed*, *updateNode*, *updateArc*);
4:   **else**
5:     **if** $p \neq q$ **then**
6:       $parP \leftarrow$ FIND(parent, f, parent[p]);
7:       **if** *changed* and $i > 1$ **then**
8:         $updateNode \leftarrow updateNode \cup \{p\}$; $updateArc \leftarrow updateArc \cup \{p\}$;
9:       **if** $parP \neq q$ **then**
10:         **if** $f(parP) \geq f(q)$ **then**
11:           UNION(parent, f, $parP$, q, i, *changed*, *updateNode*, *updateArc*);
12:         **else**
13:           parent[p] $\leftarrow q$;
14:           **if** $i > 1$ **then**
15:             $updateArc \leftarrow updateArc \cup \{p\}$;
16:           UNION(parent, f, q, $parP$, i, True, *updateNode*, *updateArc*);

straightforward: for all marked arcs (the ones in *updateArc*), we add them (linking compact nodes) in the graph (see Algorithm 5). In both Algorithms 4 and 5, arc addition depends on the gray-levels and neighborhood indices of their nodes. If they have different gray-levels (resp., indices), a parent (resp., composite) arc is added. Note that these two conditions are not exclusive: it is possible to add both arcs, like the ones from the green node (with pixel p_{17}) to the gray node in Fig. 4 (right).

Algorithm 4. Allocation of new nodes from the set *updateNode*.

1: **procedure** ALLOCATENODES(*nodes*, compactNode, *updateNode*, parent, f, i)
2:   **for** $p \in updateNode$ **do**
3:     $rP \leftarrow$ FIND(parent, f, p);
4:     **if** $p = rP$ **then**
5:       Allocate node $N' \leftarrow (rP, f(rP), i)$;
6:       **if** compactNode[rP] $\neq \emptyset$ **then**
7:         $N \leftarrow$ compactNode[rP];
8:         Add arc (N, N');
9:       compactNode[rP] $\leftarrow N'$;
10:       $nodes \leftarrow nodes \cup \{N'\}$;

Algorithm 5. Updating arcs involving new nodes.

1: **procedure** UPDATENEWARCS(compactNode, *updateArc*, parent);
2:   **for** $p \in updateArc$ **do**
3:     Add arc (compactNode[p], compactNode[parent[p]]);

4 Experiments

In this section, we analyze how much memory is saved by using our minimal representation of component-hypertrees, compared to the complete representation and the naive strategy explained in Sect. 3.2. On one hand, in the complete representation, all CCs for all component trees from $\mathcal{A}_1$ to $\mathcal{A}_n$ are stored in memory. On the other hand, in the naive representation, max-trees (the compact representation for component trees) are used.

For our tests, we used images from ICDAR 2011 [2]. For each image of this set, we used a sequence of square neighborhoods $\mathbb{A} = (\mathcal{A}_1, \ldots, \mathcal{A}_n)$, with $n = 50$, where $\mathcal{A}_i$ is defined using a SE of size $(2i + 1) \times (2i + 1)$. Then, we computed the average of the number of nodes and arcs for each i for the 3 structures: the complete hypertree, the naive implementation and our minimal representation. The results in Fig. 6 shows that the minimal representation can save a considerable amount of memory compared to the other ones. For example, in average, for 10 neighborhoods, we have a saving of about 50% compared to the naive implementation and 80% compared to the complete hypertree, both in terms of number of nodes and number of arcs.

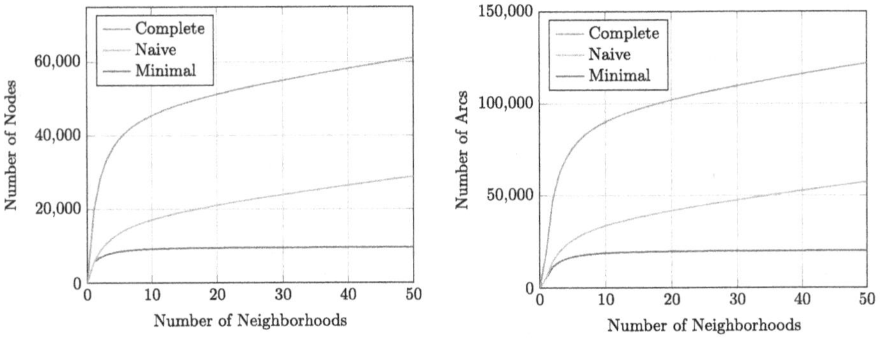

Fig. 6. Left: average number of nodes for each representation for up to 50 neighborhoods. Right: the same experiments but for the average number of arcs.

For more complex images from ICDAR 2017 [5], our method still saves between 50% and 80% of memory when compared to the naive approach and about 60% to 85% when compared to the complete representation (considering $n = 50$). In terms of time consumption, using the optimized approach given in [3] to update the array and our approach for node allocation resulted in total time ranging from 1 s (for images with 0.25 mega-pixels) to about a minute (for images with 8 mega-pixels).

5 Conclusion

In this paper, we presented algorithms and data structures behind the construction of minimal component-hypertrees that efficiently store them without redundancy and without loss of information about the inclusion relation of CCs.

Experiments show that our approach saves a considerable amount of memory compared to the complete representation and the strategy of independently building the max-trees for each neighborhood. As a future work, we plan to study how to efficiently compute attributes in these structures.

Acknowledgements. This study was financed in part by the CAPES - Coordenação de Aperfeiçoamento de Pessoal de Nível Superior (Finance Code 001); FAPESP - Fundação de Amparo a Pesquisa do Estado de São Paulo (Proc. 2018/15652-7); CNPq - Conselho Nacional de Desenvolvimento Científico e Tecnológico (Proc. 428720/2018-8).

References

1. Carlinet, E., Géraud, T.: A comparative review of component tree computation algorithms. IEEE Trans. Image Process. **23**(9), 3885–3895 (2014)
2. Karatzas, D., Mestre, S.R., Mas, J., Nourbakhsh, F., Roy, P.P.: ICDAR 2011 robust reading competition-challenge 1: reading text in born-digital images (web and email). In: 2011 International Conference on Document Analysis and Recognition (ICDAR), pp. 1485–1490. IEEE (2011)
3. Morimitsu, A., Alves, W.A.L., Hashimoto, R.F.: Incremental and efficient computation of families of component trees. In: Benediktsson, J.A., Chanussot, J., Najman, L., Talbot, H. (eds.) ISMM 2015. LNCS, vol. 9082, pp. 681–692. Springer, Cham (2015). https://doi.org/10.1007/978-3-319-18720-4_57
4. Najman, L., Couprie, M.: Building the component tree in quasi-linear time. IEEE Trans. Image Process. **15**(11), 3531–3539 (2006)
5. Nayef, N., et al.: ICDAR 2017 robust reading challenge on multi-lingual scene text detection and script identification-RRC-MLT. In: 2017 14th IAPR International Conference on Document Analysis and Recognition (ICDAR), vol. 1, pp. 1454–1459. IEEE (2017)
6. Ouzounis, G.K., Wilkinson, M.H.: Mask-based second-generation connectivity and attribute filters. IEEE Trans. Pattern Anal. Mach. Intell. **29**(6), 990–1004 (2007)
7. Ouzounis, G.K., Wilkinson, M.H.: Hyperconnected attribute filters based on k-flat zones. IEEE Trans. Pattern Anal. Mach. Intell. **33**(2), 224–239 (2011)
8. Passat, N., Naegel, B.: Component-hypertrees for image segmentation. In: Soille, P., Pesaresi, M., Ouzounis, G.K. (eds.) ISMM 2011. LNCS, vol. 6671, pp. 284–295. Springer, Heidelberg (2011). https://doi.org/10.1007/978-3-642-21569-8_25
9. Salembier, P., Oliveras, A., Garrido, L.: Antiextensive connected operators for image and sequence processing. IEEE Trans. Image Process. **7**(4), 555–570 (1998)
10. Serra, J.: Connectivity on complete lattices. J. Math. Imaging Vis. **9**(3), 231–251 (1998)
11. Silva, D.J., Alves, W.A., Morimitsu, A., Hashimoto, R.F.: Efficient incremental computation of attributes based on locally countable patterns in component trees. In: 2016 IEEE International Conference on Image Processing (ICIP), pp. 3738–3742. IEEE (2016)
12. Tarjan, R.E.: Efficiency of a good but not linear set union algorithm. J. ACM (JACM) **22**(2), 215–225 (1975)
13. Wilkinson, M.H., Gao, H., Hesselink, W.H., Jonker, J.E., Meijster, A.: Concurrent computation of attribute filters on shared memory parallel machines. IEEE Trans. Pattern Anal. Mach. Intell. **30**(10), 1800–1813 (2008)

Single Scan Granulometry Estimation from an Asymmetric Distance Map

Nicolas Normand(✉)

LS2N UMR 6004, Polytech Nantes, Université de Nantes, Nantes, France
Nicolas.Normand@univ-nantes.fr

Abstract. Granulometry, by characterizing the distribution of object sizes, is a powerful tool for the analysis of binary images. It may, for example, extract pertinent features in the context of texture classification. The granulometry is similar to a sieving process that filters image details of increasing sizes. Its computation classically relies on morphological openings, i.e. the sequence of an erosion followed by a dilation for each ball radius.

It is well known that a distance map can be described as an "erosion transform" that summarizes the erosions with the balls of all radii. Using a Steiner formula, we show how a vector of parameters measured on an eroded contour can be extrapolated to the measures of the dilation. Instead of completing the openings from the distance map by computing a dilation for each ball size, we can estimate their cardinality from the area, perimeter and Euler-Poincaré characteristic of each erosion. We extract these measures for all radii at once from an asymmetric distance map. The result is a fast streaming algorithm that provides an estimate of the granulometry distribution, in a single scan of the image and with very limited memory footprint.

Keywords: Granulometry · Discrete distance · Pattern spectrum

1 Preliminaries

In this paper, we consider images defined on the discrete square grid $\mathbb{Z}^2$. A binary image A, or simply, a point set, is a subset of $\mathbb{Z}^2$ and its complement is the set of grid points not in A: $\overline{A} = \mathbb{Z}^2 \setminus A$. Either A or $\overline{A}$ must be finite. A grayscale image is a mapping from a finite rectangular domain of $\mathbb{Z}^2$ to $\mathbb{N}$.

1.1 Morphology Operators on Binary Sets

Definition 1 (Erosion, dilation). *The* erosion *and* dilation *of the set A by B, referred to as the* structuring element, *respectively denoted by $\varepsilon_B(A)$ and $\delta_B(A)$, are the sets:*

$$\varepsilon_B(A) = \{p : (B)_p \subseteq A\}, \qquad \delta_B(A) = \bigcup_{p \in B} (A)_p \tag{1}$$

where $(B)_p$ denotes the translation of B by p: $(B)_p = \{p + q : q \in B\}$.

M. Couprie et al. (Eds.): DGCI 2019, LNCS 11414, pp. 288–299, 2019.
https://doi.org/10.1007/978-3-030-14085-4_23

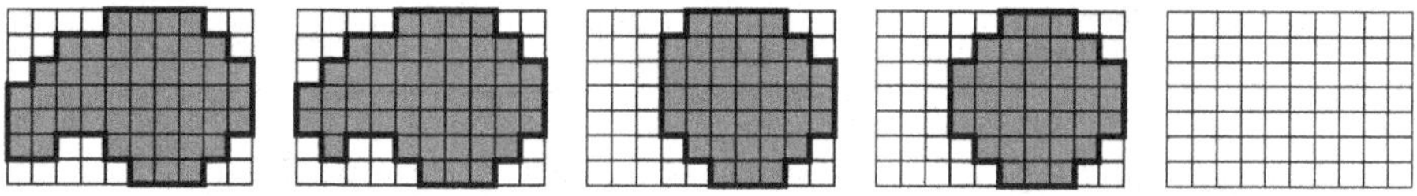

(a) $\gamma_1(X) = X$ (b) $\gamma_2(X)$ (c) $\gamma_3(X)$ (d) $\gamma_4(X)$ (e) $\gamma_5(X)$

Fig. 1. Openings with the sequence of increasing disks depicted in Fig. 2.

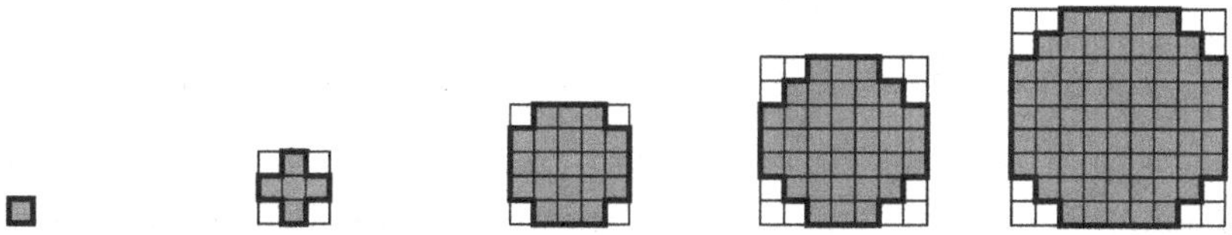

(a) $r = 1$ (b) $r = 2$ (c) $r = 3$ (d) $r = 4$ (e) $r = 5$

Fig. 2. Increasing sequence of disks of the octagonal distance used as structuring elements for morphological openings.

Definition 2 (Opening). *The* opening *of A by B, denoted by $\gamma_B(A)$, is the erosion of A with B followed by the dilation with B:*

$$\gamma_B(A) = \delta_B\left(\varepsilon_B\left(A\right)\right). \tag{2}$$

$\gamma_B(A)$ can be seen as the union of all translated images $(B)_p$ included in A, where the details of A smaller than B are deleted:

$$\gamma_B(A) = \bigcup_{p:(B)_p \subset A} (B)_p \subset A.$$

Definition 3 (Granulometry). *An opening-based* granulometry *on A is a family of openings of A with a sequence of structuring elements B_r:*

$$\left(\gamma_{B_r}(A), r \in \mathbb{N}_*\right) \tag{3}$$

with $B_1 = \{O\}$ and $\forall r \geq 0, B_{r+1} = \gamma_{B_r}(B_{r+1})$. Obviously, $\gamma_{B_1}(A) = A$.

A granulometry acts on an image like a sequence of sieves, the value r specifies the size of details to be suppressed. For the sake of simplicity, we write γ_r instead of γ_{B_r} when the sequence of B_r is clear from the context. Figure 1 pictures a family of openings with the increasing sequence of disks shown in Fig. 2.

Definition 4 (Granulometry function, pattern spectrum). *Given a sequence of sets B_r, the* granulometry function *on A, G_A, maps a positive r with the cardinality of $\gamma_r(A)$ and the* pattern spectrum *maps r with the cardinality of $\gamma_r(A) \setminus \gamma_{r+1}(A)$:*

$$\forall r > 1, \mathrm{G}_A(r) = \left|\gamma_r(X))\right|, \tag{4}$$
$$\forall r > 1, \mathrm{PS}_A(r) = \left|\gamma_r(X))\right| - \left|\gamma_{r+1}(X))\right|. \tag{5}$$

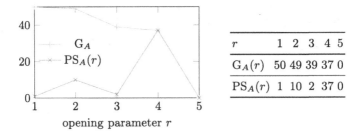

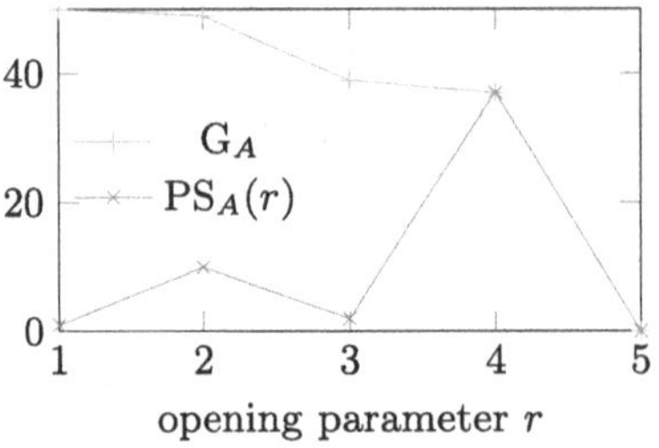

r	1	2	3	4	5
$G_A(r)$	50	49	39	37	0
$PS_A(r)$	1	10	2	37	0

Fig. 3. Granulometric spectrum of the sample image shown in Fig. 1a

Figure 3 depicts the granulometric spectrum corresponding to the sequence of morphological openings pictured in Fig. 1.

In the following, we will use distance disks $B_<(p,r) = \{q : d(q,p) < r\}$ of some distance d as structuring elements for the opening. This definition and properties of distances imply that $B_<(p,0) = \emptyset$ and $B_<(p,1) = \{p\}$.

Definition 5 (Distance transform). *The* distance transform DT_A *of the set A is the function that maps each point p to its distance from the complement of A and, equivalently, to the radius of the largest disk of center p included in A:*

$$\mathrm{DT}_A(p) = \min\{d(q,p) : q \in \overline{A}\} \tag{6}$$
$$= \max\{r : B_<(p,r) \subset A\}. \tag{7}$$

It is clear from Eq. (7) that $\mathrm{DT}_A(p) \geq r \iff B_<(p,r) \subset A \iff p \in \varepsilon_{B(O,r)}(A)$. Consequently, the erosion of A with $B_<(O,r)$ can be obtained by thresholding DT_A, *i.e.*, DT_A summarizes all erosions of A with $B_<(O,r), \forall r$.

A Neighborhood Sequence (NS) distance is a distance which disks are produced by dilations with a few neighborhoods, typically the 4-neighborhood $\mathcal{N}_4$ and the 8-neighborhood $\mathcal{N}_8$, arranged in a sequence $N(r)$: $B_<(O,r+1) = \delta_{\mathcal{N}_{\mathrm{N}(r)}}(B_<(O,r))$. The 2D NS distance produced by the strict alternation between $\mathcal{N}_4$ and $\mathcal{N}_8$ is called the octagonal distance (see Fig. 2). A so-called "translated" NS distance transform with asymmetric disks produced by translated neighborhoods $\mathcal{N}'_4$ and $\mathcal{N}'_8$, as shown in Fig. 4c, was described previously [6,7]. It was proven that those distance transforms are equivalent in terms of included disks as in Eq. (7) [7]. An efficient implementation of the translated NS distance transform that requires a single scan of the image and whose memory requirements are limited to a single line of image was described in [6].

Definition 6 (Opening function). *Consider d, a distance or pseudo-distance,* e.g., *asymmetric. The* opening function, *or* opening transform, OT_A *of the set X is a function that maps each point p to the radius of the largest opening disk of d included in X that contains p:*

$$\mathrm{OT}_A(p) = \max\{r : \exists q, B()< (q,r) \subset A, p \in B()< (q,r)\}. \tag{8}$$

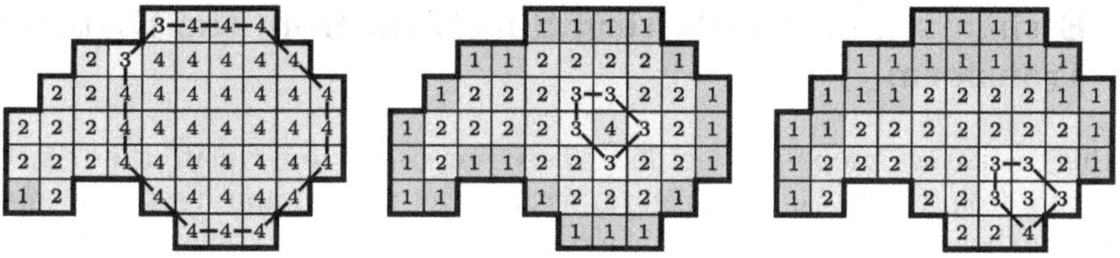

(a) OT_A (b) DT_A (c) DT'_A

Fig. 4. Opening transform OT_A, distance transform DT_A and translated distance transform DT'_A of a binary image A. All use the octagonal distance. Thresholding DT_A and DT'_A results in equal sets up to a translation. Delimited shapes correspond to pixels whose value is 3 or more. Section 2 describes how the area and perimeter of these shapes are linked.

Equivalently, $\mathrm{OT}_A(p)$ is the smallest parameter r for which p is filtered out by the morphological opening of the image:

$$\mathrm{OT}_A(p) = \min\{r : p \notin \gamma_r(X)\}. \tag{9}$$

1.2 Computation of the Granulometry Function

Morphological Openings. A simple algorithm consists in computing morphological openings for each structuring element B_r. Even with an efficient implementation of the opening operator, say linear with the size of the image domain N, the order of complexity is $\mathcal{O}(RN)$, also linear with the maximal radius R. If the structuring elements are chosen to be the disks of a distance, a speedup is achieved by computing the distance transform, that provides all the erosions at once. However, a dilation still has to be performed for each r and the order of complexity remains the same.

Opening Function. The granulometry function can be easily deduced from the opening function which it is simply the histogram.

$$\forall r > 0, f_A(r) = \mathrm{Card}\left(\{p : \mathrm{OT}_A(p) = r\}\right) \tag{10}$$

However, despite the similarity between the distance and opening transforms DT_A and OT_A, as in Eqs. (7) and (8), there is no linear time algorithm to compute the latter (except for special cases [11]).

In principle, it suffices to fill each disk $B(_{)}< (p, r)$ with its radius (given in the distance map) while keeping the maximal radius when several disks overlap. Because disks overlap, each pixel is visited more than once which increases the computational cost. In order to reduce the computation, we can observe that non maximal disks do not need to be filled since they are, by definition, included in larger disks. It is then enough to fill the maximal disks obtained in the medial axis MAT_A (or CMB_A, center of maximal balls) [1]. The number of overlaps decreases without reaching 0, in general.

2 Estimation of the Pattern Spectrum from the Distance Transform

In this section we propose a method whose purpose is to estimate the pattern spectrum from the distance transform without actually computing the opening transform nor dilations of the distance transform. For a given r, the eroded image by the ball of radius r, $\varepsilon_r(A)$, is simply the result of binarizing the distance map with threshold r. Instead of actually dilating $\varepsilon_r(A)$, the measures on the opening $\gamma_r(A)$, *i.e.* the dilated of $\varepsilon_r(A)$, are extrapolated from those of $\varepsilon_r(A)$.

After first exposing the linear behavior of the Steiner's formula in Sect. 2.1, we present a continuous interpolation of binary images based on 2×2 cells where a version of Steiner's formula can be adapted to our problem in Sect. 2.2. Section 2.3 shows how measures in the distance transform can be extracted from 2×2 local configurations (as in [3,4]), taking advantage of both the additive property and the continuous model. Finally, Sect. 2.4 presents the algorithm.

2.1 Rationale

Steiner's formula for a convex K and a Euclidean ball of radius $B(r)$ links the area and perimeter of the dilated convex $\delta_{B(r)}(K)$ with the area $\mathcal{A}$ and perimeter $\mathcal{P}$ of K and $B(r)$ [10]. In the 2D case:

$$\mathcal{A}(\delta_{B(r)}(K)) = \mathcal{A}(K) + r\mathcal{P}(K) + \mathcal{A}(B(r)), \tag{11}$$

$$\mathcal{P}(\delta_{B(r)}(K)) = \mathcal{P}(K) + P(B(r)). \tag{12}$$

Federer later extended this result to non convex compacts and introduced the notion of reach, *i.e.*, the maximal value of r for which Steiner's formula holds [2]. A further generalization to erosions was stated by Matheron: Steiner's formula holds for non convex compacts K and the erosion with $B(r)$ if and only if K is the mathematical opening of a compact with $B(r)$ [5]. This is equivalent to the notion a negative reach. If, equivalently to considering the erosion of K, we consider the dilation of its complement, $\overline{K} = \mathbb{R}^2 \setminus K$, then, by duality of the erosion and dilation:

$$\mathcal{A}(\delta_{B(r)}(\overline{K})) = \mathcal{A}(\overline{K}) + r\mathcal{P}(\overline{K}) - \mathcal{A}(B(r)), \tag{13}$$

$$\mathcal{P}(\delta_{B(r)}(\overline{K})) = \mathcal{P}(\overline{K}) - P(B(r)) \tag{14}$$

where, by convention, $\forall K, \mathcal{A}(\overline{K}) = -\mathcal{A}(K)$.

For any compact, or complement of a compact, equal to an opening with $B(r)$, we build a vector of characteristics $(\mathcal{A}, \mathcal{P}, \chi)^T$ with positive area $\mathcal{A}$, $\chi = 1$ for compacts and negative area and $\chi = -1$ for complements. Then

$$\begin{pmatrix} \mathcal{A}' \\ \mathcal{P}' \\ \chi \end{pmatrix} = \begin{pmatrix} 1 & r & \pi r^2 \\ 0 & 1 & 2\pi r \\ 0 & 0 & 1 \end{pmatrix} \begin{pmatrix} \mathcal{A} \\ \mathcal{P} \\ \chi \end{pmatrix} \tag{15}$$

where $(\mathcal{A}', \mathcal{P}', \chi)^T$ is the corresponding vector of characteristics of the set dilated by $B(r)$. Consider now a composite shape K, an opening with $B(r)$. We treat K as a mixture of its outer and inner contours and build a vector of characteristics $(\mathcal{A}, \mathcal{P}, \chi)^T$ where $\mathcal{A}$ is the area of K, *i.e.*, the sum of the areas of the interiors of outer contours minus the areas of the interiors of the inner contours, $\mathcal{P}$ the total perimeter, and χ is the count of outer minus inner contours. It is clear that $(\mathcal{A}, \mathcal{P}, \chi)^T$ is the sum of the vectors of characteristics of the contours. If the contours do not merge by dilation with K, then, by linearity, Eq. (15) holds.

2.2 Continuous Model

Let A be a binary image. Every 2×2 configuration of points of A is mapped to its convex hull $\mathfrak{C}_A(k,l) = \mathrm{conv}(\{(k,l),(k+1,l),(k,l+1),(k+1,l+1)\} \cap A)$. We define $\mathfrak{A}$, the continuous interpolation of A, as the union of all $\mathfrak{C}_A(k,l)$:

$$\mathfrak{A} = \bigcup_{k,l} \mathfrak{C}_A(k,l) = \bigcup_{k,l} \mathrm{conv}(\{(k,l),(k+1,l),(k,l+1),(k+1,l+1)\} \cap A). \quad (16)$$

It is clear that A is the Gauss discretization of $\mathfrak{A}$: $A = \mathfrak{A} \cap \mathbb{Z}^2$. Figure 5 pictures two binary images, their continuous interpolation, inner and outer contours.

Consider P, the polygon with edges $\boldsymbol{e}_i, 1 \le i \le n$, obtained by an opening with the line segment $L = [O, O + \boldsymbol{v}]$. Then

$$\mathcal{A}(\delta_L(P)) = \mathcal{A}(P) + \frac{1}{2} \sum_{i=1}^{n} |\det(\boldsymbol{e}_i, \boldsymbol{v})| \quad (17)$$

whether P is an outer or inner contour. In the former case, $\delta_L(P)$ contains 2 more edges of directions $\boldsymbol{v}$ and $-\boldsymbol{v}$ than P. In the latter case, areas are negative and the dilated contour contains two less edges of directions $\boldsymbol{v}$ and $-\boldsymbol{v}$ than P.

Practically, for neighborhood sequence distances, each structuring element, namely the 4- and 8-neighborhood, combine two dilations with orthogonal vectors, respectively $(-1,1)$ and $(1,1)$ on one hand and $(0,2)$ and $(2,0)$ on the other hand. Let $\nu_0(\mathfrak{A})$ (resp. $\nu_1(\mathfrak{A})$) be the number of horizontal and vertical (resp. diagonal) vectors. For each continuous analogue of a contour (outer or inner), we build a vector of characteristics $V(\mathfrak{A}) = (\mathcal{A}(\mathfrak{A}), \nu_0(\mathfrak{A}), \nu_1(\mathfrak{A}), \chi(\mathfrak{A}))^T$ where $A(\mathfrak{A}) > 0$ and $\chi S(\mathfrak{A}) = 1$ for outer contours, $A(\mathfrak{A}) < 0$ and $\chi S(\mathfrak{A}) = -1$ for inner contours. Then, a dilation of $\mathfrak{A}$ with $\mathcal{N}_4$ (resp. $\mathcal{N}_8$) corresponds to the matrix product of $M_{\mathcal{N}_4}$ (resp. $M_{\mathcal{N}_8}$) with $V(\mathfrak{A})$ where:

$$M_{\mathcal{N}_4} = \begin{pmatrix} 1 & 1 & 1 & 2 \\ 0 & 1 & 0 & 0 \\ 0 & 0 & 1 & 4 \\ 0 & 0 & 0 & 1 \end{pmatrix}, \qquad M_{\mathcal{N}_8} = \begin{pmatrix} 1 & 1 & 2 & 4 \\ 0 & 1 & 0 & 8 \\ 0 & 0 & 1 & 0 \\ 0 & 0 & 0 & 1 \end{pmatrix}. \quad (18)$$

Let B be a disk produced by m 4-neighborhoods and n 8-neighborhoods, then for each contour K of $\varepsilon_B(A)$:

$$V(\delta_B(K)) = (M_{\mathcal{N}_4})^m (M_{\mathcal{N}_8})^n V(K) \quad (19)$$

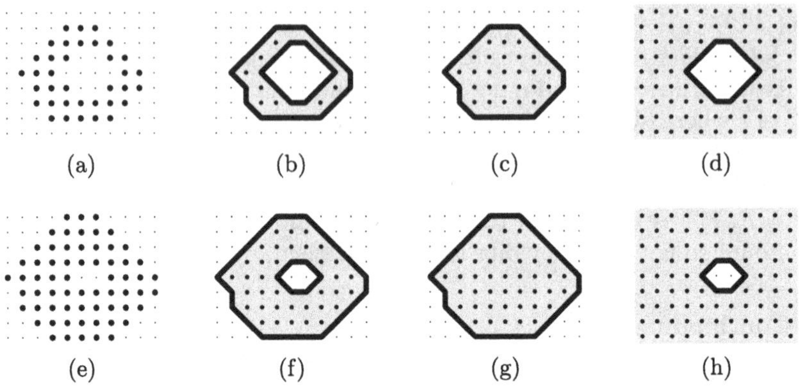

Fig. 5. Each line presents a binary image (left), its continuous interpolation (second column), its outer (third column) and inner (right) contours. Their vectors of characteristics are, from (b) to (d) and (f) to (h): $V_a = (22, 10, 18, 0)^T$, $V_b = (34, 8, 10, 1)^T$, $V_c = (-12, 2, 8, -1)^T$, $V_d = (50, 10, 18, 0)^T$, $V_e = (54, 8, 14, 1)^T$ and $V_f = (-4, 2, 4, -1)^T$ with $V_a = V_b + V_c$ and $V_d = V_e + V_f$. Each shape on the first row is the eroded of the shape underneath with the 4-neighborhood or its continuous interpolation, and conversely, each shape on the second row is the dilated of the shape above with the 4-neighborhood or its continuous interpolation. Thus, $V_d = M_{\mathcal{N}_4} V_a$, $V_e = M_{\mathcal{N}_4} V_b$ and $V_f = M_{\mathcal{N}_4} V_c$, with $M_{\mathcal{N}_4}$ from Eq. (18).

because the dilated contour $\delta_B(K)$ is an opening with B. Like in the previous section, Eq. (19) holds for a mixture of contours if the contours do not merge during the dilation, otherwise some quantities may be counted multiple times. For example, parameter vectors measured on the opening transform (Fig. 4a and Table 2 *left*) are correctly extrapolated from parameter vectors measured on the distance transform (Fig. 4c and Table 2 *right*). For instance, for radius 3:

$$M_{\mathcal{N}_4} \cdot M_{\mathcal{N}_8} \cdot (5/2\ 2\ 3\ 1)^T = (59/2\ 10\ 7\ 1)^T .$$

2.3 Local Configurations

For an arbitrary image A, the following quantities are measured on it continuous analogue $\mathfrak{A}$: the area $\mathcal{A}(\mathfrak{A})$, the number of horizontal and vertical elementary edges $\nu_0(\mathfrak{A})$, the count of diagonal elementary edges $\nu_1(\mathfrak{A})$ and the difference between the number of outer and inner contours, *i.e.*, the Euler-Poincaré characteristic $\chi(\mathfrak{A})$. For these four measures, the additivity property hold:

$$\forall \mu \in \{\mathcal{A}, \nu_0, \nu_1, \chi\}, \forall \mathfrak{A}, \mathfrak{B}, \mu(\mathfrak{A} \cup \mathfrak{B}) = \mu(\mathfrak{A}) + \mu(\mathfrak{B}) - \mu(\mathfrak{A} \cap \mathfrak{B}). \qquad (20)$$

As a consequence, $\mathcal{A}$, ν_0, ν_1 and χ can be evaluated on $\mathfrak{A}$ from the examination of subsets of $\mathfrak{A}$ and their intersections. We divide $\mathfrak{A}$ along elementary grid cells, according to its definition in Eq. (16). Then

$$\begin{aligned}\mu(\mathfrak{A}) = &+\sum_{k,l} \mu(\mathfrak{C}_A(k,l)) \\ &-\sum_{k,l} \mu(\mathfrak{C}_A(k,l) \cap \mathfrak{C}_A(k+1,l)) + \mu(\mathfrak{C}_A(k,l) \cap \mathfrak{C}_A(k,l+1)) \\ &+\sum_{k,l} \mu(\mathfrak{C}_A(k,l) \cap \mathfrak{C}_A(k+1,l) \cap \mathfrak{C}_A(k,l+1) \cap \mathfrak{C}_A(k+1,l+1)).\end{aligned}$$

Note that each $\mathfrak{C}_A(k,l)$ determines the intersections with its neighbors, *e.g.*, $\mathfrak{C}_A(k,l) \cap \mathfrak{C}_A(k+1,l) = \mathfrak{C}_A(k,l) \cap \mathrm{conv}(\{(k+1,l),(k+1,l+1)\})$. We chose to include in the measures associated to each $\mathfrak{C}_A(k,l)$, the shares of the intersections with its neighbors, equally distributed (see Fig. 6). This avoids a separate count for measures on $\mathfrak{C}_A(k,l)$ and on their intersections.

By convention, $\mathfrak{C}_A(k,l)$ is associated with the configuration index $2^0\mathbf{1}_A(k,l) + 2^1\mathbf{1}_A(k+1,l) + 2^2\mathbf{1}_A(k,l+1) + 2^3\mathbf{1}_A(k+1,l+1)$, where $\mathbf{1}_A$ is the indicator function of the set A. Table 1 presents the 16 possible configurations and their contributions to measures.

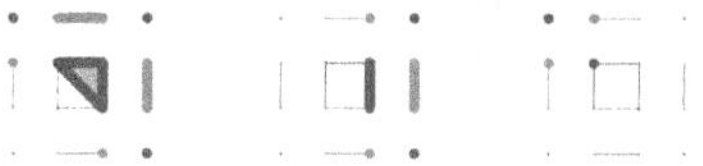

Fig. 6. Total contribution to the Euler-Poincaré characteristic in a grid cell including the shares of intersections with neighbor cells. *Left.* the cell contribution to the Euler-Poincaré characteristic for itself is 1. The intersection is non-empty in the four edges. The contribution of each one is -1, shared by two cells. The intersection is non-empty in three vertices. The contribution of each one is 1, shared by four cells. The global measure for the cell is then $\chi = 1 - 4 \times \frac{1}{2} + 3 \times \frac{1}{4} = -\frac{1}{4}$. *Middle.* $\chi = 1 - 3 \times \frac{1}{2} + 2 \times \frac{1}{4} = 0$. *Right.* $\chi = 1 - 2 \times \frac{1}{2} + 1 \times \frac{1}{4} = \frac{1}{4}$.

Table 1. Contributions of the 16 local configurations to image measures.

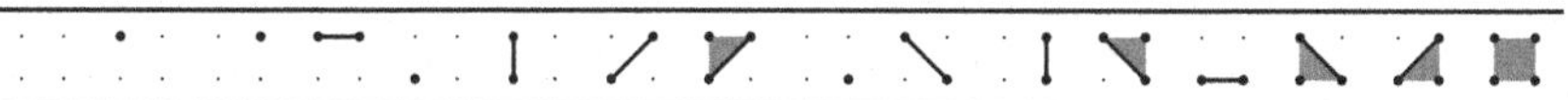

	0	1	2	3	4	5	6	7	8	9	10	11	12	13	14	15
$2\mathcal{A}$	0	0	0	0	0	0	0	1	0	0	0	1	0	1	1	2
ν_0	0	0	0	1	0	1	0	0	0	0	1	0	1	0	0	0
ν_1	0	0	0	0	0	0	2	1	0	2	0	1	0	1	1	0
4χ	0	1	1	0	1	0	-2	-1	1	-2	0	-1	0	-1	-1	0

2.4 Algorithm

The measures $\boldsymbol{\mu}_\varepsilon = (\mathcal{A}, \nu_0, \nu_1, \chi)$ have now to be computed for each erosion of A with $B_<(O,r)$, *i.e.*, in each binarization of DT'_A (or DT_A) with threshold r.

Each local configuration $\mathcal{C}_A(k,l)$ is determined by a 2×2 patch in DT'_A and the threshold r. The multiple choices of r generate at most 5 configurations according to how r compares to the values of the 4 pixels. To avoid iterating over all values of r, Algorithm 1 enumerates the 5 possible values of the configuration index $c(r)$ and update the first differences of $\boldsymbol{\mu}_\varepsilon$, $d\boldsymbol{\mu}_\varepsilon : r \mapsto \boldsymbol{\mu}_\varepsilon(r) - \boldsymbol{\mu}_\varepsilon(r-1)$. Let v contains the 4 values of the patch in increasing order of weights of the configuration index and σ be the reciprocal permutation of the ranks of elements of v in increasing order *i.e.*, $\forall i \in [0,2], v(\sigma_i) \leq v(\sigma_{i+1})$. Starting with $r = 0$ and $c(r) = 15$, $c(r)$ is decreased by 2^{σ_i} each time r exceeds some $v(\sigma_i)$. For each interval $[r_1, r_2)$ of r, $d\boldsymbol{\mu}_\varepsilon(r1)$ (resp. $d\boldsymbol{\mu}_\varepsilon(r2)$) has to be increased (resp. decreased) with the value of μ_c. These computation are performed in the first loop of Algorithm 1. An algorithm similar in principle was presented by Snidaro and Foresti [9], although limited to measuring the Euler-Poincaré characteristic.

The second loop of Algorithm 1 simultaneously accumulates $d\boldsymbol{\mu}_\varepsilon(r)$ in $\boldsymbol{\mu}_\varepsilon(r)$, computes the matrix for the dilation with the disk $B_<(O,r)$ using either $M_{\mathcal{N}_4}$ or

Algorithm 1. Simultaneous extraction of measures for all binarization thresholds.

Input: $K \times L$ asymmetric distance transform DT'_A
Output: Estimated measures on morphological openings $\hat{\mu}(r)$

begin
  for $(k,l) \in K \times L$ **do**
    $v \leftarrow (\mathrm{DT}'_A(k-1,l-1), \mathrm{DT}'_A(k,l-1), \mathrm{DT}'_A(k-1,l), \mathrm{DT}'_A(k,l))$
    determine σ, such that $v(\sigma_i) \leq v(\sigma_{i+1}), \forall i \in [0,2]$
    $c \leftarrow 15$; $r \leftarrow 0$ `// initial configuration index and threshold`
    for $i \leftarrow 0$ **to** 3 **do**
      `//` μ_c`: contribution of` v `binarized at threshold` r
      $d\boldsymbol{\mu}_\varepsilon(r) \leftarrow d\boldsymbol{\mu}_\varepsilon(r) + \mu_c$
      $r \leftarrow v(\sigma_i) + 1$ `// next threshold`
      $d\boldsymbol{\mu}_\varepsilon(r) \leftarrow d\boldsymbol{\mu}_\varepsilon(r) - \mu_c$
      $c \leftarrow c - 2^{\sigma_t}$ `// update configuration index`
    end
  end
  $M = I_4$ `// identity matrix`
  for $r = 1$ **to** $\max(\mathrm{DT}'_A)$ **do**
    $\boldsymbol{\mu}_\varepsilon(r) \leftarrow \boldsymbol{\mu}_\varepsilon(r-1) + d\boldsymbol{\mu}_\varepsilon(r)$
    $\hat{\mu}(r) \leftarrow M \cdot \boldsymbol{\mu}_\varepsilon(\{p : \mathrm{DT}'_A(p) \geq r\})$
    $M = M \cdot M_{N(r)}$
  end
end

Table 2. Parameter vectors measured (*left*) on the opening transform depicted in Fig. 4a, (*right*) on the distance transforms shown in Figs. 4b and c. The number of pixels # is deduced from the other four values and Pick's theorem [8]: $\# = \mathcal{A} + \frac{\nu_1+\nu_2}{2} + \chi$.

r	#	$\mathcal{A}$	ν_0	ν_1	χ
1	50	38	12	10	1
2	49	37.5	10	11	1
3	**39**	**29.5**	**10**	**7**	**1**
4	37	28	8	8	1
5	0	0	0	0	0

r	#	$\mathcal{A}$	ν_0	ν_1	χ
1	50	38	12	10	1
2	28	18.5	10	7	1
3	**6**	**2.5**	**2**	**3**	**1**
4	1	0	0	0	1
5	0	0	0	0	0

$M_{\mathcal{N}_8}$ according to which neighborhood is used for a specific radius and, finally, extrapolates the result vector of measures $\hat{\boldsymbol{\mu}}_\gamma(r)$ from $\boldsymbol{\mu}_\varepsilon(r)$.

The overall time complexity of the method, including the asymmetric distance transform [6], is of order $\mathcal{O}(KL + R)$, where KL is the number of pixels and R the maximal radius. The space complexity is of order $\mathcal{O}(K + R)$, as its only needs one line of K pixels for the distance transform, two lines of pixels and an array of size R for the extraction of vectors of characteristics. Moreover, the examination of 2×2 image patches can be merged with the single scan of the distance transform, leading to a streaming algorithm where computations are performed continuously while the data is input.

3 Results and Discussion

The method was tested on sections of tomographic reconstructions of composite materials and compared to the results of a morphological granulometry. Figure 8 shows an example of such an image. Numerical results are presented in Table 3 and pictured in Fig. 7. On this class of images, discrepancies on the pixel count are low (at most 2.25 %). They attain at most 13.33 % on the count of vertical and horizontal segments. Higher disparities appear on the Euler-Poincaré characteristic. This quantity measures the difference between the number of components and holes. It is highly sensitive to merging contours, which occur frequently in images with a high count of touching components like the one displayed in Fig. 8. Note that the pixel count is the only relevant measure here for granulometry. The other quantities are only displayed for completeness.

The source code of the complete method is available at https://github.com/nnormand/DGtalTools-contrib.git.

Table 3. Values $\#$, ν_0, ν_1 and χ measured on the actual openings of the image in Fig. 8 and the corresponding estimated values, $\hat{\#}$, $\hat{\nu}_0$, $\hat{\nu}_1$ and $\hat{\chi}$, obtained from a translated distance map.

r	$\#$	$\hat{\#}$	ν_0	$\hat{\nu}_0$	μ_1	$\hat{\mu}_1$	χ	$\hat{\chi}$
1	430033	430033	77442	77442	60168	60168	−333	−333
2	424083	424443	67596	67680	63763	62646	122	1386
3	408971	414309	80018	90682	52185	49706	1401	5288
4	382211	384196	67154	72648	59853	56763	2342	5545
5	133233	133896	29324	30332	15325	14700	1216	1790
6	1774	1814	270	296	209	219	14	18
7	0	0	0	0	0	0	0	0

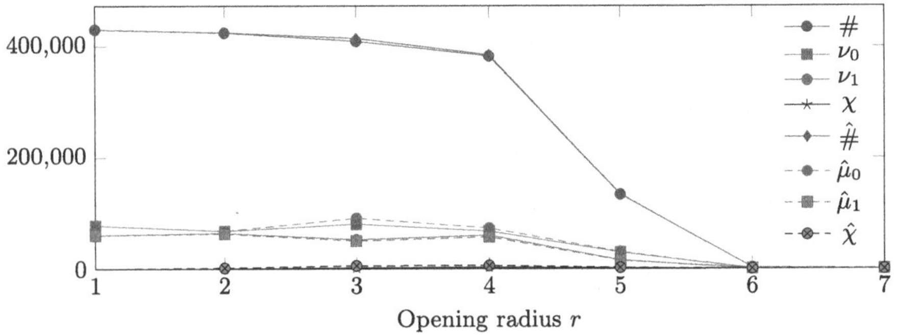

Fig. 7. Measured and estimated granulometry parameters

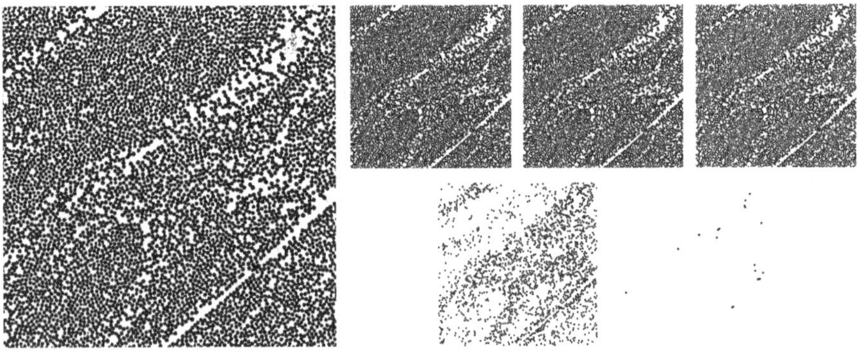

Fig. 8. Tomographic section of a composite material (orthogonal to fibers) and its openings ($r = 2$ to 6).

4 Conclusion and Future Works

This article describes a method for computing an estimation of the granulometric spectrum of a binary image by extrapolating parameters measured on the distance transform with a neighborhood-sequence distance. The algorithmic structure makes the method fit for very large amounts of data, or continuously acquired data. The time complexity of the algorithm is linear with the number of pixels for the distance transform, linear with the maximal disk radius for the measure extrapolation. Its space complexity is very low: not more that the result size (the vector of estimated parameters) and two lines of image.

The collision of contours during dilation is the sole reason why the number of pixels can be overestimated. A preliminary analysis shows that such collisions can be identified by saddle points in the distance transform. Further investigation is needed both to upper bound the pixel count error and conclude if and how the effects of contour merging can be neutralized or at least mitigated.

References

1. Coeurjolly, D.: Fast and accurate approximation of digital shape thickness distribution in arbitrary dimension. Comput. Vis. Image Underst. **116**(12), 1159–1167 (2012). ISSN 1077-3142
2. Federer, H.: Curvature measures. Trans. Am. Math. Soc. **93**(3), 418–491 (1959)
3. Guderlei, R., Klenk, S., Mayer, J., Schmidt, V., Spodarev, E.: Algorithms for the computation of the Minkowski functionals of deterministic and random polyconvex sets. Image Vis. Comput. **25**(4), 464–474 (2007). International Symposium on Mathematical Morphology 2005. ISSN 0262-8856
4. Klenk, S., Schmidt, V., Spodarev, E.: A new algorithmic approach to the computation of Minkowski functionals of polyconvex sets. Comput. Geom. **34**(3), 127–148 (2006). ISSN 0925-7721
5. Matheron, G.: La formule de Steiner pour les érosions. J. Appl. Probab. **15**(1), 126–135 (1978). French. ISSN 00219002
6. Normand, N., Strand, R., Evenou, P., Arlicot, A.: A streaming distance transform algorithm for neighborhood-sequence distances. Image Process. Line **4**, 196–203 (2014)
7. Normand, N., Strand, R., Evenou, P., Arlicot, A.: Minimal-delay distance transform for neighborhood-sequence distances in 2D and 3D. Comput. Vis. Image Underst. **117**(4), 409–417 (2013)
8. Pick, G.A.: Geometrisches zur Zahlenlehre. Sitzungsberichte des Deutschen Naturwissenschaftlich-Medicinischen Vereines für Böhmen “Lotos” in Prag **47**, 311–319 (1899). German
9. Snidaro, L., Foresti, G.L.: Real-time thresholding with Euler numbers. Pattern Recogn. Lett. **24**(9–10), 1533–1544 (2003). ISSN 0167-8655
10. Steiner, J.: Über parallele Flächen. In: Monatsbericht der Akademie der Wissenschaften zu Berlin, pp. 114–118 (1840, German)
11. Vincent, L.M.: Fast opening functions and morphological granulometries. In: Proceedings of SPIE Image Algebra and Morphological Image Processing V, vol. 2300, no. 1, pp. 253–267, July 1994

Recognizing Hierarchical Watersheds

Deise Santana Maia(✉), Jean Cousty, Laurent Najman, and Benjamin Perret

Université Paris-Est, LIGM (UMR 8049), CNRS, ENPC, ESIEE Paris, UPEM, 93162 Noisy-le-Grand, France
deisesantanamaia@gmail.com

Abstract. Combining hierarchical watersheds has proven to be a good alternative method to outperform individual hierarchical watersheds. Consequently, this raises the question of whether the resulting combinations are hierarchical watersheds themselves. Since the naive algorithm to answer this question has a factorial time complexity, we propose a new characterization of hierarchical watersheds which leads to a quasi-linear time algorithm to determine if a hierarchy is a hierarchical watershed.

1 Introduction

Hierarchical watersheds [2,5,10,13] are hierarchies of partitions obtained from an initial watershed segmentation [3,4]. They can be represented thanks to their saliency maps [1,6,13]. Departing from the watershed segmentation of an image, hierarchical watersheds are constructed by merging the regions of this initial segmentation according to regional attributes. As shown in [14], the performance of hierarchical watersheds is competitive compared to other hierarchical segmentation methods while being fast to compute.

Despite their good global performance, hierarchical watersheds based on a single regional attribute can fail at merging the adequate regions at all levels of the hierarchy. This problem is illustrated in Fig. 1, where we present the saliency maps of two hierarchical watersheds based on area [11] and dynamics [13], and two segmentation levels extracted from each hierarchy. In this representation of saliency maps, the darkest contours are the ones that persist at the highest levels of the hierarchies. From the saliency map of the hierarchical watershed based on area shown in Fig. 1, we can see that the sky region is oversegmented at high levels of this hierarchy. On the other hand, the beach is oversegmented at high levels of the hierarchy based on dynamics. To counter this problem, a method to combine hierarchical watersheds has been proposed and evaluated in [6,9]. As illustrated in Fig. 1, the combination of hierarchical watersheds can provide better segmentations than their individual counterparts.

Combining hierarchical watersheds raises the question whether the hierarchies resulting from combinations are hierarchical watersheds themselves. If so,

This research is partly funded by the Bézout Labex, funded by ANR, reference ANR-10-LABX-58.

M. Couprie et al. (Eds.): DGCI 2019, LNCS 11414, pp. 300–313, 2019.
https://doi.org/10.1007/978-3-030-14085-4_24

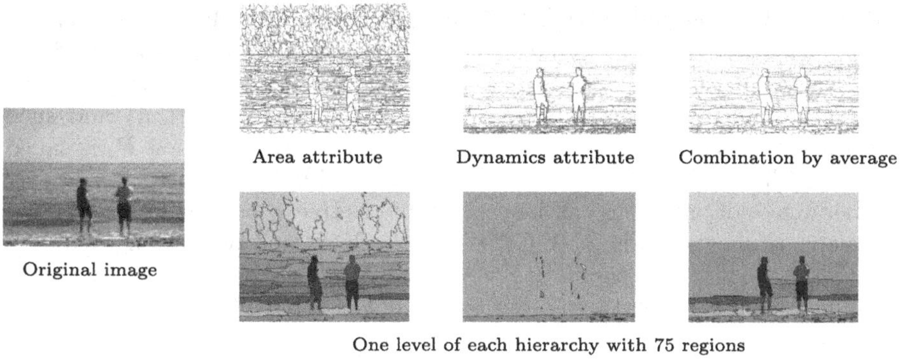

Fig. 1. Hierarchical watersheds based on area and dynamics and their combination by average.

then there might be an attribute A such that the combinations of hierarchical watersheds are precisely the hierarchical watersheds based on A. Therefore, we could simply compute hierarchical watersheds based on A, which is more efficient than combining hierarchies. Otherwise, it follows that the range of combinations of hierarchical watersheds is larger than the domain of hierarchical watersheds, which is an invitation for studying this new category of hierarchies. More generally, combining hierarchical watersheds raises the problem of recognizing hierarchical watersheds: given a hierarchy $\mathcal{H}$, decide if $\mathcal{H}$ is a hierarchical watershed.

The contributions of this article are twofold: (1) a new characterization of hierarchical watersheds; and (2) a quasi-linear time algorithm for solving the problem of recognizing hierarchical watersheds. Therefore, we have an efficient algorithm to determine if a given combination of hierarchical watersheds is a hierarchical watershed. It is noteworthy that the naive approach to recognize hierarchical watersheds has a factorial time complexity, which is explained later.

This article is organized as follows. In Sect. 2, we review basic notions on graphs, hierarchies and saliency maps. In Sect. 3, we formally state the problem of recognizing hierarchical watersheds, we introduce a characterization of hierarchical watersheds, and we present our quasi-linear time algorithm to recognize hierarchical watersheds.

2 Background Notions

In this section, we first introduce hierarchies of partitions. Then, we review the definition of graphs, connected hierarchies and saliency maps. Subsequently, we define hierarchical watersheds.

2.1 Hierarchies of Partitions

Let V be a set. A *partition* of V is a set $\mathbf{P}$ of non empty disjoint subsets of V whose union is V. If $\mathbf{P}$ is a partition of V, any element of $\mathbf{P}$ is called a *region of* $\mathbf{P}$. Let V be a set and let $\mathbf{P}_1$ and $\mathbf{P}_2$ be two partitions of V. We say that $\mathbf{P}_1$ is a *refinement* of $\mathbf{P}_2$ if every element of $\mathbf{P}_1$ is included in an element of $\mathbf{P}_2$. A *hierarchy (of partitions on* V*)* is a sequence $\mathcal{H} = (\mathbf{P}_0, \ldots, \mathbf{P}_n)$ of partitions of V such that $\mathbf{P}_{i-1}$ is a refinement of $\mathbf{P}_i$, for any $i \in \{1, \ldots, n\}$ and such that $\mathbf{P}_n = \{V\}$. For any i in $\{0, \ldots, n\}$, any region of the partition $\boldsymbol{P}_i$ is called a *region of* $\mathcal{H}$. The set of all regions of $\mathcal{H}$ is denoted by $\mathscr{R}(\mathcal{H})$.

Hierarchies of partitions can be represented as trees whose nodes correspond to regions, as shown in Fig. 2(a). Given a hierarchy $\mathcal{H}$ and two regions X and Y of $\mathcal{H}$, we say that X *is a parent of* Y *(or that* Y *is a child of* X*)* if $Y \subset X$ and X is minimal for this property. In other words, if X is a parent of Y and if there is a region Z such that $Y \subseteq Z \subset X$, then $Y = Z$.

In Fig. 2(a), the regions of the hierarchy $\mathcal{H}$ are linked to their parents (and to their children) by straight lines. It can be seen that any region X of $\mathcal{H}$ such that $X \neq V$ has exactly one parent. Thus, for any region X such that $X \neq V$, we write $parent(X) = Y$ where Y is the unique parent of X. For any region R of $\mathcal{H}$, if R is not the parent of any region of $\mathcal{H}$, we say that R *is a leaf region of* $\mathcal{H}$. Otherwise, we say that R *is a non-leaf region of* $\mathcal{H}$. The set of all non-leaf regions of $\mathcal{H}$ is denoted by $\mathscr{R}^*(\mathcal{H})$.

In the hierarchy of Fig. 2(a), we have $parent(M_1) = parent(M_2) = X_5$. The set of non-leaf regions of $\mathcal{H}$ is $\mathscr{R}^* = \{X_1, X_2, X_3, X_4, X_5, X_6, X_7\}$.

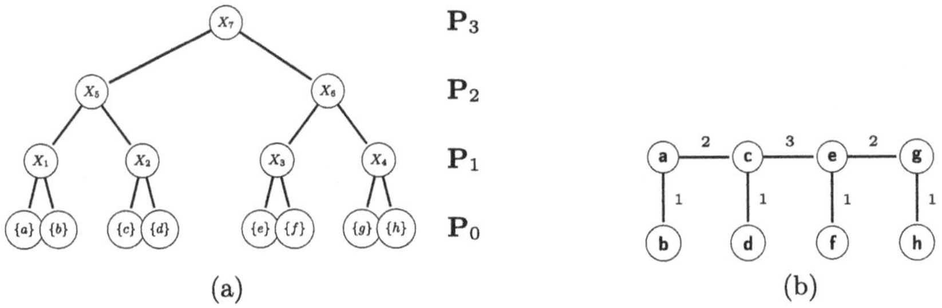

Fig. 2. (a) A representation of a hierarchy of partitions $\mathcal{H} = (\mathbf{P}_0, \mathbf{P}_1, \mathbf{P}_2, \mathbf{P}_3)$ on the set $\{a, b, c, d, e, f, g, h\}$. (b) A weighted graph (G, w).

2.2 Graphs, Connected Hierarchies and Saliency Maps

A *graph* is a pair $G = (V, E)$, where V is a finite set and E is a set of pairs of distinct elements of V, *i.e.*, $E \subseteq \{\{x, y\} \subseteq V | x \neq y\}$. Each element of V is called a *vertex (of* G*)*, and each element of E is called an *edge (of* G*)*. To simplify the notations, the set of vertices and edges of a graph G will be also denoted by $V(G)$ and $E(G)$, respectively.

Let $G = (V, E)$ be a graph and let X be a subset of V. A sequence $\pi = (x_0, \ldots, x_n)$ of elements of X is a *path (in* X*) from* x_0 *to* x_n if $\{x_{i-1}, x_i\}$ is an edge of G for any i in $\{1, \ldots, n\}$. The subset X of V is said to be *connected* if, for any x and y in X, there exists a path from x to y. The subset X is a *connected component* of G if X is connected and if, for any connected subset Y of V, if $X \subseteq Y$, then we have $X = Y$. In the following, we denote by $CC(G)$ the set of all connected components of G. It is well known that this set $CC(G)$ of all connected components of G is a partition of the set V.

Let $G = (V, E)$ be a graph. *A partition of V is connected for G* if each of its regions is connected and *a hierarchy on V is connected (for G)* if every one of its partitions is connected. For example, the hierarchy of Fig. 2(a) is connected for the graph of Fig. 2(b).

Let G be a graph. If w is a map from the edge set of G to the set $\mathbb{R}^+$ of positive real numbers, then the pair (G, w) is called an *(edge) weighted graph.* If (G, w) is a weighted graph, for any edge u of G, the value $w(u)$ is called the *weight of u (for w).*

As established in [6], a connected hierarchy can be equivalently treated by means of a weighted graph through the notion of a saliency map. Given a weighted graph (G, w) and a hierarchy $\mathcal{H} = (\mathbf{P}_0, \ldots, \mathbf{P}_n)$ connected for G, the *saliency map of $\mathcal{H}$* is the map from $E(G)$ to $\{0, \ldots, n\}$, denoted by $\Phi(\mathcal{H})$, such that, for any edge $u = \{x, y\}$ in $E(G)$, the value $\Phi(\mathcal{H})(u)$ is the smallest value i in $\{0, \ldots, n\}$ such that x and y belong to a same region of $\mathbf{P}_i$. It follows that any connected hierarchy has a unique saliency map. Moreover, from any saliency map, we can recover the departing connected hierarchy. Consequently, the bijection between connected hierarchies and saliency maps allows us to work indifferently with any of those notions during our study. For instance, the map depicted in Fig. 2(b) is the saliency map of the hierarchy of Fig. 2(a).

2.3 Hierarchical Minimum Spanning Forests and Watersheds

The watershed segmentation [3,4] derives from the topographic definition of watersheds lines and catchment basins. A catchment basin is a region whose collected precipitation drains to the same body of water, as a sea, and the watershed lines are the separating lines between neighbouring catchment basins. In [4], the authors formalize watersheds in the framework of weighted graphs and show the optimality of watersheds in the sense of minimum spanning forests. In this section, we present hierarchical watersheds following the definition of hierarchical minimum spanning forests (see efficient algorithm in [5,12]).

Let G be a graph. We say that *G is a forest* if, for any edge u in $E(G)$, the number of connected components of $(V(G), E(G) \setminus \{u\})$ is larger than the number of connected components of G. Given another graph G', we say that *G' is a subgraph of G*, denoted by $G' \sqsubseteq G$, if $V(G') \subseteq V(G)$ and $E(G') \subseteq E(G)$. Let (G, w) be a weighted graph and let G' be a subgraph of G. A graph G'' is a *Minimum Spanning Forest (MSF) of G rooted in G'* if:

1. the graphs G and G'' have the same set of vertices, *i.e.*, $V(G'') = V(G)$; and
2. the graph G' is a subgraph of G''; and
3. each connected component of G'' includes exactly one connected component of G'; and

4. the sum of the weight of the edges of G'' is minimal among all graphs for which the above conditions 1, 2 and 3 hold true.

Intuitively, a drop of water on a topographic surface drains in the direction of a local minimum. Indeed, there is a bijection between the catchment basins of a surface and its local minima. As established in [4], in the framework of weighted graphs, watershed-cuts can be induced by the minimum spanning forest rooted in the minima of this graph. Let (G, w) be a weighted graph and let k be a value in $\mathbb{R}^+$. A subgraph G' of G is a *minimum* (of w) at level k if:

1. $V(G')$ is connected for G; and
2. for any edge u in $E(G')$, the weight of u is equal to k; and
3. for any edge $\{x, y\}$ in $E(G) \setminus E(G')$ such that $|\{x, y\} \cap V(G')| \geq 1$, the weight of $\{x, y\}$ is strictly greater than k.

In the following, we define hierarchical watersheds which are optimal in the sense of minimum spanning forests [5].

Definition 1 (hierarchical watershed). *Let (G, w) be a weighted graph and let $\mathcal{S} = (M_1, \ldots, M_n)$ be a sequence of pairwise distinct minima of w such that $\{M_1, \ldots, M_n\}$ is the set of all minima of w. Let $(G_0, \ldots, G_{n-1})$ be a sequence of subgraphs of G such that:*

1. for any $i \in \{0, \ldots, n-1\}$, the graph G_i is a MSF of G rooted in $(\cup\{V(M_j) \mid j \in \{i+1, \ldots, n\}\}, \cup\{E(M_j) \mid j \in \{i+1, \ldots, n\}\})$; and
2. for any $i \in \{1, \ldots, n-1\}$, we have $G_{i-1} \sqsubseteq G_i$.

The sequence $\mathcal{T} = (CC(G_0), \ldots, CC(G_{n-1}))$ is called a hierarchical watershed *of (G, w) for $\mathcal{S}$. Given a hierarchy $\mathcal{H}$, we say that $\mathcal{H}$* is a hierarchical watershed *of (G, w) if there exists a sequence $\mathcal{S} = (M_1, \ldots, M_n)$ of pairwise distinct minima of w such that $\{M_1, \ldots, M_n\}$ is the set of all minima of w and such that $\mathcal{H}$ is a hierarchical watershed for $\mathcal{S}$.*

A weighted graph (G, w) and a hierarchical watershed $\mathcal{H}$ of (G, w) are illustrated in Fig. 3(a) and (b), respectively. We can see that $\mathcal{H}$ is the hierarchical watershed of (G, w) for the sequence $\mathcal{S} = (M_1, M_2, M_3, M_4)$.

Important Notations and Notions: In the sequel of this article, the symbol G denotes a tree, *i.e.*, a forest with a unique connected component. To shorten the notation, the vertex set of G is denoted by V and its edge set is denoted by E. The symbol w denotes a map from E into $\mathbb{R}^+$ such that, for any pair of distinct edges u and v in E, we have $w(u) \neq w(v)$. Thus, the pair (G, w) is a weighted graph. The number of minima of w is denoted by n. Every hierarchy considered in this article is connected for G. Therefore, for the sake of simplicity, we use the term *hierarchy* instead of *hierarchy which is connected for G*.

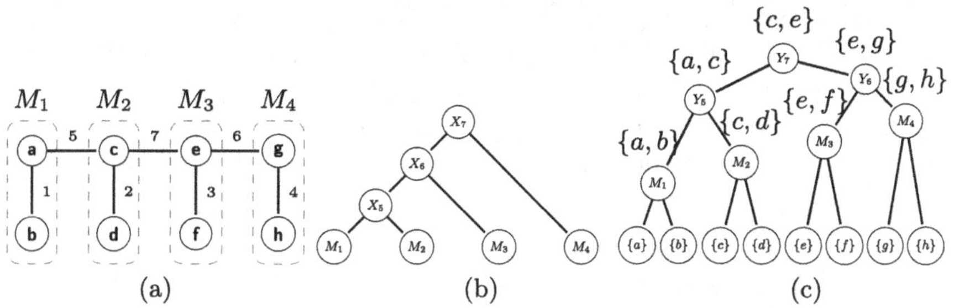

Fig. 3. (a) A weighted graph (G, w) with four minima delimited by the dashed lines. (b) The hierarchical watershed of (G, w) for the sequence (M_1, M_2, M_3, M_4). (c) The binary partition hierarchy $\mathcal{B}$ of (G, w).

3 Recognition of Hierarchical Watersheds

In this study, we tackle the following recognition problem:

(P) given a weighted graph (G, w) and a hierarchy of partitions $\mathcal{H}$, determine if $\mathcal{H}$ is a hierarchical watershed of (G, w).

Problem (P) can be related to the problem studied in [8]. In [8], the authors search for a minimum set of markers which lead to a given watershed segmentation. In our case, we are interested in ordering a predefined set of markers (the set of all minima of w) that allows to solve the minimum set of markers problem for the series of all watershed segmentations (partitions) of a given hierarchy.

A naive approach to solve Problem (P) is to verify if there is a sequence $\mathcal{S} = (M_1, \ldots, M_n)$ of pairwise distinct minima of w such that $\mathcal{H}$ is the hierarchical watershed of (G, w) for $\mathcal{S}$. However, there exist $n!$ sequences of n pairwise minima of w, which leads to an algorithm of factorial time complexity.

To solve Problem (P) more efficiently, we propose in Sect. 3.2 a characterization of hierarchical watersheds based on the binary partition hierarchy by altitude ordering (Sect. 3.1) which, as stated in [7], is known to be closely related to hierarchical watersheds. This characterization leads to a quasi-linear time algorithm to recognize hierarchical watersheds (Sect. 3.3).

3.1 Binary Partition Hierarchies by Altitude Ordering

Given any set X, we denote by $|X|$ the cardinality of X. Let k be any element in $\{1, \ldots, |E|\}$. We denote by u_k the edge in E such that there are $k-1$ edges in E whose weights are strictly smaller than $w(u_k)$. We set $\mathbf{B}_0 = \{\{x\} \mid x \in V\}$. The *$k$-partition of V (by the map w)* is defined by $\mathbf{B}_k = \{\mathbf{B}^y_{k-1} \cup \mathbf{B}^x_{k-1}\} \cup \mathbf{B}_{k-1} \setminus \{\mathbf{B}^x_{k-1}, \mathbf{B}^y_{k-1}\}$ where $u_k = \{x, y\}$ and $\mathbf{B}^x_{k-1}$ and $\mathbf{B}^y_{k-1}$ are the regions of $\mathbf{B}_{k-1}$ that contain x and y, respectively. The edge u_k is called a *building edge of the region* $\{\mathbf{B}^y_{k-1} \cup \mathbf{B}^x_{k-1}\}$. The *binary partition hierarchy by altitude ordering (of (G, w))*, denoted by $\mathcal{B}$, is the hierarchy $(\mathbf{B}_0, \ldots, \mathbf{B}_{|E|})$. Since G is a tree, we have $\mathbf{B}_i \neq \mathbf{B}_{i-1}$ for any i in $\{1, \dots, |E|\}$, so each region of $\mathcal{B}$ has a unique building edge and each edge of G is the building edge of a region of $\mathcal{B}$. Given an edge u in E, we denote by R_u the region of $\mathcal{B}$ whose building edge is u.

In Fig. 3(c), we present the binary partition hierarchy $\mathcal{B}$ of the graph (G, w) (shown in Fig. 3(a)). The building edges are shown above the regions in $\mathscr{R}^*(\mathcal{B})$.

Important Notation: In the sequel of this article, the binary partition hierarchy by altitude ordering of (G, w) is denoted by $\mathcal{B}$.

Let X and Y be two distinct regions of $\mathcal{B} \setminus \{V\}$. If the parents of X and of Y are the same, we say that X is a sibling of Y, that Y is a sibling of X or that X and Y are siblings. It can be seen that X has exactly one sibling and we denote this unique sibling of X by $sibling(X)$.

3.2 Characterization of Hierarchical Watersheds

In this section, we first introduce one-side increasing maps, which, as shown later, are closely related to hierarchical watersheds. Then, we state the main result of this article (Theorem 3), which characterizes the hierarchical watersheds as the hierarchies whose saliency maps are one-side increasing maps. Finally, we present a sketch of the proof of Theorem 3.[1]

To define one-side increasing maps, we first introduce watershed-cut edges for a map. In a topographic surface, the watershed lines are the set of separating lines between neighbouring catchment basins. In the framework of weighted graphs, *the catchment basins of the map* w are the connected components of the minimum spanning forest rooted in the minima of w. In other words, the catchment basins of w are the leaf regions of any hierarchical watershed of (G, w). Thus, given any edge $u = \{x, y\}$ in E, we say that u *is a watershed-cut edge for* w if x and y are in distinct catchment basins of w.

Important Notation: In the sequel of this article, we denote by $WS(w)$ the set of watershed-cut edges of w.

Definition 2 (one-side increasing map). *We say that a map f from E into $\mathbb{R}^+$ is a* one-side increasing map for $\mathcal{B}$ *if:*

1. *$range(f) = \{0, \dots, n-1\}$;*
2. *for any u in E, $f(u) > 0$ if and only if $u \in WS(w)$; and*
3. *for any u in E, there exists a child R of R_u such that $f(u) \geq \vee\{f(v)$ such that R_v is included in $R\}$, where $\vee\{\} = 0$.*

where $range(f) = \{f(u) \mid u \in E\}$

The next theorem establishes that hierarchical watersheds can be characterized as the hierarchies whose saliency maps are one-side increasing maps.

[1] The proofs of the lemmas, properties and theorem presented in this article can be found in https://perso.esiee.fr/~santanad/proofs.pdf.

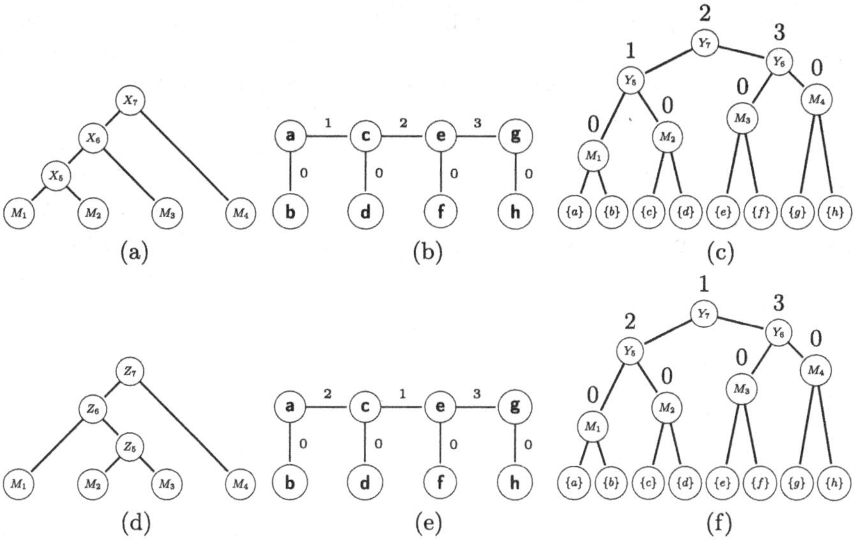

Fig. 4. (a) and (d) The hierarchies $\mathcal{H}$ and $\mathcal{H}'$, respectively. (b) and (e) the weighted graphs $(G, \Phi(\mathcal{H}))$ and $(G, \Phi(\mathcal{H}'))$, respectively. (c) and (f) The maps $\Phi(\mathcal{H})$ and $\Phi(\mathcal{H}')$ represented on the hierarchy $\mathcal{B}$ of Fig. 3(c). For any edge u, the values $\Phi(\mathcal{H})(u)$ and $\Phi(\mathcal{H}')(u)$ are shown above the region R_u of $\mathcal{B}$.

Theorem 3 (characterization of hierarchical watersheds). *Let $\mathcal{H}$ be a hierarchy and let $\Phi(\mathcal{H})$ be the saliency map of $\mathcal{H}$. The hierarchy $\mathcal{H}$ is a hierarchical watershed of (G, w) if and only if $\Phi(\mathcal{H})$ is a one-side increasing map for $\mathcal{B}$.*

Let us consider the hierarchy $\mathcal{H}$, the saliency map $\Phi(\mathcal{H})$ of $\mathcal{H}$ and the binary partition hierarchy by altitude ordering $\mathcal{B}$ (of (G, w) of Fig. 3(a)) shown in Fig. 4(a), (b) and (c), respectively. It can be verified that $\Phi(\mathcal{H})$ is one-side increasing for $\mathcal{B}$. Thus, by Theorem 3, we may affirm that $\Phi(\mathcal{H})$ is the saliency map of a hierarchical watershed of (G, w) and that $\mathcal{H}$ is a hierarchical watershed of (G, w). On the other hand, the saliency map $\Phi(\mathcal{H}')$ shown in Fig. 4(e), is not one-side increasing for $\mathcal{B}$. Indeed, the weight $\Phi(\mathcal{H}')(\{c, e\})$ of the building edge of the region Y_7 is 1, which is smaller than both $\vee\{\Phi(\mathcal{H}')(v) \mid R_v \subseteq Y_5\} = 2$ and $\vee\{\Phi(\mathcal{H}')(v) \mid R_v \subseteq Y_6\} = 3$. Hence, the condition 3 of Definition 2 is not satisfied by $\Phi(\mathcal{H}')$. Thus, by Theorem 3, we may deduce that $\Phi(\mathcal{H}')$ is not the saliency map of a hierarchical watershed of (G, w) and that $\mathcal{H}'$ is not a hierarchical watershed of (G, w).

In the remaining of this section, we present a sketch of the proof of Theorem 3. To prove the forward implication of Theorem 3, we first present the definition of extinction values and extinction maps.

Let $S = (M_1, \ldots, M_n)$ be a sequence of n pairwise distinct minima of w. Let R be a region of $\mathcal{B}$. As defined in [7], the *extinction value of R for S* is zero if there is no minimum of w included in R and, otherwise, it is the maximum value i in $\{1, \ldots, n\}$ such that the minimum M_i is included in R. Given a map P from the regions of $\mathcal{B}$ to $\mathbb{R}^+$, we say that P *is an extinction map of* w if there exists a sequence S of n pairwise distinct minima of w such that, for any region R of $\mathcal{B}$, the value $P(R)$ is the extinction value of R for $\mathcal{S}$.

We provide an example of an extinction map in Fig. 5(a). We can see that the map P is the extinction map of w for the sequence $\mathcal{S} = (M_2, M_1, M_4, M_3)$.

The next property clarifies the relation between hierarchical watersheds and extinction maps. It can be deduced from the results of [7], which makes a correspondence between extinction values for a given sequence of minima of w and the hierarchical watershed for this sequence of minima.

Property 4. *Let f be a map from E into $\mathbb{R}^+$. The map f is the saliency map of a hierarchical watershed of (G, w) if and only if there exists an extinction map P of w such that, for any u in E, we have*

$$f(u) = \min\{P(R)\; such\, that\, R\, is\, a\, child\, of\; R_u\}.$$

The forward implication of Theorem 3 can be obtained from Property 4. Given a hierarchy $\mathcal{H}$, if $\mathcal{H}$ is a hierarchical watershed, then there exists an extinction map P of w such that, for any u in E, we have $\Phi(\mathcal{H})(u) = \min\{P(R)$ such that R is a child of $R_u\}$, which can be proven to be a one-side increasing map. In order to establish the backward implication of Theorem 3, we introduce the notion of estimated extinction maps. Given the saliency map $\Phi(\mathcal{H})$ of a hierarchy $\mathcal{H}$, an estimated extinction map of $\Phi(\mathcal{H})$ is a map P such that, for any u in E, we have $\Phi(\mathcal{H})(u) = \min\{P(R)$ such that R is a child of $R_u\}$. Indeed, if P is an extinction map, then $\Phi(\mathcal{H})$ is a one-side increasing map.

Important Notation: In the sequel, we consider a total ordering $\prec$ on the regions of $\mathcal{B}$ such that, given any two edges u and v in E, if $w(u) < w(v)$ then $R_u \prec R_v$. Since the leaf regions of $\mathcal{B}$ do not have building edges, there are several orderings on the regions of $\mathcal{B}$ with this property. However, the lemmas stated here hold for any arbitrary choice of $\prec$.

Definition 5 (estimated extinction map). *Let f be a one-side increasing map. The* estimated extinction map of f *is the map ξ_f from $\mathscr{R}(\mathcal{B})$ into $\mathbb{R}^+$ such that:*

1. $\xi_f(R) = n$ *if* $R = V$;
2. $\xi_f(R) = 0$ *if there is no minimum M of w such that $M \subseteq R$;*
3. $\xi_f(R) = f(u)$, *where R_u is the parent of R, if there is a minimum M of w such that $M \subseteq R$ and $sibling(R) \in \mathscr{R}^*(\mathcal{B})$ and*
   - *if* $\vee^f(R) < \vee^f(sibling(R))$
   - *or if* $\vee^f(R) = \vee^f(sibling(R))$ *and* $R \prec sibling(R)$; *and*
4. $\xi_f(R) = \xi_f(S)$, *where S is the parent of R otherwise.*

where $\vee^f(R) = \vee\{f(v) \mid v \in E, R_v \subseteq R\}$

The next lemma establishes that the estimated extinction map of any one-side increasing map is indeed an extinction map.

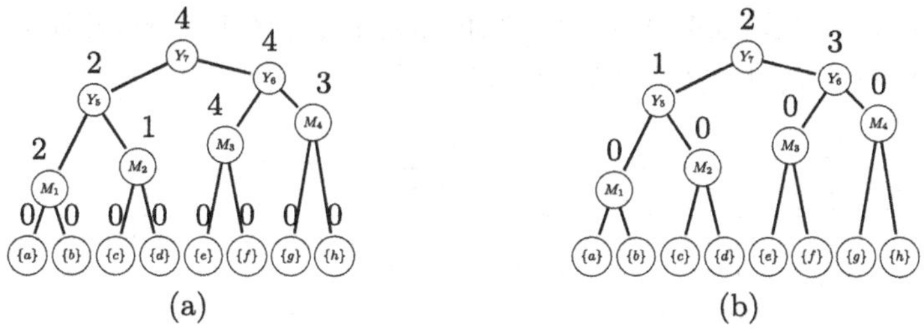

Fig. 5. (a) An estimated extinction map P of w (Fig. 3(a)). (b) The saliency map of Fig. 4(b) represented on the binary partition hierarchy by altitude ordering $\mathcal{B}$ of the graph (G, w) (Fig. 3(a)).

Lemma 6. *Let f be a one-side increasing map for $\mathcal{B}$. The estimated extinction map ξ_f of f is an extinction map of w.*

For instance, let $\prec$ be a total ordering on the regions of the hierarchy $\mathcal{B}$ of Fig. 3(c) such that $\{a\} \prec \{b\} \prec \{c\} \prec \{d\} \prec \{e\} \prec \{f\} \prec \{g\} \prec \{h\} \prec M_2 \prec M_1 \prec M_4 \prec M_3 \prec Y_5 \prec Y_6 \prec Y_7$. Then, the extinction map P of Fig. 5(a) is precisely the estimated extinction map $\xi_{\Phi(\mathcal{H})}$ of the map $\Phi(\mathcal{H})$ of Fig. 5(b).

The next lemma is the key result for establishing the backward implication of Theorem 3.

Lemma 7. *Let f be a one-side increasing map. Then, for any u in E, we have*

$$f(u) = min\{\xi_f(R)\, such\, that\, R\, is\, a\, child\, of\, R_u\}.$$

The backward implication of Theorem 3 is a consequence of Lemmas 6 and 7 and the backward implication of Property 4. Let $\mathcal{H}$ be a hierarchy. If $\Phi(\mathcal{H})$ is a one-side increasing map, then the estimated extinction map $\xi_{\Phi(\mathcal{H})}$ of $\Phi(\mathcal{H})$ is an extinction map of w by Lemma 6. Thus, for any u in E, we have $\Phi(\mathcal{H})(u) = min\{\xi_{\Phi(\mathcal{H})}(R)$ such that R is a child of $R_u\}$ by Lemma 7. Then, by the backward implication of Property 4, we conclude that $\Phi(\mathcal{H})$ is the saliency map of a hierarchical watershed of (G, w) and that $\mathcal{H}$ is a hierarchical watershed of (G, w).

To illustrate Lemma 7, in Fig. 5, we can verify that $\Phi(\mathcal{H})(u) = \min\{P(R)$ such that R is a child of $R_u\}$ for any edge u in $E(G)$ where P (shown in Fig. 5(a)) is precisely the estimated extinction map of $\Phi(\mathcal{H})$.

3.3 Recognition Algorithm for Hierarchical Watersheds

In this section, we present a quasi-linear time algorithm to recognize hierarchical watersheds based on Theorem 3. Given any hierarchy $\mathcal{H}$, to test if $\mathcal{H}$ is a hierarchical watershed of (G, w), it is sufficient to verify that the saliency map of $\mathcal{H}$ is a one-side increasing map for $\mathcal{B}$.

Algorithm 1 provides a description of our algorithm to recognize hierarchical watersheds. The inputs are a weighted graph $((V, E), w)$ whose edge weights are already sorted and the saliency map f of a hierarchy $\mathcal{H}$. In this implementation, the edges in E are represented by unique indexes ranging from 1 to $|E|$.

Algorithm 1. Recognition of hierarchical watersheds

Data: $((V, E), w)$: a weighted graph whose edges are sorted in increasing order of weights for w
f: the saliency map of a hierarchy $\mathcal{H}$
Result: *true* if $\mathcal{H}$ is a hierarchical watershed of (G, w) and *false* otherwise

// In this algorithm, we consider that the default value of any array position is zero
Compute the binary partition hierarchy $\mathcal{B}$ of $((V, E), w)$
// Computation of the array WS of watershed edges of w and of their number $k = |WS(w)|$
Declare WS as an array of $|E|$ integers
$k := 0$
for each edge u in E **do**
  if none of the children of R_u is a leaf node of $\mathcal{B}$ **then**
    $WS[u] := 1$
    $k := k + 1$
// Computation of the array Max such that, for any region R of $\mathcal{B}$ whose building edge is u, we have $Max[u] = \vee^f(R)$, where $\vee^f$ is given in Definition 5
Declare Max as an array of $|E|$ real numbers
for each edge u in increasing order of weights for w **do**
  $Max[u] := f[u]$
  for each child X of R_u **do**
    if X is not a leaf node of $\mathcal{B}$ **then**
      $v :=$ the building edge of X
      $Max[u] := max(Max[u], Max[v])$
// Testing of the conditions 1, 2 and 3 of Definition 2 for f to be a one-side increasing map for $\mathcal{B}$
Declare *range* as an array of $|E|$ integers
for each edge u in E **do**
  if $f[u] \notin \{0, 1, \dots, k\}$ **then return** *false*
  if $f[u] \neq 0$ and $range[f[u]] \neq 0$ **then return** *false*
  $range[f[u]] := 1$
  if $(WS[u] = 0$ and $f[u] \neq 0)$ or $(WS[u] = 1$ and $f[u] = 0)$ **then return** *false*
  if both children of R_u are in $\mathscr{R}^*(\mathcal{B})$ **then**
    $v_1 :=$ *the building edge of a child of* u
    $v_2 :=$ *the building edge of* $sibling(R_{v_1})$
    if $f[u] \leq Max[v_1]$ and $f[u] \leq Max[v_2]$ **then return** *false*
return *true*

The first step of Algorithm 1 is to compute the binary partition hierarchy by altitude ordering $\mathcal{B}$ of $((V, E), w)$. As established in [12], any binary partition hierarchy can be computed in quasi-linear time with respect to $|E|$ provided that

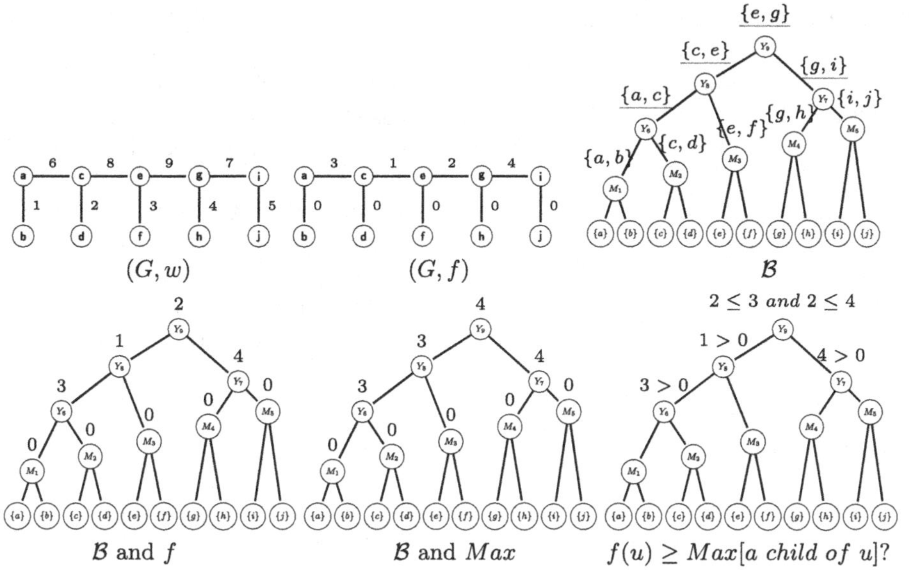

Fig. 6. Toy example of Algorithm 1.

the edges in E are already sorted or can be sorted in linear time. Subsequently, the computation of the watershed edges of w and the number of watershed edges at lines 2–7 is based on the quasi-linear time algorithm proposed in [12]. At lines 8–14, for each edge u in E, we compute $Max[u]$, which is the maximal value $f(v)$ such that $R_v \subseteq R_u$. Since each region of $\mathcal{B}$ has at most two children, the time complexity to compute the array Max is also linear with respect to $|E|$. The last **for** loop (lines 15–24) verifies that the three conditions of Definition 2 for f to be a one-side increasing map hold true. Each instruction between lines 17 and 24 can be performed in constant time. Therefore, the overall time complexity of Algorithm 1 is quasi-linear with respect to $|E|$.

We illustrate Algorithm 1 in Fig. 6. The inputs are a weighted graph (G, w) and a saliency map f. First, we obtain the binary partition hierarchy by altitude ordering $\mathcal{B}$ of (G, w) and the four watershed-cut edges of w (underlined). Then, we compute the array Max. For each edge u of G, the value $Max[u]$ is the greatest value in the set $\{f(v) \mid R_v \subseteq R_u\}$. We can verify that the range of f is $\{0, 1, 2, 3, 4\}$ and that only the watershed-cut edges of w have non-zero weights for f. Therefore, the conditions 1 and 2 of Definition 2 for f to be a one-side increasing map for $\mathcal{B}$ hold true. Finally, we test the condition 3 of Definition 2. For each watershed-cut edge u of G, we test if $f(u) > Max[v]$ for an edge v such that R_v is a child of R_u. For the building edges of the regions Y_6, Y_7 and Y_8 the condition 3 hold true, but this is not the case for Y_9. Thus, f is not a one-side increasing map for $\mathcal{B}$ and Algorithm 1 returns *false*.

4 Conclusion

We introduced a new characterization of hierarchical watersheds based on binary partition hierarchies by altitude ordering. Based on this characterization, we designed a quasi-linear time algorithm to determine if a hierarchy is a hierarchical watershed.

In future work, we will extend the recognition of hierarchical watersheds to arbitrary graphs, *i.e.*, graphs which are not trees and whose edge weights are not pairwise distinct. We are also interested in the frequency study of hierarchical watersheds, namely investigating which hierarchies are more likely to be the hierarchical watersheds of a weighted graph (G, w) for arbitrary sequences of minima of w.

References

1. Arbelaez, P., Maire, M., Fowlkes, C., Malik, J.: Contour detection and hierarchical image segmentation. IEEE PAMI **33**(5), 898–916 (2011)
2. Beucher, S.: Watershed, hierarchical segmentation and waterfall algorithm. In: Serra, J., Soille, P. (eds.) ISMM, pp. 69–76. Kluwer, Dordrecht (1994)
3. Beucher, S., Meyer, F.: The morphological approach to segmentation: the watershed transformation. Opt. Eng. **34**, 433 (1992). Marcel Dekker Incorporated, New York
4. Cousty, J., Bertrand, G., Najman, L., Couprie, M.: Watershed cuts: minimum spanning forests and the drop of water principle. IEEE PAMI **31**(8), 1362–1374 (2009)
5. Cousty, J., Najman, L.: Incremental algorithm for hierarchical minimum spanning forests and saliency of watershed cuts. In: Soille, P., Pesaresi, M., Ouzounis, G.K. (eds.) ISMM 2011. LNCS, vol. 6671, pp. 272–283. Springer, Heidelberg (2011). https://doi.org/10.1007/978-3-642-21569-8_24
6. Cousty, J., Najman, L., Kenmochi, Y., Guimarães, S.: Hierarchical segmentations with graphs: quasi-flat zones, minimum spanning trees, and saliency maps. JMIV **60**(4), 479–502 (2018)
7. Cousty, J., Najman, L., Perret, B.: Constructive links between some morphological hierarchies on edge-weighted graphs. In: Hendriks, C.L.L., Borgefors, G., Strand, R. (eds.) ISMM 2013. LNCS, vol. 7883, pp. 86–97. Springer, Heidelberg (2013). https://doi.org/10.1007/978-3-642-38294-9_8
8. Lotufo, R., Silva, W.: Minimal set of markers for the watershed transform. In: Proceedings of ISMM 2002, pp. 359–368 (2002)
9. Santana Maia, D., de Albuquerque Araujo, A., Cousty, J., Najman, L., Perret, B., Talbot, H.: Evaluation of combinations of watershed hierarchies. In: Angulo, J., Velasco-Forero, S., Meyer, F. (eds.) ISMM 2017. LNCS, vol. 10225, pp. 133–145. Springer, Cham (2017). https://doi.org/10.1007/978-3-319-57240-6_11
10. Meyer, F.: The dynamics of minima and contours. In: Maragos, P., Schafer, R., Butt, M. (eds.) ISMM, pp. 329–336. Kluwer, Dordrecht (1996)
11. Meyer, F., Vachier, C., Oliveras, A., Salembier, P.: Morphological tools for segmentation: connected filters and watersheds. Ann. Télécommun. **52**, 367–379 (1997)

12. Najman, L., Cousty, J., Perret, B.: Playing with Kruskal: algorithms for morphological trees in edge-weighted graphs. In: Hendriks, C.L.L., Borgefors, G., Strand, R. (eds.) ISMM 2013. LNCS, vol. 7883, pp. 135–146. Springer, Heidelberg (2013). https://doi.org/10.1007/978-3-642-38294-9_12
13. Najman, L., Schmitt, M.: Geodesic saliency of watershed contours and hierarchical segmentation. IEEE PAMI **18**(12), 1163–1173 (1996)
14. Perret, B., Cousty, J., Guimaraes, S.J.F., Maia, D.S.: Evaluation of hierarchical watersheds. IEEE TIP **27**(4), 1676–1688 (2018)

Shape Representation, Recognition and Analysis

Straight Line Reconstruction for Fully Materialized Table Extraction in Degraded Document Images

Héloïse Alhéritière[1,2](✉), Walid Amaïeur[1](✉), Florence Cloppet[1](✉), Camille Kurtz[1](✉), Jean-Marc Ogier[2](✉), and Nicole Vincent[1](✉)

[1] LIPADE, Université Paris Descartes, Paris, France
{heloise.alheritiere,walid.amaieur,florence.cloppet,
camille.kurtz,nicole.vincent}@parisdescartes.fr
[2] Université de La Rochelle, L3i, La Rochelle, France
{heloise.alheritiere,jean-marc.ogier}@univ-lr.fr

Abstract. Tables are one of the best ways to synthesize information such as statistical results, key figures in documents. In this article we focus on the extraction of materialized tables in document images, in the particular case where acquisition noise can disrupt the recovering of the table structures. The sequential printings/scannings of a document and its deterioration can lead to "broken" lines among the materialized segments of the tables. We propose a method based on the search for straight line segments in documents, relying on a new image transform that locally defines primitives well suited for pattern recognition and on a proposed theoretical model of lines in order to confirm their presence among a set of confident potential line parts. The extracted straight line segments are then used to reconstruct the table structures. Our approach has been evaluated both from quality and stability points of view.

Keywords: Table extraction · Straight line features · Local diameter transforms · Degraded lines · Document images

1 Introduction

A good image analysis procedure is largely depending on relevant choices for the image representation and the features used. In our case, we are concerned with images of documents, whose quantity exchanged world-widely increases exponentially for multiple usages. The semantic aspect of such images is mainly contained in the text that can be recovered by optical character recognition (OCR) softwares, but also by other media such as graphics or images. Besides the layout of the image helps the reader to interpret the message carried by the document. Unfortunately, in most OCR softwares, the presence of a table may have a negative impact on text recognition. If tables are extracted first, text

This work was supported by the French ANR under Grant ANR-14-CE28-0022.

M. Couprie et al. (Eds.): DGCI 2019, LNCS 11414, pp. 317–329, 2019.
https://doi.org/10.1007/978-3-030-14085-4_25

within each cell can be processed separately leading to better results. Thus the table extraction presents a two-fold interest, first to improve the text recognition rate and second to interpret the information contained in the table, yielding the improvement of the global document understanding.

Tables are one of the most specific structure of the image layout and they are present in a wide variety of documents such as administrative documents, invoices, scientific articles, question forms, etc. By their nature, the semantic aspect of the tables may be lost by the raster representation of the image. Thanks to recent office softwares, several table displays are used, linked to materialized cells surrounded by straight line segments or linked to different colors indicating columns or rows or to well chosen alignments. In this paper we limit ourselves to fully materialized tables, composed and surrounded by straight lines that intersect at right angles and are completely closed. This kind of table is most prevalent in commercial classic text editors. The acquisition noise due to sequential printings/scannings, deteriorations of the document may damage the table structure in raster images by generating "broken" segments. This lead to a proposal of straight line recovery that we experiment on the task of table extraction in degraded documents where the lines are no longer visible.

In this article, we address the problem of table extraction in binary images based on a model of straight lines representation of images. Indeed, we have chosen to replace the pixel classical representation of images by the use of a unique feature, the straight line segments. They can cover both the foreground and the background of the document. A pixel is one pixel long straight line segment and we will see in the proposed method that an area will be considered as a unique union of straight line segments characterized by their lengths, their orientations and their positions. Such image representation allows us to reason in a novel space representing differently the image content. This space is particularly suited for table structure extraction.

The paper is organized as follows. Section 2 recalls the main approaches from the literature for table extraction. Section 3 presents the proposed approach for table extraction based on a transform previously defined [1] to extract straight lines, that will be involved in the global process for table reconstruction. As experimental study in Sect. 4, the approach is evaluated both from quality and stability points of view on two datasets. Finally, conclusions and future works are provided in Sect. 5.

2 Related Works

Several approaches have been proposed in the literature for table extraction. They vary according to the input document format (html, image, pdf) [5,10]. We focus on approaches dedicated to extract tables from raster images of documents. Two categories of methods co-exist: the ones based on the presence of lines (i.e. table separators) considered as primitives, and those not based on their presence.

Main Approaches for Table Extraction. As pioneer work among the first category, the approach proposed in [19] looks for blocks of individual elements delimited by vertical and horizontal line segments. Line segments are first detected and the points that form the corners are determined. The connectivity relationships between the extracted corners and thus the individual blocks are interpreted using global and local tree structures. The strategy in [7] also relies on detecting lines by considering all their intersections. In [14] the arrangement of the detected lines is compared with that of the text blocks in the same area. However the results showed a sensibility to the resolution of the scanned page and noise. A top-down approach is proposed in [9], based on a hierarchical characterization of the physical cells. Horizontal/vertical lines and spaces are used as features to extract the regions of the table. A recursive partitioning produces the regions of interest, and the hierarchy is used to determine the cells. The authors of [4] presented a table localization system using a recursive analysis of a modified X-Y tree to identify regions delimited by horizontal and vertical lines. The search is refined by looking for parallel lines that can be found in the deepest levels of the tree. This requires that at least two parallel lines are present.

A second category of methods for table extraction do not rely on the presence of lines but use textual information. In [15], tables are supposed to have distinct columns, with the hypothesis that the spaces between the fields are larger than the spaces between words in normal text lines. The table extraction relies on the formation of word blobs in text lines and then on finding the set of consecutive text lines which may form a table. The system in [13] takes the information from the bounding box of the input word, and outputs the corresponding logical text block units, for example the cells in a table environment. Starting with an arbitrary word as block seed, the algorithm recursively extends this block to all the words that interlace with their vertical neighbors. Since even the smallest gaps in the table columns prevent their words from interlacing each other, this segmentation is able to isolate such columns, though sensitive to the fit of the columns. The method[1] presented in [17] is based on the concept of tab-stops in documents. Tab-stops inform about text aligns (left, right, center, etc.) and they are used to extract the blocks composing the layout of a document. Extracted blocks are then evaluated as table candidates using a specific strategy.

Despite of their interest, methods from this last category strongly depend on the textual information generally previously extracted using an OCR. However the presence of a table may have a negative impact on the character recognition rate. In this context, it appears that methods based on straight lines are more adapted to deal with fully materialized tables in potentially noisy documents with complex layouts. In the next section, we will focus on methods allowing straight line detection that we consider as primitives to extract tables.

Main Approaches for Straight Line Detection. Classical approaches derive from the Hough or Radon Transforms. The Hough Transform [2,16,20] can be applied on binary raster images in order to detect lines by considering a parametric representation. The approach uses an accumulator array, where each cell

[1] Included in the Tesseract software: https://opensource.google.com/projects/tesseract.

models a physical line. For every point, all the lines the point belongs to are considered and the associated values are incremented. The highest values in the array, representing the most salient lines, are then extracted.

Another way of detecting straight lines in image content is to consider a more local point of view than global Hough methods. The patch-based approach proposed in [18] uses a mono-dimensional accumulator array. A density map is locally computed from the input image and used to segment the image into high and low contour density areas.

In discrete geometry, the concept of discrete (blurred) segment has also been studied [6,11] but these works are more focused on building novel representations, decomposition or analysis of discrete curves and more generally shapes. The approach presented in [3] relies on the definition of specific regions, named line support regions, which contain line segment. The input image is a gradient image. Edges detected are first combined in regions using both gradient magnitude and direction of neighboring pixels. The main line of each region is finally obtained from a statistical study of gradient magnitude in the region.

Some of the methods mentioned previously make the assumption that the lines present in the images are correctly and completely materialized, which makes it difficult to apply them in noisy documents where portions of lines materializing tables may be disrupted. On the other hand, Hough-based methods generally consider globally the image content, while local approaches do not take into account the geometry and topology of the image regions. In this work, we propose a new method for extracting tables based on a new Transform (called RLDT [1]) that we employ for line detection. This transform operates at a local "region" level, adapting the region of interest to the local context, and gives more information about the spatial organization of the segments than traditional global transforms. The "broken" lines are reconstructed using line seeds obtained *via* this transform. Finally the obtained straight line segments are used to reconstruct potential table structures.

3 Table Extraction

As we consider only fully materialized tables, our approach relies on a line extraction strategy. These lines extracted from the document content can be part of a table or only separators or part of a graphic or any image media. The global flowchart (Fig. 1) contains two main stages: (1) the extraction of the horizontal/vertical lines coupled to their (potential) reconstruction, and (2) the table reconstruction based on the lines limiting the table cells.

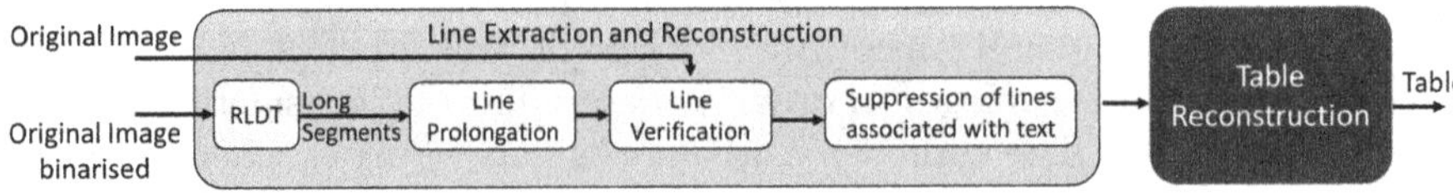

Fig. 1. Flowchart of our approach for table extraction in images of documents.

The first step of the line extraction is based on a recent transform defined in [1] that enables the definition of some primitives, in a novel image representation space, that are useful for the extraction of text, image and separators. To deal with "broken" lines and to ensure a better stability, the line seeds extracted *via* this transform are then prolongated. Then in the line verification step, a theoretical model of a line confirms or infirms the presence of segments among line segment candidates, especially in degraded conditions.

3.1 Relative Local Diameter Transform [1]

Classical methods in document analysis are based on binary images, as they yield in a fast and efficient way to the extraction of the foreground and background. These dual sources of information can be exploited to extract lines. The following notations will be used: an image of a document page D is a function that associates each point $(x, y) \in \mathbb{Z}^2$ with a value of the set V. We assume that for the image I, $V = [0, 1]$ and for the binary version I_b, $V = \{0, 1\}$ (white is marked with 0, black with 1). The foreground is assumed to be black. Both images I and I_b share the same image definition set Δ_I.

We define in this section some transforms (originally introduced in [1]) that give more local information about the spatial organization of the segments contained in a document content than traditional global transforms. We propose to adapt the neighborhood level around each pixel according to the image content. As we focus on the straight lines contained in the image, we refer to the Radon Transform. This transform was modified in order to obtain more localized information, and we defined in [1] the local Radon Transform (LR).

At each point, we define a *local* Radon Transform by

$$\begin{aligned} LR(I_b)(\theta, x_0, y_0) = & \\ & \int_{-\infty}^{+\infty} I_b(\rho(\theta)\cos(\theta) - t\sin(\theta), \rho(\theta)\sin(\theta) + t\cos(\theta)) \times \\ & \delta_0\Big(\int_{t_0}^{t} I_b(\rho(\theta)\cos(\theta) - u\sin(\theta), \rho(\theta)\sin(\theta) + u\cos(\theta)) - 1 \, du\Big) \, dt \end{aligned} \tag{1}$$

where δ_0 is the Dirac distribution in 0. LR gives the maximum length of the segment passing through the point (x_0, y_0) in the direction θ.

Local Diameter Transform. The length of a segment is not relevant here in an absolute way but only relatively to the size of the considered binary image I_b. Then, based on the LR, the local diameter at a point is measured by evaluating the length of the largest segment passing through this point and contained in the set of (foreground) pixels labeled 1. Thus only one value at each point corresponding to the maximum length of the segment is kept, independent of its direction. This is what we call the Local Diameter Transform (LDT) defined as $LDT(I_b)(x, y) = \max_{\theta \in [0,\pi]} LR(I_b)(\theta, x, y)$. Obviously, depending on the application, the importance of the segment must be estimated relative to the dimensions of the document D. By denoting $diam(\Delta_I, \theta)$ the diameter of Δ_I in the direction θ, we then define in each pixel of I_b, the relative local diameter by:

$$RLDT(I_b)(x, y) = \max_{\theta \in [0,\pi]} \frac{LR(I_b)(\theta, x, y)}{diam(\Delta_I, \theta)} \quad (2)$$

and the Relative Local Diameter Transformation of I_b is noted $RLDT(I_b)$. An example of applying the RLDT to a toy-case image can be seen in Fig. 2.

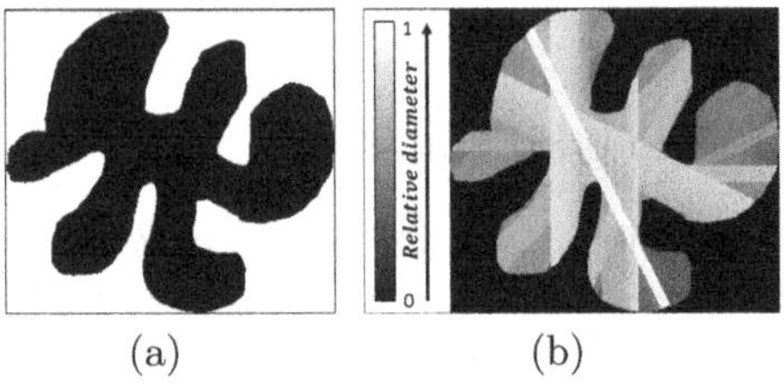

Fig. 2. Illustration of the Relative Local Diameter Transform: (a) binary image, (b) RLDT calculated on the image (a) by taking into account 8 orientations.

The union of all the lines is covering the image content. Here, we are interested in only the long lines composing the table structure. (A threshold has been fixed to 2 according to another study that has shown that lines less than 2% of the size of the document are associated with text zones [1].) We then do consider only the points whose maximum length of the segment is horizontal or vertical ($\theta = 0$ or $\theta = \frac{\pi}{2}$). The connected components (CC) contained in this image correspond to the set of straight line segments marked $L(I_b)$. Figure 3(c) illustrates this step. Note that the union of small straight lines can allow to support a slight angle to capture non-purely horizontal or vertical straight lines.

The set $L(I_b)$ comprises the separators belonging to tables, some other separators and also some short straight lines that may correspond to configuration of text or logos for example that need to be eliminated from the final set of straight lines belonging to materialized tables. The problem to be faced is the robustness of the straight line detection in degraded contexts, addressed in next section.

3.2 Line Extraction and Reconstruction

Line Prolongation. As the previous step is based on a binary image I_b, the segment detection is sensitive to the binarization method and to the quality of the initial image. One assumption of our work is that the first step of the process, even if it presents weak points, gives some seeds of real straight lines that may be incomplete with respect to the content of the document. The final goal is to find cells involved in tables. These cells are limited by four segments defining a rectangle. Thus, the seeds of the real straight lines contained in $L(I_b)$ are extended on the whole image (Fig. 3(d)) and the presence of a segment on each portion limited by consecutive orthogonal segments will be tested.

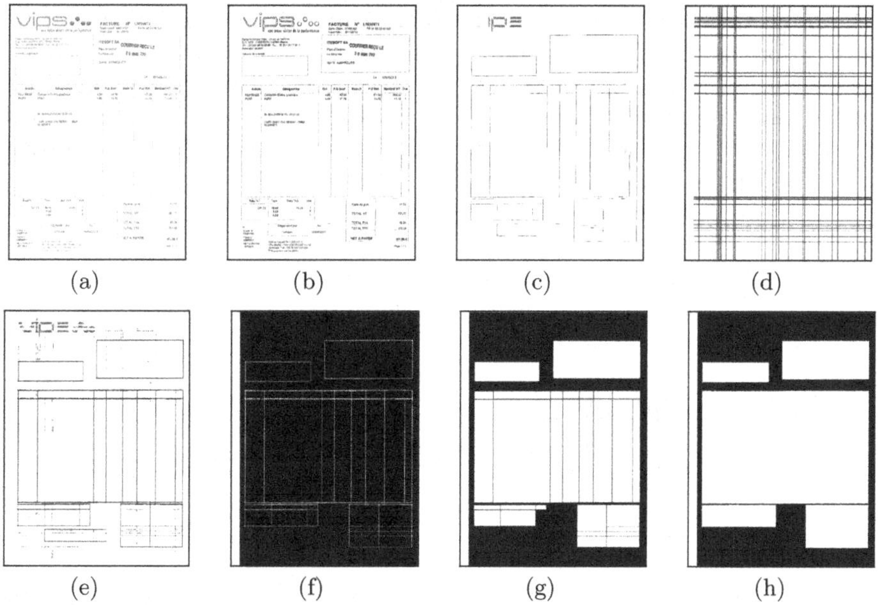

Fig. 3. Results of the different steps of the table extraction process: (a) initial image I, (b) binarized image I_b, (c) straight lines $L(I_b)$ from the RLDT, (d) candidate straight segments, (e) positively tested long segments, (f) long segments I_s (in white), (g) potential cells R^c, (h) potential tables.

Line Verification. Lines extracted at the previous step do not always fit with borders of cell tables. Some information in the original gray level image I can be useful to take a non ambiguous decision. Along the candidate straight lines, a local working zone is defined, where a theoretical model of a line is applied in order to confirm or infirm the presence of a line. This model is based on a local behavior of 1D sets of pixels and a global behavior on the working zone. The evolution of the gray levels along the sections of the zone (black line in Fig. 4(a)) are compared to the theoretical behavior in case of a line, where from left to right the gray levels should begin to increase then (potentially) stabilize and decrease. Otherwise, if there is no line at this place, the evolution is negligible. On Fig. 4(a), red zone corresponds to an increase of gray levels, green to a decrease and blue to a stability in gray levels relatively to I. A global vision of the row processing leads to the segmentation of the working zone according to three different color level behaviors Zi, Zs and Zd. This enables to eliminate the hypothesis of a straight line presence in some parts.

The result is depicted on Fig. 4(b) where the working zones are colored. As illustrated, a vertical black line well corresponds to the adjacency of red, blue and green zones. On a hypothetical segment, on each section where there is no seed point, the presence of adjacent pixels of (Zi, Zs and Zd) or (Zi and Zd) enables to maintain the hypothesis of a straight line and then a line is rebuilt as the projection of the seed on the section. A dilation in the direction of the line is performed. From seeds, i.e. the elements of $L(I_b)$, we thus built a set of longer segments where even some new segments may be added $Lr(I, L(I_b))$. It happens sometimes that $Lr(I, L(I_b))$ comprises small segments associated with text that are in the same vertical position as an horizontal segment in an other part of the page as illustrated in Fig. 3(e).

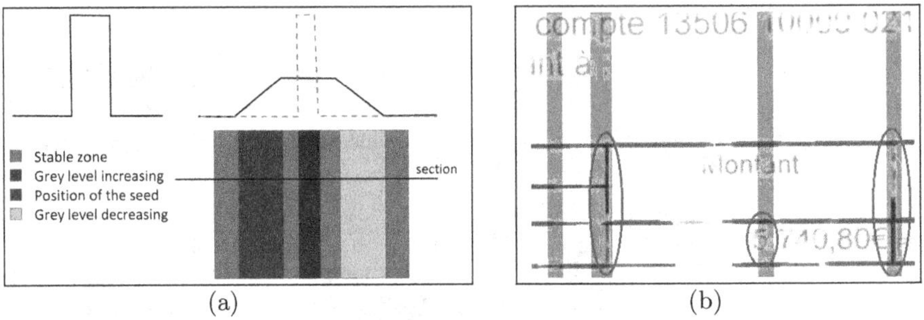

Fig. 4. Behavior of gray levels along a segment hypothesis: (a) proposed line model; (b) illustration of the different zones Zi in red, Zd in green, Zs in blue (segments of $L(I_b)$ are in black). (Color figure online)

Suppression of Lines Associated with Text. The last step is to discriminate between isolated lines and virtual lines associated with text. Let an element of $Lr(I, L(I_b))$ be noted s. In a real straight line, $Zi(s)$ is a long ribbon along the line. In text context the $Zi(s)$ zone is more complex. Such a zone may be actually composed of several connected components denoted as $\{z_{s,a}\}_{a=1}^{n_z}$. Let $proj_s(x)$ be the orthogonal projection of a set x on the principal direction of s. It is assumed that the $z_{s,a}$ have been sorted according to the lengths of their projection. Let n_0 be defined by $0.1 * length(proj_s(s))$.

The segment s is then considered as a straight line segment if:

$$\exists n \leq n_0 / length(\cup_{a=1}^{n} proj_s(z_{s,a})) \geqslant 0.9 * length(proj_s(s)) \tag{3}$$

enabling to discard some non significant straight lines. We only retain these selected long segments in a final set I_s comprising segments that may be part of a table. An example of result is illustrated on Fig. 3(f). For simplicity, we now assume the same name for the set of segments and the raster associated image.

3.3 Table Reconstruction

As we have considered the connected components in the binary image containing long segments, the components of I_s can be noisy if some text characters are touching a straight line, which has to be taken into account. Let E^c be the complementary image content (background) of binary image content E (foreground).

The non bounded CC of I_s^c cannot be part of a table and make a region R in the image domain. The complement of R, R^c comprises pixels of lines S and background pixels B (Fig. 3(g)). The CC of R^c are labeled as table if they contain pixels of B (Fig. 3(h)). Otherwise, they correspond to isolated separator lines.

4 Experimental Study

The evaluation of our approach is achieved from two points of view:

- Quality: in a classical way, the recall and precision in the predicted tables give an information on the accuracy of the process;
- Stability: nowadays more and more documents are hybrid documents, that is to say after a document is created, along its life, it can alternatively be printed and scanned. Document images are associated with the same document content but may differ in resolution, in capture noise or any degradation due to a natural use. We want to test the stability of the method on several images representing a same initial document at different dates.

To achieve this we have considered two different datasets presented in the next section before describing the obtained results.

4.1 Datasets

In order to have quantitative evaluations, two datasets were considered: one to evaluate quality and the other for stability. For quality we have considered the dataset proposed in an ICDAR 2013 competition [8]. As a matter of fact, on the one hand, the data are not images but are in pdf format and on the other hand the tables proposed are not all materialized. We have chosen to raster the images at 150 dpi and have also removed the images containing non-fully materialized tables. The datasubset contains 179 images of documents, among them 69 are containing tables. Then the results we obtain cannot be compared to those of the competition participants, focusing on the whole dataset.

For stability evaluation, no public dataset was available. We designed one annotated dataset called SETSTABLE, containing 293 document images associated with 14 hybrid documents (see composition in Table 2 (lines 1–3)).

For each hybrid document, the results associated with all the couples of the document occurrences are analyzed in order to check if they are similar. The identity of two results can be binary in particular when quality of the extraction is measured, then an extracted table is compared to the ground truth. For two occurrences D_1 and D_2 of a hybrid document D, $\{T_1^i\}_{i=1}^{N_1}$ and $\{T_2^i\}_{i=1}^{N_2}$ are the tables respectively extracted in the two document images. The covering level between two tables is computed as:

$$CL(T_1^i, T_2^j) = \frac{A(T_1^i \cap T_2^j)}{A(T_1^i \cup T_2^j)} \tag{4}$$

where A(X) is the area of X. Accuracy is measured using the binary version of Eq. 4, with risk parameter p:

$$CL_p(T_1^i, T_2^j) = \begin{cases} 1 & \text{if } CL(T_1^i, T_2^j) \geqslant p \\ 0 & \text{otherwise} \end{cases} \quad (5)$$

The stability of a table extraction in D_1 is then measured with respect to the process of a second document D_2 by:

$$ST(T_1^i, D_2) = \begin{cases} 0 & \text{if } N_2 = 0 \\ \max_{j \in [1, N_2]} \{CL(T_1^i, T_2^j)\} & \text{otherwise} \end{cases} \quad (6)$$

Finally the stability of the processing of D_1 and D_2 is measured by:

$$Stab(D_1, D_2) = \begin{cases} 1 & \text{if } N_1 = N_2 = 0 \\ 0 & \text{if } N_1 \cdot N_2 = 0 \text{ and } N_1 + N_2 \neq 0 \\ \frac{1}{2}\left(\frac{1}{N_1} \sum_{i=1}^{N_1} ST(T_1^i, D_2) + \frac{1}{N_2} \sum_{i=1}^{N_2} ST(T_2^i, D_1)\right) & \text{otherwise} \end{cases} \quad (7)$$

For an hybrid document D for which we have N ($N > 1$) different images we can compute a global stability score $SC(D)$ defined by:

$$SC(D) = \frac{2}{N^2 - N} \sum_{i=1}^{N} \sum_{j=1}^{i-1} Stab(D_i, D_j) \quad (8)$$

SC takes value in $[0, 1]$; full stability is obtained when $SC = 1$. Replacing CL by CL_p, a more binary stability score denoted $SC_p(D)$ can be computed.

4.2 Results

Quality of Table Extraction. The NICK method [12] was used to binarize all the document images. The table extraction quality is computed at the object (table) level with classical precision (P), recall (R) and F_1-measure (F) indexes. An extracted table is considered as true positive if Eq. 5 is satisfied with risk parameter $p = 0.80$, relatively to ground truth. The ICDAR 2013 document ground truth contains 88 tables bounding boxes. Among this set, our method identified correctly 86 tables and 18 extra tables have been found. This happens for example when a graph comprises histogram and horizontal lines. A further process should fix these problems when hypothesizing a table (Fig. 5(b)).

As comparative study, we compared our results to the ones obtained with the method proposed in [17], relying on a OCR to extract tables. We used the open source implementation of this algorithm provided as part of the Tesseract OCR engine, with the recommended parameter values.

In order to evaluate the efficiency of our line reconstruction step (LRS), we compared the results with or without this step on the SETSTABLE dataset.

Table 1 presents the quantitative results obtained on the datasets, while Fig. 5(a) and (b) illustrate some qualitative results.

From Table 1 (left), one can note than our proposal outperforms the results provided by the Tesseract [17] method on the ICDAR 2013 datasubset. The high precision of our method can be explained by the pdf nature of the initial data leading to high quality images. Experiments showed that in this condition the LRS step does not improve the results.

The lower precision values on the SETSTABLE dataset (Table 1 (right)) can be explained by logos detected as tables. We can see that the global quality has improved using the LRS step, especially the recall value has increased of 43%.

Table 1. Quality of fully materialized table extraction.

ICDAR 2013 datasubset				SETSTABLE dataset			
Method	P	R	F	Method	P	R	F
Tesseract [17]	0.25	0.33	0.28	Our proposal without LRS	0.64	0.46	0.54
Our proposal	0.98	0.83	0.90	Our proposal with LRS	0.73	0.66	0.70

Stability of the Method. We compared our results to the ones obtained with Tesseract [17] according to the accuracy criterion $CL_{0.80}$ of Eq. 5 (leading to $SC_{0.80}$). This does not take into account the small variations due to image acquisition. Then, we have also used $CL_{0.70}$. This relaxes the constraints and the score value should increase since the offset due to document rotations, translations in the different versions of the hybrid documents are absorbed. To be independent of the accuracy measurement, we have also used the pure cover level CL of Eq. 4 to compute SC.

The obtained results are presented in Table 2 and illustrated on Fig. 5(c) and (d). It has to be noticed that the stability results from this table cannot be directly compared to the quality results (Table 1). The lack of stability reflects an inconstancy in the errors and not the number of errors.

Results from Table 2 suggest that our strategy is more stable than Tesseract [17]. Documents with full stability are documents numbered 6, 7 and 13 that contain no table. Unfortunately, we found 15 tables in the other images without table, they correspond to logo zones.

As mentioned earlier the use of $SC_{0.70}$ shows a more optimistic point of view of the stability of the results (and improves the overall scores) than $SC_{0.80}$ where the identity criterion between two tables is stricter (i.e. smaller shifts, rotations are allowed). This is particularly penalizing for small tables. Figure 5(c, d) shows the projection of all extracted tables from the instances of an hybrid document. They seem all correct, but are slightly differing. The final stability result will depend on the choice of the risk parameter in the covering level definition.

Table 2. Evaluation of the stability of our method from the SETSTABLE dataset.

Document		1	2	3	4	5	6	7	8	9	10	11	12	13	14	Summary
# of occurrences		21	21	21	21	21	21	21	21	21	21	21	21	20	21	$\sum = 293$
# of tables		6	6	1	6	2	0	0	3	0	0	0	7	0	0	$\sum = 31$
Tesseract [17]	sc (in %)	65	65	76	49	92	29	32	54	39	27	58	7	100	61	$\bar{x} = 54$
	$sc_{0.70}$ (in %)	52	57	77	28	90	13	10	47	27	7	45	2	100	63	$\bar{x} = 44$
	$sc_{0.80}$ (in %)	46	56	67	20	90	4	3	40	16	4	29	0	100	63	$\bar{x} = 38$
Our proposal	sc (in %)	80	80	87	82	86	100	100	88	59	57	81	49	100	73	$\bar{x} = 80$
	$sc_{0.70}$ (in %)	88	87	90	87	86	100	100	97	58	56	81	47	100	73	$\bar{x} = 82$
	$sc_{0.80}$ (in %)	62	73	84	78	74	100	100	91	57	56	81	36	100	73	$\bar{x} = 76$

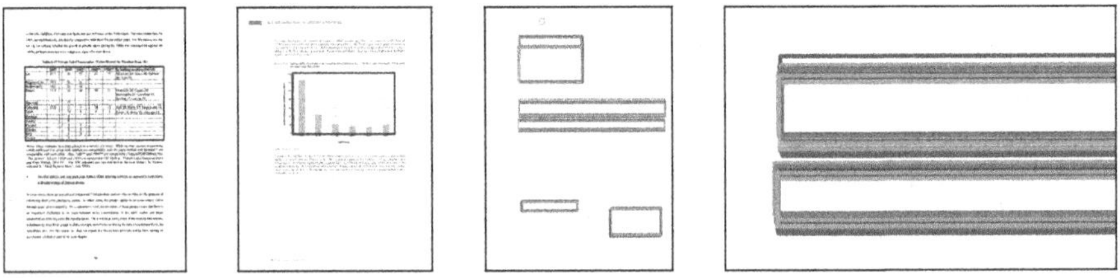

(a) (b) (c) (d)

Fig. 5. Table extraction results: (a) and (b) documents from ICDAR 2013 dataset; (c) illustrates a stability result where each color represents a result of the table extraction in a version of the hybrid document ((d) is a zoom). (Color figure online)

5 Conclusion

Table recognition in documents is still a hot topic and we proposed here an approach for their extraction in document images, in particular when acquisition noise can disrupt the recovering of the table structures. It relies on the search for straight line segments in documents. As the sequential printings and scannings of a document and its deterioration can lead to "broken" lines, the cornerstone of our approach is to first reconstruct the degraded lines given a new image transform, before to extract potential table structures. The proposed approach has been evaluated on two datasets given two points of view: the stability of the extraction, by analyzing several images of the same (hybrid) document, and the accuracy of the results. The obtained results are encouraging and highlight the ability of our approach to extract tables, even in quite noisy documents.

This work opens up several perspectives. As a limit of this approach, the output of our method strongly depends on the initial binarization of the page which can lead to potentially different results for many instances of a hybrid document, decreasing the overall stability. We plan to consider (and to couple) different methods of binarization to design a more stable process. An other way could be to generalize the proposed image transforms for gray level or color images to avoid binarization.

References

1. Alhéritière, H., Cloppet, F., Kurtz, C., Ogier, J.M., Vincent, N.: A document straight line based segmentation for complex layout extraction. In: ICDAR, Proceedings, pp. 1126–1131 (2017)
2. Ben-Tzvi, D., Sandler, M.B.: A combinatorial Hough transform. Patt. Rec. Lett. **11**(3), 167–174 (1990)
3. Burns, J.B., Hanson, A.R., Riseman, E.M.: Extracting straight lines. IEEE Trans. Patt. Anal. Mach. Intell. **8**(4), 425–455 (1986)
4. Cesarini, F., Marinai, S., Sarti, L., Soda, G.: Trainable table location in document images. In: ICPR, Proceedings, pp. 236–240 (2002)
5. Coüasnon, B., Lemaitre, A.: Recognition of tables and forms. In: Doermann, D., Tombre, K. (eds.) Handbook of Document Image Processing and Recognition, pp. 647–677. Springer, London (2014). https://doi.org/10.1007/978-0-85729-859-1_20
6. Debled-Rennesson, I., Feschet, F., Rouyer-Degli, J.: Optimal blurred segments decomposition of noisy shapes in linear time. Comput. Graph. **30**(1), 30–36 (2006)
7. Gatos, B., Danatsas, D., Pratikakis, I., Perantonis, S.J.: Automatic table detection in document images. In: ICAPR, Proceedings, pp. 609–618 (2005)
8. Göbel, M., Hassan, T., Oro, E., Orsi, G.: ICDAR 2013 table competition. In: ICDAR, Proceedings, pp. 1449–1453 (2013)
9. Green, E., Krishnamoorthy, M.: Model-based analysis of printed tables. In: ICDAR, Proceedings, pp. 214–217 (1995)
10. Kasar, T., Barlas, P., Adam, S., Chatelain, C., Paquet, T.: Learning to detect tables in scanned document images using line information. In: ICDAR, Proceedings, pp. 1185–1189 (2013)
11. Kerautret, B., Even, P.: Blurred segments in gray level images for interactive line extraction. In: IWCIA, Proceedings, pp. 176–186 (2009)
12. Khurshid, K., Siddiqi, I., Faure, C., Vincent, N.: Comparison of Niblack inspired binarization methods for ancient documents. In: DRR, Proceedings, pp. 724–732 (2009)
13. Kieninger, T.: Table structure recognition based on robust block segmentation. In: DRR, Proceedings, pp. 22–32 (1998)
14. Laurentini, A., Viada, P.: Identifying and understanding tabular material in compound documents. In: ICPR, Proceedings, pp. 405–409 (1992)
15. Mandal, S., Chowdhury, S.P., Das, A.K., Chanda, B.: A simple and effective table detection system from document images. Int. J. Doc. Anal. Recognit. **8**(2), 172–182 (2006)
16. Mukhopadhyay, P., Chaudhuri, B.B.: A survey of Hough transform. Patt. Rec. **48**(3), 993–1010 (2015)
17. Shafait, F., Smith, R.: Table detection in heterogeneous documents. In: DAS, Proceedings, pp. 65–72 (2010)
18. Shpilman, R., Brailovsky, V.: Fast and robust techniques for detecting straight line segments using local models. Patt. Rec. Lett. **20**(9), 865–877 (1999)
19. Watanabe, T., Naruse, H., Luo, Q., Sugie, N.: Structure analysis of table-form documents on the basis of the recognition of vertical and horizontal line segments. In: ICDAR, Proceedings, pp. 638–646 (1991)
20. Xu, Z., Shin, B.S., Klette, R.: Determination of length and width of a line-segment by using a Hough transform. In: DGCI, Proceedings, pp. 190–201 (2014)

A Spatial Convexity Descriptor for Object Enlacement

Sara Brunetti[1], Péter Balázs[2(✉)], Péter Bodnár[2], and Judit Szűcs[2]

[1] Dipartimento di Ingegneria dell'Informazione e Scienze Matematiche, Via Roma, 56, 53100 Siena, Italy
sara.brunetti@unisi.it

[2] Department of Image Processing and Computer Graphics, University of Szeged, Árpád tér 2, Szeged 6720, Hungary
{pbalazs,bodnaar,jszucs}@inf.u-szeged.hu

Abstract. In (Brunetti et al.: Extension of a one-dimensional convexity measure to two dimensions, LNCS 10256 (2017) 105–116) a spatial convexity descriptor is designed which provides a quantitative representation of an object by means of relative positions of its points. The descriptor uses so-called Quadrant-convexity and therefore, it is an immediate two-dimensional convexity descriptor. In this paper we extend the definition to spatial relations between objects and consider complex spatial relations like enlacement and interlacement. This approach permits to easily model these kinds of configurations as highlighted by the examples, and it allows us to define two interlacement descriptors which differ in the normalization. Experiments show a good behavior of them in the studied cases, and compare their performances.

Keywords: Shape descriptor · Spatial relations · Q-convexity

1 Introduction

Shape representation is a current topic in digital image analysis, for example, for object recognition and classification issues. Suitable approaches for handling the problems consist in the design of new shape descriptors and measures for descriptors sensitive to distinguish the shapes but robust to noise. Some methods provide a unified approach that can be applied to determine a variety of shape measures, but more often they are specific to a single aspect of shape.

Over the years, measures for descriptors based on convexity have been developed. Area based measures form one popular category [6,23,25], while boundary-based ones [28] are also frequently used. Other methods use simplification of the

This research was supported by the project "Integrated program for training new generation of scientists in the fields of computer science", no EFOP-3.6.3-VEKOP-16-2017-0002. The project has been supported by the European Union and co-funded by the European Social Fund. Ministry of Human Capacities, Hungary grant 20391-3/2018/FEKUSTRAT is acknowledged.

M. Couprie et al. (Eds.): DGCI 2019, LNCS 11414, pp. 330–342, 2019.
https://doi.org/10.1007/978-3-030-14085-4_26

contour [18] or a probabilistic approach [21,22] to solve the problem. Recently, measures based on directional (line) convexity have been defined and investigated, independently, in [3,26] and in [16,17], to use the degree of directional convexity as a shape prior in image segmentation. These methods cannot be extended easily to a two-dimensional convexity measure. A different approach to this aim is to employ the concept of so-called Quadrant-convexity [8,9] which is inherently two-dimensional. In this way, an extension of the directional convexity in [3] which uses quantitative information was introduced in [7], whereas in [1,2] a different 2D convexity measure based on salient points [12,13] was presented.

In this paper, we study more deeply the measure we derived the shape descriptors from, in [7]. The concepts on the basis of the measure definition have a counterpart in the framework of fuzzy sets for spatial relationships since the measure gives a quantitative representation of the object by means of relative positions of its points. Thus, the derived descriptors provide a model for studying the spatial relative position concepts in connection with unary relationships (to a reference object) and binary relationships (between two objects) [19].

Spatial relations have been studied in many disciplines (see Sect. 8 of [19] for a review of the related literature), and constitute an important part of the knowledge for image processing and understanding [4]. Two types of questions are raised when dealing with spatial relations: (1) given one reference object, define the region of space in which a relation to this reference is satisfied (to some degree); (2) given two objects (possibly fuzzy), assess the degree to which a relation is satisfied. Concerning the latter ones, they can be categorized into directional (relations like "to the left"), distance (relations like "far"), and topological. In particular, questions related to measuring complex spatial relations like enlacement, interlacement, and surrounding have been studied in [10,11]. The term *enlacement* between F and G does indicate that object G is somehow between object F (the reference). Given a direction, a straight line parallel to the direction intersects the object in a one-dimensional finite slice (possibly empty), called *longitudinal cut.* Roughly speaking, the *directional enlacement* of G with regards to F along an oriented line is given by summing up the contributions of longitudinal cuts to count the number of triplet of points in $F \times G \times F$, as G is enlaced by F if their points alternate in this order [10]. If we consider the unary relationship, given the reference object F, the *directional enlacement landscape* of F along an oriented line consists in quantifying the region of space that is enlaced by F by longitudinal cuts. It is worth mentioning that objects can be imbricated each other only when the objects are concave, or constituted by multiple connected components, and indeed the use of some kind of convexity to deal with these spatial relationships has been already investigated also in [5]. Here we propose our Quadrant-convexity measure to tackle these issues directly in the two-dimensional space, i.e., to define a shape descriptor based on spatial relative positions of points which permits to define enlacement and interlacement between two objects. This approach is evaluated by suitable examples. In particular, we develop the idea to use our descriptor for enlacement landscape, i.e., taking the foreground as one object and the background as the other object. This permits to deal with the classification problem, and the experiments show a good performance in the studied cases.

In Sect. 2 we recall some information on Quadrant-convexity. In Sect. 3 we introduce the enlacement descriptor by normalizing the Quadrant-convexity measure. Then, in Sect. 4 we define the object enlacement and interlacement descriptors. Section 5 is for the experiments and Sect. 6 is for the conclusion.

2 Quadrant-Convexity

In the sequel, consider a two-dimensional object F to be represented by a non-empty *lattice set*, i.e. a finite subset of $\mathbb{Z}^2$, or equivalently, a function $f : \mathbb{Z}^2 \to \{0, 1\}$. Let $\mathcal{R}$ be the smallest rectangle containing F and suppose it is of size $m \times n$. We illustrate F as a union of white unit squares (foreground pixels), up to translations, and $\mathcal{R} \setminus F$ as a union of black unit squares (background pixels).

Each position (i, j) in the rectangle together with the horizontal and vertical directions determines the following four quadrants:

$$\begin{aligned}
Z_0(i, j) &= \{(l, k) \in \mathcal{R} :\ 0 \le l \le i,\ 0 \le k \le j\},\\
Z_1(i, j) &= \{(l, k) \in \mathcal{R} :\ i \le l \le m-1,\ 0 \le k \le j\},\\
Z_2(i, j) &= \{(l, k) \in \mathcal{R} :\ i \le l \le m-1,\ j \le k \le n-1\},\\
Z_3(i, j) &= \{(l, k) \in \mathcal{R} :\ 0 \le l \le i,\ j \le k \le n-1\}.
\end{aligned}$$

Let us denote the number of object points (foreground pixels) of F in $Z_p(i, j)$ by $n_p(i, j)$, for $p = 0, \ldots, 3$, i.e.,

$$n_p(i, j) = card(Z_p(i, j) \cap F\})\ (p = 0, \ldots, 3). \tag{1}$$

Definition 1. *A lattice set F is Quadrant-convex (shortly, Q-convex) if for each (i, j) $(n_0(i, j) > 0 \wedge n_1(i, j) > 0 \wedge n_2(i, j) > 0 \wedge n_3(i, j) > 0)$ implies $(i, j) \in F$.*

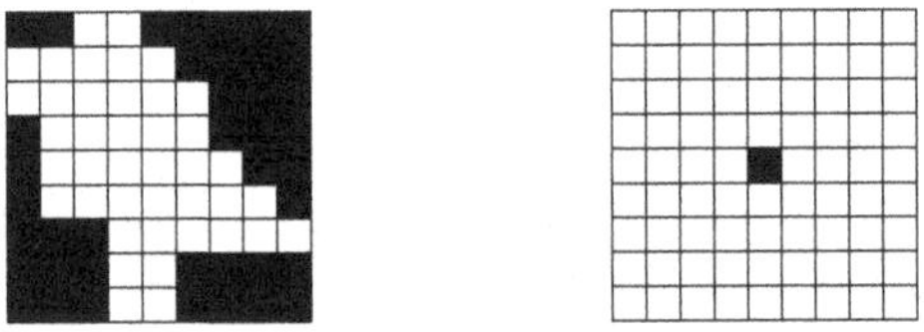

Fig. 1. A Q-convex image (left) and a non Q-convex image (right). The four quadrants around the point (5, 5) are: $Z_0(5, 5)$ left-bottom, $Z_1(5, 5)$ right-bottom, $Z_2(5, 5)$ right-top, $Z_3(5, 5)$ left-top.

If F is not Q-convex, then there exists a position (i, j) violating the Q-convexity property, i.e. $n_p(i, j) > 0$ for all $p = 0, \ldots, 3$ and $(i, j) \notin F$. Figure 1 illustrates the definition of Q-convexity: the lattice set on the right is Q-*concave* (not Q-convex) because $(5,5) \notin F$ but $Z_p(5,5)$ contains object points, for all $p = 0,1,2,3$. In the figure we have $n_0(5,5) = n_1(5,5) = n_2(5,5) = n_3(5,5) = 24$. Let us notice that Definition 1 is based on the relative positions of the object points: the quantification of the object points in the quadrants with respect to the considered point gives rise to a quantitative representation of Q-concavity.

We define the Q-concavity measure of F as the sum of the contributions of non-Q-convexity for each point in $\mathcal{R}$. Formally,

$$\varphi_F(i,j) = n_0(i,j)n_1(i,j)n_2(i,j)n_3(i,j)(1 - f(i,j)), \tag{2}$$

where (i,j) is an arbitrary point of $\mathcal{R}$, and $f(i,j) = 1$ if the point in position (i,j) belongs to the object, otherwise $f(i,j) = 0$. Moreover,

$$\varphi_F = \varphi_F(F) = \sum_{(i,j)\in\mathcal{R}} \varphi_F(i,j). \tag{3}$$

For example, for the image on the right of Fig. 1, we have $\varphi_F(5,5) = 24 \cdot 24 \cdot 24 \cdot 24 \cdot 1 = 331776$.

Remark 1. If $f(i,j) = 1$, then $\varphi_F(i,j) = 0$. Moreover, if $f(i,j) = 0$ and there exists $n_p(i,j) = 0$, then $\varphi_F(i,j) = 0$. Thus, F is Q-convex if and only if $\varphi_F = 0$.

Remark 2. By definition, φ is invariant by reflection and by point symmetry.

Remark 3. In [7] the authors showed that the Quadrant-concavity measure φ extends the directional convexity in [3].

Remark 4. Previous definitions can be viewed in a slightly different way to provide a relationship to a reference object. Since, the latter one is the discrete version of the directional enlacement landscape of F along one oriented line, the Q-concavity measure extends fuzzy directional enlacement: The intersections of F with the four quadrants Z_0, Z_1, Z_2, Z_3 are an extension of the concept of longitudinal cut to two dimensions, and so relation (2) gives a quantification of the enlacement by the reference object F for the (landscape) point (i,j). Indeed, Q-convexity evaluates if a landscape point is among points of the reference object by looking at the presence of points of the reference object in each quadrant around the landscape point into consideration.

3 Obtaining Enlacement Descriptor by Normalization

In order to measure the degree of Q-concavity, or equivalently the degree of landscape enlacement for a given object F, we normalize φ so that it ranges in $[0,1]$ (fuzzy enlacement landscape). We propose two possible normalizations gained by normalizing each contribution. A global one is obtained following [11]:

$$\mathcal{E}_F^{(1)}(i,j) = \frac{\varphi_F(i,j)}{\max_{(i',j')\in\mathcal{R}} \varphi_F(i',j')}. \tag{4}$$

Second (local) one is based on [7]. There the authors proved

Proposition 1. *Let $f(i,j)=0$, and h_i^F and v_j^F be the i-th row and j-th column sums. Then,* $\varphi_F(i,j) \leq ((card(F)+h_i^F+v_j^F)/4)^4$.

As a consequence we may normalize each contribution as

$$\mathcal{E}_F^{(2)}(i,j) = \frac{\varphi_F(i,j)}{((card(F)+h_i^F+v_j^F)/4)^4}. \tag{5}$$

In order to obtain a normalization for the global enlacement for both (4) and (5), we sum up each single contribution and then we divide by the number of non-zero contributions.

Definition 2. *For a given binary image F, its global enlacement landscape* $\mathcal{E}_F^{(\cdot)}$ *is defined by*

$$\mathcal{E}_F^{(\cdot)} = \sum_{(i,j)\in\bar{F}} \frac{\mathcal{E}_F^{(\cdot)}(i,j)}{card(\bar{F})},$$

where $\bar{F}$ denotes the subset of (landscape) points in $\mathcal{R}\setminus F$ for which the contribution is not null.

Figure 2 illustrates the values of the descriptors for four images. The first image is Q-convex and thus the descriptor gives 0. The second image receives a greater value than the third one since the background region inside is deeper. In the fourth image part of the background is entirely closed into the object, thus this image receives a high value.

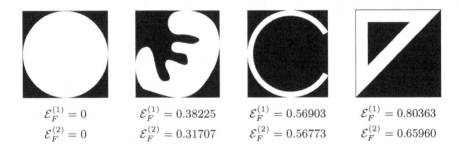

$\mathcal{E}_F^{(1)} = 0$	$\mathcal{E}_F^{(1)} = 0.38225$	$\mathcal{E}_F^{(1)} = 0.56903$	$\mathcal{E}_F^{(1)} = 0.80363$
$\mathcal{E}_F^{(2)} = 0$	$\mathcal{E}_F^{(2)} = 0.31707$	$\mathcal{E}_F^{(2)} = 0.56773$	$\mathcal{E}_F^{(2)} = 0.65960$

Fig. 2. Example images with their enlacement values. In case of the last image a thin black frame is added just for visibility, it does not belong to the image.

Figure 3 shows the contribution of each pixel according to (4): the gray-scale levels correspond to the degrees of fuzzy enlacement of each pixel in the landscape. We can see that pixels inside the concavities have lighter colors (representing higher values) than pixels outside. Similar images can be given according to (5). Of course, in case of the first image of Fig. 2, the enlacement landscape is a completely black image. We point out that normalization plays a key role in Definition 2, as we will show in the experiments.

Fig. 3. Enlacement landscapes of the last three images of Fig. 2.

4 Object Enlacement and Interlacement

In the previous sections we have defined a shape measure φ_F based on the concept of Q-convexity which provides a quantitative measure for studying relationships to a reference object. Now, we modify it into a spatial relationship (a relationship between two objects). Let F and G be two objects. How much is G enlaced by F? The idea is to capture how many occurrences of points of G are somehow between points of F. This is obtained in a straightforward way, by restricting the points in $\mathcal{R} \setminus F$ taken into account to the points in G. Therefore,

$$\varphi_{FG}(i,j) = \begin{cases} \varphi_F(i,j) & \text{if } (i,j) \in G, \\ 0 & \text{otherwise.} \end{cases} \tag{6}$$

Trivially note that if $G = \mathcal{R} \setminus F$, then $\varphi_{FG}(i,j) = \varphi_F(i,j)$. The enlacement descriptors of G by F are thus

$$\mathcal{E}^{(1)}_{FG}(i,j) = \frac{\varphi_{FG}(i,j)}{\max_{(i,j)\in G}\ \varphi_{FG}(i,j)}, \tag{7}$$

and

$$\mathcal{E}^{(2)}_{FG}(i,j) = \frac{\varphi_{FG}(i,j)}{((card(F) + h^F_i + v^F_j)/4)^4}. \tag{8}$$

For clarity, we observe that, by definition, $\varphi_{FG}(i,j) = 0$ if $(i,j) \notin \bar{F}$, thus the maximum of $\varphi_{FG}(i,j)$ in G is equal to the maximum in $G \cap \bar{F}$. In order to obtain a normalization for the global enlacement, it is enough to get.

Definition 3. *Let F and G be two objects. The enlacement of G by F is*

$$\mathcal{E}^{(\cdot)}_{FG} = \sum_{(i,j)\in G} \frac{\mathcal{E}^{(\cdot)}_{FG}(i,j)}{card(G \cap \bar{F})}.$$

Remark 5. Let us notice that the definition of enlacement of two objects can be extended as follows:

$$\frac{\sum_{(i,j)\in\mathcal{R}} \min(\mathcal{E}^{(1)}_{FG}(i,j), g(i,j))}{\sum_{(i,j)\in\mathcal{R}} g(i,j)},$$

$\mathcal{E}_{FG}^{(1)} = 0,\ \mathcal{E}_{GF}^{(1)} = 0.93259,\ \mathcal{I}_{FG}^{(1)} = 0$
$\mathcal{E}_{FG}^{(2)} = 0,\ \mathcal{E}_{GF}^{(2)} = 0.93332,\ \mathcal{I}_{FG}^{(2)} = 0$

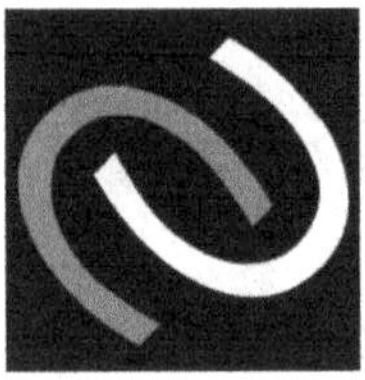

$\mathcal{E}_{FG}^{(1)} = 0.54401,\ \mathcal{E}_{GF}^{(1)} = 0.54914,\ \mathcal{I}_{FG}^{(1)} = 0.54656$
$\mathcal{E}_{FG}^{(2)} = 0.52357,\ \mathcal{E}_{GF}^{(2)} = 0.52856,\ \mathcal{I}_{FG}^{(2)} = 0.52605$

Fig. 4. Example images [10] with their interlacement values, where F and G is represented by white and gray pixels, respectively.

where g is the membership function representing G. If G is a fuzzy set, we may generalize the formula by substituting g with the membership function, and it corresponds to a satisfiability measure such as a normalized intersection [27].

Of course, the enlacement of two objects is an asymmetric relation so that the enlacement of G by F and the enlacement of F by G provide different "views". We may combine both by their harmonic mean to give a symmetric relation which is a description of mutual enlacement.

Definition 4. *Let F and G be two objects. The interlacement of F and G is*

$$\mathcal{I}_{FG}^{(\cdot)} = \frac{2\mathcal{E}_{FG}^{(\cdot)}\mathcal{E}_{GF}^{(\cdot)}}{\mathcal{E}_{FG}^{(\cdot)} + \mathcal{E}_{GF}^{(\cdot)}}, \tag{9}$$

where $\mathcal{E}_{FG}^{(\cdot)}$ and $\mathcal{E}_{GF}^{(\cdot)}$ are the enlacement of G by F and the enlacement of F by G, respectively.

Figure 4 shows two example images with their interlacement values.

The measures can be efficiently implemented in linear time in the size of the image. By definition, $Z_0(l,k) \subseteq Z_0(i,j)$ if $l \le i$ and $k \le j$, and hence $n_0(l,k) \le n_0(i,j)$ with $l \le i$ and $k \le j$. Analogous relations hold for Z_1, Z_2, Z_3 and for n_1, n_2, n_3 accordingly. Exploiting this property, and proceeding line by line, we can count the number of points in F for $Z_p(i,j)$, for each (i,j) in linear time, and store them in a matrix for any $p = \{1,2,3,4\}$. Then, $\varphi(i,j)$ can be computed in constant time for any (i,j). Normalization is straightforward.

5 Experiments

We first conducted an experiment to investigate scale invariance on a variety of different shapes (Fig. 5) from [22]. After digitalizing the original 14 images on different scales ($32 \times 32, 64 \times 64, 128 \times 128, 256 \times 256, 512 \times 512$) we computed the two different interlacement values for each and compared them (being F the foreground and G the background). Table 1 shows the average of the measured interlacement differences over the 14 pairs of consecutive images. From the small values we can deduce that scaling has no significant impact on these measures, from the practical point of view. Obviously, in lower resolutions the smaller parts of the shapes may disappear, therefore the differences can be higher.

Fig. 5. Variety of different shapes

Table 1. Average difference of interlacement values between consecutive image pairs

Size	$\mathcal{E}^{(1)}$	$\mathcal{E}^{(2)}$
$32 \rightarrow 64$	0.0513	0.0163
$64 \rightarrow 128$	0.0164	0.0077
$128 \rightarrow 256$	0.0069	0.0027
$256 \rightarrow 512$	0.0027	0.0017

In the second and third experiments we used public datasets of fundus photographs of the retina for classification issues. We decided to compare our descriptor with its counterpart in [10] since they model a similar idea based on a quantitative concept of convexity. The main difference is that we provide a fully 2D approach, whereas the directional enlacement landscape in [10] is one-dimensional. The CHASEDB1 [15] dataset is composed of 20 binary images with centered optic disks, while the DRIVE [24] dataset contains 20 images where the optic disk is shifted from the center (Fig. 6).

Following the strategy of [10] we gradually added different types of random noise to the images of size 1000×1000 (which can be interpreted as increasingly strong segmentation errors). Gaussian and Speckle noise were added with 10 increasing variances $\sigma^2 \in [0, 2]$, while salt & pepper noise was added with 10 increasing amounts in $[0, 0.1]$. Some example images are shown in Fig. 7. Then, we tried to classify the images into two classes (CHASEDB1 and DRIVE) based on their interlacement values (being F the object and G the background, again), by the 5 nearest neighbor classifier with inverse Euclidean distance (5NN). For

Fig. 6. Examples of the CHASEDB1 (left) and DRIVE datasets (right)

the implementation we chose the WEKA Toolbox [14] and we used leave-one-out cross validation to evaluate accuracy. The results are presented in Table 2. In [10] the authors reported to reach an average accuracy of 97.75%, 99.25%, and 98.75% for Speckle, Gaussian, and salt & pepper noise, respectively, on the same dataset with 5NN. In comparison, $\mathcal{I}_{FG}^{(1)}$ shows a worse performance. Fortunately, the way of normalization in (5) solves the problem, thus the results using $\mathcal{I}_{FG}^{(2)}$ are just slightly worse in case of Speckle and salt & pepper noise. This is especially promising, since our descriptor uses just two directions (four quadrants) being two-dimensional based whereas that of [10] uses 180 directions being one-dimensional based, and thus, it takes also more time to compute. Nevertheless, even a moderate amount of Gaussian noise distorts drastically the structures (see, again, Fig. 7), thus two-directional enlacement can no longer ensure a trustable classification.

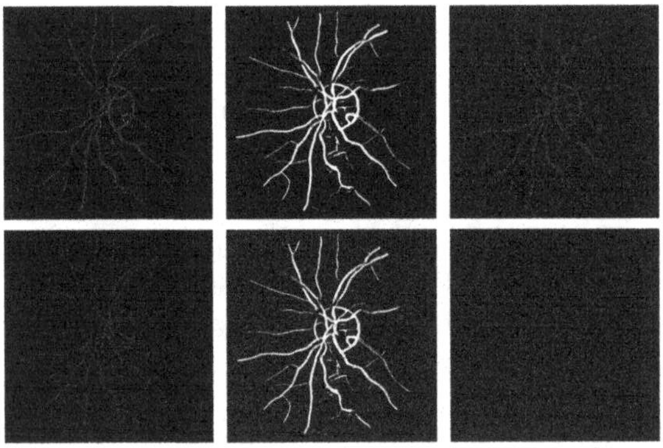

Fig. 7. Retina images with moderate (top row) and high (bottom row) amount of noise. Speckle, salt & pepper, and Gaussian noise, from left to right, respectively.

Finally, for a more complex classification problem, we used the High-Resolution Fundus (HRF) dataset [20], which is composed of 45 images of fundus: 15 healthy, 15 with glaucoma symptoms and 15 with diabetic retinopathy symptoms (Fig. 8). Using the same classifier as before we tried to separate the 15 healthy images from the 30 diseased cases. Figure 9 shows the precision-recall

Table 2. 5NN classification accuracy (in percentage) of CHASEDB1 and DRIVE images for different types and levels of noise.

	Speckle		Salt & pepper		Gaussian	
	$\mathcal{I}_{FG}^{(1)}$	$\mathcal{I}_{FG}^{(2)}$	$\mathcal{I}_{FG}^{(1)}$	$\mathcal{I}_{FG}^{(2)}$	$\mathcal{I}_{FG}^{(1)}$	$\mathcal{I}_{FG}^{(2)}$
Level 1	60.0	95.0	87.5	92.5	65.0	95.0
Level 2	85.0	95.0	85.0	95.0	72.5	87.5
Level 3	60.0	95.0	85.0	95.0	95.0	80.0
Level 4	67.5	95.0	87.5	95.0	82.5	67.5
Level 5	82.5	95.0	85.0	95.0	67.5	47.5
Level 6	70.0	95.0	87.5	95.0	67.5	52.5
Level 7	47.5	95.0	85.0	92.5	52.5	50.0
Level 8	57.5	95.0	85.0	90.0	72.5	57.5
Level 9	55.0	95.0	85.0	90.0	32.5	55.0
Level 10	67.5	95.0	87.5	90.0	57.5	55.0

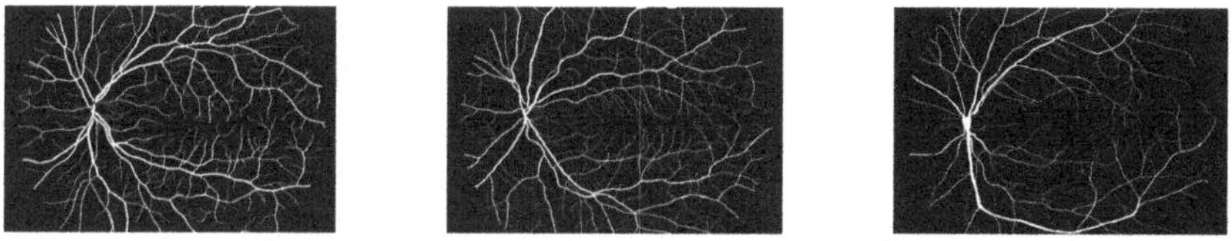

Fig. 8. Examples of the HRF dataset (healthy case, glaucoma, diabetic retinopathy, from left to right, respectively)

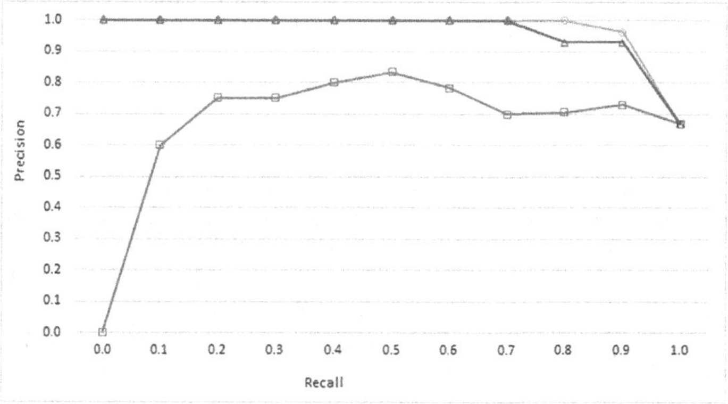

Fig. 9. Precision-recall curves obtained for classifying healthy and diseased cases of the HRF images. Green: $\mathcal{I}_{FG}^{(1)}$ Blue: $\mathcal{I}_{FG}^{(2)}$ Red: curve of [10] (Color figure online)

curves obtained for this classification problem by the two different interlacement descriptors. For comparison we present also the curve based on the interlacement descriptor introduced in [10]. The performance of $\mathcal{I}_{FG}^{(1)}$ seems to be worse, again, in this issue. On the other hand, we observe that descriptor $\mathcal{I}_{FG}^{(2)}$ performs as well as the one presented in [10] (or even slightly better). We stress again that our descriptor uses just two directions.

6 Conclusion

In this paper, we extended the Q-convexity measure of [7] to describe spatial relations between objects and consider complex spatial relations like enlacement and interlacement. We proposed two possible normalizations to obtain enlacement measures. The idea of the first one comes from [11], while the second one is based on our former work in [7]. As testified by the experiments, normalization plays a crucial role: the local normalization $\mathcal{E}^2$ performed generally better than the global normalization $\mathcal{E}^1$ in all the studied cases. In the experiments for classification issues, we developed the idea to assume the foreground being one object and the background being the other. This method is similar to the one in [10] so that the two methods aim at modelling the same idea. However, our method is based on a fully two-dimensional approach whereas the other is based on the one-dimensional directional enlacement. We believe that the two-dimensional approach is more powerful, and indeed in the experiments, just using two directions (so four quadrants), we reached comparable and even better results than that obtained by [10] employing many directions. This is the price to pay for the one-dimensional approach. Nevertheless, starting from the definition of Q-convexity, a further developement could consist in the employment of more directions. Another point is that the descriptors are not rotation invariant for rotations different from 90°. A possibility is to preprocess the image by computing the principal axes and rotate the image before computing enlacement so that the principal axes will be aligned in the horizontal and vertical directions.

References

1. Balázs, P., Brunetti, S.: A measure of Q-convexity. In: Normand, N., Guédon, J., Autrusseau, F. (eds.) DGCI 2016. LNCS, vol. 9647, pp. 219–230. Springer, Cham (2016). https://doi.org/10.1007/978-3-319-32360-2_17
2. Balázs, P., Brunetti, S.: A new shape descriptor based on a Q-convexity measure. In: Kropatsch, W.G., Artner, N.M., Janusch, I. (eds.) DGCI 2017. LNCS, vol. 10502, pp. 267–278. Springer, Cham (2017). https://doi.org/10.1007/978-3-319-66272-5_22
3. Balázs, P., Ozsvár, Z., Tasi, T.S., Nyúl, L.G.: A measure of directional convexity inspired by binary tomography. Fundam. Inform. **141**(2–3), 151–167 (2015)
4. Bloch, I.: Fuzzy sets for image processing and understanding. Fuzzy Sets Syst. **281**, 280–291 (2015)
5. Bloch, I., Colliot, O., Cesar Jr., R.M.: On the ternary spatial relation "between". IEEE Trans. Syst. Man Cybern. B Cybern. **36**(2), 312–327 (2006)
6. Boxer, L.: Computing deviations from convexity in polygons. Pattern Recogn. Lett. **14**, 163–167 (1993)

7. Brunetti, S., Balázs, P., Bodnár, P.: Extension of a one-dimensional convexity measure to two dimensions. In: Brimkov, V.E., Barneva, R.P. (eds.) IWCIA 2017. LNCS, vol. 10256, pp. 105–116. Springer, Cham (2017). https://doi.org/10.1007/978-3-319-59108-7_9
8. Brunetti, S., Daurat, A.: An algorithm reconstructing convex lattice sets. Theor. Comput. Sci. **304**(1–3), 35–57 (2003)
9. Brunetti, S., Daurat, A.: Reconstruction of convex lattice sets from tomographic projections in quartic time. Theor. Comput. Sci. **406**(1–2), 55–62 (2008)
10. Clement, M., Poulenard, A., Kurtz, C., Wendling, L.: Directional enlacement histograms for the description of complex spatial configurations between objects. IEEE Trans. Pattern Anal. Mach. Intell. **39**, 2366–2380 (2017)
11. Clément, M., Kurtz, C., Wendling, L.: Fuzzy directional enlacement landscapes. In: Kropatsch, W.G., Artner, N.M., Janusch, I. (eds.) DGCI 2017. LNCS, vol. 10502, pp. 171–182. Springer, Cham (2017). https://doi.org/10.1007/978-3-319-66272-5_15
12. Daurat, A.: Salient points of Q-convex sets. Int. J. Pattern Recognit. Artif. Intell. **15**, 1023–1030 (2001)
13. Daurat, A., Nivat, M.: Salient and reentrant points of discrete sets. Electron. Notes Discret. Math. **12**, 208–219 (2003)
14. Frank, E., Hall, M.A., Witten, I.H.: The WEKA Workbench. Online Appendix for "Data Mining: Practical Machine Learning Tools and Techniques", 4th edn. Morgan Kaufmann, Burlington (2016)
15. Fraz, M.M., et al.: An ensemble classification-based approach applied to retinal blood vessel segmentation. IEEE Trans. Biomed. Eng. **59**(9), 2538–2548 (2012)
16. Gorelick, L., Veksler, O., Boykov, Y., Nieuwenhuis, C.: Convexity shape prior for segmentation. In: Fleet, D., Pajdla, T., Schiele, B., Tuytelaars, T. (eds.) ECCV 2014. LNCS, vol. 8693, pp. 675–690. Springer, Cham (2014). https://doi.org/10.1007/978-3-319-10602-1_44
17. Gorelick, L., Veksler, O., Boykov, Y., Nieuwenhuis, C.: Convexity shape prior for binary segmentation. IEEE Trans. Pattern Anal. Mach. Intell. **39**, 258–271 (2017)
18. Latecki, L.J., Lakamper, R.: Convexity rule for shape decomposition based on discrete contour evolution. Comput. Vis. Image Underst. **73**(3), 441–454 (1999)
19. Matsakis, P., Wendling, L., Ni, J.: A general approach to the fuzzy modeling of spatial relationships. In: Jeansoulin, R., Papini, O., Prade, H., Schockaert, S. (eds.) Methods for Handling Imperfect Spatial Information, vol. 256, pp. 49–74. Springer, Heidelberg (2010). https://doi.org/10.1007/978-3-642-14755-5_3
20. Odstrcilik, J., et al.: Retinal vessel segmentation by improved matched filtering: evaluation on a new high-resolution fundus image database. IET Image Process. **7**(4), 373–383 (2013)
21. Rahtu, E., Salo, M., Heikkila, J.: A new convexity measure based on a probabilistic interpretation of images. IEEE Trans. Pattern Anal. **28**(9), 1501–1512 (2006)
22. Rosin, P.L., Zunic, J.: Probabilistic convexity measure. IET Image Process. **1**(2), 182–188 (2007)
23. Sonka, M., Hlavac, V., Boyle, R.: Image Processing, Analysis, and Machine Vision, 3rd edn. Thomson Learning, Toronto (2008)
24. Staal, J.J., Abramoff, M.D., Niemeijer, M., Viergever, M.A., Van Ginneken, B.: Ridge based vessel segmentation in color images of the retina. IEEE Trans. Med. Imaging **23**(4), 501–509 (2004)
25. Stern, H.: Polygonal entropy: a convexity measure. Pattern Recogn. Lett. **10**, 229–235 (1998)

26. Tasi, T.S., Nyúl, L.G., Balázs, P.: Directional convexity measure for binary tomography. In: Ruiz-Shulcloper, J., Sanniti di Baja, G. (eds.) CIARP 2013. LNCS, vol. 8259, pp. 9–16. Springer, Heidelberg (2013). https://doi.org/10.1007/978-3-642-41827-3_2
27. Vanegas, M.C., Bloch, I., Inglada, J.: A fuzzy definition of the spatial relation "surround" - application to complex shapes. In: Proceedings of EUSFLAT, pp. 844–851 (2011)
28. Zunic, J., Rosin, P.L.: A new convexity measure for polygons. IEEE Trans. Pattern Anal. **26**(7), 923–934 (2004)

The Propagated Skeleton: A Robust Detail-Preserving Approach

Bastien Durix[1(✉)], Sylvie Chambon[1], Kathryn Leonard[2], Jean-Luc Mari[3], and Géraldine Morin[1]

[1] IRIT-University of Toulouse, CNRS, Toulouse, France
bastien.durix@enseeiht.fr
[2] Occidental College, Los Angeles, CA, USA
[3] Aix Marseille Univ, Université de Toulon, CNRS, LIS, Marseille, France

Abstract. A skeleton is a centered geometric representation of a shape that describes the shape in a simple and intuitive way, typically reducing the dimension by at least one. Skeletons are useful in shape analysis and recognition since they provide a framework for part decomposition, are stable under topology preserving deformation, and supply information about the topology and connectivity of the shape. The main drawback to skeletonization algorithms is their sensitivity to small boundary perturbations: noise on a shape boundary, such as pixelation, will produce many spurious branches within a skeleton. As a result, skeletonizations often require a second pruning step. In this article, we propose a new 2D skeletonization algorithm that directly produces a clean skeleton for a shape, avoiding the creation of noisy branches. The approach propagates a circle inside the shape, maintaining neighborhood-based contact with the boundary and bypassing boundary oscillations below a chosen threshold. By explicitly modeling the scale of noise via two parameters that are shape-independent, the algorithm is robust to noise while preserving important shape details. Neither preprocessing of the shape boundary nor pruning of the skeleton is required. Our method produces skeletons with fewer spurious branches than other state-of-the-art methods, while outperforming them visually and according to error measures such as Hausdorff distance and symmetric difference, as evaluated on the MPEG-7 database (1033 images).

1 Introduction

A skeleton is an internal structure that describes a shape, defined by the centers of the maximally inscribed balls enclosed within the shape. The set of circle centers and radii gives the medial axis, which completely describes the shape [5]. Because the skeleton can be represented by a graph where each edge represents a feature of the shape, it is useful in applications such as shape analysis and pattern recognition [4,12,18,20,22]. To be useful in such applications, the skeleton should possess certain properties: centeredness within the shape, thinness, topological equivalence with the shape, and completeness. It should also be robust to noise.

M. Couprie et al. (Eds.): DGCI 2019, LNCS 11414, pp. 343–354, 2019.
https://doi.org/10.1007/978-3-030-14085-4_27

The noise property is one of most difficult to respect, requiring a distinction between noise and details of the shape. We propose a skeletonization method that possesses the above properties, and that is robust to noise on the shape boundary.

Many skeletonization methods have been proposed [16]. Three common categories of computing a skeleton from a discrete structure are:

- Thinning methods that use a morphological operator on a binary shape,
- Distance map methods that consider the distance to the boundary,
- Methods based on the Voronoï diagram of points on the shape boundary.

Thinning methods erode shape pixels in 2D [9] or voxels in 3D [6] according to specific rules and criteria. The resulting skeleton is a 1-pixel-wide structure. We refer to [21] for a survey of these methods. Unfortunately, the resulting skeleton lies on a 1-pixel-wide grid centered only up to a pixel size, thereby inducing a loss of precision. Furthermore, the radius is not directly computed. In contrast, our proposed skeletonization produces a structure not constrained by the grid, and gives the radius for each skeletal point.

Distance maps are defined as the minimal distance between a point and the shape boundary. The singularities of the distance map generate a skeleton. One distance map skeleton uses an active contour evolving inside the shape, following the gradients of the distance transform [13] until the fronts from opposite sides of the shape meet at the skeleton. One drawback to this method is that the structure of the skeleton depends on the local speed of the active contour, which is a function of the local curvature. Another distance map approach computes singularities of the distance transform using image based-methods, then estimates connectivity of the medial axis [15]. Because the distance map is discrete on a discrete image, its singularities can only be computed up to a convolution operator. The choice of this operator is crucial, and can create arbitrarily large branches due to noise. Although our method also uses a distance map, its main principle is to follow the boundary of the shape, which, consequently, makes it more robust to noise.

Voronoï skeletonization is the most popular method for computing a skeleton. Introduced by Ogniewicz [14] *et al.*, the interior edges of the Voronoï diagram of a discrete boundary represent the skeleton. This method is popular because it is efficient and precise. Unfortunately, the resulting skeleton usually includes many uninformative branches and is therefore unnecessarily complex (*cf.* Fig. 1(a)). For more on stability, see [3]. In 3D, the complexity is compounded, which has led to methods such as the power crust [2] to extract a simpler structure. Once a skeleton is computed, most approaches require a second step to prune noise. We next present pruning methods, and then describe approaches to generate a clean skeleton without pruning.

Pruning methods remove noisy portions of a skeleton after computation. Usually, a criterion is evaluated at each skeletal point to determine if that point should be removed. We present three popular pruning approaches. The λ-medial axis [7], with parameter λ, removes a skeletal point if the point's circle intersects the boundary at tangency points that are contained in a circle of radius λ. The

θ-homotopy medial axis [19], with parameter θ, evaluates the angle between the center of the skeletal circle and the tangency points on the boundary. If the angle is less than θ, then the associated skeletal point is removed. Finally, the scale-axis-transform (SAT) [8] grows each skeletal circle by a multiplicative factor s, then computes a skeleton from the new shape. The circles generating the new skeleton are then reduced using the multiplicative factor $1/s$. A variant of SAT that ensures homotopy equivalence with the boundary grows all circles by the factor s and removes grown circles that are contained inside another grown circle. These three methods have the similar drawbacks. First, the homotopy of the pruned skeleton can not be guaranteed, except for SAT variant and the λ-medial axis, for which the connectivity of the skeleton is ensured only if λ is lower than the smallest radius of its skeleton. Second, the choice of the parameter is difficult and often must be tuned to each shape. Finally, no distinction can be made between noise and small shape details, which means that desirable features of the shape may be pruned away (cf. Fig. 1(c) and (d)).

Directly computing a clean skeleton is an appealing alternative, with two main approaches. The first approach includes a pruning criterion within the skeletonization. For example, Digital Euclidean Connected Skeleton (DECS) [11] combines the ridges of the distance map and the centers of the maximal balls to obtain a skeleton on a grid of pixels (*cf.* Fig. 1(e)). As with thinning methods, the resulting skeleton has the width of one pixel. Furthermore, this method returns a skeleton that tends to be irregular. The second approach modifies the boundary itself, as in the circular boundary representation [1]. The resulting skeleton avoids many noisy branches, but the construction is complicated: the conversion of the boundary into arcs requires that the boundary be represented by polynomial splines which are then converted into circular arcs based on a chosen parameter. In contrast, our method uses the information of the discrete boundary directly without preprocessing.

We propose a new skeletonization method that directly produces a clean skeleton (*cf.* Fig. 1(f)) from boundaries affected by noise. Two common types of noise are pixelation noise resulting from the rasterization of the shape, and Gaussian noise on the boundary. In 2D, because most shapes are extracted from binary shape images, we focus here on rasterization noise. Our method propagates a circle inside the shape, ensuring continuous contact along the boundary. This "growing approach" has previously been applied only for grid-constrained skeletons [17]. By taking into account the scale of the rasterization noise, we can avoid creating noisy branches. Our method draws from distance map skeletons, searching for singularities of a distance map, and Voronoï skeletons, computing the connectivity of the skeleton directly from the boundary. The method generates a clean skeleton a priori and does not require a pruning step. Contributions of our method are twofold: we explicitly avoid noise on the boundary, and we capture salient geometric information with fewer branches.

Section 2 describes our method, and Sect. 3 compares our method with the most common skeletonization approach, the Voronoï skeletonization [14], pruned by scale-axis-transform [8], λ-medial axis [7] and θ-homotopy medial axis [19]. We also compare with DECS [11].

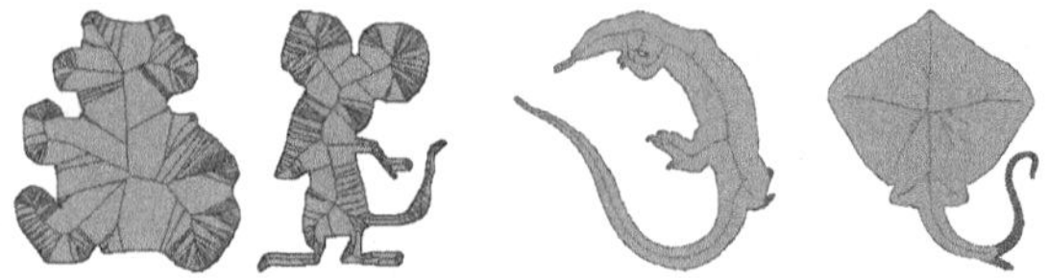

(a) Skeletons computed by a Voronoï method (b) $s = 1.7$ (c) $\lambda = 9$

(d) $\theta = 90^o$ (e) Skeletons computed with DECS

(f) Skeletons computed by our method, without changing the parameters

Fig. 1. Illustration of the limits of Voronoï skeletonization (a), the pruning methods (scale-axis-transform (b), λ-medial axis (c), θ-homotopy medial axis (d)), and DECS (e), and the advantages of ours (f). Green parts correspond to the reconstructed shape, and red parts show what has been lost. Voronoï (a) skeletons are noisy, requiring pruning. Some pruning methods struggle to differentiate between noise and shape details, and parts can be lost while noisy branches are preserved (b),(c). Others have shape dependent parameters (d). The DECS method returns a skeleton that is simple, but it tends to miss some parts, and be irregular. Our method (f) distinguishes between noise and details with parameters depending only on the scale of the noise. (Color figure online)

2 Propagating Skeletons

We consider a shape characterized by a matrix of pixels (from a binary image, for example), $\mathbf{S} \in \mathcal{M}_{h,l}(\{0;1\})$, with h the height of the image, and l its width. We extract a piecewise linear contour $\mathcal{B}$ from the shape pixels. From the boundary contour, we construct a skeleton represented by a graph whose vertices correspond to centers of skeletal balls and whose edges give connectivity information, consistent with boundary connectivity, without branches generated by pixelation noise. Following the philosophy of the Voronoï skeleton, where vertices are centers of maximally inscribed bitangent balls, we propagate circles inside the shape ensuring continuous contact with the boundary on at least two sides.

Before describing the complete algorithm in Sect. 2.4, we present some necessary tools. Section 2.1 defines a smooth distance estimation (SDE), used to compute a smoothed radius of skeletal circles. Section 2.2 explains the notion of a contact set, a key construction for robustness to noise. Finally, Sect. 2.3 describes the circle propagation principle that determines neighbors of a circle given its contact sets and the SDE.

2.1 Smooth Distance Estimation (SDE)

In this section, we consider a rasterized boundary. Typically, radius estimation measures the minimal Euclidean distance to a point on the contour for each point in the interior of the shape. This leads to poor radius estimates for rasterized boundaries, as shown in Fig. 2(a).

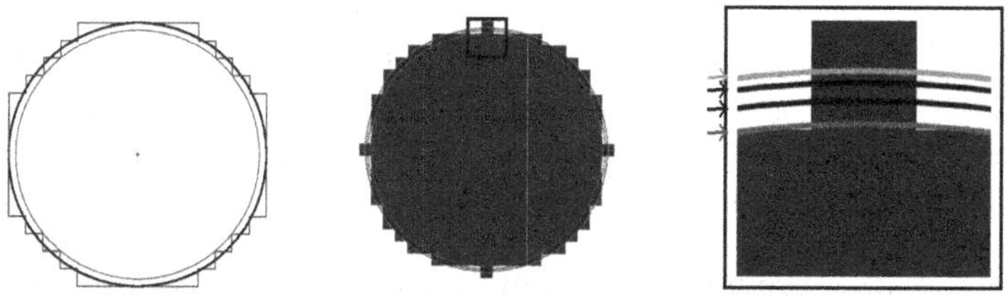

(a) (b) (c)

Fig. 2. The limitations of the Euclidean distance function compared to our proposed distance function. (a) The black circle, with center represented by a cross, is rasterized, giving the blue boundary: each pixel is considered inside the binary shape if its center is in the black circle. The red circle has same center as the black circle with radius given by the minimum Euclidean distance to a point of the contour. As the radius of the blue circle is less than that of the black circle, the Euclidean estimation of the radius is incorrect. (b) Estimated circle from the SDE, for several values of σ. In (c), the arrows indicate from top to bottom: • the original circle ($r = 10$), • the circle associated with $\sigma = 1$ ($r = 9.89$), • the circle associated with $\sigma = 0.5$ ($r = 9.55$), • the circle associated with the classic distance function ($r = 9.51$). When σ approaches 0, the radius of the circle converges to the radius from the Euclidean distance function. (Color figure online)

To improve radius estimation, we consider more than one boundary point to construct a smoothed version of the Euclidean distance:

$$d_{\mathcal{B}}(P) = \min_{B \in \mathcal{B}} \|P - B\|.$$

The proposed distance adds a weight to points on the boundary to capture their relative importance to the particular circle being computed:

$$f_{\mathcal{B},\sigma}(P) = \frac{\sum_{B\in\mathcal{B}} p_{\mathcal{B},\sigma}(P,B)\|P-B\|}{\sum_{B\in\mathcal{B}} p_{\mathcal{B},\sigma}(P,B)},$$

where $p_{\mathcal{B},\sigma}(P,B)$ is a weight assigned to the point B of the boundary. To determine the weight, we consider the distance between B and the circle $\mathcal{C}(P,\rho)$ with $\rho = d_{\mathcal{B}}(P)$. The larger the distance, the less influential the point should be. In other words, the weight function should decrease rapidly as distance to the circle increases. A semi-normal distribution with parameter σ fits these requirements:

$$P_\sigma(t_0 \geq 0) = \frac{2}{\sigma\sqrt{2\pi}} \int_{t_0}^{+\infty} e^{-\frac{t^2}{2\sigma^2}}\, dt.$$

The influence of each boundary point B on P is then given by:

$$p_{\mathcal{B},\sigma}(P,B) = P_\sigma(d_{\mathcal{C}}(P,B)), \text{with} d_{\mathcal{C}}(P,B) = \|P-B\| - d_{\mathcal{B}}(p).$$

The behavior of the function, illustrated in Fig. 2, is related to the values of the parameter σ. For $\sigma = 0$, $f_{\mathcal{B},\sigma}$ is equal to the Euclidean distance. This function is also used to identify an initial point of the skeleton. To estimate a first point P_0, we randomly choose a point Q_0 inside the shape, then maximize the function $f_{\mathcal{B},\sigma}(P,Q_0)$. Any local maximum is a singular point of the function $f_{\mathcal{B},\sigma}$ and will therefore be a skeletal point. Once P_0 is estimated, we identify its contact set and begin propagation. We define contact set in the next section.

2.2 Contact Between the Circle and the Shape

We now define the contact set between a circle $\mathcal{C}(P,\rho)$, with center P and radius $\rho = f_{\mathcal{B},\sigma}(P)$, and the contour $\mathcal{B}$. We consider the parameter σ from $f_{\mathcal{B},\sigma}$, and introduce a new parameter, α, which defines the maximal distance from a point to its associated circle. The contact set is the topological closure of boundary points at a distance less than σ to the circle together with points at a distance less than α to the circle, as illustrated in Fig. 3. A contact set corresponds to consecutive boundary points where the first and last points are at most σ, and all points are at most α, from the skeleton point. The points in the contact set are analogous to the bitangent contact points in the continuous model, denoting a set of points "closest" to the skeletal point. The contact sets also determine the directions in which the circle $\mathcal{C}(P,\rho)$ can propagate. The parameter α allows us to avoid noise on the boundary while simultaneously maintaining topological consistency of the skeleton with the boundary (*cf.* Fig. 3).

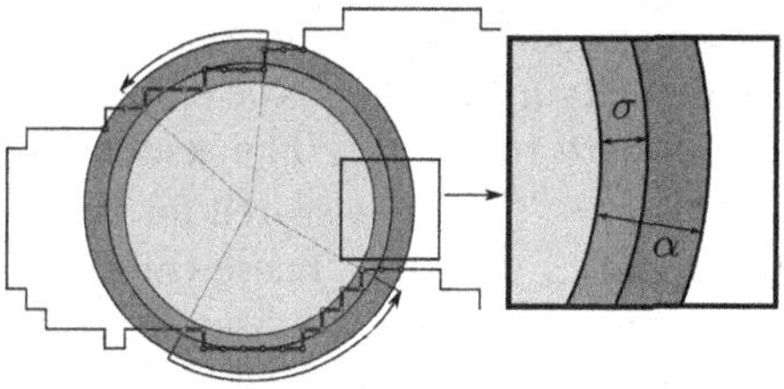

Fig. 3. Contact sets for a skeletal point. In (a), we divide the area around the estimated SDE circle into three parts, see (b) for details: the thick green part $r + \sigma$ (where r is the radius of the original circle), the thick orange part $\alpha - \sigma$, and the outer part. If only the green is taken into account, we find 6 contact sets (in green, 3 at the top, 3 at the bottom). Adding the uncertainty zone using α merges these into 2 contact sets (blue lines), which prevents the creation of four noisy branches. (Color figure online)

The definition of contact sets in our algorithm serves two purposes: reducing noise and maintaining topological equivalence to the shape. First, the contact sets for a point designate the suitable directions for propagation without noise. The value of α, chosen to distinguish between noise on the contour and details of the shape, defines the maximum distance between a point of the contour and the closest skeletal circle, which defines the Hausdorff distance between the true boundary and the approximated boundary produced by the skeleton. Second, as two neighboring circles must have overlapping contact sets, we are guaranteed to consider all points of the boundary, which guarantees the topological equivalence between skeleton and boundary.

2.3 Circle Propagation

After estimating contact sets associated to a point P of the skeleton, we propagate the circle between successive pairs of contact sets. To propagate between two successive contact sets, $\mathcal{T}_1$ and $\mathcal{T}_2$, we search for the farthest circle that maintains contact with points of $\mathcal{T}_1$ and $\mathcal{T}_2$ (*cf.* Fig. 4). This produces a nested loop, with one loop searching for the distance, *i.e.* the circular arc on which there is the next circle, and the other searching for the angle, *i.e.* the position of the center on the given circular arc.

The search region for the next point of the skeleton is the portion of the circle between the last point A_1 of $\mathcal{T}_1$ and the first point A_2 of $\mathcal{T}_2$. For a given distance d, less than the radius $\rho = f_{\mathcal{B},\sigma}(P)$ of the circle, we search for an angle θ_d where $f_{\mathcal{B},\sigma}(Q_d)$ is local maxima (with $Q_d = P + (d\cos(\theta_d), d\sin(\theta_d))^t$). Not every value of d can describe a valid neighbor of P, since contact sets of the neighbors must contain A_1 and A_2. Using this property, we can validate, for each d, the possibility of Q_d as a neighbor of P. We select the highest value of d so that Q_d is a valid neighbor of P.

To estimate the value of θ_d, we take an initial interval $\left[\theta_0^1, \theta_0^2\right]$, where θ_0^i designates the angle of A_i on the circle, and, then, we decrease it with search by dichotomy. We compute the middle angle θ_i between θ_i^1 and θ_i^2. If the derivative of $g_d(\theta) = f_{\mathcal{B},\sigma}(P + (d\cos(\theta), d\sin(\theta))^t)$ is positive on θ_i, we update the interval to $\left[\theta_i, \theta_i^2\right]$, otherwise we update to $\left[\theta_i^0, \theta_i\right]$. The search stops when $(\theta_i^2 - \theta_i^1)$ is less than a given threshold $\epsilon_\theta = \frac{0{,}01\sigma}{\rho}$. The search for the distance d is done in a similar way, up to a threshold $\epsilon_d = 0{,}01\sigma$. These thresholds could be chosen as machine precision, but because the scale of noise on the boundary is σ, we can stop the search when the distance given by each interval has less influence than the noise.

Using this algorithm, we find an approximation of a local skeleton around the point P. The number of contact sets gives the degree of connectivity of the skeleton at P. By construction, only the variations of the boundary more than α from the circle are considered as new propagation directions. Finally, forcing contact sets of any two successive circles to intersect guarantees that all the points of the boundary are associated with at least one circle, and we produce a connected skeleton. We summarize the complete algorithm in the next section.

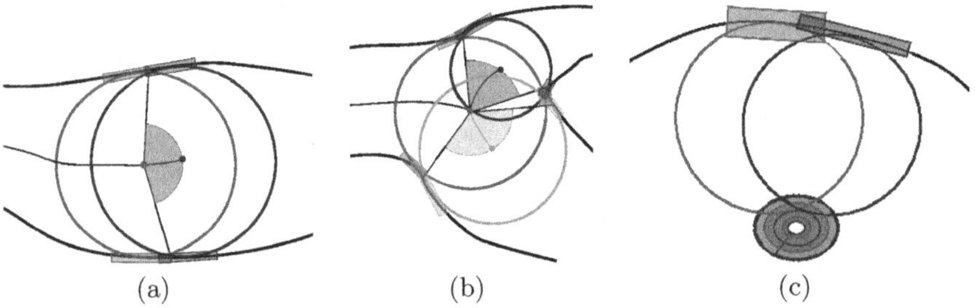

Fig. 4. Propagation of a circle, for 2 contact sets in (a) and 3 contact sets in (b). In (a), as the propagated circle comes from the left, we can propagate the red circle to the right. The blue circle is validated when its contact sets (blue rectangles) intersect the contact sets of the red circle (red rectangles). In the case of 3 contact sets (b), we have two possible directions for the propagation, and both are explored. If there is a small hole in the shape (c) that is completely covered by a contact set, the algorithm still considers that the contact set has a beginning and end and will execute correctly. (Color figure online)

2.4 Main Algorithm

Algorithm 1 summarizes our approach, which uses two parameters, σ and α where σ models the noise on the boundary and α gives a threshold on the Hausdorff distance between the shape modeled by the skeleton and the input shape, in order to discriminate between noise and details on the shape. We have chosen $\sigma = \sqrt{2}/2 \simeq 0.7$ and $\alpha = 3\sqrt{2}/2 \simeq 2.1$ and these choices are explained in Fig. 5.

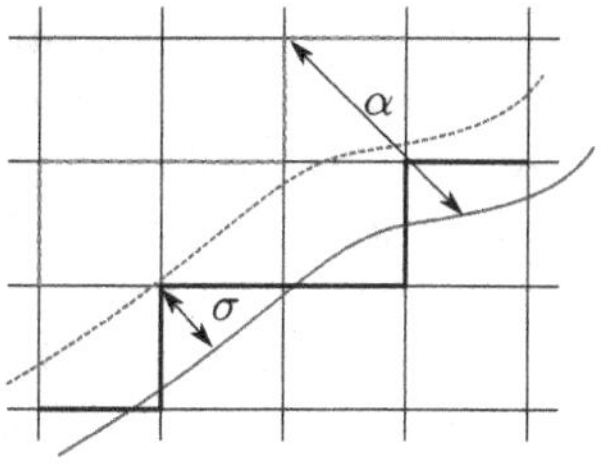

Fig. 5. Justification of the choices of σ and α. σ models the rasterization of the boundary, so it can vary from $-\sqrt{2}/2$ to $+\sqrt{2}/2$. Thus $\sigma = \sqrt{2}/2$. α models random noise: if there is a 1-pixel error on the boundary (like for the red line), the maximal distance between the boundary and the rasterized boundary is $3\sqrt{2}/2$. Thus $\alpha = 3\sqrt{2}/2$. (Color figure online)

Algorithm 1. Propagating skeleton algorithm

Parameters: σ: Noise on the boundary
$\alpha > \sigma$: Maximum distance to the shape generated by the skeleton
Data: **S**: Binary grid representing the shape
$\mathcal{B}$: Boundary of the shape (with 4-adjacency connectivity)
Result: S : Shape skeleton
Estimation of the point P_0, local maxima of $f_{\mathcal{B},\sigma}$ (*cf.* Section 2.1)
$l_c \leftarrow \{P_0\}$
while $l_c \neq \emptyset$ **do**
  Removing the first element P of l_c
  Construction of $\mathcal{C}(P, f_{\mathcal{B},\sigma}(P))$, maximal circle with center P and radius $f_{\mathcal{B},\sigma}(P)$
  Estimation of the tangency points between $\mathcal{C}(P, f_{\mathcal{B},\sigma}(P))$ and $\mathcal{B}$, with noise parameter σ and maximal distance to the skeleton α (*cf.* Section 2.2)
  Estimation by propagation of $Q_1, \ldots, Q_n$ the neighbor circles of $\mathcal{C}(P, f_{\mathcal{B},\sigma}(P))$ (*cf.* Section 2.3)
  Checking for junctions between $Q_1, \ldots, Q_n$ and l_c
  Adding $Q_1, \ldots, Q_n$ (not used in junction) in l_c
end

The algorithm ends when the list l_c is empty. The algorithm can account for loops in the skeleton: For each Q_i that results from propagation of P, we check if there is a neighbor in the current center list l_c. If so, we interrupt the propagation from Q_i and add a loop-closing edge.

The complexity is on the order of $N^2 \log_2(S/\epsilon_d) \log_2(2\pi/\epsilon_\theta)$, where N is the number of boundary points and S their maximum pairwise distance. Note that the number of skeletal points is of the same order as the boundary points, and that the cost of propagation is the cost of the nested loop and the distance function. The complexity of the loop to search for the distance to the next center is $\log_2(d_m/\epsilon_d)$, where d_m is the maximal radius of a skeletal ball which is bounded above by S. The complexity of the loop searching for the angle is $\log_2(\theta_m/\epsilon_\theta)$, where θ_m is the maximal angle between the extremities of two contact sets, bounded above by 2π. The complexity of the SDE is of order N.

We have described a skeletonization algorithm obtained by propagating centers of skeletal balls under the constraint that each circle maintains contact with the boundary. The resulting skeleton is centered, thin, topologically equivalent to the original shape, and complete. It is also robust to noise. In the next section, we show skeletonization results and quantify the strengths of this method.

3 Results and Evaluation

To evaluate our algorithm, we compare it to DECS [11], a grid-based method, and to pruned versions of the classical Voronoï skeletonization algorithm [14]. For pruning, we choose the three most common methods: scale-axis-transform [8], λ-medial axis [7] and θ-homotopy medial axis [19].

We use three evaluation criteria. The first is the Hausdorff distance between the initial shape and the shape reconstructed by the skeleton. This measures the influence of the parameter α, which sets a maximum distance to the original boundary. The second criterion measures the difference between the initial surface $\mathbf{S}$ and the reconstructed surface $\mathbf{S}'$:

$$\frac{\mathcal{A}(\mathbf{S}\Delta\mathbf{S}')}{\mathcal{A}(\mathbf{S})},$$

where $\mathcal{A}(\mathbf{S}\Delta\mathbf{S}')$ is the area of the symmetric shape difference, and $\mathcal{A}(\mathbf{S})$ is the area of the original shape. These criteria do not assess noise reduction, so our third criterion counts the number of branches of the skeleton. Taken together, we evaluate the simplicity of the considered skeleton in providing a given level of shape fidelity. We perform tests on 1033 images from the MPEG-7 images [10]. See Table 1 for the averages obtained for each of the criteria and methods.

3.1 Analysis of the Results

Our method produces a simpler skeleton than the Voronoï skeleton while maintaining good shape fidelity. The scale-axis-transform pruning and λ-medial axis, in comparison, require many more branches to maintain a similar or lower fidelity. The θ-homotopy medial axis maintains fidelity with a reasonable number of branches, but our method outperforms it in the Hausdorff distance and produces a similar area difference with comparable number of branches. This shows that our method prioritizes the most informative branches both globally and locally when compared to other Voronoï-based methods. For certain shapes, such as the stingray, this results in a significant difference in shape quality. With similar numbers of branches, DECS gives better precision in terms of area difference, but ours is better for Hausdorff distance.

Table 1. Comparison between propagation skeletonization (with $\sigma = 0.7$ and $\alpha = 2.1$), Voronoï skeletonization with three different pruning methods and DECS [11] on 1033 images from [10]. Evaluation is done with three criteria: the Hausdorff distance in pixels, the area difference in percentages, and the number of branches. Red scores indicate where our algorithm is better, and blue where others are, showing the relation between the number of kept branches and the information lost.

	Voronoï skeletonization										DECS	Ours
	All	s			λ			θ				
		1.1	1.2	1.5	1	2	4	60°	80°	90°		
Mean Hausdorff (px)	0	2.2	3.3	8.3	1.0	2.1	6.8	2.5	4.9	7.5	8.1	2.0
Median Hausdorff (px)	0	1.9	2.7	6.3	1.0	1.5	3.1	2.1	3.9	6.0	4.6	2.1
Mean area (%)	3.0	3.1	3.4	4.4	3.0	2.9	3.7	3.3	3.9	4.7	2.6	4.1
Median area (%)	2.2	2.5	2.8	4.1	2.1	2.1	2.5	2.7	3.5	4.2	2.1	3.2
Mean branches	655	196	96	26	432	192	60	36	16	10	9	16
Median branches	537	115	49	17	355	157	49	19	11	7	7	12

Computation time of our algorithm averages 982 ms for the MPEG-7 dataset on a i7-4800MQ CPU at 2.70 GHz. By comparison, the mean computation for the full Voronoï is 47 ms and filtering with the scale-axis-transform takes between 451 ms and 1118 ms. The λ-medial axis takes around 86 ms, and θ-medial axis takes between 788 ms and 1171 ms, depending on choices for their respective parameters. The execution time of the DECS is 16 ms.

We note the influence of the parameters σ and α. Lower values for these parameters produce more spurious branches, while higher values produce simpler skeletons that are less accurate. These values can then be tuned appropriately to image resolution or application.

4 Conclusion

We present a new skeletonization method that avoids creating uninformative branches by explicitly modeling boundary noise. The method relies on two parameters that are straightforward to choose and do not require hand-tuning for each shape: σ, the scale of noise on the boundary, and α, distinguishing between noise and detail on the shape. We compare our method with the Voronoï skeletonization pruned with three popular methods, and DECS. We provide quantitative evaluation showing that our method models the shape better in fewer branches in terms of Hausdorff distance.

This work could be extended quite naturally. We focus here on rasterized shapes, but we can generalize our approach to boundary representations with other types of noise. We can also accommodate lack of connectivity on the boundary, such as for point clouds, by redefining contact sets to recognize the directions in which we can propagate. Finally, we can use a similar method to compute 3D skeletons by propagation.

References

1. Aichholzer, O., Aigner, W., Aurenhammer, F., Hackl, T., Jüttler, B., Rabl, M.: Medial axis computation for planar free-form shapes. Comput. Aided Des. **41**(5), 339–349 (2009)
2. Amenta, N., Choi, S., Kolluri, R.K.: The power crust, unions of balls, and the medial axis transform. Comput. Geom. **19**(2–3), 127–153 (2001)
3. Attali, D., Montanvert, A.: Computing and simplifying 2D and 3D continuous skeletons. Comput. Vis. Image Underst. **67**(3), 261–273 (1997)
4. Bai, X., Latecki, L.J.: Path similarity skeleton graph matching. IEEE Trans. PAMI **30**(7), 1282–1292 (2008)
5. Blum, H.: A transformation for extracting new descriptors of shape. In: Symposium on Models for the Perception of Speech and Visual Form (1967)
6. Borgefors, G., Nyström, I., di Baja, G.S.: Computing skeletons in three dimensions. Pattern Recognit. **32**(7), 1225–1236 (1999)
7. Chazal, F., Lieutier, A.: The λ-medial axis. Graph. Models **67**(4), 304–331 (2005)
8. Giesen, J., Miklos, B., Pauly, M., Wormser, C.: The scale axis transform. In Annual Symposium on Computational Geometry (2009)
9. Lam, L., Lee, S.-W., Suen, C.Y.: Thinning methodologies - a comprehensive survey. IEEE Trans. PAMI **14**(9), 869–885 (1992)
10. Latecki, L.J., Lakamper, R., Eckhardt, T.: Shape descriptors for non-rigid shapes with a single closed contour. In: IEEE Conference on CVPR (2000)
11. Leborgne, A., Mille, J., Tougne, L.: Noise-resistant digital euclidean connected skeleton for graph-based shape matching. J. Vis. Commun. Image Represent. **31**, 165–176 (2015)
12. Leonard, K., Morin, G., Hahmann, S., Carlier, A.: A 2D shape structure for decomposition and part similarity. In: ICPR (2016)
13. Leymarie, F., Levine, M.D.: Simulating the grassfire transform using an active contour model. IEEE Trans. PAMI **14**(1), 56–75 (1992)
14. Ogniewicz, R., Ilg, M.: Voronoi skeletons: theory and applications. In: IEEE CVPR (1992)
15. Rumpf, M., Telea, A.: A continuous skeletonization method based on level sets. In: Proceedings of the symposium on Data Visualisation (2002)
16. Saha, P.K., Borgefors, G., di Baja, G.S.: A survey on skeletonization algorithms and their applications. Pattern Recognit. Lett. **76**(Suppl. C), 3–12 (2015)
17. Shen, W., Bai, X., Hu, R., Wang, H., Latecki, L.J.: Skeleton growing and pruning with bending potential ratio. Pattern Recognit. **44**(2), 196–209 (2011)
18. Siddiqi, K., Shokoufandeh, A., Dickinson, S.J., Zucker, S.W.: Shock graphs and shape matching. Int. J. Comput. Vis. **35**(1), 13–32 (1999)
19. Sud, A., Foskey, M., Manocha, D.: Homotopy-preserving medial axis simplification. In: ACM Symposium on Solid and Physical Modeling (2005)
20. Sundar, H., Silver, D., Gagvani, N., Dickinson, S.: Skeleton based shape matching and retrieval. In: Shape Modeling International (2003)
21. Vincze, M., Kővári, B.A.: Comparative survey of thinning algorithms. In: International Symposium of Hungarian Researchers on Computational Intelligence and Informatics (2009)
22. Xu, Y., Wang, B., Liu, W., Bai, X.: Skeleton graph matching based on critical points using path similarity. In: Zha, H., Taniguchi, R., Maybank, S. (eds.) ACCV 2009. LNCS, vol. 5996, pp. 456–465. Springer, Heidelberg (2010). https://doi.org/10.1007/978-3-642-12297-2_44

Non-centered Voronoi Skeletons

Maximilian Langer(✉), Aysylu Gabdulkhakova, and Walter G. Kropatsch

Pattern Recognition and Image Processing Group,
193-03 Institute of Visual Computing and Human-Centered Technology,
Technische Universität Wien, Favoritenstrasse 9-11, Vienna, Austria
{mlanger,aysylu,krw}@prip.tuwien.ac.at

Abstract. We propose a novel Voronoi Diagram based skeletonization algorithm that produces non-centered skeletons. The first strategy considers utilizing Elliptical Line Voronoi Diagrams with varied density based sampling of the polygonal shapes. The second strategy applies a weighting scheme on Elliptical Line Voronoi Diagrams and Line Voronoi Diagrams. The proposed skeletonization algorithm uses precomputed distance fields and basic element-wise operations, thus can be easily adapted for parallel execution. Non-centered Voronoi Skeletons give a representation that is more similar to real world skeletons and retain many of the desirable properties of skeletons.

Keywords: Generalized Voronoi Diagram · Line Voronoi · Voronoi Skeleton · Elliptical distance · Weighted hausdorff

1 Introduction

Skeletonization plays an important role in the area of shape representation and description, since it decreases the dimensionality of the problem. By definition, a *skeleton* is a compact one-point wide representation of the shape that is associated with a locus of the points that are equidistant to two or more shape boundary points. This notion was originally introduced by Blum [1] as a *medial axis transform (MAT)*. Skeletons provide useful properties for computer vision applications: invariance to translation, rotation and scaling; preservation of the topology of the shape; it can be computed for any 2D shape; incorporation of adjacency and neighborhood information; it is possible to completely reconstruct the original shape [13].

In order to represent the anatomical skeleton of an animal more closely, it is desirable to have a shape representation where the skeleton does not lie on the medial axis. In particular, MAT-based representation introduces wrong points to shape part connections (or joints) and results in deformations of rigid skeletal parts during articulated movement. This paper presents two approaches to obtain a non-centered skeleton. First, based on the properties of the Elliptical Line Voronoi Diagram we investigate the effect of the varied density based sampling of the polygonal shape approximation. Second, we propose a multiplicative weighting of shape boundary lines that enables modification of the medial axis

M. Couprie et al. (Eds.): DGCI 2019, LNCS 11414, pp. 355–366, 2019.
https://doi.org/10.1007/978-3-030-14085-4_28

position. Non-centered Voronoi skeletons fulfill all properties of skeletons except the reconstruction property. In the proposed algorithm, the shift of the medial axis towards the desired part of the shape is achieved by using varied density of the polygonal approximation of the shape and multiplicative weighting of shape boundary lines.

The remainder of the paper is organized as follows: Sect. 2 provides an overview of existing Voronoi Diagram types and presents a novel algorithm for computing a Generalized Voronoi Diagram using the Distance Transform. Section 3 discusses Voronoi Skeletons with an emphasis on Line Voronoi Skeletons, since dealing with polygonal shape approximations. Section 4 introduces Non-Centered Voronoi skeletons that can be obtained by (1) varied density-based sampling of the shape boundary and (2) weighting scheme for the distance maps. Section 5 concludes the paper.

2 Voronoi Diagram

The general idea behind the *Voronoi Diagram (VD)* is to associate each element in a set S with the closest element in a target set T (called *sites*) based on a distance metric $d(s, t), s \in S, t \in T$. In other words, VD is a partition of the set S with regard to proximity of the elements to the target set. The *Voronoi cell* is the subset of S, where each element is closest to a single element in the target set. Points with identical proximity to two elements in the target set form *Voronoi edges.*

2.1 Point Voronoi Diagram

In classical VD, or *Point Voronoi Diagram (PVD)*, a target set contains points in $\mathbb{R}^2$, whereas S is the 2D plane. When the distance is Euclidean, the Voronoi edges are bisectors. PVD with 8 sites is illustrated in Fig. 1a.

2.2 Line Voronoi Diagram

In a *Line Voronoi Diagram* (LVD), sites in the target set T are line segments. In contrast to the Point Voronoi Diagrams (PVD), the Voronoi edges contain lines and parabolic arcs [12]. Voronoi cells might not be connected if the corresponding elements in the target set are crossing each other (see Fig. 1b). In classical LVD, distance between a point and a line segment is defined with the regard to the Hausdorff distance $d_h(l, P) = \inf\{\delta(P, L) | L \in l\}$.

2.3 Elliptical Line Voronoi Diagram

Gabdulkhakova and Kropatsch [3] proposed a different metric - *Confocal Ellipse-based Distance* (CED), $d_e(P, l)$ - that defines a distance from the point P to

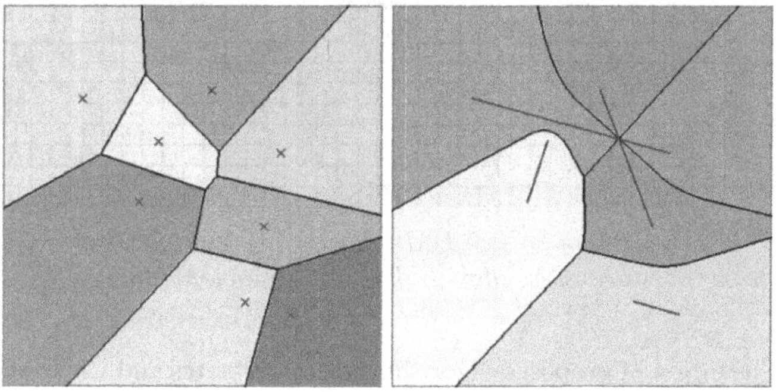

(a) (b)

Fig. 1. Examples of the (a) Point Voronoi Diagram and (b) Line Voronoi Diagram

the line segment l. As opposed to the Hausdorff distance, CED uses only the endpoints F_1, F_2 of a line segment [4]:

$$d_e(P,l) = \delta(P,F_1) + \delta(P,F_2) - \delta(F_1,F_2), \tag{1}$$

where δ denotes a Euclidean distance between two points.

CED is highly dependent on the length of the line segment. This enables longer line segments to have a greater influence on the Voronoi boundary and push the boundary away from them. This property is discussed in Sect. 4.1. Using the CED Gabdulkhakova et al. [4] introduced the Elliptical Line Voronoi Diagram (ELVD). This variation of the VD enables spatial control over the Voronoi edges without introducing weights.

Figure 2 shows this property. ELVD (black) has smaller regions associated with shorter line segments than the LVD (gray) using the Hausdorff distance.

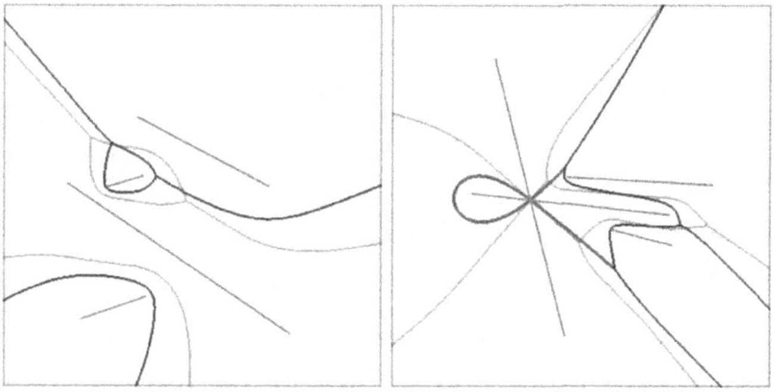

(a) Separate line segments (b) Crossing line segments

Fig. 2. ELVD (black) and LVD (gray). Note the loop (green) generated by intersecting line-sites with ELVD in (b). (Color figure online)

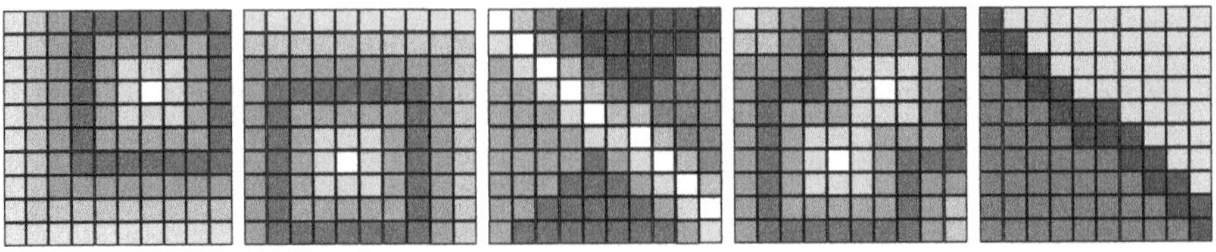

(a) $dm_1 = DM$ (b) dm_2 (c) $DM - dm_2 = dm_1 - dm_2$ (d) New $DM = \min(DM, dm_2)$ (e) IDM

Fig. 3. Single iteration of proposed algorithm with point sites and Chebyshev distance. Cyan area in (d) belongs to first site, magenta to second, dark cells are included in Voronoi edges. (Color figure online)

2.4 Generalized Voronoi Diagrams from Distance Transform

Computational complexity of Generalized Voronoi Diagrams increases with the size of the target set. Especially explicit computations are difficult, since they rely on computing complex higher-order algebraic curves (see the green curve in Fig. 2b). In many cases an explicit computation [5,7,12,14] of Voronoi edges is not necessary and algorithms [2,6,16,17] have been proposed to implicitly calculate the boundaries based on grids.

Strzodka and Telea [15] introduced an approach to calculate Distance Transforms (DT) and Voronoi Skeletons on graphics hardware using different point distance functions. They propose a technique called *Distance Splatting*, where they go through every boundary point and then every image point and compare its distance to the minimum distance of the previous boundary points, iteratively building up the DT. They also store the closest site in a separate map by propagating it from a initialization on the boundary points. Additionally they provide implementation details on a GPU with a fixed graphics pipeline.

Since ELVD requires distances to a line segment, we propose an approach similar to the one from Strzodka and Telea [15], that uses arbitrary sites instead of boundary pixels and uses precomputed DT of these primitive sites. This allows for arbitrary distance measures and weighting schemes. Additionally, it is computationally efficient for point sites and line sites using CED.

The algorithm takes a list of distance maps $L_D = \{dm_t \in \mathbb{R}^{N \times M} | t \in 0..T\}$ of T sites as input and an image resolution of N by M. Here, dm_t is the t-th distance map defining for each pixel the distance to the t-th site. It then loops through all sites building up a combined distance map DM and an ID-map IDM iteratively. Every pixel where $DM - dm_t > 0$ the IDM is assigned t. Figure 3 shows the first iteration of this loop. In the end, the ID-map is used to create the Voronoi edges by comparing neighboring pixels. The algorithm can be found in Algorithm 1.

Distance Map Creation. For PVD and LVD the speed of distance map creation can be significantly increased. The distance maps are sampled from a bigger

Algorithm 1. Generalized Voronoi Diagram from Distance Transform

Data: Target set T, map size (N, M)
Result: Map IDM containing indices of the nearest elements in T; Distance Map DM

```
for t = 1 to size_of(T) do
    dm_t ⟵ compute_distance_field(t, N, M);
IDM ⟵ ∅; DM ⟵ dm_1;
for t = 2 to size_of(T) do
    for p ∈ Grid(N, M) do
        if DM[p] − dm_t[p] > 0 then
            IDM[p] ⟵ t;
            DM[p] ⟵ min(DM[p], dm_t[p]) ;
```

point distance map of size $2N \times 2M$. Each pixel value represents the Euclidean distance to the center pixel. Therefore, it needs only one distance map creation, that can also be stored on the disk rather than being re-computed.

For ELVD the distance maps of the endpoints are summed up, then the distance between the two endpoints is subtracted. The above algorithm has to process a distance map dm_i only once, it can be generated on-the-fly by the single bigger point distance map.

Complexity and Error Estimation. Considering $N \times M$ is the grid size and T is the number of sites (samples/line segments), sequential time complexity lies in $O(T \cdot N \cdot M)$, leading to parallel complexity of $O(T)$ on graphics hardware. For point sites and lines using CED the distance transform can be loaded from disc (or generated once), leading to linear time, otherwise, the time complexity of the distance transform times the number of sites must be taken into account. Memory complexity is $O(N \cdot M)$, consisting of the $N \cdot M$ maps DM, IDM and either another for on-the-fly generated distance maps, or one $2N \times 2M$ map.

Since the algorithm considers only values on a grid, there is an accuracy error e. It is composed of the error of the distance transform e_d and the error of the boundary extraction e_b. The distance transform error for the point distance described above is $e_d = \frac{\sqrt{2}}{2}$ based on half the maximum distance between two pixel centers. The boundary extraction error is $e_b = \sqrt{2}$, resulting from a maximum of one wrong pixel distance diagonally. This gives a combined error of $e = \frac{3}{2}\sqrt{2}$. Additionally, the true Voronoi boundary after accounting for the distance transform error lies in the two pixel wide boundary found by the algorithm.

3 Voronoi Skeleton

A skeleton is a compact shape representation, which elements are equidistant from at least two points of the shape boundary. VD-based skeletonization is a continuous approach that preserves both geometrical and topological information of the shape. According to the definition, the boundaries of the Voronoi cells are equidistant to two or more elements in a target set. Ogniewicz and Ilg [10] use this property to obtain a *Voronoi Skeleton (VS)* of a shape by including all boundary pixels into the target set T. The resultant ridges are part of the skeleton that go through the edges connecting pixels along the boundary, but not through the pixels. Furthermore, they assign a residual value to each boundary between the two Voronoi cells that indicates its importance for the whole skeleton.

As can be seen throughout this paper, VD-based approaches construct the skeleton of the object (*Endoskeleton*), as well as the skeleton of the background (*Exoskeleton*) simultaneously.

3.1 Line Voronoi Skeletons

As VS edges by Ogniewicz and Ilg [10] lie between pixels their algorithm is extended to LVD with Voronoi edges passing through polygon points. The VD is computed on a polygonal shape using the DT approach described above.

After construction of the VD, each Voronoi edge between two cells must be evaluated for being a part of the skeleton. Lee [8] shows that all Voronoi edges that are not going through concave points are part of the skeleton. To avoid spurious branches (i.e. skeleton branches generated by noise) pruning is applied. In particular, densely sampled shapes have many spurious branches. Ogniewicz and Ilg propose four different residual values for each ridge that are based on the difference of the distance of the two adjacent points on the boundary and the distance through the shape. Ogniewicz [11] proves that their residual functions are monotonic and, thus, do not break topology. By using the midpoints of lines in the Line Voronoi Skeleton as adjacent points the same residual functions can be used. Mayya and Rajan [9] propose a similar procedure, but take only the number of intermediate object boundary segments into account. They show that topology is preserved by deleting only Voronoi ridges with no intermediate boundary segments in their adjacent sites.

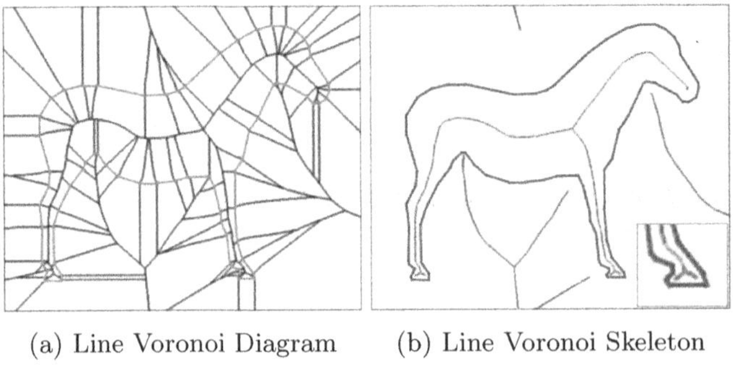

(a) Line Voronoi Diagram (b) Line Voronoi Skeleton

Fig. 4. Line Voronoi Diagram and skeleton with circular (purple), bicircular (green) and chord (blue) residual. (Color figure online)

Figure 4 shows the VD for a simplified horse shape and the Line Voronoi Skeleton. All three residuals are depicted with circular being purple, bicircular being green and chord being blue. The residuals follow the relation: circular $\subseteq$ bicircular $\subseteq$ chord.

4 Non-centered Skeletons

Using Euclidean distance as a distance measure has two advantages from the point of skeletonization: invariance to affine transformations and bending. The resultant skeleton is visually in the middle of the shape. For most cases the medial axis does not correspond to the physical skeleton of objects (e.g. animals, clothed objects). We introduce skeletons based on the medial axis using different distance measures to obtain a non-centered skeleton.

Definition 1. *A non-centered skeleton of a 2D shape S is the medial axis of S given a different distance measure than a p-Norm. The medial axis using this distance must have the same topology as S.*

4.1 Varied Density Based Sampling for Elliptic Voronoi Skeletons

It can be observed that the position of the VS produced by ELVD can be influenced by the length of the lines in the polygonal approximation of the shape. In our experiments we discovered that the influence gained solely by varied density based sampling of the shape is minor. Figure 6 gives three different approximations of the horse shape and the resulting skeletons.

In the following we study a special case, where the influence of varied density based sampling can be quantified. Let $l = (F_1, F_2)$ be the long line segment with length $2f$ and (P, P) be the shortest possible line segment coinciding with the single point $P \in \mathbb{R}^2$ (Fig. 5).

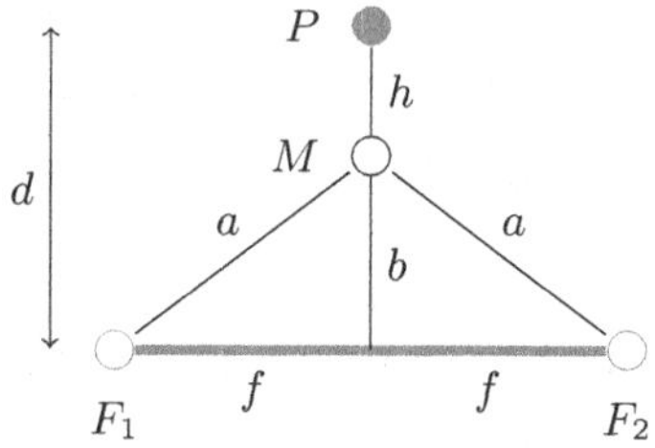

Fig. 5. Midpoint between line segment and point

Let a, b, f be the parameters of the confocal ellipses that have foci at the end points of the line segment l with $a^2 = b^2 + f^2$. Then, we know that the CED from the midpoint M to l is $d_e(M, (F_1, F_2)) = 2(a - f)$. The distance between

P and M is $2h$ and the normal distance of the point P from the line segment is $d = h + b$. Given the requirement that $d_e(M,(P,P)) = d_e(M,(F_1,F_2))$ we get $h = a - f$.

Lemma 1. *The ratio between the distance $h = \delta(P, M)$ and the normal distance d of P to l ranges between 0.5 (the symmetric case $f = 0$) to smaller values with increasing length of $l = 2f$ with ratio*

$$\frac{h}{d} = \frac{1}{2(1 + f/d)}. \tag{2}$$

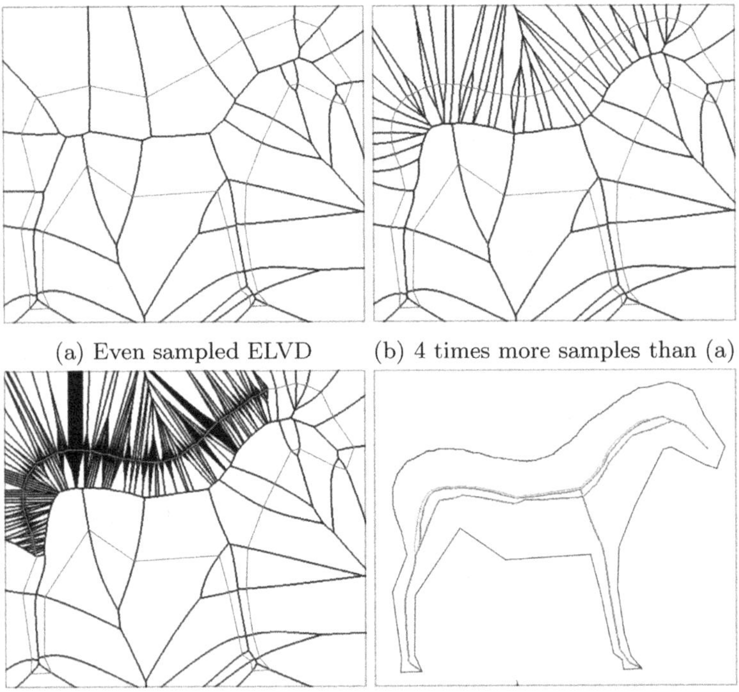

(a) Even sampled ELVD (b) 4 times more samples than (a)

(c) 40 times more samples than (a) (d) Resulting Skeletons from (a-c)

Fig. 6. ELVDs of different samples on the back of the horse. (d) skeletons (green: evenly sampled, cyan: 4 times more samples, magenta: 40 times more samples). (Color figure online)

Proof. The symmetric case would require $b = h = d/2$ and $b = a - f$. From the eccentricity formula $a^2 = b^2 + f^2$ with f being the linear eccentricity (or focal length) we derive $2bf = 0$ which implies either

- that $b = 0$, $M \in l$, $a = f$, and $h = 0 : P = M \in l$ or
- that $f = 0$, $F_1 = F_2$, $a = b = h$: M is the midpoint between P and F_1.

For all other cases $f > 0, b > 0$ we assume that the values for f and d are known. From the equations given above it follows that $a = d + f - b = \sqrt{b^2 + f^2}$. We solve for b, $b = \frac{d}{2}\left(1 + \frac{f}{d+f}\right)$ and, finally, with $b = d - h$ get the ratio $\frac{h}{d}$. □

Strength of influence is not only determined by the length of the line segment, but also by the distance to the opposite point (line). In order to off-center the medial axis, it is unpractical to rely on varied density-based sampling alone - the difference in density has to be great to handle it in a reasonable image size.

4.2 Weighted Line Voronois

In order to have a greater shift of the non-centered skeleton from the medial axis, it is proposed to weight the distance maps. We use an adaptation of multiplicative weighting of lines [12] to move the medial axis closer to higher weighted lines by multiplying the weight instead of the reciprocal of the weight with the distance map. Figure 7 shows the horse shape with higher weights on its back for the ELVD- and LVD-based skeleton. It can be observed that the weighting on LVD has more influence than on ELVD. Figure 8 illustrates the difference between LVD and ELVD weighting.

For LVD the skeleton will move in one direction according to the weight, i.e. the Euclidean distance from line l_1 to a point s on the skeleton ridge will be indirect proportional to the Euclidean distance from line l_2 to S: $\frac{d_h(l_1,s)}{d_h(l_2,s)} = \frac{w_2}{w_1}$, with $w_1, w_2 \in \mathbb{R}$ being the weights of l_1 and l_2 respectively and $d_h(l,p)$ being the Hausdorff distance between a line segment and a point.

For ELVD the task of estimating the skeleton position is harder. The distance is dependent on the focal length f and weighting is effecting the distance, as well as the elliptical parameter a.

For LVD and ELVD it is clear to see that no amount of weighting can push the medial axis outside of the object, because their values along the boundary lines are zero and can therefore never be closer to a different boundary segment.

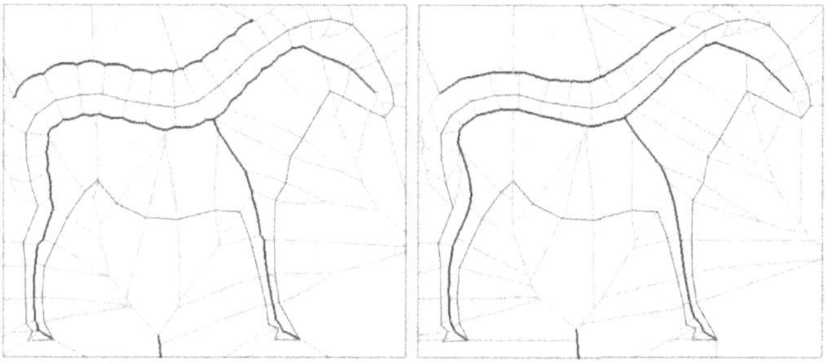

(a) ELVD (b) LVD

Fig. 7. Horse weighted on back with higher weights. ELVD and LVD are shown in gray, whereas skeleton - in black.

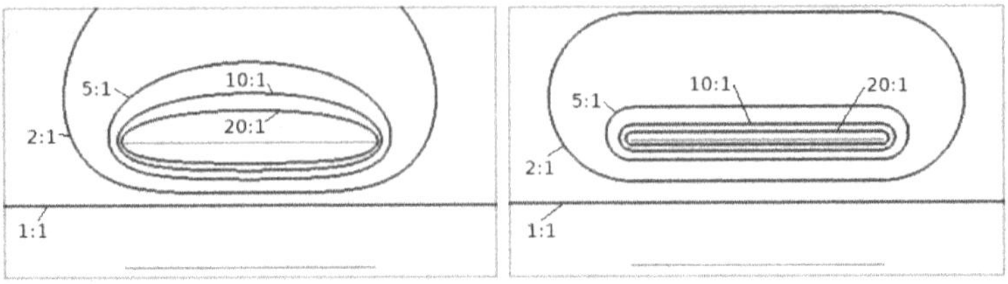

(a) ELVD (b) LVD

Fig. 8. Offset of Voronoi boundary, when using weights 1, 2, 5, 10 and 20.

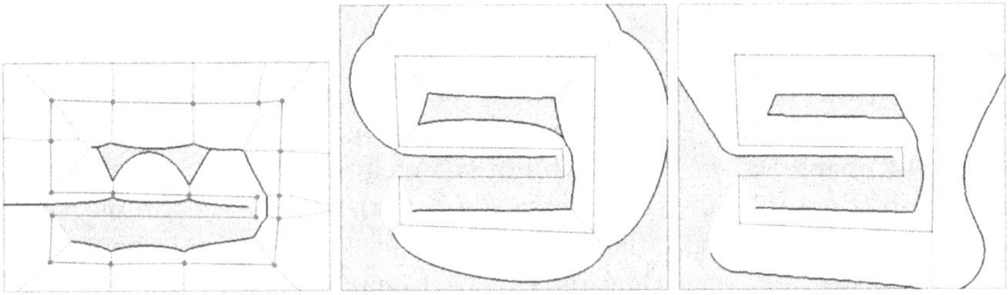

(a) Skeleton using Varied Density Based Sampling (b) Weighted ELVD (c) Weighted LVD

Fig. 9. Possible configurations for Voronoi cells with non-connected parts and resulting topology changes in the skeleton. (E)LVD in gray, skeletons in black. Gray regions are non-connected Voronoi cells.

4.3 Topology Preservation

In any of the skeletonization approaches presented above there exists the possibility of topology breakage, because the underlying VD may have cells that consist of multiple non-connected parts. This happens to skeletons that considered varied density based sampling, if line segment length proportions get too big. Analogically, for weighted skeletons it can happen, if weight proportions get too big. Figure 9 shows the problem. The gray areas belong to the same Voronoi cell, but are not connected and, hence, change topology in the medial axis.

Lemma 2. *The medial axis obtained by weighted (E)LVD or ELVD with Varied Density Based Sampling of a convex object preserves connectivity.*

Proof. This follows directly from the statement that the medial axis cannot be pushed outside the object. The influence region of a line can be behind another line, but since the object is convex, all additional regions are outside the object. □

Lemma 3. *The medial axis of a concave object constructed by weighted LVD does not break topology under the following constraint: if two lines l_1, l_2 in a concavity are weighted with w_1, w_2 $(w_1 > w_2)$, points in a distance greater than*

d times the distance from $P_1 \in l_1$ to $P_2 \in l_2$ in direction P_2 to P_1 from l_1 must be closest to another line than l_2 for all $P_1 \in l_1$ and $P_2 \in l_2$. The distance multiplier d is given as

$$d = \frac{1}{w_1}\frac{w_1 + w_2}{w_1 - w_2}. \tag{3}$$

Proof. Only considering two points $P_1 \in l_1$ and $P_2 \in l_2$, because the Hausdorff distance can be separated to a point level. Okabe et al. [12] show that multiplicative weighting results in *Apollonius circles* with two intersection points B_i, B_a on the line spanned by P_1, P_2 (Fig. 10). B_i and B_a lie on the Voronoi edge of line l_1. It follows, $\frac{\delta(P_1,B_i)}{\delta(P_2,B_i)} = \frac{w_2}{w_1}$ and $\frac{\delta(P_1,B_a)}{\delta(P_2,B_i)} = \frac{w_2}{w_1}$. The cross-ratio of the points $A = P_2, B = P_1, C = B_i, D = B_a$ is 1 as a property of Apollonius circles. By setting $\delta(B,C) = \frac{1}{w_1}$, $\delta(A,C) = \frac{1}{w_2}$, $\delta(B,D) = d$ and $\delta(A,D) = \frac{1}{w_1} + \frac{1}{w_2} + d$ the cross-ratio equation ($\frac{\delta(A,C)\delta(B,D)}{\delta(B,C)\delta(A,D)} = 1$) can be solved for d resulting in (3). □

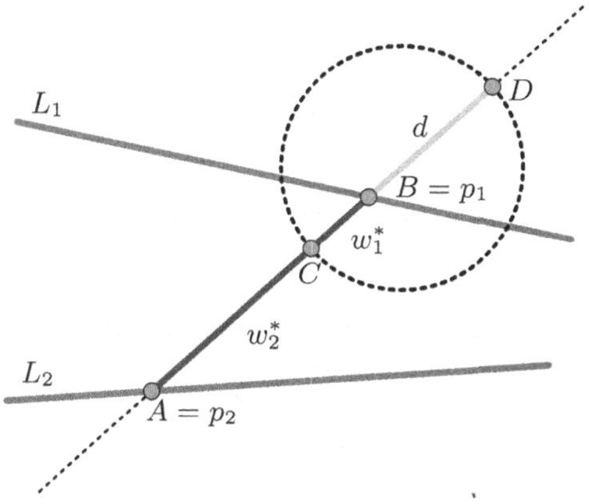

Fig. 10. Illustration of distance and *Apollonois circle* ($\frac{w_1^*}{w_2^*} = \frac{w_1}{w_2}$).

As with weighting influence, deriving rules for topologically correct skeletonization using ELVD is hard and remains an open problem.

5 Conclusion and Future Work

This paper presents a novel algorithm for building the Generalized Voronoi Diagram based on Distance Transform and uses it to create Line Voronoi Skeletons. It further explores the possibilities to shift the skeleton from the center using (1) ELVD with varied density based sampling and (2) multiplicative weighting strategy for Elliptical Line Voronoi Diagrams and Line Voronoi Diagrams.

The experimental results prove the applicability of the proposed approach for the problem of skeletonization. Non-centered skeleton opens new possibilities in the area of shape representation and description. Possible directions are thin skeleton extraction by boundary following with the help of the combined distance map and learning weights and samples to obtain the medial axis following anatomical skeletons of various shapes.

References

1. Blum, H.: A transformation for extracting descriptors of shape. In: Models for the Perception of Speech and Visual Form (1967)
2. Boada, I., Coll, N., Madern, N., Antoni Sellares, J.: Approximations of 2D and 3D generalized voronoi diagrams. Int. J. Comput. Math. **85**(7), 1003–1022 (2008)
3. Gabdulkhakova, A., Kropatsch, W.G.: Confocal ellipse-based distance and confocal elliptical field for polygonal shapes. In: Proceedings of the 24th International Conference on Pattern Recognition, ICPR (2018)
4. Gabdulkhakova, A., Langer, M., Langer, B.W., Kropatsch, W.G.: Line Voronoi diagrams using elliptical distances. In: Bai, X., Hancock, E.R., Ho, T.K., Wilson, R.C., Biggio, B., Robles-Kelly, A. (eds.) S+SSPR 2018. LNCS, vol. 11004, pp. 258–267. Springer, Cham (2018). https://doi.org/10.1007/978-3-319-97785-0_25
5. Held, M.: Voronoi diagrams and offset curves of curvilinear polygons. Comput.-Aided Des. **30**(4), 287–300 (1998)
6. Kalra, N., Ferguson, D., Stentz, A.: Incremental reconstruction of generalized voronoi diagrams on grids. Robot. Auton. Syst. **57**(2), 123–128 (2009)
7. Klein, R., Langetepe, E., Nilforoushan, Z.: Abstract Voronoi diagrams revisited. Comput. Geom. **42**(9), 885–902 (2009)
8. Lee, D.T.: Medial axis transformation of a planar shape. IEEE Trans. Pattern Anal. Mach. Intell. **4**(4), 363–369 (1982)
9. Mayya, N., Rajan, V.: Voronoi diagrams of polygons: a framework for shape representation. J. Math. Imaging Vis. **6**(4), 355–378 (1996)
10. Ogniewicz, R., Ilg, M.: Voronoi skeletons: theory and applications. In: Proceedings of 1992 IEEE Computer Society Conference on Computer Vision and Pattern Recognition, CVPR 1992 (1992)
11. Ogniewicz, R.L.: Discrete Voronoi skeletons. Ph.D. thesis, ETH Zurich (1992)
12. Okabe, A., Boots, B., Sugihara, K., Chiu, S.N.: Spatial Tessellations: Concepts and Applications of Voronoi Diagrams. Wiley, Hoboken (2009)
13. Saha, P., Borgefors, G., Sanniti di Baja, G.: Skeletonization: Theory, Methods, and Applications (2017)
14. Setter, O., Sharir, M., Halperin, D.: Constructing two-dimensional Voronoi diagrams via divide-and-conquer of envelopes in space. In: Gavrilova, M.L., Tan, C.J.K., Anton, F. (eds.) Transactions on Computational Science IX. LNCS, vol. 6290, pp. 1–27. Springer, Heidelberg (2010). https://doi.org/10.1007/978-3-642-16007-3_1
15. Strzodka, R., Telea, A.: Generalized distance transforms and skeletons in graphics hardware. In: Proceedings of the Sixth Joint Eurographics-IEEE TCVG Conference on Visualization. Eurographics Association (2004)
16. Vleugels, J.M., Overmars, M.H.: Approximating generalized Voronoi diagrams in any dimension. Utrecht University (1995)
17. Yuan, Z., Rong, G., Guo, X., Wang, W.: Generalized Voronoi diagram computation on GPU. In: 2011 Eighth International Symposium on Voronoi Diagrams in Science and Engineering (ISVD) (2011)

Dual Approaches for Elliptic Hough Transform: Eccentricity/Orientation vs Center Based

Philippe Latour(✉) and Marc Van Droogenbroeck

Montefiore Institute, University of Liège, Liège, Belgium
{philippe.latour,m.vandroogenbroeck}@uliege.be

Abstract. Ellipse matching is the process of extracting (detecting and fitting) elliptic shapes from digital images. This process typically requires the determination of 5 parameters, which can be obtained by using an Elliptic Hough Transform (EHT) algorithm.

In this paper, we focus on Elliptic Hough Transform (EHT) algorithms based on two edge points and their associated image gradients. For this set-up, it is common to first reduce the dimension of the 5D EHT by means of some geometrical observations, and then apply a simpler HT. We present an alternative approach, with its corresponding algebraic framework, based on the pencil of bi-tangent conics, expressed in two dual forms: the point or the tangential forms. We show that, for both forms, the locus of the ellipse parameters is a line in a 5D space.

With our framework, we can split the EHT into two steps. The first step accumulates 2D lines, which are computed from planar projections of the parameter locus (5D line). The second part back-projects the peak of the 2D accumulator into the 5D space, to obtain the three remaining parameters that we then accumulate in a 3D histogram, possibly represented as three separated 1D histograms.

For the point equation, the first step extracts parameters related to the ellipse orientation and eccentricity, while the remaining parameters are related to the center and a sizing parameter of the ellipse. For the tangential equation, the first step is the known center extraction algorithm, while the remaining parameters are related to the ellipse half-axes and orientation.

Keywords: Ellipse detection · Ellipse matching · Hough Transform · Pencil of conics · Tangential equation

1 Introduction

Ellipses are planar closed curves described by five independent geometric parameters (or six algebraic coefficients, defined up to a scaling factor). They are ubiquitous in images, especially those representing human environments, and they help in describing a scene. Ellipse extraction in images is a particular case of template matching, generally solved either by area- or feature-based approaches.

M. Couprie et al. (Eds.): DGCI 2019, LNCS 11414, pp. 367–379, 2019.
https://doi.org/10.1007/978-3-030-14085-4_29

This paper investigates specific feature-based approaches, where the chosen features are the edge points (or contour points) extracted from the image and their associated image gradients. This is a meaningful combination as, if most of an ellipse is visible in the image, we expect edge points to be localizable on the ellipse border and the image gradient at these edge points to be directed along the normal to the ellipse contour (and then also orthogonal to the tangent to the ellipse). Therefore, edge points with their associated gradients are possible evidences of the presence of an ellipse in the image.

Formally, ellipse matching comprises two interdependent parts: detection (selection of the ellipse in the image) and fitting (estimation of the parameters). Ellipse fitting algorithms (such as in [1]) estimate the parameters of an unknown ellipse from a set of points originating from this ellipse reasonably well, but they are not robust to outliers. So, in practical applications, we first need to select or detect the points belonging to an ellipse before being able to fit it.

There exist two main approaches for ellipse detection: RANSAC-like methods (see [1], Chap. 4) and Hough Transform (HT) algorithms (see [2,3]). Both approaches may even be intermixed in applications, but this paper will only consider the case of HT. So, we focus on a specific contour-based *ellipse matching* method named *Elliptic Hough Transform* (EHT), where the contour is represented by a list of edge points extracted from the image.

In HT, the unknown template (ellipse) is represented by a point in its parameter space (five-dimensional, or 5D, space for ellipses), and all the evidences of the presence of the template in the image (edge points) vote for some subset of this parameter space. For instance, an edge point and its associated gradient impose two constraints to the unknown ellipse (passing by the point and touching the line orthogonal to the gradient at this point). Accordingly, these two evidences reduce the compatible ellipse parameter set to a three-dimensional (3D) space (5D parameter space minus two constraints). We say that these two evidences votes for the compatible 3D parameter subset. Then, in the parameter space, we accumulate the votes of all the evidences (edge points and gradients) extracted from the image. At the end of the accumulation process, the algorithm chooses (extracts) the ellipse that has accumulated the largest number of votes.

EHT methods first aim at detecting the presence of any number of elliptic structures in an image and, incidentally, at computing an approximation of their geometric parameters or algebraic coefficients. They are not intended to provide a precise value for the parameters or coefficients, which is the role of a subsequent ellipse fitting algorithm.

In the literature, authors use different types of evidences for EHT. In this paper, the evidences are *pairs of edge points with their associated gradients,* and we impose no conditions on the gradients. In the following, the availability of associated gradients is assumed implicitly. A pair of edge point thus represents four evidences and imposes four constraints on the compatible ellipses (passing by two points and touching two lines). Therefore, a pair of edge points votes for a one-dimensional (1D) ellipse parameter subset.

In EHT, we would need to accumulate the votes in a 5D space, which is complicated in practice. So, authors separate the problem in two steps. In [4], Tsuji *et al.* presented an EHT algorithm based on pairs of edge points with parallel gradients, which uses a kind of 2D histogram for computing the center of the ellipse. Likewise, Yuen *et al.* [5] were the first to use pairs of edge points with arbitrary gradients. They accumulate lines in the 2D Hough space of the ellipse centers and then accumulate planes in the 3D Hough space of the remaining normalized algebraic coefficients. Other authors (see [6,7]) use the same first step but resort to specific geometric parameters for the three remaining parameters, which they determine in two steps rather than one.

It is interesting to note that two articles, more specifically [8] and [9], presented, independently and without explicitly naming it, the pencil of bi-tangent conics as the right mathematical framework for analyzing the family of ellipses built on two edge points. In fact, Yoo *et al.* [8] use the framework only for the first part of the EHT algorithm. Besides, Benett *et al.* [9] noticed that the framework enables to compute, as a first step, any pair or triplet of parameters (not only the center) and to easily incorporate any prior geometrical constraints. None of these works consider the framework for the second step of the EHT.

A key point of the EHT is the choice of the set of (algebraic or geometric) parameters used for the accumulation process. In the following, we show that the usual method of center accumulation is well described within the framework of the dual equation. We also show that the set of parameters of the direct equation of the framework, introduced in [10] for fitting applications (see [11,12]), has never been used explicitly in the accumulation process of the EHT. This direct form accumulates a pair of parameters, which are equivalent to the eccentricity and orientation of the ellipse.

2 Problem Statement

In this section, we introduce the main notation and equations. In the following, a specific point, named P, is represented by the $2D$ vector of its Cartesian coordinates $\bar{\boldsymbol{P}} = \begin{pmatrix} X_P \; Y_P \end{pmatrix}^T$ and/or by the $3D$ vector of its homogeneous coordinates $\boldsymbol{P} = \begin{pmatrix} x_P \; y_P \; z_P \end{pmatrix}^T$ or $\boldsymbol{P} = \begin{pmatrix} X_P \; Y_P \; 1 \end{pmatrix}^T$. Likewise, an unknown point is represented by its Cartesian coordinate $\bar{\boldsymbol{X}} = \begin{pmatrix} X \; Y \end{pmatrix}^T$ and also by its homogeneous coordinates $\boldsymbol{X} = \begin{pmatrix} x \; y \; z \end{pmatrix}^T$ or $\boldsymbol{X} = \begin{pmatrix} X \; Y \; 1 \end{pmatrix}^T$. A line, named l, then satisfies an equation of type $\boldsymbol{l}^T\boldsymbol{X} = 0$, where $\boldsymbol{l} = \begin{pmatrix} u_l \; v_l \; w_l \end{pmatrix}^T$ is the $3D$ vector of the homogeneous coefficients of its equation. Also, an unknown line satisfies an equation $\boldsymbol{u}^T\boldsymbol{X} = 0$, where $\boldsymbol{u} = \begin{pmatrix} u \; v \; w \end{pmatrix}^T$. Most of the time, we use the homogeneous coordinates of the projective plane $\mathbb{P}^2 = P^2(\mathbb{R})$ and, for the sake of conciseness, we often identify the geometric elements with their representation.

2.1 Ellipses and Conics

Geometric Parameters of Ellipses. Geometrically speaking, an ellipse is commonly defined by (the Cartesian coordinates of) its center (C, C'), its half-axes (r, r'), and its orientation θ. In the following, we will also use the eccentricity e, defined as $e^2 = 1 - {r'^2}/{r^2}$.

Point (Direct) Equation. An ellipse may be defined algebraically as a special kind of conics, which is a second degree algebraic curve. A conic $\mathcal{C}$ is any curve whose point coordinates satisfy the following equation

$$\mathcal{C} : aX^2 + 2b''XY + a'Y^2 + 2b'X + 2bY + a'' = \boldsymbol{X}^T \mathtt{F} \boldsymbol{X} = 0\,, \tag{1}$$

where the symmetric matrix $\mathtt{F}$ is easily obtained by identification of the two members of Eq. (1).

The conic is also fully determined by the vector of its six homogeneous coefficients $\boldsymbol{f} = (a, b'', a', b', b, a'')^T$, defined up to a scaling factor. This vector belongs to the projective space of the conic coefficient vectors $\boldsymbol{f} \in \mathbb{P}^5 = P^5(\mathbb{R})$. In the following, we use the operator *flatten* that operates on the symmetric matrix $\mathtt{F}$ and returns the vector $\boldsymbol{f}$; so we write $\boldsymbol{f} = \text{flatten}(\mathtt{F})$.

We later show (see Eq. (8)) that, when the conic is an ellipse, we always have $a + a' \neq 0$, and we may also use the following alternative form (introduced in [10]) for the algebraic equation of an ellipse:

$$\left(X^2 + Y^2\right) - d\left(X^2 - Y^2\right) - 2d'XY - 2cX - 2c'Y - c'' = 0\,, \tag{2}$$

where the five *direct* coefficients $\bar{\boldsymbol{f}} = (d, d', c, c', c'')$ are related to the homogeneous direct coefficients in Eq. (1) and to the geometric parameters by

$$\begin{cases} d = \frac{a'-a}{a+a'} = \frac{e^2 \cos 2\theta}{2-e^2} \\ d' = \frac{-2b''}{a+a'} = \frac{e^2 \sin 2\theta}{2-e^2} \end{cases}, \begin{cases} c = \frac{-2b'}{a+a'} = (1-d)C - d'C' \\ c' = \frac{-2b}{a+a'} = -d'C + (1+d)C' \\ c'' = \frac{-2a''}{a+a'} = \frac{2r^2r'^2}{r^2+r'^2} - cC - c'C'\,. \end{cases} \tag{3}$$

The notation $\bar{\boldsymbol{f}}$ for the $5D$ vector and $\boldsymbol{f}$ for the $6D$ homogeneous vector is similar to the notation $\bar{\boldsymbol{P}}$ for the $2D$ Cartesian coordinates and $\boldsymbol{P}$ for the $3D$ homogeneous coordinates. In some way, the vectors $\boldsymbol{f}$ and $\bar{\boldsymbol{f}}$ may be respectively considered as the projective and Cartesian direct coordinates of an ellipse.

Tangential (Dual) Equation. A line $\boldsymbol{u}^T \boldsymbol{X} = 0$ is a tangent to the conic described by Eq. (1) if and only if its triplet of coefficients $\boldsymbol{u}$ satisfies the dual equation (also called *tangential equation*) of the conic

$$\mathcal{C}^* : Au^2 + 2B''uv + A'v^2 + 2B'uw + 2Bvw + A''w^2 = \boldsymbol{u}^T \mathtt{F}^* \boldsymbol{u} = 0\,, \tag{4}$$

where $\mathtt{F}^*$ is the matrix of the cofactors of $\mathtt{F}$. We can also represent the dual conic by the vector of its six homogeneous coefficients $\boldsymbol{F} = (A, B'', A', B', B, A'')^T = \text{flatten}(\mathtt{F}^*)$.

For an ellipse, $A'' = aa' - b''^2 \neq 0$ and we also have an alternative form for the dual equation:

$$D''\left(u^2 + v^2\right) + 2D'uv - D\left(u^2 - v^2\right) + 2Cuw + 2C'vw + w^2 = 0\,, \tag{5}$$

where the 5 *dual* coefficients $\bar{\boldsymbol{F}} = (C, C', D, D', D'')$ are related to the dual homogeneous coefficients of Eq. (4) and to the geometric parameters by

$$\begin{cases} C = \frac{B'}{A''} \\ C' = \frac{B}{A''} \end{cases}, \begin{cases} D = \frac{A'-A}{2A''} = \frac{1}{2}(C^2 - C'^2) + \frac{1}{2}(r^2 - r'^2)\cos 2\theta \\ D' = \frac{B''}{A''} = CC' - \frac{1}{2}(r^2 - r'^2)\sin 2\theta \\ D'' = \frac{A+A'}{2A''} = \frac{1}{2}(C^2 + C'^2 - r^2 - r'^2). \end{cases} \tag{6}$$

We highlight the fact that the two coefficients (C, C') in Eq. (6) effectively represents the Cartesian coordinates of the ellipse center and thus have the same notation. The vectors $\boldsymbol{F}$ and $\bar{\boldsymbol{F}}$ may be respectively considered as the projective and Cartesian dual coordinates of an ellipse.

Type of a Conic. It is known that a conic is an ellipse if and only if

$$A'' = aa' - b''^2 > 0 \quad \text{or} \quad 0 \leq e < 1\,. \tag{7}$$

This condition implies that $aa' > b''^2 \geq 0$. Subsequently, a and a' must have the same sign and may not be null. As a consequence, we have the following relations:

$$\text{sign}\,(a) = \text{sign}\,(a') = \text{sign}\,(a + a') \quad \text{and} \quad a + a' \neq 0\,. \tag{8}$$

With the notation $\Delta = \det \mathtt{F}$, an ellipse is a real ellipse if, in addition,

$$a\Delta < 0 \Longleftrightarrow (a + a')\,\Delta < 0\,. \tag{9}$$

2.2 The Triangle of Edge Points and Associated Gradients

We now consider two edge points P and Q and their associated gradients $\overrightarrow{\mathrm{G}}$ and $\overrightarrow{\mathrm{H}}$; this configuration is illustrated in Fig. 1.

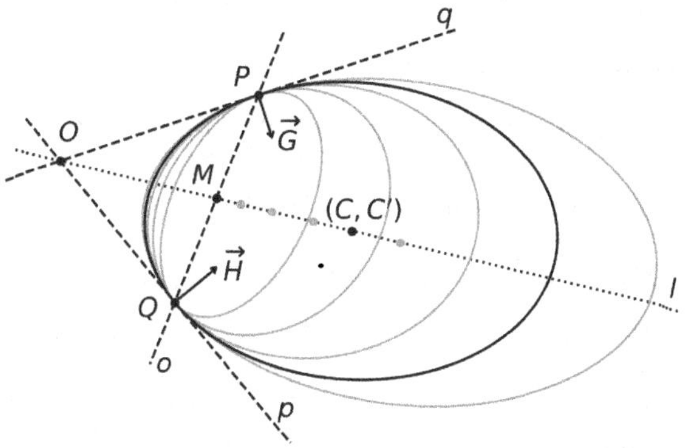

Fig. 1. The edge points P, Q, and their associated gradients $\overrightarrow{\mathrm{G}}$, $\overrightarrow{\mathrm{H}}$ are evidences of the presence of an ellipse in the image.

The mid-point of P and Q is represented by $\boldsymbol{M} = \frac{1}{2}(\boldsymbol{P}+\boldsymbol{Q})$ and their difference by $\boldsymbol{K} = \boldsymbol{Q} - \boldsymbol{P}$. The line o (with $\boldsymbol{o} = \boldsymbol{P} \wedge \boldsymbol{Q}$, where $\wedge$ is the cross-product), joining the points $\boldsymbol{P}$ and $\boldsymbol{Q}$, is a chord secant to the unknown ellipse.

The vectors $\overrightarrow{\mathrm{G}}$ and $\overrightarrow{\mathrm{H}}$ are the image gradients at the points P and Q, they are represented by the directions (points at infinity) $\boldsymbol{G} = \begin{pmatrix} X_G \; Y_G \; 0 \end{pmatrix}^T$ and $\boldsymbol{H} = \begin{pmatrix} X_H \; Y_H \; 0 \end{pmatrix}^T$. We also introduce the dot products $\lambda_G = \boldsymbol{G} \cdot \boldsymbol{K}$ and $\lambda_H = \boldsymbol{H} \cdot \boldsymbol{K}$; we later show that they cannot be null in the practical situations of interest.

The line q, defined by $\boldsymbol{q} = \frac{1}{\lambda_G} \boldsymbol{P} \wedge (\boldsymbol{G} \wedge \boldsymbol{e_z})$ with $\boldsymbol{e_z} = \begin{pmatrix} 0 \; 0 \; 1 \end{pmatrix}^T$, passing by P and orthogonal to $\overrightarrow{\mathrm{G}}$, is tangent to the unknown ellipse. Likewise, the line p (with $\boldsymbol{p} = \frac{1}{\lambda_H} (\boldsymbol{H} \wedge \boldsymbol{e_z}) \wedge \boldsymbol{Q}$) passing by Q and orthogonal to $\overrightarrow{\mathrm{H}}$, is tangent to the unknown ellipse. In the expression of $\boldsymbol{p}$ and $\boldsymbol{q}$, we choose the arbitrary scaling factors λ_G and λ_H to simplify the expression of some of the following equations. The intersection point O of the two lines q and p is represented by $\boldsymbol{O} = \boldsymbol{p} \wedge \boldsymbol{q}$.

We consider the triangle $\triangle$OPQ defined by the edge points P, Q, and the intersection point O. The sides of this triangle are thus the lines o, q, and p. The matrix of the triangle vertices is named $\mathtt{V}$. Thanks to the choice of the scaling factors λ_G and λ_H, it can be shown that the matrix $\mathtt{S} = \mathtt{V}^*$ of the cofactors of $\mathtt{V}$ is also the matrix of the coefficients of the side lines of the triangle. Therefore, we have

$$\mathtt{V} = \begin{pmatrix} \boldsymbol{O} \; \boldsymbol{P} \; \boldsymbol{Q} \end{pmatrix} \quad \text{and} \quad \mathtt{V}^* = \mathtt{S} = \begin{pmatrix} \boldsymbol{o} \; \boldsymbol{p} \; \boldsymbol{q} \end{pmatrix}, \tag{10}$$

with $\det \mathtt{S} = \det \mathtt{V} = 1$, $\mathtt{V}^{-1} = \mathtt{S}^T$, and $\mathtt{S}^{-1} = \mathtt{V}^T$.

Conditions on the Edge Points. From Fig. 1, we see that it is only possible to build an ellipse from a pair of edge points P and Q if they are distinct, if their gradients $\overrightarrow{\mathrm{G}}$ and $\overrightarrow{\mathrm{H}}$ are not null and if the edge point Q (resp. P) do not belongs to q (resp. p). So, in practice, we have to discard the pairs of edge points that do not verify the corresponding conditions given hereafter:

$$\begin{cases} \boldsymbol{K} \neq \boldsymbol{0}, \boldsymbol{G} \neq \boldsymbol{0}, \boldsymbol{H} \neq \boldsymbol{0} \\ \lambda_G = \boldsymbol{G} \cdot \boldsymbol{K} \neq 0, \lambda_H = \boldsymbol{H} \cdot \boldsymbol{K} \neq 0. \end{cases} \tag{11}$$

2.3 Pencil of Bi-Tangent Conics

The unknown ellipse belongs to the family of conics passing through the two points P and Q, and tangent to the lines q and p. Mathematically speaking, this family is the pencil of conic bi-tangent to the lines q and p at their respective intersection points (P and Q) with the line o.

The general equation of a conic $\mathcal{C}_k$ belonging to this pencil is

$$\mathcal{C}_k : \left(\boldsymbol{p}^T \boldsymbol{X}\right)\left(\boldsymbol{q}^T \boldsymbol{X}\right) - \frac{k}{2}\left(\boldsymbol{o}^T \boldsymbol{X}\right)^2 = 0, \tag{12}$$

where k is any real number, which is the *index* of the conic $\mathcal{C}_k$ in the pencil.

As said before, a conic is fully described by five parameters. As all the conics in the pencil have to verify four constraints (to pass through two given points and to touch two given lines), we obtain an equation for a conic in the pencil that only depends on one parameter, namely the index k. The notion of index of the conic is an essential component of our approach in Sect. 3.

Point (Direct) Equation of the Pencil. From Eq. (12), we compute the matrix of a conic in the pencil $\mathtt{F_k} = \boldsymbol{p}\boldsymbol{q}^T + \boldsymbol{q}\boldsymbol{p}^T - k\boldsymbol{o}\boldsymbol{o}^T$ and also $\mathtt{F_k} = \mathtt{S}\left(-k\boldsymbol{e}_x\ \boldsymbol{e}_z\ \boldsymbol{e}_y\right)\mathtt{S}^T$, where $\mathtt{S}$ is the matrix defined in Eq. (10). By identification with Eq. (1), we obtain the corresponding vector of direct homogeneous coefficients $\boldsymbol{f}_k = \boldsymbol{f}_{pq} - k\boldsymbol{f}_{oo}$, where $\boldsymbol{f}_{pq} = \text{flatten}\left(\boldsymbol{p}\boldsymbol{q}^T + \boldsymbol{q}\boldsymbol{p}^T\right)$ and $\boldsymbol{f}_{oo} = \text{flatten}\left(\boldsymbol{o}\boldsymbol{o}^T\right)$.

If we take $\alpha_{pq} = u_p u_q + v_p v_q$ and $\alpha_{oo} = u_o^2 + v_o^2$, we obtain the direct Cartesian coefficients of the ellipses in the pencil, where $\bar{\boldsymbol{f}}_{pq}$ and $\bar{\boldsymbol{f}}_{oo}$ may easily be computed from Eq. (3),

$$\bar{\boldsymbol{f}}_k = \xi_k \bar{\boldsymbol{f}}_{pq} - \left(1 + \alpha_{pq}\xi_k\right)\bar{\boldsymbol{f}}_{oo}\,,\ \text{with}\ \xi_k = \frac{2}{-2\alpha_{pq} + k\alpha_{oo}}\,. \tag{13}$$

Tangential (Dual) Equation of a Conic in the Pencil. Likewise, the dual or tangential equation is determined by the matrix of the cofactors of the matrix of the conic and may be expressed (when the conic is not degenerated) in two ways $\mathtt{F}^*_\mathtt{k} = \mathtt{V}\left(-\boldsymbol{e}_x\ k\boldsymbol{e}_z\ k\boldsymbol{e}_y\right)\mathtt{V}^T$ and $\mathtt{F}^*_\mathtt{k} = -\boldsymbol{O}\boldsymbol{O}^T + k\left[\boldsymbol{P}\boldsymbol{Q}^T + \boldsymbol{Q}\boldsymbol{P}^T\right]$. By identification with Eq. (4), we obtain the corresponding vector of dual homogeneous coefficients $\boldsymbol{F}_k = -\boldsymbol{F}_{OO} + k\boldsymbol{F}_{PQ}$, where $\boldsymbol{F}_{PQ} = \text{flatten}\left(\boldsymbol{P}\boldsymbol{Q}^T + \boldsymbol{Q}\boldsymbol{P}^T\right)$ and $\boldsymbol{F}_{OO} = \text{flatten}\left(\boldsymbol{O}\boldsymbol{O}^T\right)$.

Finally, we obtain the dual Cartesian coefficients of the ellipses in the pencil, where $\bar{\boldsymbol{F}}_{PQ}$ and $\bar{\boldsymbol{F}}_{OO}$ may easily be computed from Eq. (6),

$$\bar{\boldsymbol{F}}_k = \left(1 + \Xi_k {z_O}^2\right)\bar{\boldsymbol{F}}_{PQ} - \Xi_k \bar{\boldsymbol{F}}_{OO}\,,\ \text{with}\ \Xi_k = \frac{1}{2k - z_O^2}\,. \tag{14}$$

Type of a Conic in the Pencil. Because we are looking for a real ellipse, based on Eqs. (7) and (9), we should have $2k > {z_O}^2 \Leftrightarrow \Xi_k > 0$ and $k > 2\frac{\alpha_{pq}}{\alpha_{oo}} \Leftrightarrow \xi_k > 0$, and it is possible to prove that the second condition is always verified when the first one is true.

3 An Algorithm for Obtaining the Matching Ellipse

3.1 Parameters of an Ellipse Built upon Two Edge Points and Their Gradients

We may identify the set of all the conics in the plane with the projective space $\mathbb{P}^5 = P^5(\mathbb{R})$ of their algebraic coefficients. The subset of the (algebraic coefficients of the) conics compatible with four evidences (a pair of edge points, plus their gradients) is a pencil of bi-tangent conics. From Eq. (12), we derive that such a pencil of conics is represented by a line in $\mathbb{P}^5$.

Unfortunately, the scale and the domain of the algebraic coefficients are not easy to determine because these coefficients are only defined up to a scaling factor. So, it would be more appropriate to accumulate them in the space of the geometric parameters. But, in such a space, the pencil of conics is represented by a much more complicated curve, which is intractable.

Possible solution consist to use Eq. (13) in the $5D$ space of the parameters (d, d', c, c', c'') or Eq. (14) in the dual $5D$ space of the parameters (C, C', D, D', D''). In both spaces, the pencil of bi-tangent conics is again a line and the domain of variation of the parameters is easy to determine from a practical perspective. Indeed, these parameters may be easily expressed in terms of the geometric parameters of the ellipse.

3.2 A General Algorithm

The general principles and steps of a typical algorithm for the computation of the Elliptic Hough Transform (EHT) are:

1. We define a $5D$ discrete accumulator corresponding to the $5D$ space of the parameters (d, d', c, c', c'') or (C, C', D, D', D''). In this accumulator, each cell is indexed by a $5D$ vector whose components are the quantized values of the corresponding parameters. All cells are initialized to 0.
2. For all pairs of edge points which verify the conditions mentioned in (11), we draw the line corresponding to the pencil of bi-tangent conics in the $5D$ accumulator.
3. When all the evidences has been processed and the corresponding lines drawn, we look for the peak (maximum) values in the accumulator.
4. If the value of this peak, which represents the number of votes for a vector of parameters, is not large enough, then the process stops with no ellipse detected nor matched.
5. To a contrary, an ellipse is detected if a peak is validated and we look back for all the evidences that voted for this peak. These evidences are then removed from the initial list of evidences and we repeat the process from the beginning until no ellipses are detected anymore.

Obviously, managing a $5D$ accumulator may be very time and memory consuming in practice. Fortunately, solutions exist to reduce the dimension of the accumulator.

3.3 First Step: Reduce the Problem Dimension

Projection of the Dual Parameters to (C, C'). In the literature, without always explicitly noting it, authors reduce the dimension of the problem by considering a projection (of the process explained in the previous section) to a lower dimension space. For example, some geometric observations lead to the solution chosen by all authors that consists in accumulating the line l joining the mid-point M of the edge points P and Q from one side, and the intersection point O of the two tangents q and p from the other side (see Fig. 1). This line is always a diameter of the unknown ellipse and passes through the ellipse center. The accumulation of this line in a plane, whose dimensions are similar to those of the image, highlights the possible ellipse centers.

In the framework of the pencil of bi-tangent conics, this method is equivalent to the projection of the line representing the pencil in the $5D$ space to a line in the $2D$ space. Figure 2 illustrates the process for the dual approach; the blue line in the $5D$ space (represented in $3D$ in the figure) represents the locus of the parameters (C, C', D, D', D'') of the conics in the pencil defined by a specific pair of edge points, and the blue line in the plane (C, C') represents its projection.

From Eq. (14), we see that a degenerate case occurs when the gradients are parallel. In this situation, the $5D$ line is exactly orthogonal to the (C, C') plane and its projection reduces to a point. This means that, when the gradients are parallel, the position of the ellipse center is known perfectly.

Projection of the Direct Parameters to (d, d'). However, there are other possibilities to reduce the dimensions of the problem. For instance, we could also project the line in the $5D$ space (d, d', c, c', c'') into a line in the $2D$ space (d, d'). Indeed, from Eq. (3), we know that these (d, d') parameters are equivalent to the eccentricity e (aspect ratio) and orientation θ of the ellipse. An interesting property of these parameters is that they are always comprised within the unit circle of their plane. The domain of the accumulator is then well defined and bounded. Also, based on Eq. (13), it is possible to demonstrate that, in the case of this new projection, there is no degenerated case, and that the $5D$ line is never exactly orthogonal to the (d, d') plane.

Accumulation in $2D$. Regardless of the chosen (direct or dual) $2D$ projection, the first step of the algorithm is to iterate on all pairs of edge points which verify the conditions expressed in Eq. (11), and to draw a line in the corresponding plane ($2D$ accumulator). In Fig. 2, the gray lines in the plane (C, C') are instances, in the dual framework, of these $2D$ lines (diameters OM) for several pairs of edge points.

When the accumulation or voting process is complete, we look for the maximum value (or the most prominent cluster). If this value (or the sum of the cluster values) is not large enough, we stop the process with no ellipse detected. Otherwise, the position (a $2D$ index) of this peak (or cluster center) is chosen as the estimated value (Δ, Δ') of the two ellipse parameters (the big red point in Fig. 2).

Edge Points Selection. In the dual framework, when the detected center is close, in the $2D$ space (C, C'), to a line (diameter OM) related to a given pair of edge points, we consider that this pair of edge points has voted for the detected center and we select this pair for the second step of the algorithm. In Fig. 2, the blue line of the plane (C, C') is a $2D$ line which is related to a specific pair of edge points. It passes near the detected center and is then selected as having voted for this center.

3.4 Second Step: Back-Projection

State of the Art Methods. In the literature, authors switch to other methods or algorithms to find the remaining parameters. Most of the time, they use the information about the center location to obtain the equation of the centered ellipse. Then, all the edge points (and their gradients) that previously voted for the detected center are used to provide two conditions on the three remaining parameters. So, by a similar process of line accumulation in a three dimension space, they obtain an estimate of the remaining parameters.

Our Proposed Algorithm. In the framework of the pencil of bi-tangent conics, the three remaining parameters can advantageously be computed by back-projecting the points obtained during the $2D$ accumulation process onto the line in the full $5D$ space.

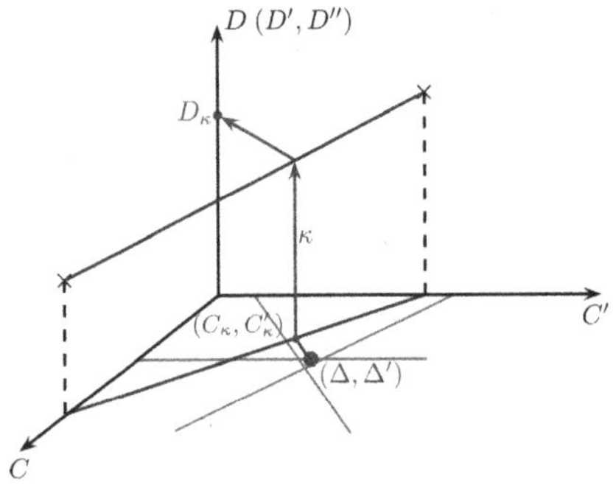

Fig. 2. Back-projection of the estimated center (Δ, Δ') into a $3D$ space (illustration of the back-projection into a $5D$ space). (Color figure online)

Index of the Closest Ellipse. In Fig. 2, the detected center (Δ, Δ') (big red point) is orthogonally projected onto the $2D$ line (blue line in the (C, C') plane) relating to a specific pair of edge points. As a point of the line, this orthogonal projection should be of the form (C_κ, C'_κ) (given by the two first equations in (12) and (14)), where κ is a specific value for the index k. In the following, κ will be named the index of the closest conic in the pencil defined by this specific pair of edge points.

Likewise, for the direct approach, the same reasoning holds for finding the index of the couple of parameters that is the closest to the estimated values in the (d, d') plane.

However, there is a major difference between the two approaches. In the dual approach, when the gradients are parallel, the $5D$ line is orthogonal to the plane (C, C'). So, in this situation, it is not possible to derive the value of κ from the estimated center and, for this step, we must discard the pairs of edge points whose gradients are parallel. In the direct approach, there is no such situation; the $5D$ line is never orthogonal to the plane (d, d') and all pairs of edge points may be used for the back-projection step.

The Three Remaining Parameters. In order to compute the three remaining ellipse parameters (D, D', D''), we use, in Eq. (14), this index κ of the closest ellipse whose the parameters (C_κ, C'_κ) are the closest to the estimated center (Δ, Δ'). This means that a pair of edge points, which voted for a detected center, also provides (or votes for) one $3D$ vector of the remaining parameters. In Fig. 2, a point (C_κ, C'_κ) on the $2D$ line corresponds to one $5D$ point $(C_\kappa, C'_\kappa, D_\kappa, D'_\kappa, D''_\kappa)$.

The last step consists to iterate on all the pairs of edge points supporting the estimated center and to fill a $3D$ histogram (or three $1D$ histograms) with the provided vectors of the remaining parameters. If the peak of (or the most prominent cluster in) the histogram is not large enough, no ellipse is detected nor matched. Otherwise, the $3D$ vector of indices of the peak (or cluster center) is chosen as the estimated value for the three remaining parameters.

Likewise, the same reasoning also holds in the direct approach for finding the three remaining parameters (c, c', c'').

Finally, we remove from the list of pairs of edge points those supporting the chosen vector of remaining parameters, and we start the whole process again until no ellipses are detected anymore.

4 Results

To illustrate the feasibility of the methods introduced in this paper, we present some preliminary comparative results between our dual algorithm and the State-of-the-Art (abbreviated SoA in the following) algorithm outlined in Sect. 3.4. Both algorithms have been implemented with the same list of edge points, the same center detection first step, the same accumulator cell size and without any post-processing. Obviously, in a practical set-up, both algorithms could benefit from additional processing step, but we are interested here in the bare properties of the methods.

We built a custom dataset with 100 synthetic images (of size 450 by 600 pixels) containing one random black ellipse on a white background (see on the left of Fig. 3) and we added salt-and-pepper noise and a randomly (Gaussian) textured image.

For comparing the SoA and dual algorithms, we computed the Hausdorff distance between the detected and the true ellipse for each test image; these statistics are given in Table 1. We also provided the numbers of detected ellipses whose Hausdorff distance to the true ellipse are respectively less than 2, 5, 10 and 15 pixels (Table 2).

In this experiment, we observe that the dual approach provides a better estimate (in terms of Hausdorff distance) of the ellipse than our implementation of the SoA approach. In addition, if we consider that an ellipse is not (correctly) detected when its Hausdorff distance to the true ellipse is larger than 10, from the Table 2, we see that, in our experiment, 1% (resp. 15%) of the ellipse were not detected by the dual (resp. SoA) approach.

Figure 3 shows a typical example of one test image (left) with the true ellipse delineated in red. The background of the middle and the right images represents

Table 1. Statistics of the Hausdorff distance between the detected and true ellipse.

	SoA	Dual
Mean	6.1	2.3
Std dev	8.4	1.5
Min	0.8	0.6
Max	59	10.6

Table 2. Number of detected ellipses whose Hausdorff distance to the true ellipse is less than the threshold given in the first column.

Threshold	SoA	Dual
2	30	56
5	67	94
10	85	99
15	90	100

the same edge points (in white) extracted from the left image. In the middle (resp. right) image, the green curve is the detected ellipse and the red points are the selected edge points, both by our implementation of the SoA (resp. dual) algorithm. The selected edge points (in red) are chosen as being closer than 10 pixels to the ellipse detected and may be used as the input of an additional ellipse fitting step, if necessary. On this example, the Hausdorff distance between the detected and the true ellipse is 11.8 for the SoA algorithm and 3.4 for the dual approach. This illustration, for the SoA algorithm, how a poor precision for the ellipse parameter estimation negatively impacts on the capacity of the algorithm to correctly extract the edge points belonging to an ellipse.

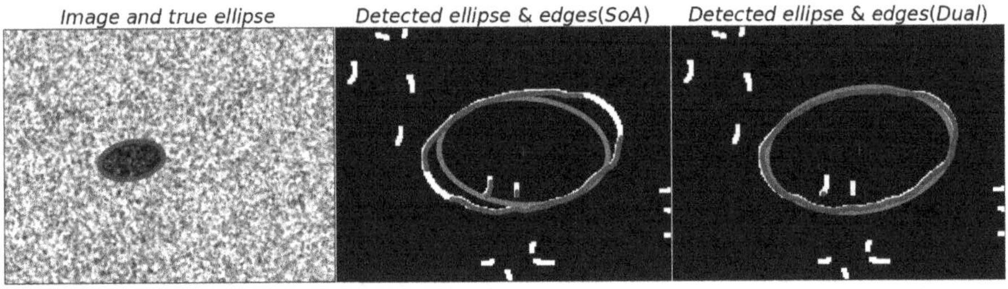

Fig. 3. A sample test image (left), the results of the SoA algorithm (middle), and of our method (right). The detected ellipse is drawn in green and the selected edge points are drawn in red. (Color figure online)

5 Conclusions

In this paper, we show that the concept of pencil of bi-tangent conics, either in its direct or dual form, is an adequate framework to discuss the theoretical aspects of the EHT based on pairs of edge points and their associated gradients. The usual process of splitting the full $5D$ problems into two (or three sometimes) steps of reduced dimension is well modeled by the notion of projection of the problem to a $2D$ sub-space followed by a back-projection of the results into the full $5D$ space.

The algebraic coherence allows us to unify both parts of the EHT into a single framework, highlights the eligibility conditions of the pairs of edge points (to be possible evidences of the presence of an ellipse), and is capable to manage special cases for the pairs of edges points or their gradients.

The dual form of the framework copies the first step of the state-of-the-art methods that consists in finding the ellipse center, but it innovates for finding the remaining parameters. Our preliminary results suggest that our new approach could have a better accuracy in the estimation of the ellipse parameters.

The direct form of the framework is, as far as we now, a new approach to reduce the dimension and solve the problem. This direct approach first looks for a couple of ellipse parameters similar to the eccentricity and the orientation. Theoretically, both approaches have benefits and drawbacks; the choice between them will be a matter of practical considerations.

In the future, we expect that the use of direct and dual equations of the pencil of bi-tangent conics for the EHT will pave the way for new approaches.

References

1. Kanatani, K., Sugaya, Y., Kanazawa, Y.: Ellipse Fitting for Computer Vision: Implementation and Applications. Synthesis Lectures on Computer Vision. Morgan & Claypool, San Rafael (2016)
2. Duda, R.O., Hart, P.E.: Use of the Hough transformation to detect lines and curves in pictures. Commun. ACM **15**(1), 11–15 (1972)
3. Mukhopadhyay, P., Chaudhuri, B.: A survey of Hough transform. Pattern Recogn. **48**(3), 993–1010 (2015)
4. Tsuji, S., Matsumoto, F.: Detection of ellipses by a modified Hough transformation. IEEE Trans. Comput. **C-27**(8), 777–781 (1978)
5. Yuen, H.K., Illingworth, J., Kittler, J.: Ellipse detection using the Hough transform. In: Proceedings of the Alvey Vision Conference, pp. 41.1–41.8. Alvey Vision Club (1988)
6. Guil, N., Zapata, E.L.: Lower order circle and ellipse Hough transform. Pattern Recogn. **30**(10), 1729–1744 (1997)
7. Muammar, H.K., Nixon, M.: Tristage Hough transform for multiple ellipse extraction. IEE Proc. E Comput. Digit. Tech. **138**(1), 27–35 (1991)
8. Yoo, J.H., Sethi, I.K.: An ellipse detection method from the polar and pole definition of conics. Pattern Recogn. **26**(2), 307–315 (1993)
9. Bennett, N., Burridge, R., Saito, N.: A method to detect and characterize ellipses using the Hough transform. IEEE Trans. Pattern Anal. Mach. Intell. **21**(7), 652–657 (1999)
10. Forbes, A.B.: Fitting an Ellipse to Data. NPL Report DITC 95/87, National Physical Laboratory (1987)
11. Leavers, V.F.: The dynamic generalized Hough transform: its relationship to the probabilistic Hough transforms and an application to the concurrent detection of circles and ellipses. Comput. Vis. Graph. Image Process.: Image Underst. **56**(3), 381–398 (1992)
12. Lu, W., Yu, J., Tan, J.: Direct inverse randomized Hough transform for incomplete ellipse detection in noisy images. J. Pattern Recogn. Res. **1**, 13–24 (2014)

Digital Plane Recognition with Fewer Probes

Tristan Roussillon[1(✉)] and Jacques-Olivier Lachaud[2]

[1] Université de Lyon, INSA Lyon, LIRIS, UMR CNRS 5205, 69622 Villeurbanne, France
tristan.roussillon@liris.cnrs.fr

[2] Université Savoie Mont Blanc, LAMA, UMR CNRS 5127, 73376 Le Bourget-du-Lac, France
jacques-olivier.lachaud@univ-smb.fr

Abstract. We present a new plane-probing algorithm, i.e., an algorithm that computes the normal vector of a digital plane from a starting point and a predicate "Is a point $\boldsymbol{x}$ in the digital plane?". This predicate is used to probe the digital plane as locally as possible and decide on-the-fly the next points to consider. We show that this algorithm returns the same vector as another plane-probing algorithm proposed in Lachaud et al. (J. Math. Imaging Vis., 59, 1, 23–39, 2017), but requires fewer probes. The theoretical upper bound is indeed lowered from $O(\omega \log \omega)$ to $O(\omega)$ calls to the predicate, where ω is the arithmetical thickness of the digital plane, and far fewer calls are experimentally observed on average. This reduction is made possible by a study that shows how to avoid computations that do not contribute to the final solution. In the context of digital surface analysis, this new algorithm is expected to be of great interest for normal estimation and shape reconstruction.

Keywords: Digital plane recognition · Normal estimation · Plane-probing algorithm

1 Introduction

Analyzing the geometry of digital surfaces is a challenging task, since the local geometry is very poor (only six possible normal directions). A classical approach is to estimate the geometry by observing a larger neighborhood, whose size is given rather arbitrarily by the user. However, this approach comes at the cost of blurring sharp features. The trade-off between a sufficiently large neighborhood to get a relevant normal direction and a sufficiently small neighborhood to preserve sharp features is hard to find and may vary across the digital shape. Purely digital methods have thus emerged and try to perform digital surface analysis without any external parameter.

This work has been partly funded by CoMeDiC ANR-15-CE40-0006 and PARADIS ANR-18-CE23-0007-01 research grants.

M. Couprie et al. (Eds.): DGCI 2019, LNCS 11414, pp. 380–393, 2019.
https://doi.org/10.1007/978-3-030-14085-4_30

A natural approach, e.g., [3], consists in computing a set of digital plane segments (DPSs) that locally fit the digital surface. This strategy has also been used for surface area estimation [7] or reversible polyhedrization [12]. However, finding how to scan the digital surface to efficiently recognize DPSs whose size and shape reveal the local geometry is difficult. There are numerous algorithms for the recognition of DPSs (to quote a few: [1,2,4–6,11,13,14]). All these algorithms take a point set as input (possibly in an incremental way), determine whether this set can be a DPS or not, and if so, provide its geometric characteristics. But the most difficult part consists in determining which input points should be taken into account during the recognition process in order to guarantee that the obtained DPSs are indeed *tangent* to the digital surface.

Therefore, recently, another category of recognition algorithms have been developped [8–10]. These algorithms, called *plane-probing* algorithms in [10], decide on-the-fly where to probe next the digital surface while growing a triangular facet, which is tangent and separating by construction. The growth direction is given by both arithmetic and geometric properties.

Two of these algorithms proposed in [10], called H- and R-algorithm, are local in the sense that the returned triangular facet is guaranteed to stay around the starting point (but this is not the case for the one proposed in [8]). The R-algorithm is even more local in the sense that the final triangular facet has experimentally always acute angles and is less elongated.

In this paper, we present a new plane-probing algorithm that returns the same triangular facet (and normal vector) as the R-algorithm, but requires fewer probes. For comparison, the R-algorithm requires $O(\omega \log \omega)$ calls to the predicate "Is $\boldsymbol{x}$ in the digital plane?" and the exhibited worst-case example implies $\Theta(\omega)$ calls. We present here an improvement that achieves the tight bound of $O(\omega)$ calls. Furthermore, far fewer calls are observed in practice.

In Sect. 2, we give an overview of our new algorithm. In Sects. 3 and 4, we go into details and show how to avoid computations that do not contribute to the final solution. Some experimental results are discussed in Sect. 5.

2 A New Plane-Probing Algorithm

We keep definitions and notations introduced in [10]. We wish to extract the parameters of a standard digital plane $\mathbf{P}$, defined as the set

$$\mathbf{P} = \{\boldsymbol{x} \in \mathbb{Z}^3 \mid 0 \leq \boldsymbol{x} \cdot \mathbf{N} < \omega\},$$

where $\mathbf{N} \in \mathbb{N}^3$ is the *normal vector* whose components (a, b, c) are such that $0 < a \leq b \leq c$, $\gcd(a, b, c) = 1$ and $\omega := (1, 1, 1) \cdot \mathbf{N}$ is the *thickness*. Our approach can be straightforwardly extended to digital plane with arbitrary intercept and with a normal vector in any orthant (see [10]).

As in [9,10], we propose an algorithm that, given a predicate $P(\boldsymbol{x})$:="Is $\boldsymbol{x} \in \mathbf{P}$?" and a starting point $\boldsymbol{p}$ at the base of a reentrant corner of $\mathbf{P}$, computes the *normal vector* of a piece of digital plane surrounding $\boldsymbol{p}$. Moreover, if $\boldsymbol{p}$ is a *lower leaning point*, i.e., $\boldsymbol{p} \cdot \mathbf{N} = 0$, the algorithm extracts the exact normal $\mathbf{N}$ of

P and a *basis of* **P**, that is a pair of vectors that forms a basis of the 2D lattice $\{\boldsymbol{x} \in \mathbb{Z}^3 \mid \boldsymbol{x} \cdot \mathbf{N} = 0\}$. This basis is returned as three *upper leaning points* of **P**, i.e., points $\boldsymbol{x}$ such that $\boldsymbol{x} \cdot \mathbf{N} = \omega - 1$.

Initialization. Given a starting point $\boldsymbol{p} \in \mathbf{P}$, the algorithm places an initial triplet of points $\mathbf{T}^{(0)} := (\boldsymbol{v}_k^{(0)})_{k \in \{0,1,2\}}$ such that $\forall k, \boldsymbol{v}_k^{(0)} := \boldsymbol{p} + \boldsymbol{e}_k + \boldsymbol{e}_{k+1}$, where $(\boldsymbol{e}_0, \boldsymbol{e}_1, \boldsymbol{e}_2)$ is the canonical basis of $\mathbb{R}^3$, and "$\forall k$" stands for "$\forall k \in \mathbb{Z}/3\mathbb{Z}$" for clarity. The algorithm requires $\mathbf{T}^{(0)} \subset \mathbf{P}$ which is the case when $\boldsymbol{p}$ is the base of a reentrant corner (see inset figure). We also denote by $\boldsymbol{q}$ the point $\boldsymbol{p} + (1, 1, 1)$, which is not in **P**.

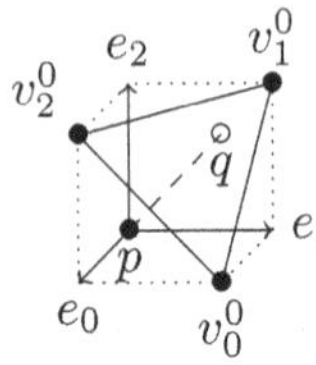

Evolution. At each step $i \in \mathbb{Z}_{\geq 0}$, the triangle $\mathbf{T}^{(i)}$ represents the current approximation of the plane **P**. The algorithm updates one vertex of $\mathbf{T}^{(i)}$ per iteration, while $\boldsymbol{q}$ does not move and stays above the triangle. The new vertex $\boldsymbol{x}^\star$ in $\mathbf{T}^{(i+1)} \setminus \mathbf{T}^{(i)}$, is a point both in **P** and in a specific neighborhood $\mathcal{N}^{(i)}$ yet to be defined, such that the circumsphere of $\mathbf{T}^{(i)} \cup \{\boldsymbol{x}^\star\}$ does not include any point $\boldsymbol{x} \in (\mathcal{N}^{(i)} \cap \mathbf{P})$ in its interior. Denoting by Σ the set of permutations over $\{0,\ 1,\ 2\}$, we introduce the following notations (illustrated in Fig. 1) in order to define $\mathcal{N}^{(i)}$:

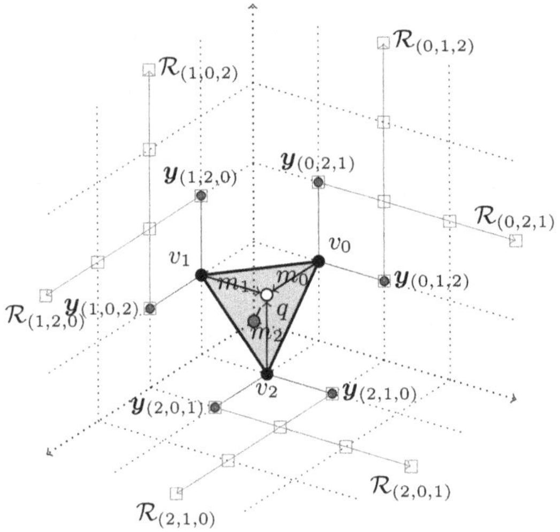

Fig. 1. The triangle $conv(\mathbf{T}^{(i)})$ is depicted in grey. The H-neighborhood $\mathcal{N}_H^{(i)} := (\boldsymbol{y}_\sigma^{(i)})_{\sigma \in \Sigma}$ is depicted with red disks, whereas rays $(\mathcal{R}_\sigma^{(i)})_{\sigma \in \Sigma}$ are depicted with green squares and include the H-neighborhood (iteration number dropped). (Color figure online)

$$\forall k,\ \boldsymbol{m}_k^{(i)} := \boldsymbol{q} - \boldsymbol{v}_k^{(i)} \tag{1}$$

$$\forall \sigma \in \Sigma, \boldsymbol{y}_\sigma^{(i)} := \boldsymbol{v}_{\sigma(0)}^{(i)} + \boldsymbol{m}_{\sigma(1)}^{(i)} \tag{2}$$

$$\forall \sigma \in \Sigma, \forall \lambda \in \mathbb{Z}_{\geq 0}, \mathcal{R}_\sigma^{(i)}[\lambda] := \boldsymbol{y}_\sigma^{(i)} + \lambda \boldsymbol{m}_{\sigma(2)}^{(i)} \tag{3}$$

$$\forall \sigma \in \Sigma, \mathcal{R}_\sigma^{(i)} := (\mathcal{R}_\sigma^{(i)}[\lambda])_{\lambda \in \mathbb{Z}_{\geq 0}} \tag{4}$$

At step i, the six points $\mathcal{N}_H^{(i)} := (\boldsymbol{y}_\sigma^{(i)})_{\sigma \in \Sigma}$ forms an hexagon, called *H-neighborhood*, while the six rays $\mathcal{N}_R^{(i)} := (\mathcal{R}_\sigma^{(i)})_{\sigma \in \Sigma}$ forms the *R-neighborhood*. In [10], the H-(resp. R-)algorithm uses the H-(resp. R-)neighborhood (see Fig. 2 for an illustration of the running of the two algorithms). While the update procedure of the H-algorithm is trivial and constant-time, since $\boldsymbol{x}^\star$ is one of the six points of the H-neighborhood, the update procedure of the R-algorithm computes a candidate point *for each ray* having a non-empty intersection with $\mathbf{P}$, before choosing one of them as $\boldsymbol{x}^\star$. This strategy is not optimal since a candidate point belonging to a ray may not be chosen finally.

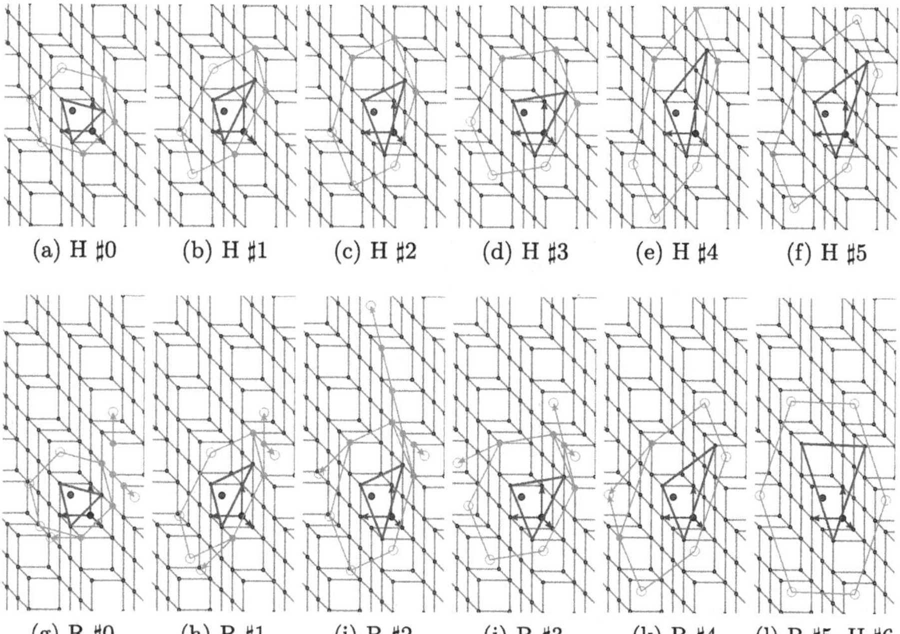

(a) H ♯0 (b) H ♯1 (c) H ♯2 (d) H ♯3 (e) H ♯4 (f) H ♯5

(g) R ♯0 (h) R ♯1 (i) R ♯2 (j) R ♯3 (k) R ♯4 (l) R ♯5, H ♯6

Fig. 2. The H and R algorithms are applied on a digital plane of normal vector $\mathbf{N}(9, 2, 3)$. Images (a) to (f) and (l) show the six iterations of the H-algorithm, whereas images (g) to (l) show the five iterations of the R-algorithm. In each image, the current triangle is depicted in blue, whereas the neighborhood is depicted in red – disks (resp. circles) for points lying inside (resp. outside) the digital plane. Note that, in this example, the output of the two algorithms only differ at step ♯4. (Color figure online)

On the contrary, in this paper, we propose to only probe the H-neighborhood and *one* ray to determine triangle $\mathbf{T}^{(i+1)}$. In addition, we may probe fewer points onto the ray. We call this new algorithm R^1 since it tests at most one ray per iteration. It can be coarsely described as repetitive calls to the function UPDATETRIANGLE (see Algorithm 1).

Algorithm 1. R^1-ALGORITHM: it extracts a triplet of upper leaning points by probing the H-neighborhood and one ray.

Input: The predicate $P(\boldsymbol{x})$ "Is $\boldsymbol{x} \in \mathbf{P}$?", the exterior point $\boldsymbol{q}$ and triangle $\mathbf{T}^{(0)}$
Output: A triangle $\mathbf{T}^{(n)}$
$i \leftarrow 0$;
while $\mathcal{N}_H^{(i)} \cap \mathbf{P} \neq \emptyset$ **do**
  $\mathbf{T}^{(i+1)} \leftarrow$ UPDATETRIANGLE$(P, \mathbf{T}^{(i)}, \boldsymbol{q})$;
  $i \leftarrow i + 1$;
return $\mathbf{T}^{(i)}$

Table 1. Function UPDATETRIANGLE$(P, \mathbf{T}, \boldsymbol{q})$, with $\mathbf{T} = (\boldsymbol{v}_0, \boldsymbol{v}_1, \boldsymbol{v}_2)$. After all cases, the triangle $\mathbf{T}$ is returned.

$\|\mathcal{N}_H \cap \mathbf{P}\|$	$\mathcal{N}_H \cap \mathbf{P}$	Output
0	$()$	Algorithm termination
1	$(\boldsymbol{y}_\sigma)$	$\boldsymbol{v}_{\sigma(0)} \leftarrow \boldsymbol{v}_{\sigma(0)} + \boldsymbol{m}_{\sigma(1)}$
2	$(\boldsymbol{y}_\sigma, \boldsymbol{y}_{\sigma'})$ with $\sigma(0) \neq \sigma'(0), \sigma(1) = \sigma'(1)$	**if** $\boldsymbol{y}_{\sigma'} \leq_{\mathbf{T}} \boldsymbol{y}_\sigma$ **then** $\sigma \leftarrow \sigma'$; $\boldsymbol{v}_{\sigma(0)} \leftarrow \boldsymbol{v}_{\sigma(0)} + \boldsymbol{m}_{\sigma(1)}$
2	$(\boldsymbol{y}_\sigma, \boldsymbol{y}_{\sigma'})$ with $\sigma(0) = \sigma'(0), \sigma(1) \neq \sigma'(1)$	**if** $\boldsymbol{m}_{\sigma(1)} \geq \boldsymbol{m}_{\sigma'(1)}$ **then** $(\tau, \tau') \leftarrow (\sigma, \sigma')$; **else** $(\tau, \tau') \leftarrow (\sigma', \sigma)$; $(\pi, \alpha) \leftarrow$ CLOSESTAMONGPOINTANDRAY$(P, \mathbf{T}, \boldsymbol{q}, \tau', \tau)$ $\boldsymbol{v}_{\pi(0)} \leftarrow \boldsymbol{v}_{\pi(0)} + \boldsymbol{m}_{\pi(1)} + \alpha \boldsymbol{m}_{\pi(2)}$
3	$(\boldsymbol{y}_\sigma, \boldsymbol{y}_{\sigma'}, \boldsymbol{y}_{\sigma''})$ with $\sigma(0) = \sigma'(0)$, $\sigma(0) \neq \sigma''(0)$, $\sigma(1) \neq \sigma'(1)$	**if** $\boldsymbol{m}_{\sigma(1)} \geq \boldsymbol{m}_{\sigma'(1)}$ **then** $(\tau, \tau') \leftarrow (\sigma, \sigma')$; **else** $(\tau, \tau') \leftarrow (\sigma', \sigma)$; **if** $\boldsymbol{y}_{\sigma''} \leq_{\mathbf{T}} \boldsymbol{y}_{\tau'}$ **then** $\tau' \leftarrow \sigma''$; $(\pi, \alpha) \leftarrow$ CLOSESTAMONGPOINTANDRAY$(P, \mathbf{T}, \boldsymbol{q}, \tau', \tau)$ $\boldsymbol{v}_{\pi(0)} \leftarrow \boldsymbol{v}_{\pi(0)} + \boldsymbol{m}_{\pi(1)} + \alpha \boldsymbol{m}_{\pi(2)}$
Else		Error, $\mathbf{P}$ is not a plane.

Function UPDATETRIANGLE is detailed in Table 1 and performs a case analysis on the cardinal and the composition of the H-neighborhood.

Order Induced by Circumspheres to $\mathbf{T}$*:* Let I^+ be the half-plane delimited by $\mathbf{T}$ and containing $\boldsymbol{q}$. We claim that the ball $C(\mathbf{T}, \boldsymbol{z})$ circumscribing $\mathbf{T}$ and some $\boldsymbol{z} \in I^+$, induces a total pre-order on I^+ through the inclusion relation. Indeed, if $\boldsymbol{z}' \in I^+$, whenever $\boldsymbol{z}' \in C(\mathbf{T}, \boldsymbol{z})$, then $(C(T, \boldsymbol{z}') \cap I^+) \subset (C(T, \boldsymbol{z}) \cap I^+)$. Clearly, this relation is reflexive, transitive and connex. We shall say that $\boldsymbol{z}'$ is *closer* to $\mathbf{T}$ than $\boldsymbol{z}$ and we write $\boldsymbol{z}' \leq_{\mathbf{T}} \boldsymbol{z}$. We can use this relation because the R-neighborhood is included in I^+.

Correctness: To prove that R^1-algorithm extracts the normal vector of $\mathbf{P}$ when starting from a lower leaning point $\boldsymbol{p}$, it is enough to show that UPDATETRIANGLE outputs in all cases the same triangle $\mathbf{T}$ as in the R-algorithm, i.e., it updates a vertex of $\mathbf{T}$ by a point $\boldsymbol{x}^\star$ in $\mathcal{N}_R \cap \mathbf{P}$ such that $\forall \boldsymbol{x} \in (\mathcal{N}_R \cap \mathbf{P}), \boldsymbol{x}^\star \leq_{\mathbf{T}} \boldsymbol{x}$. In the next sections, we show that it does so, by appropriate calls to function CLOSESTAMONGPOINTANDRAY, which is detailed in Algorithm 4.

3 Local Configuration

Informally, function UPDATETRIANGLE updates $\mathbf{T}$ as follows:

$|\mathcal{N}_H \cap \mathbf{P}| = 0$: the algorithm stops,
$|\mathcal{N}_H \cap \mathbf{P}| = 1$: the algorithm updates the associated vertex of $\mathbf{T}$ with this point,
$|\mathcal{N}_H \cap \mathbf{P}| = 2$, **the two points are linked to distinct vertices of T:** the algorithm picks the closest according to $\leq_{\mathbf{T}}$ and updates the associated vertex,
$|\mathcal{N}_H \cap \mathbf{P}| = 2$, **the two points are linked to the same vertex of T:** the algorithm determines which ray originating from these points may possibly contain the target point, then picks the closest according to $\leq_{\mathbf{T}}$ and updates the associated vertex,
$|\mathcal{N}_H \cap \mathbf{P}| = 3$, **two of the points are linked to the same vertex of T:** the algorithm determines which ray originating from these points may possibly contain the target point, then picks the closest according to $\leq_{\mathbf{T}}$ and updates the associated vertex.

Even if there are exactly six points in $\mathcal{N}_H$ by definition, there are fewer points in $\mathcal{N}_H \cap \mathbf{P}$:

Lemma 1. *There are no more than three points in $\mathcal{N}_H \cap \mathbf{P}$. In addition, they are consecutive when counter-clockwise ordered around $\boldsymbol{q}$.*

For the proof, we introduce the edge vectors defined as $\boldsymbol{d}_k := \boldsymbol{v}_{k+1} - \boldsymbol{v}_k$ for all $k \in \{0, 1, 2\}$.

Proof. Since $\mathcal{N}_H = \{\boldsymbol{q} \pm \boldsymbol{d}_k\}_{k \in \{0,1,2\}}$ and $\boldsymbol{q} \notin \mathbf{P}$, it is clear that $\boldsymbol{q} - \boldsymbol{d}_k$ and $\boldsymbol{q} + \boldsymbol{d}_k$ cannot be both in $\mathbf{P}$ by linearity, which means that there are no more than three points in $\mathcal{N}_H \cap \mathbf{P}$.

For the second part, it is enough to see that for any k, $\boldsymbol{q} - \boldsymbol{d}_{k+1}$ is necessarily in $\mathbf{P}$ by linearity if both $\boldsymbol{q} + \boldsymbol{d}_k$ and $\boldsymbol{q} + \boldsymbol{d}_{k+2}$ are in $\mathbf{P}$. □

Lemma 1 shows that all possible cardinalities are taken into account in function UPDATETRIANGLE. Now, we explain why we can only probe the H-neighborhood and one ray at each step. To do so, we use the following lemma:

Lemma 2 ([10], Lemma 7). *For any permutation $\sigma \in \Sigma$, if there is a point $\boldsymbol{x}$ of ray $\mathcal{R}_\sigma$ that is not in $\mathbf{P}$, then no point further than $\boldsymbol{x}$ on the ray is in $\mathbf{P}$.*

Due to Lemma 2, if the starting point of a ray is not in $\mathbf{P}$, then no other ray point has to be considered, which explains the first two lines of Table 1. The following result explains the third line:

Lemma 3. *For any k, let $\mathcal{R}_\sigma$ and $\mathcal{R}_{\sigma'}$ be two distinct rays such that $\sigma(0) = \sigma'(0)$. If the starting point of one ray is not in $\mathbf{P}$, then only the starting point of the other ray may be in $\mathbf{P}$ among all the points of $\mathcal{R}_\sigma \cup \mathcal{R}_{\sigma'}$.*

Proof. The key point is to notice that the two rays cross at a point $\boldsymbol{v}_{\sigma(0)} + \boldsymbol{m}_{\sigma(1)} + \boldsymbol{m}_{\sigma(2)}$ (because $\sigma'(1) = \sigma(2)$ and $\sigma'(2) = \sigma(1)$, see Fig. 1). Due to Lemma 2, we conclude that if the starting point of a ray, let us say $\mathcal{R}_\sigma$, is not in $\mathbf{P}$, then neither the crossing point, nor any further point in $\mathcal{R}_{\sigma'}$ (and obviously in $\mathcal{R}_\sigma$) is in $\mathbf{P}$. □

In other words, only one point, instead of two rays, has to be considered in this case. If only two such points belong to $\mathcal{N}_H \cap \mathbf{P}$, it is enough to determine which one is the closest according to $\leq_\mathbf{T}$, hence the third line of Table 1. The following result provides the rationale for the fourth and fifth lines:

Lemma 4. *Let $\mathcal{R}_\sigma$ and $\mathcal{R}_{\sigma'}$ be two distinct rays such that $\sigma(0) = \sigma'(0)$ and such that $\boldsymbol{y}_\sigma, \boldsymbol{y}_{\sigma'} \in \mathbf{P}$. If $\boldsymbol{m}_{\sigma(1)} \geq \boldsymbol{m}_{\sigma'(1)}$ (resp. $\boldsymbol{m}_{\sigma(1)} \leq \boldsymbol{m}_{\sigma'(1)}$), a closest point $\boldsymbol{x}^\star$ according to $\leq_\mathbf{T}$, among all the points of $\mathcal{R}_\sigma \cup \mathcal{R}_{\sigma'}$, belongs to $\mathcal{R}_\sigma \cup \boldsymbol{y}_{\sigma'}$ (resp. $\mathcal{R}_{\sigma'} \cup \boldsymbol{y}_\sigma$).*

The proof of Lemma 4 requires this result in the case of an acute angle:

Lemma 5. *Let $\mathcal{R}_\sigma$ and $\mathcal{R}_{\sigma'}$ be two distinct rays such that $\sigma(0) = \sigma'(0)$ and such that $\boldsymbol{y}_\sigma, \boldsymbol{y}_{\sigma'} \in \mathbf{P}$. If $\boldsymbol{m}_{\sigma(1)} \cdot \boldsymbol{m}_{\sigma'(1)} \geq 0$, a closest point $\boldsymbol{x}^\star$ according to $\leq_\mathbf{T}$, among all the points of $\mathcal{R}_\sigma \cup \mathcal{R}_{\sigma'}$, is either $\boldsymbol{y}_\sigma$ or $\boldsymbol{y}_{\sigma'}$.*

Proof. Let us focus on $\mathcal{R}_\sigma$ because the same is true for $\mathcal{R}_{\sigma'}$. To show that either $\boldsymbol{y}_\sigma$ or $\boldsymbol{y}_{\sigma'}$ is closer than any point of $\mathcal{R}_\sigma$, let us consider the parallelogram whose first diagonal links $\boldsymbol{v}_{\sigma(0)}$ to a ray point $\mathcal{R}_\sigma[\lambda]$, with $\lambda \in \mathbb{Z}_{>0}$, and the second one links $\boldsymbol{v}_{\sigma(0)} + \boldsymbol{m}_{\sigma(1)} = \boldsymbol{y}_\sigma$ to $\mathcal{R}_\sigma[\lambda] - \boldsymbol{m}_{\sigma(1)} = \boldsymbol{v}_{\sigma(0)} + \lambda\boldsymbol{m}_{\sigma(2)}$. We prove below that the first diagonal is always strictly longer than the second one, which means that a sphere passing by $\boldsymbol{v}_{\sigma(0)}$ and $\mathcal{R}_\sigma[\lambda]$ contains either $\boldsymbol{y}_\sigma$ or $\boldsymbol{v}_{\sigma(0)} + \lambda\boldsymbol{m}_{\sigma(2)}$, i.e., $\boldsymbol{y}_\sigma \leq_\mathbf{T} \mathcal{R}_\sigma[\lambda]$ or $\boldsymbol{y}_{\sigma'} = \boldsymbol{v}_{\sigma(0)} + \boldsymbol{m}_{\sigma(2)} \leq_\mathbf{T} \boldsymbol{v}_{\sigma(0)} + \lambda\boldsymbol{m}_{\sigma(2)} \leq_\mathbf{T} \mathcal{R}_\sigma[\lambda]$.

Indeed, we have:

$$(\boldsymbol{m}_{\sigma(1)} + \lambda\boldsymbol{m}_{\sigma(2)})^2 - (-\boldsymbol{m}_{\sigma(1)} + \lambda\boldsymbol{m}_{\sigma(2)})^2 = 4\lambda(\boldsymbol{m}_{\sigma(1)} \cdot \boldsymbol{m}_{\sigma(2)}) > 0,$$

because $\lambda > 0$ and $\boldsymbol{m}_{\sigma(1)} \cdot \boldsymbol{m}_{\sigma(2)} = \boldsymbol{m}_{\sigma(1)} \cdot \boldsymbol{m}_{\sigma'(1)} \geq 0$ by hypothesis.

Proof (Lemma 4). The proof is divided into two cases. If $\boldsymbol{m}_{\sigma(1)} \cdot \boldsymbol{m}_{\sigma'(1)} \geq 0$, $\boldsymbol{x}^\star$ is the starting point of $\mathcal{R}_\sigma$ or $\mathcal{R}_{\sigma'}$ by Lemma 5.

Otherwise, $\boldsymbol{m}_{\sigma(1)} \cdot \boldsymbol{m}_{\sigma'(1)} < 0$ and we have:

$$-\max\{\boldsymbol{m}^2_{\sigma(1)}, \boldsymbol{m}^2_{\sigma'(1)}\} < \boldsymbol{m}_{\sigma(1)} \cdot \boldsymbol{m}_{\sigma'(1)} < 0,$$

which is equivalent to

$$0 < \boldsymbol{m}_{\sigma(1)} \cdot \boldsymbol{m}_{\sigma'(1)} + \max\{\boldsymbol{m}^2_{\sigma(1)}, \boldsymbol{m}^2_{\sigma'(1)}\} < \max\{\boldsymbol{m}^2_{\sigma(1)}, \boldsymbol{m}^2_{\sigma'(1)}\}.$$

Let us assume w.l.o.g. that $\boldsymbol{m}_{\sigma(1)} \geq \boldsymbol{m}_{\sigma'(1)}$ so that $\boldsymbol{m}_{\sigma(1)} \cdot (\boldsymbol{m}_{\sigma(1)} + \boldsymbol{m}_{\sigma'(1)}) \geq 0$. Let us consider the following linear transform: $\widetilde{\boldsymbol{m}}_{\sigma(1)} := \boldsymbol{m}_{\sigma(1)}$ and $\widetilde{\boldsymbol{m}}_{\sigma'(1)} := \boldsymbol{m}_{\sigma(1)} + \boldsymbol{m}_{\sigma'(1)}$. Since $\widetilde{\boldsymbol{m}}_{\sigma(1)} \cdot \widetilde{\boldsymbol{m}}_{\sigma'(1)} \geq 0$, due to Lemma 5, $\boldsymbol{v}_{\sigma(0)} + \widetilde{\boldsymbol{m}}_{\sigma(1)}$ or $\boldsymbol{v}_{\sigma(0)} + \widetilde{\boldsymbol{m}}_{\sigma'(1)}$ is closer according to $\leq_{\mathbf{T}}$, than any point $\boldsymbol{v}_{\sigma(0)} + \widetilde{\boldsymbol{m}}_{\sigma'(1)} + \lambda \widetilde{\boldsymbol{m}}_{\sigma(1)}$, with $\lambda \in \mathbb{Z}_{>0}$. In other words, a closest point $\boldsymbol{x}^\star$ cannot belong to $\{\mathcal{R}_{\sigma'}[\lambda+1], \lambda \in \mathbb{Z}_{>0}\} \subset \mathcal{R}_{\sigma'} \backslash \boldsymbol{y}_{\sigma'}$, because either $\boldsymbol{v}_{\sigma(0)} + \widetilde{\boldsymbol{m}}_{\sigma(1)} = \mathcal{R}_\sigma[0]$ or $\boldsymbol{v}_{\sigma(0)} + \widetilde{\boldsymbol{m}}_{\sigma'(1)} = \mathcal{R}_\sigma[1]$, are closer according to $\leq_{\mathbf{T}}$. □

According to Lemma 4, it is enough to call the function ClosestAmongPointAndRay on an appropriate point and on an appropriate ray (lines 4 and 5 of Table 1). The body of the function is given in the next section.

4 One Point and One Ray $\mathcal{R}_\sigma$

It is well known that the implicit equation of the sphere can be written as a determinant (e.g., see MathWorld). More precisely, the algebraic distance of $\boldsymbol{x}'$ to the circumsphere of $\mathbf{T} \cup \{\boldsymbol{x}\}$ is given by the following 5×5 matrix determinant:

$$\delta_{\mathbf{T}}(\boldsymbol{x}, \boldsymbol{x}') := \begin{vmatrix} \boldsymbol{v}_0 & \boldsymbol{v}_1 & \boldsymbol{v}_2 & \boldsymbol{x} & \boldsymbol{x}' \\ {\boldsymbol{v}_0}^2 & {\boldsymbol{v}_1}^2 & {\boldsymbol{v}_2}^2 & \boldsymbol{x}^2 & {\boldsymbol{x}'}^2 \\ 1 & 1 & 1 & 1 & 1 \end{vmatrix}.$$

Note that $\delta_{\mathbf{T}}(\boldsymbol{x}, \boldsymbol{x}') \leq 0 \Leftrightarrow \boldsymbol{x}' \leq_{\mathbf{T}} \boldsymbol{x}$ means that $\boldsymbol{x}'$ is inside or on the circumsphere of $\mathbf{T} \cup \{\boldsymbol{x}\}$.

From now on, we assume w.l.o.g. that $\sigma(0) = 0$, $\sigma(1) = 1$, $\sigma(2) = 2$ so that we can take $\boldsymbol{v}_0$ as the origin. In order to shorten notations, we set $\boldsymbol{y} := \boldsymbol{x} - \boldsymbol{v}_0$, $\boldsymbol{y}' := \boldsymbol{x}' - \boldsymbol{v}_0$ and using $\boldsymbol{d}_k = \boldsymbol{v}_{k+1} - \boldsymbol{v}_k = \boldsymbol{m}_k - \boldsymbol{m}_{k+1}$ for all k, we have:

$$\delta^0_{\mathbf{T}}(\boldsymbol{y}, \boldsymbol{y}') := \delta_{\mathbf{T}}(\boldsymbol{v}_0 + \boldsymbol{y}, \boldsymbol{v}_0 + \boldsymbol{y}') = \begin{vmatrix} \boldsymbol{d}_0 & -\boldsymbol{d}_2 & \boldsymbol{y} & \boldsymbol{y}' \\ {\boldsymbol{d}_0}^2 & {\boldsymbol{d}_2}^2 & \boldsymbol{y}^2 & {\boldsymbol{y}'}^2 \end{vmatrix}. \tag{5}$$

Let us denote by $[\boldsymbol{z}, \boldsymbol{z}', \boldsymbol{z}'']$ the 3×3 matrix composed of columns $\boldsymbol{z}, \boldsymbol{z}', \boldsymbol{z}''$. We give below a formula for $\delta^0_{\mathbf{T}}(\boldsymbol{z}, \boldsymbol{z}' + \alpha \boldsymbol{z}'')$ for any $\boldsymbol{z}, \boldsymbol{z}', \boldsymbol{z}'' \in \mathbb{R}^3$ using the cofactor expansion of the determinant (5):

$$\begin{aligned} \delta^0_{\mathbf{T}}(\boldsymbol{z}, \boldsymbol{z}' + \alpha \boldsymbol{z}'') = & -{\boldsymbol{d}_0}^2 \det[-\boldsymbol{d}_2, \boldsymbol{z}, \boldsymbol{z}' + \alpha \boldsymbol{z}''] + {\boldsymbol{d}_2}^2 \det[\boldsymbol{d}_0, \boldsymbol{z}, \boldsymbol{z}' + \alpha \boldsymbol{z}''] \\ & - \boldsymbol{z}^2 \det[\boldsymbol{d}_0, -\boldsymbol{d}_2, \boldsymbol{z}' + \alpha \boldsymbol{z}''] + (\boldsymbol{z}' + \alpha \boldsymbol{z}'')^2 \det[\boldsymbol{d}_0, -\boldsymbol{d}_2, \boldsymbol{z}]. \end{aligned}$$

Since the determinant is multilinear, the following identity can be obtained:

$$\begin{aligned} \delta^0_{\mathbf{T}}(\boldsymbol{z}, \boldsymbol{z}' + \alpha \boldsymbol{z}'') = & \ \delta^0_{\mathbf{T}}(\boldsymbol{z}, \boldsymbol{z}') + \alpha \delta^0_{\mathbf{T}}(\boldsymbol{z}, \boldsymbol{z}'') \\ & + \Big(\alpha^2 ({\boldsymbol{z}''}^2) + \alpha\big(-{\boldsymbol{z}''}^2 + 2\boldsymbol{z}' \cdot \boldsymbol{z}''\big)\Big) \det[\boldsymbol{d}_0, -\boldsymbol{d}_2, \boldsymbol{z}]. \end{aligned} \tag{6}$$

We now use (6) in order to find an implementation of Algorithms 2 and 3 in constant time, which are used in Algorithm 4.

SphereRayIntersection. In order to implement Algorithm 2, we consider the circumsphere of $\mathbf{T} \cup \{\boldsymbol{v}_0 + \boldsymbol{z}\}$, where $\boldsymbol{z} \in \{\boldsymbol{m}_2, \boldsymbol{d}_0 + \boldsymbol{m}_2, -\boldsymbol{d}_2 + \boldsymbol{m}_1\}$, and its intersection with the ray points $\boldsymbol{v}_0 + \boldsymbol{m}_1 + \lambda \boldsymbol{m}_2$, $\lambda \in \mathbb{Z}_{\geq 0}$.

First, $\det[\boldsymbol{d}_0, -\boldsymbol{d}_2, \boldsymbol{z}] = 1$ for all $\boldsymbol{z} \in \{\boldsymbol{m}_2, \boldsymbol{d}_0 + \boldsymbol{m}_2, -\boldsymbol{d}_2 + \boldsymbol{m}_1\}$, because the determinant is multilinear and $\det[\boldsymbol{m}_0, \boldsymbol{m}_1, \boldsymbol{m}_2] = 1$ [10, Lemma 3]. Consequently, replacing $\boldsymbol{z}'$ with $\boldsymbol{m}_1$, $\boldsymbol{z}''$ with $\boldsymbol{m}_2$ and α with λ in (6), we get:

$$\delta^0_{\mathbf{T}}(\boldsymbol{z}, \boldsymbol{m}_1 + \lambda \boldsymbol{m}_2) = \lambda^2 (\boldsymbol{m}_2^2) + \lambda\big(2\boldsymbol{z} \cdot \boldsymbol{m}_1 - \boldsymbol{m}_2^2 + \delta^0_{\mathbf{T}}(\boldsymbol{z}, \boldsymbol{m}_2)\big) + \delta^0_{\mathbf{T}}(\boldsymbol{z}, \boldsymbol{m}_1). \quad (7)$$

The ray points $\boldsymbol{v}_0 + \boldsymbol{m}_1 + \lambda \boldsymbol{m}_2$ are in the circumsphere of $\mathbf{T} \cup \{\boldsymbol{v}_0 + \boldsymbol{z}\}$ if and only if $\delta^0_{\mathbf{T}}(\boldsymbol{z}, \boldsymbol{m}_1 + \lambda \boldsymbol{m}_2) \leq 0$. Since $\delta^0_{\mathbf{T}}(\boldsymbol{z}, \boldsymbol{m}_1 + \lambda \boldsymbol{m}_2)$ is a quadratic function in λ, there are either zero or two (possibly equal) real roots $\lambda_1 \leq \lambda_2$. In the first case, there is no intersection between the circumsphere of $\mathbf{T} \cup \{\boldsymbol{v}_0 + \boldsymbol{z}\}$ and the ray, whereas in the second case, the ray points such that $\lambda \in [\lambda_1; \lambda_2] \cap \mathbb{Z}_{\geq 0}$, lie in the circumsphere of $\mathbf{T} \cup \{\boldsymbol{v}_0 + \boldsymbol{z}\}$ and are closer than $\boldsymbol{z}$ according to $\leq_{\mathbf{T}}$.

In Algorithm 2, we check the sign of the discriminant and either return an empty list if it is strictly negative or return the (possibly empty) range of ray points as the list of the (possibly equal) lower and upper bounds.

Algorithm 2. SPHERERAYINTERSECTION($\mathbf{T}, \boldsymbol{q}, \sigma', \sigma$)

Input: the base triangle $\mathbf{T}$, the exterior point $\boldsymbol{q}$, a point $\boldsymbol{y}_{\sigma'}$, a ray $\mathcal{R}_\sigma$
Output: either empty () or the bounds $(\lambda_1, \lambda_2) \in \mathbb{Z}^2_{\geq 0}$ of the greatest interval of points $\{\mathcal{R}_\sigma[\lambda], \lambda_1 \leq \lambda \leq \lambda_2, \lambda \in \mathbb{Z}_{\geq 0}\}$, such that $\boldsymbol{y}_{\sigma'} \leq_{\mathbf{T}} \mathcal{R}_\sigma[\lambda]$

$(\boldsymbol{z}, \boldsymbol{m}_2, \boldsymbol{m}_1) \leftarrow (\boldsymbol{y}_{\sigma'} - \boldsymbol{v}_{\sigma(0)}, \boldsymbol{q} - \boldsymbol{v}_{\sigma(2)}, \boldsymbol{q} - \boldsymbol{v}_{\sigma(1)})$;
$(a, b, c) \leftarrow ({\boldsymbol{m}_2}^2, -{\boldsymbol{m}_2}^2 + 2\boldsymbol{z} \cdot \boldsymbol{m}_1 + \delta^0_{\mathbf{T}}(\boldsymbol{z}, \boldsymbol{m}_2), \delta^0_{\mathbf{T}}(\boldsymbol{z}, \boldsymbol{m}_1))$;
$d \leftarrow b^2 - 4ac$;
if $d \geq 0$ **then**
  $(\lambda_1, \lambda_2) \leftarrow \big(\lceil(-b - \sqrt{d})/(2a)\rceil, \lfloor(-b + \sqrt{d})/(2a)\rfloor\big)$;
  if $\lambda_1 \leq \lambda_2$ *and* $0 \leq \lambda_2$ **then return** $(\max(0, \lambda_1), \lambda_2)$;
return ()

ClosestOnRay. In order to implement Algorithm 3, we consider the family of spheres passing by the vertices of $\mathbf{T}$ and a ray point $\boldsymbol{v}_0 + \boldsymbol{m}_1 + \lambda \boldsymbol{m}_2$, for any $\lambda \in \mathbb{Z}_{\geq 0}$. Given a sphere, we want to check whether the next ray point, i.e., $\boldsymbol{v}_0 + \boldsymbol{m}_1 + (\lambda + 1)\boldsymbol{m}_2$, is located inside it or not.

Replacing both $\boldsymbol{z}, \boldsymbol{z}'$ with $\boldsymbol{m}_1 + \lambda \boldsymbol{m}_2$, $\boldsymbol{z}''$ with $\boldsymbol{m}_2$, α with 1 in (6), we get:

$$\delta^0_{\mathbf{T}}(\boldsymbol{m}_1 + \lambda \boldsymbol{m}_2, \boldsymbol{m}_1 + (\lambda + 1)\boldsymbol{m}_2) = \delta^0_{\mathbf{T}}(\boldsymbol{m}_1 + \lambda \boldsymbol{m}_2, \boldsymbol{m}_2) \\ + 2(\boldsymbol{m}_1 + \lambda \boldsymbol{m}_2) \cdot \boldsymbol{m}_2 \det[\boldsymbol{d}_0, -\boldsymbol{d}_2, \boldsymbol{m}_1 + \lambda \boldsymbol{m}_2].$$

Using (7) (with $\boldsymbol{z} = \boldsymbol{m}_2$) and $\det[\boldsymbol{d}_0, -\boldsymbol{d}_2, \boldsymbol{m}_1 + \lambda \boldsymbol{m}_2] = \lambda + 1$ (again from [10, Lemma 3]), this expression can be simplified into:

$$\delta^0_{\mathbf{T}}(\boldsymbol{m}_1 + \lambda \boldsymbol{m}_2, \boldsymbol{m}_1 + (\lambda+1)\boldsymbol{m}_2) = \lambda^2(\boldsymbol{m}_2{}^2) + \lambda(3\boldsymbol{m}_2{}^2) + 2(\boldsymbol{m}_1 \cdot \boldsymbol{m}_2) + \delta^0_{\mathbf{T}}(\boldsymbol{m}_1, \boldsymbol{m}_2). \quad (8)$$

Clearly the determinant $\delta^0_{\mathbf{T}}(\boldsymbol{m}_1 + \lambda \boldsymbol{m}_2, \boldsymbol{m}_1 + (\lambda+1)\boldsymbol{m}_2)$ is a quadratic function in λ, whose minimum is reached at $\lambda = -3/2$. It is therefore monotonically increasing over $[0; \infty)$. Let $\lambda^\star$ be the smallest integer $\lambda \in \mathbb{Z}_{\geq 0}$ such that $\delta^0_{\mathbf{T}}(\boldsymbol{m}_1 + \lambda \boldsymbol{m}_2, \boldsymbol{m}_1 + (\lambda+1)\boldsymbol{m}_2) > 0$. By definition, $\delta^0_{\mathbf{T}}(\boldsymbol{m}_1 + \lambda^\star \boldsymbol{m}_2, \boldsymbol{m}_1 + (\lambda^\star+1)\boldsymbol{m}_2) > 0$, which means that the sphere passing by the vertices of $\mathbf{T}$ and $\boldsymbol{v}_0 + \boldsymbol{m}_1 + \lambda^\star \boldsymbol{m}_2$ contains neither $\boldsymbol{v}_0 + \boldsymbol{m}_1 + (\lambda^\star + 1)\boldsymbol{m}_2$ nor the following ray points by transitivity, because $\delta^0_{\mathbf{T}}(\boldsymbol{m}_1 + \lambda \boldsymbol{m}_2, \boldsymbol{m}_1 + (\lambda+1)\boldsymbol{m}_2) > 0$ also for all $\lambda \geq \lambda^\star$. In other words, $\forall \lambda \geq \lambda^\star, \boldsymbol{v}_0 + \boldsymbol{m}_1 + \lambda^\star \boldsymbol{m}_2 \leq_{\mathbf{T}} \boldsymbol{v}_0 + \boldsymbol{m}_1 + \lambda \boldsymbol{m}_2$. In addition, if $\lambda^\star \geq 1$, $\delta^0_{\mathbf{T}}(\boldsymbol{m}_1 + (\lambda^\star - 1)\boldsymbol{m}_2, \boldsymbol{m}_1 + \lambda^\star \boldsymbol{m}_2) \leq 0$, which is equivalent to $\delta^0_{\mathbf{T}}(\boldsymbol{m}_1 + \lambda^\star \boldsymbol{m}_2, \boldsymbol{m}_1 + (\lambda^\star - 1)\boldsymbol{m}_2) > 0$. This means that $\forall \lambda \in 0, \ldots, \lambda^\star, \boldsymbol{v}_0 + \boldsymbol{m}_1 + \lambda^\star \boldsymbol{m}_2 \leq_{\mathbf{T}} \boldsymbol{v}_0 + \boldsymbol{m}_1 + \lambda \boldsymbol{m}_2$. As a consequence, $\lambda^\star$ provides the closest ray point. Algorithm 3 just computes $\lambda^\star$ and returns the corresponding ray point.

Algorithm 3. ClosestOnRay($\mathbf{T}, \boldsymbol{q}, \sigma$). Searches for the closest point according to $\leq_{\mathbf{T}}$ on ray $\mathcal{R}_\sigma$.

Input: the base triangle $\mathbf{T}$, the exterior point $\boldsymbol{q}$, a ray $\mathcal{R}_\sigma$
Output: the smallest $\lambda \in \mathbb{Z}_{\geq 0}$ such that $\mathcal{R}_\sigma[\lambda] \leq_{\mathbf{T}} \mathcal{R}_\sigma[\lambda + 1]$
$(\boldsymbol{m}_1, \boldsymbol{m}_2) \leftarrow (\boldsymbol{q} - \boldsymbol{v}_{\sigma(1)}, \boldsymbol{q} - \boldsymbol{v}_{\sigma(2)})$;
$(a, b, c) \leftarrow (\boldsymbol{m}_2{}^2, 3\boldsymbol{m}_2{}^2, 2\boldsymbol{m}_1 \cdot \boldsymbol{m}_2 + \delta^0_{\mathbf{T}}(\boldsymbol{m}_1, \boldsymbol{m}_2))$;
$d \leftarrow b^2 - 4ac$;
if $d < 0$ **then return** 0;
$\lambda^\star \leftarrow \lceil (-b + \sqrt{d})/(2a) \rceil$;
return $\max(0, \lambda^\star)$

ClosestAmongPointAndRay. Once the ray that can hold a candidate point on $\mathbf{P}$ has been identified, it remains to determine whether the closest point to the current triangle $\mathbf{T}$ (according to $\leq_{\mathbf{T}}$) lies on the ray or is another point $\boldsymbol{y}$ of the H-neighborhood. Algorithm 4 performs this operation with the following steps. First it calls Algorithm 2 to find if there may be an interval of points on the ray that are closer than $\boldsymbol{y}$. If it is the case, it checks if at least one belongs to $\mathbf{P}$. If this is the case it calls Algorithm 3 to determine the one that is closest. If it belongs to $\mathbf{P}$, we are done. Otherwise, we have to find the closest point on the ray that belongs to $\mathbf{P}$ by a call to Algorithm 5. This routine simply performs an exponential march followed by a binary search.

5 Complexity Analysis and Experimental Results

Upper Bound on the Number of Calls to Predicate. Given the previous functions that are used to update the triangle at each step, we can prove:

Algorithm 4. ClosestAmongPointAndRay($P, \mathbf{T}, \boldsymbol{q}, \sigma', \sigma$)

Input: the predicate P, the base triangle $\mathbf{T}$, the exterior point $\boldsymbol{q}$, a candidate point $\boldsymbol{y}_{\sigma'}$ with $P(\boldsymbol{y}_{\sigma'})$, a ray $\mathcal{R}_\sigma$ with $P(\boldsymbol{y}_\sigma)$
Output: A couple (τ, α), such that $\tau \in \{\sigma, \sigma'\}$, $\alpha \in \mathbb{Z}_{\geq 0}$ and $P(\mathcal{R}_\tau[\alpha])$

```
L ← SphereRayIntersection(T, q, y_σ', σ);
if L ≠ ∅ then
    α1, α2 ← L;
    if P(R_σ[α1]) then
        α ← ClosestOnRay(T, q, σ);
        if α1 ≤ α ≤ α2 then
            if P(R_σ[α]) then return (σ, α);
            else return (σ, FindLast(P, T, q, σ, α1))
return (σ', 0)
```

Algorithm 5. FindLast($P, \mathbf{T}, \boldsymbol{q}, \sigma, \alpha_1$) Use exponential march then binary search to find the last point in $\mathcal{R}_\sigma$ that is in $\mathbf{P}$.

Input: the predicate P, the base triangle $\mathbf{T}$, the exterior point $\boldsymbol{q}$, a ray $\mathcal{R}_\sigma$, an integer α_1 with $P(\mathcal{R}_\sigma[\alpha_1])$
Output: the integer λ such that $P(\mathcal{R}_\sigma[\lambda])$ and $\neg P(\mathcal{R}_\sigma[\lambda+1])$

```
J ← 1;
while P(R_σ[α1 + J]) do J ← 2J;
λ1 ← α1 + ⌊J/2⌋; λ2 ← α1 + J;
while λ2 ≠ λ1 + 1 do
    λ ← ⌊(λ1+λ2)/2⌋ ;
    if P(R_σ[λ]) then λ1 ← λ;
    else λ2 ← λ
return λ1
```

Theorem 1. *The number of calls to predicate $P(\boldsymbol{x}) :=$ "Is $\boldsymbol{x}$ in $\mathbf{P}$" in R^1-algorithm (Algorithm 1) is upper bounded by $O(\omega)$.*

Proof. Let $A^{(i)} := \boldsymbol{v}_0^{(i)} \cdot \mathbf{N} + \boldsymbol{v}_1^{(i)} \cdot \mathbf{N} + \boldsymbol{v}_2^{(i)} \cdot \mathbf{N}$ be the height of the current triangle. Let us denote $\lambda^{(i)}$ the integer related to the update of a vertex at iteration i, i.e., $\exists \pi \in \Sigma, \boldsymbol{v}_{\pi(0)}^{(i+1)} \leftarrow \boldsymbol{v}_{\pi(0)}^{(i)} + \boldsymbol{m}_{\pi(1)}^{(i)} + \lambda^{(i)} \boldsymbol{m}_{\pi(2)}^{(i)}$.

Since $\forall k, \boldsymbol{m}_k^{(i)} \cdot \mathbf{N} \geq 1$ [10, Lemma 5] then $A^{(i+1)} - A^{(i)} \geq 1 + \lambda^{(i)}$. Noticing that $A^{(0)} \geq 2\omega$, $A^{(n)} \leq 3\omega - 3$ [10, Theorem 2], suming over all iterations gives

$$\omega - 3 \geq A^{(n)} - A^{(0)} \geq \sum_{i=0}^{n-1} (1 + \lambda^{(i)}). \tag{9}$$

If we look now at the number $B^{(i)}$ of calls to P at iteration i, we must count the 6 calls for determining the H-neighborhood, 2 more possible calls in ClosestAmongPointAndRay, and possibly $2 \log_2 J$ calls in FindLast. But $\alpha_1 + J/2 \leq \lambda^{(i)}$, thus $2 \log_2 J \leq 2 \log_2(1 + \lambda^{(i)})$. We get straightforwardly that $B^{(i)} \leq 8 + 2\log_2(1 + \lambda^{(i)})$. Recalling that $x \geq \log_2(1+x)$ and suming the $B^{(i)}$, we derive from (9) that

$$\omega - 3 \geq \sum_{i=0}^{n-1}(1 + \lambda^{(i)}) \geq \frac{1}{8}\sum_{i=0}^{n-1}(8 + 8\log_2(1 + \lambda^{(i)})) \geq \frac{1}{8}\sum_{i=0}^{n-1} B^{(i)}. \tag{10}$$

Noticing that $\sum_{i=0}^{n-1} B^{(i)}$ is the total number of calls to P concludes. □

It is worthy to note that the R^1-algorithm performs also $O(\omega)$ arithmetic, square root and rounding operations.

Furthermore, plane-probing algorithms are run in this work on a digital plane, i.e., an infinite point set. However, since they are local and stop after a finite number of steps (Theorem 1), there is some finite subsets for which processing the whole digital plane or only one of such subsets would be equivalent. Characterizing these subsets and bounding their size or diameter is not easy and may involve geometrical arguments based on the empty-circumsphere criterion. Then it will be possible to express the time complexity of plane-probing algorithms relatively to this bound, but it is still an open question.

Experimental Evaluation. We ran all three algorithms (H, R and R^1) on planes whose normal vector is ranging from (1, 1, 1) to (200, 200, 200). There are 6578833 vectors with relatively prime components in this range. Results are reported in the inset table. For the number of steps, i.e., n, and the total number of calls to predicate P, i.e., $\sum_i B^i$, we computed the average over the 6578833 runs. However, for the number of calls to P at step i, i.e., B^i, we computed the average and maximum over all steps and runs.

	n	B^i		$\sum_{i=0}^{n-1} B^i$
Alg.	Avg.	Avg.	Max.	Avg.
H	24.84	6.00	6	149.01
R	17.59	14.49	25	254.95
R^1	17.32	7.06	14	122.36

First, the number of steps is 70% lower on average for the R- and R^1-algorithms than for the H-algorithm. Note that the number of steps is not exactly the same for R and R^1, because the arbitrary choice of a closest point in case of several closest co-spheric points is not the same. All three algorithms have however the same worst case: there are indeed always $n = 2r - 1$ steps for planes with normal $\mathbf{N}(1, r, r)$.

Second, the number of calls to P at each step is exactly 6 for the H-algorithm, close to 14 on average for the R-algorithm, but between 6 and 8 in most cases for the R^1-algorithm – a greater number of calls happens only occasionally, when function FINDLAST is called in Algorithm 4.

The total number of calls to P is the lowest on average for the R^1-algorithm, which is a good trade-off between the number of steps and the number of calls at each step. On small-size vectors, i.e., ranging from (1, 1, 1) to (200, 200, 200), the R^1-algorithm is approximately twice faster than the R-algorithm, because the total number of calls to P is close to 122 on average for the R^1-algorithm while it is close to 255 for the R-algorithm.

Alg.\k	10	20	30
R	18.50	25.39	36.33
R^1	6.80	6.20	6.11

For larger vectors, we can observe that the number of calls to P per step is still constant on average for the R^1-algorithm, but may be arbitrary large for the R-algorithm. Indeed, on digital planes of normal $\mathbf{N}(1, F_k, F_{k+1})$, where F_k is the k-th term of the Fibonacci sequence, the average number of predicate calls at each step increases as k increases for R, while it remains close to 6 for R^1 (see inset table).

6 Conlusion and Perspectives

We have presented a new plane-probing algorithm that outperforms both in theory and practice the state-of-the-art R-algorithm. It avoids computations that do not contribute to the final solution and performs fewer calls to the predicate "Is $\boldsymbol{x} \in \mathbf{P}$?": between 6 and 8 calls most of the time at each step and $O(\omega)$ calls in total. This algorithm is expected to be an efficient tool for digital surface analysis.

It remains to understand why these algorithms perform so much better on average than the worst-case bound $O(\omega)$. Therefore, we will investigate in the future their link with multi-dimensional continued fractions and dynamic number theory.

References

1. Buzer, L.: A linear incremental algorithm for naive and standard digital lines and planes recognition. Graph. Models **65**(1–3), 61–76 (2003)
2. Charrier, E., Buzer, L.: An efficient and quasi linear worst-case time algorithm for digital plane recognition. In: Coeurjolly, D., Sivignon, I., Tougne, L., Dupont, F. (eds.) DGCI 2008. LNCS, vol. 4992, pp. 346–357. Springer, Heidelberg (2008). https://doi.org/10.1007/978-3-540-79126-3_31
3. Charrier, E., Lachaud, J.-O.: Maximal planes and multiscale tangential cover of 3D digital objects. In: Aggarwal, J.K., Barneva, R.P., Brimkov, V.E., Koroutchev, K.N., Korutcheva, E.R. (eds.) IWCIA 2011. LNCS, vol. 6636, pp. 132–143. Springer, Heidelberg (2011). https://doi.org/10.1007/978-3-642-21073-0_14
4. Debled-Rennesson, I., Reveillès, J.: An incremental algorithm for digital plane recognition. In: Discrete Geometry for Computer Imagery, pp. 194–205 (1994)
5. Fernique, T.: Generation and recognition of digital planes using multi-dimensional continued fractions. Pattern Recogn. **42**(10), 2229–2238 (2009)
6. Gérard, Y., Debled-Rennesson, I., Zimmermann, P.: An elementary digital plane recognition algorithm. Discrete Appl. Math. **151**(1), 169–183 (2005)
7. Klette, R., Sun, H.J.: Digital planar segment based polyhedrization for surface area estimation. In: Arcelli, C., Cordella, L.P., di Baja, G.S. (eds.) IWVF 2001. LNCS, vol. 2059, pp. 356–366. Springer, Heidelberg (2001). https://doi.org/10.1007/3-540-45129-3_32
8. Lachaud, J.O., Provençal, X., Roussillon, T.: An output-sensitive algorithm to compute the normal vector of a digital plane. J. Theor. Comput. Sci. (TCS) **624**, 73–88 (2016)

9. Lachaud, J.-O., Provençal, X., Roussillon, T.: Computation of the normal vector to a digital plane by sampling significant points. In: Normand, N., Guédon, J., Autrusseau, F. (eds.) DGCI 2016. LNCS, vol. 9647, pp. 194–205. Springer, Cham (2016). https://doi.org/10.1007/978-3-319-32360-2_15
10. Lachaud, J.O., Provençal, X., Roussillon, T.: Two plane-probing algorithms for the computation of the normal vector to a digital plane. J. Math. Imaging Vis. **59**(1), 23–39 (2017)
11. Mesmoudi, M.M.: A simplified recognition algorithm of digital planes pieces. In: Braquelaire, A., Lachaud, J.-O., Vialard, A. (eds.) DGCI 2002. LNCS, vol. 2301, pp. 404–416. Springer, Heidelberg (2002). https://doi.org/10.1007/3-540-45986-3_36
12. Sivignon, I., Dupont, F., Chassery, J.M.: Decomposition of a three-dimensional discrete object surface into discrete plane pieces. Algorithmica **38**(1), 25–43 (2004)
13. Veelaert, P.: Digital planarity of rectangular surface segments. IEEE Trans. Pattern Anal. Mach. Intell. **16**(6), 647–652 (1994)
14. Veelaert, P.: Fast combinatorial algorithm for tightly separating hyperplanes. In: Barneva, R.P., Brimkov, V.E., Aggarwal, J.K. (eds.) IWCIA 2012. LNCS, vol. 7655, pp. 31–44. Springer, Heidelberg (2012). https://doi.org/10.1007/978-3-642-34732-0_3

Geometric Computation

Convex and Concave Vertices on a Simple Closed Curve in the Triangular Grid

Lidija Čomić(✉)

Faculty of Technical Sciences, University of Novi Sad, Novi Sad, Serbia
comic@uns.ac.rs

Abstract. We propose a classification of convex and concave vertices on a simple closed curve in the triangular grid into two types, based on the angle turn the curve makes at the vertex. We prove a combinatorial property for the number of convex and concave vertices of different types.

Keywords: Digital topology · Triangular grid · Convex and concave vertices · Salient and reentrant vertices · Simple closed curve

1 Introduction

There are three regular grids in the plane, inducing the tiling of the plane into regular triangles, squares or hexagons, called collectively pixels. Although the square grid received the most attention in the literature, the two alternative grids were also widely investigated in different frameworks, such as topology-preserving transformations [14,15,21,24,25,34], computation of the Euler characteristic [4,22,33], analytical [1,5,16,20,30] or computational [31,32] geometry, tomography [29], topological/combinatorial coordinate systems [23,28], distance transform [2,3] and neighborhood sequences [17], to name just a few.

Chain codes were designed to represent digital curves composed of pixels in the square grid, by specifying the direction in which the next pixel on the curve lies. They were extended to vertex chain codes [4], to represent one-dimensional interpixel boundaries of well-composed [27] binary objects, i.e., simple closed curves composed of grid edges. Vertex chain codes were defined in all three regular grids. The code of a vertex is equal to the number of interior (black or object) pixels incident to the vertex.

In the square grid, if the curve is traced in the counterclockwise direction, at each vertex with code 1 the curve makes a left turn, and at each vertex with code 3 the curve makes a right turn. Such vertices are called salient and reentrant, [12,13], respectively. In the hexagonal and the triangular grids, salient and reentrant vertices were also defined as those vertices at which the curve makes a left and a right turn, respectively [9]. It was shown [12,13] that every simple closed curve in the square grid has four salient vertices more than it has reentrant ones. This result was generalized to arbitrary curves (not necessarily

M. Couprie et al. (Eds.): DGCI 2019, LNCS 11414, pp. 397–408, 2019.
https://doi.org/10.1007/978-3-030-14085-4_31

simple) in the square grid, and to curves in the hexagonal grid [8,9]. In the latter grid, each curve has six more salient vertices than reentrant ones. This relationship between the number of salient and reentrant vertices was used in the study of the tiling problem on the square and hexagonal grids [10]. It was also shown [9] that there is no similar relationship between the number of salient and reentrant vertices in the triangular grid.

Here, we propose a finer classification of the vertices on a simple closed curve in the triangular grid, based on the number of incident interior triangles, i.e., based on the angle the curve makes at the vertex. We call such vertices (type 1 and type 2) convex and concave vertices. We prove a relationship between the number of different types of convex and concave vertices that every simple closed curve in the triangular grid satisfies.

2 Preliminaries

We introduce some basic notions on the regular grids in 2D [15,26], on simple closed curves [4,19] and on salient and reentrant vertices in these grids [9,13].

2.1 Regular Grids in the Plane

There are three regular grids in the plane: the triangular, square and hexagonal grids. They induce the tessellations of the plane into regular triangles, squares and hexagons, respectively, called pixels. Each triangle, square and hexagon is incident to three, four and six edges and vertices, respectively. Each edge (in all three grids) is incident to two vertices and two pixels. Two vertices incident to the same edge are called neighbors. Each vertex in the triangular, square and hexagonal grids is incident to six, four and three pixels and edges, and has six, four and three neighbors, respectively.

Different types of adjacency relations are defined between the pixels in these grids, depending on their intersection. Edge-adjacent pixels share an entire edge; (strictly) vertex-adjacent pixels share (only) a vertex. Each triangle is edge-adjacent to three triangles, one across each of its edges. It is strictly vertex-adjacent to another nine triangles, three across each of its vertices. Each square is edge-adjacent to four squares, and is strictly vertex-adjacent to other four squares. Each hexagon is edge-adjacent to six hexagons. Two hexagons that share a vertex, share also an entire edge. Thus, there is no strict vertex-adjacency in the hexagonal grid.

A (binary) object O in these grids is a finite set of pixels in the grid. The pixels in O are called black (object), those in the complement O^c of O are called white (background). Two pixels p and q in an object O are edge-connected (vertex-connected) in O, if there is a sequence of pixels in O, starting at p and ending in q, such that any two consecutive pixels in the sequence are edge-adjacent (vertex-adjacent).

2.2 Simple Closed Curves in the Regular Grids

A simple closed curve C of length k in a grid is a cyclic sequence $v_1, e_1, v_2, e_2, ..., v_k, e_k, v_1$, of grid vertices v_i and grid edges e_i, $1 \leq i \leq k$, if each vertex in the sequence C is shared by the previous and the next edge in C and there is no repetition of vertices (or of edges) in C. Thus, C is a discrete 1-surface [18] and it defines a (polygonal) Jordan curve. Two consecutive vertices in the sequence are called adjacent on C. We will assume that C is oriented counterclockwise.

The interior of the simple closed curve C is composed of a finite set S of grid pixels, called interior pixels of C. Such sets S in the square [27] and in the triangular grid [11] are called well-composed or gapless [6,7]. In the hexagonal grid, each such set of pixels (hexagons) is well-composed, due to the restricted number of pixels incident to a vertex in the hexagonal grid (three, as opposed to four in the square and six in the triangular grid). The pixels in the complement S^c of S are called exterior pixels. The set of pixels incident to a vertex v on a simple closed curve C can be divided in two edge-connected subsets: the edge-connected subset of interior pixels, and the edge-connected subset of exterior pixels of C.

2.3 Salient and Reentrant Vertices in the Regular Grids

For the counterclockwise orientation of a simple closed curve C, the set S of interior pixels is on the left side of C. At each vertex v on C, the curve C makes a left turn, a right turn, or continues straight ahead. Salient and reentrant vertices in the three regular grids are defined as vertices at which the curve C makes a left and right turn, respectively.

In the square grid, if the curve C makes a left turn at v, the salient vertex v is incident to exactly one interior square (and exactly three exterior squares). If C makes a right turn at v, the reentrant vertex v is incident to exactly three interior squares (and one exterior square). If C continues straight at v, then v is incident to exactly two edge-adjacent interior squares (and two edge-adjacent exterior ones). As C is simple, the vertex v cannot be incident to exactly two strictly vertex-adjacent interior squares (and two strictly vertex-adjacent exterior ones).

It was shown [9,13,35] that for a simple closed curve C in the square grid, the number n^+ of salient vertices is larger by 4 than the number n^- of reentrant vertices, i.e.,

$$n^+ - n^- = 4.$$

In the hexagonal grid, a salient vertex v on a curve C is incident to exactly one interior hexagon, and exactly two exterior ones. A reentrant vertex is incident to exactly two interior hexagons, and exactly one exterior hexagon. The number n^+ of salient vertices is larger by 6 than the number n^- of reentrant vertices [9], i.e.,

$$n^+ - n^- = 6.$$

In the triangular grid, a salient vertex is incident to one or two interior triangles and to five or four edge-connected exterior triangles, respectively. A reentrant vertex is incident to five or four edge-connected interior triangles and to one or two exterior triangles, respectively. There is no (nontrivial) combinatorial relationship between the number of salient and reentrant vertices on C of the form

$$kn^+ + ln^- = m$$

for some integers k, l and m. For example [9], the boundary curve of a single triangle has three salient and no reentrant vertices, while the boundary curve of two edge-adjacent triangles has four salient and no reentrant vertices.

3 Convex and Concave Vertices in the Triangular Grid

We obtain a combinatorial relationship between different types of vertices on a simple closed curve C in the triangular grid by making a distinction between different types of salient and reentrant vertices, based on the number of incident interior and exterior triangles, i.e., based on the angle at which the curve C turns at the vertex.

At a vertex v on a simple closed curve C in the triangular grid, the curve C can make a turn left at the angle $\pi/3$ or $2\pi/3$, a turn right at the same angles, or it can continue straight ahead.

Definition 1. *A vertex v on a counterclockwise oriented simple closed curve C in the triangular grid is called convex (concave) if C turns left (right) at v. If the turning angle is $\pi/3$, the vertex is of type 1; if the angle is $2\pi/3$ it is of type 2. If C makes no turn at v, then v is called straight.*

The different types (up to rotation) of vertices are illustrated in Fig. 1.

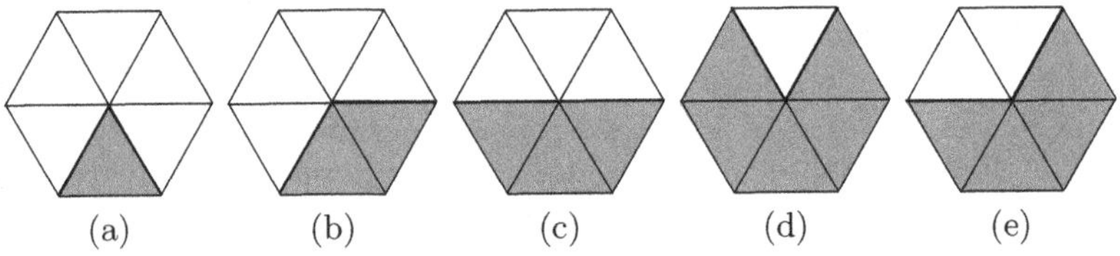

Fig. 1. The possible configurations (up to rotation) of interior (dark) and exterior (white) triangles around a vertex on a curve in the triangular grid: (a) type 1 convex; (b) type 2 convex; (c) straight; (d) type 1 concave; (e) type 2 concave.

Each type 1 convex vertex is incident to one interior triangle (and five exterior ones), and each type 2 convex vertex is incident to two edge-adjacent interior triangles (and four edge-connected exterior ones). Each type 1 concave vertex is incident to one exterior triangle (and five interior ones), and each type 2 concave vertex is incident to two edge-adjacent exterior triangles (and four edge-connected interior ones).

We denote as n_1^+, n_2^+, n_1^- and n_2^- the number of type 1 convex, type 2 convex, type 1 concave and type 2 concave vertices on a simple closed curve C, respectively. We will show that

$$2n_1^+ + n_2^+ - 2n_1^- - n_2^- = 6,$$

i.e.,

$$A(C) = n_1^+ + n_2^+/2 - n_1^- - n_2^-/2 = 3.$$

The explanation follows from considering some simple cases.

- If C is the boundary curve of a single triangle, it has three type 1 convex vertices ($n_1^+ = 3$), and $A(C) = 3$.
- If C is the boundary curve of two edge-adjacent triangles, it has two type 1 convex vertices and two type 2 convex ones ($n_1^+ = n_2^+ = 2$), and $A(C) = 2 + 2/2 = 3$.
- If C is the boundary curve of five triangles incident to a vertex v, it has two type 1 convex, four type 2 convex and one type 1 concave vertex ($n_1^+ = 2$, $n_2^+ = 4$, $n_1^- = 1$), and $A(C) = 2 + 4/2 - 1 = 3$.
- If C is the boundary curve of four edge-connected triangles incident to a vertex v, it has two type 1 convex, three type 2 convex and one type 2 concave vertex ($n_1^+ = 2$, $n_2^+ = 3$, $n_2^- = 1$), and $A(C) = 2 + 3/2 - 1/2 = 3$.

In the following, we give a formal proof of this combinatorial relationship.

Proposition 1. *For a simple closed curve C in the triangular grid,*

$$A(C) = n_1^+ + n_2^+/2 - n_1^- - n_2^-/2 = 3.$$

Proof. The proof is by induction on the length n of C.

By the above discussion for convex vertices, the claim is true for the base cases $n = 3, 4$.

For the inductive step, let the proposition be true for all curves of length smaller than n and let C be a curve of length $n \geq 5$. Let us assume, without loss of generality, that the triangular grid is oriented so that it has $\bigtriangledown$ and $\bigtriangleup$ shaped triangles. In this orientation, each vertex has an east, west, south-east, south-west, north-east and north-west neighbor. Let v be the vertex on C with the smallest x Cartesian coordinate among those with the largest y coordinate (v is the leftmost of the topmost vertices on C). The vertex v is necessarily a convex vertex. We denote by e_1 and e_2 the previous and the next edge of the vertex v on C, and by v_1 and v_2 the other endpoints of the edges e_1 and e_2, respectively.

We will distinguish several cases, depending on the type of the vertex v (v is type 1 convex (case 1) or type 2 convex (case 2)) and, in case 2, depending on the south-east neighbor v' of the vertex v (v' is adjacent on C to v_1 or v_2 (case 2a); v' is not adjacent on C to either v_1 or v_2 and v' is a type 1 concave vertex on C (case 2b); v' is not adjacent on C to either v_1 or v_2 and v' is a type 2 concave vertex on C (case 2c); v' is not on C (case 2d)). The different configurations are illustrated in Fig. 2.

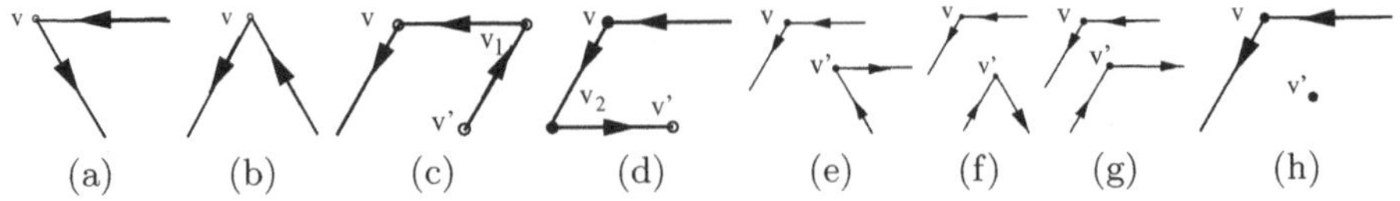

Fig. 2. The different cases in the proof of Proposition 1. The topmost leftmost vertex v on the curve C is: (a) and (b) a type 1 convex vertex; (c) and (d) a type 2 concave vertex with the south-east neighbor v' adjacent on C to the vertex v_1 or the vertex v_2, respectively; (e) and (f) a type 2 convex vertex with type 1 concave south-east neighbor v' on C not adjacent on C to v_1 or v_2; (g) a type 2 convex vertex with type 2 concave south-east neighbor v' on C not adjacent on C to v_1 or v_2; (h) a type 2 convex vertex with south-east neighbor v' not on C.

1. If v is a type 1 convex vertex, then the vertices v_1 and v_2 are neighbors, and they determine the edge $e = v_1v_2$ in the triangular grid. Since there are no repetitions of vertices or edges in C, the edge e does not belong to C. Replacement of the vertex v and edges e_1 and e_2 with the edge e produces a curve of length $n - 1$, for which the proposition is true. This replacement
   - removes from C one type 1 convex vertex (the vertex v), decreasing the quantity $A(C)$ by 1;
   - changes the type of the vertices v_1 and v_2 by changing the angle the curve makes at them, increasing the quantity $A(C)$ by 1/2 for each of the two vertices v_1 and v_2.

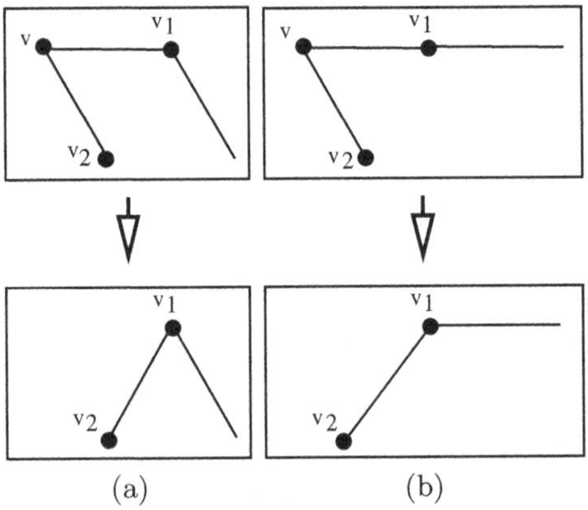

Fig. 3. The possible configurations around the vertex v_1 if the vertex v is type 1 convex and the edge $e_1 = v_1v$ is horizontal (top) before and (bottom) after the replacement. (a) A type 2 convex vertex v_1 becomes type 1 convex. (b) A straight vertex v_1 becomes type 2 convex.

Neither of the two vertices v_1 and v_2 is a type 1 convex vertex, because the length n of C is greater than 4 and C is a simple closed curve. The possible

cases for the vertex v_1 before and after the replacement are illustrated in Figs. 3 and 4, and for the vertex v_2 in Figs. 5 and 6. We give details for the vertex v_1, the analysis for the vertex v_2 being similar.
The possible types for the vertex v_1 depend on the edge e_1 being horizontal or not.

- If the edge e_1 connecting the vertices v_1 and v is a horizontal edge, then the vertex v_1 is not a concave vertex, because of the choice of the vertex v. In this case, the vertex v_1 can be a straight vertex, or a type 2 convex vertex (see Fig. 3 (top)). The replacement of the vertex v and the edges $e_1 = v_1v$ and $e_2 = vv_2$ with the edge $e = v_1v_2$ decreases by one the number of interior triangles incident to v_1 (see Fig. 3 (bottom)). It thus transforms
  - a type 2 convex vertex to a type 1 convex vertex, decreasing n_2^+ by 1 and increasing n_1^+ by 1;
  - a straight vertex to a type 2 convex vertex, increasing n_2^+ by 1;

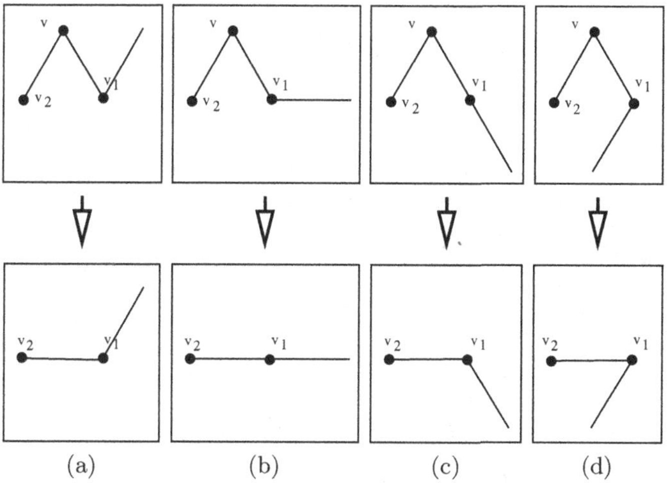

Fig. 4. The possible configurations around the vertex v_1 if the vertex v is type 1 convex and the edge $e_1 = v_1v$ is not horizontal (top) before and (bottom) after the replacement. (a) A type 1 concave vertex v_1 becomes type 2 concave. (b) A type 2 concave vertex v_1 becomes straight. (c) A straight vertex v_1 becomes type 2 convex. (d) A type 2 convex vertex v_1 becomes type 1 convex.

- If the edge e_1 joining the vertices v_1 and v is not horizontal, then the vertex v_1 can also be a type 1 or type 2 concave vertex. The replacement in these two cases transforms
  - a type 1 concave vertex to a type 2 concave vertex, decreasing n_1^- by 1 and increasing n_2^- by 1.
  - a type 2 concave vertex to a straight vertex, decreasing n_2^- by 1;

Each of these changes increases the quantity $A(C)$ by $1/2$. The possible configurations around the vertex v_1 before and after the replacement in the case when the edge e_1 is not horizontal are illustrated in Fig. 4.

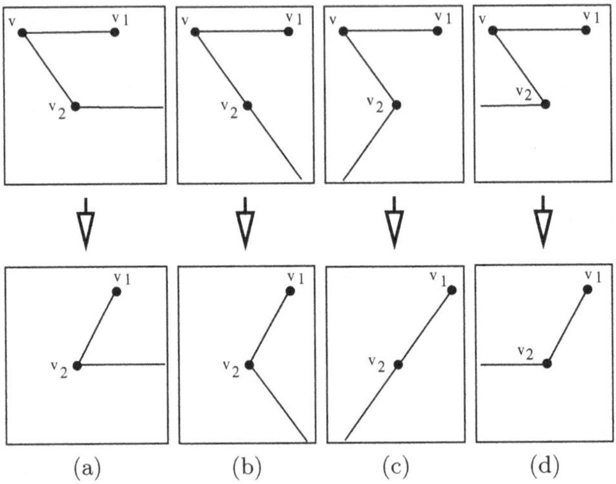

Fig. 5. The possible configurations around the vertex v_2 if the vertex v is type 1 convex and the edge $e_1 = v_1 v$ is horizontal (top) before and (bottom) after the replacement. (a) A type 2 convex vertex v_2 becomes type 1 convex. (b) A straight vertex v_2 becomes type 2 convex. (c) A type 2 concave vertex v_2 becomes straight. (d) A type 1 concave vertex v_2 becomes type 2 concave.

The type of the vertex v_2 also depends on the edge e_1 being horizontal or not. Similarly as for the vertex v_1, the change of the type of the vertex v_2 induced by the replacement increases the quantity $A(C)$ by $1/2$, as illustrated in Figs. 5 and 6.

Thus, the total change in the quantity $A(C)$ induced by this replacement is equal to $-1 + 1/2 + 1/2 = 0$, i.e., the proposition is true for the curve C.

2. If v is a type 2 convex vertex, then v is incident to two edge-adjacent interior triangles. Let e' be the common edge of the two triangles. The other endpoint of e' is the south-east neighbor v' of v. We distinguish several cases depending on the vertex v'.
   2a. If the vertex v' is adjacent on C to v_1, then v_1 is a type 1 convex vertex (see Fig. 2 (c)) and it can be replaced, together with its two incident edges $v'v_1$ and v_1v, with the edge $e' = v'v$. In the same way as for the type 1 convex vertex v above it can be shown that the replacement decreases the length of the curve by 1 and it does not change the quantity $A(C)$. A similar argument holds if the vertex v' is adjacent on C to the vertex v_2 (see Fig. 2 (d)). The vertex v' cannot be adjacent on C to both v_1 and v_2 because the length of the (simple closed) curve C is greater than 4.

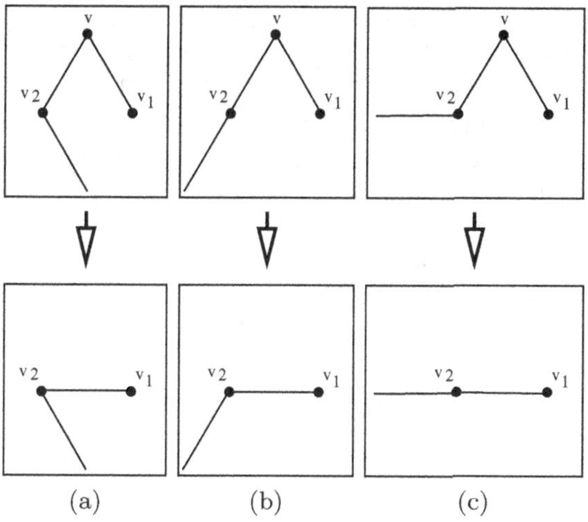

Fig. 6. The possible configurations around the vertex v_2 if the vertex v is type 1 convex and the edge $e_1 = v_1v$ is not horizontal (top) before and (bottom) after the replacement. (a) A type 2 convex vertex v_2 becomes type 1 convex. (b) A straight vertex v_2 becomes type 2 convex. (c) A type 2 concave vertex v_2 becomes straight.

2b. If the vertex v' is not adjacent on C to either v_1 or v_2 and v' is a type 1 concave vertex on the curve C (see Figs. 2 (e) and (f)), we denote e'_1 and e'_2 the previous and the next edge on C incident to v', respectively. We can bypass v' and its two incident edges e'_1 and e'_2, as we did with the type 1 convex vertex v above, replacing them with the edge connecting the other two endpoints of the edges e'_1 and e'_2. A similar analysis as above shows that the quantity $A(C)$ remains unchanged, and the length of the curve decreases by 1.

2c. If the vertex v' is not adjacent on C to either v_1 or v_2 and v' is a type 2 concave vertex on C (see Fig. 2 (g)), then the edges on C incident to v and those incident to v' are pairwise parallel. We split the curve C into two curves C_1 and C_2. The curve C_1 is obtained by connecting the vertex v to the vertex v', and taking the part of the curve C from the vertex v' and the edge e'_2 to the edge e_1 and the vertex v. The curve C_2 is obtained in a similar fashion by connecting the vertex v' to the vertex v, and taking the part of the curve C from the vertex v and the edge e_2 to the edge e'_1 and the vertex v'. Both curves C_1 and C_2 are simple closed counterclockwise oriented curves. They are both shorter than the curve C (at least by 2), and proposition is valid for the two curves. The sum $A(C_1) + A(C_2)$ of the quantities A of the two curves is 6. The split of the curve C into C_1 and C_2 induces the following changes:

- The type 2 convex vertex v is replaced by two type 1 convex vertices, one on the curve C_1 and the other on C_2, increasing the total count of the quantity $A(C)$ by $-1/2 + 1 + 1 = 3/2$.

- The type 2 concave vertex v' is replaced by two type 2 convex vertices, one on each of the curves C_1 and C_2, increasing the total count of the quantity $A(C)$ by $-(-1/2) + 1/2 + 1/2 = 3/2$.

Thus, the total increase in the quantity $A(C_1) + A(C_2)$ obtained by the vertex count is $3/2 + 3/2 = 3$, which exactly accounts for the increased number of curves (two curves C_1 and C_2 instead of one curve C) after the split.

2d. If the vertex v' is not on the curve C, we replace the vertex v and its two incident edges e_1 and e_2 in C with the vertex v' and the edges $v'v_1$ and $v'v_2$, respectively. The new curve is of the same length as C.
- The type 2 convex vertex v is replaced with the type 2 concave vertex v', decreasing the quantity $A(C)$ by 1.
- The vertices v_1 and v_2 loose one incident interior triangle each, increasing the quantity $A(C)$ in total by 1.

Thus, the total change in the quantity $A(C)$ induced by this replacement is $-1 + 1 = 0$.

The vertex v_1 has the same y coordinate as the vertex v, and it is the vertex with the smallest x coordinate among the vertices with that y coordinate. The vertex v_1 now takes the role of the vertex v: it is either a type 1 or a type 2 convex vertex. We process it in the same way as the vertex v. We either
- decrease the length of the curve (if the vertex v_1 is a type 1 convex vertex; if v_1 is a type 2 convex vertex with the south-east neighbor adjacent on C to one of the neighbors of v_1 on C; if v_1 is a type 2 convex vertex with a type 1 concave south-east neighbor not adjacent on C to either of the two neighbors of v_1 on C),
- split the curve in two (if v_1 is a type 2 convex vertex with the type 2 concave south-east neighbor), or
- eliminate the vertex v_1, keeping the length of the curve, making its east neighbor the vertex with the smallest x coordinate among those vertices with the largest y coordinate.

This process will continue until the curve is shortened or split in two shorter curves, or until the last vertex v'' on C with the same y coordinate as the vertices v and v_1 is reached. The vertex v'' is either a type 1 convex vertex, or it becomes a type 1 convex vertex after the final replacement of two edges in C with another two edges at the vertex following v'' on C. At this point, the vertex v'' is eliminated decreasing the length of C and maintaining the quantity $A(C)$.

4 Summary and Future Work

We gave a classification of convex and concave vertices on a simple closed curve in the triangular grid, based on the angle at which the curve turns at the vertex, i.e., based on the number of interior triangles incident to the vertex, thus refining the existing classification into salient and reentrant vertices. We showed a combinatorial relationship between the number of convex and concave vertices of different types.

We plan to extend this work to general (non-simple) closed curves and to curves in semiregular grids.

Acknowledgement. We thank the reviewers for reading the paper carefully, spotting some gaps in the first version of the proof and making constructive suggestions. This work has been partially supported by the Ministry of Education and Science of the Republic of Serbia within the Project No. 34014.

References

1. Bell, S.B.M., Holroyd, F.C., Mason, D.C.: A digital geometry for hexagonal pixels. Image Vis. Comput. **7**(3), 194–204 (1989)
2. Borgefors, G.: Distance transformations on hexagonal grids. Pattern Recogn. Lett. **9**(2), 97–105 (1989)
3. Borgefors, G., Sanniti di Baja, G.: Skeletonizing the distance transform on the hexagonal grid. In: 9th International Conference on Pattern Recognition, ICPR, pp. 504–507 (1988)
4. Bribiesca, E.: A new chain code. Pattern Recogn. **32**(2), 235–251 (1999)
5. Brimkov, V.E., Barneva, R.P.: Analytical honeycomb geometry for raster and volume graphics. Comput. J. **48**, 180–199 (2005)
6. Brimkov, V.E., Maimone, A., Nordo, G.: Counting gaps in binary pictures. In: Reulke, R., Eckardt, U., Flach, B., Knauer, U., Polthier, K. (eds.) IWCIA 2006. LNCS, vol. 4040, pp. 16–24. Springer, Heidelberg (2006). https://doi.org/10.1007/11774938_2
7. Brimkov, V.E., Maimone, A., Nordo, G., Barneva, R.P., Klette, R.: The number of gaps in binary pictures. In: Bebis, G., Boyle, R., Koracin, D., Parvin, B. (eds.) ISVC 2005. LNCS, vol. 3804, pp. 35–42. Springer, Heidelberg (2005). https://doi.org/10.1007/11595755_5
8. Brlek, S., Labelle, G., Lacasse, A.: A note on a result of Daurat and Nivat. In: De Felice, C., Restivo, A. (eds.) DLT 2005. LNCS, vol. 3572, pp. 189–198. Springer, Heidelberg (2005). https://doi.org/10.1007/11505877_17
9. Brlek, S., Labelle, G., Lacasse, A.: Properties of the contour path of discrete sets. Int. J. Found. Comput. Sci. **17**(3), 543–556 (2006)
10. Brlek, S., Provençal, X., Fedou, J.: On the tiling by translation problem. Discrete Appl. Math. **157**(3), 464–475 (2009)
11. Čomić, L.: Gaps and well-composed objects in the triangular grid. In: Marfil, R., Calderón, M., Díaz del Río, F., Real, P., Bandera, A. (eds.) CTIC 2019. LNCS, vol. 11382, pp. 54–67. Springer, Cham (2019). https://doi.org/10.1007/978-3-030-10828-1_5
12. Daurat, A., Nivat, M.: Salient and reentrant points of discrete sets. Electron. Notes Discrete Math. **12**, 208–219 (2003)
13. Daurat, A., Nivat, M.: Salient and reentrant points of discrete sets. Discrete Appl. Math. **151**(1–3), 106–121 (2005)
14. Deutsch, E.S.: On parallel operations on hexagonal arrays. IEEE Trans. Comput. **19**(10), 982–983 (1970)
15. Deutsch, E.S.: Thinning algorithms on rectangular, hexagonal, and triangular arrays. Commun. ACM **15**(9), 827–837 (1972)

16. Dutt, M., Andres, E., Largeteau-Skapin, G.: Characterization and generation of straight line segments on triangular cell grid. Pattern Recogn. Lett. **103**, 68–74 (2018)
17. Dutt, M., Biswas, A., Nagy, B.: Number of shortest paths in triangular grid for 1- and 2-neighborhoods. In: Barneva, R.P., Bhattacharya, B.B., Brimkov, V.E. (eds.) IWCIA 2015. LNCS, vol. 9448, pp. 115–124. Springer, Cham (2015). https://doi.org/10.1007/978-3-319-26145-4_9
18. Evako, A.V., Kopperman, R., Mukhin, Y.V.: Dimensional properties of graphs and digital spaces. J. Math. Imaging Vis. **6**(2–3), 109–119 (1996)
19. Freeman, H.: Computer processing of line-drawing images. ACM Comput. Surv. **6**(1), 57–97 (1974)
20. Freeman, H.: Algorithm for generating a digital straight line on a triangular grid. IEEE Trans. Comput. **28**(2), 150–152 (1979)
21. Golay, M.J.E.: Hexagonal parallel pattern transformations. IEEE Trans. Comput. **18**(8), 733–740 (1969)
22. Gray, S.: Local properties of binary images in two dimensions. IEEE Trans. Comput. **20**, 551–561 (1971)
23. Her, I.: Geometric transformations on the hexagonal grid. IEEE Trans. Image Process. **4**(9), 1213–1222 (1995)
24. Kardos, P., Palágyi, K.: Topology preservation on the triangular grid. Ann. Math. Artif. Intell. **75**(1–2), 53–68 (2015)
25. Kardos, P., Palágyi, K.: On topology preservation of mixed operators in triangular, square, and hexagonal grids. Discrete Appl. Math. **216**, 441–448 (2017)
26. Klette, R., Rosenfeld, A.: Digital Geometry. Geometric Methods for Digital Picture Analysis. Morgan Kaufmann Publishers, San Francisco, Amsterdam (2004)
27. Latecki, L.J., Eckhardt, U., Rosenfeld, A.: Well-composed sets. Comput. Vis. Image Underst. **61**(1), 70–83 (1995)
28. Nagy, B.: Cellular topology and topological coordinate systems on the hexagonal and on the triangular grids. Ann. Math. Artif. Intell. **75**(1–2), 117–134 (2015)
29. Nagy, B., Lukić, T.: Dense projection tomography on the triangular tiling. Fundam. Inform. **145**(2), 125–141 (2016)
30. Nagy, B., Strand, R.: Approximating Euclidean circles by neighbourhood sequences in a hexagonal grid. Theor. Comput. Sci. **412**(15), 1364–1377 (2011)
31. Sarkar, A., Biswas, A., Dutt, M., Bhowmick, P., Bhattacharya, B.B.: A linear-time algorithm to compute the triangular hull of a digital object. Discrete Appl. Math. **216**, 408–423 (2017)
32. Sarkar, A., Biswas, A., Dutt, M., Mondal, S.: Finding shortest triangular path and its family inside a digital object. Fundam. Inform. **159**(3), 297–325 (2018)
33. Sossa-Azuela, J.H., Cuevas-Jiménez, E.V., Zaldivar-Navarro, D.: Computation of the Euler number of a binary image composed of hexagonal cells. J. Appl. Res. Technol. **8**, 340–350 (2010)
34. Wiederhold, P., Morales, S.: Thinning on quadratic, triangular, and hexagonal cell complexes. In: Brimkov, V.E., Barneva, R.P., Hauptman, H.A. (eds.) IWCIA 2008. LNCS, vol. 4958, pp. 13–25. Springer, Heidelberg (2008). https://doi.org/10.1007/978-3-540-78275-9_2
35. Yip, B., Klette, R.: Angle counts for isothetic polygons and polyhedra. Pattern Recogn. Lett. **24**(9–10), 1275–1278 (2003)

Efficient Algorithms to Test Digital Convexity

Loïc Crombez(✉), Guilheme D. da Fonseca, and Yan Gérard

Université Clermont Auvergne and LIMOS, Clermont-Ferrand, France
lcrombez@isima.fr

Abstract. A set $S \subset \mathbb{Z}^d$ is *digital convex* if $\text{conv}(S) \cap \mathbb{Z}^d = S$, where $\text{conv}(S)$ denotes the convex hull of S. In this paper, we consider the algorithmic problem of testing whether a given set S of n lattice points is digital convex. Although convex hull computation requires $\Omega(n \log n)$ time even for dimension $d = 2$, we provide an algorithm for testing the digital convexity of $S \subset \mathbb{Z}^2$ in $O(n + h \log r)$ time, where h is the number of edges of the convex hull and r is the diameter of S. This main result is obtained by proving that if S is digital convex, then the well-known quickhull algorithm computes the convex hull of S in linear time. In fixed dimension d, we present the first polynomial algorithm to test digital convexity, as well as a simpler and more practical algorithm whose running time may not be polynomial in n for certain inputs.

Keywords: Convexity · Digital geometry

1 Introduction

Digital geometry is the field of mathematics that studies the geometry of points with integer coordinates, also known as *lattice points* [1]. Convexity is a fundamental concept in digital geometry, as well as in continuous geometry [2]. From a historical perspective, the study of digital convexity dates back to the works of Minkowski [3] and it is the main subject of the mathematical field of geometry of numbers.

While convexity has a unique well stated definition in any linear space, different definitions have been investigated in $\mathbb{Z}^2$ and $\mathbb{Z}^3$ [4–8]. In two dimensions, we encounter at least five different approaches, called respectively digital line, triangle, line [4], HV (for Horizontal and Vertical [9]), and Q (for Quadrant [10]) convexities. These definitions were created in order to guarantee that a digital convex set is connected (in terms of the induced grid subgraph), which simplifies several algorithmic problems.

The original definition of digital convexity in the geometry of number does not guarantee connectivity of the grid subgraph, but provides several other important mathematical properties, such as being preserved under certain affine

M. Couprie et al. (Eds.): DGCI 2019, LNCS 11414, pp. 409–419, 2019.
https://doi.org/10.1007/978-3-030-14085-4_32

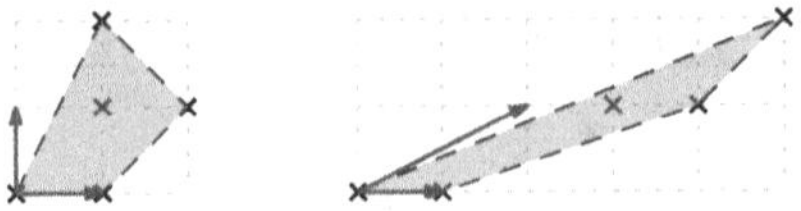

Fig. 1. Shearing a digital convex set. Example of a set whose connectivity is lost after a linear shear.

transformations (Fig. 1). The definition is the following. A set of lattice points $S \subset \mathbb{Z}^d$ is *digital convex* if $\text{conv}(S) \cap \mathbb{Z}^d = S$, where $\text{conv}(S)$ denotes the convex hull of S.

Herein, we consider the fundamental problem of verifying whether a given set of lattice points is digital convex.

Problem `TestConvexity`(S)
Input: Set $S \subset \mathbb{Z}^d$ of n lattice points given by their coordinates.
Output: Determine whether S is digital convex or not.

The input of `TestConvexity`(S) is an unstructured finite lattice set (without repeating elements). Related work considered more structured data in dimension 2, in which S is assumed to be connected. The *contour* of a connected set S of lattice points is the ordered list of the points of S having a grid neighbor outside S. When S is connected, it is possible to represent S by its contour, either directly as in [11] or encoded as binary word [12]. The algorithms presented in [11,12] test digital convexity in linear time on the respective input representations.

Our work, however, does not make any assumption on S being connected, or any particular ordering of the input. In this setting, a naive approach to test the digital convexity is:

1. Compute the convex hull $\text{conv}(S)$ of the n lattice points of S.
2. Compute the number n' of lattice points inside the convex hull of S.
3. If $n = n'$, then S is convex. Otherwise, it is not.

Step 1 consists of computing the convex hull of n points. The field of computational geometry provides a plethora of algorithms to compute the convex hull of a finite set $S \subset \mathbb{R}^d$ of n points [13]. The fastest algorithms for dimensions 2 and 3 take $O(n \log n)$ time [14], which matches the lower bound in the algebraic decision tree model of computation [15]. In dimension $d \leq 3$, if we also take into consideration the output size h, i.e. the number of vertices of the convex hull, the fastest algorithms take $O(n \log h)$ time [16,17]. Some polytopes with n vertices (e.g., the cyclic polytope) have $\Theta(n^{\lfloor (d-1)/2 \rfloor})$ facets. Therefore, any algorithm that outputs this facet description of the convex hull requires $\Omega(n^{\lfloor (d-1)/2 \rfloor})$ time. Optimal algorithms to compute the convex hull in dimension $d \geq 4$ match this lower bound [18].

Step 2 consists of computing the number of lattice points inside a polytope (represented by its vertices), which is a well studied problem. In dimension 2, it can be solved using Pick's formula [19]. The question has been widely investigated in the framework of the geometry of numbers, from Ehrhart theory [20] to Barvinok's algorithm [21]. Currently best known algorithms have a complexity of $O(n^{O(d)})$ for fixed dimension d [22]. As conclusion, the time complexity of this naive approach is at least the one of the computation of the convex hull.

1.1 Results

In Sect. 2, we consider the 2-dimensional version of the problem and show that the convex hull of digital convex sets can be computed in linear time. Our main result is an algorithm for dimension $d = 2$ to solve `TestConvexity`(S) in $O(n + h \log r)$ time, where h is the number of edges of the convex hull and r is the diameter of S.

In Sect. 3, we consider the problem in fixed dimension d. We present the first polynomial-time algorithm to test digital convexity, as well as a simpler and more practical algorithm whose running time may not be polynomial in n for certain inputs.

2 Digital Convexity in 2 Dimensions

The purpose of this section is to provide an algorithm to test the convexity of a finite lattice $S \subset \mathbb{Z}^2$ in linear time in n. To this endeavour, we show that the convex hull of a digital convex set S can be computed in linear time. In fact, we show that this linear running time is achieved by the well-known quickhull algorithm [23].

Quickhull is one the many early algorithms to compute the convex hull in dimension 2. Its worst case time is $O(n^2)$, which makes it generally less attractive than the $O(n \log n)$ algorithm. However for certain inputs and variations of the algorithm, the average time complexity is reduced to $O(n \log n)$ or $O(n)$ [13,24].

The quickhull algorithm starts by initializing a convex polygon in the following manner. First it computes the top-most and bottom-most points of the set. Then it computes the two extreme points in the normal direction of the line supported by the top-most and bottom-most points. Those four points describe a convex polygon that we call a *partial hull*, which is contained inside the convex hull of S. The points contained in the interior of the partial hull are discarded. Furthermore, horizontal lines and lines parallel to the top-most to bottom-most line passing through these points describe an outlying bounding box in which the convex hull lies (Fig. 2).

The algorithm adds vertices to the partial hull until it obtains the actual convex hull. This is done by inserting new vertices in the partial hull one by one. Given an edge of the partial hull, let v denote its outwards normal vector. The algorithm searches for the extreme point in direction v. If this point is already an edge point, then the edge is part of the convex hull. Otherwise, we insert the farthest point found between the two edge vertices, discarding the points that are inside the new partial hull. Throughout this paper, we call a *step* of the quickhull algorithm the computation of the farthest point of every edge for a given partial hull. When adding new vertices to the partial hull, the region inside the partial hull expands. Points inside that expansion are discarded by

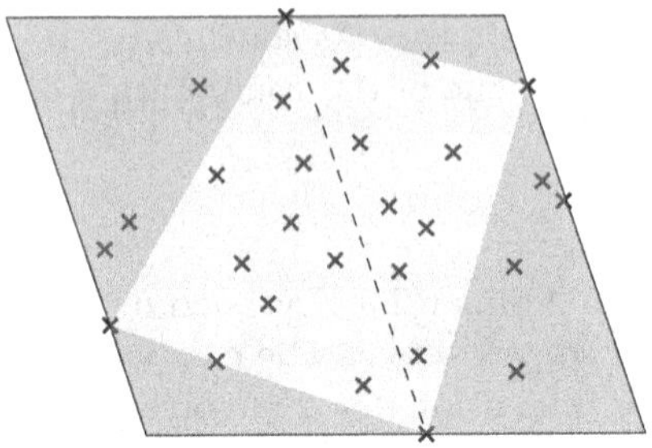

Fig. 2. Quickhull initialization. Points inside the partial hull (light brown) are discarded. The remaining points are potentially part of the hull. (Color figure online)

quickhull and herein we name this region *discarded region*. The points that still lie outside the partial hull are preserved, and we call the region within which points might still lies *preserved region* (Fig. 3).

We show that quickhull steps takes linear time and that at each step half of the remaining input points of the convex hull is discarded. Therefore, as in standard decimation algorithms, the total running time remains linear. In Sect. 2.2, we explain how to use this algorithm to test the digital convexity of any lattice set in linear time in n.

Theorem 1. *If the input is a digital convex set of* n *points, then QuickHull has* $O(n)$ *time and space complexities.*

2.1 Proof of Theorem 1

We prove Theorem 1 with the help of the following lemma.

Lemma 1. *The area of the discarded region is larger than the area of the preserved region.*

Proof. Consider one step of the algorithm: Let ab be the edge associated to the step. When a was added to the hull, it was as the farthest point in a given direction. Hence, there is no point behind the line orthogonal to this direction going through a. (Fig. 3b). The same can be said for b. Let c be the intersection point of those two lines. Every point that lies within $\triangle abc$ will be fed to the following steps. At this step, we are looking for the point that is the farthest from the supporting line of ab and outside the partial hull (let that point be d) (Fig. 4). Let e and f be the intersections between the line parallel to ab going through d, and respectively ac and bc. There are no points from S inside the triangle $\triangle cef$. Adding d to the partial hull creates two other edges to further be treated: one with ad as an edge that will be fed the points inside $\triangle ade$ and one with bd as the edge that will be fed the points inside $\triangle bdf$. The triangle $\triangle abd$ lies within the partial hull, therefore $\triangle abd$ is the region in which points are discarded (Fig. 4).

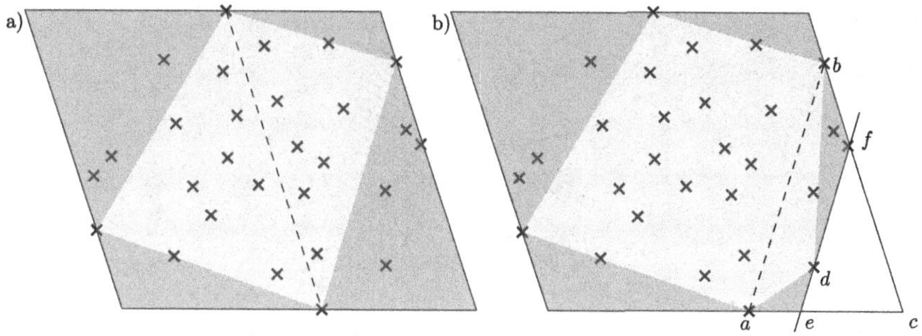

Fig. 3. Quickhull regions. The preserved region (region in which we look for the next vertex to be added to the partial hull) is a triangle. This stays true when adding new vertices to the hull (as shown here in the bottom right corner). The partial hull (whose interior is shown in light brown) grows at each vertex insertions to the partial hull. The new region added to the partial hull is called discarded region. (Color figure online)

We established that the preserved lattice points are the lattice points within $\triangle ade$ and $\triangle bdf$. Also the discarded lattice points are those within $\triangle abd$. Let c_1 be the middle of ad and c_2 be the middle of bd. As shown in Fig. 4, the symmetrical of $\triangle ade$ and $\triangle bdf$ through respectively c_1 and c_2 both lie inside $\triangle abd$ and do not intersect each other. Hence $\triangle abd$ is larger in terms of area than $\triangle aed \cup \triangle bdf$. □

Remark 1. Pick's formula does not apply here since all vertices of the triangle (namely c in Fig. 4) are not necessarily lattice points.

Remark 2. As there is no direct relation between the area of a triangle and the number of lattice points inside it, this result is not sufficient to conclude that a constant proportion of points are discarded at each step.

Corollary 1. *The reflection of lattice points inside $\triangle aed$ and $\triangle bdf$ across respectively c_1 and c_2 are lattice points.*

Proof. The points a, b, d are lattice points so c_1 and c_2 (middle of respectively ad and bd) have their coordinates in multiple of half integers. Hence the reflection of a lattice point across c_1 or c_2 is a lattice point. Therefore, every lattice point within $\triangle aed$ has a lattice point reflection across c_1 within $\triangle ae_1d$ and every lattice point within $\triangle bfd$ has a lattice point reflection across c_2 within $\triangle bf_2d$. □

Remark 3. This previous result would prove that half the points are discarded at each step if it were not for the lattice points on the diagonals ad and bd.

We will now show that quickhull discards at least half of the remaining points at each step, hence proving Theorem 1

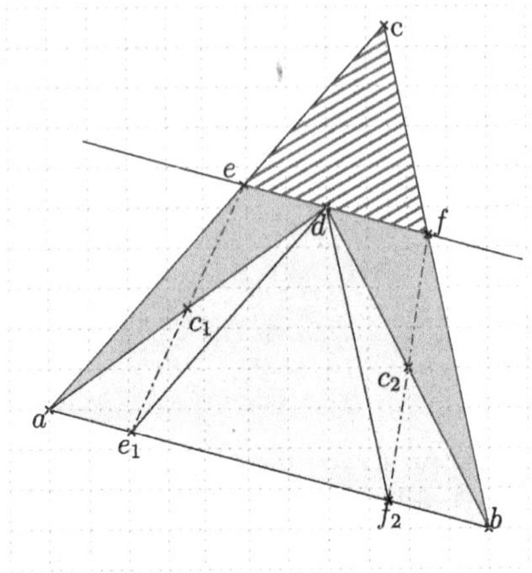

Fig. 4. Symmetrical regions. The next step of the algorithm will only be fed the points inside the dark brown regions (search regions). Each lattice points inside the light brown region (discarded region) is inside the partial hull and is therefore discarded. Each search region (in dark brown) has a symmetrical region (either through c_1 or c_2) that lies inside the discarded region. Furthermore, this symmetrical transformation also preserve lattice points. (Color figure online)

Proof. We established in Corollary 1 that lattice points inside the search regions ($\triangle aed$ and $\triangle bfd$) have symmetrical counterparts inside the discarded region (more precisely inside $\triangle ae_1d$ and $\triangle bf_2d$) (Fig. 4). By preserving each points inside $\triangle aed$ and $\triangle bfd$ at each step, we do not have a discarded symmetrical counterpart for the lattice points lying on ad and bd. But we do not need to preserve those points, since ad and bd are at this step edges of the partial hull. Removing lattice points from ad and bd implies that in the following step there will be no lattice points on ab, leaving lattice points on ef without a discarded symmetrical counterparts (Fig. 5).

Let actually discard every points on ef, since they all are equally farthest from ab in the outer direction, they all belong to the hull. Hence we can add the first and last lattice point on ef to the partial hull (Fig. 5). Note that this only takes linear time and does not change the time complexity of each individual step. Hence, at each step of quickhull, for every preserved points there is at least a discarded point. Consequently, the number of operations is proportional to $n \sum_{i=0}^{\infty} (\frac{1}{2})^i = 2n$ and quickhull takes linear time for digital convex sets. □

2.2 Determining the Digital Convexity of a Set

We showed in Theorem 1 that the quickhull algorithm computes the convex hull of digital convex sets in linear time thanks to the fact that at each step quickhull discards at least half of the remaining points. By running quickhull on any given set S, and stopping the computation if any step of the algorithm discards less

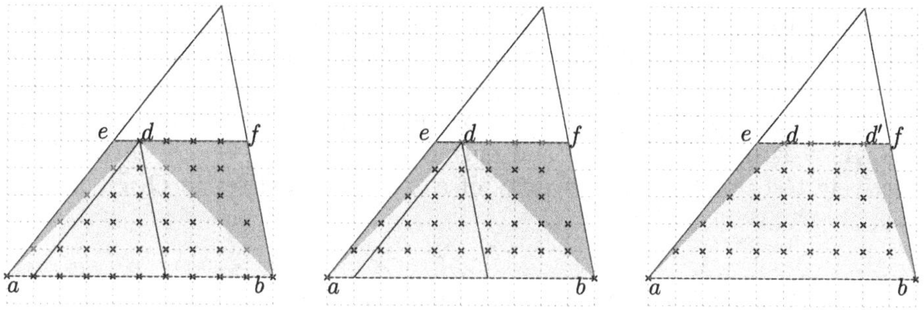

Fig. 5. Lonely points. The lattice points without discarded symmetrical counterparts are shown in red. On the left: if every points inside the triangle is preserved, and in the center: if the points on the edges of the partial hull are discarded. Finally on the right a visualization of what happens if we discard all the farthest points and update the partial hull accordingly. (Color figure online)

than half of the remaining points, we ensure both that the running time is linear, and that if S is digital convex, quickhull finishes and we get the convex hull of S. If the computation finishes for S, we still need to test its digital convexity. To do so, we use the previously computed convex hull and compute $|conv(S) \cap \mathbb{Z}^2|$ using Pick's formula [19]. The set S is digital convex if $|conv(S) \cap \mathbb{Z}^2| = |S|$. Hence the resulting Algorithm 1.

Algorithm 1. isDigitalConvex(S)

Input: S a set of points
Output: true if S is digital convex, false if not.

```
while S is not empty do
    Run one step of the quickhull algorithm on S
    if quickhull discarded less than half the remaning points of S then
        return false
Compute |conv(S) ∩ Z^2|
if |conv(S) ∩ Z^2| > |S| then
    return false
return true
```

Theorem 2. *Algorithm 1 tests digital convexity of any 2 dimensional set S, and runs in $O(n + h \log r)$ time, where h is the number of edges of $conv(S)$ and r is the diameter of S.*

Proof. As Algorithm 1 runs quickhull, but stops as soon as less than half the remaining points have been removed, the running time of the quickhull part is bounded by the series $n \sum_{i=0}^{\infty} (\frac{1}{2})^i = 2n$, and is hence linear. Thanks to Theorem 1 we know that the computation of quickhull will not stop for any digital convex sets. Computing $|conv(S) \cap \mathbb{Z}^2|$ using Pick's formula requires the computation of the area of conv(S) and of the number of lattice points lying on its boundary, which requires the computation of a greatest common divisor. Hence this takes $O(h \log r)$ time where h is the number of edge of $conv(S)$ and r is the diameter of S. As S is digital convex if and only if $|S| = |conv(S) \cap \mathbb{Z}^2|$, Algorithm 1 effectively tests the digital convexity of a 2 dimensional set in $O(n + h \log r)$ time. □

3 Test Digital Convexity in Dimension *d*

We provide two algorithms for verifying the digital convexity in any fixed dimension.

3.1 Naive Algorithm

The naive algorithm mentioned in the Introduction is based on the following equivalence: the set $S \subset \mathbb{Z}^d$ is digital convex if and only if its cardinality is equal to the cardinality of $conv(S) \cap \mathbb{Z}^d$. In Step 1, we compute the convex hull of S (in $O(n \log n + n^{\lfloor \frac{d}{2} \rfloor})$ time [18]). In Step 2, we need to count the number of integer points inside conv(S). The classical algorithm to achieve this goal is known as Barvinok algorithm [21]. This approach determines only the number of missing points. If we want to enumerate the points, it is possible to do so through a formal computation of the generating functions used in Barvinok algorithm.

Theorem 3. *The naive algorithm tests digital convexity in any fixed dimension d and runs in polynomial time.*

Proof. Computing the convex hull of any set can be done in $O(n \log n + n^{\lfloor \frac{d}{2} \rfloor})$ time [18]). Counting lattice points inside a convex lattice polytope can be done in polynomial time [22]. A direct consequence of the digital convexity definition is that a set $S \subset \mathbb{Z}^d$ is digital convex if and only if $|S| = |conv(S) \cap \mathbb{Z}^d|$, hence the naive algorithm tests digital convexity in any fixed dimension d and runs in polynomial time. □

3.2 Alternative Algorithm

This new algorithm computes all integer points in the convex hull of S with a more direct approach. Its principle is to enumerate the points x of a finite lattice set $S' \subset \mathbb{Z}^d$ surrounding $conv(S) \cap \mathbb{Z}^d$ ($conv(S) \cap \mathbb{Z}^d \subset S'$). In a first variant, we count the number of points of S' belonging to $conv(S)$. At the end, the set S is convex if and only if $|conv(S) \cap \mathbb{Z}^d|$ is equal to the cardinality of S. In a second variant, for each point of S', we test whether it belongs to S and in the negative case, we test whether it belongs to the convex hull of S. If a point of $S' \setminus S \cap conv(S)$ is found, then S is not convex (Fig. 6).

We define the set S' as the set of points $x \in \mathbb{Z}^d$ such that the cube $x + [-\frac{1}{2}, \frac{1}{2}]^d$ has a nonempty intersection with the convex hull of S, where $+$ denotes the

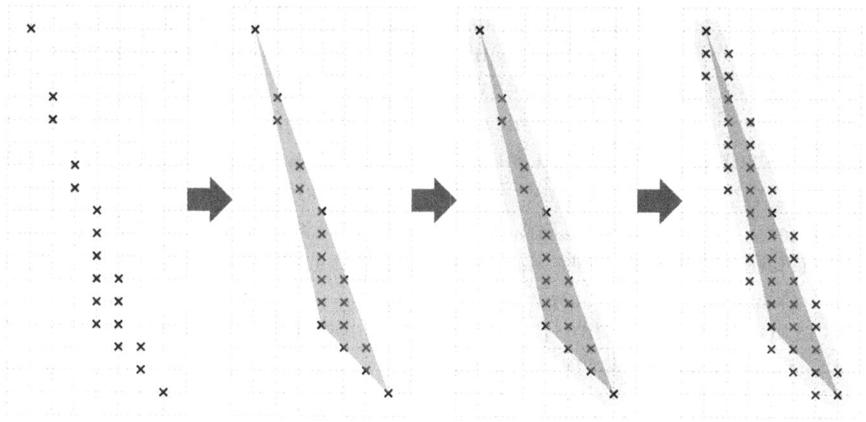

Fig. 6. Practical algorithm. A lattice set S, its convex hull and its dilation by a centered cube of side 1. The intersection of $\text{conv}(S) + [-\frac{1}{2}, \frac{1}{2}]^d$ with the lattice is the set S'. It is $2d$-connected and contains the convex hull of S. The principle of the algorithm is either to count the points of S' in $\text{conv}(S)$ (variant 1) or to search for a point of $S' \setminus S$ (blue points) in the convex hull of S (variant 2). (Color figure online)

Minkowski sum. It can be easily proved that S' is $2d$-connected (the $2d$ neighbors of a lattice point $x \in \mathbb{Z}^d$ are the $2d$ integer points at Euclidean distance 1) and by construction, it contains S. The graph structure induced by the $2d$-connectivity on S' allows to visit all the points of S' efficiently: for each point $x \in S'$, we consider its $2d$ neighbors and test whether they belong to S'. If they do, we add them to the stack of the remaining points of S'. The goal is to test whether a point of $S' \setminus S$ is in the convex hull of S.

Then the algorithm has two main routines:

- $\texttt{InConvexHull}_S$ tests whether a given point $x \in \mathbb{R}^d$ belongs to the convex hull of S. It is equivalent with testing whether there exists a hyperplane separating x from the points of S. It can be done by linear programming with a worst-case time complexity of $O(n)$ for fixed dimension d [13].
- $\texttt{InConvexHull}_{S+[-\frac{1}{2},\frac{1}{2}]^d}$ tests whether a given point x belongs to the convex hull of $S+[-\frac{1}{2}, \frac{1}{2}]^d$. It follows the same principle as $\texttt{InConvexHull}_S$ with $2^d n$ points. The time complexity remains linear in fixed dimension. This routine is used to test whether an integer point belongs to S'.

The algorithm is the following. First, we create a stack T of the points of S' to visit and initialize it with the set S. For each point x in T, we remove it from the stack T and label it as already visited. Then, we consider its $2d$ neighbors x'. If x' belongs to S' and has not been visited previously, we add it in the stack T. We test whether x belongs to $conv(S)$ and increment the cardinality of $conv(S) \cap \mathbb{Z}^d$ accordingly (variant 1) or test whether x is in S and $conv(S)$ and return `S not convex` if $x \in conv(S) \setminus S$ (variant 2).

The running time is strongly dependent on the cardinality of S'. It is $O(n|S'|)$. If the size of S' is of the same magnitude as the initial set, the algorithm runs in $O(n^2)$ time. It is unfortunately not possible to bound $|S'|$ as a function of n. The ratio $\frac{|S'|}{|S|}$ can go to infinity. It is easy to build such an example with a set S consisting of only two lattice points, for instance for any $k \in \mathbb{Z}$ the set $S = \{(0,0); (1,2k)\}$ induces $\frac{|S'|}{|S|} \geq k$. A direction of improvement could be to consider a linear transformation of the lattice $\mathbb{Z}^d$ in order to obtain a more compact lattice set and then a lower ratio $\frac{|S'|}{|S|}$. LLL algorithm [25] could be useful to achieve this goal in future work.

As in the naive algorithm, a variant of this approach can be easily developed in order to enumerate the missing points.

4 Perspectives

In this paper, we presented an algorithm to test digital convexity in time linear in n for dimension $d = 2$. In higher dimensions, our running time depends on the complexity of general convex hull algorithms. The questions of whether digital convexity can be tested in linear time in 3 dimensions, or faster than convex hull computation in arbitrary dimensions remain open. A tentative approach consists of changing the lattice base, in order to obtain certain connectivity properties.

We showed that the convex hull of a digital convex set in dimension 2 can be computed in linear time. Can the convex hull of digital convex sets be computed in linear time in dimension 3, or more generally, what is the complexity of convex hull computation of a digital convex set in any fixed dimension? We note that the number of faces of any digital convex set in d dimensions is $O(V^{(d-1)/(d+1)})$, where V is the volume of the polytope [26,27]. Therefore, the lower bound of $\Omega(n^{\lfloor (d-1)/2 \rfloor})$ for the complexity of the convex hull of arbitrary polytopes does not hold for digital convex sets.

Acknowledgement. This work has been sponsored by the French government research program "Investissements d'Avenir" through the IDEX-ISITE initiative 16-IDEX-0001 (CAP 20-25).

References

1. Klette, R., Rosenfeld, A.: Digital Geometry: Geometric Methods for Digital Picture Analysis. Elsevier, Amsterdam (2004)
2. Ronse, C.: A bibliography on digital and computational convexity (1961–1988). IEEE Trans. Pattern Anal. Mach. Intell. **11**(2), 181–190 (1989)
3. Minkowski, H.: Geometrie der Zahlen, vol. 2. B.G. Teubner, Leipzig (1910)
4. Kim, C.E., Rosenfeld, A.: Digital straight lines and convexity of digital regions. IEEE Trans. Pattern Anal. Mach. Intell. **4**(2), 149–153 (1982)
5. Kim, C.E., Rosenfeld, A.: Convex digital solids. IEEE Trans. Pattern Anal. Mach. Intell. **4**(6), 612–618 (1982)

6. Chassery, J.-M.: Discrete convexity: definition, parametrization, and compatibility with continuous convexity. Comput. Vis. Graph. Image Process. **21**(3), 326–344 (1983)
7. Kishimoto, K.: Characterizing digital convexity and straightness in terms of length and total absolute curvature. Comput. Vis. Image Underst. **63**(2), 326–333 (1996)
8. Chaudhuri, B.B., Rosenfeld, A.: On the computation of the digital convex hull and circular hull of a digital region. Pattern Recognit. **31**(12), 2007–2016 (1998)
9. Barcucci, E., Del Lungo, A., Nivat, M., Pinzani, R.: Reconstructing convex polyominoes from horizontal and vertical projections. Theoret. Comput. Sci. **155**(2), 321–347 (1996)
10. Daurat, A.: Salient points of Q-convex sets. Int. J. Pattern Recognit. Artif. Intell. **15**(7), 1023–1030 (2001)
11. Debled-Rennesson, I., Rémy, J.-L., Rouyer-Degli, J.: Detection of the discrete convexity of polyominoes. Discrete Appl. Math. **125**(1), 115–133 (2003). 9th International Conference on Discrete Geometry for Computer Im agery (DGCI 2000)
12. Provençal, X., Reutenauer, C., Brlek, S., Lachaud, J.-O.: Lyndon + Christoffel = digitally convex. Pattern Recognit. **42**(10), 2239–2246 (2009). Selected papers from the 14th IAPR International Conference on Discrete Geometry for Computer Imagery (2008)
13. de Berg, M., Cheong, O., van Kreveld, M., Overmars, M.: Computational Geometry: Algorithms and Applications, 3rd edn. Springer, Santa Clara (2008). https://doi.org/10.1007/978-3-540-77974-2
14. Yao, A.C.-C.: A lower bound to finding convex hulls. J. ACM **28**(4), 780–787 (1981)
15. Preparata, F.P., Hong, S.J.: Convex hulls of finite sets of points in two and three dimensions. Commun. ACM **20**(2), 87–93 (1977)
16. Kirkpatrick, D., Seidel, R.: The ultimate planar convex hull algorithm? SIAM J. Comput. **15**(1), 287–299 (1986)
17. Chan, T.M.: Optimal output-sensitive convex hull algorithms in two and three dimensions. Discrete Comput. Geom. **16**(4), 361–368 (1996)
18. Chazelle, B.: An optimal convex hull algorithm in any fixed dimension. Discrete Comput. Geom. **10**(4), 377–409 (1993)
19. Pick, G.: Geometrisches zur zahlenlehre. Sitzungsberichte des Deutschen Naturwissenschaftlich-Medicinischen Vereines für Böhmen "Lotos" in Prag, vol. 47–48, pp. 1899–1900 (1899)
20. Ehrhart, E.: Sur les polyèdres rationnels homothétiques à n dimensions. Technical report, académie des sciences, Paris (1962)
21. Barvinok, A.I.: A polynomial time algorithm for counting integral points in polyhedra when the dimension is fixed. Math. Oper. Res. **19**(4), 769–779 (1994)
22. Barvinok, A.I.: Computing the Ehrhart polynomial of a convex lattice polytope. Discrete Comput. Geom. **12**(1), 35–48 (1994)
23. Barber, C.B., Dobkin, D.P., Huhdanpaa, H.: The quickhull algorithm for convex hulls. ACM Trans. Math. Softw. **22**, 469–483 (1996)
24. Greenfield, J.S.: A proof for a quickhull algorithm. Technical report, Syracuse University (1990)
25. Lenstra, A.K., Lenstra, H.W., Lovasz, L.: Factoring polynomials with rational coefficients. Math. Ann. **261**(4), 515–534 (1982)
26. Andrews, G.E.: A lower bound for the volumes of strictly convex bodies with many boundary points. Trans. Am. Math. Soc. **106**, 270–279 (1963)
27. Bárány, I.: Extremal problems for convex lattice polytopes: a survey. Contemp. Math. **453**, 87–103 (2008)

Compact Packings of the Plane with Three Sizes of Discs

Thomas Fernique[1(✉)], Amir Hashemi[2], and Olga Sizova[3,4]

[1] Univ. Paris 13, CNRS, Sorbonne Paris Cité, UMR 7030, 93430 Villetaneuse, France
thomas.fernique@lipn.univ-paris13.fr
[2] Department of Mathematical Sciences, Isfahan University of Technology, Isfahan, Iran
[3] Faculty of Mathematics, Higher School of Economics, 119048 Moscow, Russia
[4] Semenov Institute of Chemical Physics, 119991 Moscow, Russia

Abstract. A compact packing is a set of non-overlapping discs where all the holes between discs are curvilinear triangles. There is only one compact packing by discs of radius 1. There are exactly 9 values of r which allow a compact packing with discs of radius 1 and r. It has been proven that at most 11462 pairs (r, s) allow a compact packing with discs of radius 1, r and s. We prove that there are exactly 164 such pairs.

Keywords: Compact packing · Disc packing · Tiling

1 Introduction

A *packing* of the plane by discs is a set of interior-disjoint discs. Packings are of special interest to model the structure of materials, *e.g.*, crystals or granular materials, and the goal in this context is to understand which typical or extremal properties have the packings. In 1964, Tóth coined the notion of *compact packing* [4]: this is a packing such that the graph which connects the center of mutually tangent discs is triangulated. Equivalently, a packing is compact when all its holes are curvilinear triangles.

There is only one compact packing by one disc, called the *hexagonal compact packing*, where the disc centers are located on the triangular grid. In [9], it has been proven that there are exactly 9 values of r which allow a compact packing with discs of radius 1 and r (Fig. 1 – all already appeared in [4], except the fifth which later appeared in [11] and the second which was new at this time). Recently, it was proven in [12] that there are at most 11462 pairs (r, s) which allow a compact packing by discs of radii 1, r and s. The author provided several examples and suggested that a complete characterization of all the possible pairs

The work of O. S was supported within frameworks of the state task for ICP RAS 0082-2014-0001 (state registration AAAA-A17-117040610310-6). The work of Th. F and A. H was supported by the Partenariat Hubert Curien (PHC) Gundishapur.

M. Couprie et al. (Eds.): DGCI 2019, LNCS 11414, pp. 420–431, 2019.
https://doi.org/10.1007/978-3-030-14085-4_33

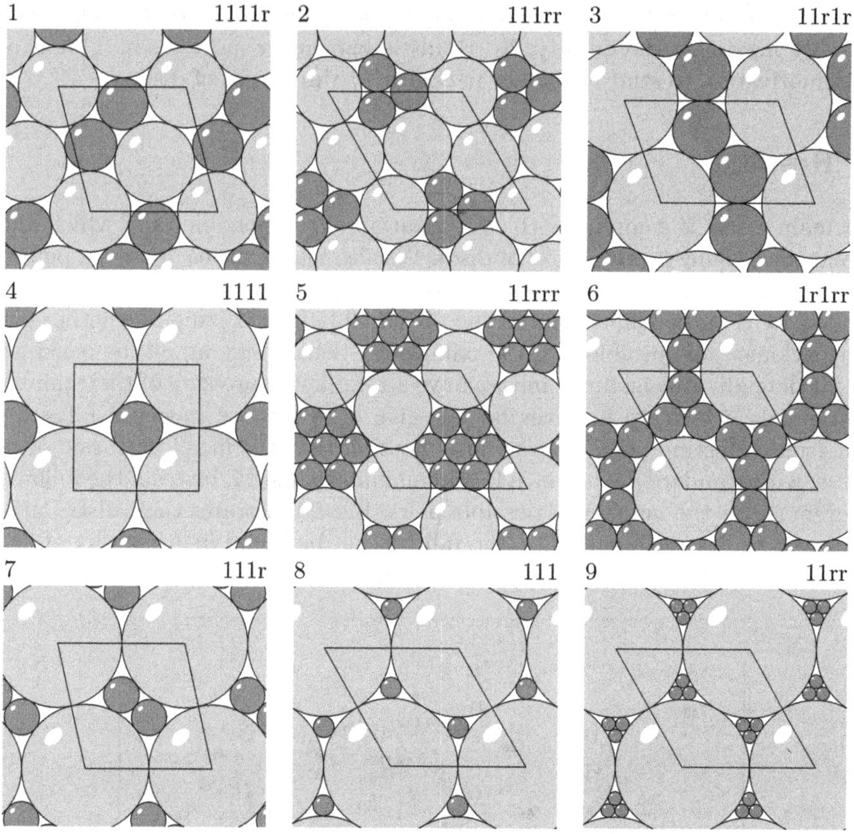

Fig. 1. Compact packing with discs of radii 1 and $r < 1$ for each of the 9 possible values of r. They are all periodic, with the parallelogram showing a fundamental domain. The top-right word over $\{1, r\}$ gives the sequence of radii of the discs around a small disc.

is probably beyond the actual capacity of computers. Little is know for more discs, in particular it is still open whether there are always finitely many possible radii.

A classic problem in packing theory is to find the maximal density, defined by

$$\delta := \sup_{\text{packings}} \limsup_{k\to\infty} \frac{\text{covered area the } k\times k \text{ square}}{k^2}.$$

For packings with only one disc, the maximal density has been proven in [3] to be $\delta_1 := \pi/\sqrt{12} \simeq 0.9069$, attained for the unique compact packing. For packings with discs of size 1 and r, the maximal density $\delta(r)$ is arbitrarily close to $\delta_1 + (1-\delta_1)\delta_1 \simeq 0.9913$ for r small enough, and it has been proven in [1] to be still equal to δ_1 for $r \geq 0.7429$. An upper bound on $\delta(r)$ has been proven in [6] for any r, but the only exact results correspond to radii which allow a compact packing, namely those on Fig. 1, except the fifth and the last ones for which no proof has been given, see [7,8,10]. Compact packings thus seem to be good candidates to provably maximize the density. No result is known for more discs. This is our main motivation to study compact packings by three sizes of discs.

2 Results

Our main result is a complete characterization of the possible radii which allow a compact packing by three size of discs. We also found, for each case, a periodic compact packing. The characterization of the compact packings which maximize the density is still open, as well as the presumably harder issue of whether compact packings (for suitable radii) maximize the density among *all* the packings[1].

All the radii are algebraic and we have an explicit expression of their minimal polynomials, which we however do not give here because it would fill several rather uninteresting pages. Let us just mention that the mean algebraic degree is 6.08, with standard deviation 4.45 and maximal value 24. Instead, the following theorem states the number of possible pairs, Fig. 2 illustrates their distribution, and Fig. 3 shows some packings (the full list can be found in Appendix of [5]).

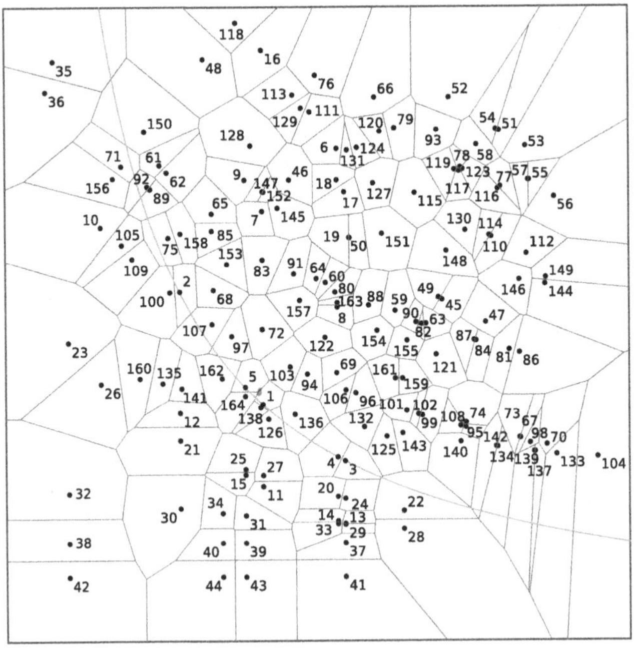

Fig. 2. Distribution of the 164 pairs (r, s), with abscissa r and ordinate $\frac{s}{r}$. Pairs below the hyperbola are such that a small disc fits in the hole between three large discs. Voronoï cells aim to give an idea of how close are two pairs.

[1] This does not hold when small discs can be added in holes between larger discs, breaking the compacity but increasing the density, see, *e.g.*, packing 33 on Fig. 3.

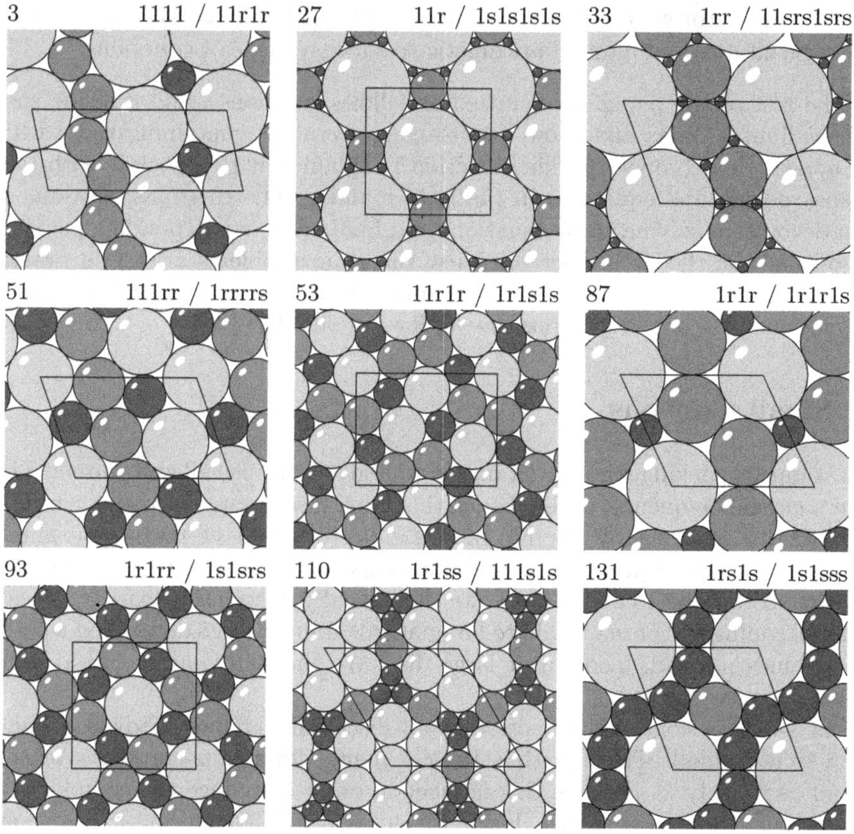

Fig. 3. Some compact packings with three sizes of discs. Those on the first line are easily derived from compact packing with two sizes of discs. The other ones are more original. The top-left number refers to numbers on Fig. 2, and the two top-right words over $\{1, r, s\}$ give a possible sequence of radii of discs around, respectively, a disc of radius s and one of radius r (small and medium corona, see Sects. 3 and 4).

Theorem 1. *There are exactly* 164 *pairs* (r, s) *which allow a compact packing by discs of radii* 1, r *and* s, *with* $0 < s < r < 1$.

The examples given in [12] correspond on Fig. 2 to numbers 143, 145, 146, 144, 104 and 99. Number 107 also already appeared in [4] (p. 187). Number 51 (depicted on Fig. 3) can be seen on the floor in front of the library of Weggis, Switzerland (4 Luzernerstrasse). Other examples probably appear elsewhere. The challenge is to find them all.

This is rather simple in theory: as we shall see, we just need basic trigonometry and solving systems of polynomial equations. The challenge is computational: there are many cases to consider and the equations are very complicated. To get over the barrier encountered in [12], the main new ingredient we introduced are:

- use combinatorics to reduce the number of cases before computing;
- use resultants and interval arithmetic to solve systems of equations.

The rest of the paper is organized as follows. Sections 3 and 4 define small and medium coronas and show how to compute a list containing those which can appear in a compact packing. Section 5 explain how to associate with these coronas polynomial equations in the radii r and s. The three last sections are then devoted to solving these equations and finding the corresponding packings. We outline the theory in Sect. 6, review the main problems arising in practice in Sect. 7 and detail how we solved them in Sect. 8. The presentation is a bit sketchy due to space limitations, see [5] for full details.

3 Small Coronas

In a compact packing, the discs tangent to a given disc is called a *corona*. They form a circular sequence, ordered such that each disc is tangent to the next one. We *code* this sequence by the disc radii, *i.e.*, a word over $\{1, \mathrm{r}, \mathrm{s}\}$ (see examples on Figs. 1 and 3). Since circular permutations or reversal of this word encode the same corona up to an isometry, we shall usually choose the lexicographically minimal coding. A corona is said to be small, medium or large depending whether it surrounds a small, medium or large disc. We shall also call x-disc a disc of radius x, and x-corona

An s-corona contains at most 6 small discs, and at most 5 discs if one of them is not small. There is thus finitely many different s-coronas. We shall bound as sharply as possible this number in order to reduce the complexity of the further exhaustive search. For this, let us define for $0 < s < r < 1$ and $\boldsymbol{k} = (k_1, \ldots, k_6) \in \mathbb{N}^6$ the function

$$S_{\boldsymbol{k}}(r,s) := k_1 \widehat{1s1} + k_2 \widehat{1sr} + k_3 \widehat{1ss} + k_4 \widehat{rsr} + k_5 \widehat{rss} + k_6 \widehat{sss},$$

where $\widehat{xyz}$ denotes, in the triangle which connects the centers of mutually tangent discs of radii x, y and z, the angle at the center of the disc of radius y. The function $S_{\boldsymbol{k}}$ counts the angles to pass from disc to disc in an s-corona. To each s-corona corresponds a vector $\boldsymbol{k}$, called its *angle vector*, such that $S_{\boldsymbol{k}}(r,s) = 2\pi$. For example, to the s-corona 1rsrs corresponds the angle vector $(0,1,1,0,3,0)$. We have to find the possible values $\boldsymbol{k}$ for which the equation $S_{\boldsymbol{k}}(r,s) = 2\pi$ admits a solution $0 < s < r < 1$.

The angles which occur in $S_{\boldsymbol{k}}(r,s)$ decrease with s, except $\widehat{sss}$, and increase with r. This yields the following inequalities, strict except for the s-corona ssssss:

$$S_{\boldsymbol{k}}(r,s) \leq \lim_{\substack{r \to 1 \\ s \to 0}} S_{\boldsymbol{k}}(r,s) = k_1\pi + k_2\pi + k_3\frac{\pi}{2} + k_4\pi + k_5\frac{\pi}{2} + k_6\frac{\pi}{3},$$

$$S_{\boldsymbol{k}}(r,s) \geq \inf_r \lim_{s \to r} S_{\boldsymbol{k}}(r,s) = \lim_{\substack{r \to 1 \\ s \to 1}} S_{\boldsymbol{k}}(r,s) = k_1\frac{\pi}{3} + k_2\frac{\pi}{3} + k_3\frac{\pi}{3} + k_4\frac{\pi}{3} + k_5\frac{\pi}{3} + k_6\frac{\pi}{3}.$$

The existence of (r, s) such that $S_{\boldsymbol{k}}(r, s) = 2\pi$ thus yields the inequalities

$$k_1 + k_2 + k_3 + k_4 + k_5 + k_6 < 6 < 3k_1 + 3k_2 + \frac{3}{2}k_3 + 3k_4 + \frac{3}{2}k_5 + k_6,$$

except for the s-corona ssssss ($k_1 = \ldots = k_5 = 0$ and $k_6 = 6$). An exhaustive search on computer yields 383 possible values for $\boldsymbol{k}$. For each of them, one shall check that there indeed exists a coding over $\{1, r, s\}$ with this angle vector[2]. This is also checked by computer and eventually yields 56 s-coronas, listed in Table 1.

Table 1. The 55 possible s-coronas, besides ssssss. Those on the first line have no 1-disc and those on the second line no r-disc. All the codings in a given column are identical up to the replacement $1 \to r$ (this is the keypoint of Lemma 1).

rrrrr	rrrrs	rrrss	rrsrs	rrrr	rrsss	rsrss	rrrs	rrr	rrss
11111	1111s	111ss	11s1s	1111	11sss	1s1ss	111s	111	11ss
1111r	111rs	11rss	11srs	111r	1rsss	1srss	11rs	11r	1rss
111rr	11r1s	1r1ss	1rs1s	11rr			1r1s	1rr	
11r1r	11rrs	1rrss	1rsrs	1r1r			1rrs		
11rrr	1r1rs	r1rss	rrs1s	1rrr			r1rs		
1r1rr	1rr1s								
1rrrr	1rrrs								
	r11rs								
	r1rrs								

4 Medium Coronas

Since s can be arbitrarily smaller than r, there can be infinitely many s-discs in an r-corona. Let us however see that this cannot happen in a compact packing.

Lemma 1. *The ratio $\frac{s}{r}$ is uniformly bounded from below in compact packings with three sizes of discs.*

Proof. Consider a compact packings with three sizes of discs. It contains an s-disc and not only s-discs, thus an s-corona other than ssssss. By replacing each 1 by r in this s-corona, Table 1 shows that we still get an s-corona. In this new corona, the ratio $\frac{s}{r}$ is smaller. Indeed, the 1-discs have been "deflated" in r-discs, so that the perimeter of the corona decreased, whence the size of the surrounded small disc too. But there is at most 10 possible ratios $\frac{s}{r}$ for an s-corona without large discs: they correspond to the values computed in [9] for compact packing with two sizes of discs (the smallest is $5 - 2\sqrt{6} \simeq 0.101$).

[2] For example, $\boldsymbol{k} = (0, 3, 0, 0, 0, 0)$ corresponds to three angles $\widehat{1sr}$ around an s-disc: this is combinatorially impossible.

This lemma ensures that the number of s-discs in an r-corona is uniformly bounded in compact packings. There is thus only finitely many different r-coronas in compact packings. To find them all, we proceed similarly as for s-coronas. We define for $0 < s < r < 1$ and $\boldsymbol{l} = (l_1, \ldots, l_6) \in \mathbb{N}^6$ the function

$$M_{\boldsymbol{l}}(r,s) := l_1\widehat{1r1} + l_2\widehat{1rr} + l_3\widehat{1rs} + l_4\widehat{rrr} + l_5\widehat{rrs} + l_6\widehat{srs}.$$

We have to find the possible values $\boldsymbol{l}$ for which the equation $M_{\boldsymbol{l}}(r,s) = 2\pi$ admits a solution $0 < s < r < 1$. Actually, since the solution should correspond to an r-corona which occur in a packing, we can assume that it satisfies $\frac{s}{r} \geq \alpha$, where α is the lower bound on $\frac{s}{r}$ given by the s-coronas which occur in this packing. The angles which occur in $M_{\boldsymbol{l}}(r,s)$ decrease with r (except $\widehat{rrr}$) and increase with s. This yields the following inequalities, strict except for the r-corona rrrrrr:

$$M_{\boldsymbol{l}}(r,s) \leq \sup_r \lim_{s \to r} M_{\boldsymbol{l}}(r,s) = l_1\pi + l_2\frac{\pi}{2} + l_3\frac{\pi}{2} + l_4\frac{\pi}{3} + l_5\frac{\pi}{3} + l_6\frac{\pi}{3},$$

$$M_{\boldsymbol{l}}(r,s) \geq \inf_r \lim_{\frac{s}{r} \to \alpha} M_{\boldsymbol{l}}(r,s) = l_1\frac{\pi}{3} + l_2\frac{\pi}{3} + l_3 u_\alpha + l_4\frac{\pi}{3} + l_5 u_\alpha + l_6 v_\alpha,$$

where $\widehat{1rs}$ has been bounded from below by $\widehat{rrs}$ for any r, and the limits u_α and v_α of $\widehat{rrs}$ and $\widehat{srs}$ when $\frac{s}{r} \to \alpha$ are obtained via the cosine law:

$$u_\alpha := \arccos\left(\frac{1}{1+\alpha}\right) \quad \text{and} \quad v_\alpha := \arccos\left(1 - \frac{2\alpha^2}{(1+\alpha)^2}\right).$$

The existence of (r,s) such that $M_{\boldsymbol{l}}(r,s) = 2\pi$ thus yields the inequalities

$$l_1 + l_2 + l_4 + \frac{3}{\pi}(l_3 u_\alpha + l_5 u_\alpha + l_6 v_\alpha) < 6 < 3l_1 + \frac{3}{2}l_2 + \frac{3}{2}l_3 + l_4 + l_5 + l_6,$$

except for the r-corona rrrrrr ($l_1 = \ldots = l_5 = 0$ and $l_6 = 6$). We also impose

$$l_1 + l_2 + l_4 + \frac{1}{2}(l_3 + l_5) < 6,$$

which tells that an r-corona (other than rrrrrr) contains at most 5 r- or 1-discs. An exhaustive search on computer, which also checks whether there indeed exists a coding for each possible $\boldsymbol{l}$, eventually yields the possible r-coronas for each value of α. Table 2 gives the numbers of such coronas.

Table 2. Each s-corona whose 1-discs have been deflated in r-discs (first line) yields a lower bound α on $\frac{s}{r}$ in any compact packing which contains it (second line), and thus an upper bound on the number of possible r-coronas in this packing (third line).

rrrrr	rrrrs	rrrss	rrsrs	rrrr	rrsss	rsrss	rrrs	rrr	rrss
0.701	0.637	0.545	0.533	0.414	0.386	0.349	0.280	0.154	0.101
84	94	130	143	197	241	272	386	889	1654

Since any r-corona for some lower bound α also appears for a smaller α, there are at most 1654 different r-coronas. Combining Tables 1 and 2 shows that there is at most 16805 pairs formed by an s-corona and an r-corona which can appear in the same compact packing with three sizes of discs.

5 Equations

We here describe an algorithm to associate with an s-corona with angle vector $\boldsymbol{k}$ a polynomial equation in r and s whose solutions contain those of $S_{\boldsymbol{k}}(r, s) = 2\pi$.

Start from the equation $S_{\boldsymbol{k}}(r, s) = 2\pi$. Take the cosine of both sides and fully expand the left-hand side. Substract $\cos(2\pi) = 1$ to both sides. This yields a polynomial equation in cosines and sines of the angles occuring in $S_{\boldsymbol{k}}$.

The power of each sine can be decreased by 2 by using $\sin^2 a = 1 - \cos^2 a$. Doing this as much as possible yields a polynomial where no sines is raised to a power greater than 1. The cosine law then allows to replace each cosine by a rational fraction in r and s. This yields a rational function in r, s and sines. Multiplying by the least common divisors of the denominators yields a polynomial in r, s and the sines which is equal to zero.

We shall now "remove" the sines one by one. Whenever the polynomial writes $A \sin a + B$, where A and B are polynomials without $\sin a$, we multiply by $A \sin a - B$ to get $A^2 \sin^2 a - B^2 = 0$. We can then replace $\sin^2 a$ by $1 - \cos^2 a$, use the cosine law to replace $\cos^2 a$ by a rational function in r and s, and multiply by the least common divisors of the denominators. By iterating this for each sine, we eventually get a polynomial equation in r and s. The final polynomial can however be quite large since each sine removal roughly doubles its degree.

The solutions (r, s) of $S_{\boldsymbol{k}}(r, s) = 2\pi$ are still solution of this polynomial equation, but each multiplication by $A \sin a - B$ could have added new solutions which shall eventually be ruled out (we shall come back to this later).

This algorithm has been implemented on computer. For example, it associates with the s-corona $11rs$ the polynomial equation

$$r^2s^4 - 2r^2s^3 - 2rs^4 - 23r^2s^2 - 28rs^3 + s^4 - 24r^2s - 58rs^2 - 2s^3 + 16r^2 - 8rs + s^2.$$

The same algorithm also associates with each r-corona a polynomial equation.

We cannot write here all the equations. Let us just provide some statistics. The equations associated with the 55 s-coronas are computed on our laptop[3] in less that 5 s and yield a 17Ko file. The mean degree of these polynomial equations is 6.71 (standard deviation 5.77), with maximum 28 for the s-corona 11rrs. The equations associated with the 1654 r-coronas are computed on our laptop in 2 h 21 min and yield a 35Mo file. The mean degree of these polynomial equations is 57.88 (standard deviation 50.16), with maximum 416 for the s-corona 11rrs^{12}.

[3] Intel Core i5-7300U with 4 cores at 2.60 GHz and 15,6 Go RAM.

6 Computating Radii: The Theory

The path to compute the possible radii is now marked:

1. Compute the solutions (r, s) with $0 < s < r < 1$ of each system of two polynomial equations in r and s associated with a possible pair formed by an s-corona and an r-corona.
2. Use the computed values of r and s to compute the angles appearing in $S_{\boldsymbol{k}}$ and $M_{\boldsymbol{l}}$, as well as in the function defined for $0 < s < r < 1$ and $\boldsymbol{n} \in \mathbb{N}^6$ by

$$L_{\boldsymbol{n}}(r,s) := n_1\widehat{111} + n_2\widehat{11r} + n_3\widehat{11s} + n_4\widehat{r1r} + n_5\widehat{r1s} + n_6\widehat{s1s}.$$

   Perform an exhaustive search for all the possible values $\boldsymbol{k}$, $\boldsymbol{l}$ and $\boldsymbol{n}$ which satisfy the equations $S_{\boldsymbol{k}}(r,s) = 2\pi$, $M_{\boldsymbol{l}}(r,s) = 2\pi$ and $L_{\boldsymbol{n}}(r,s) = 2\pi$. This yields all the possible sets of s-, r- and 1-coronas.
3. Check the existence of a compact packing for each set of coronas.

Consider, for example, the pair s/r-coronas 1r1r/1r1r1s. The associated equations are respectively $rs+s^2-r+s$ and $r^3+r^2s-rs-s$. There is only one solution (r, s) with $0 < s < r < 1$, namely $r \simeq 0.751$ is a root of $X^4-4X^3-2X^2+2X+1$ and $s \simeq 0.356$ is a root of X^4+6X^3+2X-1. Searching for all the other coronas compatible with these values yields only large coronas, namely $\mathrm{r^3sr^3s}$, $\mathrm{r^4sr^2s}$ and $\mathrm{r^5srs}$ (same angle vector). On then easily finds a periodic compact packing, namely the number 87 on Fig. 3.

7 Computating Radii: Problems

The above approach however suffers from a number of problems:

1. The example 1r1r/1r1r1s in the previous section has been chosen because it yields one of the most simple system to solve. The coronas 11rs/111srrss yield equations of degree 6 and 56 - close to the mean degree of equations associated with s- and r-coronas - and we need 22 min 44 s with SageMath [14] on our laptop to solve the system. For $\mathrm{11rrs/11rrs}^{12}$, the equations have degree 28 and 416 and solving the system seems out or reach of our laptop[4].
2. Algebraic values of r and s yield non-algebraic angles: how to perform exact computation to find all the compatible coronas and only them?
3. Not only systems can be highly complicated, but there are 16805 of them!
4. Checking the existence of a compact packing for a given set of coronas could be hard. This indeed amounts do decide whether a set of ten triangular *tiles*, namely the triangles between the centers of three mutually adjacent discs, does tile the plane. This is close to the *domino problem*, which is known to be undecidable [13].

The next section sketches how we got over this (full details can be found in [5]).

4 It required 6 min just to compute the equations. Trying to solve the system exhausted the memory of our laptop after around 30 h and crashed thereafter.

8 Computating Radii: The Practice

8.1 Solving Equations

We use the *hidden variable* method (see, *e.g.*, Sect. 3.5 of [2] or Sect. 9.2.4 of [15]). Recall that the *resultant* of two univariate polynomials is a scalar which is equal to zero iff the two polynomials have a common root. Now, if P and Q are two polynomials of $\mathbb{Z}[r, s]$, we can see them as polynomials in r with coefficients in $\mathbb{Z}[s]$, *i.e.*, we *hide* the variable s. Their resultant is thus a polynomial in s which has a root s_0 iff $P(r, s_0)$ and $Q(r, s_0)$ have a common root r. In other words, we can compute the second coordinates of the roots (r, s) of P and Q by computing the roots of an univariate polynomial (the above resultant). We can similarly compute the first coordinates by exchanging r and s. The cartesian product of these first and second coordinates then contains the roots (r, s) of P and Q.

We shall apply this to the two polynomial equations in r and s associated with each pair of s/r-coronas. The point is that resultants are easily computed (as determinants of Sylvester matrices), as well as the roots of an univariate polynomial (it amounts to find a proper interval for each root). Of course, the obtained cartesian product generally contains, besides the solutions, many "false pairs" (r, s) which have to be ruled out. Actually, even among the solutions there are also false pairs which must be ruled out: those introduced when we remove the sines to get a polynomial equation (Sect. 5). But too much is better than not enough: we shall now focus on finding the pairs which indeed allow coronas.

8.2 Finding Coronas

Assume two algebraic real numbers $0 < s < r < 1$ are given. We want to find all the values $\boldsymbol{k}$, $\boldsymbol{l}$ and $\boldsymbol{n}$ which satisfy the equations $S_{\boldsymbol{k}}(r, s) = 2\pi$, $M_{\boldsymbol{l}}(r, s) = 2\pi$ and $L_{\boldsymbol{n}}(r, s) = 2\pi$. If they are no possible value for $\boldsymbol{k}$, $\boldsymbol{l}$ or $\boldsymbol{n}$, then we rule out the pair (r, s).

We rely on *interval arithmetic* as much as possible. We compute intervals for r and s, then for the angles appearing in $S_{\boldsymbol{k}}$, $M_{\boldsymbol{l}}$, and $L_{\boldsymbol{n}}$ via arccos. We then compute the $\boldsymbol{k}$, $\boldsymbol{l}$ and $\boldsymbol{n}$ such that the intervals $S_{\boldsymbol{k}}(r, s)$, $M_{\boldsymbol{l}}(r, s)$ and $L_{\boldsymbol{n}}(r, s)$ contain 2π. As we shall see, this actually suffices to rule out all the false pairs (a hint is that the remaining set is stable when the precision used for intervals is increased).

To check that the found coronas are indeed valid, we check the associated polynomial equations. We compute them as in Sect. 5, except that each time we multiply by $A \sin a - B$ to remove $\sin a$, we compute the interval for this expression and check that it does not contain zero. If it does[5], we check exactly whether $A \sin a + B = 0$ and rule out the case if it does not. We finally put the algebraic values of r and s in the polynomial equation to check it exactly[6].

[5] In all the cases we eventually considered, it occured only for 1r1r/11r1s.

[6] There were eventually only 213 cases to check; it tooks less than 5 min on our laptop.

8.3 Reducing the Number of Cases

As seen in Sect. 4, there are 16805 pairs of s- and r-coronas, *i.e.*, 16805 systems of polynomials equations to solve. Many of them are moreover too hard to solve. However, as we shall see, combinatorial arguments can be used to show that most of them do not allow compact packing, without even computing r and s. For this purpose, we distinguish three classes of compact packings:

1. those with no s- and r-discs in contact;
2. those with two different s-coronas (besides ssssss);
3. all the other ones.

The first class is very simple. There is an s-corona without r-disc which thus characterizes one of the 10 values of s for a compact packings with discs of radii 1 and s. Similarly, there is an r-corona without s-disc which yield 10 possibles values for r. With moreover $s < r$, this immediately yields $\binom{10}{2} = 45$ pairs (r, s).

The second class relies only on s-coronas. There are 55 of them, moreover associated with simpler equations than r-coronas. This yields $\binom{55}{2} = 1485$ systems. The hidden variable method gives 1573 pairs (r, s) where r and s are real algebraic numbers such that $0 < s < r < 1$. Only 37 of them admit coronas of all sizes[7]. All these computations required no more than a couple of minutes.

The last class is the main one. The fact that there is only one s-corona (besides ssssss) however yields a strong combinatorial constraint on the possible r-coronas. Since these compact packings contain an s-corona and a r-corona which intersect, *i.e.*, the surrounded s- and r-discs are tangent, we can assume that the considered pair of s- and r-coronas intersect. Then, each pattern xsy in the coding of the r-corona force the pattern xry in the coding of the s-corona (the discs of radii x and y are those which are tangent to the centers of both coronas). This simple constraint reduces to only 803 the number of pairs of s- and r-coronas to be considered. Moreover, the remaining r-coronas are among the simplest ones[8]. The hidden variable method gives 62892 pairs (r, s) where r and s are real algebraic numbers such that $0 < s < r < 1$. Only 176 of them admit coronas of all sizes. All these computations required around 13 min[9].

8.4 Finding Packings

The three above combinatorial classes now respectively contain 45, 37 and 176 pairs (r, s) which allow coronas of all sizes. It was a bit long but not too difficult to find, by hands, an example of a periodic compact packing with three size of discs for respectively 18, 1 and 145 of these pairs[10]. We then had to combinatorially

[7] That is, the values of r and s allows s-, r- and 1-coronnas, which is necessary to have a packing with the three size of discs.

[8] There are 192 different r-coronas (among the 1654 initial ones), and the mean algebraic degree of the associated equations is 14, with maximum 80 for 11rrsrss.

[9] Except for 1rr1s/11rrs: the hidden variable method with SageMath inexplicably crashed and solving the system via Gröbner basis tooks 45 min on our laptop.

[10] The undecidability of the domino problem seems to need more discs to appear....

prove that the other pairs do not allow any compact packing with three size of discs. Since $18 + 1 + 145 = 164$, Theorem 1 follows.

The 27 cases ruled out in the first combinatorial class were exactly those with no 1-corona containing both s- and r-discs. They do admit compact packings but not with all the three size of discs. All the cases in the second combinatorial class turned out to have no 1-corona containing an r-disc, and the only case which allows three sizes of discs is the one with an s-corona which contains both r- and 1-discs (namely 1srrs). In the last class, we proved that an s-corona 1rss, 11rss, 1rrss or 1srss in a compact packing actually implies another s-corona in this packing, which thus falls into the second combinatorial class. This ruled out 24 cases. The 7 remaining ones were ruled out one by one, each with a short combinatorial argument. They do not allow any compact packing.

References

1. Blind, G.: Über Unterdeckungen der Ebene durch Kreise. J. reine angew. Math. **236**, 145–173 (1969)
2. Cox, D.A., Little, J., O'Shea, D.: Using Algebraic Geometry. Graduate Texts in Mathematics, vol. 185. Springer, New York (2005). https://doi.org/10.1007/b138611
3. Fejes Tóth, L.: Über die dichteste Kugellagerung. Math. Z. **48**, 676–684 (1943)
4. Fejes Tóth, L.: Regular Figures. Pergamon Press, Oxford (1964)
5. Fernique, Th., Hashemi, A., Sizova, O.: Empilements compacts avec trois tailles de disque. arxiv:1808.10677 (2018)
6. Florian, A.: Ausfüllung der Ebene durch Kreise. Rend. Circ. Mat. Palermo **9**, 300–312 (1960)
7. Heppes, A.: On the densest packing of discs of radius 1 and $\sqrt{2} - 1$. Stud. Sci. Math. Hung. **36**, 433–454 (2000)
8. Heppes, A.: Some densest two-size disc packings in the plane. Discrete Compuat. Geom. **30**, 241–262 (2003)
9. Kennedy, T.: Compact packings of the plane with two sizes of discs. Discrete Comput. Geom. **35**, 255–267 (2006)
10. Kennedy, T.: A densest compact planar packing with two sizes of discs. arxiv:0412418 (2004)
11. Likos, C.N., Henley, C.L.: Complex alloy phases for binary hard-disc mixtures. Philos. Mag. B **68**, 85–113 (1993)
12. Messerschmidt, M.: On compact packings of the plane with circles of three radii. arxiv:1709.03487 (2017)
13. Robinson, R.: Undecidability and nonperiodicity for tilings of the plane. Invent. Math. **12**, 177–209 (1971)
14. The Sage Developers: SageMath, the Sage Mathematics Software System (Version 7.4) (2016). http://www.sagemath.org
15. Zimmerman, P., et al.: Calcul Mathématique avec Sage. CreateSpace Independent Publishing Platform (2013)

Convex Aggregation Problems in $\mathbb{Z}^2$

Yan Gérard(✉)

Université Clermont Auvergne - LIMOS, Clermont-Ferrand, France
yan.gerard@uca.fr

Abstract. We introduce a family of combinatorial problems of digital geometry that we call *convex aggregation* problems. Two variants are considered. In *Unary* convex aggregation problems, a first lattice set $A \subseteq \mathbb{Z}^d$ called *support* and a family of lattice sets $B^i \subseteq \mathbb{Z}^d$ called *pads* are given. The question to determine whether there exists a non-empty subset of pads (the set of their indices is denoted I) whose union $A \cup_{i \in I} B^i$ with the support is convex. In the *binary* convex aggregation problem, the input contains the support set $A \subseteq \mathbb{Z}^2$ and pairs of pads B^i and $\overline{B}^i$. The question is to aggregate to the support either a pad B^i or its correspondent $\overline{B}^i$ so that the union $A \cup_{i \in I} B^i \cup_{i \notin I} \overline{B}^i$ is convex.

We provide a first classification of the classes of complexities of these two problems in dimension 2 under different assumptions: if the support is 8-connected and the pads included in its enclosing rectangle, if the pads are all disjoint, if they intersect and at least according to the chosen kind of convexity. In the polynomial cases, the algorithms are based on a reduction to Horn-SAT while in the NP-complete cases, we reduce 3-SAT to an instance of convex aggregation.

Keywords: Digital Geometry · Discrete Tomography · Convexity · Complexity · Horn-SAT

1 Introduction

The first convex aggregation problem has been considered in 1996 in Discrete Tomography where it has been used for the reconstruction of $\mathcal{HV}$-convex polyominoes i.e $\mathcal{HV}$-convex 4-connected lattice sets [1]. More generally, convex aggregation problems are the bottleneck of the reconstruction of convex lattice sets in different problems of Discrete Tomography [2–4]. Although they might be of major interest for the field, this class of combinatorial problems have never been neither formulated independently nor investigated in a rather exhaustive manner. The first parameter of this family of problems is the chosen kind of convexity. In the lattice $\mathbb{Z}^2$, the two main ones are denoted here $\mathcal{HV}$ and $\mathcal{C}$-convexity. A lattice set $S \subseteq \mathbb{Z}^2$ is $\mathcal{HV}$-convex if its intersection with any horizontal or vertical line is a set of consecutive points and $S \subset \mathbb{Z}^d$ is $\mathcal{C}$-convex if it is equal to its intersection with its real convex hull $S = \mathrm{conv}_{\mathbb{R}^d}(S) \cap \mathbb{Z}^d$ (Fig. 1). The $\mathcal{HV}$-convexity is a directional convexity reduced to the horizontal and vertical

M. Couprie et al. (Eds.): DGCI 2019, LNCS 11414, pp. 432–443, 2019.
https://doi.org/10.1007/978-3-030-14085-4_34

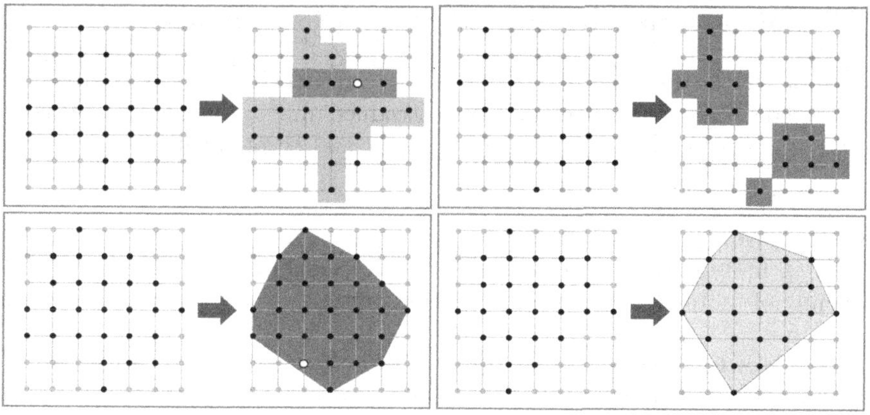

Fig. 1. $\mathcal{HV}$ **and** $\mathcal{C}$**-convexities.** Top left, the set is not $\mathcal{HV}$-convex while the top-right, the lattice set is $\mathcal{HV}$-convex. Below, the left lattice set is not $\mathcal{C}$-convex while the right set is $\mathcal{C}$-convex. Notice that neither $\mathcal{HV}$-convexity, nor $\mathcal{C}$-convexity imply digital connectivity.

directions. The $\mathcal{C}$-convexity is the classical convexity of geometry of numbers. The $\mathcal{C}$-convex lattice sets are the lattice polytopes.

In Discrete Tomography, the problem of convex aggregation appears as follows: a partial convex solution $A \subseteq \mathbb{Z}^2$ and a list of conjugate subsets $B^i, \overline{B}^i$ of $\mathbb{Z}^2$ with an index i from 1 to n have been computed. In the sequel, the subset A is called the *support* while the subsets $B^i \subseteq \mathbb{Z}^2$ and $\overline{B}^i \subseteq \mathbb{Z}^2$ are called *pads*. The problem is to aggregate either the pad B^i or its conjugate $\overline{B}^i$ to the support so that the union of the support with all chosen pad is convex. In other words, the problem is to find a list of indices I so that the union $A \cup_{i \in I} B^i \cup_{i \notin I} \overline{B}^i$ is convex. These problems of convex aggregation are called *binary*. We choose this term to highlight the difference with another class of problems that we call *unary*. In unary convex aggregation problems, the input is a support set $A \subseteq \mathbb{Z}^2$ and a family of pads $B^i \subseteq \mathbb{Z}^2$. The problem is to find a non-empty set of pads whose union $A \cup_{i \in I} B^i$ with the support is convex.

The purpose of the paper is to determine the classes of complexities of these two problems by considering either the $\mathcal{HV}$ or the $\mathcal{C}$-convexity in the general case and under several assumptions on the pads and the support.

In the next section (Sect. 2), we start by stating formally the problems. Section 3 provides a very brief state of the art and the results. Some complexities are shown to be polynomial and others NP-complete. The polynomial-time algorithms are given in Sect. 4 while Sect. 5 provides sketches of proofs of NP-completeness.

2 Problem Statement

For reducing the number of notations, we consider the chosen convexity as a parameter denoted convexity. We introduce first the problem of unary convex aggregation $\mathtt{1CA}_{\text{convexity}}(A, (B^i)_{1 \leq i \leq n})$ and secondly its binary variant $\mathtt{2CA}_{\text{convexity}}(A, (B^i)_{1 \leq i \leq n}, (\overline{B}^i)_{1 \leq i \leq n})$.

Problem 1. $\mathtt{1CA}_{\text{convexity}}(A, (B^i)_{1 \leq i \leq n})$
Input: - A finite lattice set $A \subseteq \mathbb{Z}^d$ that we call *support*,
- a family of finite lattice sets $B^i \subseteq \mathbb{Z}^d$ that we call *pads*, with an index i going from 1 to n.
Output: Does there exist a non-empty set of indices $I \subseteq [1 \,..\, n]$ such that the union $A \cup_{i \in I} B^i$ of the pads B^i with $i \in I$ and A is convex (Fig. 2)?

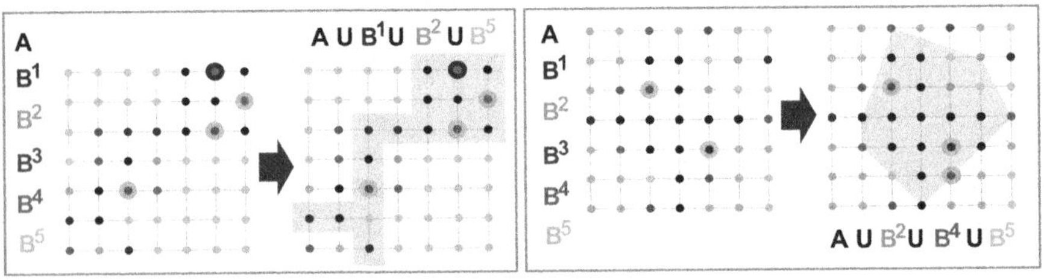

Fig. 2. Two instances of $\mathtt{1CA}_{\text{convexity}}(A, (B^i)_{1 \leq i \leq n})$ **and their solutions.** The input is a set A (black points) with pads B^i (each pad is a set of colored points) while a solution is a non-empty set of indices I such that the union $A \cup_{i \in I} B^i$ is convex. On the left, we consider the $\mathcal{HV}$-convexity and the $\mathcal{C}$-convexity on the right.

Notice that the support A from an instance can be removed by adding it to all the pads: the instance $\mathtt{1CA}_{\text{convexity}}(\emptyset, (A \cup B^i)_{1 \leq i \leq n})$ is equivalent with $\mathtt{1CA}_{\text{convexity}}(A, (B^i)_{1 \leq i \leq n})$.

We introduce now the second class of problems of convex aggregation occuring in Discrete Tomography and we call them *binary*.

Problem 2. $\mathtt{2CA}_{\text{convexity}}(A, (B^i)_{1 \leq i \leq n}, (\overline{B}^i)_{1 \leq i \leq n})$
Input: - A finite lattice set $A \subseteq \mathbb{Z}^d$ that we call *support*,
- a family of pairs of finite lattice sets $B^i \subseteq \mathbb{Z}^2$ and $\overline{B}^i \subseteq \mathbb{Z}^2$ that we call *pads*, with an index i going from 1 to n.
Output: Does there exist a set of indices $I \subseteq [1 \,..\, n]$ such that the union $A \cup_{i \in I} B^i \cup_{i \notin I} \overline{B}^i$ is convex (Fig. 3)?

Note that with a non-convex support A, the unary problem $\mathtt{1CA}_{\text{convexity}}(A, (B^i)_{1 \leq i \leq n})$ is a particular case of $\mathtt{2CA}_{\text{convexity}}(A, (B^i)_{1 \leq i \leq n}), (\overline{B}^i)_{1 \leq i \leq n})$ with empty conjugate pads $\overline{B}^i$. It follows that the complexity of the binary problems is necessarily at least the complexity of the unary cases.

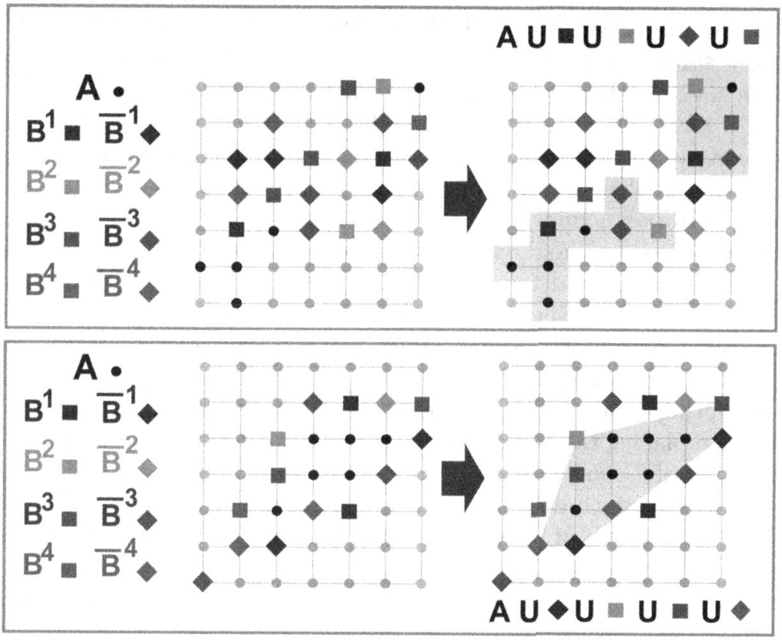

Fig. 3. Two instances of $2\mathtt{CA}_{\text{convexity}}(A, (B^i)_{1\leq i\leq n}, (\overline{B}^i)_{1\leq i\leq n})$ **and their solutions.** The input is a set A (black points) and for each index i, a pair of conjugate pads B^i and $\overline{B}^i$. The points of the pads B^i are represented by squares while the points of their conjugate $\overline{B}^i$ are represented by diamonds, with a different color for each index. The problem of binary convex aggregation is to aggregate to the support A either the squares or the diamonds of each color, so that the union $A \cup_{i\in I} B^i \cup_{i\notin I} \overline{B}^i$ becomes convex. The instances are drawn on the left (above with $\mathcal{HV}$-convexity, below with $\mathcal{C}$-convexity) and a solution is given on the right.

3 State of the Art and Results

The state of the art is mainly reduced to the following result known in Discrete Tomography. It requires to define the *$\mathcal{HV}$-grid* of a finite lattice set $A \subseteq \mathbb{Z}^2$ as the finite grid $[\min(x(A)) \mathinner{..} \max(x(A))] \times [\min(y(A)) \mathinner{..} \max(y(A))] \subseteq \mathbb{Z}^2$ where x and y are the two coordinates of $\mathbb{Z}^2$. We recall also that a lattice set A of $\mathbb{Z}^2$ is *8-connected* if any pair of points $(a, a') \in A^2$ is connected by a discrete path $(a_i)_{1\leq i\leq n}$ of points of A verifying $a_1 = a$, $a_n = a'$ and $\forall i \in [1 \mathinner{..} n-1], d_2(a_i, a_{i+1}) \leq 1$ where d_2 denotes the Euclidean distance.

Theorem 1. *If A is 8-connected, and all the pads are disjoint and included in the $\mathcal{HV}$-grid of A, $2\mathit{CA}_{\mathcal{HV}}(A, (B^i)_{1\leq i\leq n}, (\overline{B}^i)_{1\leq i\leq n})$ can be solved in polynomial time* [1].

If A is 8-connected, and all the pads are included in the $\mathcal{HV}$-grid of A, $1\mathit{CA}_{\mathcal{HV}}(A, (B^i)_{1\leq i\leq n})$ can be solved in polynomial time.

The original proof is stated in the binary case with disjoint pads [1]. It is based on a simple idea: encode the choice to add either the pad B^i or $\overline{B}^i$ by a Boolean variable that we denote $\mathfrak{b}_i$ (with the convention that $\mathfrak{b}_i = 1$ if B^i is added to A and then $i \in I$ while $\mathfrak{b}_i = 0$ if $i \notin I$). Then, with disjoint pads, the $\mathcal{HV}$-convexity of the union of the pads is encoded by a conjunction of 2-clauses. It reduces the problem of convex aggregation to an instance of 2-SAT and allows us to solve it in polynomial time [5].

The result does not hold without the assumption of disjoint pads. The approach of [1] remains valid for unary $\mathcal{HV}$-convex aggregation with disjoint and non-disjoint pads under the assumption that the support A is 8-connected and the pads included in the $\mathcal{HV}$-grid of A. The $\mathcal{HV}$-convexity of a solution is encoded by a Horn-SAT instance and can be solved in polynomial time [6].

On the other hand, different authors have noticed that the same problem obtained by replacing the $\mathcal{HV}$-convexity by the $\mathcal{C}$-convexity is more difficult [3, 7]. The constraint of $\mathcal{C}$-convexity can be formulated with 3-clauses. 3-SAT being NP-complete, this approach does not provide a solution in polynomial time. Does it mean that the problem of convex aggregation is itself NP-complete? It is one of the questions that we investigate in what follows.

We determine in the paper a first classification of the complexities of the problems of convex aggregation.

Theorem 2. *The classes of complexities of the problems of convex aggregation under several assumptions are given in Table 1.*

Table 1. Complexities of unary and binary convex aggregation problems with either $\mathcal{HV}$ or $\mathcal{C}$-convexity, with or without two different assumptions. We can assume that the support is 8-connected and the pads included in the $\mathcal{HV}$-grid of A and secondly, we can assume that the pads are disjoint. In each cell, *P* means polynomial-time and *NP* is written for NP-complete.

Problem	Unary				Binary			
	8-connected support with pads in the $\mathcal{HV}$-grid of A				8-connected support with pads in the $\mathcal{HV}$-grid of A			
	Disjoint pads		Disjoint pads		Disjoint pads		Disjoint pads	
$\mathcal{HV}$-convexity	P (Theorem 1)	P (Theorem 1)	P	NP	P (Theorem 1)	NP	NP	NP
$\mathcal{C}$-convexity	P	NP	P	NP	NP	NP	NP	NP

In the case of binary convex aggregation, the only polynomial case among the configurations that we investigate is the one of Theorem 1. All the others are NP-complete. In particular, the condition that the pads are disjoint or the 8-connectivity of the support is not sufficient to provide a polynomial time algorithm.

We do not provide here polynomial time algorithms for binary $\mathcal{C}$-convex aggregation but only results of NP-completeness. With specific properties of the pads, such polynomial-time algorithms might be obtained [7] but these assumptions used in Discrete Tomography are too specific to fit into the scope of this general classification. The reader should just keep in mind that some assumptions on the pads might lead to polynomial time algorithms. We conjecture in particular that the problem $\mathtt{2CA}_{\text{convexity}}(A, (B^i)_{1\leq i\leq n}, (\overline{B}^i)_{1\leq i\leq n})$ is polynomial with convex disjoint pads.

4 Polynomial Time Algorithms

Theorem 2 states that with disjoint pads, $\mathtt{1CA}_{\text{convexity}}$ $(A, (B^i)_{1\leq i\leq n})$ can be solved in polynomial in the two cases of convexity (convexity $= \mathcal{HV}$ is the case 1 and convexity $= \mathcal{C}$ in the case 2). The purpose of this section is to describe the two algorithms.

According to the same strategy as the proof of Theorem 1, we encode the choice to add or not the pad B^i to A by a Boolean variable. By convention, $\mathfrak{b}_i$ is equal to 1 if i is in the set I (in other words, if the pad B^i is added to the union) and $\mathfrak{b}_i = 0$ if it is excluded. The main idea of the two algorithms is to express the convexity of the union $A \cup_{I\in I} B^i$ by a conjunction of clauses.

In the following, a point of the lattice $\mathbb{Z}^2$ is said *covered* if it is in the union $A \cup_{1\leq i\leq n} B^i$ of the pads and the support and *uncovered* otherwise. In terms of data structure, it is practical to build two tables. The first one contains the abscissas of the covered points and for each abscissa, the ordered list of their ordinates with the index of the pad containing the point. The second table uses first the ordinates and secondly, the abscissas. Denoting the number of pads n and the total number of points $N = |A| + \sum_{i=1}^{n} |B^i|$, the two tables can be computed in $O(N\log(N))$ time, with a storage in $O(N)$.

4.1 Case 1 - with $\mathcal{HV}$-Convexity and Disjoint Pads

We consider the problem $\mathtt{1CA}_{\mathcal{HV}}(A, (B^i)_{1\leq i\leq n})$ with disjoint pads. We provide an algorithm in $O(N^3 + n^4)$. Its strategy is to build a conjunction of clauses characterizing the $\mathcal{HV}$-convexity of the union $A \cup_{I\in I} B^i$:

1. For all pairs of points $(a, b) \in \cup_{1\leq i\leq n} A \times B^i$ with a and b on the same row or on the same column, we consider the integer points in the segment of end points a and b. If there is an uncovered point, the pad B^i can be directly excluded since its union with the support A cannot be convex (there is a hole in between). Otherwise, we build a clause $\mathfrak{b}_i \implies \mathfrak{b}_j$ for all the pads b^j of the points on the segment. It expresses the constraint that the points in-between have to be included in the union.

2. For all pairs of points $(b, b') \in \cup_{1 \leq i \leq j \leq n} B^i \times B^j$ with b and b' on the same row or on the same column, we consider the integer points in-between. If there is an uncovered point, we build the clause $\mathfrak{b}_i \implies \overline{\mathfrak{b}}_j$. It expresses that the two pads cannot be chosen simultaneously. Otherwise, for all the pads B^k having points between b and b', we add the clause $\mathfrak{b}_i \wedge \mathfrak{b}_j \implies \mathfrak{b}_k$ (notice that we use here the assumption of disjoint pads. Otherwise the clause could be $\mathfrak{b}_i \wedge \mathfrak{b}_j \implies \mathfrak{b}_k \vee \mathfrak{b}_{k'}$ if the pads B^k and $B^{k'}$ have for instance a common point between b and b').
3. We remove the redundant clauses. The number of clauses becomes $O(n^3)$.

The clauses built with this algorithm guarantee the $\mathcal{HV}$-convexity of the corresponding union $A \cup_{1 \leq i \leq n} B^i$. Conversely, an $\mathcal{HV}$-convex union $A \cup_{i \in I} B^i$ provides a solution of the instance of satisfiability. In the case where the support A is $\mathcal{HV}$-convex, a null solution can satisfy the clauses but the empty set of indices $I = \emptyset$ is not a valid solution of $1\mathtt{CA}_{\text{convexity}}(A, (B^i)_{1 \leq i \leq n})$. This case apart (we consider it afterwards), finding a non-null solution of the conjunction of clauses is equivalent to the problem of convex aggregation $1\mathtt{CA}_{\text{convexity}}(A, (B^i)_{1 \leq i \leq n})$. The clauses are either of the form $\mathfrak{b}_i \implies \mathfrak{b}_j$, $\mathfrak{b}_i \implies \overline{\mathfrak{b}}_j$ or $\mathfrak{b}_i \wedge \mathfrak{b}_j \implies \mathfrak{b}_k$. They can respectively be rewritten as $\overline{\mathfrak{b}}_i \vee \mathfrak{b}_j$, $\overline{\mathfrak{b}}_i \vee \overline{\mathfrak{b}}_i$ and $\overline{\mathfrak{b}}_i \vee \overline{\mathfrak{b}}_j \vee \mathfrak{b}_k$. They have at most one positive literal. It makes them Horn clauses and Horn-SAT can be solved in linear time [6]. The condition of non-nullity can be written as $\vee_{1 \leq i \leq n} \mathfrak{b}_i$ and it is not a Horn clause. A naive way to search for a non-null solution of the Horn-SAT instance is to fix successively $\mathfrak{b}_i = 1$ and to solve the new SAT instance, now in linear time. If $\mathfrak{b}_1 = 1$ does not provide a solution, try $\mathfrak{b}_2 = 1$ and so on until $\mathfrak{b}_n = 1$ if no solution has been found previously.

Complexity

The number of pairs of points considered in the algorithm is bounded by N^2. Then it generates at most a cubic number of clauses $O(N^3)$ in a cubic time $O(N^3)$. Their number is reduced to $O(n^3)$ by removing the redundant clauses in $O(N^3)$ time. The resolution of the n instances of the main Horn-SAT instance obtained by fixing $\mathfrak{b}_i = 1$ requires at most n times $O(n^3)$, namely $O(n^4)$. It provides an algorithm with an overall complexity in $O(N^3 + n^4)$.

4.2 Case 2 - $\mathcal{C}$-Convexity with Disjoint Pads

We consider now the problem $1\mathtt{CA}_{\mathcal{C}}(A, (B^i)_{1 \leq i \leq n})$ with disjoint pads and provide an algorithm in time $O(N^4 + n^5)$. Its strategy is similar to the algorithm given in the previous subsection. In order to reduce the number of cases, we introduce a supplementary Boolean variable $\mathfrak{b}_0$ associated with the support A (its value will be fixed to 1 after the generation of all clauses). We build an instance of satisfiability expressing the $\mathcal{C}$-convexity of $A \cup_{i \in I} B^i$ with the following process:

For all triplets of points $(a, b, c) \in (A \cup_{1 \leq i \leq n} B^i)^3$, we consider the triangle $T_{a,b,c}$ with vertices a, b and c. The indices of the pads of the points a, b and c are denoted i, j and k. If one of the points of $T_{a,b,c}$ is uncovered, then the three variables cannot be true in the same time. It is expressed by the conjunction $\overline{\mathfrak{b}}_i \vee \overline{\mathfrak{b}}_j \vee \overline{\mathfrak{b}}_k$. If all the points of the triangle are covered, we add the clause $\mathfrak{b}_i \wedge \mathfrak{b}_j \wedge \mathfrak{b}_k \implies \mathfrak{b}_q$ for all the indices q of the pads of the points in the triangle $T_{a,b,c}$. This clause can be rewritten as $\overline{\mathfrak{b}}_i \vee \overline{\mathfrak{b}}_j \vee \overline{\mathfrak{b}}_k \vee \mathfrak{b}_q$. They are both Horn clauses. Once all triangles have been considered, we fix $b_0 = 1$ and remove the redundant clauses. Their final number becomes bounded by $O(n^4)$.

According to Carathéodory's theorem in dimension 2, the convex hull of a finite set S is the union of the triangles with vertices in S [8]. It follows that S is convex iff no triangle with vertices in S has a point outside S. It explains that the 3-clauses and 4-clauses obtained from all the triangles with vertices in $A \cup_{1 \leq i \leq n} B^i$ guarantee the $\mathcal{C}$-convexity of the union $A \cup_{\in I} B^i$. The $\mathcal{C}$-convexity is expressed by the clauses. As in the previous case, the constraint to find a non-empty union of pads $\cup_{i \in I} B^i$ has to taken into account. If the support A is convex, we consider n sub-instances of the Horn-SAT instance with $\mathfrak{b}_i = 1$ for all indices from 1 to n. By construction, this sequence of problems is equivalent with $1\mathtt{CA}_{\mathcal{C}}(A, (B^i)_{1 \leq i \leq n})$ and it can be solved in polynomial time.

Complexity

The number of triplets of points considered by the algorithm is bounded by N^3. We have at most a quartic number of clauses $O(N^4)$ generated in a quartic time $O(N^4)$. Their number is reduced to $O(n^4)$ by removing the redundant clauses in $O(N^4)$ time. The resolution of the n instances of the main Horn-SAT instance with $\mathfrak{b}_i = 1$ requires at most n times $O(n^4)$, namely $O(n^5)$. It provides an algorithm with an overall complexity in $O(N^4 + n^5)$.

5 Proofs of NP-Completeness

We prove here that the convex aggregation problems are NP-complete in the five following cases:

1. $1\mathtt{CA}_{\mathcal{HV}}(A, (B^i)_{1 \leq i \leq n})$ where convexity $= \mathcal{HV}$.
2. $1\mathtt{CA}_{\mathcal{C}}(A, (B^i)_{1 \leq i \leq n})$ where convexity $= \mathcal{C}$.
3. $2\mathtt{CA}_{\mathcal{HV}}(A, (B^i)_{1 \leq i \leq n}, (\overline{B}^i)_{1 \leq i \leq n})$ with a 8-connected support and pads included in the $\mathcal{HV}$-grid of A.
4. $2\mathtt{CA}_{\mathcal{HV}}(A, (B^i)_{1 \leq i \leq n}, (\overline{B}^i)_{1 \leq i \leq n})$ with disjoint pads.
5. $2\mathtt{CA}_{\mathcal{C}}(A, (B^i)_{1 \leq i \leq n}, (\overline{B}^i)_{1 \leq i \leq n})$ with disjoint pads.

For any of these five problems, given a set of indices I, we can test in polynomial time whether I is a solution of the instance. It follows that all these problems are in NP. Secondly, we have to provide a polynomial time reduction of an instance of a known NP-complete problem to each one of these problems. The five proofs follow the same strategy: reduce an instance of 3-SAT.

We assume that we have an instance of 3-SAT with n Boolean variables $(\mathfrak{b}_i)_{1\leq i\leq n} \in \{0,1\}^n$ and N 3-clauses. The goal is to build an instance of convex aggregation which admits a solution if and only if the 3-SAT instance is feasible. The strategy is to use pads B^i to represent literals. In the case of unary convex aggregation, we use the pad B^i of index i to represent the literal $\mathfrak{b}_i$ and the pad B^{i+n} of index $i+n$ to represent the literal $\overline{\mathfrak{b}}_i$. Therefore, we use $2n$ pads for representing n variables. In the case of binary convex aggregation, the pad B^i represents the literal $\mathfrak{b}_i$ while the conjugate pad $\overline{B}^i$ represents its negation $\overline{\mathfrak{b}}_i$. According to this strategy, the purpose of each proof is to provide a geometrical structure where the constraint of convex aggregation encodes, first, the relations of negation between two literals and secondly, the relations given by the 3-clauses.

5.1 Case 3 - $\mathcal{HV}$-Convex Binary Aggregation with Non-disjoint Pads

We show how to reduce a 3-SAT instance with n Boolean variables and c clauses to an equivalent instance $2\mathtt{CA}_{\mathcal{HV}}(A,(B^i)_{1\leq i\leq n},(\overline{B}^i)_{1\leq i\leq n})$ of polynomial size. The structure is made by a support of size $(2c+1)\times(c+1)$. Its upper points can be used to encode the clauses with Horizontal convexity (Fig. 4). The equivalence between the instance of 3-SAT and the corresponding instance of $2\mathtt{CA}_{\mathcal{HV}}(A,(B^i)_{1\leq i\leq n},(\overline{B}^i)_{1\leq i\leq n})$ is a consequence of the rewriting of the clauses as implications before their encoding with pads. Then, the NP-completeness of the problem is obvious.

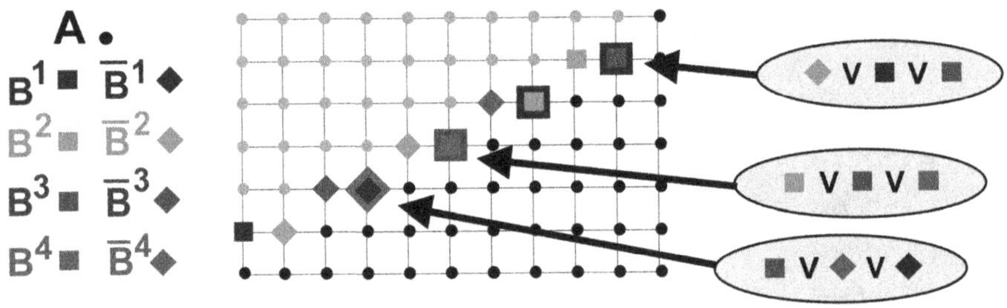

Fig. 4. Reduction of 3-SAT instance to $2\mathtt{CA}_{\mathcal{HV}}(A,(B^i)_{1\leq i\leq n})$. We encode the 3-SAT instance $(\overline{\mathfrak{b}}_1\vee\overline{\mathfrak{b}}_2)\wedge(\mathfrak{b}_3\vee\overline{\mathfrak{b}}_1\vee\overline{\mathfrak{b}}_4)\wedge(e\mathfrak{b}_2\vee\mathfrak{b}_3\vee\mathfrak{b}_4)\wedge(\mathfrak{b}_4\vee\mathfrak{b}_1\vee\mathfrak{b}_2)\wedge(\overline{\mathfrak{b}}_2\vee\mathfrak{b}_1\vee\mathfrak{b}_3)$. The clauses can be rewritten as $\mathfrak{b}_1 \implies \overline{\mathfrak{b}}_2$, $\overline{\mathfrak{b}}_3 \implies \overline{\mathfrak{b}}_1\vee\overline{\mathfrak{b}}_4$, $\overline{\mathfrak{b}}_2 \implies \mathfrak{b}_3\vee\mathfrak{b}_4$, $\overline{\mathfrak{b}}_4 \implies \mathfrak{b}_1\vee\mathfrak{b}_2$, $\mathfrak{b}_2 \implies \mathfrak{b}_1\vee\mathfrak{b}_3$ before being encoded on each row.

5.2 Cases 1 and 4 - $\mathcal{HV}$-Convex Aggregation

We provide now a sketch of reduction of any 3-SAT instance with n Boolean variables and c clauses to an equivalent instance $1\mathtt{CA}_{\mathcal{HV}}(A,(B^i)_{1\leq i\leq n})$ of polynomial size. The strategy is illustrated in Fig. 5. Two rectangular regions are used to encode a relation of negation between B^i and B^{n+i}: the pad B^i is joined with

A $(i \in I)$ if and only if the pad B^{i+n} is not joined with A $(i+n \notin I)$. Therefore the pad B^i represents the literal $\mathfrak{b}_i$ while B_{n+i} represents its negation $\overline{\mathfrak{b}}_i$. Then each clause is encoded on an upper row. The size of the geometrical structure is $(3+3n+c)\times(2n+c)$. The equivalence between the instance of 3-SAT and the corresponding instance of $1\mathtt{CA}_{\mathcal{HV}}(A,(B^i)_{1\le i\le n})$ provides the wanted result.

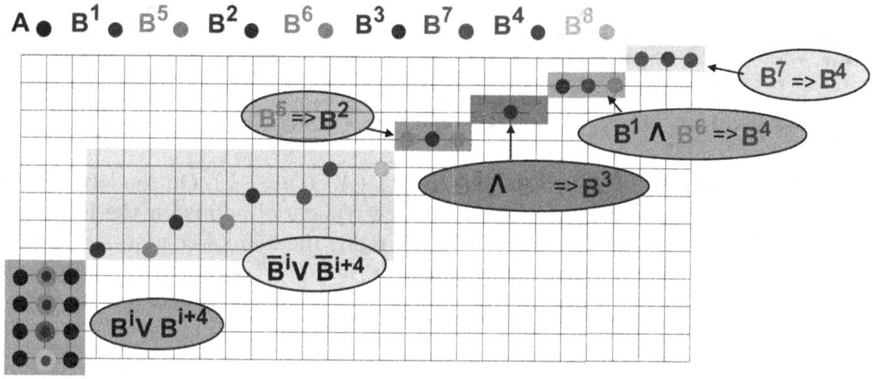

Fig. 5. Reduction of 3-SAT instance to $1\mathtt{CA}_{\mathcal{HV}}(A,(B^i)_{1\le i\le n})$. We encode the 3-SAT instance $(\mathfrak{b}_1 \vee \mathfrak{b}_2)\wedge(\mathfrak{b}_1 \vee \mathfrak{b}_2 \vee \mathfrak{b}_3)\wedge(\overline{\mathfrak{b}}_1 \vee \mathfrak{b}_2 \vee \mathfrak{b}_4)\wedge(\mathfrak{b}_3 \vee \mathfrak{b}_4)$. The clauses can be rewritten as $\overline{\mathfrak{b}}_1 \implies \mathfrak{b}_2$, $\overline{\mathfrak{b}}_1 \wedge \overline{\mathfrak{b}}_2 \implies \mathfrak{b}_3$, $\mathfrak{b}_1 \wedge \overline{\mathfrak{b}}_2 \implies \mathfrak{b}_4$ and $\overline{\mathfrak{b}}_3 \implies \mathfrak{b}_4$. The $\mathcal{HV}$-convexity in the grey and yellow zones guarantee that we have to aggregate either B^i or B^{i+n} but not both. Then, as the pad B^i encodes the literal $\mathfrak{b}_i$, the pad B^{i+n} encodes its negation $\overline{\mathfrak{b}}_i$. In the upper part, each clause is easily encoded by a row.

The NP-completeness of the case 4 is obtained with the same strategy and a more simple structure. The pad B^i represents the literal $\mathfrak{b}_i$ while its conjugate $\overline{B}^i$ represents directly its negation $\overline{\mathfrak{b}}_i$. It allows us to encode any 3-SAT instance with a row for each clause (as in the upper part of Fig. 5). It makes the NP-completeness of $2\mathtt{CA}_{\mathcal{HV}}(A,(B^i)_{1\le i\le n},(\overline{B}^i)_{1\le i\le n})$ with disjoint pads straightforward.

5.3 Cases 2 and 5 - $\mathcal{C}$-Convex Aggregation

We build a new geometrical structure with two goals: encode a negation relation between the pads B^i and B^{i+n}, and encode the c clauses of the 3-SAT instance. The structure has a support made by a polygon of consecutive edges of coordinates $(4k, 4)$ with k going from 0 to $2n+c$ as drawn in Fig. 6. On each border, the polygon has 3 integer points which can be added or removed without any interference with the other points of the same kind. Each triplet of points is called a *nest*. The strategy of the algorithm is to use each nest to encode either the negation between the literals, or the clauses. We need 2 nests for each Boolean variable to encode the negation relation between the pads B^i and B^{i+n} and 1 nest per 3-clause.

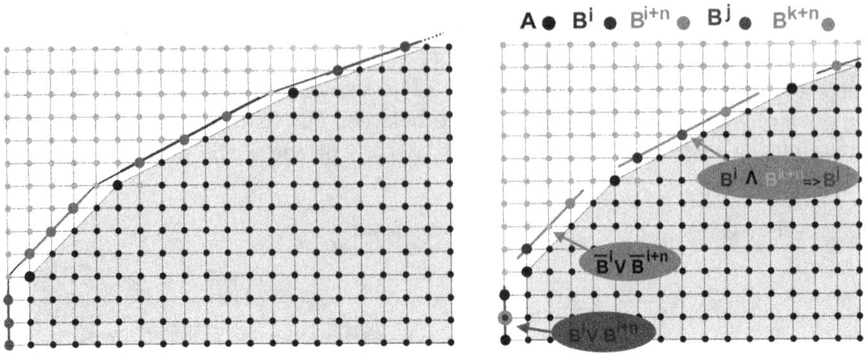

Fig. 6. Reduction of 3-SAT instance to $2\mathtt{CA}_{\mathcal{C}}(A,(B^i)_{1\leq i\leq n})$. On the left, we provide the construction of the nests i.e the colored triplets of points on the left of the polygon of edges of coordinates $(4k, 4)$. On the right, the two lower nests are used to encode the negation relation between the pads B^i and B^{i+n} so that B^{k+n} represents the literal $\overline{\mathfrak{b}}_i$. The third nest is used to encode $\mathfrak{b}_i \wedge \overline{\mathfrak{b}}_k \implies \mathfrak{b}_j$, namely $\overline{\mathfrak{b}}_i \vee \mathfrak{b}_k \vee \mathfrak{b}_j$.

This construction allows to prove that $1\mathtt{CA}_{\mathcal{C}}(A,(B^i)_{1\leq i\leq n})$ is NP-complete. It proves of course that $2\mathtt{CA}_{\mathcal{C}}(A,(B^i)_{1\leq i\leq n},(\overline{B}^i)_{1\leq i\leq n})$ is NP-complete since it is true with empty pads $\overline{B}^i$. For proving the case 5, the NP-completeness of $2\mathtt{CA}_{\mathcal{C}}(A,(B^i)_{1\leq i\leq n},(\overline{B}^i)_{1\leq i\leq n})$ with disjoint pads, we have to notice that the intersections between the pads B^i and B^{i+n} used for encoding the negation of $\mathfrak{b}_i$ are now useless. The pads $\overline{B}^i$ encode directly the negative literals $\overline{\mathfrak{b}}_i$. Then, only one nest per clause is sufficient to encode geometrically the 3-SAT instance (Fig. 7). As the construction has a polynomial size and can be done in polynomial time, it proves the NP-completeness of $2\mathtt{CA}_{\mathcal{C}}(A,(B^i)_{1\leq i\leq n},(\overline{B}^i)_{1\leq i\leq n})$ with disjoint pads. We can at last notice that the support A is 8-connected and that there is no difficulty to have pads included in the $\mathcal{HV}$-grid of A. It follows that $2\mathtt{CA}_{\mathcal{C}}(A,(B^i)_{1\leq i\leq n},(\overline{B}^i)_{1\leq i\leq n})$ remains NP-complete in this subcase.

6 Perspectives

Binary convex aggregation is the bottleneck of several problems of Discrete Tomography. The problem of binary convex aggregation $2\mathtt{CA}_{\mathcal{HV}}(A,(B^i)_{1\leq i\leq n},(\overline{B}^i)_{1\leq i\leq n})$ with an 8-connected support and disjoint pads is the last step of a possible strategy for reconstructing $\mathcal{C}$-convex lattice sets from their horizontal and vertical X-rays [3,7]. Unless $P = NP$, its NP-completeness shown in previous section lets no hope to tackle this last step with a general polynomial time algorithm. It shows that structural properties of the pads are necessary in this framework to achieve this strategy.

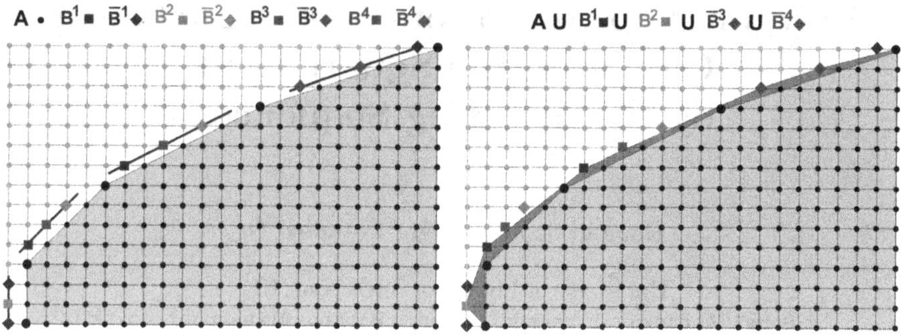

Fig. 7. Reduction of 3-SAT instance to $2\mathtt{CA}_{\mathcal{C}}(A, (B^i)_{1\leq i\leq n}, (\overline{B}^i)_{1\leq i\leq n})$ with disjoint pads. We encode the 3-SAT instance $(\mathfrak{b}_1 \vee \mathfrak{b}_2) \wedge (\overline{\mathfrak{b}}_1 \vee \mathfrak{b}_2 \vee \mathfrak{b}_3) \wedge (\overline{\mathfrak{b}}_1 \vee \mathfrak{b}_2 \vee \mathfrak{b}_4) \wedge (\mathfrak{b}_3 \vee \mathfrak{b}_1 \vee \overline{\mathfrak{b}}_4)$. By rewriting the clauses $\overline{\mathfrak{b}}_1 \implies \mathfrak{b}_2$, $\mathfrak{b}_1 \wedge \overline{\mathfrak{b}}_2 \implies \mathfrak{b}_3$, $\mathfrak{b}_1 \wedge \overline{\mathfrak{b}}_2 \implies \mathfrak{b}_4$ and $\overline{\mathfrak{b}}_3 \wedge \overline{\mathfrak{b}}_1 \implies \overline{\mathfrak{b}}_4$, each one is encoded in a nest. A solution of the convex aggregation problem $(A \cup B^1 \cup B^2 \cup \overline{B}^3 \cup \overline{B}^4)$ provides a solution of the 3-SAT instance ($\mathfrak{b}_1 = 1$, $\mathfrak{b}_2 = 1$, $\mathfrak{b}_3 = 0$ and $\mathfrak{b}_4 = 0$).

Acknowledgement. This work has been sponsored by the French government research program "Investissements d'Avenir" through the IDEX-ISITE initiative 16-IDEX-0001 (CAP 20-25).

References

1. Barcucci, E., Del Lungo, A., Nivat, M., Pinzani, R.: Reconstructing convex polyominoes from horizontal and vertical projections. Theor. Comput. Sci. **155**(2), 321–347 (1996)
2. Brunetti, S., Daurat, A.: Reconstruction of convex lattice sets from tomographic projections in quartic time. Theor. Comput. Sci. **406**(1), 55–62 (2008). Discrete Tomography and Digital Geometry: In memory of Attila Kuba
3. Dulio, P., Frosini, A., Rinaldi, S., Tarsissi, L., Vuillon, L.: First steps in the algorithmic reconstruction of digital convex sets. In: Brlek, S., Dolce, F., Reutenauer, C., Vandomme, É. (eds.) WORDS 2017. LNCS, vol. 10432, pp. 164–176. Springer, Cham (2017). https://doi.org/10.1007/978-3-319-66396-8_16
4. Gerard, Y.: Regular switching components. Theoretical Computer Science, Special Issue in the memory of Maurice Nivat (2019, to appear)
5. Aspvall, B., Plass, M.F., Tarjan, R.E.: A linear-time algorithm for testing the truth of certain quantified boolean formulas. Inf. Process. Lett. **8**(3), 121–123 (1979)
6. Dowling, W.F., Gallier, J.H.: Linear-time algorithms for testing the satisfiability of propositional horn formulae. J. Logic Program. **1**(3), 267–284 (1984)
7. Gerard, Y.: Polynomial time reconstruction of regular convex lattice sets from their horizontal and vertical X-rays. preprint hal-01854636, August 2018
8. Carathéodory, C.: Über den variabilitätsbereich der koeffizienten von potenzreihen, die gegebene werte nicht annehmen. Math. Ann. **64**(1), 95–115 (1907)

Polygon Approximations of the Euclidean Circles on the Square Grid by Broadcasting Sequences

Haomin Song and Igor Potapov(✉)

University of Liverpool, Liverpool, UK
H.Song12@student.liverpool.ac.uk, Potapov@liverpool.ac.uk

Abstract. Euclidean circle approximation on the square grid is an important problem in digital geometry. Recently several schemes have been proposed for approximation of Euclidean circles based on Neighbourhood Sequences, which correspond to repeated application of the von Neumann and Moore neighbourhoods on a square grid. In this paper we study polygon approximations of the Euclidean circles on the square grid with Broadcasting Sequences which can be seen as a generalization of Neighbourhood Sequences. The polygons generated by Broadcasting Sequences are the Minkowski sums of digital disks defined by a given set of broadcasting radii. We propose a polynomial time algorithm that can generate Broadcasting Sequences which are providing flexible and accurate approximation of Euclidean circles.

1 Introduction

In the area of digital image processing the Distance Transform (DT) computation is computationally expensive if it is based on explicit calculation of Euclidean distances. In order to overcome this problem, other distance metrics (such as Manhattan, Chessboard, Neighbourhood, Chamfer, Broadcasting) have been proposed to calculate the distance considering only small neighbourhoods. The transforms that apply these metrics can be seen as an approximation of Euclidean distance with considerably less complex algorithms and which is suitable for parallel implementation on cellular processor arrays. In this context approximation of Euclidean distance leads to the problem of Euclidean circle approximation on regular grids [1–3]. Polygonal approximation is one of the efficient approaches to approximate a circle in a discrete domain [4]. The main objective is to approximate a given circle by a polygonal chain satisfying certain optimality criteria, such as minimising global approximation or restricting local error in some predefined threshold. Moreover, such approximations usually require to reduce storage spaces that needed to represent a shape of some circle as well as to improve the extraction and properties of required features from a given set of digital curves [5]. The simpler and more efficient approximations has been constructed then more likely it can fit to a larger class of applied problems.

M. Couprie et al. (Eds.): DGCI 2019, LNCS 11414, pp. 444–456, 2019.
https://doi.org/10.1007/978-3-030-14085-4_35

Several schemes have been proposed for an approximation of circles based on Neighbourhood Sequences, which is an application of the von Neumann and Moore neighbourhoods on the square lattice [6–8], triangular [9] or hexagon lattices [2]. In case of the neighbourhood sequences the best approximation for a circle on the square lattice is the regular octagon and the problem of generating such shape by neighbourhood sequences has been analysed in [10]. The main downsides for a circle approximation by this approach are high error rate and low expansibility due to restricted polygonal shapes.

This paper provides another methodology, using broadcasting automata as a tool, to implement polygonal approximation. Broadcasting Automata can be defined on some form of grid or lattice structure and have a simple computational primitives comparative to finite state automata with the ability to receive and send messages both from and to those automata which are within its transmission radius [11]. In this work, we only consider the square grid structure, but due to flexibility of the model similar broadcasting processes can be defined on other lattices or other distance measured methods. However in connection with simple agregation functions it can be used to generate more complex non-convex shapes, the approximation of L^p metric or even arbitrary patterns, see [11].

Originally the broadcasting automata has been studied in the context of the distributed algorithms and pattern formation problems. Each state in the grid or lattice is regarded as a cell, which only has limited sensor range and restricted common knowledge such as positions [12]. Therefore, all those cells within a certain transmission radius, receive the message from the sender. As a result, the communication between cells works like a Multi Agent System, which increases the robustness and ability to tackle failures and abnormal conditions [13]. In analogy to the neighbourhood sequences the similar broadcasting sequences can be defined and in the case of 2D grid it covers classical Moore and Von Neumann neighbourhoods as well as other neighbourhoods corresponding to digital disks.

By changing broadcasting radii we are changing sizes and shapes of transmission polygons on a grid (i.e. digital disks) [12]. Obviously using larger number of radii corresponds to transmission polygons with larger number of line segments which in its turn leads to better approximation of the circles. So along this line we are focusing on the problem of finding the best approximation of the Euclidean circles for a given fixed set of transmission radii. Following recent results from [12] about the composition properties of their Minkowski sum and methods for efficient description of the polygons by the chain codes we design an algorithm to find a Broadcasting Sequence that provides polygon approximation of Euclidean circles. The error rate between polygon and circle has been used as criterion to evaluate efficiency. We provide the analysis of the time complexity for the proposed algorithm and experimental results to compare generated approximations with optimal solution for a given set of radii and the distance to the Euclidean circles.

2 Approximation with Neighbourhood Sequences

The circle approximation using neighbourhood sequences on a square grid has been studied in [10]. The neighbourhood sequences can generate octagon and the more regular octagon can be generated the better approximation can be achieved. The area of a polygon is defined by the Eq. (1) in [10]

$$S = a^2 + 2\sqrt{2}ab + b^2, \tag{1}$$

where a and b are side-lengths of the sequence-generated octagon. So the best approximation can be achieved when the side-lengths are the same, i.e. when $a = b$. In this case, the area of the ideal polygon is equal to $S_{polygon} = (2 + 2\sqrt{2})a^2$. The relationship between a and the radius of circumscribed circle R is represented by the equation $a = \sqrt{2-\sqrt{2}}R$. In that case, the area of the circumscribed circle is $S_{circumscribed} = \pi R^2 = (1 + \frac{\sqrt{2}}{2})\pi a^2$. To evaluate the approximation between polygon and the circumscribed circle the following Error Rate (ER) defined by Eq. (2) has been used in [10]

$$ER = 1 - \frac{S_{polygon}}{S_{circumscribed}}. \tag{2}$$

It is obvious that the better approximation corresponds to lower error rate and the error rate of neighbourhood sequences approximation is about 9.97% as shown in [10]. The objective for this paper is to provide more accurate polygonal approximation and to design an algorithm for its construction.

3 Broadcasting Sequences

As the alternative to the neighbourhood sequences we consider the problem of approximation of Euclidean circles by polygons generated by a set of neighbourhoods which correspond to a finite set of digital disks. Such sequences are known as *Broadcasting sequences* [11]. They are naturally generalizing neighbourhood sequences and enriching the model by a variety of generated polygons. However finding the optimal approximation of Euclidean circles by broadcasting sequences for a given finite set of digital disks is quite nontrivial computational problem, since it is hard to provide simple canonical forms corresponding to the optimal solutions for a set of digital disks.

In this section we introduce basic concepts of broadcasting sequences and digitized circles, which will be used as neighbourhoods for approximation of Euclidean circles. Originally broadcasting automata model has been defined in [11,12] as a network of finite automata on the integer grid. The mechanism for communication between automata is message passing. The messages generated by an automaton (sender) transfer to its transmission neighbourhood (receiver), which is a subset of its neighbors within the transmission range. Messages are generated and transferred instantaneously at a discrete time step, regarding as synchronous steps [12]. In this paper we study the concept of transmission neighbourhoods in the context of the Euclidean circle approximation problem.

Definition 1. *The **Euclidean distance** between two cells c_1 and c_2 on a square grid with integer coordinates (x_1, y_1) and (x_2, y_2) is defined as*

$$d(c_1, c_2) = \sqrt{(x_1 - x_2)^2 + (y_1 - y_2)^2}.$$

Definition 2. ***The square transmission radius** r, which is used to quantify the transmission range, is the square of the max cell distance of the sender and its receivers. **The Broadcasting Neighbourhood** of the cell c_0 on the distance $\sqrt{r}$ is defined as a set $\{c|d(c, c_0)^2 \leq r\}$ which we also call as **primitive transmission polygon (PTP)**.*

Varying the transmission radius we construct different digital disks, i.e. different sizes and shapes of PTP. We also use a notion of the square radius to identify a PTP for simplicity to stay with integer numbers. Let us illustrate a variety of PTPs corresponding to different square transmission radii on Fig. 1 [11].

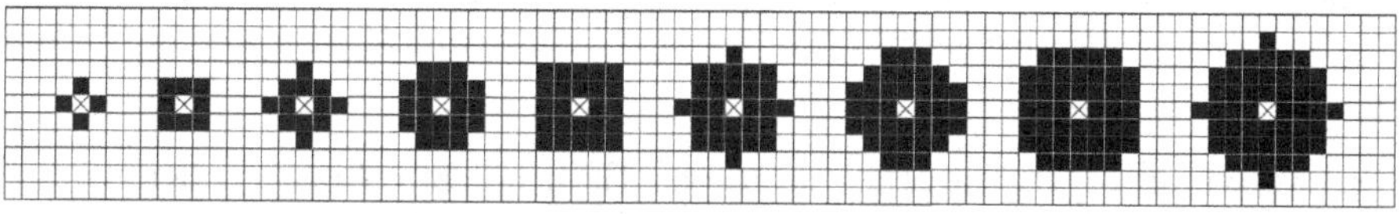

Fig. 1. Examples of PTPs with square transmission radii in order $\mathcal{R} = \{1, 2, 4, 5, 8, 9, 10, 13, 16\}$

Definition 3. *A **Broadcasting sequence** is a sequence of radii*

$$\mathcal{R} = \{r_1, r_2, r_3 ... r_n\},$$

where r_i is a square transmission radius and n is the number of elements in the sequence. The polygon generated by broadcasting sequence is the Minkowski sum of all primitive transmission polygons (digital disks) with radii from $\mathcal{R}$.[1]

In order to represent symmetrical polygons generated by broadcasting sequences we use the chain code [14]. The polygon is divided into eight octants (Fig. 2) and it is sufficient and necessary to depict the first octant of a polygon because the rest of the part is "mirrored" from the first octant [11].

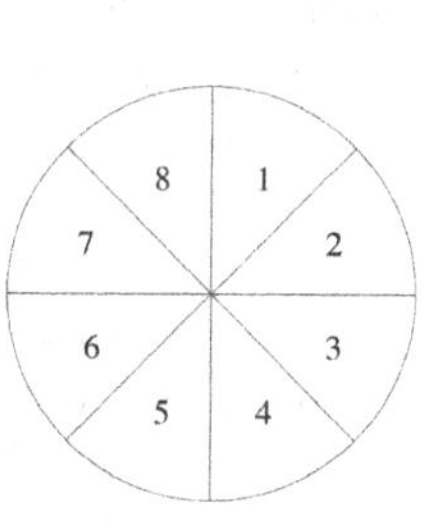

Fig. 2. Eight octants of digital polygons

[1] Minkowski sum gives the same polygon for different order of the composition as it is commutative operation.

Definition 4 [11]. *A* ***chain code*** *for a digitized circle in the first octant is a word in* $\{0,1\}^*$ *where* 0 *is positive motion along the* x*-axis,* 1 *is positive motion along the* x*-axis and negative motion along the* y*-axis. The* 9th *image in Fig. 1 is an example of chain code "101".*

It has been proven in [11] that the edges of primitive transmission polygon in the first octant can be represented by chain code the formal definition of the line segments has been introduced to describe the sides of these polygons.

Definition 5 [11]. ***Line segments****, sub-word of chain code, represent a straight edge of a circle in one of the following forms:* 0^*, $(10^n)^*$,$(10^n)(10^{n+1})^*$, $(10^n)^*(10^{n+1})$. *For example, the line segments of* 9th *polygon in Fig. 1 are:* $\{10,1\}$.

Definition 6 [11]. *A* ***gradient****,* $G(s)$*, of a line segment* s*, is* $G(s) = \frac{|s|_1}{|s|}$*, where the function* $|s|_1$ *returns the number of* 1*'s in the chain code and* $|s|$ *is the length of* s*. It is also called tangent of line segment. For example, the gradient of line segment "1000" is 0.25.*

It was shown in [11] that for these polygons, the gradients of line segments in the first octant are strictly increasing and the following composition theorem holds:

Theorem 1 [11]. ***Composition theorem:*** *Given two chain codes* u *and* v *which contain line segment* $l_1^u l_2^u ... l_t^u$ *and* $l_1^v l_2^v ... l_{t'}^v$ *with strictly increasing gradients, the chain code of a composition of* u *and* v *can be constructed by combining line segments of* u *and* v*, then ordering them by increasing gradient. The composition of two chain codes* u *and* v *is* ***commutative*** *and corresponds to the chain code of Minkowski sum of digital disks in the first octant.*

Definition 7. *A* ***length****,* $L(s)$*, of a line segment* s*, is* $L(s) = |s|\sqrt{1+G(s)^2}$*, where* $|s|$ *is the length of line segment* s*,* $G(s)$ *is the gradient of* s*.*

4 Polygon Approximation of Circle

In this paper we design the algorithm to find a polygon approximation of a circle using broadcasting sequences. Single primitive transmission polygons may have a very high error rate for circle approximation and clearly polygons with more segments can achieve better approximation rate circle. Following Theorem 1 it is clear that the composition of multiple PTPs contains more types of line segments, which results in a better approximation for circle. However when the total number of radii is fixed the main question is to find the right ratios of radii that should be used together for the best approximation of a circle.

The problem of finding the best **approximation of Euclidean circle** can be formulated as follows: Given a set of square transmission radii $\{r_1, r_2, \ldots, r_m\}$ and the maximal length of broadcasting sequence h find a broadcasting sequence

$$\mathcal{R} = \{r_1...r_1, r_2...r_2, \ldots, r_m...r_m\},$$

where $|\mathcal{R}| \leq h$ such that the polygon generated by a sequence $\mathcal{R}$ (i.e. Minkowski sum of digital disks with radii from $\mathcal{R}$) has the smaller error rate in relation to Euclidean circle.

In order to solve the above problem we represent polygons in the vector form, look for the ideal polygon based on the line segments and finally approximate the ideal polygon by given PTPs. So the algorithm contains three parts: vector system setup, ideal polygon and broadcasting sequence construction.

4.1 Vector System Setup

The vector system is used to represent polygons by line segments. Each line segment in a polygon has two associated values: the gradient and the length. These line segments can form a polygon by strict increasing gradients in the first octant. In that case, **gradient set** records gradients of line segments and **length vector** records lengths of corresponding line segments.

Definition 8. ***Gradient set****, $GV(p) = \{g_1, g_2, ...g_n\}$, of the polygon p, denotes a list of gradients of line segments of p. For convenience, gradients in the set are in ascending order ($g_i < g_{i+1}$).*

Definition 9. ***Length vector****, $LV(p) = [l_1, l_2, ...l_n]$, of the polygon p, denotes a list of length of line segments of p. The elements of the same index in $GV(p)$ and $LV(p)$ indicate the gradient and length of one line segment respectively.*

In order to operate with several polygons we will introduce extensive length vector. It represents a length vector of p in the context of larger gradient set of p_0, where $GV(p)$ is the subset of $GV(p_0)$.

Definition 10. ***Extensive length vector****, $LV(p, p_0) = [l_1, l_2, ...l_n]$, of the polygon p, denotes a list of length of line segments of p corresponding to $GV(p_0)$ and if some g exist in $GV(p_0)$ but not in $GV(p)$, the corresponding element in this vector is 0. The size of this vector is the same as $GV(p_0)$.*

Later we use p_0 to represent the objective polygon and $PTPs$=$\{p_1, p_2...p_m\}$ represents all PTPs according to $\mathcal{R}$. Since the objective polygon is generated by the given PTPs, gradient set of objective polygon is identified as $GV(p_0) = \bigcup_{i=1}^{m} GV(p_i)$. In that case, the gradient set of each PTP is the subset of $GV(p_0)$.

Definition 11. ***Length matrix*** *of a list of $PTPs = (p_1, p_2 \cdots p_m)$ is $LM(PTPs, p_0) = [LV(p_1, p_0)^T, LV(p_2, p_0)^T \ldots LV(p_m, p_0)^T]$, which is a matrix of lengths of line segments of $PTPs$ corresponding to $GV(p_0)$. The size of this matrix is the product of the size of $PTPs$ and the size of $GV(p_0)$.*

Example: Let us consider a set of PTPs $P = \{p_1, p_2, p_3\}$ with corresponding radii $\{r_1 = 16, r_2 = 18, r_3 = 25\}$. Let p_0 be the objective polygon, then the Gradient set of p_1, p_2 p_3 and p_0 will be defined as $GV(p_1) = \{\frac{1}{2}, 1\}$, $GV(p_2) = \{0, \frac{1}{2}\}$, $GV(p_3) = \{\frac{1}{3}, 1\}$ and $GV(p_0) = \{0, \frac{1}{3}, \frac{1}{2}, 1\}$. The Extensive Length vectors of

PTPs in this case are: $LV(p_1, p_0) = \{0, 0, \sqrt{5}, \sqrt{2}\}$, $LV(p_2, p_0) = \{2, 0, \sqrt{5}, 0\}$ and $LV(p_3, p_0) = \{0, \sqrt{10}, 0, \sqrt{2}\}$ and the Length matrix is:

$$LM(P, p_0) = \begin{bmatrix} 0 & 2 & 0 \\ 0 & 0 & \sqrt{10} \\ \sqrt{5} & \sqrt{5} & 0 \\ \sqrt{2} & 0 & \sqrt{2} \end{bmatrix}.$$

Thus the length matrix $X = LM(PTPs, GV(p_0))$ represents PTPs and the length vector $Y = LV(p_0)$ represents the length vector of objective polygon. We use the notion of the Parikh vector to combine these two objects X and Y.

Definition 12 [15]. *Let $\sum = \{a_1, a_2 ... a_k\}$ be an alphabet. The **Parikh vector** of a word w is defined as the function $p(w) = (|w|_{a_1}, |w|_{a_2}, |w|_{a_3} ... |w|_{a_k})$, where $|w|_{a_i}$ denotes the number of occurrences of the letter a_i in the word w.*

In other words, the elements in Parikh vector are the number of occurrences of corresponding radii that appear in the broadcasting sequence. Finally, the relationship of length vector of objective polygon Y, the length matrix of PTPs X and the Parikh vector P can be defined as Eq. (3).

$$XP = Y \tag{3}$$

The value of X can be derived from PTPs, however the values of Y and P are unknown. In Subsect. 4.2 we show how to find the value of Y and in the Subsect. 4.3 we will calculate P from X and Y.

4.2 Ideal Polygon Construction

Let us start with finding the length vector of the ideal polygon Y according to the gradient set $GV(p_0)$, which contains all gradients of line segments that PTPs provide. Since the polygon is an inscribed polygon with max area, each vertex of polygon should be on the circle with the same distance to the center of circle and the polygon occupies the circle as large proposition as possible. To calculate the area of polygon, the polygon is divided into isosceles triangles whose point angles are the center of the circle and the base edges are the edges of polygon. In that case, the area of polygon is the sum of areas of all triangles, see Fig. 3. In addition, since the gradients of edges are fixed, there are constraints (Eq. (4)) among point angles and base angles, i.e θ_i and α_i, where θ_i is known as outer angle of polygon and α_i is as the inner angle of a divided triangle of the polygon, both shown on Fig. 3:

$$2\pi - 2\theta_i - \alpha_i - \alpha_{i+1} = 0 \tag{4}$$

Due to the symmetry of constructed polygons, we only need to count up triangles in the first octant, which occupies $\frac{1}{8}$ of the total polygon. However, due to the double line segments, the triangles with gradient of 0 or 1 only half

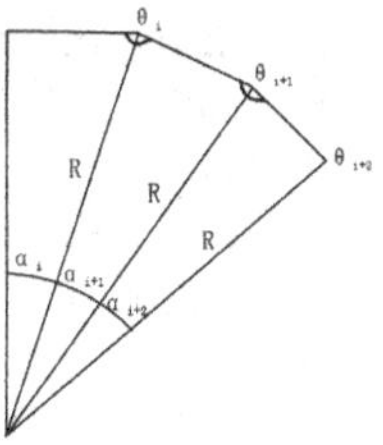

Fig. 3. Triangle division of polygon

occupy the first octant. In that case, when calculating the area, the area should be half for these triangles. The area function of a polygon in the first octant can be defined as Eq. (5) and the objective is to find the maximum of S_P.

$$S_P(\alpha) = \begin{cases} \frac{1}{2}R^2\sum_1^n \sin\alpha_i & \text{no double line segment} \\ \frac{1}{2}R^2(\sum_2^n \sin\alpha_i + \frac{1}{2}\sin\alpha_1) & G(\alpha_1)=0, G(\alpha_n)\neq 1 \\ \frac{1}{2}R^2(\sum_1^{n-1} \sin\alpha_i + \frac{1}{2}\sin\alpha_n) & G(\alpha_1)\neq 0, G(\alpha_n)=1 \\ \frac{1}{2}R^2(\sum_2^{n-1} \sin\alpha_i + \frac{1}{2}(\sin\alpha_1+\sin\alpha_n)) & G(\alpha_1)=0, G(\alpha_n)=1 \end{cases} \tag{5}$$

From the criterion equation, since the circumscribed circle is fixed, it is clear that the larger area the polygon corresponds to the lower error rate of an approximation. Therefore, we need to find the maxima of this function subject to equality constraints. We use Lagrange Multiplier method, which is the proper mathematical optimization technique to find the solutions of the function. Therefore, for constraints, we define scalars $\lambda_1, \lambda_2 ... \lambda_{n-1}$ and an auxiliary function shown in Eq. (6), then we solve the Equation set (7). There are $2n-1$ equations and $2n-1$ variables, which are solvable.

$$F = S_P(\alpha) + \sum_{i=1}^{n-1} \lambda_i(2\pi - 2\theta_i - \alpha_i - \alpha_{i+1}) \tag{6}$$

$$F(\alpha;\lambda) = \begin{cases} \nabla S_P(\alpha) - \sum_{k=1}^{n-1}\lambda_k \nabla(2\pi - 2\theta_i - \alpha_i - \alpha_{i+1}) = 0 \\ 2\pi - 2\theta_i - \alpha_i - \alpha_{i+1} = 0 \end{cases} \tag{7}$$

Since the above equations are not linear, we utilized Newton-Raphson method to solve the equation set, which is the method of finding successively better approximations to the roots of a real-valued function. All variables are combined in a vector $x = [\alpha; \lambda]$. The iterative equation can be defined as Eq. (8)

$$x^{(k+1)} = x^{(k)} - J^{-1}(x^{(k)})S_P(x^{(k)}) \tag{8}$$

where $x^{(k)}$ is previous point, $x^{(k+1)}$ is the derivative point, S_P is the column vector of equations at $x^{(k)}$ and J is Jacobian matrix of S_P at $x^{(k)}$, a matrix of all first-order partial derivatives of S_P.

We create an iterative algorithm to implement this method. In the algorithm, we define the max iterative time max_itr and deviation range err_range. The algorithm repeats until deviation between calculated function and objective values is smaller than the deviation range. The founded x is result of equation set and k is total iterative step. If the iterative time overtakes the predefined value max_itr, the algorithm returns no results.

Consequently, the resulted vector $x = [\alpha; \lambda]$ of Newton-Raphson method contains the sine values of ideal angles of triangles in the first octant. Based on the equation $l_i = 2R \sin \alpha_i$, the ratio of lengths of line segment is the same as ratio of sine values of inner angles since that all divided triangles are isosceles with equal side length R. In addition, only the shape of polygon is considered. Therefore, the relative lengths of line segments should be adopted.

4.3 Broadcasting Sequence Construction

In this section we will show how to calculate P and find the broadcasting sequence which can be close to the ideal polygon. In the Sect. 4.1 we show that the ideal result of Parikh vector should satisfies to the Eq. 3, however, for length matrix X, we still need to consider several cases:

1. If the number of rows is smaller than the number of columns, which means the number of PTPs is larger than the number of gradient types, the result of P is not unique and infinite.
2. If the number of rows equal to the number of columns, which means the number of PTPs is the same as the number of gradient types, the result of P is unique.
3. If the number of rows is larger than the number of columns, which means the number of PTPs is larger than the number of gradient types, the result of P is not exist.

For the first case, we can define some elements of P in the admissible domain until the number of undefined elements is equal to the number of columns. Then the result of P is unique and computable. For the second case, P is derived directly by the Eq. (9)

$$P = X^{-1}Y. \tag{9}$$

For the third case, assuming that there is no P that satisfy the equation, we aim to find the best fitting that minimizes the deviation. The Least Square Fitting is the most common method which solves the Eq. (10)

$$S(P) = \arg \sin(\|XP - Y\|)^2. \tag{10}$$

We also need to assume that all elements in P must be positive. In that case, a constraint is added in the equation and another extensive Least Square Fitting method— Non-negative least squares (NNLS) with the constraint $x > 0$ for equation. This method is based on the observation that only a small subset of constraints is usually active at the solution [16], which is given by Lawson and Hanson [17]. The basic principle uses iterative method to calculate the equation until Kuhn-Tucker conditions are satisfied. Currently, MathWorks [18] provides the function for the algorithm NNLS, which is called to "lsqnonneg". Therefore, the Eq. (11) can be used directly for a given X and Y

$$P = (X^T X)^{-1} X^T Y. \tag{11}$$

After generating P we can derive the broadcasting sequence. In the broadcasting sequence, the number of radii is the values of corresponding elements in P. Note that the absolute values of elements in P are not important. The main focus are the relative values among elements. The values of elements in P can be shrunk by same proposition according to the accuracy requirements. Considering that the number of elements in broadcasting sequence should be integers, all elements in P shrink and approximate an integer within the requirement range.

Proposition 1. *For a given set of radii $r_1, \ldots, r_m$ with corresponding number of line segments $n_1, \ldots, n_m$ the constructed algorithm can find the broadcasting sequence approximating Euclidean circle in polynomial time $O(nm^2 + n^3 k)$ where n is the sum of all n_i for $i = 1..m$ and assuming that $k \propto \log t$ where t is approximation precision.*

Proof. According to the Sect. 4, which describes the whole process of broadcasting sequence construction, there are 3 steps needed to be considered.

In the first step "Vector system setup", based on the given set of square radii where the size of set is m and max value of square radius is r_{max}, the max number of line segments n for each PTP satisfies the equation n, which is a linear relationship to square root of r_{max}: $n \propto \sqrt{r_{max}}$. In this case, the time complexity for deriving the set of line segments of objective polygon p_0 is $O(nm)$. According to the definition (10), the time complexity to derive the length matrix X of PTPs is $O(nm^2)$.

In the second step "Ideal Polygon Construction", a non-linear equation set is solved by Newton method, which is quadratic convergent [19]. Assuming that the iterative time is k, based on the results discussing complexity of Newton method [20], the iterative times depends on the initial points of Newton method and approximation precision. Normally k is linear relative to $\log t$ where t is approximation precision. On the other hand, for each iterative step, several components are calculated with the time complexity listed below:

1. Equation vector S_P at the iterative point: $O(n^2)$.
2. Jacobian matrix at the iterative point: $O(n^3)$.
3. Matrix transposition: $O(n^2)$.
4. Matrix inversion and matrix multiplication: $O(n^3)$.

To sum up, the time complexity of Newton Method is $O(n^3 logt)$, where n is the total number of line segments and t is approximation precision. In the third step "Broadcasting Sequence Construction", the time complexity of Eqs. (9)

and (11) are $O(n^3)$ since the calculations only includes matrix transposition, inversion and multiplication. As a result, the time complexity of this method is $O(nm^2 + n^3k)$ and normally $k \propto \log t$ where t is approximation precision.

5 Construction and Comparison

Let us illustrate the results of experiments and a few constructions for specific set of radii. Let us consider the first example for the radius set $\mathcal{R}_1 = \{r_1 = 2, r_2 = 4\}$ which corresponds to application of Von Neumann neighbourhood (PTP with $r = 2$) and Moore neighbourhood (PTP with $r = 4$). Table 1 shows all intermediate values as well as the optimal broadcasting sequence in this case.

Table 1. Immediate results of construction process

Gradient set	Length vector (X)	Length vector (Y)	Parikh vector (P)	Broadcasting sequence of length 60
$\{0, 1\}$	$\begin{bmatrix} 2 & 0 \\ 0 & 2\sqrt{2} \end{bmatrix}$	$\begin{bmatrix} 0.78 \\ 0.79 \end{bmatrix}$	$\begin{bmatrix} 0.39 \\ 0.28 \end{bmatrix}$	$\{35, 25\}$

The image of transmission polygons for the radius set $\mathcal{R}_1$ is shown on Fig. 4 (Left). For this example, the transmission polygons of broadcasting sequence are the same as in case of the neighbourhood sequence. Our broadcasting sequence is optimal and it is matching with optimal error rate of 9.97% in this case.

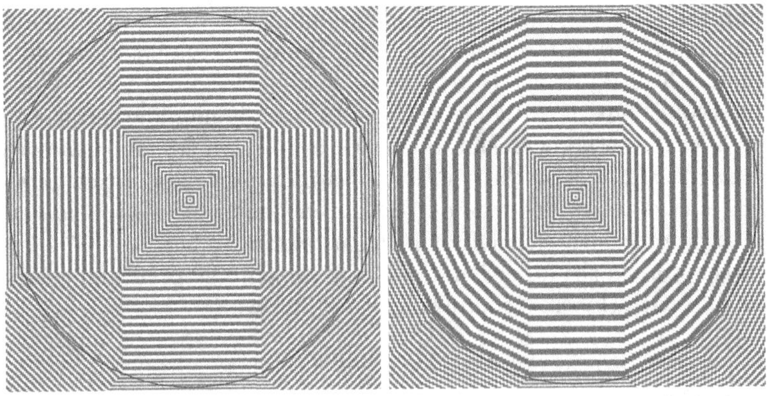

Fig. 4. Constructed polygons for $\mathcal{R}_1 = \{r_1 = 2, r_2 = 4\}$ (Left) and for $\mathcal{R}_2 = \{r_1 = 2, r_2 = 4, r_3 = 9\}$ (Right)

The second example uses three types of PTPs for the design of broadcasting sequence. Let the radius set $\mathcal{R}_2$ be $\{r_1 = 2, r_2 = 4, r_3 = 9\}$. Table 2 shows the intermediate results of each step and Fig. 4 (Right) shows the generated image of transmission polygon. The error rate in this case is 3.03% and the solution is also optimal.

Table 2. Euclidean distance method result

Gradient set	Length vector (X)	Length vector (Y)	Parikh vector (P)	Broadcasting sequence of length 60
$\{0, \frac{1}{2}, 1\}$	$\begin{bmatrix} 2 & 0 & 0 \\ 0 & 0 & \sqrt{5} \\ 0 & 2\sqrt{2} & 0 \end{bmatrix}$	$\begin{bmatrix} 0.52 \\ 0.41 \\ 0.24 \end{bmatrix}$	$\begin{bmatrix} 0.26 \\ 0.08 \\ 0.18 \end{bmatrix}$	$\{29, 10, 21\}$

In the Table 3 we provide comparison between the error rate for optimal solution approximating Euclidean circle by broadcasting sequences of length 60 and the sequence generated by our algorithm.

Table 3. Experimental results comparing error rates between the optimal and approximate solution

Gradient set	Radius set	Approximated broadcasting sequence	Error rate of approximated solution	Optimal broadcasting sequence	Error rate of optimal solution
$0, 1$	{2,4}	{35,25}	9.97%	{35,25}	9.97%
$0, \frac{1}{2}, 1$	{2,4,9}	{29,10,21}	3.03%	{29,10,21}	3.03%
$0, \frac{1}{2}, 1$	{5,16,18}	{24,9,27}	3.03%	{24,9,27}	3.03%
$0, \frac{1}{3}, 1$	{2,25}	{25,35}	7.55%	{0,60}	5.78%
$0, \frac{1}{3}, 1$	{5,25}	{31,29}	4.51%	{31,29}	4.51%
$0, \frac{1}{3}, \frac{1}{2}, 1$	{2,9,36}	{36,9,15}	5.79%	{22,18,20}	2.31%
$0, \frac{1}{3}, \frac{1}{2}, 1$	{16,18,25}	{7,31,22}	3.77%	{11,28,21}	2.98%
$0, \frac{1}{4}, \frac{1}{3}, \frac{1}{2}, 1$	{171,239}	{34,26}	1.68%	{35,25}	1.68%
$0, \frac{1}{4}, \frac{1}{3}, \frac{1}{2}, 1$	{233,239}	{40,20}	2.15%	{11,49}	1.93%

Conclusion. The main contribution of the paper is polynomial time algorithm that can find for a fixed set of digital disks the broadcasting sequence which Minkowski sum (i.e. covered area) can approximate Euclidean circles with a small error rate. The efficiency of the method is limited by the ideal polygon (see Subsect. 4.2) with the same list of gradients as in a given set of digital disks. The error difference between an optimal polygon that can be constructed for a given set of digital disks and a polygon produced by our algorithms can be zero in some cases and also within 3.5% for all tested cases. Our approximation algorithm can calculate the optimal solution (in some cases) or a solution with a small error rate when gradients are repeated for several disks. From the structure of our algorithm follows that the larger set of digital disks we take, the more common gradients they have leading to smaller error rate between optimal and approximate solutions.

References

1. Worring, M., Smeulders, A.W.M.: Digitized circular arcs: characterization and parameter estimation. IEEE Trans. Pattern Anal. Mach. Intell. **17**(6), 587–598 (1995)
2. Nagy, B., Strand, R.: Approximating Euclidean circles by neighbourhood sequences in a hexagonal grid. Theor. Comput. Sci. **412**(15), 1364–1377 (2011)
3. Mukherjee, J.: On approximating Euclidean metrics by weighted t-cost distances in arbitrary dimension. Pattern Recogn. Lett. **32**, 824–831 (2011)
4. Debledrennesson, I., Tabbone, S., Wendling, L.: Fast polygonal approximation of digital curves. In: ICPR 2004, vol. 1, pp. 465–468 (2004)
5. Bhowmick, P., Bhattacharya, B.B.: Approximation of digital circles by regular polygons. In: Singh, S., Singh, M., Apte, C., Perner, P. (eds.) ICAPR 2005. LNCS, vol. 3686, pp. 257–267. Springer, Heidelberg (2005). https://doi.org/10.1007/11551188_28
6. Das, P.P., Chakrabarti, P.P., Chatterji, B.N.: Distance functions in digital geometry. Inf. Sci. **42**(2), 113–136 (1987)
7. Das, P.P., Chatterji, B.N.: Octagonal distances for digital pictures. Inf. Sci. **50**(2), 123–150 (1990)
8. Nagy, B., Strand, R., Normand, N.: Distance functions based on multiple types of weighted steps combined with neighborhood sequences. J. Math. Imaging Vis. **60**, 1209–1219 (2018)
9. Mir-Mohammad-Sadeghi, H., Nagy, B.: On the chamfer polygons on the triangular grid. In: Brimkov, V.E., Barneva, R.P. (eds.) IWCIA 2017. LNCS, vol. 10256, pp. 53–65. Springer, Cham (2017). https://doi.org/10.1007/978-3-319-59108-7_5
10. Farkas, J., Bajak, S., Nagy, B.: Approximating the Euclidean circle in the square grid using neighbourhood sequences. Pure Math. Appl. **17**(3–4), 309–322 (2010)
11. Nickson, T., Potapov, I.: Broadcasting automata and patterns on $\mathbb{Z}^2$. In: Adamatzky, A. (ed.) Automata, Universality, Computation. ECC, vol. 12, pp. 297–340. Springer, Cham (2015). https://doi.org/10.1007/978-3-319-09039-9_14
12. Martin, R., Nickson, T., Potapov, I.: Geometric computations by broadcasting automata. Nat. Comput.: Int. J. **11**, 623–635 (2012)
13. Efrima, A., Peleg, D.: Distributed algorithms for partitioning a swarm of autonomous mobile robots. Theor. Comput. Sci. **410**, 1355–1368 (2009)
14. Klette, R., Rosenfeld, A.: Digital straightness-a review. Discrete Appl. Math. **139**(1), 197–230 (2004)
15. Kozen, D.C.: Automata and Computability. Springer, New York (1997). https://doi.org/10.1007/978-1-4612-1844-9
16. Chen, D., Plemmons, R.J.: Nonnegativity constraints in numerical analysis. In: Symposium on the Birth of Numerical Analysis, pp. 109–139 (2009)
17. Lawson, C.L., Hanson, R.J.: Solving Least Squares Problems. Prentice-Hall/SIAM, Englewood Cliffs/Philadelphia (1995)
18. Shure, L.: Brief history of nonnegative least squares in MATLAB (2006). http://blogs.mathworks.com/loren/2006/
19. Ferreira, O.: Local convergence of Newton's method under majorant condition. J. Comput. Appl. Math. **235**(5), 1515–1522 (2011)
20. Batra, P.: Newton's method and the computational complexity of the fundamental theorem of algebra. Electron. Notes Theor. Comput. Sci. **202**, 201–218 (2008)

Unfolding Level 1 Menger Polycubes of Arbitrary Size With Help of Outer Faces

Lydie Richaume(✉), Eric Andres, Gaëlle Largeteau-Skapin, and Rita Zrour

Laboratory XLIM, ASALI, UMR CNRS 7252, University of Poitiers, H1, Tel. 2, Bld. Marie et Pierre Curie, BP 30179, 86962 Futuroscope-Chasseneuil Cedex, France
lydie.richaume@univ-poitiers.fr

Abstract. In this article, we suggest a grid-unfolding of level 1 Menger polycubes of arbitrary size with L holes along the x-axis, M the y-axis and N the z-axis. These polycubes can have a high genus, and most vertices are of degree 6. The unfolding is based mainly on the inner faces (that do not lie on the outer most envelope) except for some outer faces that are needed to connect lines or planes in the object. It is worth noticing that this grid-unfolding algorithm is deterministic and without refinement.

Keywords: Polycube · Orthogonal polyhedra · Unfolding · High genus object

1 Introduction

Unfolding a geometric three-dimensional shape consists in providing a two-dimensional flat non-overlapping and connected representation of it. This problem is as old as cartography and is still widely open with many potential applications including puzzle design [1], 3D paper craft [12], manufacturing [14,16]...

Our interest lies in the grid edge-unfolding without refinement of a very specific class of polyhedra: the level 1 Menger polycube of arbitrary size (a generalisation of the Menger sponge [11]). A polycube is a set of face-connected voxels without inner cavity. An edge unfolding is an unfolding along the original edges of the polyhedron. A grid unfolding of a polycube is an unfolding that consists in creating a 2D net of all the faces of the polycube (faces that are face-connected to the complement) in a 2D orthogonal grid. Many grid unfolding algorithms found in the literature require refinements. A $n \times m$ refinement means that a polycube face is cut into $n \times m$ rectangular subfaces. Oftentimes, algorithms deal with orthogonal polyhedra rather than polycubes. An orthogonal polyhedron is a polyhedron where all the faces are parallel to Cartesian coordinate planes and can be seen as a generalisation of polycubes.

Edge-unfolding of a general polyhedron is not always possible [8]. There are several examples of non-convex polyhedra that cannot be edge-unfolded. As for

M. Couprie et al. (Eds.): DGCI 2019, LNCS 11414, pp. 457–468, 2019.
https://doi.org/10.1007/978-3-030-14085-4_36

convex polyhedra, it is still an open question whether they always admit an edge unfolding. These last couple of years, the community tried to make some progress by focusing on various subclasses of orthogonal polyhedra that admit a grid edge-unfolding without refinement such as orthotubes [2], well-separated orthotrees [6], orthoterrains [13], H-convex Manhattan towers... [15] For some other subclasses, the known methods require the grid to be refined further by adding cutting planes in a constant or linear number of times: for example, 4×5 grid-unfolding for Manhattan towers [7], 1×2 grid-unfolding for orthostacks... [2]

More recently, the community has begun to focus on orthogonal polyhedra of genus greater than zero. In [10], edge-unfolding of several classes of one-layered orthogonal polyhedra (one layer of voxels) with cubic holes are considered and it was shown that several special classes of them admit grid unfoldings. In [4], authors generalised the work of [10] and offered a grid unfolding method suitable for a wider class of one layer polyhedra of genus greater than zero using a (2×1)-refinement. In [5], an algorithm for unfolding all orthogonal polyhedra of genus 1 or 2 is proposed; it extends the idea of [3] to deal with holes, and both require a linear refinement. In [9], Ho *et al.* treat two subclasses of orthogonal polyhedra with arbitrary genus with 2×1 refinement. Our Menger polycube is a subclass of their well-separated orthographs, but our algorithm does not need any refinement.

The general question whether there always exists a grid unfolding without refinement for any polycube is an open question. A grid unfolding without refinement means that each polycube voxel face can be unfolded as a square 2D grid face (a pixel if you prefer). The final 2D net should be connected and it should be noted that two faces in the net are considered as neighbours iff the original corresponding faces in the 3D polyhedra share a common edge. Two 2D grid faces may be neighbours in the classical 4-connected sense and not be neighbours in the net.

One way of looking at unfolding problems is to consider 3D vertices. In three dimensions, a polycube vertex may have a degree from 3 to 7, meaning that there is a cycle of 3 to 7 faces around a vertex. However, once unfolded, a face can have a maximum degree of 4. All the faces around a vertex of degree over 4 cannot be placed together on the unfolding. Does the proportion of vertices of degree greater than 4 limit the unfoldability of a polycube? This question led us to the level 1 Menger polycube of arbitrary size with $L \times M \times N$ holes [11]. The dimensions correspond to the number of holes in a direction: L holes along the x-axis, M along the y-axis and N along the z-axis. This represents a polycube with a potentially high genus and almost all the vertices in such a polycube are of degree 6 except for the 8 corner vertices that are of degree 3, some outer faces that are of degree 4 and the hole vertices on the outer envelope faces that are of degree 5.

In this paper we propose a deterministic algorithm for grid unfolding without refinement (edge unfolding) of level 1 Menger polycubes of arbitrary size. The idea is to focus on the "inner" faces (those that are not on the outer envelope). The outer envelope is defined as all the faces that have minimum or maximum

x, y or z centroid coordinates. The inner faces is where the vertices of degree 6 can be found. Our unfolding method requires the help of some "outer" faces only at specific places. Once all the inner faces are unfolded with help of some outer faces, all the other outer faces can then be unfolded very simply.

The paper is organized as follows: Sect. 2 presents some preliminaries as well as the particular polycubes we are interested in. The unfolding method is then detailed. We finish the paper with a conclusion and prospects.

2 Unfolding Level 1 Menger Polycubes

2.1 Introducing Level 1 Menger Polycubes of Arbitrary Size

A Menger sponge is a fractal object [11]. In this paper, we only consider the level 1 of this construction: a simple level 1 Menger polycube is made of 20 unit cubes forming a greater $3 \times 3 \times 3$ cube where the cubes in the centre of each face are missing as well as the cube in the very centre of the greater cube (Fig. 1a). This is what we call here a level 1 $(1, 1, 1)$-Menger polycube. The genus of this polycube is 5.

A level 1 (L, M, N)-Menger polycube is the Minkowski addition between a Cartesian (L,M,N) subset of points of $(2Z)^3$ and a structuring element that is simply the $(1, 1, 1)$-Menger sponge. This polycube has L holes along the x-axis, M holes along the y-axis and N holes along the z-axis. Each hole is the size of one unit cube and is separated from the holes around it by one unit cube. Such a polycube has a size of $(2L+1) \times (2M+1) \times (2N+1)$. An example is presented on Fig. 2a for the level 1 $(1, 3, 1)$-Menger polycube. This polycube is of genus 13. The general formulas for a level 1 (L, M, N)-Menger are:

- number of voxels: $1 + 2(L + M + N) + 3(LM + LN + MN) + 4LMN$
- number of faces: $6 + 8(L + M + N) + 10(LM + LN + MN) + 12LMN$
- number of vertices: $8(1 + L)(1 + M)(1 + N)$. From which:
  - 8 vertices of degree 3;
  - $8(L + M + N)$ vertices of degree 4
  - $8(LM + MN + LN)$ vertices of degree 5
  - and $8LMN$ vertices of degree 6.
- genus: $LM + LN + MN + 2LMN$.

For instance, a $(3, 3, 3)$-Menger polycube (see Fig. 6) has a genus of 81, is composed of 208 voxels, 672 faces with 8 vertices of degree 3, 72 vertices of degree 4, 216 vertices of degree 5 and 216 vertices of degree 6. The ratio of vertices of degree 6 to other vertices climbs rapidly. For a $(100, 100, 100)$-Menger we have 8 million vertices of degree 6 and 242408 vertices of other degrees (and a genus of 2030000).

In the following, we start with handling the inner faces of such polycubes. These faces form a volume that, to unfold, we will divide in simpler structures: we can unfold a volume if we know how to unfold planes. In the same way, each plane is divided into lines and each line into crosses.

The outer faces (forming the outer envelope of the polycube) can be unfolded in a similar way as a simple cube as we will see in Sect. 2.5.

2.2 Unfolding an Inner Line

We call inner cross (or simply cross) the inner faces of a (1, 1, 1)-Menger polycube, i.e. the faces around the central “hole”. A complete cross has 6 branches (see Fig. 1a and b). We can merge several crosses to form an inner line: the back branch of one cross is also the front branch of the next one (see Fig. 2b). An inner line is defined as the inner faces of a $(1, M, 1)$-Menger.

We first explain how to unfold a simple cross and then how to go on with another connected one and finally how to unfold the whole inner line.

Unfolding a Cross: First we start with the simple cross which is the object obtained with the (1, 1, 1)-Menger polycube (i.e. the Menger sponge) inner faces (see Fig. 1b).

As we can see on Fig. 1d, unfolding one branch leads to four aligned faces. In Fig. 1d, “u, b, d, f, l, r” stands for “up, back, down, front, left, right” faces as they appear visually in Fig. 1b. We can navigate from one branch to the next by reversing the faces cycle: this way, the unfolding keeps progressing in the same direction, ensuring the non-overlapping property we seek. On Fig. 1c and d we propose an order to browse the branches and show the corresponding unfolding.

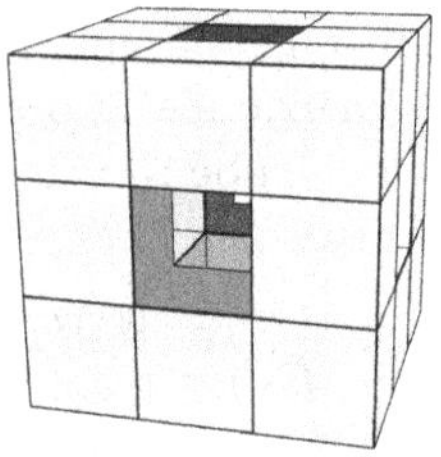

(a) (1, 1, 1)-Menger polycube.

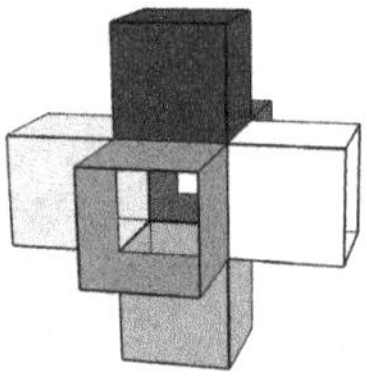

(b) A cross: the inside of a (1, 1, 1)-Menger.

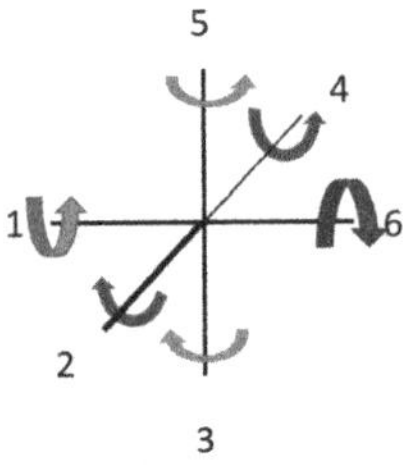

(c) The order and rotating direction to unfold the branches.

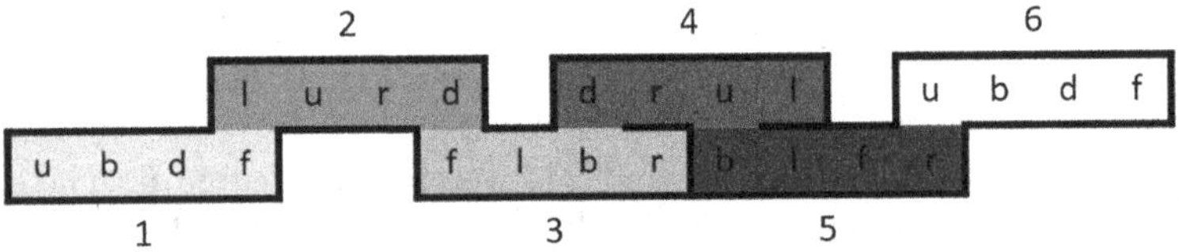

(d) Unfolded inner faces of a (1, 1, 1)-Menger polycube.

Fig. 1. The (1, 1, 1)-Menger polycube. The inner faces of this polycube and their deterministic edge-unfolding.

Linking Two Crosses: As we can see on Fig. 2b, between two consecutive crosses, there is a common branch. On the pattern we used to unfold a single cross, the last and the first branch are opposite to each other. If we apply this pattern to each cross of the inner line, the first and last branches are the shared ones. This allows us to simply concatenate the unfoldings by merging the shared branches (see Fig. 2c).

(a) A Line $(1,3,1)$-Menger Polycube

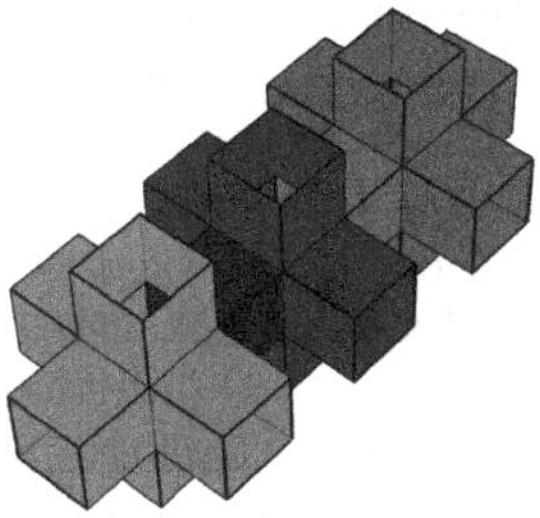

(b) The faces of an inner line Menger. The green cross and the red cross each share a branch with the blue cross.

(c) Unfolded Inner Faces of an Inner Line $(1,3,1)$-Menger

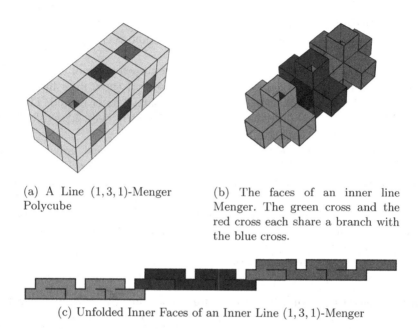

Fig. 2. The line $(1,3,1)$-Menger polycube. The inner faces of this polycube and its deterministic edge-unfolding. (Color figure online)

Unfolding Inner Line Results: We now have all we need to unfold several crosses in a row: we simply have to concatenate the cross pattern M times to get a M sized line unfolding.

We can notice (see Fig. 2c) that the unfolding of each cross is on the right and slightly above the previous one, preventing any two faces from different crosses to overlap each other. The unfolding of the back branch of a cross is also the unfolding of the front branch of the next cross (presented by mixed colour on the figure). It allows linking every consecutive cross together.

2.3 Unfolding an Inner Plane

In the same way we formed a line by merging several crosses, we can now merge several lines side by side to form the inside of a $(L,M,1)$-Menger plane, called an inner plane for what follows. As previously, some branches will be shared between two lines (see Fig. 3a). We will first see how to unfold an inner line without the side branches and then how to connect the nets of two consecutive inner lines to finally unfold an inner plane.

Unfolding a Partial Inner Line: Between two consecutive inner lines, each cross shares a lateral branch with a cross from the other inner line. This time, the shared branches will not be used to link the unfoldings of the different inner lines. Since we cannot unfold the same faces twice, we choose to ignore the shared branches while unfolding the first inner line and unfold them with the following inner line and so forth until the last.

This means that, when we have M inner lines, each cross of the first $M-1$ inner lines will miss the left branch. The crosses of these lines have only 5 branches (see Fig. 3a and b) instead of (these are partial inner lines)6. Therefore a new pattern is needed. Indeed, the pattern we used in Sect. 2.2 works for an inner line with crosses that have 6 branches.

A suitable pattern can be found in Fig. 3c. Once again, each two consecutive crosses from a same partial inner line share the front and back branches (in orange and red on the figures), and we will be using them to concatenate the unfoldings. We can see that, this time, the shared branch is *not* the last branch of the first cross, but we can nonetheless link the unfoldings of two consecutive partial crosses without overlap. The top branch is unfolded under the shared branch and under the next partial cross. We can thus concatenate any number of 5 branch crosses to create a partial inner line of any length.

Linking Two Lines in an Inner Plane: To link two partial inner lines together, we use an outer face. Indeed, we chose to unfold in such a way that the last visited branch of a partial inner line is a branch connected to the outer envelope of the polycube. We use the outer face as a bridge to reach the cross to the left of the previous one (see the orange faces on Fig. 3).

In a partial inner line, each cross unfolding is placed on the top of the preceding one. Following the same idea, each partial inner line unfolding is also placed strictly on top of the preceding one. This way, two different line-unfoldings cannot overlap. Each partial inner line is visited in a reverse order from the preceding one; this is why each inner line unfolding mirrors the previous one.

Unfolding Inner Plane Results: A full $L \times M$ inner plane is composed of $L-1$ partial inner lines and one complete inner line. We just provided a way to unfold a partial inner line and the unfolding of an inner line (with 6 branches crosses) is treated in Sect. 2.2. We simply have to link these different line-unfoldings as previously described to get a full inner plane unfolding.

At each step, the unfolding of the different crosses and lines are placed on the top of the previous ones. This way we can ensure that there is no unwanted overlapping.

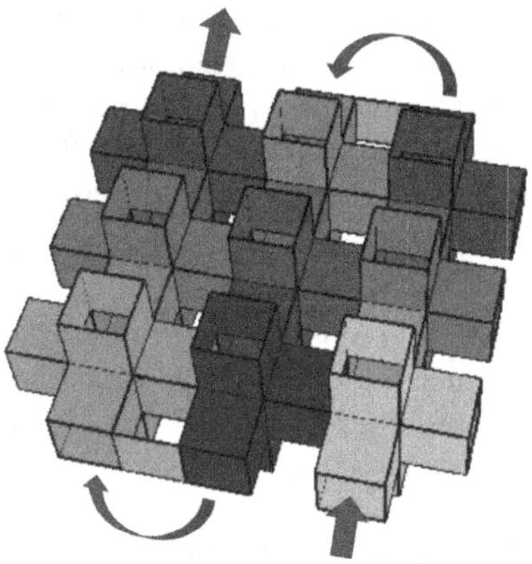

(a) A Menger inner plane and, in orange, the exterior faces used to link two lines.

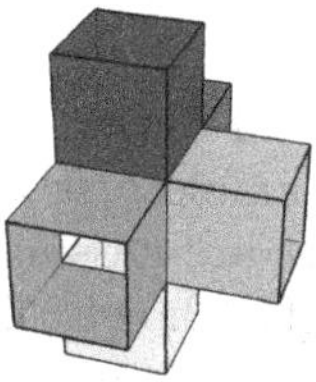

(b) A cross missing its side branch

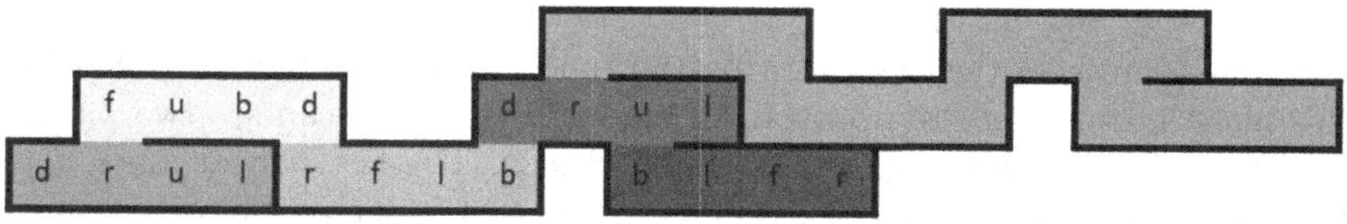

(c) The unfolding of a 5-branch cross inner line. The red branch is used to link this 5-branch cross to the next one (in grey).

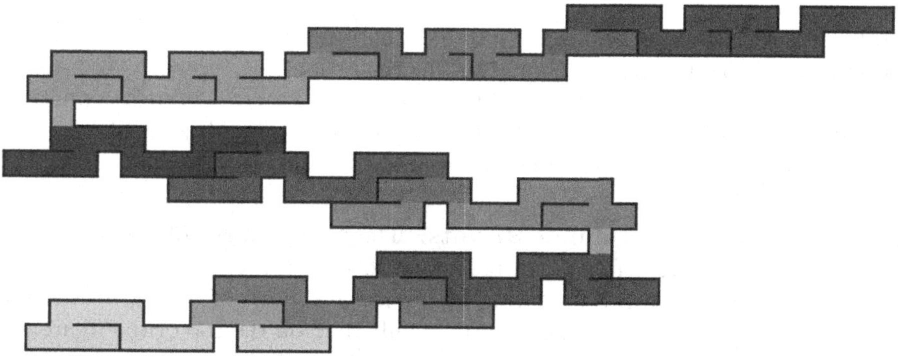

(d) The unfolding of an inner $(3,3,1)$-Menger plane. Faces of a same colour are from the same cross. Orange faces are outer faces.

Fig. 3. The inner faces of a $(3,3,1)$-Menger polycube and their unfolding. (Color figure online)

2.4 Unfolding an Inner Volume

We can now merge inner planes to build the inner faces of a (L, M, N)-Menger polycube volume. As for the previous case, some branches will be shared by two consecutive inner planes. To unfold a volume, we therefore consider that the lower planes miss their top branches (see Fig. 4a). This defines partial inner planes. The top plane is a full inner plane as discussed in Sect. 2.3. Now, we explain how to unfold the partial inner planes. Then, we explain how to link two consecutive inner plane-unfoldings together.

Unfolding a Partial Inner Plane: To unfold a partial inner plane (see Fig. 4), we have to unfold new types of partial inner lines where this time, the top and left branches are missing. This leads to unfolding crosses having 4 (for the first lines) or 5 (for the last line) branches.

When considering Fig. 3c, the pattern for 4 branch crosses is the same pattern as the one with 5 branches except that we drop the blue branch.

The last line of such planes has 5 branches, but this time it is the top branch that is missing. Globally the idea of the pattern is similar as for the 5 branch pattern in Sect. 2.3 but the last branch is linked to the "**left**" of the red branch instead of the "**up**" (see Fig. 3c).

Figure 4 presents the faces of a partial inner plane of a Menger polycube. This plane misses every top branch. In blue and green, we can see two partial inner lines missing their top and one side branch. The last partial inner line, that misses only its top branches, is in red. To unfold such a partial inner plane, we use a face of the outer envelope (in orange) to go from one line to the next (similarly to what was described previously in Sect. 2.3).

Linking Two Planes: As for the link from an inner line to the next inner line (see Sect. 2.3), we use an outer face to link a partial inner plane to the above located inner plane. We exit a partial inner plane using a cross branch connected to a side of the polycube and re-enter it by the corresponding branch of the cross located above. The outer face used is coloured in yellow on Fig. 4.

Unfolding an Inner Volume Results: The whole inner volume unfolding is obtained by adding a full last inner plane to the $N - 1$ partial inner planes unfoldings. We can see the unfolding of the inner volume of a $(3, 3, 3)$-Menger polycube on Fig. 6. It is the zigzag pattern unfolding on the top of the figure (in green and blue we have the two partial inner planes and in red, the full inner plane).

2.5 Unfolding the Remaining Outer Faces

The remaining faces all belong to the outer envelope of the polycube. We will now see that all of these faces remain connected, can be unfolded in the same fashion as a simple cube (see the bottom part of Fig. 6) and can be added to the unfolding of the inner faces without overlap.

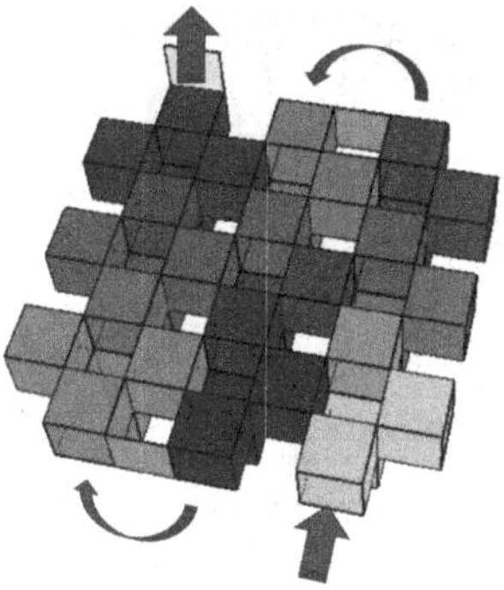

(a) A Partial Inner Plane Missing Every Top Branches.

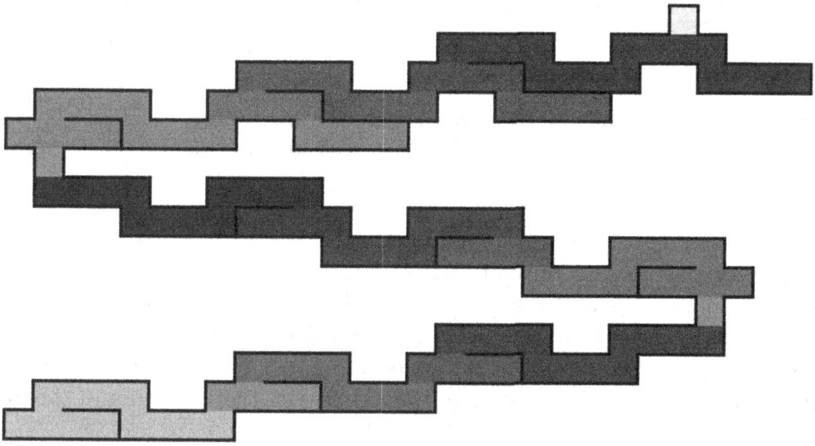

(b) The unfolding of a partial plane of a $(3, 3, N)$-Menger inner volume. The yellow face will be used to link this partial inner plane to one above it.

Fig. 4. The inner faces of a partial inner plane from a $(3, 3, N)$-Menger polycube and their unfolding. (Color figure online)

Connectedness of Outer Faces: While unfolding the inner parts of the polycube we use outer faces to link two inner lines either horizontally (for two consecutive inner lines of an inner plane) or vertically (for two inner lines of two consecutive inner planes). The way we unfold all the inner lines is similar to a sewing pattern starting on the bottom-front of the polycube, passing through to exit on the opposite side and coming back through the next inner line either on the left or on the top of the first one.

The outer faces used to link the inner unfoldings belong only to the front and the back sides of the polycube, the other four sides are untouched.

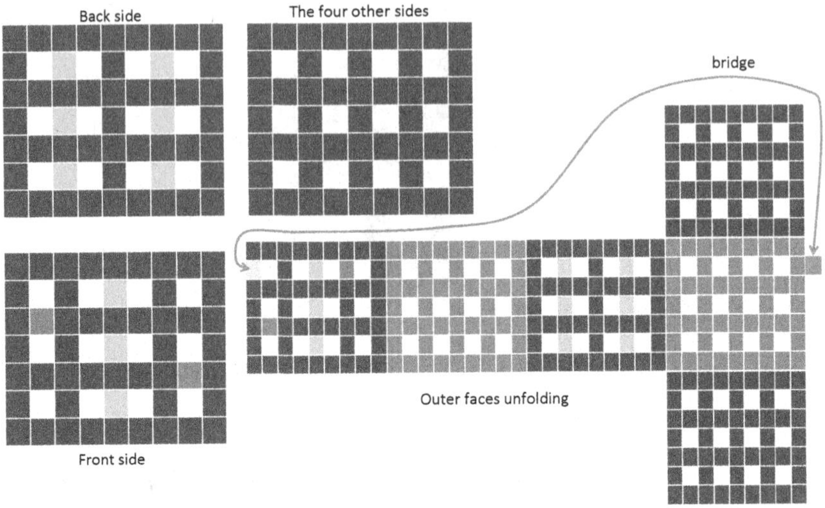

Fig. 5. The outer face unfolding in a cube unfolding fashion. The light-blue faces are used to link 2 lines, and the green faces are used to link 2 planes. (Color figure online)

We can see on Fig. 5 that the unfolding of the outer faces remains connected even with the missing outer faces (in light blue and green) that were used to link the inner lines and planes together. Despite having used some outer faces to unfold the inner part, the outer faces remain connected and can be unfolded in the same way as a simple cube (see Fig. 6). A last face, coloured in orange, has been removed to be used as a bridge between the unfolding of the outer and inner part.

Linking the Outer Unfolding to the Inner Unfolding: When we unfold the inner faces, the first unfolded branch is placed to have the lowest ordinate value of the inner faces unfolding. This branch is linked to the outer envelope of the polycube. During the unfolding process, faces are always added either at the same level as the previous ones or above. Therefore, the outer faces can be unfolded under the inner unfolding.

If we keep strictly the unfolding of a cube, we cannot connect the outer faces to the inner faces. There is no free edge in the unfolded outer faces that is common with an inner face. To create this common free edge, we move one of the two faces of the front side of the polycube that are connected to both the front branch of the starting cross and a face from another outer side of the polycube. This face makes the bridge between the inner unfolding part and the outer unfolding part.

2.6 Final Result

Figure 6 shows the unfolding of a (3, 3, 3)-Menger polycube and gathers all the previously presented unfolding elements. We can see the unfolding of most of the outer faces on the bottom. We can see the unfolding results of the inner part: the zigzag unfolding on the upper part. Each inner line unfolding corresponds to a linear part. Each zigzag composed of three linear parts belongs to the unfolding of an inner plane. Each linear part is connected to the next using an outer face either to connect two inner lines in the same inner plane (orange faces) or to connect two inner planes (yellow faces). On the front of the (3, 3, 3)-Menger polycube in Fig. 6, one can see some of the outer faces (in orange and yellow respectively) that are used to link the inner lines and the inner planes. A similar set of outer faces is located on the back of the Menger polycube.

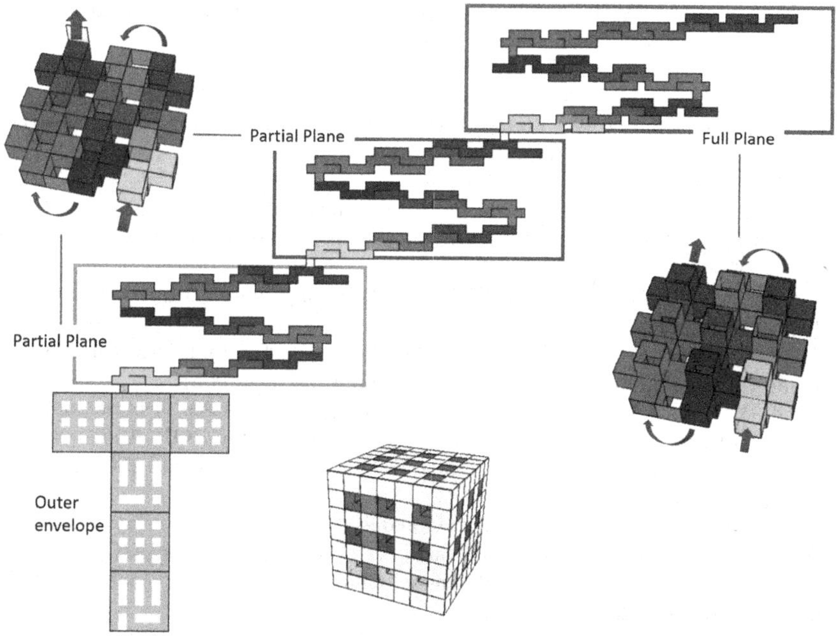

Fig. 6. The unfolding of the level 1 (3, 3, 3)-Menger polycube. (Color figure online)

3 Conclusion

In this paper, we have proposed a deterministic method to unfold without refinement high genus polycubes made out of level 1 Menger sponges. This proves that despite the high rate of 6-degree vertices, there exists a non-overlapping flat unfolding for such a polycube. It is the first unfolding example of such a complex polycube object with a high genus without refinement.

As a short time prospect, we will try to unfold the inner part of the Menger polycubes without the help of outer faces. The goal is to prove that it is possible to unfold an object composed only of degree 6 vertices. In the long term, we want to study if our method can be adapted to other types of high genus polycubes and maybe come up with a deterministic unfolding for any type of polycube.

References

1. Aloupis, G., et al.: Common unfoldings of polyominoes and polycubes. In: Akiyama, J., Bo, J., Kano, M., Tan, X. (eds.) CGGA 2010. LNCS, vol. 7033, pp. 44–54. Springer, Heidelberg (2011). https://doi.org/10.1007/978-3-642-24983-9_5
2. Biedl, T., et al.: Unfolding some classes of orthogonal polyhedra. In: Proceedings of 10th Canadian Conference on Computer Geometry, January 1998
3. Chang, Y.-J., Yen, H.-C.: Unfolding orthogonal polyhedra with linear refinement. In: Elbassioni, K., Makino, K. (eds.) ISAAC 2015. LNCS, vol. 9472, pp. 415–425. Springer, Heidelberg (2015). https://doi.org/10.1007/978-3-662-48971-0_36
4. Chang, Y.-J., Yen, H.-C.: Improved algorithms for grid-unfolding orthogonal polyhedra. Int. J. Comput. Geom. Appl. **27**(01n02), 33–56 (2017)
5. Damian, M., Demaine, E., Flatland, R., O'Rourke, J.: Unfolding genus-2 orthogonal polyhedra with linear refinement. Graphs and Combinatorics **33**(5), 1357–1379 (2017)
6. Damian, M., Flatland, R., Meijer, H., O'rourke, J.: Unfolding well-separated orthotrees. In: 15th Annual Fall Workshop Computer Geometry, pp. 23–25. Citeseer (2005)
7. Damian, M., Flatland, R., O'Rourke, J.: Unfolding manhattan towers. Comput. Geom. **40**(2), 102–114 (2008)
8. Demaine, E.D., O'Rourke, J.: A survey of folding and unfolding in computational geometry. Combinatorial Comput. Geom. **52**, 167–211 (2005)
9. Ho, K.-Y., Chang, Y.-J., Yen, H.-C.: Unfolding some classes of orthogonal polyhedra of arbitrary genus. In: Cao, Y., Chen, J. (eds.) COCOON 2017. LNCS, vol. 10392, pp. 275–286. Springer, Cham (2017). https://doi.org/10.1007/978-3-319-62389-4_23
10. Liou, M.-H., Poon, S.-H., Wei, Y.-J.: On edge-unfolding one-layer lattice polyhedra with cubic holes. In: Cai, Z., Zelikovsky, A., Bourgeois, A. (eds.) COCOON 2014. LNCS, vol. 8591, pp. 251–262. Springer, Cham (2014). https://doi.org/10.1007/978-3-319-08783-2_22
11. Menger, K.: Dimensionstheorie. Springer, Heidelberg (1928). https://doi.org/10.1007/978-3-663-16056-4
12. Mitani, J., Suzuki, H.: Making papercraft toys from meshes using strip-based approximate unfolding. ACM Trans. Graph. (TOG) **23**(3), 259–263 (2004)
13. O'Rourke, J.: Unfolding orthogonal terrains. arXiv preprint arXiv:0707.0610 (2007)
14. O'Rourke, J.: Folding and unfolding in computational geometry. In: Akiyama, J., Kano, M., Urabe, M. (eds.) JCDCG 1998. LNCS, vol. 1763, pp. 258–266. Springer, Heidelberg (2000). https://doi.org/10.1007/978-3-540-46515-7_22
15. Richaume, L., Andres, E., Largeteau-Skapin, G., Zrour, R.: Unfolding H-convex Manhattan Towers. HAL preprint, December 2018. https://hal.archives-ouvertes.fr/hal-01963257
16. Wang, C.-H.: Manufacturability-driven decomposition of sheet metal products. Ph.D. thesis, Carnegie Mellon University (1997)

A Discrete Bisector Function Based on Annulus

Sangbé Sidibe, Rita Zrour(✉), Eric Andres, and Gaelle Largeteau-Skapin

Laboratoire XLIM UMR CNRS 7252, ASALI, Université de Poitiers,
Bld Marie et Pierre Curie, BP 30179, 86962 Futuroscope Chasseneuil Cedex, France
rita.zrour@univ-poitiers.fr

Abstract. In this paper we are proposing a new way to compute a discrete bisector function, which is an important tool for analyzing and filtering Euclidean skeletons. From a continuous point of view, a point that belongs to the medial axis is the center of a maximal ball that hits the background in more than one point. The maximal angle between those points is expected to be high for most of the object points and corresponds to the bisector angle. This logic is not really applicable in the discrete space since in some configurations we miss some background points leading sometimes to small bisector angles. In this work we use annuli to find the background points in order to compute the bisector angle. The main advantage of this approach is the possibility to change the thickness and therefore to be more flexible while computing the bisector angle.

Keywords: Bisector function · Euclidean distance transform · Skeleton · Digital geometry · Digital annulus

1 Introduction

The medial axis is a shape representation tool that has been used in a wide variety of applications like pattern recognition, robotic motion planning [23], skinning for animation [14] and other domains as well. The notion of medial axis was originally proposed by Blum in 1967 [7] where it was defined as the set of points where different fire fronts meet.

In the continuous Euclidean space two definitions can be used to describe the medial axis [8]:

(a) the medial axis is formed by the center of balls included in the shape and not included in any other ball in the shape
(b) the medial axis consists of points that have more than one nearest points on the boundary of the shape.

In the state of art many methods were used to filter medial axis from some undesirable or spurious points or some branches [4,5,8,10,16]. Among those

M. Couprie et al. (Eds.): DGCI 2019, LNCS 11414, pp. 469–480, 2019.
https://doi.org/10.1007/978-3-030-14085-4_37

methods, we considered the method that uses filtering based on the bisector function [10].

The bisector function was introduced by Talbot and Vincent [22] where they generalize a notion proposed by Meyer [18]. Informally the bisector function associates to each point p of an object, the maximal angle formed by p and the points of the background that are nearest to p (Fig. 1 gives an example of the bisector function at two distinct point x and y). The first algorithms proposed to compute bisector function used vectors obtained from the distance transform [12]. One vector indicates the location of the closest background point. The main drawback of such algorithms [15,22] is that points from a same distance may be ignored. In [11] authors proposed a new definition and algorithm to compute a discrete bisector function which was revisited and improved in [10] based on the use of Voronoi diagram of a shape. Authors noticed that the projection of the point is not sufficient due to some configurations in the discrete space and extended the notion of projection of a point to the extended projection (see Sect. 2).

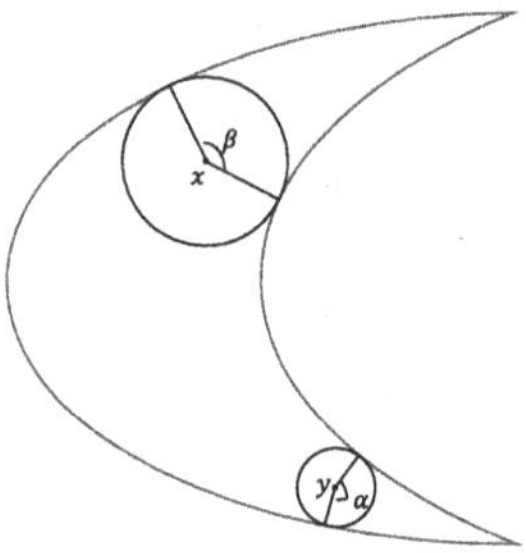

Fig. 1. Example of the bisector function. β is the angle at point x and α the angle at point y.

This paper contains one major contribution which is a new way to compute the bisector function based on an annulus and an algorithm to compute such bisector function based on digital annulus generation algorithm [2,3]. We are mainly interested in this work at points having a high bisector angles since points belonging to a skeleton or important points of the medial axis are supposed to be center of a maximal ball that must hit the background in more than one point. Different filtering results are conducted on the bisector function and showed that as the filtering increases important point of the medial axis are detected. This article presents the method in 2D however it should be stated that this work can be extended to 3D and higher dimensions.

The paper is organized as follows: in Sect. 2 we provides some basic notions and definitions. In Sect. 3 we details our method of bisector function based on annulus, present the algorithm and provide some experimental results. The final sections state some conclusion and perspectives.

2 Basic Notions and Definitions

In this section we present some basic notions and definitions. Let us denote by $\mathbb{Z}$ the set of integers, by $\mathbb{N}$ the set of non-negative integers and by E the discrete plane $\mathbb{Z}^2$. A point x in E is defined by (x_1, x_2). Let $x, y \in E$ be two points, we denote by $d^2(x, y)$ the squared Euclidean distance between x and y. Let $Y \subset E$, we denote by $d^2(x, Y)$ the square of the Euclidean distance between x and the set Y, that is $d^2(x, Y) = min\{d^2(x, y); y \in Y\}$. Let $X \subset E$ (the object), we denote by D^2_X the map from E to $\mathbb{N}$ which associates, to each point x of E, the value $D^2_X(x) = d^2(x, \overline{X})$, where $\overline{X}$ denotes the complementary of X (the background). The map D^2_X is called the squared Euclidean distance map of X. Let $x \in E$, $r \in \mathbb{N}$, we denote by $B_r(x)$ the ball of (squared) radius r centered on x, defined by $B_r(x) = \{y \in E, d^2(x,y) < r\}$. Note that the value $D^2_X(x)$ is the radius of a ball centered on the point $x \in X$ and not included in any other ball centered on x and included in X.

The considered discrete neighborhood N_4 is defined as follows :

$$N_4(x) = \{y \in E; |y_1 - x_1| + |y_2 - x_2| \leq 1\}$$

Let X be a non empty subset of E and let $x \in X$, the ***projection*** of x on $\overline{X}$ [10], denoted by $\Pi_{\overline{X}}(x) = \{y \in \overline{X}, \forall z \in \overline{X}, d(y,x) \leq d(z,x)\}$. For example, in Fig. 2, $\Pi_{\overline{X}}(x) = \{a, b\}$ and $\Pi_{\overline{X}}(y) = \{c\}$.

Let X be a non empty subset of E and let $x \in X$, the ***extended projection*** of x on $\overline{X}$ [10], denoted by $\Pi^e_{\overline{X}}(x)$ is the union of the sets $\Pi_{\overline{X}}(y)$, for all $y \in N_4(x)$ such that $d^2(y, \overline{X}) \leq d^2(x, \overline{X})$.

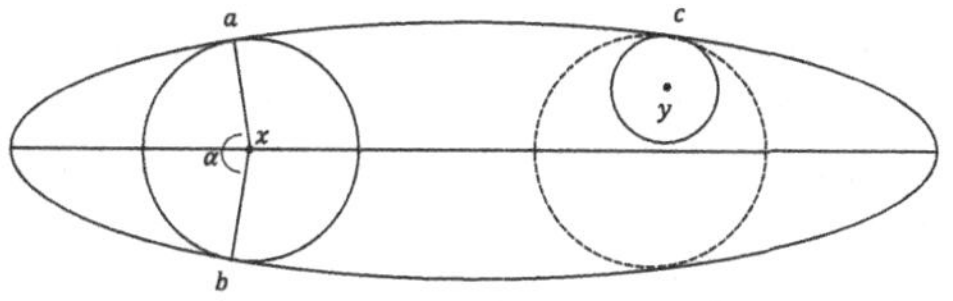

Fig. 2. An object X (represented by the ellipse) and its medial axis. The projection of the point x is $\{a, b\}$ and the projection of the point y is $\{c\}$. The ball centered on y is included in another ball (the dashed ball) and thus y does not belong to the medial axis of X.

3 Bisector Function Based on Digital Annulus

In this section we detail our method for computing the bisector function using an annulus, describe the algorithm used and present the results.

3.1 Digital Annulus

A digital annulus $\mathbb{A}(C, r, w)$ is defined by an offset of thickness ω on a 2D regular square grid [2,3]:

$$\mathbb{A}(C, r, \omega) = \{(i,j) \in \mathbb{Z}^2 : r^2 \leq (i - C_x)^2 + (j - C_y)^2 < (r + \omega)^2\} \quad (1)$$

Where $C(C_x, C_y) \in \mathbb{R}^2$ is the center and $r + \frac{\omega}{2} \in \mathbb{R}$ is the radius. In this work we refer to r as the inner radius.

In this work $C(C_x, C_y) \in \mathbb{Z}^2$ since we are considering integer coordinates of the image as X. We also need to include the equality from both sides of (2) for practical reasons. This gives us the equation of an annulus of thickness ω represented by the following equation:

$$\mathbb{A}_w(C, r, \omega) = \{(i,j) \in \mathbb{Z}^2 : r^2 \leq (i - C_x)^2 + (j - C_y)^2 \leq (r + \omega)^2\} \quad (2)$$

3.2 Bisector Function: New Definition and Exact Algorithm

Definition 1. *Annulus projection : Let X be a non empty subset of E and let $x \in X$, the annulus projection of x on $\overline{X}$, is given by:*

$$A\Pi_{\overline{X}}^{\omega}(x) = \left\{ y \in \overline{X}, D_X^2(x) \leq d^2(x,y) \leq (D_X(x) + \omega)^2 \right\} \tag{3}$$

where ω is the thickness of the annulus and $D_X(x)$, as stated in Sect. 2, is the value of the distance map at point x. (See Fig. 3c).

Now we can propose a definition of the bisector function.

Definition 2. *Bisector angle and bisector function: Let $X \subset E$, and let $x \in X$. The bisector angle, denoted by $\varphi_X^{\omega}(x)$ of x in X, denoted is the maximal unsigned angle between the vectors $\overrightarrow{xy}$, $\overrightarrow{xz}$, for all y, z in $A\Pi_{\overline{X}}^{\omega}(x)$. The bisector function of X, denoted by φ_X^{ω}, is the function which associates to each point x of X, its bisector angle in X.*

The last step to obtain the bisector angle is the computation of the maximum unsigned angle between all the pairs of vectors $\overrightarrow{xy}, \overrightarrow{xz}$ for all y, z in $A\Pi_{\overline{X}}^{\omega}(x)$. As stated in [10,11] this problem reduces to the problem of finding the maximum diameter of a convex polygon in 2D which has been solved in [21] by a linear time algorithm.

Figure 3a shows an original set X, Fig. 3b shows the extended projection [10,11] of the point x encircled in black on $\overline{X}$; the resulting points are encircled in dashed red. Figure 3c shows an example of the annulus projection of the point

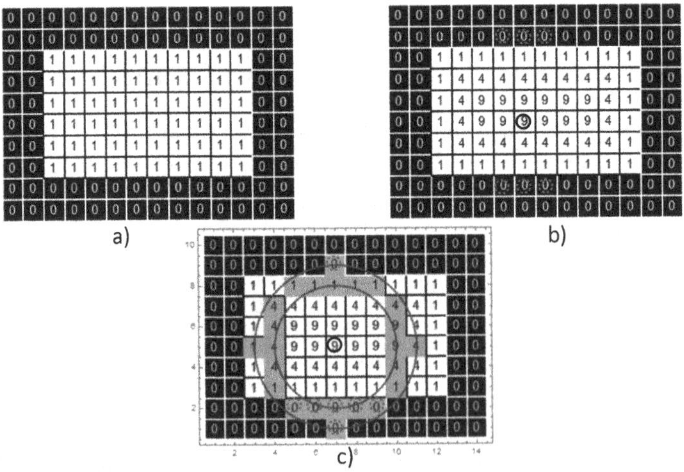

Fig. 3. (a) A set X. (b) Distance transform D_X^2 of X. A point x encircled in black and its extended projection [10], points encircled in red; the bisector angle $\theta_X(x)$ at point x is π (c) Annulus projection of x on $\overline{X}$, $A\Pi_{\overline{X}}^{\omega}(x)$ represented by the points encircled in dashed red; the bisector angle φ_X^{ω} at x is also π. (Color figure online)

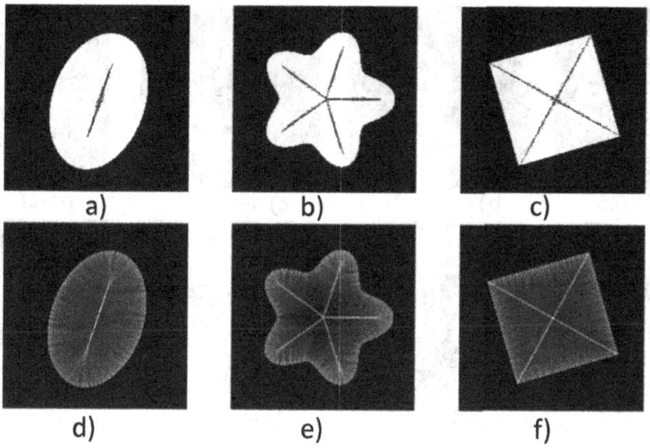

Fig. 4. a, b and c show three original images of an ellipse, a flower and a square with their medial axis colored blue. d, e and f show the bisector functions φ_X^ω obtained using an annulus of width $\omega = 1$. (Color figure online)

x (encircled in black) on $\overline{X}$ using a thickness $\omega = 1$; the resulting points are encircled in dashed red.

Figure 4a, b and c show three original images of an ellipse, a flower and a square extracted form the digital library DGtal [1] with their medial axis colored blue computed using the exact euclidean medial axis algorithm of [19]. Figure 4d, e and f show the bisector function φ_X^ω computed using a thickness ω of 1.

Algorithm 1. Annulus projection $A\Pi_{\overline{X}}^{\omega}(x)$

```
input : D²_X, x and w
output: Annulus projection of a point
        x(x1, x2), on X̄
begin
    w*=IntegerPart[w];
    AΠx=∅;
    r=D_X(x);
    if w <= 1 then
        AΠx=AΠx ∪
        Annulus(D²_X,r,x,w);
    else
        for i = 0; i <w*; i = i + 1 do
            AΠx=AΠx ∪
            Annulus(D²_X,r + i,x,1);
        w_new=w-w*;
        if w_new > 0 then
            AΠx=AΠx ∪
            Annulus(D²_X,r+w*,x,w_new);
return AΠx;
```

Algorithm 2. Annulus(D_X^2,r,x,w)

```
input : D²_X, r, x and w
output: Points of a digital Annulus that
        belong to X̄
begin
    da = ∅;
    Initialize xd = 0,
    upper = 2 * r * w + w * w, yd = ⌈r⌉,
    Δ = xd * xd * yd * yd − r²;
    while yd ≥ xd do
        if Δ ≥ 0 and Δ ≤ upper then
            v = eightsymmetry(xd, yd);
            foreach z ∈ v do
                if D²_X(x + z) = 0 then
                    da = da ∪ {x + z};
        if Δ <= 2 * (r − xd) then
            Δ = Δ + 2 * xd + 1;
            xd = xd + 1;
        else
            if Δ ≥ 2 * yd − 1 then
                Δ = Δ − 2 * yd + 1;
                yd = yd − 1;
            else
                Δ = Δ+2*(xd − yd+1);
                xd = xd + 1;
                yd = yd − 1;
    return da;
```

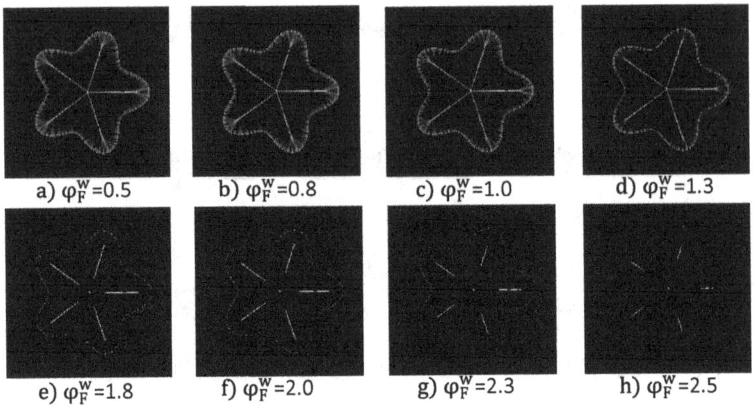

Fig. 5. The result of different filtering angle φ_F^ω applied on the bisector function φ_X^ω of the flower of Fig. 4b the filtering value is φ_F^ω in radians. We can notice that when φ_F^ω is too small we have noisy contour that disappear when φ_F^ω increases.

3.3 Algorithm

This section presents two different algorithms that allow to compute the bisector function based on annuli. The algorithms to compute the bisector function are based on the information of the distance map generated by a distance map algorithm [13,17,20]. Let us first start by the naive algorithm that is obvious but has a high time complexity and then present a second algorithm based on a slightly modified digital annulus generation algorithm adapted to our annuli definition with a "'$\leq$"' on both side of the equation as stated in Sect. 3.1.

Naive Algorithm. The naive algorithm consists in taking for every point $x(x_1, x_2)$ of the set X a square window of size $(D_X(x) - \omega, D_X(x) + \omega)$, centered at the point x and computing $A\Pi_{\overline{X}}^{\omega}(x)$ which means all the points of $\overline{X}$ that are inside the annulus $\mathbb{A}_\omega(C, D_X(x), \omega)$. This algorithm has a high time complexity that depends directly on the size of the annulus.

Algorithm Based on Incremental Digital Annulus Generation. We can decrease the time complexity by looking only into points inside the annulus using the incremental digital annulus generation algorithm detailed in [2,3]. It should be noted that in our case we use double equality from both side of the equation and therefore the algorithm of [2,3] should be slightly adapted to our case; also the thickness can vary however this is not a problem since one can think of generating digital annulus one after another when the thickness increases. For example for thickness $\omega = 2.7$ and inner radius r we must generate points inside digital annulus of inner radius r and $\omega = 1$, then those inside digital annulus of inner radius $r + 1$ and $\omega = 1$ (Lines 7 and 8 of Algorithm 1) and then those inside annulus of inner radius $r + 2$ and $w = 0.7$ (Lines 9–11 of Algorithm 1).

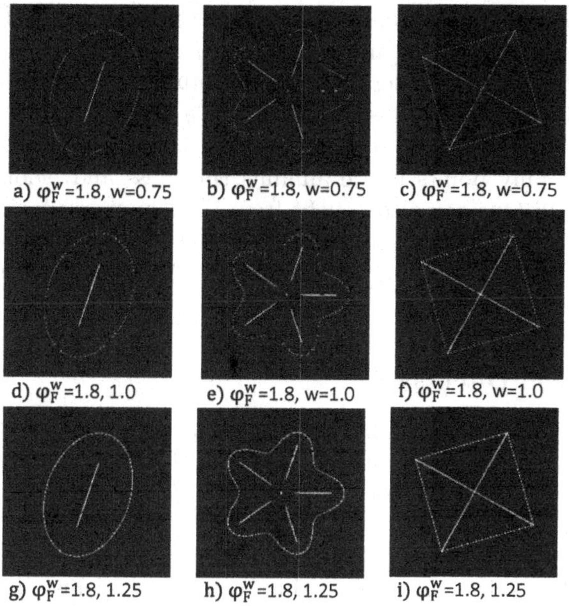

Fig. 6. The result of fixing the filtering angle φ_F^ω to 1.8 and varying the thickness from $\omega = 0.75$, $\omega = 1$, $\omega = 1.25$. As we can see that a width $\omega = 0.75$ is not sufficient to compute a correct angle and some important information are missed.

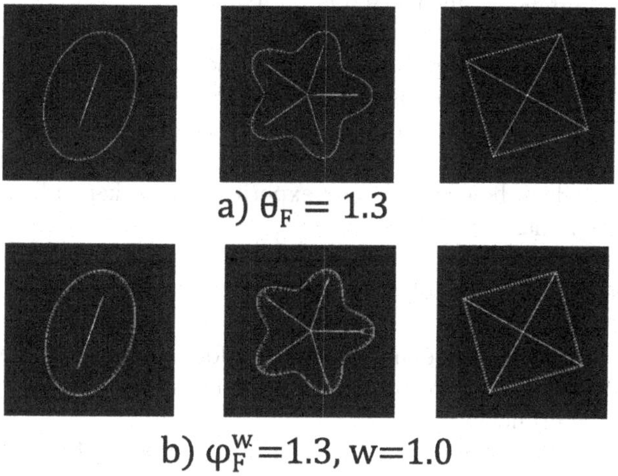

Fig. 7. (a) The result of [10,11] with an angle $\theta_F = 1.3$. (b) our approach with $\varphi_F^\omega = 1.3$ and $\omega = 1$.

Algorithm 1 shows the detailed algorithm for computing annulus projection of a point x on $\overline{X}$ the inner radius of the digital annulus at point x is determined by $D_X^2(x)$. As we can see when w is less or equal to 1 one annulus is generated and

Algorithm 2 is called only once. When w increases, Algorithm 2 may be called many times depending on the integer part of w denoted by $w*$. Algorithm 2 details the digital annulus generation of [2,3] adapted to our application. The starting point of the algorithm is the point $(x_d, y_d)=(0,\lceil r \rceil)$, the algorithm generates all points inside annulus of radius 1 and that belong to $\overline{X}$, however when w is less than 1 only points inside annulus thickness and belonging to $\overline{X}$ are added (Lines 5–9). One can also benefit from the symmetry condition of digital annulus and generate points only in the first octant and then apply the eight symmetry (Line 6 of Algorithm 2 *eightsymmetry*) to points to generate points in other octants.

3.4 Results and Discussion

This section details the results obtained using our algorithm. We tested different filterings based on the bisector angle φ_X^ω and studied also the effect of changing the width of the annulus. The original images used for the testing are shown in Fig. 4 a, b and c.

For experiments, let us denote by φ_F^ω the filtering angle applied on the bisector function φ_X^ω which consists in setting to black all points $x \in X$ having an angle $\varphi_X^\omega(x)$ less than φ_F^ω (the display of the images takes into account the value of the angles). Let us denote also by θ_F the filtering angle applied on the bisector function of [10,11].

Figure 5 shows the effect of the filtering applied on the bisector function of the flower for different φ_F^ω and a fixed annulus width $\omega = 1.0$. We can see in this image that for small angles we have some noisy contours and as φ_F^ω increases, this noisy effect disappear. For every image the angle is increased to see the filtering effect. When the angle is too high we lose some important information that we would like to preserve. It should be noted also that one could imagine to omit points with small distances to have only the points inside the image and not close to boundary however for our experiments we kept all the points and didn't omit any point.

Figure 6 shows the effect of changing the thickness while fixing φ_F^ω to 1.8 applied on the three images of ellipse, flower and square. The three thickness considered are 0.75, 1.0 and 1.25. We can see that a small thickness, less than 1, is not sufficient for most of the images since we lose some important information or branches. When the thickness increases we could capture more points which give us some nice branches.

Figure 7a and b shows a comparison of the result of [10,11] with the annulus approach proposed in this paper. The filtering angle θ_F and φ_F^ω are equal to 1.3 and the annulus thickness is fixed to 1. We can notice that our approach produced some noisy contours compared to [10,11]. When θ_F and φ_F^ω increases to 2.0 while keeping the same thickness of 1, we can notice that noise almost disappears as shown in Fig. 8.

Figure 8a shows the result of [10,11] with an angle $\theta_F = 2.0$ and Fig. 8b and c shows our result for $\varphi_F^\omega = 2.0$ and a thickness of respectively $\omega = 1$ and $\omega = 1.25$. We can see that as the filtering angle increases we have better result

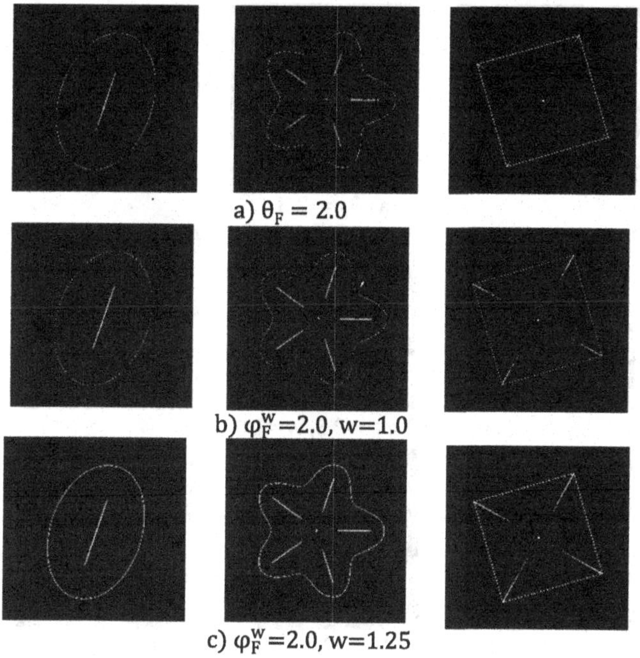

Fig. 8. (a) The result of [10,11] with an angle $\theta_F = 2.0$. (b) our approach with $\varphi_F^\omega = 2.0$ and $\omega = 1$. (c) our approach with $\varphi_F^\omega = 2.0$ and $\omega = 1.25$.

than [10,11] and increasing the thickness can compensate for some information loss. Figure 9 shows the same result as Fig. 8 but with $\theta_F = 2.3$ and $\varphi_F^\omega = 2.3$.

Figure 10b and c shows a comparison of our approach with the approach of [10,11] with filtering angle of 2.0 applied on Fig. 10a extracted from [1]. The points detected after filtering are the red ones and constitute a subset of the medial axis points; blue points represent the other points of the medial axis. Figure 11 shows the results of applying our method for filtering only medial axis points. The bisector function is thus applied only on the medial axis points. Blue points are points of the medial axis that are removed and red points are points kept.

3.5 Complexity and Extensions

In Sect. 3.3 we have proposed an incremental algorithm for digital annulus generation, that is less complex than the naive algorithm. The proposed algorithm is not linear in time: for each point x of the image, we are considering an annulus whose number of points is linear to the distance map $D_X(x)$. Let us state that the thickness does not have a large influence on the complexity since it is expected to fluctuate between 0.5 to 2 so the range is limited and this changes the overall complexity only by a constant factor. As for higher dimensions the generation of a digital hypersphere is well known and is dimensions dependent, we expect the

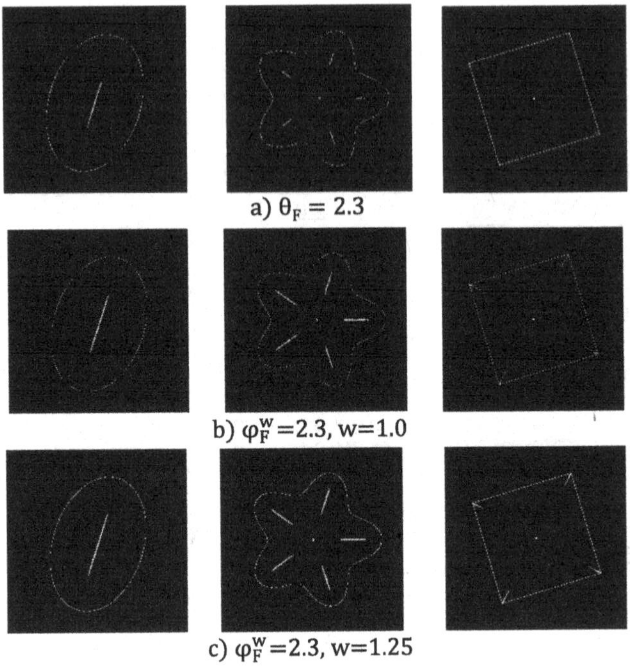

Fig. 9. (a) The result of [10,11] with an angle $\theta_F = 2.3$. (b) our approach with $\varphi_F^\omega = 2.3$ and $\omega = 1$. (c) our approach with $\varphi_F^\omega = 2.3$ and $\omega = 1.25$.

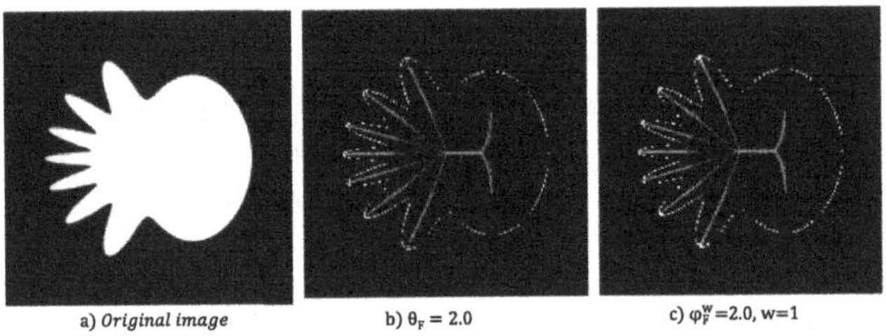

Fig. 10. (a) original image (b) results of [10,11] with an angle $\theta_F = 2.0$ (c) our approach with $\varphi_F^\omega = 2.0$ and $\omega = 1$. The points detected after filtering are the red ones; blue points represent medial axis points of the image. (Color figure online)

approach to be feasible for moderate image size. We would like also to state the possibility to improve the time complexity of the algorithm when the filtering angle is known in advance. This idea remains theoretical; more tests are needed to prove its results and convergence. The idea is to memorize for each point x of the object the closest background point when computing the distance map. Once this information is memorized, filtering the image according to a known bisector angle φ_X^w can be done while looking only at a small subsection of the annulus and neglecting the other parts making thus the computation quasi linear.

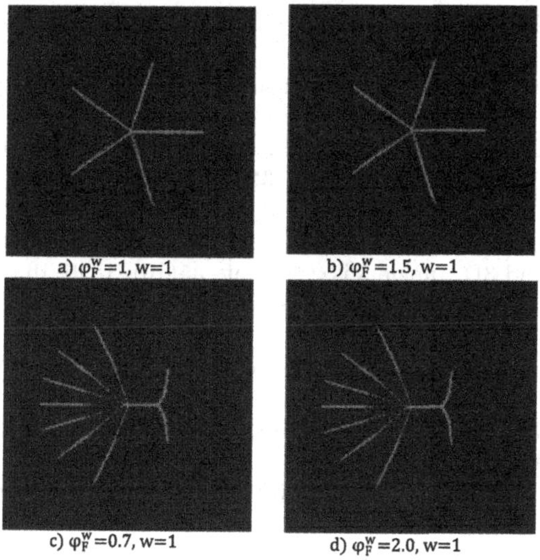

Fig. 11. Applying our method only on the medial axis points: red points are points kept and blue points are points removed. (Color figure online)

4 Conclusion

We introduced a way to compute the discrete bisector function based on an annulus. This approach shows some promising results for filtering skeletons. The results were most promising when the angles become bigger. This method can be extended to 3D and higher dimensions since both the notion of bisector function as well as digital annulus are defined in arbitrary dimensions. One of the perspectives of this work is to use homotopic thinning process and compute skeleton in 2D and higher dimension based on the bisector function information. It should be noted also that filtering using only the angle criteria is most of the time not sufficient [6] and some works combining angle and distance criteria such as [8,9] would be also a future perspective of this work.

References

1. DGtal: Digital geometry tools & algorithms library. http://dgtal.org/
2. Andres, E.: Discrete circles, rings and spheres. Comput. Graph. **18**(5), 695–706 (1994)
3. Andres, E., Jacob, M.: Discrete analytical hyperspheres. IEEE Trans. Vis. Comput. Graph **3**, 75–86 (1997)
4. Attali, D., di Baja, G.S., Thiel, E.: Pruning discrete and semicontinuous skeletons. In: Braccini, C., DeFloriani, L., Vernazza, G. (eds.) ICIAP 1995. LNCS, vol. 974, pp. 488–493. Springer, Heidelberg (1995). https://doi.org/10.1007/3-540-60298-4_303
5. Attali, D., Lachaud, J.O.: Delaunay conforming iso-surface, skeleton extraction and noise removal. Comput. Geom.: Theory Appl. **19**, 175–189 (2001)

6. Attali, D., Montanvert, A.: Modeling noise for a better simplification of skeletons. In: ICIP, vol. 3, pp. 13–16 (1996)
7. Blum, H.: A transformation for extracting new descriptors of shape. In: Models for the Perception of Speech and Visual Form, pp. 362–380 (1967)
8. Chaussard, J., Couprie, M., Talbot, H.: Robust skeletonization using the discrete λ-medial axis. Pattern Recognit. Lett. **32**(9), 1384–1394 (2011)
9. Chazal, F., Lieutier, A.: The λ-medial axis. Graph. Model. **67**(4), 304–331 (2005)
10. Couprie, M., Coeurjolly, D., Zrour, R.: Discrete bisector function and Euclidean skeleton in 2D and 3D. Image Vis. Comput. **25**(10), 1519–1698 (2007)
11. Couprie, M., Zrour, R.: Discrete bisector function and Euclidean skeleton. In: Andres, E., Damiand, G., Lienhardt, P. (eds.) DGCI 2005. LNCS, vol. 3429, pp. 216–227. Springer, Heidelberg (2005). https://doi.org/10.1007/978-3-540-31965-8_21
12. Danielsson, P.E.: Euclidean distance mapping. Comput. Graph. Image Process. **14**, 227–248 (1980)
13. Hirata, T.: A unified linear-time algorithm for computing distance maps. Inf. Process. Lett. **58**(3), 129–133 (1996)
14. Le, B.H., Hodgins, J.K.: Real-time skeletal skinning with optimized centers of rotation. ACM Trans. Graph. **35**(4), 37:1–37:10 (2016)
15. Malandain, G., Fernandez-Vidal, S.: Euclidean skeletons. Image Vis. Comput. **16**(5), 317–328 (1998)
16. Marie, R., Labbani-Igbida, O., Mouaddib, E.M.: The delta medial axis: a fast and robust algorithm for filtered skeleton extraction. Pattern Recognit. **56**, 26–39 (2016)
17. Meijster, A., Roerdink, J.B.T.M., Hesselink, W.H.: A general algorithm for computing distance transforms in linear time. In: Goutsias, J., Vincent, L., Bloomberg, D.S. (eds.) Mathematical Morphology and its Applications to Image and Signal Processing. CIVI, vol. 18, pp. 331–340. Springer, Heidelberg (2002). https://doi.org/10.1007/0-306-47025-X_36
18. Meyer, F.: Cytologie quantitative et morphologie mathématique. Ph.D. thesis, Ecole des mines de Paris (1979)
19. Remy, E., Thiel, E.: Exact medial axis with Euclidean distance. Image Vis. Comput. **23**(2), 167–175 (2005)
20. Saito, T., Toriwaki, J.I.: New algorithms for Euclidean distance transformation of an n-dimensional digitized picture with applications. Pattern Recognit. **27**(11), 1551–1565 (1994)
21. Shamos, M.I.: Computational geometry. Ph.D. thesis, Yale University (1978)
22. Talbot, H., Vincent, L.M.: Euclidean skeletons and conditional bisectors. In: Proceedings of Visual Communications and Image Processing, vol. 1818, pp. 862–877. International Society for Optics and Photonics (1992)
23. Xu, Y., Mattikalli, R., Khosla, P.: Motion planning using medial axis. IFAC Proc. **25**(28), 135–140 (1992)

Average Curve of n Digital Curves

Isabelle Sivignon(✉)

Univ. Grenoble Alpes, CNRS, Grenoble INP, GIPSA-Lab, 38000 Grenoble, France
isabelle.sivignon@gipsa-lab.grenoble-inp.fr

Abstract. Building on [23], we investigate the problem of defining and computing the average of digital curves. Given n digital curves that satisfy compatibility conditions, a set - called the *gap* - in which the average curve is looked for, is defined. The proposed definition rewrites the classical arithmetic mean for curves by (i) defining the distance between each point of the gap and its projection on each curve and (ii) computing the points that minimize the sum of squared deviations. We show that, algorithmically speaking, computing such projections comes down to computing a distance transform with visibility constraints. We propose a fast algorithm to compute a good approximation of these distance maps and finally show that the average curve can be obtained using classical watershed algorithm.

Keywords: Digital curves · Arithmetic mean · Distance transform · Visibility

1 Introduction

Context. Shape averaging is an important task in many applications in order to take into account fluctuation of data, sensibility to the choice of parameters or to compute a morphing between several shapes. To cite only a few examples, it may be used in geological studies to define a shoreline despite short-lived deformations, or as a post-treatment in image segmentation to average the effect of the choice of algorithm parameters. Many works have been conducted on this subject, especially when input data is a set of n polygonal curves. When $n = 2$, medial axis provides a solution. When $n > 2$ the problem is much more complex, and a two-steps process is usually followed: first establish a correspondence between input curves, and then construct a "centroid" from this correspondence. Correspondence between input curves can be achieved using for instance the Fréchet distance [5], or normal map computation [15] for instance. However, in the examples cited above, shapes are defined as objects in digital images, so that defining the average of digital shapes is particularly relevant for many applications. On this matter, most previous works focused on morphing-based approaches [3].

M. Couprie et al. (Eds.): DGCI 2019, LNCS 11414, pp. 481–493, 2019.
https://doi.org/10.1007/978-3-030-14085-4_38

Arithmetic Mean. Going back to basics, the arithmetic mean $\bar{x}$ is the best way to select a "typical" value for a set of input values x_i, in the sense that it minimizes the sum of squared deviations. Indeed, if we denote $f(x) = \sum_i (x - x_i)^2$, $\bar{x} = argmin_x f(x)$. Minimizing f can be done either by looking for local minima of f, or by looking for the zero set of its derivative $f'(x) = \sum_i (x - x_i)$. Anyway, the arithmetic mean $\bar{x}$ lies in the interval $[x_m, x_M]$, where x_m and x_M denote the minimal and maximal value of the x_i respectively.

Outline. In [23], the authors present a framework to extend the notion of arithmetic mean to "values" that are not numbers but smooth curves. In this article, we investigate the extension and relevance of this definition of average curve in the case of digital curves. We also discuss efficient algorithmic solutions to compute it. In Sect. 2 we recall the main results of [23] and revisit them for digital curves, which reveals a new constrained distance transform problem. Section 3 focuses on algorithmic solutions to solve this problem. Then, in Sect. 4 we show how to obtain an average curve from the constrained distance map and give some experimental results. Last, Sect. 5 raises the question of necessary conditions on the input curves for this approach to be valid, and provides tentative answers.

2 Which Definition of Average Curve?

2.1 "Arithmetic" Mean Curve

Given a set of n smooth Jordan curves $\{\mathscr{C}_i\}$, the authors of [23] define the Valley Average Curve as the valley of the scalar height field $Q(p) = \sum_i d(p, \pi_i(p))^2$, and the Zero Average Curve as the zero set of the scalar height field $D(p) = \sum_i d(p, \pi_i(p))$, where d denotes the Euclidean distance and $\pi_i(p)$ the "projection" of p onto $\mathscr{C}_i$. Since all curves are Jordan, each $\mathscr{C}_i$ is the boundary of a bounded shape $\mathscr{S}_i$. As the arithmetic mean of a set of numbers lies somewhere in between the minimum and maximum values, the average curve is looked for somewhere in the symmetric difference of the $\mathscr{S}_i$: no point of the average curve may lie in the interior or in the exterior of all the $\mathscr{C}_i$. Formally, in [23] the authors define the *gap* of $\{\mathscr{C}_i\}$ as $\Delta(\{\mathscr{C}_i\}) = (\bigcup \overline{\mathscr{S}}_i) \backslash (\bigcap \mathring{\mathscr{S}}_i)$ (see Fig. 1 - we use the notation Δ for short when there is no ambiguity). Note that, from this definition, an interesting property is that the average curve goes through all the points shared by all input curves.

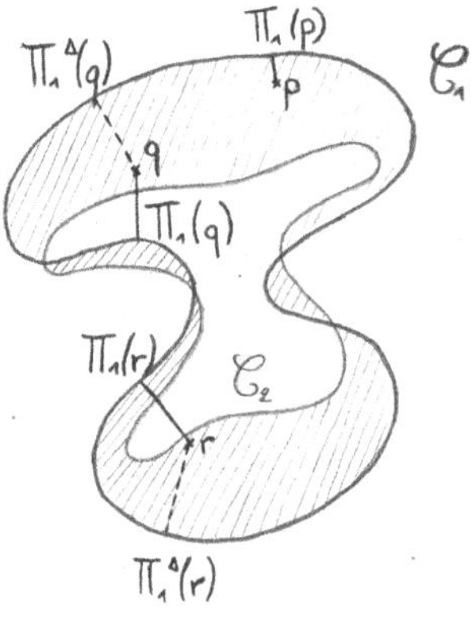

Fig. 1. Gap (dashed) of two curves $\mathscr{C}_1$ and $\mathscr{C}_2$. The closest projections $\pi_1(q)$ and $\pi_1(r)$ of q and r on $\mathscr{C}_1$ yield erroneous mean curve. Gap relative projections π_1^Δ are defined instead.

The authors remarked that if the projection $\pi_i(p)$ of a point p is defined as the point of $\mathscr{C}_i$ closest to p, some parts of the average curve may be missed in the case of the Valley Average Curve, or misplaced for the Zero Average Curve (see Fig. 1). Indeed, the idea behind the definition of average curve is to suppose that projections of each point of the average curve $\mathscr{C}$ on each $\mathscr{C}_i$ provide a continuous and bijective mapping between $\mathscr{C}$ and each $\mathscr{C}_i$: if the segment $[p, \pi_i(p)]$ crosses Δ, this property fails. Thus they introduce the *gap relative projection* $\pi_i^\Delta(p)$ as the closest point on $\mathscr{C}_i$ such that $[p, \pi_i^\Delta(p)] \subseteq \Delta$.

2.2 Arithmetic Mean Digital Curve

Consider a set of n digital 4-connected simple and closed curves $\{C_i, i = 1 \dots n\}$. Each C_i being a Jordan curve, $\mathbb{Z}^2 \backslash C_i$ has exactly two 8-connected components, the unbounded one being called its exterior, and the other its interior. We denote by O_i the digital object defined as the union of C_i and its interior, that we denote $\mathring{O}_i$. Similarly to [23] we define the *gap* of $\{C_i\}$ as $\Delta(\{C_i\}) = (\bigcup O_i) \backslash (\bigcap \mathring{O}_i)$ (or Δ for short). Note that the Zero Average Curve definition involves non-integer computation, while the Valley Average Curve can be computed using integers when all points are in $\mathbb{Z}^2$. That is why in the following, we focus on the Valley Average Curve definition, that we call arithmetic mean digital curve.

For any point $p \in \mathbb{Z}^2$, we can define $H(p) = \sum_i d(p, \pi_i^\Delta(p))^2$ (H is called height map) as before. However, the gap relative projection π_i^Δ has to be redefined for digital points and objects. Given a model of digital straight segment (DSS for short), we say that two points p and q are *visible* in a digital set A and denote $Vis_A(p, q)$ if there exists a DSS $s \subseteq A$ such that $p, q \in s$. We also define $V_A(p)$ as the visibility region of p in A: $V_A(p) = \{q \in A | Vis_A(p, q)\}$. Thus we have $\pi_i^\Delta(p) = \arg\min_{q \in V_\Delta(p) \cap C_i} \{d(p, q)\}$.

Hence, a main issue in the computation of the arithmetic mean digital curve is to compute gap relative projections efficiently for all points in the gap, which is the topic of the next section.

3 Computation of Gap Relative Projections

We consider here the most general problem of computing the visible projection of any point of a set A with respect to a subset $B \subseteq A$. In the following, A^c is the complement of A in $\mathbb{Z}^2$, and points of A^c are called *obstacles*.

Problem 1 (Visible Voronoi Map). Given two digital sets A and B with $B \subseteq A$, compute for each point $p \in A$ the projection $q \in B$ of p such that $Vis_A(p, q)$.

3.1 Brute Force Algorithm and Discussion

Algorithm. Efficient separable algorithms to compute distance transforms and Voronoi Maps have been proposed [8], and can be used to design a first straightforward brute-force algorithm. Given two digital sets A and B such that $B \subseteq A$, Algorithm 1 solves Problem 1. Function VISIBLE is the implementation of the Vis predicate defined in Sect. 2.2.

Algorithm 1. BruteForceVisibleVoronoiMap(DIGITAL SET A, DIGITAL SET B, $B \subseteq A$)

Initialization: VISIVOROMAP(p) = VORONOIMAP$(p) = \arg\min_{r \in B} d(p, r)$;
for $p \in A$ **do**
  if $\neg$VISIBLE$(A, p,$ VISIVOROMAP$(p))$ **then**
    Loop on points of B to find the visible projection q of p in A;
    VISIVOROMAP$(p) = q$
return VISIVOROMAP

Proposition 1. *If n and m are the cardinality of A and B respectively ($n > m$), if V denotes the computational cost of* VISIBLE, *and k is the number of points of A for which the Voronoi Map differs from the Visible Voronoi Map then Algorithm 1 is in $O(n + V.(n + km))$.*

The complexity V of the VISIBLE predicate is discussed in Sect. 3.3. However, even if V is neglected, this algorithm is not satisfying since, when k and m are in the order of n, the complexity becomes quadratic in n.

Discussion. Several options may be considered to design a more efficient algorithm. The first one would be to adapt the optimal separable distance transform algorithm of [8] to take into account obstacles. Following the rewriting proposed in [7] using two predicates (CLOSEST and HIDDENBY), it would be a matter of modifying the HIDDENBY predicate since a site hidden by others in the Voronoi Map may very well be the projection of a point if other sites are not visible from this point. It appears that, with no further assumption on obstacles, the list of sites cannot be shortened as efficiently as for the Voronoi Map.

Another option would be to first compute the visibility region of each point $p \in A$ before computing the distance transform of p with respect to $V_A(p) \cap B$. Computing the visibility region of a point has ben widely studied in computational geometry [16], where obstacles are polygons. In the case of digital sets, the problem has been tackled in [6] but the difficulty here is to efficiently compute the visibility region for all points of A.

Yet another alternative would be to build a tailored data structure to perform efficient visible nearest neighbour queries, as proposed in [19] for instance. Here again, obstacles are supposed to be represented as polygons, and a fast computation of a predicate returning the minimum visible distance from a point to an obstacle is key. In our context, an important point is the definition of obstacles, which could range from an obstacle per point in B (a lot of obstacles with a very fast predicate) to connected components of B (few obstacles but more convoluted predicate) for instance. The best solution would highly depend on the geometry of B with respect to A.

The last option we discuss here is to propagate projections locally, from point to point. Even though it has been reported in the literature that such local approaches fail to compute exact Voronoi Maps [12], in Sect. 3.2 we provide

Algorithm 2. FMMLOOP(ACCEPTED POINT SET Λ, CANDIDATE LIST Γ)

```
Extract from Γ the point p with smallest distance to its projection π^A(p);
if p ∉ Λ then
    Λ = Λ ∪ {p, π^A(p)};
    foreach q ∈ N(p) ∩ A do  ADDCANDIDATEPOINT(Λ,Γ,A,q) ;
```

new results on local approaches accuracy, and show that a strong point of this approach is that visibility can be checked on the fly. In the next section, we propose to adapt the Fast-Marching Method [24] (FMM for short) algorithm to compute an approximation of the solution to Problem 1, but other local schemes may also be considered.

3.2 Fast Marching Method-Like Algorithm Using Projections

Basic Loop. The algorithm follows the propagation principle of the Fast Marching Method (which in turns is very similar to Dijkstra's shortest path algorithm). It was designed primarily to describe the evolution of an interface as a function of time and at a given speed in the normal direction to the surface. But more generally, given a domain A and a set $B \subseteq A$, the algorithm computes for each point p of A an approximation of the length of the shortest geodesic from B to p in A. From the list of points for which the distance is known (called *accepted points*), a list of *candidate points* that lie in the neighbourhood of accepted points is maintained. For each candidate point, a tentative value of distance is computed from its neighbours belonging to the accepted point set. Each step of the algorithm consists in adding the candidate point with the minimum distance. We adopt the same simplified FMM loop as presented in [17], which is sketched in Algorithm 2, and implemented in DGtal library [1]. Initially, accepted point set Λ is set to B, candidate list Γ is empty.

Algorithm 3 details the ADDCANDIDATEPOINT function. Parts in orange are to be disregarded at this point and will be explained later in this section. In order to be able to include a visibility constraint in the course of the algorithm, instead of propagating distance information from point to point, visible projection point is propagated. Note that a point may appear several times in the candidate list, with different projections, and thus distance values.

About the Accuracy of the Algorithm. Propagating closest point information in FMM, without the visibility constraint, was already done in [4]. Accuracy relies on the assumption that for any point $p \in A$, its visible projection in B is also the visible projection of one of its neighbours. We formalize this condition in Proposition 2 when $\mathcal{N}$ is the classical 8-neighbourhood $\mathcal{N}_8$.

Proposition 2. *Let $p \in A$, and $\pi^A(p)$ its ground truth visible projection on B. If there exists an 8-connected path of points $\{p_i\}_{1..k}$ in A such that (i) $p_1 = p$, $p_k = \pi^A(p)$, for all $i \in \{1 \dots k\}$ $\pi^A(p_i) = \pi^A(p)$ (ii) the distance to $\pi^A(p)$ is decreasing, then Algorithm 2 computes the exact visible projection $\pi^A(p)$ of p.*

Algorithm 3. ADDCANDIDATEPOINT(ACCEPTED POINT SET Λ, CANDIDATE LIST Γ, DIGITAL SET A, DIGITAL SET B, POINT p)

1 $L = \emptyset$;
2 **foreach** $q \in \mathcal{N}(p) \cap \Lambda$ **do**
3   **if** VISIBLE$(A,\ p, \pi^A(q))$ **then** $L = L \cup \pi^A(q)$;
4   else
5         $\tau(p) =$ COMPUTEPROJTANGENT(p, q, A, B);
6   **if** $\tau(p)$ *exists* **then** $L = L \cup \tau(p)$;
7     $\pi^A(p) = \arg\min_{x \in L}\{d(p, x)\}$; $\Gamma = \Gamma \cup \{p, \pi^A(p)\}$;

Proof. We proceed by induction on the length of the path.

[Initialization] If $k = 2$, then p is 8-neighbour to $\pi^A(p) \in B$. If $\pi^A(p)$ is the only point of B in $\mathcal{N}_8(p)$, we are done. Otherwise, any point of B wad added to Λ before p, and p was added in the candidate point set Γ, using its 8-neighbours, in particular $\pi^A(p)$. Then p may have been added several times in Γ with different tentative projections, but only the one with the closest tentative projection, here $\pi^A(p)$, is actually added in Λ thanks to line 2 of Algorithm 2.

[Induction step] Consider a point $p \in A$ for which there exists a path $\{p_i\}$ of length $j + 1$, with $p_1 = p$ and such that the two conditions of Proposition 2 are satisfied. Point p_2 is such that $\pi^A(p_2) = \pi^A(p)$ and there exists a path of length j between p_2 and $\pi^A(p_2)$. By induction hypothesis, Algorithm 2 correctly computes $\pi^A(p_2)$. Consider now the moment when p_2 is included in the accepted point set Λ (Algorithm 2 line 4). Several cases may arise:

- if p_1 is not in Λ, and whether it is already in the list of candidates Γ or not, it is added to the list, with the smallest distance possible since $\pi^A(p_2)$ is equal to the true visible projection $\pi^A(p)$ of p;
- if p_1 is already in Λ with an erroneous projection, this means that p_1 was added before p_2 (otherwise p_1, as a 8-neighbour of p_2, would have been added to Γ line 7 of Algorithm 3, since $\pi^A(p_2) = \pi^A(p)$). Let q be the projection associated to p when it was added to Λ. We have $d(p, q) \leq d(p_2, \pi^A(p_2))$. But by hypothesis we have $d(p, \pi^A(p)) \geq d(p_2, \pi^A(p)) = d(p_2, \pi^A(p_2)) \geq d(p, q)$, thus contradicting the fact that $\pi^A(p)$ is the ground truth projection.

The sufficient conditions of Proposition 2 may be broken in at least two situations. Firstly, even if there is no obstacle around p and $\pi^A(p)$, $\pi^A(p)$ may not be the projection of any of p's neighbours. This issue is inherent in any approach based on local propagation because digital Voronoi cells may be not connected. This has been discussed in the literature, and we extend here the results of [12] by showing that erroneous values cannot occur close to B.

Proposition 3. *For any digital set B, for all p such that $d(p, B) < 13$, there exists a point $q \in \mathcal{N}_8(p)$ such that $\pi(q) = \pi(p)$, where $\mathcal{N}_8$ denotes the 8 neighbourhood.*

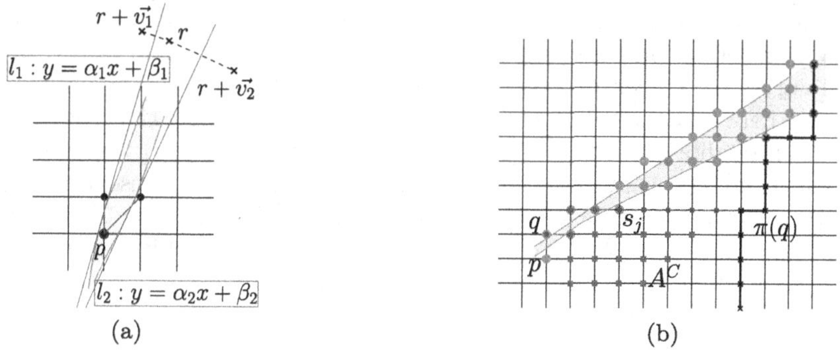

Fig. 2. (a) Illustration of the proof of Proposition 3. V is (incompletely) depicted in green. r is the site associated to V. (b) Computation of a tangent from a point p using the DSS from one of its neighbours q to $\pi^A(q)$ (in blue). Red points are obstacles. Points in B are marked by black crosses. (Color figure online)

Proof. Suppose wlog that $p(0,0)$. See Fig. 2(a) for an illustration of the notations. If there is no point in $\mathcal{N}_8(p)$ such that $\pi(q) = \pi(p)$, this implies that the Voronoi cell V of $\pi(p)$ does not contain any point of $\mathcal{N}_8(p)$. Up to symmetries, this means that the boundary of V contains two segments, one cutting the segment $[p, p+(0,1)]$ and the other cutting the segment $[p, p+(1,1)]$. If we call $l1$ and $l2$ respectively the two lines (of positive slopes) supporting these two segments, this translates into the following constraints: $\beta_1 < 1$, $\beta_1 > 0$, $\beta_2 + \alpha_2 > 1$ and $\beta 2 < 0$. Let r be the site associated to p (its closest point in B). The two lines l_1 and l_2 are bisectors of r with points $r + \boldsymbol{v_1} \in B$ and $r + \boldsymbol{v_2} \in B$ respectively. Since B is a digital set, r, $r + \boldsymbol{v_1}$ and $r + \boldsymbol{v_2}$ are points of $\mathbb{Z}^2$. The values β_1, β_2, $\beta_2 + \alpha_2$ can be written as functions of r, $\boldsymbol{v_1}$ and $\boldsymbol{v_2}$. Combining the constraints stated above with a study of the variations of these functions to derive necessary conditions on r, $\boldsymbol{v_1}$ and $\boldsymbol{v_2}$ is quite tedious. Experimentally, it is however pretty straightforward to find that the triplet of points defined by $r(5,12)$, $\boldsymbol{v_1} = (-4,1)$ and $\boldsymbol{v_2} = (2,-1)$ fulfills the constraints. This gives an upper bound on the minimal distance ($d(p,r) = 13$) for which there exists a point fulfilling the constraints. We verify that this is also a lower bound by exhaustively checking the constraints for all digital points at distance less than 13[1].

Secondly, visible projections of p's neighbours may not be visible from p, and p's projection may be somewhere else. In the next paragraph, we propose an improved version of the function ADDCANDIDATEPOINT, using the parts in orange, in order to take into account such cases.

Algorithm Using Tangents. If, for a given neighbour q of p, $\pi^A(q)$ is not visible from p, there is a connected component C of A^C between p and $\pi^A(q)$. The idea is to compute a tangent to C passing through p. Such a construction echoes the

[1] Python code to make this test is available at www.gipsa-lab.grenoble-inp.fr/~isabelle.sivignon/Code.

computation of visibility regions in polygons [16]. The returned point, if there is one, is the intersection between this tangent and B. Many tangent estimators for digital curves have been proposed in the literature [22,26], but the problem we are facing here is slightly different. Indeed, we need to compute the tangent to an obstacle through a given point (not on the obstacle's boundary) with no apriori information on the obstacle. In order to solve this problem, we present here a fast heuristic. Figure 2(b) illustrates how COMPUTEPROJTANGENT is implemented. Let us suppose that a DSS $S(q) = \{s_1 = q \dots s_k = \pi^A(q)\}$ between q and $\pi^A(q)$ is known (in blue on Fig. 2(b) and see Sect. 3.3 for a discussion). Since $\pi^A(q)$ is not visible from p, $p \cup S(q)$ is not a DSS. Let $j \in [2, k-1]$ be the index such that $p \cup \{s_1, \dots s_j\}$ is a DSS while $p \cup \{s_1, \dots s_{j+1}\}$ is not. Part $\{s_1, \dots s_j\}$ is depicted as blue and green dots in Fig. 2(b). The set $\mathcal{L}$ of all digital straight lines that contain the DSS $p \cup \{s_1, \dots s_j\}$ defines a family of lines close to C through p. In Fig. 2(b), the green region represents the parameters of this set of digital straight lines. Thus, there is not one unique intersection point between B and a line of $\mathcal{L}$, but a set of points $P = \bigcup_{l \in \mathcal{L}, l \cap A^c = \emptyset} l \cap B$ (green points overlayed by black crosses in Fig. 2(b)). Computing P to find the point of P closest to p may be computationally expensive in general. If only one intersection point between B and a specific digital straight line $l \in \mathcal{L}$, $l \cap A^C = \emptyset$ is computed, then committed error (in comparison with the exhaustive computation of P) is upper bounded by $\max_{r,s \in P} |d(p,r) - d(p,s)|$.

Computational Complexity

Proposition 4. *If A is a digital set of n points, computational complexity of Algorithm 2 is $O(n.|\mathcal{N}|.(|\mathcal{N}|.V + \log(|\Gamma|)))$ when Algorithm 3 is called without the tangent computation. $|\Gamma|$ is the maximal size of the list of candidates, and V is the computational complexity of the predicate* VISIBLE.

Complexity of Algorithm 2 is $O(n.|\mathcal{N}|.(|\mathcal{N}|.(V+T) + \log(|\Gamma|)))$ when Algorithm 3 with tangent computation is used instead. T is the computational complexity of the function COMPUTEPROJTANGENT.

Proof. The complexity analysis is pretty straightforward. The $\log(|\Gamma|)$ term comes from the insertion of a new candidate in Γ line 7 of Algorithm 3. This complexity is achieved by using an ordered set, where the elements $\{p, \pi^A(p)\}$ are ordered with respect to the value of $d(p, \pi^A(p))$.

3.3 Visibility Test

The VISIBLE predicate is called many times in the algorithm, so that its implementation is key to the algorithm efficiency. We propose below a visibility test in three steps. Each step consists in checking sufficient visibility conditions of increasing precision and complexity: VISIBLE(A,p,q) = (VISIBLEADD(A,p,q) or VISIBLEUPLOW(A,p,q) or VISIBLEPREIMAGE(A,p,q)). Note that the *or* short-circuiting ensures that the least computationally expensive tests handle most common cases.

VISIBLEADD. For each accepted point $q \in \Lambda$, we also store a DSS $S(q) \subseteq A$ such that $q, \pi^A(q) \in S(q)$. Maintaining and propagating this information is actually very easy using a classical incremental DSS recognition algorithm [13] and the fact that Algorithm 2 computes the projection of a point from the projections of its neighbours. The VISIBLEADD function returns true if $p \cup S(q)$ is still a DSS, false otherwise. If true, we set $S(p) = p \cup S(q)$ and store it with p and $\pi^A(q)$ in the list of candidates Γ. This test runs in constant time.

VISIBLEUPLOW. This predicates verifies whether the DSS having p and q as upper (resp. lower) leaning points is included in A. This test runs in $O(\max(abs(x_p - x_q), abs(y_p - y_q)))$, which is upper bounded by the maximal distance between two points of A.

VISIBLEPREIMAGE. The last test finally enables to thoroughly check whether there exists a DSS included in A among all possible DSSs between p and q. This problem was tackled in [6] and the solution proposed can be implemented using the linear-in-time stabbing lines algorithm of [20] for which an implementation is available in DGtal [1]. The general idea of the algorithm is the following: given a set of digital points that must belong to the DSS, and another set of digital points that must not, saying that a given digital point must or must not belong to a DSS is translated into a couple of constraints on the DSS parameters. When all the points have been added, it is enough to check whether the set of possible DSS parameters is empty or not. In our test, the only two points we want in the DSS are p and q. The points that must not belong to the DSS are obstacles points, *i.e.* points of A^C. However, in general, few points of A^C actually interfere with a DSS between p and q, and such points can be detected while tracking the upper and lower DSS in function VISIBLEUPLOW.

3.4 Experimental Comparison of Brute Force and FMM-Like Algorithms

Algorithms 1 and 2 were implemented using DGtal library [1]. Code is available at www.gipsa-lab.grenoble-inp.fr/~isabelle.sivignon/Code. Figure 3 gives elements of comparison in terms of computational speed and accuracy. The result of Algorithm 1 is considered as the ground truth since an exhaustive search of visible projections is performed when necessary. We test our algorithms on three different sets A (one obstacle in (d), two large obstacles in (e), many obstacles in (f)), with the same set B. Figure 3(a-c) show experimental computation times in the three cases, at different resolutions and for each algorithm. Unsurprisingly, Algorithm 1 is much faster when $\pi(p) = \pi^A(p)$ for all p as in case (d). However, when visibility issues arise for many points p, the trend is reversed. Figures (d-h) provide elements to evaluate the quality of the approximation computed by Algorithm 2. The results of this algorithm are depicted in (d) and (e): we do not show the results of Algorithm 1 since it is the same apart for very few points where the error is very small (equal to 1 on these examples, see Proposition 3). However, the results tend to differ more for the shape with many small obstacles

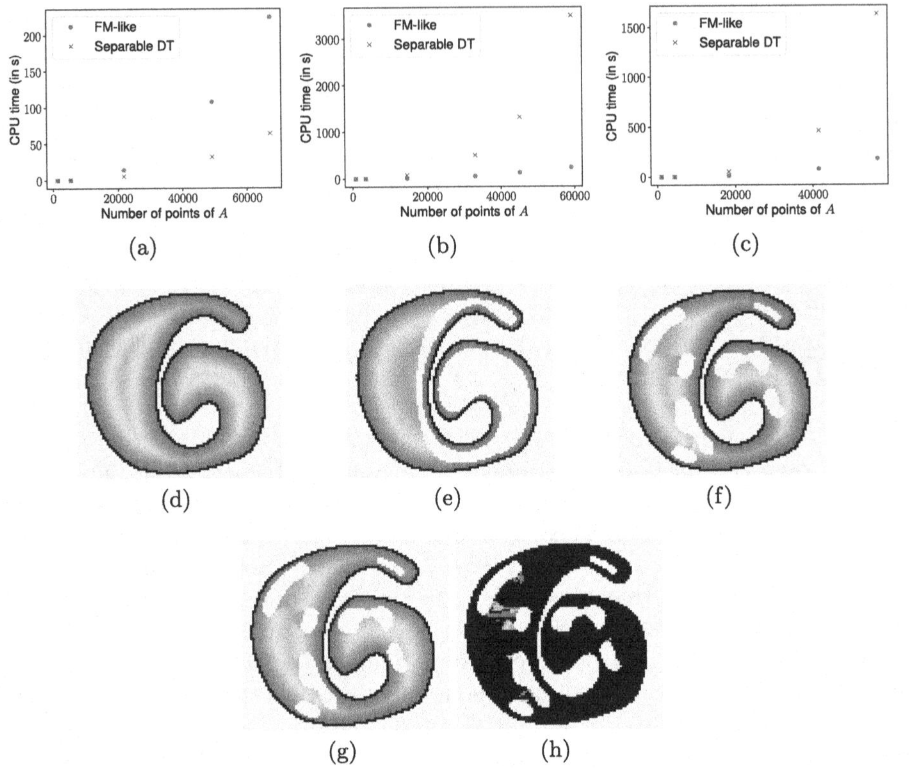

Fig. 3. In (d-g), A^C is depicted in light grey, B in black. Figures (a) to (c) report computation times when running Algorithms 1 and 2 on input data depicted in (d) to (f) respectively. In (d-g) a color map from red (low values) to green (large values) is used to represent the distance between a point and its visible projection computed thanks to Algorithm 2, except for (g) where Algorithm 1 is used. In (h), the same color map is used to depict the difference between (f) and (g). (Color figure online)

(see (f), (g) and (h)), especially in parts where the heuristic tangent estimator was called.

4 Computation of the Arithmetic Mean Digital Curve

Coming back to the initial problem, recall that the goal is the following: given a set of digital curves $\{C_i\}$, (i) compute the height map $H : \Delta(\{C_i\}) \to \mathbb{R}$, $p \mapsto \sum_i d(p, \pi_i^{\Delta}(p))^2$ (ii) compute the valley of the height field H. Section 3 was dedicated to the design of an algorithm to solve step (i). The concepts of ridges and valleys involved in step (ii) have been widely studied in computational imaging, with applications in image segmentation or shape recognition (see [14] for an overview). As stated in [21], there are two main approaches to define these

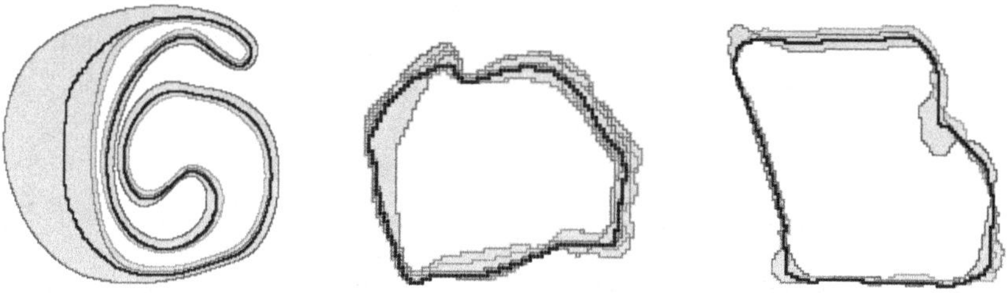

(a) (b) (c)

Fig. 4. Results of arithmetic mean digital curve computation: input curves are in colors, the result mean curve is in black, and the gap in gray: (a) three curves issued from [23]; in (b) and (c), input curves are the results of a morphological snake image segmentation algorithm [2] with different parameters - eight curves in (b), three in (c).

concepts. The first one is to define ridges as solution manifolds of algebraic equations using the height field and its derivatives (see [14,21]). The second approach is the well-known concept of watershed, for which many efficient algorithms have been designed for the last 30 years [9,11,25]. Experimental results presented in Fig. 4 use the implementation of [25] in the ImageJ open source software [18], which is widely used in biomedical image analysis for instance.

5 About Compatibility Conditions of Input Curves

In this last part, we give tentative elements about conditions input curves must fulfill for this approach to be correct, and leave some open questions.

Conditions of [23]. To define compatibility conditions of input curves, the authors of [23] define *spikes* for each point $p \in \mathscr{C}_i$: if l is the line l normal to $\mathscr{C}_i$ at p, the spike of $\mathscr{C}_i$ at p is the segment of $l \cap \Delta$ containing p. The assumption they make on the set of input curves is that for any curve $\mathscr{C}_i$, the spikes of $\mathscr{C}_i$ are pairwise distinct and their union is the gap. Under this assumption, they conjecture that the height map f has exactly one local minimum on each spike, for any curve $\mathscr{C}_i$. They also provide a sufficient condition to ensure that the assumption is true: all input curves should be pairwise normal-compatible [15] (meaning that the closest point maps between any pair of curves must be an homeomorphism). Finally, they also enounce a necessary condition (called gap compatibility), used to fastly detect families of curves for which the approach fails: for any point $p \in \Delta$ and any curve $\mathscr{C}_i$, $\pi_i^{\Delta}(p)$ must be unique. This can be rewritten as: if $p \in \Delta$ belongs to the medial axis of $\mathscr{S}_i$, then only one of its closest points is visible in Δ.

Which Conditions for a Set of Digital Curves? The properties and conditions above do not translate straightforwardly for digital curves. First, the gap compatibility condition is not relevant in our context since many points that do not belong to the medial axis (which is the set of maximal balls) may have several

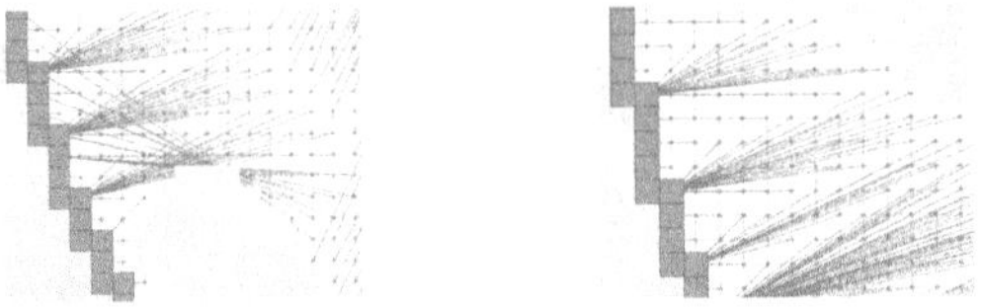

(a) (b)

Fig. 5. C_i is in dark grey, Δ^C in light grey and arrows are depicted between points $p \in \Delta$ and $\pi_i^\Delta(p)$. (a) Intersecting spikes, (b) no spikes intersect.

closest points. We propose to rewrite this condition as follows: if for a point p, $\pi_i^\Delta(p)$ is not unique, say $\pi_i^\Delta(p) = \{s_j\}$, if (i) the digital ball $b(p, d(p, s_j))$ is not maximal in O_i and (ii) let s_1 and s_k be the projections such that $C_i(s_0..s_k)$ (the part of C_i between s_0 and s_k) is of minimal length and for all j, $s_j \in C_i(s_0..s_k)$; then any point in $C_i(s_0..s_k)$ is visible from p. Verifying that this condition is fulfilled requires to compute all closest points instead of one, and a solution was proposed in [10]. Last, the definition of spike is key to ensure that the minima of the height field define a curve. Again, the definition does not extend to digital curves. The spike of a point $p \in \mathscr{C}_i$ is actually the Voronoi Cell of p in Δ. Thus we propose to define the spike of a point $p \in C_i$ as $\text{spike}(p, C_i) = \text{ConvexHull}(VoronoiCell(p) \cap \Delta)$. The assumption on spikes translates as follows: for any point x in the interior of $\text{spike}(p, C_i)$, $p = \pi_i^\Delta(x)$. This assumption is violated is many cases in the example of the shape with many holes of Fig. 3(f), while it is always fulfilled for shape (e). Figure 5 shows a close-up on spikes in that two cases. We end up with an open question: what are the properties of the height field minima with respect to spikes when they are pairwise distinct and their union is the gap? May we prove that minima define a curve?

References

1. DGtal: Digital Geometry Tools and Algorithms Library. http://dgtal.org
2. Alvarez, L., Baumela, L., Márquez-Neila, P., Henríquez, P.: A real time morphological snakes algorithm. Image Process. Line **2**, 1–7 (2012)
3. Boukhriss, I., Miguet, S., Tougne, L.: Discrete average of two-dimensional shapes. In: Gagalowicz, A., Philips, W. (eds.) CAIP 2005. LNCS, vol. 3691, pp. 145–152. Springer, Heidelberg (2005). https://doi.org/10.1007/11556121_19
4. Breen, D.E., Mauch, S., Whitaker, R.T.: 3D scan conversion of CSG models into distance volumes. In: Proceedings of the 1998 IEEE Symposium on Volume Visualization, pp. 7–14. ACM (1998)
5. Buchin, K., et al.: Median trajectories. Algorithmica **66**(3), 595–614 (2013)
6. Coeurjolly, D., Miguet, S., Tougne, L.: 2D and 3D visibility in discrete geometry: an application to discrete geodesic paths. Pattern Recognit. Lett. **25**(5), 561–570 (2004)
7. Coeurjolly, D.: 2D subquadratic separable distance transformation for path-based norms. In: Barcucci, E., Frosini, A., Rinaldi, S. (eds.) DGCI 2014. LNCS, vol. 8668, pp. 75–87. Springer, Cham (2014). https://doi.org/10.1007/978-3-319-09955-2_7

8. Coeurjolly, D., Montanvert, A.: Optimal separable algorithms to compute the reverse euclidean distance transformation and discrete medial axis in arbitrary dimension. IEEE Trans. Pattern Anal. Mach. Intell. **29**(3), 437–448 (2007)
9. Couprie, M., Bertrand, G.: Topological gray-scale watershed transformation (1997)
10. Couprie, M., Coeurjolly, D., Zrour, R.: Discrete bisector function and Euclidean skeleton in 2D and 3D. Image Vis. Comput. **25**(10), 1543–1556 (2007)
11. Couprie, M., Najman, L., Bertrand, G.: Quasi-linear algorithms for the topological watershed. J. Math. Imaging Vis. **22**(2), 231–249 (2005)
12. Danielsson, P.E.: Euclidean distance mapping. Comput. Graph. Image Process. **14**, 227–248 (1980)
13. Debled-Rennesson, I., Reveillès, J.P.: A linear algorithm for segmentation of digital curves. Int. J. Pattern Recognit. Artif. Intell. **09**(4), 635–662 (1995)
14. Eberly, D.: Ridges in Image and Data Analysis. Springer, Heidelberg (1996). https://doi.org/10.1007/978-94-015-8765-5
15. Chazal, F., Lieutier, A., Rossignac, J.: Normal-map between normal-compatible manifolds. Int. J. Comput. Geom. Appl. **17**(5), 403–421 (2007)
16. Goodman, J.E., O'Rourke, J. (eds.): Handbook of Discrete and Computational Geometry. CRC Press Inc., Boca Raton (1997)
17. Jones, M.W., Baerentzen, J.A., Sramek, M.: 3D distance fields: a survey of techniques and applications. IEEE Trans. Vis. Comput. Graph. **12**(4), 581–599 (2006)
18. Legland, D., Arganda-Carreras, I.: ImageJ-MorphoLibJ. http://imagej.net/MorphoLibJ
19. Nutanong, S., Tanin, E., Zhang, R.: Incremental evaluation of visible nearest neighbor queries. IEEE Trans. Knowl. Data Eng. **22**(5), 665–681 (2010)
20. O'Rourke, J.: An on-line algorithm for fitting straight lines between data ranges. Commun. ACM **24**(9), 574–578 (1981)
21. Peikert, R., Sadlo, F.: Height ridge computation and filtering for visualization. In: 2008 IEEE Pacific Visualization Symposium, pp. 119–126 (2008)
22. Prasad, D.K., Leung, M.K.H., Quek, C., Brown, M.S.: DEB: definite error bounded tangent estimator for digital curves. IEEE Trans. Image Process. **23**(10), 4297–4310 (2014)
23. Sati, M., Rossignac, J., Seidel, R., Wyvill, B., Musuvathy, S.: Average curve of n smooth planar curves. Comput.-Aided Design **70**, 46–55 (2016)
24. Sethian, J.A.: A fast marching level set method for monotonically advancing fronts. Proc. Natl. Acad. Sci. **93**(4), 1591–1595 (1996)
25. Soille, P., Vincent, L.: Determining watersheds in digital pictures via flooding simulations (1990)
26. de Vieilleville, F., Lachaud, J.O.: Comparison and improvement of tangent estimators on digital curves. Pattern Recognit. **42**(8), 1693–1707 (2009). Advances in Combinatorial Image Analysis

Author Index

Printed in the United States
By Bookmasters